S_pringer_ **M**_onographs_ in **M**_athematics_

For further volumes:
www.springer.com/series/3733

Roelof Koekoek • Peter A. Lesky •
René F. Swarttouw

Hypergeometric Orthogonal Polynomials and Their q-Analogues

With a Foreword by Tom H. Koornwinder

Springer

Roelof Koekoek
Delft Institute of Applied Mathematics
Delft University of Technology
P.O. Box 5031
2600 GA Delft
The Netherlands
r.koekoek@tudelft.nl

Peter A. Lesky (1927–2008)
University of Stuttgart, Germany

René F. Swarttouw
Department of Mathematics
Faculty of Sciences
VU University Amsterdam
De Boelelaan 1081a
1081 HV Amsterdam
The Netherlands
rene@few.vu.nl

Foreword by
Tom H. Koornwinder
Korteweg-de Vries Institute of Mathematics
Faculty of Science
University of Amsterdam
P.O. Box 94248
1090 GE Amsterdam
The Netherlands
t.h.koornwinder@uva.nl

ISSN 1439-7382
ISBN 978-3-642-05013-8 e-ISBN 978-3-642-05014-5
DOI 10.1007/978-3-642-05014-5
Springer Heidelberg Dordrecht London New York

Library of Congress Control Number: 2010923797

Mathematics Subject Classification (2000): 33C45; 33D45

Cover illustration: The Authors
Cover design: deblik

Printed on acid-free paper

Springer is part of Springer Science+Business Media (www.springer.com)

Foreword

The present book is about the Askey scheme and the q-Askey scheme, which are graphically displayed right before chapter 9 and chapter 14, respectively. The families of orthogonal polynomials in these two schemes generalize the classical orthogonal polynomials (Jacobi, Laguerre and Hermite polynomials) and they have properties similar to them. In fact, they have properties so similar that I am inclined (following Andrews & Askey [34]) to call all families in the (q-)Askey scheme classical orthogonal polynomials, and to call the Jacobi, Laguerre and Hermite polynomials *very classical* orthogonal polynomials.

These very classical orthogonal polynomials are good friends of mine since almost the beginning of my mathematical career. When I was a fresh PhD student at the Mathematical Centre (now CWI) in Amsterdam, Dick Askey spent a sabbatical there during the academic year 1969–1970. He lectured to us in a very stimulating way about hypergeometric functions and classical orthogonal polynomials. Even better, he gave us problems to solve which might be worth a PhD. He also pointed out to us that there was more than just Jacobi, Laguerre and Hermite polynomials, for instance Hahn polynomials, and that it was one of the merits of the Higher Transcendental Functions (Bateman project) that it included some newer stuff like the Hahn polynomials (see [198, §10.23]). Note that the emphasis in this section of the Bateman project is on Chebyshev's (or Tchebichef's) polynomials of a discrete variable, the special case of Hahn polynomials where we have equal weights on equidistant points. This special case is very important for applications, in particular in numerical analysis, see for instance Savitzky & Golay [468] (this paper from 1964 has now 2778 citations in Google Scholar) and Meer & Weiss [404]. Ironically, as Askey later wrote in his comments on [494], Chebyshev already published in 1875 on what we now call the Hahn polynomials of general parameters.

Of course, Askey told us during 1969–1970 also about the limit transitions Jacobi $\to$ Laguerre, Jacobi $\to$ Hermite and Laguerre $\to$ Hermite (formulas (9.8.16), (9.8.18) and (9.12.13) in this volume). During the seventies there grew a greater awareness that these three limit relations were part of a larger system of such limits, for instance also involving some discrete orthogonal polynomials like Hahn and Meixner polynomials. The idea to present these limits graphically was born at an

Oberwalfach meeting in 1977 on "Combinatorics and Special Functions" organized by George Andrews and Dominique Foata, also attended by me. In Dick Askey's words (personal communication):

"I gave a talk about many of the classical type orthogonal polynomials and it fell flat. Few there appreciated it. Later in the week Michael Hoare, a physicist then at Bedford College, talked about some very nice work [155] he had done with R.D. Cooper and Mizan Rahman. In this talk he had an overhead of the polynomials they had dealt with, starting with Hahn polynomials at the top and moving down to limiting cases with arrows illustrating the limits which they had used. The audience did not seem to care much about the probability problem, but they were very excited about the chart[1] he had shown and wanted copies. If there was that much interest in his chart, I thought that it should be extended to include all of the classical type polynomials which had been found. This was included in the Memoir [72, Appendix] Jim Wilson and I wrote. We missed one case, since we had found the symmetric continuous Hahn polynomials, but had not realized that the symmetry was not needed. That was done by Atakishiyev and Suslov [81]."

During the conference *Polynômes Orthogonaux et Applications* in Bar-le-Duc, France in 1984, Jacques Labelle presented a poster of size 89×122 cm containing what he called *Tableau d'Askey* [362]. For some years I had it hanging on the wall of my office, but it gradually faded away.

At this 1984 Bar-le-Duc conference Andrews & Askey [34] talked about the q-analogues of the polynomials in the Askey scheme, which had already been around for some seven years, starting with the work of Askey together with his PhD student Jim Wilson. This culminated into their joint memoir [72] in 1985. As Andrews & Askey wrote in [34]:

"A set of orthogonal polynomials is *classical* if it is a special case or a limiting case of the Askey-Wilson polynomials or q-Racah polynomials."

It took some time before also the q-Askey scheme was graphically displayed. As Labelle wrote in [362], one would need a 3-dimensional chart, because there are both arrows within the q-Askey scheme and from the q-Askey scheme to the Askey scheme. But if one is satisfied with just the arrows of the first type, then one can find the q-Askey scheme just before chapter 14 in the present volume.

The present book is a merger of the report *The Askey-scheme of hypergeometric orthogonal polynomials and its q-analogue* by Koekoek and Swarttouw (1994, and thoroughly revised and updated in 1998) and of a series of papers by Lesky on the classification of these polynomials. The report of Koekoek and Swarttouw had its roots in a regular seminar on orthogonal polynomials at Delft University of Technology in 1990 or so, when Koekoek and Swarttouw were PhD students there, and where I also participated. The Koekoek-Swarttouw report, in its various versions, has become very well-known, a real standard reference although it was lacking during all those years the status of a publication at a recognized publisher. At present, Google Scholar gives 686 citations for the version at arXiv (`arXiv:math/9602214v1`

[1] An extension of [155, p.285, Figure 2], called *The seven-fold way of orthogonal polynomials* and *The seven-fold way of probability distributions*

[math.CA]), 106 citations for the 1998 report version and 69 citations for the 1994 report version (all these numbers are crude lower bounds). Many citations are from applied fields. The pleasant systematics of the report, giving for each family a limited but important list of formulas, always in the same order, will have greatly contributed to the success of the report. It is used not only to look up formulas of a family of polynomials one has already in mind, but maybe even more in the "inverse way": one has a formula and one wonders if this is a formula for one of the polynomials in the (q-)Askey scheme.

Given the classification results of Bochner and Hahn which are mentioned in the Preface of this book, it is natural to look for similar results in the case of a quadratic lattice $z(s) = as^2 + bs + c$ or a q-quadratic lattice $z(s) = aq^s + bq^{-s} + c$. The task is to classify for such $z(s)$ all systems of polynomials p_n of degree n ($n = 0, 1, 2, \ldots, N$ or $n = 0, 1, 2, \ldots$) such that the functions $s \mapsto p_n(z(s))$ are eigenfunctions of a second-order difference operator in s. This was done by Grünbaum & Haine [256] under the additional assumption that the p_n are orthogonal polynomials and next by Ismail [275] without this assumption. The result was just what we already knew for the (q-) quadratic lattice cases in the (q-)Askey scheme. In an earlier elegant paper Leonard [371] had shown that a finite system of orthogonal polynomials of which the dual is again a finite system of orthogonal polynomials, must be (q-)Racah or one of its limit cases which are finite systems. Finally, the condition that $z(s)$ is a (q-)linear or (q-) quadratic lattice was dropped for finite systems in Terwilliger [499] and for infinite systems in Vinet & Zhedanov [507]. Again nothing more came out than was already in the (q-)Askey scheme. In particular in [256] and [371] the notion of *bispectrality* is important: on the one hand orthogonal polynomials $p_n(x)$ are eigenfunctions of a second-order difference operator in the n-variable with eigenvalue x, on the other hand, if they are in the (q-)Askey scheme then they are eigenfunctions of a second-order operator in the x-variable with eigenvalue depending on n.

While the classification results mentioned in the previous paragraph could be written up in relatively short papers, one might also try to structure the families in the (q-)Askey scheme in a more refined way and to list precisely for which values of the parameters there will exist a positive orthogonality measure, and to give such an orthogonality measure possibly in an explicit way. Such tasks can be quite laborious and very technical. On the one hand, as mentioned in the Preface, such work was done in the books by Nikiforov and Uvarov (Suslov being also an author for the second book), and continued subsequently by many authors, in particular belonging to the Spanish school of orthogonal polynomials. On the other hand, rather independently, this was done in a long series of papers by Lesky, which material now constitutes a large part of the present book.

An interesting aspect of Lesky's approach, also reflected in the present book, is the special consideration for finite systems of orthogonal polynomials. In particular, Romanovski introduced in 1929 two finite systems of Jacobi polynomials and one finite system of Bessel polynomials which are here included in sections 9.8 (Jacobi), 9.9 (Pseudo Jacobi) and 9.13 (Bessel). As already proposed by Lesky [384] in 1998, these families, in particular the Bessel polynomials, should be embedded in the Askey scheme, together with suitable limit arrows. This is indeed done in the Askey scheme in this book, right before chapter 9.

Why are the three finite orthogonal systems just mentioned much less known than for instance the Krawtchouk and Hahn polynomials, which are also finite systems? Are they possibly of different nature? From the point of view of Favard's theorem (see also the discussion in §3.1 of the book), whenever one has a three-term recurrence relation

$$x\,p_n(x) = p_{n+1}(x) + B_n\,p_n(x) + C_n\,p_{n-1}(x) \qquad (B_n \in \mathbb{R},\, C_n > 0)$$

for $n = 0,1,\ldots,N-1$, yielding monic polynomials p_n of degree n for $n = 0,1,\ldots,N$, then these polynomials will form a finite system of orthogonal polynomials. This finite system can be extended to an infinite system of orthogonal polynomials by extending the above recurrence relation for $n = N, N+1, \ldots$ with arbitrary choices for $B_n \in \mathbb{R}$ and $C_n > 0$. On the other hand, any infinite system of orthogonal polynomials p_n yields a finite system by only considering p_n for $n \leq N$. In this way the notion of a finite system seems quite arbitrary, but the finite systems will come up in an essential way if one requires that the p_n are eigenfunctions of some second-order operator. Still one may wonder, for a given finite system $p_0, p_1, \ldots, p_N$, if there are natural choices for p_{N+1}, $B_N \in \mathbb{R}$ and $C_N > 0$ such that

$$x\,p_N(x) = p_{N+1}(x) + B_N\,p_N(x) + C_N\,p_{N-1}(x).$$

If so, then p_{N+1} will have $N+1$ distinct real zeros interlacing the N zeros of p_N and there is a natural way of realizing the orthogonality of the finite system $p_0, p_1, \ldots, p_N$ by suitable weights on the zeros of p_{N+1}.

Let us consider this for the example of Krawtchouk polynomials given in §9.11. For N a generic complex number the normalized recurrence relation (9.11.4) is satisfied for all $n = 0,1,2,\ldots$ by

$$p_n(x) = (-N)_n\, p^n\, {}_2F_1\left(\begin{array}{c} -n, -x \\ -N \end{array}; p^{-1}\right) = \sum_{k=0}^{n} \binom{n}{k} p^{n-k}\, (-N+k)_{n-k}\, (x-k+1)_k.$$

This is continuous in $N \in \mathbb{C}$, and similarly are the coefficients in the normalized recurrence relation (9.11.4) continuous in N. If we let N a positive integer and if $n > N$ then only the terms with $N < k \leq n$ in the expression for $p_n(x)$ survive. In particular, we then find:

$$\begin{aligned} p_{N+1}(x) &= x(x-1)(x-2)\ldots(x-N), \\ p_{N+2}(x) &= x(x-1)(x-2)\ldots(x-N)\big(x-N-1+p(N+2)\big) \\ &= \big(x-N-1+p(N+2)\big)\,p_{N+1}(x), \end{aligned}$$

which is compatible with $C_{N+1} = 0$. Since $C_N = Np(1-p) > 0$, we can realize the orthogonality of the Krawtchouk polynomials $p_0, p_1, \ldots, p_N$ by weights on the zeros $\{0,1,\ldots,N\}$ of $p_{N+1}(x)$, and this gives the usual orthogonality for Krawtchouk polynomials. The Hahn and Racah polynomials and their q-analogues can be treated in a similar way.

The situation is very different with the finite system of Jacobi polynomials $P_n^{(\alpha,\beta)}$ given by (9.8.3). There $\alpha+\beta+1<0$ and $\beta>-1$. Take the integer N such that $N<-(\alpha+\beta+1)/2\le N+1$. Then the Jacobi polynomials $P_n^{(\alpha,\beta)}$ for $n=0,1,\ldots,N$ form a finite orthogonal system. Then the coefficient of $p_{n-1}(x)$ in the normalized recurrence relation (9.8.5) is positive for $n=0,1,\ldots,N-1$. If moreover $2n+\alpha+\beta+2\neq 0$ then (9.8.5) for $n=N$ has no singularities and the coefficient of $p_{N-1}(x)$ there is still positive. If also $2n+\alpha+\beta+4\neq 0$ then (9.8.5) for $n=N+1$ has no singularities and the coefficient of $p_N(x)$ there is negative. So one might realize the orthogonality of the finite system by weights on the zeros of $P_{N+1}^{(\alpha,\beta)}(x)$ (still assuming $2n+\alpha+\beta+2\neq 0$). But there is no explicit expression for these zeros and there is no evident second-order operator adapted to these zeros which has the Jacobi polynomials as eigenfunctions. Therefore the orthogonality (9.8.3) is a much better choice.

The finite orthogonal system (9.8.3) of Jacobi polynomials discussed above can also be obtained in terms of Jacobi functions $\phi_\mu^{(\alpha,\beta)}$ in the case that $\alpha+\beta+1<0$. The finitely many values of μ for which the Jacobi functions are L^2 with respect to the suitable weight function, then yield the finite system of Jacobi polynomials. See Flensted-Jensen [206, Appendix 1] for details. This example is an illustration of an important principle: If one has a finite system of orthogonal polynomials which "ends" in a somewhat unnatural way, then look for a natural completion to some infinite dimensional L^2-space, possibly using a continuous (generalized) orthogonal system like the Jacobi functions. Similarly, Koelink & Stokman [321, §7] observed finite orthogonal systems of Askey-Wilson polynomials as specializations to discrete parts of the spectrum for the Askey-Wilson function transform.

The so-called Macdonald-Koornwinder polynomials are analogues of Askey-Wilson polynomials in several variables associated with root system BC_n. Apart from q they have five parameters, one more than in the one-variable case. There are strong indications that all families and arrows in the (q-)Askey scheme have their analogues in this multi-variable case. See for instance Stokman [491].

The families in the (q-)Askey scheme have found interpretations in various settings, for instance in the representation theory of specific Lie or finite or quantum groups, in combinatorics and in probability theory. Some of the limits in the scheme could also be brought over to limit relations of the structures where the polynomials were interpreted.

In principle, the limit formulas in chapters 9 and 14 can be applied to all suitable identities there, and the resulting identities will be again present in these chapters on the suitable places. It is quite straightforward, although a little technical, to do this for the identities in the following categories: (normalized) recurrence relation, (q-) difference or differential equation, shift operators and Rodrigues-type formula. For the orthogonality relation taking the limit will not always be possible or evident. It is fun to walk downwards from one of the generating functions (14.1.13)–(14.1.15) for Askey-Wilson polynomials. An immediate question then is if it will be possible to walk upwards from all the generating functions for Jacobi polynomials listed in (9.8.11)–(9.8.15). Certainly there are formulas for Jacobi polynomials which have not yet found all their way up to the Askey-Wilson polynomials, for instance the addition formula for Jacobi polynomials, see [332].

Potentially, there are many further formulas which could be filled in for the various families in the (q-)Askey scheme. One category is formed by the structure relations and raising and lowering operators, see [351] and references given there. Related with this are the generators and relations for the *Zhedanov algebra* which one can associate with each family in the scheme, see [254]. Something which in future should also be written down for all families in the scheme are the so-called *non-symmetric* versions of the various orthogonal polynomials, to be understood in the framework of Cherednik's *double affine Hecke algebra* or some degenerate version of it. See this in Noumi & Stokman [423] for the case of Askey-Wilson polynomials.

Finally, Bessel functions, although not being polynomials, are very natural companions of Jacobi polynomials, from which they can be obtained as limit cases. Remarkably enough, several kinds of q-Bessel functions occur as limits of families in the q-Askey scheme, see for instance Koelink & Stokman [322].

Tom Koornwinder, August 2009
University of Amsterdam

Preface

In 1929 S. Bochner (see [106]) found all families of polynomials satisfying a second-order differential equation with polynomial coefficients. This led to the continuous classical orthogonal polynomials named after Jacobi, Laguerre and Hermite. The Bessel polynomials also appear in this study by S. Bochner. However, these polynomials can only be seen as continuous classical orthogonal polynomials in the case of a *finite* system (in case of positive-definite orthogonality). The continuous classical orthogonal polynomials are treated in chapter 4.

In 1949 W. Hahn (see [261]) found orthogonal polynomial solutions of second-order q-difference equations. This class of families of orthogonal polynomials is known as the Hahn class of orthogonal polynomials.

Many other families of orthogonal polynomials such as the discrete classical orthogonal polynomials have been very well known for a long time, but a classification of all of these families did not exist. A first attempt to combine both the continuous and discrete classical orthogonal polynomials was made in 1985 by R. Askey and J.A. Wilson (see [72]) by introducing the so-called Askey scheme of hypergeometric orthogonal polynomials. In [72] the continuous classical orthogonal polynomials were introduced as limiting cases of the Wilson polynomials and the discrete classical orthogonal polynomials as limiting cases of the Racah polynomials. These polynomials are treated in chapter 7.

In [72] R. Askey and J.A. Wilson also introduced q-analogues of the Wilson polynomials, which are known as the Askey-Wilson polynomials.

We also mention the books [417] by A.F. Nikiforov and V.B. Uvarov (1988) and [416] by A.F. Nikiforov, S.K. Suslov and V.B. Uvarov (1991) in this perspective.

Furthermore we refer to [256], [275] and [507] for characterizations of the Askey-Wilson polynomials.

In 1994 the first and third author (see [318]) published a preliminary version of their report on *the Askey scheme of hypergeometric orthogonal polynomials and its q-analogue*. All known q-analogues of the families of orthogonal polynomials belonging to the Askey scheme were arranged into a q-analogue of this Askey scheme. In 1998 (see [319]) a completely revised and updated version of this report appeared, containing more formulas for these families of orthogonal polynomials. However, a classification of the orthogonal polynomials was still missing.

In the meanwhile, the second author studied the classification of several kinds of both continuous and discrete orthogonal polynomials. This led to several publications on orthogonal polynomial solutions of several kinds of eigenvalue problems. In [385] four types of q-orthogonal polynomials in x and three types of q-orthogonal polynomials in q^{-x} were introduced. These seven types form a comprehensive basis for the classical q-orthogonal polynomials in x and in q^{-x}. These polynomials are treated in chapter 10 and chapter 11.

The intention of this book is to give a classification of all families of classical orthogonal polynomials and their q-analogues, the classical q-orthogonal polynomials. In order to do this we make the following observations:

- We only consider *positive-definite orthogonality* in terms of an inner product.
- We consider both *infinite* and *finite* systems of orthogonal polynomials.

The characterization of classical orthogonal polynomials can be written in terms of *eigenvalue problems* as follows: the polynomial solutions of all degrees $n=0,1,2,\ldots$ satisfy a three-term recurrence relation (with respect to the degree n) from which we can deduce necessary and sufficient conditions by use of the theorem of Favard. These conditions lead to all possible weight functions for orthogonality. It is also possible to have finite systems of orthogonal polynomials.

We consider the following cases:

1. *Second-order differential equations* (chapter 4).

The polynomial solutions lead to the continuous classical orthogonal polynomials; three infinite systems of Hermite, Laguerre and Jacobi polynomials and three finite systems of Jacobi, Bessel and pseudo Jacobi polynomials.

2. *Second-order difference equations with real coefficients* (chapter 5).

The polynomial solutions lead to the first part of the discrete classical orthogonal polynomials; two infinite systems of Charlier and Meixner polynomials and two finite systems of Krawtchouk and Hahn polynomials.

3. *Second-order difference equations with complex coefficients* (chapter 6).

The polynomial solutions lead to the second part of the discrete classical orthogonal polynomials; two infinite and finite systems of Meixner-Pollaczek and continuous Hahn polynomials.

Second-order difference equations can also be seen as three-term recurrence relations in terms of the argument x. The connection between the three-term recurrence relation with respect to the degree n and the one with respect to the argument x leads to the concept of *duality*.

4. The concept of *duality* leads to polynomial solutions in $x(x+u)$, with $x \in \mathbb{R}$ and $u \in \mathbb{R}$ a constant, of *second-order difference equations with real coefficients* (chapter 7).

The polynomial solutions lead to the third part of the discrete classical orthogonal polynomials; two infinite and finite systems of dual Hahn and Racah polynomials.

5. The concept of *duality* also leads to polynomial solutions in $z(z+u)$, with $z \in \mathbb{C}$ and $u \in \mathbb{R}$ a constant, of *second-order difference equations with complex coefficients* (chapter 8).

The polynomial solutions lead to the fourth part of the discrete classical orthogonal polynomials; two infinite and finite systems of continuous dual Hahn and Wilson polynomials.

In chapter 9 we list all families of hypergeometric orthogonal polynomials belonging to the Askey scheme. In each case we use the most common notation and we list the most important properties of the polynomials such as a representation as hypergeometric function, orthogonality relation(s), the three-term recurrence relation, the second-order differential or difference equation, the forward shift (or degree lowering) and backward shift (or degree raising) operator, a Rodrigues-type formula and some generating functions. Moreover, in each case we mention the connection between various families by given the appropriate limit relations.

6. Hahn's q-operator leads to eigenvalue problems in terms of *second-order q-difference equations* (chapter 10).

The polynomial solutions lead to the first part of the classical q-orthogonal polynomials, containing the Stieltjes-Wigert, the q-Laguerre, the little q-Jacobi, the little q-Laguerre, the q-Bessel, the big q-Jacobi, the big q-Laguerre, the Al-Salam-Carlitz and the discrete q-Hermite polynomials.

7. By changing x into q^{-x}, we obtain even more *second-order q-difference equations* (chapter 11).

The polynomial solutions lead to the second part of the classical q-orthogonal polynomials, containing the q-Meixner, the q-Krawtchouk, the quantum q-Krawtchouk, the affine q-Krawtchouk, the q-Hahn and the q-Charlier polynomials.

8. The concept of *duality* can also be applied to the case of q-orthogonal polynomials, which leads to polynomial solutions in $q^{-x}+uq^{x}$, with $x \in \mathbb{R}$ and $u \in \mathbb{R}$ a constant, of *second-order q-difference equations with real coefficients* (chapter 12).

The polynomial solutions lead to the third part of the classical q-orthogonal polynomials, containing the q-Racah, the dual q-Racah, the dual q-Hahn, the dual q-Krawtchouk and the dual q-Charlier polynomials.

9. By changing q^{x} into $\frac{z}{a}$, we also obtain polynomial solutions in $\frac{a}{z}+\frac{uz}{a}$, with $z \in \mathbb{C}$ and $u \in \mathbb{R}$ and $a \in \mathbb{C}$ constants, of *second-order q-difference equations with complex coefficients* (chapter 13).

Real polynomial solutions can only exist for $|z| = 1$. These lead to the fourth part of the classical q-orthogonal polynomials, containing the Askey-Wilson, the continuous q-Hahn, the continuous q-Jacobi, the continuous dual q-Hahn, the q-Meixner-Pollaczek, the Al-Salam-Chihara, the continuous q-Laguerre and the continuous (big) q-Hermite polynomials.

In chapter 14 we list all families of basic hypergeometric orthogonal polynomials belonging to the q-analogue of the Askey scheme. Again, in each case we use the most common notation and we list the most important properties of the polynomials such as a representation as basic hypergeometric function, orthogonality relation(s), the three-term recurrence relation, the second-order q-difference equation, the forward shift (or degree lowering) and backward shift (or degree raising) operator, a Rodrigues-type formula and some generating functions. Moreover, in each case we also indicate the limit relations between various families of q-orthogonal polynomials and the limit relations ($q \to 1$) to the classical hypergeometric orthogonal polynomials belonging to the Askey scheme.

Roelof Koekoek, Peter A. Lesky[†] and René F. Swarttouw

[†] Unfortunately, Peter Lesky died in Innsbruck (Austria) on February 12, 2008 at the age of 81.

Contents

Chapter 1
Definitions and Miscellaneous Formulas

1.1 Orthogonal Polynomials

A system of polynomials $\{p_n(x)\}_{n=0}^{\infty}$ with degree$[p_n(x)] = n$ for all $n \in \{0,1,2,\ldots\}$ is called orthogonal on an interval (a,b) with respect to the weight function $w(x) \geq 0$ if

$$\int_a^b p_m(x)p_n(x)w(x)\,dx = 0, \quad m \neq n, \quad m,n \in \{0,1,2,\ldots\}. \tag{1.1.1}$$

Here $w(x)$ is continuous or piecewise continuous or integrable, and such that

$$0 < \int_a^b x^{2n}w(x)\,dx < \infty \quad \text{for all} \quad n \in \{0,1,2,\ldots\}.$$

More generally, $w(x)\,dx$ may be replaced in this definition by a positive measure $d\alpha(x)$, where $\alpha(x)$ is a bounded nondecreasing function on $[a,b] \cap \mathbb{R}$ with an infinite number of points of increase, and such that

$$0 < \int_a^b x^{2n}\,d\alpha(x) < \infty \quad \text{for all} \quad n \in \{0,1,2,\ldots\}.$$

If this function $\alpha(x)$ is constant between its (countably many) jump points then we have the situation of positive weights w_x on a countable subset X of $\mathbb{R}$. Then the system $\{p_n(x)\}_{n=0}^{\infty}$ is orthogonal on X with respect to these weights as follows:

$$\sum_{x \in X} p_m(x)p_n(x)w_x = 0, \quad m \neq n, \quad m,n \in \{0,1,2,\ldots\}. \tag{1.1.2}$$

The case of weights w_x $(x \in X)$ on a finite set X of $N+1$ points yields orthogonality for a finite system of polynomials $\{p_n(x)\}_{n=0}^{N}$:

$$\sum_{x \in X} p_m(x)p_n(x)w_x = 0, \quad m \neq n, \quad m,n \in \{0,1,2,\ldots,N\}. \tag{1.1.3}$$

R. Koekoek et al., *Hypergeometric Orthogonal Polynomials and Their q-Analogues*, Springer Monographs in Mathematics,
DOI 10.1007/978-3-642-05014-5_1,

The orthogonality relations (1.1.1), (1.1.2) or (1.1.3) determine the polynomials $\{p_n(x)\}_{n=0}^{\infty}$ up to constant factors, which may be fixed by a suitable normalization. We set:

$$\sigma_n = \int_a^b \{p_n(x)\}^2 w(x)\,dx, \quad n = 0,1,2,\ldots, \tag{1.1.4}$$

$$\sigma_n = \sum_{x \in X} \{p_n(x)\}^2 w_x, \quad n = 0,1,2,\ldots \tag{1.1.5}$$

or

$$\sigma_n = \sum_{x \in X} \{p_n(x)\}^2 w_x, \quad n = 0,1,2,\ldots,N, \tag{1.1.6}$$

respectively and

$$p_n(x) = k_n x^n + \text{lower order terms}, \quad n = 0,1,2,\ldots. \tag{1.1.7}$$

Then we have the following normalizations:

1. $\sigma_n = 1$ for all $n = 0,1,2,\ldots$. In that case the system of polynomials is called orthonormal. If moreover $k_n > 0$, for instance, then the polynomials are uniquely determined.

2. $k_n = 1$ for all $n = 0,1,2,\ldots$. In that case the system of polynomials is called monic. Also in this case the polynomials are uniquely determined.

Polynomials $\{p_n(x)\}_{n=0}^{\infty}$ with degree$[p_n(x)] = n$ for all $n \in \{0,1,2,\ldots\}$ which are orthogonal with respect to a (piecewise) continuous or integrable weight function $w(x)$ as in (1.1.1) are called continuous orthogonal polynomials.
Polynomials $\{p_n(x)\}_{n=0}^{N}$ with degree$[p_n(x)] = n$ for all $n \in \{0,1,2,\ldots,N\}$ and possibly $N \to \infty$ which are orthogonal with respect to a countable subset X of $\mathbb{R}$ as in (1.1.2) or (1.1.3) are called discrete orthogonal polynomials.
By using the Kronecker delta, defined by

$$\delta_{mn} := \begin{cases} 0, \; m \neq n, \\ \\ 1, \; m = n, \end{cases} \quad m,n \in \{0,1,2,\ldots\}, \tag{1.1.8}$$

orthogonality relations can be written in the form

$$\int_a^b p_m(x) p_n(x) w(x)\,dx = \sigma_n\,\delta_{mn}, \quad m,n \in \{0,1,2,\ldots\} \tag{1.1.9}$$

or (with possibly $N \to \infty$)

$$\sum_{x \in X} p_m(x) p_n(x) w_x = \sigma_n\,\delta_{mn}, \quad m,n \in \{0,1,2,\ldots,N\}, \tag{1.1.10}$$

respectively.

1.2 The Gamma and Beta Function

For $\mathrm{Re}\, z > 0$ the gamma function can be defined by the gamma integral

$$\Gamma(z) := \int_0^\infty t^{z-1} \mathrm{e}^{-t}\, dt, \quad \mathrm{Re}\, z > 0. \tag{1.2.1}$$

This gamma function satisfies the well-known functional equation

$$\Gamma(z+1) = z\Gamma(z) \quad \text{with} \quad \Gamma(1) = 1, \tag{1.2.2}$$

which shows that $\Gamma(n+1) = n!$ for $n \in \{0,1,2,\ldots\}$. For non-integral values of z, the gamma function also satisfies the reflection formula

$$\Gamma(z)\Gamma(1-z) = \frac{\pi}{\sin \pi z}, \quad z \notin \mathbb{Z}. \tag{1.2.3}$$

Hence we have $\Gamma(1/2) = \sqrt{\pi}$, which implies that

$$\int_{-\infty}^\infty \mathrm{e}^{-x^2}\, dx = 2\int_0^\infty \mathrm{e}^{-x^2}\, dx = \int_0^\infty t^{-1/2}\mathrm{e}^{-t}\, dt = \Gamma(1/2) = \sqrt{\pi}. \tag{1.2.4}$$

More general we have

$$\int_{-\infty}^\infty \mathrm{e}^{-\alpha^2x^2-2\beta x}\, dx = \sqrt{\frac{\pi}{\alpha^2}}\, \mathrm{e}^{\beta^2/\alpha^2}, \quad \alpha,\beta \in \mathbb{R}, \quad \alpha \neq 0. \tag{1.2.5}$$

Further we have Legendre's duplication formula

$$\Gamma(z)\Gamma(z+1/2) = 2^{1-2z}\sqrt{\pi}\,\Gamma(2z), \quad z \in \mathbb{C}, \quad 2z \neq 0,-1,-2,\ldots \tag{1.2.6}$$

and Stirling's asymptotic formula

$$\Gamma(z) \sim \sqrt{2\pi}\, z^{z-1/2}\mathrm{e}^{-z}, \quad \mathrm{Re}\, z \to \infty. \tag{1.2.7}$$

For $z = x + iy$ with $x, y \in \mathbb{R}$ we also have

$$\Gamma(x+iy) \sim \sqrt{2\pi}\, |y|^{x-1/2}\mathrm{e}^{-|y|\pi/2}, \quad |y| \to \infty \tag{1.2.8}$$

and for the ratio of two gamma functions we have the asymptotic formula

$$\frac{\Gamma(z+a)}{\Gamma(z+b)} \sim z^{a-b}, \quad a,b \in \mathbb{C}, \quad |z| \to \infty. \tag{1.2.9}$$

For $\mathrm{Re}\, x > 0$ and $\mathrm{Re}\, y > 0$ the beta function can be defined by the integral

$$\mathrm{B}(x,y) := \int_0^1 t^{x-1}(1-t)^{y-1}\, dt, \quad \mathrm{Re}\, x > 0, \quad \mathrm{Re}\, y > 0. \tag{1.2.10}$$

The connection between the beta function and the gamma function is given by the relation

$$\mathrm{B}(x,y)=\frac{\Gamma(x)\Gamma(y)}{\Gamma(x+y)},\quad \mathrm{Re}\,x>0,\quad \mathrm{Re}\,y>0. \tag{1.2.11}$$

There is another beta integral due to Cauchy, which can be written as

$$\frac{1}{2\pi}\int_{-\infty}^{\infty}\frac{dt}{(r+it)^{\rho}(s-it)^{\sigma}}=\frac{(r+s)^{1-\rho-\sigma}\Gamma(\rho+\sigma-1)}{\Gamma(\rho)\Gamma(\sigma)} \tag{1.2.12}$$

for $\mathrm{Re}\,r>0$, $\mathrm{Re}\,s>0$ and $\mathrm{Re}(\rho+\sigma)>1$.

1.3 The Shifted Factorial and Binomial Coefficients

The shifted factorial – or Pochhammer symbol – is defined by

$$(a)_0:=1 \quad\text{and}\quad (a)_k:=\prod_{i=1}^{k}(a+i-1),\quad k=1,2,3,\ldots. \tag{1.3.1}$$

This can be seen as a generalization of the factorial since

$$(1)_n=n!,\quad n=0,1,2,\ldots.$$

The binomial coefficient can be defined by

$$\binom{\alpha}{\beta}:=\frac{\Gamma(\alpha+1)}{\Gamma(\beta+1)\Gamma(\alpha-\beta+1)}. \tag{1.3.2}$$

For integer values of the parameter β we have

$$\binom{\alpha}{k}:=\frac{(-\alpha)_k}{k!}(-1)^k,\quad k=0,1,2,\ldots$$

and when the parameter α is an integer too, we have

$$\binom{n}{k}:=\frac{n!}{k!\,(n-k)!},\quad k=0,1,2,\ldots,n,\quad n=0,1,2,\ldots.$$

The latter formula can be used to show that

$$\binom{2n}{n}=\frac{\left(\frac{1}{2}\right)_n}{n!}4^n,\quad n=0,1,2,\ldots.$$

1.4 Hypergeometric Functions

The hypergeometric function ${}_rF_s$ is defined by the series

$$ {}_rF_s\left(\begin{matrix} a_1,\ldots,a_r \\ b_1,\ldots,b_s \end{matrix};z\right) := \sum_{k=0}^{\infty} \frac{(a_1,\ldots,a_r)_k}{(b_1,\ldots,b_s)_k} \frac{z^k}{k!}, \tag{1.4.1} $$

where

$$ (a_1,\ldots,a_r)_k := (a_1)_k \cdots (a_r)_k. $$

Of course, the parameters must be such that the denominator factors in the terms of the series are never zero. When one of the numerator parameters a_i equals $-n$, where n is a nonnegative integer, this hypergeometric function is a polynomial in z. Otherwise the radius of convergence ρ of the hypergeometric series is given by

$$ \rho = \begin{cases} \infty & \text{if} \quad r < s+1 \\ 1 & \text{if} \quad r = s+1 \\ 0 & \text{if} \quad r > s+1. \end{cases} $$

A hypergeometric series of the form (1.4.1) is called balanced or Saalschützian if $r = s+1$, $z = 1$ and $a_1 + a_2 + \ldots + a_{s+1} + 1 = b_1 + b_2 + \ldots + b_s$.

Many limit relations between hypergeometric orthogonal polynomials are based on the observations that

$$ {}_rF_s\left(\begin{matrix} a_1,\ldots,a_{r-1},\mu \\ b_1,\ldots,b_{s-1},\mu \end{matrix};z\right) = {}_{r-1}F_{s-1}\left(\begin{matrix} a_1,\ldots,a_{r-1} \\ b_1,\ldots,b_{s-1} \end{matrix};z\right), \tag{1.4.2} $$

$$ \lim_{\lambda\to\infty} {}_rF_s\left(\begin{matrix} a_1,\ldots,a_{r-1},\lambda a_r \\ b_1,\ldots,b_s \end{matrix};\frac{z}{\lambda}\right) = {}_{r-1}F_s\left(\begin{matrix} a_1,\ldots,a_{r-1} \\ b_1,\ldots,b_s \end{matrix};a_r z\right), \tag{1.4.3} $$

$$ \lim_{\lambda\to\infty} {}_rF_s\left(\begin{matrix} a_1,\ldots,a_r \\ b_1,\ldots,b_{s-1},\lambda b_s \end{matrix};\lambda z\right) = {}_rF_{s-1}\left(\begin{matrix} a_1,\ldots,a_r \\ b_1,\ldots,b_{s-1} \end{matrix};\frac{z}{b_s}\right) \tag{1.4.4} $$

and

$$ \lim_{\lambda\to\infty} {}_rF_s\left(\begin{matrix} a_1,\ldots,a_{r-1},\lambda a_r \\ b_1,\ldots,b_{s-1},\lambda b_s \end{matrix};z\right) = {}_{r-1}F_{s-1}\left(\begin{matrix} a_1,\ldots,a_{r-1} \\ b_1,\ldots,b_{s-1} \end{matrix};\frac{a_r z}{b_s}\right). \tag{1.4.5} $$

Mostly, the left-hand side of (1.4.2) occurs as a limit case where some numerator parameter and some denominator parameter tend to the same value.

All families of discrete orthogonal polynomials $\{P_n(x)\}_{n=0}^N$ are defined for $n = 0,1,2,\ldots,N$, where N is a positive integer. In these cases something like (1.4.2) occurs in the hypergeometric representation when $n = N$. In these cases we have to be aware of the fact that we still have a polynomial (in that case of degree N). For instance, if we take $n = N$ in the hypergeometric representation (9.5.1) of the Hahn polynomials, we have

$$Q_N(x;\alpha,\beta,N) = \sum_{k=0}^{N} \frac{(N+\alpha+\beta+1)_k(-x)_k}{(\alpha+1)_k k!}.$$

So these cases must be understood by continuity.

In cases of discrete orthogonal polynomials, we need a special notation for some of the generating functions. We define

$$[f(t)]_N := \sum_{k=0}^{N} \frac{f^{(k)}(0)}{k!} t^k,$$

for every function f for which $f^{(k)}(0)$, $k = 0,1,2,\ldots,N$ exists. As an example of the use of this Nth partial sum of a power series in t we remark that the generating function (9.11.12) for the Krawtchouk polynomials must be understood as follows: the Nth partial sum of

$$\mathrm{e}^t \, {}_1F_1\left(\begin{matrix} -x \\ -N \end{matrix} ; -\frac{t}{p}\right) = \sum_{k=0}^{\infty} \frac{t^k}{k!} \sum_{m=0}^{x} \frac{(-x)_m}{(-N)_m m!} \left(-\frac{t}{p}\right)^m$$

equals

$$\sum_{n=0}^{N} \frac{K_n(x;p,N)}{n!} t^n$$

for $x = 0,1,2,\ldots,N$.

The classical exponential function e^z and the trigonometric functions $\cos z$ and $\sin z$ can be expressed in terms of hypergeometric functions as

$$\mathrm{e}^z = {}_0F_0\left(\begin{matrix} - \\ - \end{matrix} ; z\right), \tag{1.4.6}$$

$$\cos z = {}_0F_1\left(\begin{matrix} - \\ \frac{1}{2} \end{matrix} ; -\frac{z^2}{4}\right) \tag{1.4.7}$$

and

$$\sin z = z \, {}_0F_1\left(\begin{matrix} - \\ \frac{3}{2} \end{matrix} ; -\frac{z^2}{4}\right). \tag{1.4.8}$$

Further we have the well-known Bessel function of the first kind $J_\nu(z)$, which can be defined by

$$J_\nu(z) := \frac{\left(\frac{1}{2}z\right)^\nu}{\Gamma(\nu+1)} \, {}_0F_1\left(\begin{matrix} - \\ \nu+1 \end{matrix} ; -\frac{z^2}{4}\right). \tag{1.4.9}$$

1.5 The Binomial Theorem and Other Summation Formulas

One of the most important summation formulas for hypergeometric series is given by the binomial theorem:

$$ {}_1F_0\left(\begin{matrix} a \\ - \end{matrix} ; z\right) = \sum_{n=0}^{\infty} \frac{(a)_n}{n!} z^n = (1-z)^{-a}, \quad |z| < 1, \tag{1.5.1}$$

which is a generalization of Newton's binomium

$$ {}_1F_0\left(\begin{matrix} -n \\ - \end{matrix} ; z\right) = \sum_{k=0}^{n} \frac{(-n)_k}{k!} z^k = \sum_{k=0}^{n} \binom{n}{k} (-z)^k = (1-z)^n, \quad n = 0,1,2,\ldots. \tag{1.5.2}$$

We also have Gauss's summation formula

$$ {}_2F_1\left(\begin{matrix} a,b \\ c \end{matrix} ; 1\right) = \frac{\Gamma(c)\Gamma(c-a-b)}{\Gamma(c-a)\Gamma(c-b)}, \quad \mathrm{Re}(c-a-b) > 0 \tag{1.5.3}$$

and the Vandermonde or Chu-Vandermonde summation formula

$$ {}_2F_1\left(\begin{matrix} -n,b \\ c \end{matrix} ; 1\right) = \frac{(c-b)_n}{(c)_n}, \quad n = 0,1,2,\ldots. \tag{1.5.4}$$

On the next level we have the summation formula

$$ {}_3F_2\left(\begin{matrix} -n,a,b \\ c, 1+a+b-c-n \end{matrix} ; 1\right) = \frac{(c-a)_n(c-b)_n}{(c)_n(c-a-b)_n}, \quad n = 0,1,2,\ldots, \tag{1.5.5}$$

which is called the Saalschütz or Pfaff-Saalschütz summation formula.

For a very-well-poised ${}_5F_4$ we have the summation formula

$$ {}_5F_4\left(\begin{matrix} 1+a/2,a,b,c,d \\ a/2, 1+a-b, 1+a-c, 1+a-d \end{matrix} ; 1\right) = \frac{\Gamma(1+a-b)\Gamma(1+a-c)\Gamma(1+a-d)\Gamma(1+a-b-c-d)}{\Gamma(1+a)\Gamma(1+a-b-c)\Gamma(1+a-b-d)\Gamma(1+a-c-d)}. \tag{1.5.6}$$

The limit case $d \to -\infty$ leads to

$$ {}_4F_3\left(\begin{matrix} 1+a/2,a,b,c \\ a/2, 1+a-b, 1+a-c \end{matrix} ; -1\right) = \frac{\Gamma(1+a-b)\Gamma(1+a-c)}{\Gamma(1+a)\Gamma(1+a-b-c)}. \tag{1.5.7}$$

Finally, we mention Dougall's bilateral sum

$$ \sum_{n=-\infty}^{\infty} \frac{\Gamma(n+a)\Gamma(n+b)}{\Gamma(n+c)\Gamma(n+d)} = \frac{\Gamma(a)\Gamma(1-a)\Gamma(b)\Gamma(1-b)\Gamma(c+d-a-b-1)}{\Gamma(c-a)\Gamma(d-a)\Gamma(c-b)\Gamma(d-b)}, \tag{1.5.8}$$

which holds for $\mathrm{Re}(a+b)+1 < \mathrm{Re}(c+d)$.

By using

$$\Gamma(n+a)\Gamma(n+b) = \frac{\Gamma(a)\Gamma(1-a)\Gamma(b)\Gamma(1-b)}{\Gamma(1-a-n)\Gamma(1-b-n)}, \quad n \in \mathbb{Z},$$

Dougall's bilateral sum can also be written in the form

$$\sum_{n=-\infty}^{\infty} \frac{1}{\Gamma(n+c)\Gamma(n+d)\Gamma(1-a-n)\Gamma(1-b-n)} = \frac{\Gamma(c+d-a-b-1)}{\Gamma(c-a)\Gamma(d-a)\Gamma(c-b)\Gamma(d-b)} \tag{1.5.9}$$

for $\mathrm{Re}(a+b)+1 < \mathrm{Re}(c+d)$.

1.6 Some Integrals

For the ${}_2F_1$ hypergeometric function we have Euler's integral representation

$${}_2F_1\left(\begin{matrix} a,b \\ c \end{matrix}; z\right) = \frac{\Gamma(c)}{\Gamma(b)\Gamma(c-b)} \int_0^1 t^{b-1}(1-t)^{c-b-1}(1-zt)^{-a}\, dt, \tag{1.6.1}$$

where $\mathrm{Re}\, c > \mathrm{Re}\, b > 0$, which holds for $z \in \mathbb{C} \setminus (1,\infty)$. Here it is understood that $\arg t = \arg(1-t) = 0$ and that $(1-zt)^{-a}$ has its principal value.

We also have Barnes' integral representation

$$\frac{\Gamma(a)\Gamma(b)}{\Gamma(c)}\, {}_2F_1\left(\begin{matrix} a,b \\ c \end{matrix}; z\right) = \frac{1}{2\pi i} \int_{-i\infty}^{i\infty} \frac{\Gamma(a+s)\Gamma(b+s)\Gamma(-s)}{\Gamma(c+s)} (-z)^s\, ds \tag{1.6.2}$$

for $|z| < 1$ with $\arg(-z) < \pi$, where the path of integration is deformed if necessary, to separate the decreasing poles $s = -a-n$ and $s = -b-n$ from the increasing poles $s = n$ for $n \in \{0,1,2,\ldots\}$. Such a path always exists if $a,b \notin \{\ldots,-3,-2,-1\}$.

Secondly, we have the Mellin-Barnes integral or Barnes' first lemma

$$\frac{1}{2\pi i} \int_{-i\infty}^{i\infty} \Gamma(a+s)\Gamma(b+s)\Gamma(c-s)\Gamma(d-s)\, ds = \frac{\Gamma(a+c)\Gamma(a+d)\Gamma(b+c)\Gamma(b+d)}{\Gamma(a+b+c+d)} \tag{1.6.3}$$

for $\mathrm{Re}(a+b+c+d) < 1$, where the contour must also be taken in a such a way that the increasing poles and the decreasing poles remain separate. By using analytic continuation, one can avoid the condition $\mathrm{Re}(a+b+c+d) < 1$. The Barnes integral (1.6.3) can be seen as a continuous analogue of Gauss' summation formula (1.5.3).

We also have a continuous analogue of the Pfaff-Saalschütz summation formula (1.5.5) given by the integral, which is also called Barnes' second lemma,

$$\frac{1}{2\pi i}\int_{-i\infty}^{i\infty}\frac{\Gamma(a+s)\Gamma(b+s)\Gamma(c+s)\Gamma(1-d-s)\Gamma(-s)}{\Gamma(e+s)}ds$$
$$=\frac{\Gamma(a)\Gamma(b)\Gamma(c)\Gamma(1+a-d)\Gamma(1+b-d)\Gamma(1+c-d)}{\Gamma(e-a)\Gamma(e-b)\Gamma(e-c)} \tag{1.6.4}$$

with $d+e=a+b+c+1$, where the contour must be taken in a such a way that the increasing poles and the decreasing poles stay separated.

A continuous analogue of the summation formula (1.5.6) for a very-well-poised ${}_5F_4$ is given by Bailey's integral (see also [90])

$$\frac{1}{2\pi i}\int_{-i\infty}^{i\infty}\frac{\Gamma(1+a/2+s)\Gamma(a+s)\Gamma(b+s)\Gamma(c+s)\Gamma(d+s)\Gamma(b-a-s)\Gamma(-s)}{\Gamma(a/2+s)\Gamma(1+a-c+s)\Gamma(1+a-d+s)}ds$$
$$=\frac{\Gamma(b)\Gamma(c)\Gamma(d)\Gamma(b+c-a)\Gamma(b+d-a)}{2\Gamma(1+a-c-d)\Gamma(b+c+d-a)} \tag{1.6.5}$$

which can also be written in a more symmetric form

$$\frac{1}{2\pi i}\int_{-i\infty}^{i\infty}\frac{\Gamma(a+s)\Gamma(b+s)\Gamma(c+s)\Gamma(d+s)}{\Gamma(a+2s)}$$
$$\times\frac{\Gamma(-s)\Gamma(b-a-s)\Gamma(c-a-s)\Gamma(d-a-s)}{\Gamma(-a-2s)}ds$$
$$=\frac{2\Gamma(b)\Gamma(c)\Gamma(d)\Gamma(b+c-a)\Gamma(b+d-a)\Gamma(c+d-a)}{\Gamma(b+c+d-a)} \tag{1.6.6}$$

and also in the following form due to Wilson (see also [512])

$$\frac{1}{2\pi i}\int_{-i\infty}^{i\infty}\frac{\Gamma(a+s)\Gamma(b+s)\Gamma(c+s)\Gamma(d+s)\Gamma(a-s)\Gamma(b-s)\Gamma(c-s)\Gamma(d-s)}{\Gamma(2s)\Gamma(-2s)}ds$$
$$=\frac{2\Gamma(a+b)\Gamma(a+c)\Gamma(a+d)\Gamma(b+c)\Gamma(b+d)\Gamma(c+d)}{\Gamma(a+b+c+d)}, \tag{1.6.7}$$

where the contours must be taken in a such a way that the increasing poles and the decreasing poles remain separate.

We also mention the integral

$$\frac{1}{2\pi i}\int_{-i\infty}^{i\infty}\Gamma(a+s)\Gamma(b-s)z^{-s}\,ds=\Gamma(a+b)\frac{z^a}{(1+z)^{a+b}}. \tag{1.6.8}$$

Again the contour must be taken in such a way that the increasing poles of $\Gamma(b-s)$ and the decreasing poles of $\Gamma(a+s)$ remain separate.

The integrals (1.6.3) through (1.6.7) are all special cases of a more general formula (id est formula (4.5.1.2) in [471]), which has more interesting and useful special cases such as

$$\frac{1}{2\pi i}\int_{-i\infty}^{i\infty}\frac{\Gamma(a+s)\Gamma(b-s)}{\Gamma(c+s)\Gamma(d-s)}ds=\frac{\Gamma(a+b)\Gamma(c+d-a-b-1)}{\Gamma(c+d-1)\Gamma(c-a)\Gamma(d-b)}, \tag{1.6.9}$$

where the contour must be taken in a such a way that the increasing poles and the decreasing poles remain separate and

$$\frac{1}{2\pi i}\int_{-i\infty}^{i\infty}\frac{\Gamma(a+s)\Gamma(b+s)\Gamma(c+s)\Gamma(-s)\Gamma(b-a-s)\Gamma(c-a-s)}{\Gamma(a+2s)\Gamma(-a-2s)}ds$$
$$=\frac{1}{2}\Gamma(b)\Gamma(c)\Gamma(b+c-a), \tag{1.6.10}$$

where the contour must be taken in a such a way that the increasing poles and the decreasing poles remain separate.

If a,b,c,d are positive or $b=\overline{a}$ and/or $d=\overline{c}$ and the real parts are positive, Wilson's integral (1.6.7) can be written in the form

$$\frac{1}{2\pi}\int_0^{\infty}\left|\frac{\Gamma(a+ix)\Gamma(b+ix)\Gamma(c+ix)\Gamma(d+ix)}{\Gamma(2ix)}\right|^2 dx$$
$$=\frac{\Gamma(a+b)\Gamma(a+c)\Gamma(a+d)\Gamma(b+c)\Gamma(b+d)\Gamma(c+d)}{\Gamma(a+b+c+d)}. \tag{1.6.11}$$

The limit case $d\to\infty$ leads to

$$\frac{1}{2\pi}\int_0^{\infty}\left|\frac{\Gamma(a+ix)\Gamma(b+ix)\Gamma(c+ix)}{\Gamma(2ix)}\right|^2 dx=\Gamma(a+b)\Gamma(a+c)\Gamma(b+c). \tag{1.6.12}$$

1.7 Transformation Formulas

In this section we list a number of transformation formulas which can be used to transform hypergeometric representations and other formulas into equivalent but different forms.

First of all we have Euler's transformation formula:

$${}_2F_1\left(\begin{matrix}a,b\\c\end{matrix};z\right)=(1-z)^{c-a-b}\,{}_2F_1\left(\begin{matrix}c-a,c-b\\c\end{matrix};z\right). \tag{1.7.1}$$

Another transformation formula for the ${}_2F_1$ series, which is also due to Euler, is

$${}_2F_1\left(\begin{matrix}a,b\\c\end{matrix};z\right)=(1-z)^{-a}\,{}_2F_1\left(\begin{matrix}a,c-b\\c\end{matrix};\frac{z}{z-1}\right). \tag{1.7.2}$$

This transformation formula is also known as the Pfaff or Pfaff-Kummer transformation formula.

As a limit case of the Pfaff-Kummer transformation formula we have Kummer's transformation formula for the confluent hypergeometric series:

$${}_1F_1\left(\begin{matrix}a\\c\end{matrix};z\right)=\mathrm{e}^z\,{}_1F_1\left(\begin{matrix}c-a\\c\end{matrix};-z\right). \tag{1.7.3}$$

If we reverse the order of summation in a terminating ${}_1F_1$ series, we obtain a ${}_2F_0$ series; in fact we have

$$ {}_1F_1\left(\begin{matrix}-n\\ a\end{matrix};x\right)=\frac{(-x)^n}{(a)_n}\,{}_2F_0\left(\begin{matrix}-n,-a-n+1\\ -\end{matrix};-\frac{1}{x}\right),\quad n=0,1,2,\ldots. \tag{1.7.4} $$

If we apply this technique to a terminating ${}_2F_1$ series, we find

$$ {}_2F_1\left(\begin{matrix}-n,b\\ c\end{matrix};x\right)=\frac{(b)_n}{(c)_n}(-x)^n\,{}_2F_1\left(\begin{matrix}-n,-c-n+1\\ -b-n+1\end{matrix};\frac{1}{x}\right),\quad n=0,1,2,\ldots. \tag{1.7.5} $$

On the next level we have Whipple's transformation formula for a terminating balanced ${}_4F_3$ series:

$$ \begin{aligned} &{}_4F_3\left(\begin{matrix}-n,a,b,c\\ d,e,f\end{matrix};1\right)\\ &=\frac{(e-a)_n(f-a)_n}{(e)_n(f)_n}\,{}_4F_3\left(\begin{matrix}-n,a,d-b,d-c\\ d,a-e-n+1,a-f-n+1\end{matrix};1\right) \end{aligned} \tag{1.7.6} $$

provided that $a+b+c+1=d+e+f+n$.

1.8 The q-Shifted Factorial

The theory of q-analogues or q-extensions of classical formulas and functions is based on the observation that

$$ \lim_{q\to 1}\frac{1-q^{\alpha}}{1-q}=\alpha. $$

Therefore the number $(1-q^{\alpha})/(1-q)$ is sometimes called the *basic number* $[\alpha]$. For $q\neq 0$ and $q\neq 1$ we define

$$ [\alpha]:=\frac{1-q^{\alpha}}{1-q}, \tag{1.8.1} $$

which implies that

$$ [0]=0,\quad [n]=\frac{1-q^n}{1-q}=\sum_{k=0}^{n-1}q^k,\quad n=1,2,3,\ldots. \tag{1.8.2} $$

Now we can give a q-analogue of the Pochhammer symbol $(a)_k$ defined by (1.3.1):

$$ (a;q)_0:=1\quad\text{and}\quad (a;q)_k:=\prod_{i=1}^{k}(1-aq^{i-1}),\quad k=1,2,3,\ldots. \tag{1.8.3} $$

It is clear that

$$\lim_{q\to 1}\frac{(q^{\alpha};q)_k}{(1-q)^k}=(\alpha)_k. \tag{1.8.4}$$

The symbols $(a;q)_k$ are called q-shifted factorials. For negative subscripts we define

$$(a;q)_{-k}:=\frac{1}{\prod_{i=1}^{k}(1-aq^{-i})},\quad a\neq q,q^2,q^3,\ldots,q^k,\quad k=1,2,3,\ldots. \tag{1.8.5}$$

Now we have

$$(a;q)_{-n}=\frac{1}{(aq^{-n};q)_n}=\frac{(-qa^{-1})^n}{(qa^{-1};q)_n}q^{\binom{n}{2}},\quad n=0,1,2,\ldots. \tag{1.8.6}$$

If we replace q by q^{-1}, we obtain

$$(a;q^{-1})_n=(a^{-1};q)_n(-a)^n q^{-\binom{n}{2}},\quad a\neq 0. \tag{1.8.7}$$

We can also define

$$(a;q)_\infty=\prod_{k=0}^{\infty}(1-aq^k),\quad 0<|q|<1.$$

This implies that

$$(a;q)_n=\frac{(a;q)_\infty}{(aq^n;q)_\infty},\quad 0<|q|<1, \tag{1.8.8}$$

and, for any complex number λ,

$$(a;q)_\lambda=\frac{(a;q)_\infty}{(aq^\lambda;q)_\infty},\quad 0<|q|<1, \tag{1.8.9}$$

where the principal value of q^λ is taken.

Finally, we list a number of transformation formulas for the q-shifted factorials, where k and n are nonnegative integers:

$$(a;q)_{n+k}=(a;q)_n(aq^n;q)_k, \tag{1.8.10}$$

$$\frac{(aq^n;q)_k}{(aq^k;q)_n}=\frac{(a;q)_k}{(a;q)_n}, \tag{1.8.11}$$

$$(aq^k;q)_{n-k}=\frac{(a;q)_n}{(a;q)_k},\quad k=0,1,2,\ldots,n, \tag{1.8.12}$$

$$(a;q)_n=(a^{-1}q^{1-n};q)_n(-a)^n q^{\binom{n}{2}},\quad a\neq 0, \tag{1.8.13}$$

$$(aq^{-n};q)_n=(a^{-1}q;q)_n(-a)^n q^{-n-\binom{n}{2}},\quad a\neq 0, \tag{1.8.14}$$

$$\frac{(aq^{-n};q)_n}{(bq^{-n};q)_n} = \frac{(a^{-1}q;q)_n}{(b^{-1}q;q)_n}\left(\frac{a}{b}\right)^n, \quad a \neq 0, \quad b \neq 0, \tag{1.8.15}$$

$$(a;q)_{n-k} = \frac{(a;q)_n}{(a^{-1}q^{1-n};q)_k}\left(-\frac{q}{a}\right)^k q^{\binom{k}{2}-nk}, \quad a \neq 0, \quad k = 0,1,2,\ldots,n, \tag{1.8.16}$$

$$\frac{(a;q)_{n-k}}{(b;q)_{n-k}} = \frac{(a;q)_n}{(b;q)_n}\frac{(b^{-1}q^{1-n};q)_k}{(a^{-1}q^{1-n};q)_k}\left(\frac{b}{a}\right)^k, \\ a \neq 0, \quad b \neq 0, \quad k = 0,1,2,\ldots,n, \tag{1.8.17}$$

$$(q^{-n};q)_k = \frac{(q;q)_n}{(q;q)_{n-k}}(-1)^k q^{\binom{k}{2}-nk}, \quad k = 0,1,2,\ldots,n, \tag{1.8.18}$$

$$(aq^{-n};q)_k = \frac{(a^{-1}q;q)_n}{(a^{-1}q^{1-k};q)_n}(a;q)_k q^{-nk}, \quad a \neq 0, \tag{1.8.19}$$

$$(aq^{-n};q)_{n-k} = \frac{(a^{-1}q;q)_n}{(a^{-1}q;q)_k}\left(-\frac{a}{q}\right)^{n-k} q^{\binom{k}{2}-\binom{n}{2}}, \\ a \neq 0, \quad k = 0,1,2,\ldots,n, \tag{1.8.20}$$

$$(a;q)_{2n} = (a;q^2)_n(aq;q^2)_n, \tag{1.8.21}$$

$$(a^2;q^2)_n = (a;q)_n(-a;q)_n, \tag{1.8.22}$$

$$(a;q)_\infty = (a;q^2)_\infty(aq;q^2)_\infty, \quad 0 < |q| < 1, \tag{1.8.23}$$

$$(a^2;q^2)_\infty = (a;q)_\infty(-a;q)_\infty, \quad 0 < |q| < 1. \tag{1.8.24}$$

We remark that by using (1.8.22) we have

$$\frac{1-a^2q^{2n}}{1-a^2} = \frac{(a^2q^2;q^2)_n}{(a^2;q^2)_n} = \frac{(aq;q)_n(-aq;q)_n}{(a;q)_n(-a;q)_n}. \tag{1.8.25}$$

1.9 The q-Gamma Function and q-Binomial Coefficients

The q-gamma function is defined by

$$\Gamma_q(x) := \frac{(q;q)_\infty}{(q^x;q)_\infty}(1-q)^{1-x}, \quad 0 < q < 1. \tag{1.9.1}$$

This is a q-analogue of the gamma function given by (1.2.1). In fact we have

$$\lim_{q \to 1} \Gamma_q(x) = \Gamma(x).$$

Note that the q-gamma function satisfies the functional equation

$$\Gamma_q(z+1)=\frac{1-q^z}{1-q}\Gamma_q(z) \quad \text{with} \quad \Gamma_q(1)=1, \tag{1.9.2}$$

which is a q-extension of the functional equation (1.2.2) for the ordinary gamma function.

If we take the principal values of q^x and $(1-q)^{1-x}$ definition (1.9.1) holds for $0<|q|<1$. For $q>1$ the q-gamma function can be defined by

$$\Gamma_q(x)=\frac{(q^{-1};q^{-1})_\infty}{(q^{-x};q^{-1})_\infty}q^{\binom{x}{2}}(q-1)^{1-x}, \quad q>1. \tag{1.9.3}$$

The q-binomial coefficient is defined by

$$\begin{bmatrix} n \\ k \end{bmatrix}_q := \frac{(q;q)_n}{(q;q)_k(q;q)_{n-k}} = \begin{bmatrix} n \\ n-k \end{bmatrix}_q, \quad k=0,1,2,\ldots,n, \tag{1.9.4}$$

where n denotes a nonnegative integer.

This definition can be generalized in the following way. For arbitrary complex α we have

$$\begin{bmatrix} \alpha \\ k \end{bmatrix}_q := \frac{(q^{-\alpha};q)_k}{(q;q)_k}(-1)^k q^{k\alpha-\binom{k}{2}}. \tag{1.9.5}$$

Or more general, for all complex α and β and $0<|q|<1$ we have

$$\begin{bmatrix} \alpha \\ \beta \end{bmatrix}_q := \frac{\Gamma_q(\alpha+1)}{\Gamma_q(\beta+1)\Gamma_q(\alpha-\beta+1)} = \frac{(q^{\beta+1};q)_\infty(q^{\alpha-\beta+1};q)_\infty}{(q;q)_\infty(q^{\alpha+1};q)_\infty}. \tag{1.9.6}$$

For instance this implies that

$$\frac{(q^{\alpha+1};q)_n}{(q;q)_n} = \begin{bmatrix} n+\alpha \\ n \end{bmatrix}_q.$$

Note that

$$\lim_{q\to 1}\begin{bmatrix} \alpha \\ \beta \end{bmatrix}_q = \frac{\Gamma(\alpha+1)}{\Gamma(\beta+1)\Gamma(\alpha-\beta+1)} = \binom{\alpha}{\beta}.$$

Finally, we remark that

$$\frac{1}{(q;q)_n} = \sum_{k=0}^{n}\frac{q^k}{(q;q)_k}, \quad n=0,1,2,\ldots, \tag{1.9.7}$$

which can easily be shown by induction, and that

$$(a;q)_n = \sum_{k=0}^{n}\begin{bmatrix} n \\ k \end{bmatrix}_q q^{\binom{k}{2}}(-a)^k, \quad n=0,1,2,\ldots, \tag{1.9.8}$$

which is a special case of (1.11.2).

1.10 Basic Hypergeometric Functions

The basic hypergeometric or q-hypergeometric function ${}_r\phi_s$ is defined by the series

$$\begin{aligned}&{}_r\phi_s\left(\begin{array}{c}a_1,\ldots,a_r\\ b_1,\ldots,b_s\end{array};q,z\right)\\ &:=\sum_{k=0}^{\infty}\frac{(a_1,\ldots,a_r;q)_k}{(b_1,\ldots,b_s;q)_k}(-1)^{(1+s-r)k}q^{(1+s-r)\binom{k}{2}}\frac{z^k}{(q;q)_k},\end{aligned}\qquad(1.10.1)$$

where

$$(a_1,\ldots,a_r;q)_k:=(a_1;q)_k\cdots(a_r;q)_k.$$

Again we assume that the parameters are such that the denominator factors in the terms of the series are never zero. If one of the numerator parameters a_i equals q^{-n}, where n is a nonnegative integer, this basic hypergeometric function is a polynomial in z. Otherwise the radius of convergence ρ of the basic hypergeometric series is given by

$$\rho=\begin{cases}\infty & \text{if}\quad r<s+1\\ 1 & \text{if}\quad r=s+1\\ 0 & \text{if}\quad r>s+1.\end{cases}$$

The special case $r=s+1$ reads

$${}_{s+1}\phi_s\left(\begin{array}{c}a_1,\ldots,a_{s+1}\\ b_1,\ldots,b_s\end{array};q,z\right)=\sum_{k=0}^{\infty}\frac{(a_1,\ldots,a_{s+1};q)_k}{(b_1,\ldots,b_s;q)_k}\frac{z^k}{(q;q)_k}.$$

This basic hypergeometric series was first introduced by Heine in 1846; therefore it is sometimes called Heine's series. A basic hypergeometric series of this form is called balanced or Saalschützian if $z=q$ and $a_1a_2\cdots a_{s+1}q=b_1b_2\cdots b_s$.

The q-hypergeometric function is a q-analogue of the hypergeometric function defined by (1.4.1) since

$$\lim_{q\to 1}{}_r\phi_s\left(\begin{array}{c}q^{a_1},\ldots,q^{a_r}\\ q^{b_1},\ldots,q^{b_s}\end{array};q,(q-1)^{1+s-r}z\right)={}_rF_s\left(\begin{array}{c}a_1,\ldots,a_r\\ b_1,\ldots,b_s\end{array};z\right).$$

This limit will be used frequently in chapter 14. In all cases the hypergeometric series involved is in fact a polynomial so that convergence is guaranteed.

In the sequel of this paragraph we also assume that each (basic) hypergeometric function is in fact a polynomial. We remark that

$$\lim_{a_r\to\infty}{}_r\phi_s\left(\begin{array}{c}a_1,\ldots,a_r\\ b_1,\ldots,b_s\end{array};q,\frac{z}{a_r}\right)={}_{r-1}\phi_s\left(\begin{array}{c}a_1,\ldots,a_{r-1}\\ b_1,\ldots,b_s\end{array};q,z\right).$$

In fact this is the reason for the factors $(-1)^{(1+s-r)k}q^{(1+s-r)\binom{k}{2}}$ in the definition (1.10.1) of the basic hypergeometric function.

Many limit relations between basic hypergeometric orthogonal polynomials are based on the observations that

$$
{}_r\phi_s\left(\begin{array}{c} a_1,\ldots,a_{r-1},\mu \\ b_1,\ldots,b_{s-1},\mu \end{array} ; q, z\right) = {}_{r-1}\phi_{s-1}\left(\begin{array}{c} a_1,\ldots,a_{r-1} \\ b_1,\ldots,b_{s-1} \end{array} ; q, z\right), \tag{1.10.2}
$$

$$
\lim_{\lambda\to\infty} {}_r\phi_s\left(\begin{array}{c} a_1,\ldots,a_{r-1},\lambda a_r \\ b_1,\ldots,b_s \end{array} ; q, \frac{z}{\lambda}\right) = {}_{r-1}\phi_s\left(\begin{array}{c} a_1,\ldots,a_{r-1} \\ b_1,\ldots,b_s \end{array} ; q, a_r z\right), \tag{1.10.3}
$$

$$
\lim_{\lambda\to\infty} {}_r\phi_s\left(\begin{array}{c} a_1,\ldots,a_r \\ b_1,\ldots,b_{s-1},\lambda b_s \end{array} ; q, \lambda z\right) = {}_r\phi_{s-1}\left(\begin{array}{c} a_1,\ldots,a_r \\ b_1,\ldots,b_{s-1} \end{array} ; q, \frac{z}{b_s}\right), \tag{1.10.4}
$$

and

$$
\lim_{\lambda\to\infty} {}_r\phi_s\left(\begin{array}{c} a_1,\ldots,a_{r-1},\lambda a_r \\ b_1,\ldots,b_{s-1},\lambda b_s \end{array} ; q, z\right) = {}_{r-1}\phi_{s-1}\left(\begin{array}{c} a_1,\ldots,a_{r-1} \\ b_1,\ldots,b_{s-1} \end{array} ; q, \frac{a_r z}{b_s}\right). \tag{1.10.5}
$$

Mostly, the left-hand side of (1.10.2) occurs as a limit case when some numerator parameter and some denominator parameter tend to the same value.

All families of discrete orthogonal polynomials $\{P_n(x)\}_{n=0}^N$ are defined for $n = 0,1,2,\ldots,N$, where N is a positive integer. In these cases something like (1.10.2) occurs in the basic hypergeometric representation when $n = N$. In these cases we have to be aware of the fact that we still have a polynomial (in that case of degree N). For instance, if we take $n = N$ in the basic hypergeometric representation (14.6.1) of the q-Hahn polynomials, we have

$$
Q_N(q^{-x};\alpha,\beta,N|q) = \sum_{k=0}^{N} \frac{(\alpha\beta q^{N+1};q)_k (q^{-x};q)_k}{(\alpha q;q)_k (q;q)_k} q^k.
$$

So these cases must be understood by continuity.

1.11 The q-Binomial Theorem and Other Summation Formulas

A q-analogue of the binomial theorem (1.5.1) is called the q-binomial theorem:

$$
{}_1\phi_0\left(\begin{array}{c} a \\ - \end{array} ; q, z\right) = \sum_{n=0}^{\infty} \frac{(a;q)_n}{(q;q)_n} z^n = \frac{(az;q)_\infty}{(z;q)_\infty}, \quad 0<|q|<1, \quad |z|<1. \tag{1.11.1}
$$

For $a = q^{-n}$ with n a nonnegative integer we find

$$
{}_1\phi_0\left(\begin{array}{c} q^{-n} \\ - \end{array} ; q, z\right) = (zq^{-n};q)_n, \quad n = 0,1,2,\ldots. \tag{1.11.2}
$$

In fact this is a q-analogue of Newton's binomium (1.5.2). Note that the case $a = 0$ of (1.11.1) is the limit case ($n \to \infty$) of (1.9.7).

Gauss's summation formula (1.5.3) and the Vandermonde or Chu-Vandermonde summation formula (1.5.4) have the following q-analogues:

$$ {}_2\phi_1\left(\begin{matrix} a,b \\ c \end{matrix}; q, \frac{c}{ab}\right) = \frac{(a^{-1}c, b^{-1}c; q)_\infty}{(c, a^{-1}b^{-1}c; q)_\infty}, \quad 0 < |q| < 1, \quad \left|\frac{c}{ab}\right| < 1, \tag{1.11.3} $$

$$ {}_2\phi_1\left(\begin{matrix} q^{-n},b \\ c \end{matrix}; q, \frac{cq^n}{b}\right) = \frac{(b^{-1}c;q)_n}{(c;q)_n}, \quad n = 0,1,2,\ldots \tag{1.11.4} $$

and

$$ {}_2\phi_1\left(\begin{matrix} q^{-n},b \\ c \end{matrix}; q, q\right) = \frac{(b^{-1}c;q)_n}{(c;q)_n} b^n, \quad n = 0,1,2,\ldots. \tag{1.11.5} $$

By taking the limit $b \to \infty$ in (1.11.3) we also obtain a summation formula for the ${}_1\phi_1$ series:

$$ {}_1\phi_1\left(\begin{matrix} a \\ c \end{matrix}; q, \frac{c}{a}\right) = \frac{(a^{-1}c;q)_\infty}{(c;q)_\infty}, \quad 0 < |q| < 1. \tag{1.11.6} $$

By taking the limit $c \to \infty$ in (1.11.4) we obtain a summation formula for a terminating ${}_2\phi_0$ series:

$$ {}_2\phi_0\left(\begin{matrix} q^{-n},b \\ - \end{matrix}; q, \frac{q^n}{b}\right) = \frac{1}{b^n}, \quad n = 0,1,2,\ldots. \tag{1.11.7} $$

By taking the limit $a \to \infty$ in (1.11.6) we obtain

$$ {}_0\phi_1\left(\begin{matrix} - \\ c \end{matrix}; q, c\right) = \frac{1}{(c;q)_\infty}, \quad 0 < |q| < 1. \tag{1.11.8} $$

On the ${}_3\phi_2$ level we have the summation formula

$$ {}_3\phi_2\left(\begin{matrix} q^{-n},a,b \\ c, abc^{-1}q^{1-n} \end{matrix}; q, q\right) = \frac{(a^{-1}c, b^{-1}c; q)_n}{(c, a^{-1}b^{-1}c; q)_n}, \quad n = 0,1,2,\ldots, \tag{1.11.9} $$

which is a q-analogue of the Saalschütz or Pfaff-Saalschütz summation formula (1.5.5).

On the ${}_6\phi_5$ level we have the Jackson summation formula

$$ {}_6\phi_5\left(\begin{matrix} q\sqrt{a}, -q\sqrt{a}, a, b, c, d \\ \sqrt{a}, -\sqrt{a}, ab^{-1}q, ac^{-1}q, ad^{-1}q \end{matrix}; q, \frac{aq}{bcd}\right) $$
$$ = \frac{(aq, ab^{-1}c^{-1}q, ab^{-1}d^{-1}q, ac^{-1}d^{-1}q; q)_\infty}{(ab^{-1}q, ac^{-1}q, ad^{-1}q, ab^{-1}c^{-1}d^{-1}q; q)_\infty}, \quad \left|\frac{aq}{bcd}\right| < 1 \tag{1.11.10} $$

for a non-terminating very-well-poised ${}_6\phi_5$. We also have the summation formula

$$
\begin{aligned}
& {}_6\phi_5\left(\begin{array}{c} q\sqrt{a},-q\sqrt{a},a,b,c,q^{-n} \\ \sqrt{a},-\sqrt{a},ab^{-1}q,ac^{-1}q,aq^{n+1} \end{array}; q, \frac{aq^{n+1}}{bc}\right) \\
& \quad = \frac{(aq,ab^{-1}c^{-1}q;q)_n}{(ab^{-1}q,ac^{-1}q;q)_n}, \quad n=0,1,2,\ldots \qquad (1.11.11)
\end{aligned}
$$

for a terminating very-well-poised ${}_6\phi_5$, which is also due to Jackson. By taking the limit $c \to 0$ we obtain

$$
\begin{aligned}
& {}_6\phi_4\left(\begin{array}{c} q\sqrt{a},-q\sqrt{a},a,b,0,q^{-n} \\ \sqrt{a},-\sqrt{a},ab^{-1}q,aq^{n+1} \end{array}; q, \frac{q^n}{b}\right) \\
& \quad = \frac{(aq;q)_n}{(ab^{-1}q;q)_n} b^{-n}, \quad n=0,1,2,\ldots. \qquad (1.11.12)
\end{aligned}
$$

If we take the limit $b \to 0$ we obtain

$$
\begin{aligned}
& {}_6\phi_3\left(\begin{array}{c} q\sqrt{a},-q\sqrt{a},a,0,0,q^{-n} \\ \sqrt{a},-\sqrt{a},aq^{n+1} \end{array}; q, \frac{q^{n-1}}{a}\right) \\
& \quad = (-1)^n a^{-n} q^{-\binom{n+1}{2}} (aq;q)_n, \quad n=0,1,2,\ldots. \qquad (1.11.13)
\end{aligned}
$$

1.12 More Integrals

In this section we list some q-extensions of the beta integral. For instance, we will need the following integral:

$$
\int_0^\infty x^{c-1} \frac{(-ax,-bq/x;q)_\infty}{(-x,-q/x;q)_\infty} dx = \frac{\pi}{\sin \pi c} \frac{(ab,q^c,q^{1-c};q)_\infty}{(bq^c,aq^{-c},q;q)_\infty}, \qquad (1.12.1)
$$

which holds for $0 < q < 1$, $\mathrm{Re}\, c > 0$ and $|aq^{-c}| < 1$. If we set $b = 1$ in (1.12.1), we find that

$$
\int_0^\infty x^{c-1} \frac{(-ax;q)_\infty}{(-x;q)_\infty} dx = \frac{\pi}{\sin \pi c} \frac{(a,q^{1-c};q)_\infty}{(aq^{-c},q;q)_\infty} \qquad (1.12.2)
$$

for $0 < q < 1$, $\mathrm{Re}\, c > 0$ and $|aq^{-c}| < 1$, and by taking the limit $c \to 1$ in (1.12.1), we obtain

$$
\int_0^\infty \frac{(-ax,-bq/x;q)_\infty}{(-x,-q/x;q)_\infty} dx = -\ln q \frac{(ab,q;q)_\infty}{(bq,a/q;q)_\infty} \qquad (1.12.3)
$$

for $0 < q < 1$ and $\left|aq^{-1}\right| < 1$.

Finally, we mention the Askey-Wilson q-beta integral

$$
\frac{1}{2\pi} \int_{-1}^1 \frac{w(x)}{\sqrt{1-x^2}} dx = \frac{1}{2\pi} \int_0^\pi w(\cos\theta)\, d\theta = \frac{(abcd;q)_\infty}{(ab,ac,ad,bc,bd,cd,q;q)_\infty}, \qquad (1.12.4)
$$

where

$$w(x)=\frac{h(x,1)h(x,-1)h(x,q^{1/2})h(x,-q^{1/2})}{h(x,a)h(x,b)h(x,c)h(x,d)}=\left|\frac{(\mathrm{e}^{2i\theta};q)_\infty}{(a\mathrm{e}^{i\theta},b\mathrm{e}^{i\theta},c\mathrm{e}^{i\theta},d\mathrm{e}^{i\theta};q)_\infty}\right|^2$$
$$=\frac{(\mathrm{e}^{2i\theta},\mathrm{e}^{-2i\theta};q)_\infty}{(a\mathrm{e}^{i\theta},a\mathrm{e}^{-i\theta},b\mathrm{e}^{i\theta},b\mathrm{e}^{-i\theta},c\mathrm{e}^{i\theta},c\mathrm{e}^{-i\theta},d\mathrm{e}^{i\theta},d\mathrm{e}^{-i\theta};q)_\infty},\quad x=\cos\theta$$

with

$$h(x,\alpha)=\prod_{k=0}^{\infty}\left(1-2\alpha xq^k+\alpha^2q^{2k}\right)$$
$$=\left|(\alpha\mathrm{e}^{i\theta};q)_\infty\right|^2=\left(\alpha\mathrm{e}^{i\theta},\alpha\mathrm{e}^{-i\theta};q\right)_\infty,\quad x=\cos\theta.$$

The Askey-Wilson integral (1.12.4) is a q-analogue of Wilson's integral (1.6.11) and holds for $\max(|a|,|b|,|c|,|d|)<1$ and a,b,c,d positive or $b=\overline{a}$ and/or $d=\overline{c}$ with positive real parts.

1.13 Transformation Formulas

In this section we list a number of transformation formulas which can be used to transform basic hypergeometric representations and other formulas into equivalent but different forms.

First of all we have Heine's transformation formulas for the ${}_2\phi_1$ series:

$${}_2\phi_1\left(\begin{matrix}a,b\\c\end{matrix};q,z\right)=\frac{(az,b;q)_\infty}{(c,z;q)_\infty}\,{}_2\phi_1\left(\begin{matrix}b^{-1}c,z\\az\end{matrix};q,b\right) \tag{1.13.1}$$
$$=\frac{(b^{-1}c,bz;q)_\infty}{(c,z;q)_\infty}\,{}_2\phi_1\left(\begin{matrix}abc^{-1}z,b\\bz\end{matrix};q,\frac{c}{b}\right) \tag{1.13.2}$$
$$=\frac{(abc^{-1}z;q)_\infty}{(z;q)_\infty}\,{}_2\phi_1\left(\begin{matrix}a^{-1}c,b^{-1}c\\c\end{matrix};q,\frac{abz}{c}\right). \tag{1.13.3}$$

The latter formula is a q-analogue of Euler's transformation formula:

$${}_2F_1\left(\begin{matrix}a,b\\c\end{matrix};z\right)=(1-z)^{c-a-b}\,{}_2F_1\left(\begin{matrix}c-a,c-b\\c\end{matrix};z\right). \tag{1.13.4}$$

Another transformation formula for the ${}_2\phi_1$ series is

$${}_2\phi_1\left(\begin{matrix}a,b\\c\end{matrix};q,z\right)=\frac{(az;q)_\infty}{(z;q)_\infty}\,{}_2\phi_2\left(\begin{matrix}a,b^{-1}c\\c,az\end{matrix};q,bz\right), \tag{1.13.5}$$

which is a q-analogue of the Pfaff-Kummer transformation formula (1.7.2).

Limit cases of Heine's transformation formulas are

$$
{}_2\phi_1\left(\begin{matrix} a,0 \\ c \end{matrix}; q, z\right) = \frac{(az;q)_\infty}{(c,z;q)_\infty}\,{}_1\phi_1\left(\begin{matrix} z \\ az \end{matrix}; q, c\right) \tag{1.13.6}
$$

$$
= \frac{1}{(z;q)_\infty}\,{}_1\phi_1\left(\begin{matrix} a^{-1}c \\ c \end{matrix}; q, az\right), \tag{1.13.7}
$$

$$
{}_2\phi_1\left(\begin{matrix} 0,0 \\ c \end{matrix}; q, z\right) = \frac{1}{(c,z;q)_\infty}\,{}_1\phi_1\left(\begin{matrix} z \\ 0 \end{matrix}; q, c\right) \tag{1.13.8}
$$

$$
= \frac{1}{(z;q)_\infty}\,{}_0\phi_1\left(\begin{matrix} - \\ c \end{matrix}; q, cz\right), \tag{1.13.9}
$$

$$
{}_1\phi_1\left(\begin{matrix} a \\ c \end{matrix}; q, z\right) = \frac{(a,z;q)_\infty}{(c;q)_\infty}\,{}_2\phi_1\left(\begin{matrix} a^{-1}c,0 \\ z \end{matrix}; q, a\right) \tag{1.13.10}
$$

$$
= (ac^{-1}z;q)_\infty \cdot {}_2\phi_1\left(\begin{matrix} a^{-1}c,0 \\ c \end{matrix}; q, \frac{az}{c}\right) \tag{1.13.11}
$$

and

$$
{}_2\phi_1\left(\begin{matrix} a,b \\ 0 \end{matrix}; q, z\right) = \frac{(az,b;q)_\infty}{(z;q)_\infty}\,{}_2\phi_1\left(\begin{matrix} z,0 \\ az \end{matrix}; q, b\right) \tag{1.13.12}
$$

$$
= \frac{(bz;q)_\infty}{(z;q)_\infty}\,{}_1\phi_1\left(\begin{matrix} b \\ bz \end{matrix}; q, az\right). \tag{1.13.13}
$$

The q-analogues of (1.7.4) and (1.7.5) are

$$
{}_1\phi_1\left(\begin{matrix} q^{-n} \\ a \end{matrix}; q, z\right) = \frac{(q^{-1}z)^n}{(a;q)_n}\,{}_2\phi_1\left(\begin{matrix} q^{-n},a^{-1}q^{1-n} \\ 0 \end{matrix}; q, \frac{aq^{n+1}}{z}\right) \tag{1.13.14}
$$

for $n = 0,1,2,\ldots$, and

$$
{}_2\phi_1\left(\begin{matrix} q^{-n},b \\ c \end{matrix}; q, z\right)
$$

$$
= \frac{(b;q)_n}{(c;q)_n} q^{-n-\binom{n}{2}}(-z)^n\,{}_2\phi_1\left(\begin{matrix} q^{-n},c^{-1}q^{1-n} \\ b^{-1}q^{1-n} \end{matrix}; q, \frac{cq^{n+1}}{bz}\right) \tag{1.13.15}
$$

for $n = 0,1,2,\ldots$.

A limit case of the latter formula is

$$
{}_2\phi_0\left(\begin{matrix} q^{-n},b \\ - \end{matrix}; q, zq^n\right) = (b;q)_n z^n\,{}_2\phi_1\left(\begin{matrix} q^{-n},0 \\ b^{-1}q^{1-n} \end{matrix}; q, \frac{q}{bz}\right) \tag{1.13.16}
$$

for $n = 0,1,2,\ldots$.

The next transformation formula is due to Jackson:

$$
{}_2\phi_1\left(\begin{matrix} q^{-n},b \\ c \end{matrix};q,z\right)=\frac{(bc^{-1}q^{-n}z;q)_\infty}{(bc^{-1}z;q)_\infty}\,{}_3\phi_2\left(\begin{matrix} q^{-n},b^{-1}c,0 \\ c,b^{-1}cqz^{-1} \end{matrix};q,q\right) \tag{1.13.17}
$$

for $n=0,1,2,\ldots$. Equivalently, we have

$$
{}_3\phi_2\left(\begin{matrix} q^{-n},a,0 \\ b,c \end{matrix};q,q\right)=\frac{(b^{-1}q;q)_\infty}{(b^{-1}q^{1-n};q)_\infty}\,{}_2\phi_1\left(\begin{matrix} q^{-n},a^{-1}c \\ c \end{matrix};q,\frac{aq}{b}\right) \tag{1.13.18}
$$

for $n=0,1,2,\ldots$.

Other transformation formulas of this kind are given by:

$$
\begin{aligned}
&{}_2\phi_1\left(\begin{matrix} q^{-n},b \\ c \end{matrix};q,z\right)\\
&=\frac{(b^{-1}c;q)_n}{(c;q)_n}\left(\frac{bz}{q}\right)^n{}_3\phi_2\left(\begin{matrix} q^{-n},qz^{-1},c^{-1}q^{1-n} \\ bc^{-1}q^{1-n},0 \end{matrix};q,q\right) &(1.13.19)\\
&=\frac{(b^{-1}c;q)_n}{(c;q)_n}\,{}_3\phi_2\left(\begin{matrix} q^{-n},b,bc^{-1}q^{-n}z \\ bc^{-1}q^{1-n},0 \end{matrix};q,q\right) &(1.13.20)
\end{aligned}
$$

for $n=0,1,2,\ldots$, or equivalently

$$
\begin{aligned}
{}_3\phi_2\left(\begin{matrix} q^{-n},a,b \\ c,0 \end{matrix};q,q\right)&=\frac{(b;q)_n}{(c;q)_n}a^n\,{}_2\phi_1\left(\begin{matrix} q^{-n},b^{-1}c \\ b^{-1}q^{1-n} \end{matrix};q,\frac{q}{a}\right) &(1.13.21)\\
&=\frac{(a^{-1}c;q)_n}{(c;q)_n}a^n\,{}_2\phi_1\left(\begin{matrix} q^{-n},a \\ ac^{-1}q^{1-n} \end{matrix};q,\frac{bq}{c}\right) &(1.13.22)
\end{aligned}
$$

for $n=0,1,2,\ldots$.

Limit cases of these formulas are

$$
{}_2\phi_0\left(\begin{matrix} q^{-n},b \\ - \end{matrix};q,z\right)=b^{-n}\,{}_3\phi_2\left(\begin{matrix} q^{-n},b,bzq^{-n} \\ 0,0 \end{matrix};q,q\right),\quad n=0,1,2,\ldots, \tag{1.13.23}
$$

or equivalently

$$
\begin{aligned}
{}_3\phi_2\left(\begin{matrix} q^{-n},a,b \\ 0,0 \end{matrix};q,q\right)&=(b;q)_n a^n\,{}_2\phi_1\left(\begin{matrix} q^{-n},0 \\ b^{-1}q^{1-n} \end{matrix};q,\frac{q}{a}\right) &(1.13.24)\\
&=a^n\,{}_2\phi_0\left(\begin{matrix} q^{-n},a \\ - \end{matrix};q,\frac{bq^n}{a}\right) &(1.13.25)
\end{aligned}
$$

for $n=0,1,2,\ldots$.

On the next level we have Sears' transformation formula for a terminating balanced ${}_4\phi_3$ series:

$$
\begin{aligned}
&{}_4\phi_3\left(\begin{matrix} q^{-n},a,b,c \\ d,e,f \end{matrix};q,q\right) \\
&= \frac{(a^{-1}e,a^{-1}f;q)_n}{(e,f;q)_n} a^n \,{}_4\phi_3\left(\begin{matrix} q^{-n},a,b^{-1}d,c^{-1}d \\ d,ae^{-1}q^{1-n},af^{-1}q^{1-n} \end{matrix};q,q\right) \qquad (1.13.26)\\
&= \frac{(a,a^{-1}b^{-1}ef,a^{-1}c^{-1}ef;q)_n}{(e,f,a^{-1}b^{-1}c^{-1}ef;q)_n} \\
&\quad \times {}_4\phi_3\left(\begin{matrix} q^{-n},a^{-1}e,a^{-1}f,a^{-1}b^{-1}c^{-1}ef \\ a^{-1}b^{-1}ef,a^{-1}c^{-1}ef,a^{-1}q^{1-n} \end{matrix};q,q\right), \qquad (1.13.27)
\end{aligned}
$$

provided that $def = abcq^{1-n}$. Sears' transformation formula is a q-analogue of Whipple's transformation (1.7.6).

Finally, we have a quadratic transformation formula which is due to Singh:

$$
{}_4\phi_3\left(\begin{matrix} a^2,b^2,c,d \\ abq^{1/2},-abq^{1/2},-cd \end{matrix};q,q\right) = {}_4\phi_3\left(\begin{matrix} a^2,b^2,c^2,d^2 \\ a^2b^2q,-cd,-cdq \end{matrix};q^2,q^2\right), \qquad (1.13.28)
$$

which is valid when both sides terminate.

1.14 Some q-Analogues of Special Functions

For the exponential function we have two different natural q-extensions, denoted by $e_q(z)$ and $\mathrm{E}_q(z)$, which can be defined by

$$
\mathrm{e}_q(z) := {}_1\phi_0\left(\begin{matrix} 0 \\ - \end{matrix};q,z\right) = \sum_{n=0}^{\infty} \frac{z^n}{(q;q)_n} = \frac{1}{(z;q)_\infty}, \quad 0<|q|<1, \quad |z|<1 \qquad (1.14.1)
$$

and

$$
\mathrm{E}_q(z) := {}_0\phi_0\left(\begin{matrix} - \\ - \end{matrix};q,-z\right) = \sum_{n=0}^{\infty} \frac{q^{\binom{n}{2}}}{(q;q)_n} z^n = (-z;q)_\infty, \quad 0<|q|<1. \qquad (1.14.2)
$$

These q-analogues of the exponential function are related by

$$
\mathrm{e}_q(z)\mathrm{E}_q(-z) = 1.
$$

They are q-extensions of the exponential function since

$$
\lim_{q\to1} \mathrm{e}_q((1-q)z) = \lim_{q\to1} \mathrm{E}_q((1-q)z) = \mathrm{e}^z.
$$

Note that (1.14.1) can be seen as the special case $a=0$ of (1.11.1).

If we assume that $0<|q|<1$ and $|z|<1$, we may define

$$\cos_q(z) := \frac{e_q(iz)+e_q(-iz)}{2} = \sum_{n=0}^{\infty} \frac{(-1)^n z^{2n}}{(q;q)_{2n}} \tag{1.14.3}$$

and

$$\sin_q(z) := \frac{e_q(iz)-e_q(-iz)}{2i} = \sum_{n=0}^{\infty} \frac{(-1)^n z^{2n+1}}{(q;q)_{2n+1}}. \tag{1.14.4}$$

These are q-analogues of the trigonometric functions $\cos z$ and $\sin z$. On the other hand, we may define

$$\mathrm{Cos}_q(z) := \frac{E_q(iz)+E_q(-iz)}{2} = \sum_{n=0}^{\infty} \frac{(-1)^n q^{\binom{2n}{2}} z^{2n}}{(q;q)_{2n}} \tag{1.14.5}$$

and

$$\mathrm{Sin}_q(z) := \frac{E_q(iz)-E_q(-iz)}{2i} = \sum_{n=0}^{\infty} \frac{(-1)^n q^{\binom{2n+1}{2}} z^{2n+1}}{(q;q)_{2n+1}}. \tag{1.14.6}$$

Then we have

$$e_q(iz) = \cos_q(z) + i\sin_q(z) \quad \text{and} \quad E_q(iz) = \mathrm{Cos}_q(z) + i\,\mathrm{Sin}_q(z).$$

Further we have

$$\begin{cases} \cos_q(z)\mathrm{Cos}_q(z) + \sin_q(z)\mathrm{Sin}_q(z) = 1 \\ \sin_q(z)\mathrm{Cos}_q(z) - \cos_q(z)\mathrm{Sin}_q(z) = 0. \end{cases}$$

The q-analogues of the trigonometric functions can be used to find different forms of formulas appearing in this book, although we will not use them.

Some q-analogues of the Bessel functions are given by

$$J_\nu^{(1)}(z;q) := \frac{(q^{\nu+1};q)_\infty}{(q;q)_\infty} \left(\frac{z}{2}\right)^\nu {}_2\phi_1\left(\begin{matrix} 0,0 \\ q^{\nu+1} \end{matrix}; q, -\frac{z^2}{4}\right), \quad |z|<2 \tag{1.14.7}$$

and

$$J_\nu^{(2)}(z;q) := \frac{(q^{\nu+1};q)_\infty}{(q;q)_\infty} \left(\frac{z}{2}\right)^\nu {}_0\phi_1\left(\begin{matrix} - \\ q^{\nu+1} \end{matrix}; q, -\frac{q^{\nu+1}z^2}{4}\right). \tag{1.14.8}$$

These q-Bessel functions are connected by

$$J_\nu^{(2)}(z;q) = (-\frac{z^2}{4};q)_\infty \cdot J_\nu^{(1)}(z;q), \quad |z|<2.$$

They are q-extensions of the Bessel function of the first kind since

$$\lim_{q\to 1} J_\nu^{(k)}((1-q)z;q) = J_\nu(z), \quad k=1,2.$$

These q-Bessel functions were introduced by F.H. Jackson in 1905 and are therefore referred to as Jackson's q-Bessel functions. A third q-analogue of the Bessel function is given by

$$J_\nu^{(3)}(z;q) := \frac{(q^{\nu+1};q)_\infty}{(q;q)_\infty} z^\nu \, {}_1\phi_1\left(\begin{matrix} 0 \\ q^{\nu+1} \end{matrix} ; q, qz^2\right). \tag{1.14.9}$$

This third q-Bessel function is also known as the Hahn-Exton q-Bessel function. This is also a q-extension of the Bessel function of the first kind since

$$\lim_{q\to 1} J_\nu^{(3)}((1-q)z;q) = J_\nu(2z).$$

1.15 The q-Derivative and q-Integral

For $q \neq 1$ the q-derivative operator $\mathscr{D}_q$ is defined by

$$\mathscr{D}_q f(z) := \begin{cases} \dfrac{f(z)-f(qz)}{(1-q)z}, & z \neq 0 \\ f'(0), & z = 0. \end{cases} \tag{1.15.1}$$

Further we define

$$\mathscr{D}_q^0 f := f \quad \text{and} \quad \mathscr{D}_q^n f := \mathscr{D}_q\left(\mathscr{D}_q^{n-1} f\right), \quad n = 1,2,3,\ldots. \tag{1.15.2}$$

It is not very difficult to see that

$$\lim_{q\to 1} \mathscr{D}_q f(z) = f'(z)$$

if the function f is differentiable at z.

The q-derivative operator is a special case of Hahn's q-operator, which will be defined in section 2.1. This operator also generalizes the differentiation operator $D = d/dx$.

An easy consequence of the definition (1.15.1) is

$$\mathscr{D}_q\left[f(\gamma x)\right] = \gamma(\mathscr{D}_q f)(\gamma x) \tag{1.15.3}$$

for all $\gamma \in \mathbb{C}$, or more general

$$\mathscr{D}_q^n\left[f(\gamma x)\right] = \gamma^n \left(\mathscr{D}_q^n f\right)(\gamma x), \quad n = 0,1,2,\ldots. \tag{1.15.4}$$

Further we have

$$\mathscr{D}_q\left[f(x)g(x)\right] = f(qx)\mathscr{D}_q g(x) + g(x)\mathscr{D}_q f(x) \tag{1.15.5}$$

which is often referred to as the q-product rule. This can be generalized to a q-analogue of Leibniz' rule:

$$\mathscr{D}_q^n\left[f(x)g(x)\right] = \sum_{k=0}^{n} \begin{bmatrix} n \\ k \end{bmatrix}_q \left(\mathscr{D}_q^{n-k} f\right)(q^k x)\left(\mathscr{D}_q^k g\right)(x), \quad n = 0, 1, 2, \ldots . \tag{1.15.6}$$

The q-integral is defined by

$$\int_0^z f(t)\, d_q t := z(1-q) \sum_{n=0}^{\infty} f(q^n z) q^n, \quad 0 < q < 1. \tag{1.15.7}$$

This definition is due to J. Thomae (1869) and F.H. Jackson (1910). Jackson also defined a q-integral on $(0, \infty)$ by

$$\int_0^{\infty} f(t)\, d_q t := (1-q) \sum_{n=-\infty}^{\infty} f(q^n) q^n, \quad 0 < q < 1. \tag{1.15.8}$$

If the function f is continuous on $[0, z]$, we have

$$\lim_{q \to 1} \int_0^z f(t)\, d_q t = \int_0^z f(t)\, dt.$$

Of course, definition (1.15.7) implies that

$$\int_a^b f(t)\, d_q t = b(1-q) \sum_{n=0}^{\infty} f(bq^n) q^n - a(1-q) \sum_{n=0}^{\infty} f(aq^n) q^n, \quad 0 < q < 1$$

and definition (1.15.8) implies that

$$\int_{-\infty}^{\infty} f(t)\, d_q t = (1-q) \sum_{n=-\infty}^{\infty} \{f(q^n) + f(-q^n)\} q^n, \quad 0 < q < 1.$$

A function f is called q-integrable if the appropriate sum converges absolutely. We remark that we have

$$F(z) = \int_0^z f(t)\, d_q t \quad \Longrightarrow \quad \mathscr{D}_q F(z) = f(z)$$

and

$$\int_a^b \mathscr{D}_q f(t)\, d_q t = f(b) - f(a).$$

The q-productrule (1.15.5) implies the q-integration by parts formula

$$\int_a^b g(x) \mathscr{D}_q f(x)\, d_q x = \left[f(x)g(x)\right]_a^b - \int_a^b f(qx) \mathscr{D}_q g(x)\, d_q x \tag{1.15.9}$$

for suitable functions f and g.

Finally, we mention the following q-integral formula (see [238], formula (2.10.18))

$$\int_a^b \frac{\left(a^{-1}qt,b^{-1}qt,ct;q\right)_\infty}{(dt,et,ft;q)_\infty}\,d_qt$$
$$=(b-a)(1-q)\frac{\left(q,a^{-1}bq,ab^{-1}q,cd^{-1},ce^{-1},cf^{-1};q\right)_\infty}{(ad,ae,af,bd,be,bf;q)_\infty}, \quad (1.15.10)$$

provided that $c=abdef$. By taking the limit $f\to 0$, we obtain

$$\int_a^b \frac{\left(a^{-1}qt,b^{-1}qt;q\right)_\infty}{(dt,et;q)_\infty}\,d_qt=(b-a)(1-q)\frac{\left(q,a^{-1}bq,ab^{-1}q,abde;q\right)_\infty}{(ad,ae,bd,be;q)_\infty}. \quad (1.15.11)$$

Limit cases of (1.15.11) are

$$\int_a^\infty \frac{\left(a^{-1}qt;q\right)_\infty}{(dt,et;q)_\infty}\,d_qt=(1-q)\frac{\left(q,a,a^{-1}q,ade,a^{-1}d^{-1}e^{-1}q;q\right)_\infty}{(ad,ae,d,d^{-1}q,e,e^{-1}q;q)_\infty} \quad (1.15.12)$$

and

$$\int_{-\infty}^\infty \frac{1}{(dt,et;q)_\infty}\,d_qt$$
$$=(1-q)\frac{\left(q,-q,-1,-de,-d^{-1}e^{-1}q;q\right)_\infty}{(d,-d,d^{-1}q,-d^{-1}q,e,-e,e^{-1}q,-e^{-1}q;q)_\infty}. \quad (1.15.13)$$

1.16 Shift Operators and Rodrigues-Type Formulas

We need some more differential and difference operators to formulate the Rodrigues-type formulas. These operators can also be used to formulate the second-order differential or difference equations but this is mostly avoided. As usual, we will use the notation

$$f'(x)=\frac{d}{dx}f(x)=\frac{df}{dx}(x)=\frac{df(x)}{dx}.$$

Further we define

$$\Delta f(x)=f(x+1)-f(x), \quad (1.16.1)$$
$$\nabla f(x)=f(x)-f(x-1) \quad (1.16.2)$$

and

$$\delta f(x)=f(x+\tfrac{1}{2}i)-f(x-\tfrac{1}{2}i). \quad (1.16.3)$$

The ordinary difference operator Δ given by (1.16.1) is also a special case of Hahn's q-operator defined in section 2.1.

Note that (1.16.3) implies that

$$\delta^2 f(x)=f(x+i)-f(x)-f(x)+f(x-i)=f(x+i)-2f(x)+f(x-i),$$

$$\delta x = x + \tfrac{1}{2}i - (x - \tfrac{1}{2}i) = i \quad \text{and} \quad \delta x^2 = (x + \tfrac{1}{2}i)^2 - (x - \tfrac{1}{2}i)^2 = 2ix.$$

Further we have for $\lambda(x) := x(x + \gamma + \delta + 1)$

$$\Delta\lambda(x) = 2x + \gamma + \delta + 2 \quad \text{and} \quad \nabla\lambda(x) = 2x + \gamma + \delta.$$

In a similar way we have for $\mu(x) := q^{-x} + \gamma\delta q^{x+1}$

$$\Delta\mu(x) = q^{-x-1}(1-q)(1-\gamma\delta q^{2x+2}) \quad \text{and} \quad \nabla\mu(x) = q^{-x}(1-q)(1-\gamma\delta q^{2x})$$

and hence for $\lambda(x) := q^{-x} + cq^{x-N}$

$$\Delta\lambda(x) = q^{-x-1}(1-q)(1-cq^{2x-N+1}) \quad \text{and} \quad \nabla\lambda(x) = q^{-x}(1-q)(1-cq^{2x-N-1}).$$

Also note that

$$\Delta q^{-x} = q^{-x-1}(1-q) \quad \text{and} \quad \nabla q^{-x} = q^{-x}(1-q).$$

Finally, we define

$$D_q f(x) := \frac{\delta_q f(x)}{\delta_q x}, \quad x = \cos\theta \tag{1.16.4}$$

with

$$\delta_q f(e^{i\theta}) = f(q^{\frac{1}{2}} e^{i\theta}) - f(q^{-\frac{1}{2}} e^{i\theta}).$$

Here we have

$$\delta_q x = -\tfrac{1}{2} q^{-\frac{1}{2}}(1-q)(e^{i\theta} - e^{-i\theta}), \quad x = \cos\theta.$$

Chapter 2
Polynomial Solutions of Eigenvalue Problems

2.1 Hahn's q-Operator

Let $\mathscr{P}$ denote the space of polynomials over $\mathbb{C}$.

In [261] W. Hahn introduced the linear operator $\mathscr{A}_{q,\omega}$ defined by

$$(\mathscr{A}_{q,\omega}p)(x) := \frac{p(qx+\omega)-p(x)}{qx+\omega-x}, \quad p \in \mathscr{P}, \quad x \in \mathbb{R} \setminus \left\{\frac{\omega}{1-q}\right\} \tag{2.1.1}$$

for all $q \in \mathbb{R}\setminus\{-1,0\}$[1], $\omega \in \mathbb{R}$ and $(q,\omega) \neq (1,0)$. This class of operators includes the q-derivative operator $\mathscr{D}_q$ $(\omega = 0)$, the difference operator Δ $(q = 1$ and $\omega = 1)$ and also the differentiation operator D as a limit case $(q \to 1$ and $\omega = 0)$. In order to avoid the latter limiting process, we introduce the operator $\mathscr{A}_{q,\omega}$ in a second way. From (2.1.1) we obtain that $\mathscr{A}_{q,\omega}(1) = 0$, $\mathscr{A}_{q,\omega}(x) = 1$ and the product rule

$$(\mathscr{A}_{q,\omega}(p_1p_2))(x) = (\mathscr{A}_{q,\omega}p_1)(x)p_2(x) + p_1(qx+\omega)(\mathscr{A}_{q,\omega}p_2)(x) \tag{2.1.2}$$

for $p_1, p_2 \in \mathscr{P}$. Now we have

Theorem 2.1. *For all $q \in \mathbb{R}\setminus\{-1,0\}$ and $\omega \in \mathbb{R}$ there exists a unique linear operator $\mathscr{A}_{q,\omega}$ on $\mathscr{P}$ satisfying $\mathscr{A}_{q,\omega}(x) = 1$ and the product rule (2.1.2).*

Proof. The product rule (2.1.2) implies that $\mathscr{A}_{q,\omega}(1\cdot 1) = \mathscr{A}_{q,\omega}(1)\cdot 1 + 1\cdot\mathscr{A}_{q,\omega}(1)$. Hence we have $\mathscr{A}_{q,\omega}(1) = 0$. So $\mathscr{A}_{q,\omega}$ is uniquely defined on the basis $\{1, x, x^2, \ldots\}$ of the space $\mathscr{P}$ by

$$\mathscr{A}_{q,\omega}(x^{n+1}) = \mathscr{A}_{q,\omega}(x)x^n + (qx+\omega)\mathscr{A}_{q,\omega}(x^n), \quad n = 1,2,3,\ldots$$

and the initial values $\mathscr{A}_{q,\omega}(1) = 0$ and $\mathscr{A}_{q,\omega}(x) = 1$. □

[1] An essential property of the operator $\mathscr{A}_{q,\omega}$ is that its action on a polynomial of degree n leads to a polynomial of degree $n-1$ for all $n = 1,2,3,\ldots$. For $q = -1$ this property does not hold (for instance $\mathscr{A}_{-1,\omega}(x^2) = \omega$). That is why the case $q = -1$ is excluded.

R. Koekoek et al., *Hypergeometric Orthogonal Polynomials and Their q-Analogues*, Springer Monographs in Mathematics,
DOI 10.1007/978-3-642-05014-5_2,

As a consequence of this theorem, (2.1.2) together with $\mathscr{A}_{q,\omega}(x) = 1$ can be seen as a definition of $\mathscr{A}_{q,\omega}$. This definition holds for all $q \in \mathbb{R} \setminus \{-1, 0\}$ and $\omega \in \mathbb{R}$, and therefore includes the differentiation operator $D = \mathscr{A}_{1,0}$.

2.2 Eigenvalue Problems

We consider the eigenvalue problem[2]

$$\varphi(x)\left(\mathscr{A}_{q,\omega}^2 y_n\right)(x) + \psi(x)\left(\mathscr{A}_{q,\omega} y_n\right)(x) = \lambda_n y_n(qx+\omega) \qquad (2.2.1)$$

for polynomials y_n of degree n, where $\mathscr{A}_{q,\omega}^2 y_n = \mathscr{A}_{q,\omega}\left(\mathscr{A}_{q,\omega} y_n\right)$ with $\lambda_n \in \mathbb{C}$ and $n \in \{0,1,2,\ldots\}$. In [106], [372] and [379], for instance, it is shown that polynomial solutions of any degree n, where $n \in \{0,1,2,\ldots\}$, can only exist if φ is a polynomial of degree at most 2 and ψ is a polynomial of degree 1, say

$$\varphi(x) = ex^2 + 2fx + g, \quad \psi(x) = 2\varepsilon x + \gamma, \quad e,f,g,\varepsilon,\gamma \in \mathbb{C}, \quad \varepsilon \neq 0. \qquad (2.2.2)$$

In order to calculate the eigenvalues λ_n, we define (cf. (1.8.2))

$$[-1] = -\frac{1}{q}, \quad [0] = 0, \quad [n] = \sum_{k=0}^{n-1} q^k, \quad n = 1,2,3,\ldots.$$

Note that this definition holds for all $q \in \mathbb{R} \setminus \{-1, 0\}$. Now we obtain

$$[n] - q^{n-k}[k] = [n-k], \quad n \geq k, \quad n,k \in \{0,1,2,\ldots\} \qquad (2.2.3)$$

and

$$[n][n-1] - q^{n-k}[k][k-1] = [n-k][n+k-1], \quad n \geq k, \quad n,k \in \{0,1,2,\ldots\}. \quad (2.2.4)$$

Further we have

$$\begin{aligned}\mathscr{A}_{q,\omega}(x^n) &= \frac{(qx+\omega)^n - x^n}{qx+\omega-x} = \sum_{k=0}^{n-1}(qx+\omega)^{n-1-k}x^k \\ &= x^{n-1}\sum_{k=0}^{n-1} q^k + r(x) = [n]x^{n-1} + r(x), \quad n = 2,3,4,\ldots, \qquad (2.2.5)\end{aligned}$$

where r is a polynomial of degree at most $n-2$. The eigenvalues λ_n can now be obtained by comparing the coefficients of x^n in (2.2.1):

$$\lambda_n = \frac{[n]}{q^n}\left(e[n-1] + 2\varepsilon\right), \quad n = 0,1,2,\ldots. \qquad (2.2.6)$$

[2] It will turn out to be convenient to take $y_n(qx+\omega)$ on the right-hand side.

So we have: if (2.2.1) has a polynomial solution of degree n, then λ_n given by (2.2.6) is the corresponding eigenvalue.

Note that this result is also valid for $q=1$ and $\omega=0$.

So the eigenvalue problem (2.2.1) can be written in the form

$$\begin{aligned}&(ex^2+2fx+g)\left(\mathscr{A}_{q,\omega}^2 y_n\right)(x)+(2\varepsilon x+\gamma)\left(\mathscr{A}_{q,\omega} y_n\right)(x)\\&=\frac{[n]}{q^n}(e[n-1]+2\varepsilon)y_n(qx+\omega),\quad n=0,1,2,\ldots.\end{aligned}\tag{2.2.7}$$

In the case of the difference operator Δ (i.e. $q=1$ and $\omega=1$) this reads

$$(ex^2+2fx+g)\left(\Delta^2 y_n\right)(x)+(2\varepsilon x+\gamma)\left(\Delta y_n\right)(x)=n(e(n-1)+2\varepsilon)y_n(x+1)\tag{2.2.8}$$

for $n=0,1,2,\ldots$ and in the case of the differentiation operator D (i.e. $q=1$ and $\omega=0$)

$$(ex^2+2fx+g)y_n''(x)+(2\varepsilon x+\gamma)y_n'(x)=n(e(n-1)+2\varepsilon)y_n(x)\tag{2.2.9}$$

for $n=0,1,2,\ldots$. We remark that the operator $\mathscr{A}_{q,\omega}$ is not invariant under translations for $q\neq 1$, unlike the operators Δ and D. This can be seen as follows. If we apply the operator $\mathscr{A}_{q,\omega}$ to the polynomial $p(\cdot+c)$, where $c\in\mathbb{R}$ is a constant, we obtain

$$\left(\mathscr{A}_{q,\omega}p(\cdot+c)\right)(x)=\frac{p(qx+\omega+c)-p(x+c)}{qx+\omega-x}\tag{2.2.10}$$

and, if we apply the operator $\mathscr{A}_{q,\omega}$ to the polynomial p first and then replace the argument x by $x+c$, we have

$$\begin{aligned}\left(\mathscr{A}_{q,\omega}p(\cdot)\right)(x+c)&=\frac{p(q(x+c)+\omega)-p(x+c)}{q(x+c)+\omega-(x+c)}\\&=\frac{p(qx+\overline{\omega}+c)-p(x+c)}{qx+\overline{\omega}-x},\end{aligned}\tag{2.2.11}$$

where $\overline{\omega}=\omega+c(q-1)$. Both results coincide only if $q=1$.

Now we will give another version of the operator equation (2.2.7) in the case that $(q,\omega)\neq(1,0)$. In order to do this, we write

$$\left(\mathscr{A}_{q,\omega}y_n\right)(x)=\frac{y_n(qx+\omega)-y_n(x)}{qx+\omega-x}=:p_n(x)$$

and

$$\begin{aligned}\left(\mathscr{A}_{q,\omega}^2 y_n\right)(x)=\left(\mathscr{A}_{q,\omega}p_n\right)(x)&=\frac{p_n(qx+\omega)-p_n(x)}{qx+\omega-x}\\&=\frac{y_n(q^2x+[2]\omega)-(1+q)y_n(qx+\omega)+qy_n(x)}{q(qx+\omega-x)^2}.\end{aligned}$$

Then the operator equation (2.2.7) can be written in the form

$$C(qx+\omega)y_n(q^2x+[2]\omega)-\{C(qx+\omega)+D(qx+\omega)\}\,y_n(qx+\omega)$$
$$+D(qx+\omega)y_n(x)=\lambda_n y_n(qx+\omega),$$

where

$$C(qx+\omega)=\frac{ex^2+2fx+g}{q(qx+\omega-x)^2}\quad\text{and}\quad D(qx+\omega)=qC(qx+\omega)-\frac{2\varepsilon x+\gamma}{qx+\omega-x}.$$

If we now replace x by $(x-\omega)/q$, we obtain the so-called *symmetric* form

$$C(x)y_n(qx+\omega)-\{C(x)+D(x)\}\,y_n(x)+D(x)y_n\left((x-\omega)/q\right)=\lambda_n y_n(x)\qquad(2.2.12)$$

for $n=0,1,2,\ldots$ with

$$C(x)=\frac{e(x-\omega)^2+2fq(x-\omega)+gq^2}{q(qx+\omega-x)^2}$$
$$\text{and}\quad D(x)=qC(x)-\frac{2\varepsilon(x-\omega)+\gamma q}{qx+\omega-x}.\qquad(2.2.13)$$

Finally, we will derive a third version of the operator equation (2.2.7), which involves the operators $\mathscr{A}_{q,\omega}$ and $\mathscr{A}_{1/q,-\omega/q}$. First of all we have

$$\left(\mathscr{A}_{q,\omega}y_n\right)\left((x-\omega)/q\right)=\frac{y_n\left(q((x-\omega)/q)+\omega\right)-y_n((x-\omega)/q)}{q((x-\omega)/q)+\omega-(x-\omega)/q}$$
$$=\frac{y_n((x-\omega)/q)-y_n(x)}{(x-\omega)/q-x}=\left(\mathscr{A}_{1/q,-\omega/q}y_n\right)(x).$$

In the case of the q-derivative operator $\mathscr{D}_q$ (i.e. $\omega=0$), this reads

$$\left(\mathscr{D}_q y_n\right)(x/q)=\left(\mathscr{A}_{q,0}y_n\right)(x/q)=\left(\mathscr{A}_{1/q,0}y_n\right)(x)=\mathscr{D}_{1/q}y_n(x).$$

In the case of the difference operator Δ (i.e. $q=1$ and $\omega=1$), this reads

$$\left(\Delta y_n\right)(x-1)=\left(\mathscr{A}_{1,1}y_n\right)(x-1)=\left(\mathscr{A}_{1,-1}y_n\right)(x)=:\nabla y_n(x).$$

Now we have

$$\left(\mathscr{A}_{q,\omega}\left(\mathscr{A}_{1/q,-\omega/q}y_n\right)\right)(x)=\frac{\left(\mathscr{A}_{1/q,-\omega/q}y_n\right)(qx+\omega)-\left(\mathscr{A}_{1/q,-\omega/q}y_n\right)(x)}{qx+\omega-x}$$
$$=\frac{y_n(qx+\omega)-(1+q)y_n(x)+qy_n((x-\omega)/q)}{(qx+\omega-x)^2}$$
$$=q^{-1}\left(\mathscr{A}_{q,\omega}^2y_n\right)\left((x-\omega)/q\right).$$

Hence we obtain for $n=0,1,2,\ldots$

$$q\varphi((x-\omega)/q)\left(\mathscr{A}_{q,\omega}\left(\mathscr{A}_{1/q,-\omega/q}y_n\right)\right)(x)$$
$$+\psi((x-\omega)/q)\left(\mathscr{A}_{1/q,-\omega/q}y_n\right)(x)=\lambda_n y_n(x). \tag{2.2.14}$$

In the case of the q-derivative operator $\mathscr{D}_q$ (i.e. $\omega=0$), this reads

$$q\varphi(x/q)\left(\mathscr{D}_q\left(\mathscr{D}_{1/q}y_n\right)\right)(x)+\psi(x/q)\left(\mathscr{D}_{1/q}y_n\right)(x)=\lambda_n y_n(x) \tag{2.2.15}$$

for $n=0,1,2,\ldots$. In the case of the difference operator Δ (i.e. $q=1$ and $\omega=1$), this reads

$$\varphi(x-1)(\Delta(\nabla y_n))(x)+\psi(x-1)(\nabla y_n)(x)=\lambda_n y_n(x), \quad n=0,1,2,\ldots. \tag{2.2.16}$$

2.3 The Regularity Condition

In this section we will point out in which cases the eigenvalue problem (2.2.1) has essentially unique polynomial solutions $y_n(x)$ of degrees $n=0,1,2,\ldots,N$ for some positive integer N with possibly $N\to\infty$. Solutions are called essentially unique if they are determined up to a factor independent of x. We have

Theorem 2.2. *Let N denote a positive integer (possibly $N\to\infty$). Then the following statements are equivalent:*

1. *For each $n=0,1,2,\ldots,N$ there exists a solution y_n of the eigenvalue problem (2.2.1) and all eigenspaces are one dimensional.*
2. *For $m,n\in\{0,1,2,\ldots,N\}$ with $m\neq n$ we have $\lambda_m\neq\lambda_n$.*

Proof. Assume that $\lambda_m=\lambda_n$ for $m\neq n$. Then there is either no polynomial solution for one of the degrees m and n or the solutions y_m and y_n belong to the same eigenspace. This shows that the first statement implies the second.

Now we use induction to show that the second statement implies the first. For $n=0$ we have $\lambda_0=0$ and the one-dimensional eigenspace generated by $y_0(x)=1$. Now we assume that $n\in\{1,2,3,\ldots\}$. Suppose that the polynomials $y_\nu(x)$ are solutions of degree ν for $\nu=0,1,2,\ldots,n-1$. Then the (monic) polynomial $y_n(x)$ of degree n given by

$$y_n(x)=x^n+\sum_{\nu=0}^{n-1}\alpha_\nu y_\nu(x) \quad \text{with} \quad \alpha_\nu\in\mathbb{C} \tag{2.3.1}$$

is a solution of (2.2.1) if

$$\begin{aligned}
\varphi(x)\mathscr{A}_{q,\omega}^2(x^n)+\psi(x)\mathscr{A}_{q,\omega}(x^n)\\
+\varphi(x)\left(\mathscr{A}_{q,\omega}^2\sum_{\nu=0}^{n-1}\alpha_\nu y_\nu\right)(x)+\psi(x)\left(\mathscr{A}_{q,\omega}\sum_{\nu=0}^{n-1}\alpha_\nu y_\nu\right)(x)\\
=\lambda_n\left((qx+\omega)^n+\sum_{\nu=0}^{n-1}\alpha_\nu y_\nu(qx+\omega)\right)
\end{aligned}$$

holds. The polynomial $\varphi(x)\mathscr{A}_{q,\omega}^2(x^n)+\psi(x)\mathscr{A}_{q,\omega}(x^n)$ has degree at most n. Hence we may write

$$\varphi(x)\mathscr{A}_{q,\omega}^2(x^n)+\psi(x)\mathscr{A}_{q,\omega}(x^n)=\beta_n(qx+\omega)^n+\sum_{\nu=0}^{n-1}\beta_\nu y_\nu(qx+\omega)$$

with $\beta_n,\beta_\nu\in\mathbb{C}$. Combining the last two equations, we get

$$\beta_n(qx+\omega)^n+\sum_{\nu=0}^{n-1}(\beta_\nu+\lambda_\nu\alpha_\nu)y_\nu(qx+\omega)=\lambda_n\left((qx+\omega)^n+\sum_{\nu=0}^{n-1}\alpha_\nu y_\nu(qx+\omega)\right)$$

and therefore

$$(\beta_n-\lambda_n)(qx+\omega)^n+\sum_{\nu=0}^{n-1}(\alpha_\nu(\lambda_\nu-\lambda_n)+\beta_\nu)y_\nu(qx+\omega)=0.$$

Since $\lambda_\nu\neq\lambda_n$, this implies that the numbers α_ν are uniquely determined by this equation. So in fact this means that the (monic) polynomial solution y_n given by (2.3.1) is uniquely determined, which implies that the corresponding eigenspace is one dimensional. □

Now we use (2.2.3) and (2.2.4) to find from (2.2.6):

$$\begin{aligned}
q^n(\lambda_n-\lambda_m)&=e([n][n-1]-q^{n-m}[m][m-1])+2\varepsilon([n]-q^{n-m}[m])\\
&=e[n-m][n+m-1]+2\varepsilon[n-m]\\
&=[n-m](e[n+m-1]+2\varepsilon),\quad n\geq m,\quad m,n\in\{0,1,2,\ldots\}.
\end{aligned}$$

Hence we have

$$\lambda_n-\lambda_m=\frac{[n-m]}{q^n}(e[n+m-1]+2\varepsilon),\quad n\geq m,\quad m,n\in\{0,1,2,\ldots\}.\tag{2.3.2}$$

Since $q\neq-1$, it follows that $[n-m]\neq0$ for $m\neq n$, so $\lambda_m\neq\lambda_n$ is equivalent to $e[n+m-1]+2\varepsilon\neq0$. Therefore, theorem 2.2 leads to:

Corollary 2.3. *Let N denote a positive integer (possibly $N\to\infty$). Then the eigenvalue problem (2.2.1) has polynomial solutions y_n of degree n for all $n=0,1,2,\ldots,N$ with one-dimensional eigenspaces if and only if the regularity condition*

$$e[n]+2\varepsilon \neq 0, \quad n=0,1,2,\ldots,2N-2 \tag{2.3.3}$$

holds.

2.4 Determination of the Polynomial Solutions

We want to obtain a two-term recurrence relation for the coefficients of the polynomial solutions y_n of (2.2.1). In order to achieve this, we introduce so-called generalized binomial coefficients $\begin{bmatrix} x;c \\ n \end{bmatrix}$ with $x \in \mathbb{R}$, $c \in \mathbb{C}$ and $n \in \{0,1,2,\ldots\}$, such that

$$\left(\mathscr{A}_{q,\omega}\begin{bmatrix} \cdot;c \\ 0 \end{bmatrix}\right)(x)=0 \quad \text{and} \quad \left(\mathscr{A}_{q,\omega}\begin{bmatrix} \cdot;c \\ n \end{bmatrix}\right)(x)=\begin{bmatrix} x;c \\ n-1 \end{bmatrix} \tag{2.4.1}$$

for $n=1,2,3,\ldots$. This can be done by

$$\begin{bmatrix} x;c \\ 0 \end{bmatrix}:=1 \quad \text{and} \quad \begin{bmatrix} x;c \\ n \end{bmatrix}:=\prod_{i=1}^{n}\frac{x+cq^{i-1}-[i-1]\omega}{[i]}, \quad n=1,2,3,\ldots. \tag{2.4.2}$$

Note that these generalized binomial coefficients depend on q and ω. However, for ease of expression we omit these in the notation of the symbol $\begin{bmatrix} \; \end{bmatrix}$. In the case of the difference operator (i.e. $q=1$ and $\omega=1$) these generalized binomial coefficients reduce to the ordinary binomial coefficients and

$$\Delta\binom{x+c}{0}=0 \quad \text{and} \quad \Delta\binom{x+c}{n}=\binom{x+c}{n-1}, \quad n=1,2,3,\ldots.$$

In the case of the differentiation operator D (i.e. $q=1$ and $\omega=0$), we have

$$D(1)=0 \quad \text{and} \quad D\frac{(x+c)^n}{n!}=\frac{(x+c)^{n-1}}{(n-1)!}, \quad n=1,2,3,\ldots.$$

2.4.1 First Approach

To simplify the expressions, we write $v=\omega+c(1-q)$. Then we have

$$\begin{bmatrix} x;c \\ n \end{bmatrix}=\prod_{i=1}^{n}\frac{x+c-[i-1]v}{[i]}, \quad n=1,2,3,\ldots.$$

Now we write

$$y_n(x)=\sum_{k=0}^{n}a_{n,k}\begin{bmatrix} x;c \\ k \end{bmatrix}, \quad a_{n,n}\neq 0, \quad n=0,1,2,\ldots. \tag{2.4.3}$$

For a given $a_{n,n} \neq 0$ we want to determine the other coefficients $a_{n,k}$ in such a way that $y_n(x)$ given by (2.4.3) satisfies (2.2.7). By using (2.2.3), (2.2.4), (2.4.1),

$$x\begin{bmatrix} x\,;c \\ k-1 \end{bmatrix} = [k]\begin{bmatrix} x\,;c \\ k \end{bmatrix} + ([k-1]v - c)\begin{bmatrix} x\,;c \\ k-1 \end{bmatrix}, \quad k = 1,2,3,\ldots,$$

$$\begin{aligned} x^2\begin{bmatrix} x\,;c \\ k-2 \end{bmatrix} &= [k][k-1]\begin{bmatrix} x\,;c \\ k \end{bmatrix} + [k-1]\{([k-1]+[k-2])\,v - 2c\}\begin{bmatrix} x\,;c \\ k-1 \end{bmatrix} \\ &\quad + ([k-2]v-c)^2\begin{bmatrix} x\,;c \\ k-2 \end{bmatrix}, \quad k = 2,3,4,\ldots, \end{aligned}$$

$$\begin{bmatrix} qx+\omega\,;c \\ k \end{bmatrix} = q^k\begin{bmatrix} x\,;c \\ k \end{bmatrix} + vq^{k-1}\begin{bmatrix} x\,;c \\ k-1 \end{bmatrix}, \quad k = 1,2,3,\ldots,$$

by substituting (2.4.3) into (2.2.7) and by comparing the coefficients, we find

$$\begin{aligned} &[n-k]\,(e[n+k-1]+2\varepsilon)\,a_{n,k} + \Big\{e[n][n-1]v + 2\varepsilon[n-k]v \\ &\quad - e[k]\,(([k]+[k-1])v - 2c)\,q^{n-k} + (2\varepsilon c - 2f[k] - \gamma)q^{n-k}\Big\}a_{n,k+1} \\ &\quad - \big(e([k]v-c)^2 + 2f([k]v-c) + g\big)\,q^{n-k}a_{n,k+2} = 0 \end{aligned} \tag{2.4.4}$$

for $k = n-1, n-2, n-3, \ldots, 0$ with the convention that $a_{n,n+1} := 0$. Hence, if the regularity condition (2.3.3) holds, then (2.4.4) gives us the coefficients $a_{n,k}$ for $k = n-1, n-2, n-3, \ldots, 0$ in terms of $a_{n,n} \neq 0$.

The three-term recurrence relation (2.4.4) can be rewritten in a such a way that the possibility of a two-term recurrence relation becomes apparent:

$$\begin{aligned} &[n-k]\,(e[n+k-1]+2\varepsilon)\,a_{n,k} - (e[k-1]v - 2ec + 2f)\,[k-1]q^{n-k+1}a_{n,k+1} \\ &\quad + [n-k-1]\,(e[n+k]+2\varepsilon)\,va_{n,k+1} - (e[k]v - 2ec + 2f)\,[k]vq^{n-k}a_{n,k+2} \\ &\quad + ((e+2\varepsilon)v + \{2c(e+\varepsilon) - \gamma - 2f\}\,q)\,q^{n-k-1}a_{n,k+1} \\ &\quad - (ec^2 - 2fc + g)q^{n-k}a_{n,k+2} = 0. \end{aligned}$$

This implies that if

$$(e+2\varepsilon)v^2 + \{2c(e+\varepsilon) - \gamma - 2f\}\,vq = -(ec^2 - 2fc + g)q^2 \tag{2.4.5}$$

holds, then the recurrence relation (2.4.4) can be written in the form

$$s(k)a_{n,k} + t(k)a_{n,k+1} + v\,\big(s(k+1)a_{n,k+1} + t(k+1)a_{n,k+2}\big) = 0$$

with

$$s(k) := [n-k]\,(e[n+k-1]+2\varepsilon)$$

and

$$t(k) := \left((e+2\varepsilon)v + \left\{2\varepsilon c - \gamma + 2(ec-f)[k] - e[k-1]^2 vq\right\} q\right) q^{n-k-1}.$$

Now we consider the two-term recurrence relation

$$s(k)a_{n,k} + t(k)a_{n,k+1} = 0, \quad k = n-1, n-2, n-3, \ldots, 0. \tag{2.4.6}$$

If there exists a number c satisfying (2.4.5) and if the coefficients $a_{n,k}$ satisfy (2.4.6), then they also satisfy (2.4.4). Therefore, the coefficients $a_{n,k}$ can be determined by (2.4.6) in terms of $a_{n,n} \neq 0$ provided that there exists a number c such that (2.4.5) holds.

In the case that $v \neq 0$, and c satisfies (2.4.5), we can write the two-term recurrence relation for the coefficients $a_{n,k}$ in a different form. If we multiply (2.4.6) by v and use (2.4.5), we obtain

$$\begin{aligned}&[n-k]\left(e[n+k-1]+2\varepsilon\right) v a_{n,k}\\ &\quad - \left\{e([k-1]v-c)^2 + 2f([k-1]v-c) + g\right\} q^{n-k+1} a_{n,k+1} = 0\end{aligned} \tag{2.4.7}$$

for $k = n-1, n-2, n-3, \ldots, 0$.

With this first approach the case of the differentiation operator D (i.e. $q = 1$ and $\omega = v = 0$) can be treated completely. If there exists a c satisfying (2.4.5), id est $ec^2 - 2fc + g = 0$, then the coefficients of the polynomial solutions

$$y_n(x) = \sum_{k=0}^{n} a_{n,k} \frac{(x+c)^k}{k!}, \quad a_{n,n} \neq 0, \quad n = 0, 1, 2, \ldots \tag{2.4.8}$$

of the second-order differential equation (2.2.9) satisfy the two-term recurrence relation (2.4.6), id est

$$(n-k)\left(e(n+k-1)+2\varepsilon\right) a_{n,k} + \left(2(ec-f)k + 2\varepsilon c - \gamma\right) a_{n,k+1} = 0 \tag{2.4.9}$$

for $k = n-1, n-2, n-3, \ldots, 0$. If (2.4.5) has no solution for c, id est $e = f = 0$ and $g \neq 0$, then we find from (2.4.4) with $c = \gamma/2\varepsilon$ the two-term recurrence relation

$$2\varepsilon(n-k)a_{n,k} - g a_{n,k+2} = 0, \quad a_{n,n+1} = 0, \quad k = n-1, n-2, n-3, \ldots, 0, \tag{2.4.10}$$

which leads to the (symmetric) Hermite polynomials.

2.4.2 Second Approach

In order to deal with the cases where c cannot be determined by (2.4.5), we use a second approach. For this purpose we will modify (2.4.3) in such a way that in the numerator of the generalized binomial coefficient $x + c$ occurs as the penultimate factor. By using $v = \omega + c(1-q)$ and the definition (2.4.2), we obtain

$$\begin{aligned}\begin{bmatrix} x\,;\,c+[k-2]vq^{2-k} \\ k \end{bmatrix} &= \prod_{i=1}^{k} \frac{x+cq^{i-1}+[k-2]vq^{i+1-k}-[i-1]\omega}{[i]} \\ &= \prod_{i=1}^{k} \frac{x+c+[k-i-1]vq^{i+1-k}}{[i]} \end{aligned} \tag{2.4.11}$$

for $k=1,2,3,\ldots$. Note that

$$\left(\mathscr{A}_{q,\omega}\begin{bmatrix} \cdot\,;\,c+[k-2]vq^{2-k} \\ k \end{bmatrix}\right)(x) = \begin{bmatrix} x\,;\,c+[k-2]vq^{2-k} \\ k-1 \end{bmatrix}, \quad k=1,2,3,\ldots.$$

These alternative generalized binomial coefficients can be used to build a second form for the solutions:

$$y_n(x) = \sum_{k=0}^{n} b_{n,k} \begin{bmatrix} x\,;\,c+[k-2]vq^{2-k} \\ k \end{bmatrix}, \quad b_{n,n} \neq 0, \quad n=0,1,2,\ldots. \tag{2.4.12}$$

For a given $b_{n,n} \neq 0$ we want to determine the other coefficients $b_{n,k}$ in such a way that $y_n(x)$ given by (2.4.12) satisfies (2.2.7). By using

$$x\begin{bmatrix} x\,;\,c+[k-2]vq^{2-k} \\ k-1 \end{bmatrix} = [k]\begin{bmatrix} x\,;\,c+[k-2]vq^{2-k} \\ k \end{bmatrix} + (v-c)\begin{bmatrix} x\,;\,c+[k-2]vq^{2-k} \\ k-1 \end{bmatrix}$$

for $k=1,2,3,\ldots$,

$$\begin{aligned} x^2\begin{bmatrix} x\,;\,c+[k-2]vq^{2-k} \\ k-2 \end{bmatrix} &= [k][k-1]\begin{bmatrix} x\,;\,c+[k-2]vq^{2-k} \\ k \end{bmatrix} \\ &\quad + [k-1](v-2c)\begin{bmatrix} x\,;\,c+[k-2]vq^{2-k} \\ k-1 \end{bmatrix} + c^2\begin{bmatrix} x\,;\,c+[k-2]vq^{2-k} \\ k-2 \end{bmatrix} \end{aligned}$$

for $k=2,3,4,\ldots$,

$$x\begin{bmatrix} x\,;\,c+[k-2]vq^{2-k} \\ k-2 \end{bmatrix} = [k-1]\begin{bmatrix} x\,;\,c+[k-2]vq^{2-k} \\ k-1 \end{bmatrix} - c\begin{bmatrix} x\,;\,c+[k-2]vq^{2-k} \\ k-2 \end{bmatrix}$$

for $k=2,3,4,\ldots$ and

$$\begin{aligned} \begin{bmatrix} qx+\omega\,;\,c+[k-2]vq^{2-k} \\ k \end{bmatrix} &= q^k\begin{bmatrix} x\,;\,c+[k-1]vq^{1-k} \\ k \end{bmatrix} \\ &= q^k\begin{bmatrix} x\,;\,c+[k-2]vq^{2-k} \\ k \end{bmatrix} + vq\begin{bmatrix} x\,;\,c+[k-2]vq^{2-k} \\ k-1 \end{bmatrix} \end{aligned}$$

for $k=1,2,3,\ldots$, we find by substituting (2.4.12) into (2.2.7)

$$\sum_{k=0}^{n}[n-k]\left(e[n+k-1]+2\varepsilon\right)q^{k}\begin{bmatrix}x;c+[k-1]vq^{1-k}\\ k\end{bmatrix}b_{n,k}$$
$$+\sum_{k=1}^{n}\left\{\left(e[k-1]+2\varepsilon\right)\left([k]q^{1-k}-1\right)vq^{n}+\left(2[k-1](ec-f)+2\varepsilon c-\gamma\right)q^{n}\right\}$$
$$\times\begin{bmatrix}x;c+[k-2]vq^{2-k}\\ k-1\end{bmatrix}b_{n,k}$$
$$-\sum_{k=2}^{n}(ec^{2}-2fc+g)q^{n}\begin{bmatrix}x;c+[k-2]vq^{2-k}\\ k-2\end{bmatrix}b_{n,k}=0.$$

Hence, if $ec^2-2fc+g=0$ can be solved for c, only two sums remain. These lead to a two-term recurrence relation for the coefficients $b_{n,k}$:

$$[n-k]\left(e[n+k-1]+2\varepsilon\right)b_{n,k}+\left\{(e[k]+2\varepsilon)([k+1]q^{-k}-1)v\right.$$
$$\left.+\left(2[k](ec-f)+2\varepsilon c-\gamma\right)\right\}q^{n-k}b_{n,k+1}=0 \tag{2.4.13}$$

for $k=n-1,n-2,n-3,\dots,0$.

The case of the difference operator Δ (i.e. $q=\omega=v=1$) can be treated by using both approaches. The first approach leads to solutions of the form

$$y_n(x)=\sum_{k=0}^{n}a_{n,k}\binom{x+c}{k},\quad a_{n,n}\neq 0,\quad n=0,1,2,\dots, \tag{2.4.14}$$

where c is a solution of (2.4.5), id est

$$e(c+1)^2-2f(c+1)+g=-2\varepsilon(c+1)+\gamma$$

and the coefficients satisfy (2.4.6), which reads

$$(n-k)\left(e(n+k-1)+2\varepsilon\right)a_{n,k}$$
$$-\left(e(k-1-c)^2+2f(k-1-c)+g\right)a_{n,k+1}=0 \tag{2.4.15}$$

for $k=n-1,n-2,n-3,\dots,0$. The second approach leads to solutions of the form

$$y_n(x)=\sum_{k=0}^{n}b_{n,k}\binom{x+c+k-2}{k},\quad b_{n,n}\neq 0,\quad n=0,1,2,\dots. \tag{2.4.16}$$

If c satisfies $ec^2-2fc+g=0$, then the coefficients satisfy the two-term recurrence relation (2.4.13), which reads

$$(n-k)\left(e(n+k-1)+2\varepsilon\right)b_{n,k}$$
$$+\left(ek^2+2(ec-f+\varepsilon)k+2\varepsilon c-\gamma\right)b_{n,k+1}=0 \tag{2.4.17}$$

for $k=n-1,n-2,n-3,\dots,0$.

2.5 Existence of a Three-Term Recurrence Relation

In this section we will show that the monic polynomial solutions y_n of the operator equation (2.2.1) satisfy a three-term recurrence relation of the form

$$y_{n+1}(x) = (x - c_n)y_n(x) - d_n y_{n-1}(x), \quad c_n, d_n \in \mathbb{C}, \quad n = 1,2,3,\ldots. \tag{2.5.1}$$

For simplification we introduce the operator $\mathscr{S}_{q,\omega}$ on the space $\mathscr{P}$ of polynomials defined by

$$(\mathscr{S}_{q,\omega}p)(x) := p(qx+\omega), \quad p \in \mathscr{P}, \quad x \in \mathbb{R}. \tag{2.5.2}$$

Then the following commutation relations hold:

$$\mathscr{A}_{q,\omega}\mathscr{S}_{q,\omega} = q\mathscr{S}_{q,\omega}\mathscr{A}_{q,\omega} \quad \text{and} \quad \mathscr{S}_{q,\omega}^{-1}\mathscr{A}_{q,\omega} = q\mathscr{A}_{q,\omega}\mathscr{S}_{q,\omega}^{-1}. \tag{2.5.3}$$

Further we will use the notation

$$\widehat{p}(x) := \left(\mathscr{S}_{q,\omega}^{-1}p\right)(x) = p((x-\omega)/q), \quad p \in \mathscr{P}, \quad x \in \mathbb{R} \tag{2.5.4}$$

for convenience.

Theorem 2.4. *Let Λ be a linear functional on $\mathscr{P}$ defined by*

$$\Lambda[1] = 1 \quad \textit{and} \quad \Lambda[y_n] = 0, \quad n = 1,2,3,\ldots, \tag{2.5.5}$$

where y_n denotes a polynomial solution of the operator equation (2.2.1). Then the following distributional equation holds for every polynomial $p \in \mathscr{P}$:

$$\Lambda[q\widehat{\varphi}(\mathscr{A}_{q,\omega}p) + \widehat{\psi}p] = 0. \tag{2.5.6}$$

Furthermore, for all $m,n \in \{0,1,2,\ldots\}$ we have

$$\Lambda\left[q\widehat{\varphi}\left(\mathscr{A}_{q,\omega}y_m\right)\left(\mathscr{A}_{q,\omega}y_n\right)\right] = -\lambda_n\Lambda\left[y_m y_n\right]. \tag{2.5.7}$$

Proof. By using (2.5.2), we obtain from (2.2.1) for every polynomial

$$p^*(x) = \sum_{k=0}^{n} \alpha_k y_k(x), \quad \alpha_k \in \mathbb{C}$$

that

$$\begin{aligned}
&\varphi(x)\left(\mathscr{A}_{q,\omega}^2 p^*\right)(x) + \psi(x)\left(\mathscr{A}_{q,\omega}p^*\right)(x) \\
&= \sum_{k=0}^{n} \alpha_k \left\{\varphi(x)\left(\mathscr{A}_{q,\omega}^2 y_k\right)(x) + \psi(x)\left(\mathscr{A}_{q,\omega}y_k\right)(x)\right\} \\
&= \sum_{k=0}^{n} \alpha_k \lambda_k \left(\mathscr{S}_{q,\omega}y_k\right)(x).
\end{aligned}$$

Applying $\mathscr{S}_{q,\omega}^{-1}$ to both sides and using the commutation relation (2.5.3), we obtain by using the notation (2.5.4)

$$q\widehat{\varphi}(x)\left(\mathscr{A}_{q,\omega}\mathscr{S}_{q,\omega}^{-1}\mathscr{A}_{q,\omega}p^*\right)(x)+\widehat{\psi}(x)\left(\mathscr{S}_{q,\omega}^{-1}\mathscr{A}_{q,\omega}p^*\right)(x)=\sum_{k=0}^{n}\alpha_k\lambda_k y_k(x).$$

From (2.2.5) it can be deduced that for each polynomial $p\in\mathscr{P}$ there exists a polynomial p^* such that

$$p(x)=\left(\mathscr{S}_{q,\omega}^{-1}\mathscr{A}_{q,\omega}p^*\right)(x).$$

Now we use the fact that $\lambda_0=0$ and $\Lambda[y_k]=0$ for $k=1,2,3,\ldots$ to conclude that

$$\Lambda\left[q\widehat{\varphi}\left(\mathscr{A}_{q,\omega}p\right)+\widehat{\psi}p\right]=\sum_{k=0}^{n}\alpha_k\lambda_k\Lambda[y_k]=0,$$

which proves (2.5.6).

To prove (2.5.7), we apply $\mathscr{S}_{q,\omega}^{-1}$ to the operator equation (2.2.1), use the commutation relation (2.5.3) and multiply the result by y_m:

$$q\widehat{\varphi}(x)\left(\mathscr{A}_{q,\omega}\mathscr{S}_{q,\omega}^{-1}\mathscr{A}_{q,\omega}y_n\right)(x)y_m(x)+\widehat{\psi}(x)\left(\mathscr{S}_{q,\omega}^{-1}\mathscr{A}_{q,\omega}y_n\right)(x)y_m(x)=\lambda_n y_m(x)y_n(x).$$

If we apply the product rule (2.1.2) to $p_1(x)=\left(\mathscr{S}_{q,\omega}^{-1}\mathscr{A}_{q,\omega}y_n\right)(x)$ and $p_2(x)=y_m(x)$, then we find

$$\begin{aligned}&\left(\mathscr{A}_{q,\omega}\left(\left(\mathscr{S}_{q,\omega}^{-1}\mathscr{A}_{q,\omega}y_n\right)y_m\right)\right)(x)\\&=\left(\mathscr{A}_{q,\omega}\mathscr{S}_{q,\omega}^{-1}\mathscr{A}_{q,\omega}y_n\right)(x)y_m(x)+\left(\mathscr{A}_{q,\omega}y_n\right)(x)\left(\mathscr{A}_{q,\omega}y_m\right)(x).\end{aligned}$$

Combining the last two results, we find that

$$\begin{aligned}&q\widehat{\varphi}(x)\left\{\left(\mathscr{A}_{q,\omega}\left(\left(\mathscr{S}_{q,\omega}^{-1}\mathscr{A}_{q,\omega}y_n\right)y_m\right)\right)(x)-\left(\mathscr{A}_{q,\omega}y_n\right)(x)\left(\mathscr{A}_{q,\omega}y_m\right)(x)\right\}\\&\quad+\widehat{\psi}(x)\left(\mathscr{S}_{q,\omega}^{-1}\mathscr{A}_{q,\omega}y_n\right)(x)y_m(x)=\lambda_n y_m(x)y_n(x).\end{aligned}$$

Finally, we apply Λ on both sides of this equation to obtain

$$\begin{aligned}&\Lambda\left[q\widehat{\varphi}\left(\mathscr{A}_{q,\omega}\left(\left(\mathscr{S}_{q,\omega}^{-1}\mathscr{A}_{q,\omega}y_n\right)y_m\right)\right)+\widehat{\psi}\left(\mathscr{S}_{q,\omega}^{-1}\mathscr{A}_{q,\omega}y_n\right)y_m\right]\\&\quad-\Lambda\left[q\widehat{\varphi}\left(\mathscr{A}_{q,\omega}y_n\right)\left(\mathscr{A}_{q,\omega}y_m\right)\right]=\lambda_n\Lambda[y_m y_n].\end{aligned}$$

The first term equals zero in view of (2.5.6) with $p(x)=\left(\mathscr{S}_{q,\omega}^{-1}\mathscr{A}_{q,\omega}y_n\right)(x)y_m(x)$. This completes the proof of (2.5.7). □

The following theorem states the "weak" orthogonality of the polynomial solutions of the eigenvalue problem (2.2.1).

Theorem 2.5. *Let the regularity condition (2.3.3) hold for the operator equation (2.2.1) with polynomial solutions y_n with $n=0,1,2,\ldots$. Then the linear functional Λ given by (2.5.5) satisfies*

$$\Lambda[y_m y_n] = 0 \quad for \quad m \neq n, \quad m,n \in \{0,1,2,\ldots\}. \tag{2.5.8}$$

Proof. From (2.5.7) we get for $m,n \in \{0,1,2,\ldots\}$

$$\begin{cases} \Lambda\left[q\widehat{\varphi}\left(\mathscr{A}_{q,\omega}y_n\right)\left(\mathscr{A}_{q,\omega}y_m\right)\right] = -\lambda_n \Lambda[y_m y_n] \\ \\ \Lambda\left[q\widehat{\varphi}\left(\mathscr{A}_{q,\omega}y_m\right)\left(\mathscr{A}_{q,\omega}y_n\right)\right] = -\lambda_m \Lambda[y_m y_n]. \end{cases}$$

Subtracting these two equations, we find

$$(\lambda_n - \lambda_m)\Lambda[y_m y_n] = 0, \quad m,n \in \{0,1,2,\ldots\}.$$

The regularity condition (2.3.3) implies that $\lambda_m \neq \lambda_n$ for $m \neq n$, which implies (2.5.8). □

If the linear functional Λ in the preceding theorem is quasi-definite, it is well known that a recurrence relation of the form (2.5.1) exists for all $n = 0,1,2,\ldots$. See for instance [146]. Now we will prove:

Theorem 2.6. *With the assumptions of the preceding theorem, assume that there exists a number $N \in \{1,2,3\ldots\}$ such that*

$$\Lambda[y_n^2] \neq 0, \quad for \quad n = 0,1,2,\ldots,N-1 \quad and \quad \Lambda[y_N^2] = 0. \tag{2.5.9}$$

Then there exists a three-term recurrence relation of the form

$$y_{n+1}(x) = (x - c_n)y_n(x) - d_n y_{n-1}(x), \quad n = 1,2,3,\ldots,N+1 \tag{2.5.10}$$

with $d_N = 0$.

Proof. The monic polynomial y_{n+1} can be written as

$$y_{n+1}(x) = x y_n(x) + \sum_{k=0}^{n} \alpha_k^{(n)} y_k(x), \quad \alpha_k^{(n)} \in \mathbb{C}, \quad n = 0,1,2,\ldots. \tag{2.5.11}$$

Now we want to show that $\alpha_k^{(n)} = 0$ for $k = 0,1,2,\ldots,n-2$. To do this we multiply this equation by y_ν for $\nu \in \{0,1,2,\ldots,n-2\}$ and apply the linear functional Λ to both sides, which leads to

$$\alpha_k^{(n)} \Lambda[y_k^2] = 0, \quad k = 0,1,2,\ldots,n-2.$$

Now we use (2.5.9) to conclude that $\alpha_k^{(n)} = 0$ for $k = 0,1,2,\ldots,n-2$ for all $n = 2,3,4,\ldots,N+1$, which proves (2.5.10) with $c_n = -\alpha_n^{(n)}$ and $d_n = -\alpha_{n-1}^{(n)}$.

In order to show that $d_N = 0$, we start with (2.5.10) for $n = N$, multiply by y_{N-1} and apply the linear functional Λ to find

$$\Lambda[y_{N+1}y_{N-1}] = \Lambda[x y_N y_{N-1}] - c_N \Lambda[y_N y_{N-1}] - d_N \Lambda[y_{N-1}^2].$$

Since $\Lambda\left[xy_N y_{N-1}\right] = \Lambda\left[y_N^2\right] = 0$ and $\Lambda\left[y_{N-1}^2\right] \neq 0$, we get

$$d_N = \frac{\Lambda\left[y_N^2\right]}{\Lambda\left[y_{N-1}^2\right]} = 0.$$

This completes the proof. □

Now we will show that a finite system $\{y_n\}_{n=0}^N$ of polynomial solutions of the eigenvalue problem (2.2.1) which satisfies a three-term recurrence relation of the form (2.5.10) can be extended to an infinite system $\{y_n\}_{n=0}^{\infty}$ of polynomials which satisfies a three-term recurrence relation of the form (2.5.1). First we will prove:

Theorem 2.7. *Let the monic solutions $\{y_n\}_{n=0}^N$ of the eigenvalue problem (2.2.1) satisfy the three-term recurrence relation (2.5.10) and let the regularity condition (2.3.3) with $N \to \infty$ hold. Then we may write*

$$y_{N+k} = \widetilde{y}_k y_N, \quad k = 0, 1, 2, \ldots \tag{2.5.12}$$

with monic polynomials $\widetilde{y}_k$ of degree k which are solutions of the eigenvalue problem

$$\widetilde{\varphi}\left(\mathscr{A}_{q,\omega}^2 \widetilde{y}_k\right) + \widetilde{\psi}\left(\mathscr{A}_{q,\omega} \widetilde{y}_k\right) = \widetilde{\lambda}_k\left(\mathscr{S}_{q,\omega} \widetilde{y}_k\right), \quad k = 0, 1, 2, \ldots, \tag{2.5.13}$$

where $\widetilde{\lambda}_k = \lambda_{N+k} - \lambda_N$, $\widetilde{\varphi}$ is a polynomial of degree at most 2 with $\widetilde{\varphi} \neq 0$ and $\widetilde{\psi}$ is a polynomial of degree 1 exactly.

Proof. Since $d_N = 0$, we have from (2.5.10)

$$y_{N+1}(x) = (x - c_N) y_N(x) = \widetilde{y}_1(x) y_N(x) \quad \text{with} \quad \widetilde{y}_1(x) = x - c_N$$

and

$$\begin{aligned} y_{N+2}(x) &= (x - c_{N+1}) y_{N+1}(x) - d_{N+1} y_N(x) \\ &= \left((x - c_{N+1})(x - c_N) - d_{N+1}\right) y_N(x) = \widetilde{y}_2(x) y_N(x) \end{aligned}$$

with $\widetilde{y}_2(x) = (x - c_{N+1})(x - c_N) - d_{N+1}$. Together with $\widetilde{y}_0(x) = 1$ this proves (2.5.12) for $k = 0, 1, 2$.

Substitution of $y_{N+k} = \widetilde{y}_k y_N$ in the eigenvalue problem (2.2.1) gives

$$\varphi\left(\mathscr{A}_{q,\omega}^2\left(\widetilde{y}_k y_N\right)\right) + \psi\left(\mathscr{A}_{q,\omega}\left(\widetilde{y}_k y_N\right)\right) = \lambda_{N+k}\left(\mathscr{S}_{q,\omega}\left(\widetilde{y}_k y_N\right)\right).$$

By using the product rule (2.1.2), we find

$$\mathscr{A}_{q,\omega}\left(\widetilde{y}_k y_N\right) = \left(\mathscr{A}_{q,\omega} y_N\right)\left(\mathscr{S}_{q,\omega} \widetilde{y}_k\right) + y_N\left(\mathscr{A}_{q,\omega} \widetilde{y}_k\right),$$

$$\mathscr{A}_{q,\omega}\left(\widetilde{y}_k y_N\right) = \left(\mathscr{A}_{q,\omega} y_N\right) \widetilde{y}_k + \left(\mathscr{S}_{q,\omega} y_N\right)\left(\mathscr{A}_{q,\omega} \widetilde{y}_k\right)$$

and

$$\mathscr{A}_{q,\omega}^2(\widetilde{y}_k y_N) = \left(\mathscr{A}_{q,\omega}^2 y_N\right)(\mathscr{S}_{q,\omega}\widetilde{y}_k) + (\mathscr{A}_{q,\omega}y_N)(\mathscr{A}_{q,\omega}\widetilde{y}_k) \\ + \left(\mathscr{S}_{q,\omega}^2 y_N\right)\left(\mathscr{A}_{q,\omega}^2\widetilde{y}_k\right) + (\mathscr{A}_{q,\omega}\mathscr{S}_{q,\omega}y_N)(\mathscr{A}_{q,\omega}\widetilde{y}_k).$$

Hence

$$\varphi\left(\mathscr{S}_{q,\omega}^2 y_N\right)\left(\mathscr{A}_{q,\omega}^2\widetilde{y}_k\right) + (\varphi(\mathscr{A}_{q,\omega}\mathscr{S}_{q,\omega}y_N) + \varphi(\mathscr{A}_{q,\omega}y_N) + \psi y_N)(\mathscr{A}_{q,\omega}\widetilde{y}_k) \\ + \left(\varphi\left(\mathscr{A}_{q,\omega}^2 y_N\right) + \psi(\mathscr{A}_{q,\omega}y_N)\right)(\mathscr{S}_{q,\omega}\widetilde{y}_k) = \lambda_{N+k}(\mathscr{S}_{q,\omega}y_N)(\mathscr{S}_{q,\omega}\widetilde{y}_k).$$

Together with

$$\varphi\left(\mathscr{A}_{q,\omega}^2 y_N\right) + \psi(\mathscr{A}_{q,\omega}y_N) = \lambda_N(\mathscr{S}_{q,\omega}y_N),$$

we obtain

$$\varphi\left(\mathscr{S}_{q,\omega}^2 y_N\right)\left(\mathscr{A}_{q,\omega}^2\widetilde{y}_k\right) + (\varphi(\mathscr{A}_{q,\omega}\mathscr{S}_{q,\omega}y_N) + \varphi(\mathscr{A}_{q,\omega}y_N) + \psi y_N)(\mathscr{A}_{q,\omega}\widetilde{y}_k) \\ = (\lambda_{N+k} - \lambda_N)(\mathscr{S}_{q,\omega}y_N)(\mathscr{S}_{q,\omega}\widetilde{y}_k).$$

This equation holds for $k = 0,1,2$. For $k = 1$ this reads

$$(\varphi(\mathscr{A}_{q,\omega}\mathscr{S}_{q,\omega}y_N) + \varphi(\mathscr{A}_{q,\omega}y_N) + \psi y_N)C_1 = (\lambda_{N+1} - \lambda_N)(\mathscr{S}_{q,\omega}y_N)(\mathscr{S}_{q,\omega}\widetilde{y}_1)$$

with $C_1 = \mathscr{A}_{q,\omega}\widetilde{y}_1(\neq 0)$. Since $\lambda_{N+1} - \lambda_N \neq 0$, the polynomial on the left-hand side of this equation contains $\mathscr{S}_{q,\omega}y_N$ as a factor. Hence the polynomial $\widetilde{\psi}$ can be defined by

$$\widetilde{\psi} := \frac{\varphi(\mathscr{A}_{q,\omega}\mathscr{S}_{q,\omega}y_N) + \varphi(\mathscr{A}_{q,\omega}y_N) + \psi y_N}{\mathscr{S}_{q,\omega}y_N}.$$

For $k = 2$ we find

$$\varphi C_2\left(\mathscr{S}_{q,\omega}^2 y_N\right) + (\mathscr{A}_{q,\omega}\widetilde{y}_2)\widetilde{\psi}(\mathscr{S}_{q,\omega}y_N) = (\lambda_{N+2} - \lambda_N)(\mathscr{S}_{q,\omega}y_N)(\mathscr{S}_{q,\omega}\widetilde{y}_2)$$

with $C_2 = \mathscr{A}_{q,\omega}^2\widetilde{y}_2(\neq 0)$. Since $\lambda_{N+2} - \lambda_N \neq 0$, this implies that $\mathscr{S}_{q,\omega}y_N$ divides $\varphi\left(\mathscr{S}_{q,\omega}^2 y_N\right)$. Hence the polynomial $\widetilde{\varphi}$ can be defined by

$$\widetilde{\varphi} := \frac{\varphi\left(\mathscr{S}_{q,\omega}^2 y_N\right)}{\mathscr{S}_{q,\omega}y_N}.$$

So we have found that

$$\widetilde{\varphi}\left(\mathscr{A}_{q,\omega}^2\widetilde{y}_k\right) + \widetilde{\psi}(\mathscr{A}_{q,\omega}\widetilde{y}_k) = \widetilde{\lambda}_k(\mathscr{S}_{q,\omega}\widetilde{y}_k), \quad \widetilde{\lambda}_k = \lambda_{N+k} - \lambda_N.$$

If (2.3.2) is used, the regularity condition (2.3.3) implies that

$$\widetilde{\lambda}_k = \lambda_{N+k} - \lambda_N = \frac{[k]}{q^{N+k}}(e[2N+k-1] + 2\varepsilon)(\neq 0), \quad k = 1,2,3,\ldots.$$

This proves that for (2.5.13) polynomial solutions of all degrees exist. □

The preceding theorem allows us to prove the existence of a recurrence relation of the form (2.5.1). Let $d_n \neq 0$ for $n = 1,2,3,\ldots,N-1$ and $d_N = 0$. Then we have the recurrence relation (2.5.10). The preceding theorem shows that this recurrence relation can be continued as long as it is possible to continue the recurrence relation

$$\widetilde{y}_{k+1}(x) = (x - \widetilde{c}_k)\widetilde{y}_k(x) - \widetilde{d}_k\widetilde{y}_{k-1}(x)$$

for $\widetilde{y}_k$. This recurrence relation either holds for all $k = 0,1,2,\ldots$ or there is a number $K \in \{1,2,3,\ldots\}$ such that $\widetilde{d}_k \neq 0$ for $k = 0,1,2,\ldots,K-1$ and $\widetilde{d}_K = 0$. Then the preceding theorem can be applied again. This process can be continued until we arrive at:

Theorem 2.8. *If the regularity condition (2.3.3) with $N \to \infty$ holds, then there exist numbers $c_n, d_n \in \mathbb{C}$ such that the polynomial solutions $\{y_n\}_{n=0}^{\infty}$ of the eigenvalue problem (2.2.1) satisfy a three-term recurrence relation of the form (2.5.1).*

2.6 Explicit Form of the Three-Term Recurrence Relation

To determine the polynomial solutions in section 2.4, we needed a two-term recurrence relation for the coefficients. In this section we will not need such a two-term recurrence relation. Therefore we only need the representation (2.4.3) and the three-term recurrence relation (2.4.4). We will not need (2.4.5) here.

In the case of the differentiation operator D (i.e. $q = 1$ and $\omega = \nu = 0$), we simply use the form

$$y_n(x) = \sum_{k=0}^{n} a_{n,k}\frac{x^k}{k!}, \quad a_{n,n} \neq 0, \quad n = 0,1,2,\ldots.$$

Since we do not need a two-term recurrence relation for the coefficients, it suffices to take $c = 0$. In that case (2.4.4) reduces to the three-term recurrence relation

$$(n-k)\left(e(n+k-1)+2\varepsilon\right)a_{n,k} - (2fk+\gamma)a_{n,k+1} - ga_{n,k+2} = 0 \tag{2.6.1}$$

for $k = n-1, n-2, n-3, \ldots, 0$ with $a_{n,n+1} := 0$.

In the case of the q-derivative operator $\mathscr{D}_q$ (i.e. $q \neq 1$ and $\omega = 0$), we need an extra observation in order to deal with the lack of translation invariance. Recall that (2.2.10) and (2.2.11) imply that

$$\left(\mathscr{A}_{q,\omega}p(\cdot)\right)(x+c) = \left(\mathscr{A}_{q,\overline{\omega}}p(\cdot+c)\right)(x)$$

with $\overline{\omega} = \omega + c(q-1)$ and $c \in \mathbb{R}$. For $q \neq 1$ we can put $c = \omega/(1-q)$, which yields $\overline{\omega} = 0$. Hence, since $\mathscr{D}_q := \mathscr{A}_{q,0}$, the operator equation (2.2.1) can be written in terms of the q-derivative operator as

$$\begin{aligned}\varphi(x+c)&\left(\mathscr{D}_q^2 y_n(\cdot+c)\right)(x)\\ &+\psi(x+c)\left(\mathscr{D}_q y_n(\cdot+c)\right)(x) = \lambda_n y_n(qx+c), \quad n = 0,1,2,\ldots\end{aligned} \tag{2.6.2}$$

with $c = \omega/(1-q)$. This translation does not affect the possible orthogonality. In fact the regularity condition (2.3.3) is preserved since the leading coefficients of $\varphi(x)$ and $\psi(x)$ are equal to the leading coefficients of $\varphi(x+c)$ and $\psi(x+c)$, respectively. Similarly, the recurrence relation (2.5.1) for the polynomials y_n is transformed into the recurrence relation

$$y_{n+1}(x+c) = (x - c_n + c)y_n(x+c) - d_n y_{n-1}(x+c), \quad n = 0,1,2,\ldots \tag{2.6.3}$$

for the polynomials $y_n(x+c)$. While d_n remains unchanged, c_n is replaced by $c_n - c$. However, c_n is real if and only if $c_n - c$ is real. Together with the theorem by Favard (see the next chapter) this implies that the polynomials y_n are orthogonal in the positive-definite or quasi-definite sense if and only if this is the case for the polynomials $y_n(x+c)$. In the case of the q-derivative operator $\mathscr{D}_q$, we have $\omega = 0$ and therefore $c = \omega/(1-q) = 0$ and $v = \omega + c(1-q) = 0$. If we set $\omega = c = v = 0$ into (2.4.4), we obtain the three-term recurrence relation

$$[n-k]\,(e[n+k-1]+2\varepsilon)\,a_{n,k} - (2f[k]+\gamma)\,q^{n-k}a_{n,k+1} - g q^{n-k} a_{n,k+2} = 0 \tag{2.6.4}$$

for $k = n-1, n-2, n-3, \ldots, 0$ with $a_{n,n+1} := 0$. Note that (2.6.4) for $q = 1$ equals (2.6.1). This implies that the case of the differentiation operator D needs no special treatment.

As a generalization of the difference operator for $q = 1$ and $\omega \neq 0$, we use another observation. From the definition (2.1.1) it follows that

$$\begin{aligned}(\mathscr{A}_{q,\omega}p(\,\cdot\,))\,(\rho x) &= \frac{p(q\rho x+\omega) - p(\rho x)}{q\rho x + \omega - \rho x} \\ &= \frac{p(\rho(qx+\omega/\rho)) - p(\rho x)}{\rho(qx+\omega/\rho - x)} = \frac{1}{\rho}\left(\mathscr{A}_{q,\omega/\rho}\,p(\rho\,\cdot)\right)(x)\end{aligned}$$

for $\rho \neq 0$ and $p \in \mathscr{P}$. For $q = 1$ and $\rho = \omega (\neq 0)$ this yields $\mathscr{A}_{q,\omega/\rho} = \mathscr{A}_{1,1} = \Delta$ and the operator equation (2.2.1) reads

$$\frac{1}{\omega^2}\varphi(\omega x)\left(\Delta^2 y_n(\omega\,\cdot)\right)(x) + \frac{1}{\omega}\psi(\omega x)\left(\Delta y_n(\omega\,\cdot)\right)(x) = \lambda_n y_n(\omega x + \omega). \tag{2.6.5}$$

Similar to the case $q \neq 1$ and $\overline{\omega} = 0$ above, it can easily be seen by means of the three-term recurrence relation that the possible orthogonality is not affected by the dilatation $x \mapsto \omega x$.

In the case of the difference operator Δ (i.e. $q = 1$ and $\omega = v = 1$), we may use the form

$$y_n(x) = \sum_{k=0}^{n} a_{n,k}\binom{x}{k}, \quad a_{n,n} \neq 0, \quad n = 0,1,2,\ldots. \tag{2.6.6}$$

Since we do not need a two-term recurrence relation for the coefficients, we may simply take $c = 0$. In that case (2.4.4) reads the three-term recurrence relation

$$\begin{aligned}&(n-k)\,(e(n+k-1)+2\varepsilon)\,a_{n,k}\\&\quad+\{e\,(n(n-1)-k(2k-1))+2\varepsilon(n-k)-2fk-\gamma\}\,a_{n,k+1}\\&\quad-(ek^2+2fk+g)a_{n,k+2}=0\end{aligned}\tag{2.6.7}$$

for $k=n-1,n-2,n-3,\ldots,0$ with $a_{n,n+1}:=0$.

In order to obtain the explicit form of the three-term recurrence relation (2.5.1), we substitute the monic polynomials

$$y_n(x)=\sum_{k=0}^{n}\alpha_k^{(n)}x^k,\quad \alpha_n^{(n)}=1,\quad n=0,1,2,\ldots\tag{2.6.8}$$

in the recurrence relation (2.5.1) and $y_1(x)=x-c_0$. Comparison of the coefficients of x^n and x^{n-1} yields

$$c_0=-\alpha_0^{(1)}\quad\text{and}\quad c_n=\alpha_{n-1}^{(n)}-\alpha_n^{(n+1)},\quad n=1,2,3,\ldots\tag{2.6.9}$$

and

$$d_1=-\alpha_0^{(2)}-c_1\alpha_0^{(1)}\quad\text{and}\quad d_n=\alpha_{n-2}^{(n)}-\alpha_{n-1}^{(n+1)}-c_n\alpha_{n-1}^{(n)}\tag{2.6.10}$$

for $n=2,3,4,\ldots$. Hence it suffices to compute $\alpha_{n-1}^{(n)}$ for $n=1,2,3,\ldots$ and $\alpha_{n-2}^{(n)}$ for $n=2,3,4,\ldots$.

In the case of the q-derivative operator $\mathscr{D}_q$, we find from (2.6.4) for $k=n-1$

$$(e[2n-2]+2\varepsilon)\,a_{n,n-1}=q\,(2f[n-1]+\gamma)\,a_{n,n},\quad n=1,2,3,\ldots$$

and for $k=n-2$ and $n=2,3,4,\ldots$

$$[2]\,(e[2n-3]+2\varepsilon)\,a_{n,n-2}-q^2\,(2f[n-2]+\gamma)\,a_{n,n-1}-q^2ga_{n,n}=0.$$

Hence, if the regularity condition (2.3.3) holds for $n=1,2,3,\ldots$, we find

$$a_{n,n-1}=\frac{q\,(2f[n-1]+\gamma)}{e[2n-2]+2\varepsilon}a_{n,n},\quad n=1,2,3,\ldots$$

and

$$a_{n,n-2}=\frac{q^3\,(2f[n-1]+\gamma)\,(2f[n-2]+\gamma)+q^2g\,(e[2n-2]+2\varepsilon)}{[2]\,(e[2n-3]+2\varepsilon)\,(e[2n-2]+2\varepsilon)}a_{n,n}$$

for $n=2,3,4,\ldots$. Comparing (2.4.3) and (2.6.8), we find by using (2.4.2) that

$$a_{n,n}=[n]!,\quad \alpha_{n-1}^{(n)}=\frac{a_{n,n-1}}{[n-1]!}=[n]\frac{a_{n,n-1}}{a_{n,n}},\quad n=1,2,3\ldots$$

with

$$[0]!:=1,\quad [n]!:=\prod_{i=1}^{n}[i],\quad n=1,2,3,\ldots$$

and

$$\alpha_{n-2}^{(n)} = \frac{a_{n,n-2}}{[n-2]!} = [n][n-1]\frac{a_{n,n-2}}{a_{n,n}}, \quad n = 2,3,4,\ldots.$$

Hence we have

$$\alpha_{n-1}^{(n)} = \frac{q[n]\,(2f[n-1]+\gamma)}{e[2n-2]+2\varepsilon}, \quad n = 1,2,3,\ldots$$

and

$$\alpha_{n-2}^{(n)} = \frac{q^2[n][n-1]\,\{q\,(2f[n-1]+\gamma)\,(2f[n-2]+\gamma)+g\,(e[2n-2]+2\varepsilon)\}}{[2]\,(e[2n-3]+2\varepsilon)\,(e[2n-2]+2\varepsilon)}$$

for $n = 2,3,4,\ldots$. By using (2.6.9), we conclude that $c_0 = -\gamma q/2\varepsilon$ and

$$\begin{aligned} c_n &= \frac{q[n]\,(2f[n-1]+\gamma)}{e[2n-2]+2\varepsilon} - \frac{q[n+1]\,(2f[n]+\gamma)}{e[2n]+2\varepsilon} \\ &= \frac{q\,\{[n]\,(2f[n-1]+\gamma)\,(e[2n]+2\varepsilon)-[n+1]\,(2f[n]+\gamma)\,(e[2n-2]+2\varepsilon)\}}{(e[2n-2]+2\varepsilon)\,(e[2n]+2\varepsilon)} \\ &= -\frac{q^n\left\{(e[n-1]+2\varepsilon)\,(2f[n](1+q)+\gamma q)-e\gamma[n+1]q^{n-1}\right\}}{(e[2n-2]+2\varepsilon)\,(e[2n]+2\varepsilon)} \end{aligned} \tag{2.6.11}$$

for $n = 1,2,3,\ldots$. By using (2.6.10), we obtain

$$\begin{aligned} d_1 &= -\frac{q^2[2]\,\{\gamma q(2f+\gamma)+g\,(e[2]+2\varepsilon)\}}{[2]\,(e+2\varepsilon)\,(e[2]+2\varepsilon)} + \frac{q\{2\varepsilon\,(2f(1+q)+\gamma q)-e\gamma[2]\}}{2\varepsilon\,(e[2]+2\varepsilon)}\cdot\frac{\gamma q}{2\varepsilon} \\ &= \frac{q^2\,(e[2]+2\varepsilon)\,\left(4\varepsilon(f\gamma-g\varepsilon)-e\gamma^2\right)}{(2\varepsilon)^2\,(e+2\varepsilon)\,(e[2]+2\varepsilon)} = \frac{q^2\,\left(4\varepsilon(f\gamma-g\varepsilon)-e\gamma^2\right)}{4\varepsilon^2\,(e+2\varepsilon)} \end{aligned}$$

and for $n = 2,3,4,\ldots$

$$\begin{aligned} d_n &= \frac{q^2[n][n-1]\,\{q\,(2f[n-1]+\gamma)\,(2f[n-2]+\gamma)+g\,(e[2n-2]+2\varepsilon)\}}{[2]\,(e[2n-3]+2\varepsilon)\,(e[2n-2]+2\varepsilon)} \\ &\quad - \frac{q^2[n][n+1]\,\{q\,(2f[n]+\gamma)\,(2f[n-1]+\gamma)+g\,(e[2n]+2\varepsilon)\}}{[2]\,(e[2n-1]+2\varepsilon)\,(e[2n]+2\varepsilon)} \\ &\quad + \frac{q[n]\,(2f[n-1]+\gamma)}{e[2n-2]+2\varepsilon} \\ &\qquad \times \frac{q^n\left\{(e[n-1]+2\varepsilon)\,(2f[n](1+q)+\gamma q)-e\gamma[n+1]q^{n-1}\right\}}{(e[2n-2]+2\varepsilon)\,(e[2n]+2\varepsilon)} \\ &= \frac{q^{n+1}[n]\,(e[n-2]+2\varepsilon)}{(e[2n-3]+2\varepsilon)\,(e[2n-2]+2\varepsilon)^2\,(e[2n-1]+2\varepsilon)} \\ &\quad \times \Big\{q^{n-1}\,(2f[n-1]+\gamma)\,\left(2f\,\{e[n-1]+2\varepsilon\}-q^{n-1}e\gamma\right) \\ &\qquad - g\,(e[2n-2]+2\varepsilon)^2\Big\}. \end{aligned} \tag{2.6.12}$$

Note that the latter formula also holds for $n = 1$.

In the case of the differentiation operator D, we have $q = 1$. In that case we get

$$c_n = -\frac{2fn(e(n-1)+2\varepsilon)-\gamma(e-\varepsilon)}{2(e(n-1)+\varepsilon)(en+\varepsilon)}, \quad n = 0,1,2,\ldots \tag{2.6.13}$$

and

$$\begin{aligned} d_n = & \frac{n(e(n-2)+2\varepsilon)}{4(e(2n-3)+2\varepsilon)(e(n-1)+\varepsilon)^2(e(2n-1)+2\varepsilon)} \\ & \times\Big\{\{2f(n-1)+\gamma\}\{2f(e(n-1)+2\varepsilon)-e\gamma\} \\ & -4g(e(n-1)+\varepsilon)^2\Big\}, \quad n = 1,2,3,\ldots. \end{aligned} \tag{2.6.14}$$

In the case of the difference operator Δ, we find from (2.6.7) with $k = n-1$

$$(e(2n-2)+2\varepsilon)a_{n,n-1} = \left(e(n^2-4n+3)-2\varepsilon+2f(n-1)+\gamma\right)a_{n,n}$$

for $n = 1,2,3,\ldots$, and with $k = n-2$

$$\begin{aligned} & 2(e(2n-3)+2\varepsilon)a_{n,n-2} - \left(e(n^2-8n+10)-4\varepsilon+2f(n-2)+\gamma\right)a_{n,n-1} \\ & -\left(e(n-2)^2+2f(n-2)+g\right)a_{n,n} = 0 \end{aligned}$$

for $n = 2,3,4,\ldots$. Hence, if the regularity condition (2.3.3) holds for $n = 1,2,3,\ldots$, we find

$$a_{n,n-1} = \frac{e(n-1)(n-3)-2\varepsilon+2f(n-1)+\gamma}{2(e(n-1)+\varepsilon)}a_{n,n}, \quad n = 1,2,3,\ldots$$

and for $n = 2,3,4,\ldots$

$$\begin{aligned} a_{n,n-2} = & \left\{\frac{e(n-2)^2+2f(n-2)+g}{2(e(2n-3)+2\varepsilon)}\right. \\ & +\left(e(n^2-8n+10)-4\varepsilon+2f(n-2)+\gamma\right) \\ & \left.\times\frac{(e(n-1)(n-3)-2\varepsilon+2f(n-1)+\gamma)}{4(e(n-1)+\varepsilon)(e(2n-3)+2\varepsilon)}\right\}a_{n,n}. \end{aligned}$$

Comparing (2.6.8) and (2.6.6), we find that

$$a_{n,n} = n!, \quad \alpha_{n-1}^{(n)} = n\left\{\frac{a_{n,n-1}}{a_{n,n}} - \frac{n-1}{2}\right\} = \frac{n(2(n-1)(f-e)-(n+1)\varepsilon+\gamma)}{2(e(n-1)+\varepsilon)}$$

for $n = 1,2,3,\ldots$ and

$$\begin{aligned}
\alpha_{n-2}^{(n)} &= \binom{n}{3}\frac{3n-1}{4} - \frac{n-2}{2}\frac{a_{n,n-1}}{(n-2)!} + \frac{a_{n,n-2}}{(n-2)!} \\
&= n(n-1)\left\{\frac{(n-2)(3n-1)}{24} - \frac{n-2}{2}\frac{a_{n,n-1}}{a_{n,n}} + \frac{a_{n,n-2}}{a_{n,n}}\right\} \\
&= \frac{n(n-1)}{4\left(e(n-1)+\varepsilon\right)\left(e(2n-3)+2\varepsilon\right)} \\
&\qquad \times \left\{\frac{1}{6}(n+1)(3n+2)\left(e(n-1)+\varepsilon\right)\left(e(2n-3)+2\varepsilon\right)\right. \\
&\qquad\quad - en(n-1)^2\left(e(2n-3)+2\varepsilon\right) + e^2(n-1)^2(n-2)^2 \\
&\qquad\quad - 2\left(2f(n-1)+\gamma\right)\left(e(2n-3)+n\varepsilon\right) - \gamma\left(e(n-1)+2\varepsilon\right) \\
&\qquad\quad \left. + \left(2f(n-1)+\gamma\right)\left(2f(n-2)+\gamma\right) + 2g\left(e(n-1)+\varepsilon\right)\right\}
\end{aligned}$$

for $n = 2,3,4,\ldots$. By using (2.6.9), we conclude that $c_0 = 1 - \gamma/2\varepsilon$ and

$$\begin{aligned}
c_n &= \frac{n\left(2(n-1)(f-e)-(n+1)\varepsilon+\gamma\right)}{2\left(e(n-1)+\varepsilon\right)} \\
&\qquad - \frac{(n+1)\left(2n(f-e)-(n+2)\varepsilon+\gamma\right)}{2(en+\varepsilon)} \\
&= \frac{n\left(e(n-1)+2\varepsilon\right)\left(2(e-f)+\varepsilon\right)+(e-\varepsilon)(\gamma-2\varepsilon)}{2\left(e(n-1)+\varepsilon\right)(en+\varepsilon)}
\end{aligned} \tag{2.6.15}$$

for $n = 1,2,3,\ldots$. By using (2.6.10), we find

$$\begin{aligned}
d_1 &= -\frac{1}{2(e+\varepsilon)(e+2\varepsilon)}\left\{4(e+\varepsilon)(e+2\varepsilon) - 2e(e+2\varepsilon)\right. \\
&\qquad \left. - 2(2f+\gamma)(e+2\varepsilon) - \gamma(e+2\varepsilon) + (2f+\gamma)\gamma + 2g(e+\varepsilon)\right\} \\
&\qquad + \frac{2\varepsilon\left(2(e-f)+\varepsilon\right)+(e-\varepsilon)(\gamma-2\varepsilon)}{2\varepsilon(e+\varepsilon)}\cdot\frac{2\varepsilon-\gamma}{2\varepsilon} \\
&= \frac{4\varepsilon(f\gamma-g\varepsilon)-e\gamma^2}{4\varepsilon^2(e+2\varepsilon)}.
\end{aligned}$$

and for $n = 2,3,4,\ldots$

$$
\begin{aligned}
d_n = {} & \frac{n(n-1)}{4\,(e(n-1)+\varepsilon)\,(e(2n-3)+2\varepsilon)} \\
& \times \left\{ \frac{1}{6}(n+1)(3n+2)\,(e(n-1)+\varepsilon)\,(e(2n-3)+2\varepsilon) \right. \\
& \qquad - en(n-1)^2\,(e(2n-3)+2\varepsilon) + e^2(n-1)^2(n-2)^2 \\
& \qquad - 2\,(2f(n-1)+\gamma)\,(e(2n-3)+n\varepsilon) - \gamma(e(n-1)+2\varepsilon) \\
& \qquad \left. + (2f(n-1)+\gamma)\,(2f(n-2)+\gamma) + 2g\,(e(n-1)+\varepsilon) \right\} \\
& - \frac{n(n+1)}{4(en+\varepsilon)\,(e(2n-1)+2\varepsilon)} \\
& \times \left\{ \frac{1}{6}(n+2)(3n+5)(en+\varepsilon)\,(e(2n-1)+2\varepsilon) \right. \\
& \qquad - en^2(n+1)\,(e(2n-1)+2\varepsilon) + e^2n^2(n-1)^2 \\
& \qquad - 2(2fn+\gamma)\,(e(2n-1)+(n+1)\varepsilon) - \gamma(en+2\varepsilon) \\
& \qquad \left. + (2fn+\gamma)\,(2f(n-1)+\gamma) + 2g(en+\varepsilon) \right\} \\
& - \frac{n\,(2(e-f)+\varepsilon)\,(e(n-1)+2\varepsilon)+(e-\varepsilon)(\gamma-2\varepsilon)}{2\,(e(n-1)+\varepsilon)\,(en+\varepsilon)} \\
& \times \frac{n\,(2(n-1)(f-e)-(n+1)\varepsilon+\gamma)}{2\,(e(n-1)+\varepsilon)} \\
= {} & - \frac{n\,(e(n-2)+2\varepsilon)}{4\,(e(2n-3)+2\varepsilon)\,(e(n-1)+\varepsilon)^2\,(e(2n-1)+2\varepsilon)} \\
& \times \left\{ e(n-1)^2\,(e(n-1)+2\varepsilon)^2 \right. \\
& \qquad + 2(n-1)\,(e(n-1)+2\varepsilon)\,(2eg+2f(\varepsilon-f)-e\gamma) \\
& \qquad \left. + 4\varepsilon(g\varepsilon-f\gamma)+e\gamma^2 \right\}.
\end{aligned}
\tag{2.6.16}
$$

Note that the latter formula also holds for $n=1$.

Chapter 3
Orthogonality of the Polynomial Solutions

3.1 Favard's Theorem

In this section we consider the possible orthogonality of polynomials satisfying a three-term recurrence relation of the form (2.5.1). Hereby we use Favard's theorem (see for instance [146]):

Theorem 3.1. *Let y_n denote the monic polynomial of degree $n \in \{0,1,2,\ldots\}$ satisfying the three-term recurrence relation (2.5.1)*

$$y_{n+1}(x) = (x-c_n)y_n(x) - d_n y_{n-1}(x), \quad c_n, d_n \in \mathbb{C}, \quad n = 1,2,3,\ldots.$$

Then there exists a unique linear functional Λ with

$$\Lambda[1] = 1 \quad \textit{and} \quad \Lambda[y_m y_n] = 0 \quad \textit{for} \quad m \neq n, \quad m,n \in \{0,1,2,\ldots\}. \tag{3.1.1}$$

This linear functional Λ is quasi-definite if and only if $d_n \neq 0$ for all $n = 1,2,3,\ldots$. This linear functional Λ is positive-definite if and only if $c_n \in \mathbb{R}$ for all $n = 0,1,2,\ldots$ and $d_n > 0$ for all $n = 1,2,3,\ldots$.

Proof. For the fulfillment of (3.1.1), we must have that $\Lambda[y_0]=1$ and $\Lambda[y_n] = 0$ for $n = 1,2,3,\ldots$. Then the linear functional Λ is uniquely determined on $\mathscr{P}$. Furthermore, (2.5.1) can be written as

$$xy_n(x) = y_{n+1}(x) + c_n y_n(x) + d_n y_{n-1}(x), \quad n = 1,2,3,\ldots \tag{3.1.2}$$

and applying the linear functional Λ, we obtain

$$\Lambda[xy_n(x)] = 0, \quad n = 2,3,4,\ldots.$$

Next we use

$$x^2 y_n(x) = xy_{n+1}(x) + c_n xy_n(x) + d_n xy_{n-1}(x), \quad n = 1,2,3,\ldots$$

R. Koekoek et al., *Hypergeometric Orthogonal Polynomials and Their q-Analogues*, Springer Monographs in Mathematics,
DOI 10.1007/978-3-642-05014-5_3, © Springer-Verlag Berlin Heidelberg 2010

to conclude that $\Lambda[x^2 y_n(x)] = 0$ for $n = 3,4,5,\ldots$. Hence, by using induction, we have proved (3.1.1).

Now we multiply (3.1.2) by $y_{n-1}(x)$ to obtain

$$xy_n(x)y_{n-1}(x) = y_{n+1}(x)y_{n-1}(x) + c_n y_n(x)y_{n-1}(x) + d_n y_{n-1}^2(x)$$

for $n = 1,2,3,\ldots$. Now we use (2.5.11) to write

$$xy_{n-1}(x) = y_n(x) - \sum_{k=0}^{n-1} \alpha_k^{(n-1)} y_k(x), \quad \alpha_k^{(n-1)} \in \mathbb{C}, \quad n = 1,2,3,\ldots,$$

apply the linear functional Λ and find by using (3.1.1)

$$\Lambda[y_n^2] = d_n \Lambda[y_{n-1}^2], \quad n = 1,2,3,\ldots. \tag{3.1.3}$$

A quasi-definite linear functional Λ for a monic polynomial system $\{y_n\}_{n=0}^{\infty}$ is characterized by $\Lambda[y_n^2] \neq 0$ for all $n = 0,1,2,\ldots$. In view of (3.1.3), this is equivalent to $d_n \neq 0$ for all $n = 1,2,3,\ldots$.

A positive-definite linear functional Λ for a monic polynomial system $\{y_n\}_{n=0}^{\infty}$ is characterized by $\Lambda[y_n^2] > 0$ and each y_n has real coefficients for all $n = 0,1,2,\ldots$. In view of (2.5.1) and (3.1.3), this is equivalent to $c_n \in \mathbb{R}$ for all $n = 0,1,2,\ldots$ and $d_n > 0$ for all $n = 1,2,3,\ldots$. □

Remark. From (3.1.3) together with $\Lambda[y_0^2] = 1$ it follows that

$$\Lambda[y_n^2] = \prod_{k=1}^{n} d_k, \quad n = 1,2,3,\ldots. \tag{3.1.4}$$

Finite orthogonal polynomial systems $\{y_n\}_{n=0}^{N}$ with $N+1$ polynomials occur if $\Lambda[y_n^2] \neq 0$ for $n = 0,1,2,\ldots,N$. These polynomials satisfy a three-term recurrence relation of the form

$$y_{n+1}(x) = (x - c_n)y_n(x) - d_n y_{n-1}(x), \quad n = 0,1,2,\ldots,N-1$$

with $y_{-1}(x) := 0$. Further we define y_{N+1}^* by

$$y_{N+1}^*(x) = (x - c_N^*)y_N(x) - d_N y_{N-1}(x)$$

with c_N^* arbitrary. Then (3.1.3) also holds for $n = N$. Now we define $\{y_n^*\}_{n=0}^{\infty}$ such that

$$y_n^*(x) = y_n(x), \quad n = 0,1,2,\ldots,N.$$

The polynomials of degree higher than $N+1$ can be obtained by

$$\begin{aligned}
y^*_{N+2}(x) &= (x - c^*_{N+1})y^*_{N+1}(x) - d^*_{N+1}y_N(x),\\
y^*_{N+3}(x) &= (x - c^*_{N+2})y^*_{N+2}(x) - d^*_{N+2}y^*_{N+1}(x),\\
&\ \vdots
\end{aligned}$$

where the coefficients $\{c^*_n\}_{n=N}^{\infty}$ and $\{d^*_n\}_{n=N+1}^{\infty}$ can be chosen freely according to the following rules:

- In the quasi-definite case: $d^*_n \neq 0$ for $n = N+1, N+2, N+3, \ldots$.
- In the positive-definite case: $c^*_n \in \mathbb{R}$ for $n = N, N+1, N+2, \ldots$ and $d^*_n > 0$ for $n = N+1, N+2, N+3, \ldots$.

Let Λ^* be the linear functional given by

$$\Lambda^*[y^*_0] = 1, \quad \Lambda^*[y^*_n] = 0, \quad n = 1, 2, 3, \ldots,$$

where only the first $N+1$ polynomials $\{y^*_n\}_{n=0}^{N}$ satisfy the eigenvalue problem (2.2.1). Then Favard's theorem implies that $\{y^*_n\}_{n=0}^{\infty}$ is an orthogonal system. Then we call $\{y_n\}_{n=0}^{N}$ a finite orthogonal system with $N+1$ polynomials. See also [373].

3.2 Orthogonality and the Self-Adjoint Operator Equation

Let $q \in \mathbb{R} \setminus \{-1, 0\}$, $\omega \in \mathbb{R}$ and $(q, \omega) \neq (1, 0)$. Then the definition (2.1.1) of Hahn's q-operator $\mathscr{A}_{q,\omega}$ can be extended to

$$(\mathscr{A}_{q,\omega} f)(x) := \frac{f(qx+\omega) - f(x)}{qx + \omega - x}, \quad x \in \mathbb{C} \setminus \left\{\frac{\omega}{1-q}\right\}, \tag{3.2.1}$$

for arbitrary complex-valued functions f whose domain contains the number $qx + \omega \in \mathbb{C}$ as well as $x \in \mathbb{C}$. For two such functions, the product rule (2.1.2) is still valid:

$$(\mathscr{A}_{q,\omega}(f_1 f_2))(x) = (\mathscr{A}_{q,\omega} f_1)(x) f_2(x) + f_1(qx+\omega)(\mathscr{A}_{q,\omega} f_2)(x). \tag{3.2.2}$$

Likewise, the definition (2.5.2) of the operator $\mathscr{S}_{q,\omega}$ can be extended:

$$(\mathscr{S}_{q,\omega} f)(x) := f(qx+\omega), \quad x \in \mathbb{C} \tag{3.2.3}$$

and

$$\widehat{f}(x) := \left(\mathscr{S}_{q,\omega}^{-1} f\right)(x) = f((x-\omega)/q), \quad x \in \mathbb{C}. \tag{3.2.4}$$

Assume that w is a complex-valued function for which $w(x)$ and $(\mathscr{S}_{q,\omega} w)(x)$ are defined for infinitely many $x \in \mathbb{C}$. If we multiply the operator equation (2.2.1) by $(\mathscr{S}_{q,\omega} w)(x)$, we find by using (3.2.3)

$$\begin{aligned}&\left(\mathscr{S}_{q,\omega}w\right)(x)\varphi(x)\left(\mathscr{A}_{q,\omega}^{2}y_n\right)(x)\\&\quad+\left(\mathscr{S}_{q,\omega}w\right)(x)\psi(x)\left(\mathscr{A}_{q,\omega}y_n\right)(x)=\lambda_n\left(\mathscr{S}_{q,\omega}w\right)(x)\left(\mathscr{S}_{q,\omega}y_n\right)(x).\end{aligned}$$

By using (3.2.4), we note that this equation coincides with the self-adjoint operator equation

$$\left(\mathscr{A}_{q,\omega}\left(w\widehat{\varphi}\mathscr{A}_{q,\omega}y_n\right)\right)(x)=\lambda_n\left(\mathscr{S}_{q,\omega}w\right)(x)\left(\mathscr{S}_{q,\omega}y_n\right)(x) \tag{3.2.5}$$

if the so-called Pearson operator equation[1]

$$\left(\mathscr{A}_{q,\omega}\left(w\widehat{\varphi}\right)\right)(x)=\left(\mathscr{S}_{q,\omega}w\right)(x)\psi(x) \tag{3.2.6}$$

holds. From this Pearson operator equation the function w can be determined.

Let us state the Pearson operator equation in another form. By using the definition (3.2.1), the product rule (3.2.2) and the definition (3.2.4), we find for the left-hand side of (3.2.6)

$$\begin{aligned}\left(\mathscr{A}_{q,\omega}\left(w\widehat{\varphi}\right)\right)(x)&=\frac{w(qx+\omega)-w(x)}{qx+\omega-x}\widehat{\varphi}(x)+w(qx+\omega)\frac{\widehat{\varphi}(qx+\omega)-\widehat{\varphi}(x)}{qx+\omega-x}\\&=\frac{w(qx+\omega)\varphi(x)-w(x)\widehat{\varphi}(x)}{qx+\omega-x}.\end{aligned}$$

By using (3.2.4), we can write the right-hand side of (3.2.6) as

$$\left(\mathscr{S}_{q,\omega}w\right)(x)\psi(x)=w(qx+\omega)\psi(x).$$

Hence we have

$$w(x)\widehat{\varphi}(x)=w(qx+\omega)\left(\varphi(x)-(qx+\omega-x)\psi(x)\right). \tag{3.2.7}$$

If (2.2.2) and (2.2.13) are used, this can be written as

$$w(x)C(x)=qw(qx+\omega)D(qx+\omega). \tag{3.2.8}$$

Now we examine the relationship between the function w and the orthogonality functionals for the polynomials $\{y_n\}_{n=0}^{\infty}$. For this purpose we multiply (3.2.5) by $\left(\mathscr{S}_{q,\omega}y_m\right)(x)$ and subtract from the resulting equation the same equation with m and n exchanged. Then we apply $\mathscr{S}_{q,\omega}^{-1}$ to the result and use the commutation rules (2.5.3) to find

$$\begin{aligned}&(\lambda_n-\lambda_m)w(x)y_m(x)y_n(x)\\&=q\left(\mathscr{A}_{q,\omega}\mathscr{S}_{q,\omega}^{-1}\left(w\widehat{\varphi}\mathscr{A}_{q,\omega}y_n\right)\right)(x)y_m(x)\\&\quad-q\left(\mathscr{A}_{q,\omega}\mathscr{S}_{q,\omega}^{-1}\left(w\widehat{\varphi}\mathscr{A}_{q,\omega}y_m\right)\right)(x)y_n(x).\end{aligned} \tag{3.2.9}$$

[1] The Pearson operator equation also plays an important role in stochastics, where it is used to derive stochastic distribution functions. This shows the connection between orthogonal polynomials and stochastic distribution functions. See for instance the book [469] by W. Schoutens.

This leads to two kinds of orthogonality.

A. For $\omega = 0$, id est $\mathscr{A}_{q,0} = \mathscr{D}_q$, we have the q-integration by parts formula (1.15.9)

$$\int_a^b \left(\mathscr{A}_{q,0} f_1\right)(x) f_2(x)\, d_q x$$
$$= \Big[f_1(x) f_2(x)\Big]_a^b - \int_a^b \left(\mathscr{S}_{q,0} f_1\right)(x) \left(\mathscr{A}_{q,0} f_2\right)(x)\, d_q x \qquad (3.2.10)$$

for arbitrary complex-valued functions f_1 and f_2 which are q-integrable on the interval (a,b). Then we have for (3.2.9)

$$(\lambda_n - \lambda_m) \int_a^b w(x) y_m(x) y_n(x)\, d_q x$$
$$= \int_a^b q \Big\{ \left(\mathscr{A}_{q,0} \mathscr{S}_{q,0}^{-1} \left(w \widehat{\varphi} \mathscr{A}_{q,0} y_n\right)\right)(x) y_m(x)$$
$$- \left(\mathscr{A}_{q,0} \mathscr{S}_{q,0}^{-1} \left(w \widehat{\varphi} \mathscr{A}_{q,0} y_m\right)\right)(x) y_n(x) \Big\}\, d_q x$$
$$= q \Big(\left(\mathscr{S}_{q,0}^{-1} (w \widehat{\varphi})\right)(x) \Big\{ \left(\mathscr{S}_{q,0}^{-1} \left(\mathscr{A}_{q,0} y_n\right)\right)(x) y_m(x)$$
$$- \left(\mathscr{S}_{q,0}^{-1} \left(\mathscr{A}_{q,0} y_m\right)\right)(x) y_n(x) \Big\} \Big)_a^b.$$

In view of the regularity condition (2.3.3), we have $\lambda_m \neq \lambda_n$ for $m \neq n$. So we have

Theorem 3.2. *Let $\{y_n\}_{n=0}^{\infty}$ denote the polynomial solutions of the eigenvalue problem (2.2.1), let the regularity condition (2.3.3) hold for $n = 0, 1, 2, \ldots$ and let w denote a complex valued function which is q-integrable on the interval (a,b) on the real line and which satisfies the Pearson operator equation (3.2.6). Then we have the orthogonality relation*

$$\int_a^b w(x) y_m(x) y_n(x)\, d_q x = 0, \quad m \neq n, \quad m,n \in \{0,1,2,\ldots\} \qquad (3.2.11)$$

with weight function w if the boundary conditions

$$\left(\mathscr{S}_{q,0}^{-1}(w \widehat{\varphi})\right)(a) = 0 \quad \textit{and} \quad \left(\mathscr{S}_{q,0}^{-1}(w \widehat{\varphi})\right)(b) = 0 \qquad (3.2.12)$$

hold. Here a continuous extension of $w\widehat{\varphi}$ might be necessary.

If the necessary convergence conditions hold, the integral in (3.2.11) can also be taken over (a,∞), $(-\infty,b)$ or $(-\infty,\infty)$ with appropriate boundary conditions.

Since we used the definition (3.2.1), which is not valid for $(q,\omega) = (1,0)$, the differentiation operator D has not been covered. This case needs a separate treatment. The differential equation (2.2.9) has the self-adjoint form

$$\left(w\varphi y_n'\right)'(x) = \lambda_n w(x) y_n(x), \quad n = 0,1,2,\ldots$$

if w satisfies the Pearson differential equation

$$(w\varphi)'(x) = w(x)\psi(x), \tag{3.2.13}$$

where $\varphi(x) = ex^2 + 2fx + g$, $\psi(x) = 2\varepsilon x + \gamma$ and $\lambda_n = n(e(n-1) + 2\varepsilon)$. Partial integration over an interval (a,b) with $a,b \in \mathbb{R}$ leads to the orthogonality relation

$$\begin{aligned}(\lambda_n - \lambda_m) & \int_a^b w(x) y_m(x) y_n(x)\, dx \\ &= \left[w(x)\varphi(x)\left(y_m(x)y_n'(x) - y_n(x)y_m'(x)\right)\right]_a^b = 0 \end{aligned} \tag{3.2.14}$$

for $m \neq n$ and $m,n \in \{0,1,2,\ldots\}$ if the boundary conditions

$$w(a)\varphi(a) = 0 \quad \text{and} \quad w(b)\varphi(b) = 0 \tag{3.2.15}$$

hold. If the necessary convergence conditions hold, the integral in (3.2.14) can also be taken over (a,∞), $(-\infty,b)$ and $(-\infty,\infty)$ with appropriate boundary conditions.

B. For $N \in \{1,2,3,\ldots\}$ we consider the set of points

$$x_\nu := Aq^\nu + [\nu]\omega, \quad A \in \mathbb{C}, \quad \nu = 0,1,2,\ldots,N+1, \tag{3.2.16}$$

which implies that $x_{\nu+1} = qx_\nu + \omega$. Then we have the summation by parts formula

$$\begin{aligned}&\sum_{\nu=0}^{N} \left(\mathscr{A}_{q,\omega} f_1\right)(x_\nu) f_2(x_\nu)\left((q-1)x_\nu + \omega\right) \\ &= \left[f_1(x_\nu) f_2(x_\nu)\right]_{\nu=0}^{N+1} \\ &\quad - \sum_{\nu=0}^{N} \left(\mathscr{S}_{q,\omega} f_1\right)(x_\nu)\left(\mathscr{A}_{q,\omega} f_2\right)(x_\nu)\left((q-1)x_\nu + \omega\right)\end{aligned} \tag{3.2.17}$$

for arbitrary complex-valued functions f_1 and f_2 whose domain contains the set of points $\{x_\nu\}_{\nu=0}^{N+1}$. Hence we have for (3.2.9)

$$
\begin{aligned}
&(\lambda_n - \lambda_m) \sum_{\nu=0}^{N} w(x_\nu) y_m(x_\nu) y_n(x_\nu) \left((q-1)x_\nu + \omega\right) \\
&= \sum_{\nu=0}^{N} q \Big\{ \left(\mathscr{A}_{q,\omega} \mathscr{S}_{q,\omega}^{-1} \left(w\widehat{\varphi} \mathscr{A}_{q,\omega} y_n \right) \right)(x_\nu) y_m(x_\nu) \\
&\qquad - \left(\mathscr{A}_{q,\omega} \mathscr{S}_{q,\omega}^{-1} \left(w\widehat{\varphi} \mathscr{A}_{q,\omega} y_m \right) \right)(x_\nu) y_n(x_\nu) \Big\} \left((q-1)x_\nu + \omega\right) \\
&= q \Bigg(\left(\mathscr{S}_{q,\omega}^{-1}(w\widehat{\varphi}) \right)(x_\nu) \Big\{ \left(\mathscr{S}_{q,\omega}^{-1} \left(\mathscr{A}_{q,\omega} y_n \right) \right)(x_\nu) y_m(x_\nu) \\
&\qquad - \left(\mathscr{S}_{q,\omega}^{-1} \left(\mathscr{A}_{q,\omega} y_m \right) \right)(x_\nu) y_n(x_\nu) \Big\} \Bigg)_{\nu=0}^{N+1}.
\end{aligned}
$$

In view of the regularity condition (2.3.3), we have $\lambda_m \neq \lambda_n$ for $m \neq n$. So we have

Theorem 3.3. *Let $\{y_n\}_{n=0}^{\infty}$ denote the polynomial solutions of the eigenvalue problem (2.2.1), let the regularity condition (2.3.3) hold for $n = 0, 1, 2, \ldots$ and let w denote a complex-valued function whose domain contains the set of points $\{x_\nu\}_{\nu=0}^{N+1}$ and which satisfies the Pearson operator equation (3.2.6). Then we have the orthogonality relation*

$$
\sum_{\nu=0}^{N} w(x_\nu) y_m(x_\nu) y_n(x_\nu) \left((q-1)x_\nu + \omega\right) = 0, \qquad m \neq n, \quad m,n \in \{0,1,2,\ldots,N\} \tag{3.2.18}
$$

with weight function w if the boundary conditions

$$
\left(\mathscr{S}_{q,\omega}^{-1}(w\widehat{\varphi}) \right)(x_0) = 0 \quad \text{and} \quad \left(\mathscr{S}_{q,\omega}^{-1}(w\widehat{\varphi}) \right)(x_{N+1}) = 0 \tag{3.2.19}
$$

hold. Here a continuous extension of $w\widehat{\varphi}$ might be necessary.

Note that the boundary conditions (3.2.19) have to be satisfied for two points of the set (3.2.16). This means that two of these points must be zeros of $\left(\mathscr{S}_{q,\omega}^{-1}(w\widehat{\varphi}) \right)$. If this cannot be achieved, the above sums can be generalized by Jackson-Thomae integrals (see the next section), which also include the case where infinite sums are used instead of the finite sums (3.2.18).

3.3 The Jackson-Thomae q-Integral

In the previous section we developed a basis for the representation of orthogonality functionals for the polynomial solutions of the eigenvalue problem (2.2.1). It remains to deal with the difficulty of satisfying the boundary conditions (3.2.19) by elements of the set (3.2.16). This problem can be solved by decomposing the functional into two (infinite) sums, which can be seen as a Stieltjes integral.

It will be necessary to distinguish the cases $0 < |q| < 1$ and $|q| > 1$. Therefore we introduce the set of points

$$D(x,q,\omega) := \begin{cases} \left\{xq^k + [k]\omega\right\}_{k=0}^{\infty}, & 0 < |q| < 1 \\ \left\{q^{-k-1}\left(x - [k+1]\omega\right)\right\}_{k=0}^{\infty}, & |q| > 1. \end{cases}$$

Let $A, B \in \mathbb{C}$ and consider the two sets of points $D(A,q,\omega)$ and $D(B,q,\omega)$. For $0 < |q| < 1$ these sets of points can be described by the sequences $\{x_\nu\}_{\nu=0}^{\infty}$ and $\{x_\nu^*\}_{\nu=0}^{\infty}$, given by

$$x_\nu = Aq^\nu + [\nu]\omega \quad \text{and} \quad x_\nu^* = Bq^\nu + [\nu]\omega, \quad \nu = 0,1,2,\ldots,$$

respectively. For $|q| > 1$ we have

$$x_\nu = \frac{1}{q^{\nu+1}}\left(A - [\nu+1]\omega\right) \quad \text{and} \quad x_\nu^* = \frac{1}{q^{\nu+1}}\left(B - [\nu+1]\omega\right), \quad \nu = 0,1,2,\ldots,$$

respectively. Note that for $0 < |q| < 1$ we have

$$\lim_{\nu\to\infty}\left(Aq^\nu + [\nu]\omega\right) = \lim_{\nu\to\infty}\left(Aq^\nu + \frac{1-q^\nu}{1-q}\,\omega\right) = \frac{\omega}{1-q}$$

and for $|q| > 1$ we obtain

$$\lim_{\nu\to\infty}\frac{A - [\nu+1]\omega}{q^{\nu+1}} = \lim_{\nu\to\infty}\frac{A(1-q) - (1-q^{\nu+1})\omega}{(1-q)q^{\nu+1}} = \frac{\omega}{1-q}.$$

For a complex-valued function f with domain containing $D(A,q,\omega)$ the Jackson-Thomae integral is defined by

$$\int_{\frac{\omega}{1-q}}^{A} f(x)\,d_{q,\omega}x := \begin{cases} \sum\limits_{\nu=0}^{\infty} f(x_\nu)\left((1-q)x_\nu - \omega\right), & 0 < |q| < 1 \\ \sum\limits_{\nu=0}^{\infty} f(x_\nu)\left((q-1)x_\nu + \omega\right), & |q| > 1, \end{cases}$$

provided that the sums converge. If the domain of f also contains $D(B,q,\omega)$ and all corresponding sums converge, we define

$$\int_{A}^{B} f(x)\,d_{q,\omega}x := \int_{\frac{\omega}{1-q}}^{B} f(x)\,d_{q,\omega}x - \int_{\frac{\omega}{1-q}}^{A} f(x)\,d_{q,\omega}x. \tag{3.3.1}$$

Then we have the following analogue of the fundamental theorem of calculus:

Theorem 3.4. *Let the domain of the complex-valued function f contain $D(A,q,\omega) \cup D(B,q,\omega)$. If all corresponding sums converge and if there exists a continuous extension of f for the point $\omega/(1-q)$, then we have*

$$\int_{A}^{B} \left(\mathscr{A}_{q,\omega}f\right)(x)\,d_{q,\omega}x = f(B) - f(A). \tag{3.3.2}$$

Proof. Consider the case that $0 < |q| < 1$. First we prove the assertion for $A = \omega/(1-q)$. With $x_\nu^* = Bq^\nu + [\nu]\omega$ we get

$$\begin{aligned}
\int_{\frac{\omega}{1-q}}^{B} (\mathscr{A}_{q,\omega}f)(x)\, d_{q,\omega}x &= \sum_{\nu=0}^{\infty} (\mathscr{A}_{q,\omega}f)(x_\nu^*)\,((1-q)x_\nu^* - \omega) \\
&= \sum_{\nu=0}^{\infty} \frac{f(qx_\nu^* + \omega) - f(x_\nu^*)}{qx_\nu^* + \omega - x_\nu^*}\,((1-q)x_\nu^* - \omega) \\
&= \sum_{\nu=0}^{\infty} \left(f(x_\nu^*) - f(x_{\nu+1}^*)\right) \\
&= f(x_0^*) - \lim_{\nu\to\infty} f(x_{\nu+1}^*) = f(B) - f(\omega/(1-q)).
\end{aligned}$$

In fact since $0 < |q| < 1$, we have

$$\lim_{\nu\to\infty} \left(Bq^{\nu+1} + [\nu+1]\omega\right) = \frac{\omega}{1-q}.$$

So the continuity of f at the point $\omega/(1-q)$ implies that

$$\lim_{\nu\to\infty} f(x_{\nu+1}^*) = f(\omega/(1-q)).$$

Similarly we get

$$\int_{\frac{\omega}{1-q}}^{A} (\mathscr{A}_{q,\omega}f)(x)\, d_{q,\omega}x = f(A) - f(\omega/(1-q)).$$

By using (3.3.1), we have proved (3.3.2) in the case that $0 < |q| < 1$.

In the case that $|q| > 1$, we first set $A = \omega/(1-q)$ again.
With $x_\nu^* = \frac{1}{q^{\nu+1}}(B - [\nu+1]\omega)$ we get

$$\begin{aligned}
&\int_{\frac{\omega}{1-q}}^{B} (\mathscr{A}_{q,\omega}f)(x)\, d_{q,\omega}x \\
&= \sum_{\nu=0}^{\infty} (\mathscr{A}_{q,\omega}f)(x_\nu^*)\,((q-1)x_\nu^* + \omega) \\
&= \sum_{\nu=0}^{\infty} \frac{f(qx_\nu^* + \omega) - f(x_\nu^*)}{qx_\nu^* + \omega - x_\nu^*}\,((q-1)x_\nu^* + \omega) \\
&= f(qx_0^* + \omega) - f(x_0^*) + \sum_{\nu=1}^{\infty} \left(f(x_{\nu-1}^*) - f(x_\nu^*)\right) \\
&= f(B) - f(x_0^*) + f(x_0^*) - \lim_{\nu\to\infty} f(x_\nu^*) = f(B) - f(\omega/(1-q)).
\end{aligned}$$

In fact since $|q| > 1$, we have

$$\lim_{\nu\to\infty} \frac{1}{q^{\nu+1}}(B - [\nu+1]\omega) = \frac{\omega}{1-q}.$$

So the continuity of f at the point $\omega/(1-q)$ implies that

$$\lim_{\nu\to\infty} f(x_\nu^*) = f(\omega/(1-q)).$$

Similarly we get

$$\int_{\frac{\omega}{1-q}}^{A} \left(\mathscr{A}_{q,\omega} f\right)(x)\, d_{q,\omega}x = f(A) - f(\omega/(1-q)).$$

By using (3.3.1), we have proved (3.3.2) in the case that $|q| > 1$. □

Now the summation by parts formula (3.2.17) and the orthogonality relation (3.2.18) with the boundary conditions (3.2.19) can easily be extended to Jackson-Thomae integrals:

Theorem 3.5. *Let the function* $w : D(A,q,\omega) \cup D(B,q,\omega) \to \mathbb{C}$ *satisfy the self-adjoint operator equation (3.2.5) and the boundary conditions*

$$\left(\mathscr{S}_{q,\omega}^{-1}(w\widehat{\varphi})\right)(A) = 0 \quad \text{and} \quad \left(\mathscr{S}_{q,\omega}^{-1}(w\widehat{\varphi})\right)(B) = 0.$$

If the regularity condition (2.3.3) holds for $n = 0, 1, 2, \ldots$*, then the polynomial solutions* $\{y_n\}_{n=0}^{\infty}$ *of the eigenvalue problem (2.2.1) satisfy the orthogonality relation*

$$\int_A^B w(x) y_m(x) y_n(x)\, d_{q,\omega}x = 0, \quad m \neq n, \quad m,n \in \{0,1,2,\ldots\},$$

provided that convergence holds.

3.4 Rodrigues Formulas

In this section we will show that polynomial solutions $\{y_n\}_{n=0}^{\infty}$ of the eigenvalue problem (2.2.7) satisfy a so-called Rodrigues formula. In order to do this, we start with the differential equation (2.2.9). The procedure in this special case will give a motivation for the procedure starting from the q-operator equation (2.2.7), for which the difference equation (2.2.8) is a special case.

The differential equation (2.2.9) can be written as

$$\varphi(x) y_n''(x) + \psi(x) y_n'(x) = \lambda_n y_n(x), \quad n = 0, 1, 2, \ldots \tag{3.4.1}$$

with

$$\varphi(x) = ex^2 + 2fx + g, \quad \psi(x) = 2\varepsilon x + \gamma \quad \text{and} \quad \lambda_n = n\left(e(n-1) + 2\varepsilon\right).$$

In self-adjoint form this reads

$$\left(w(x)\varphi(x) y_n'(x)\right)' = \lambda_n w(x) y_n(x) \tag{3.4.2}$$

where w satisfies the Pearson differential equation

$$(w(x)\varphi(x))' = w(x)\psi(x). \tag{3.4.3}$$

Since $\lambda_0 = 0$, (3.4.2) can be written as

$$\left(w(x)\varphi(x)y_n'(x)\right)' = (\lambda_n - \lambda_0)w(x)y_n(x). \tag{3.4.4}$$

This can be generalized to

$$\left(w(x)\{\varphi(x)\}^k y_n^{(k)}(x)\right)' = (\lambda_n - \lambda_{k-1})w(x)\{\varphi(x)\}^{k-1} y_n^{(k-1)}(x) \tag{3.4.5}$$

for $k = 1,2,3,\ldots$. To prove this, we use induction on k. For $k = 1$ (3.4.5) equals (3.4.4). For the left-hand side of (3.4.5) we obtain by using the Pearson differential equation (3.4.3)

$$\begin{aligned}
&\left(w(x)\{\varphi(x)\}^k y_n^{(k)}(x)\right)' \\
&= w(x)\{\varphi(x)\}^k y_n^{(k+1)}(x) + \left(w(x)\varphi(x)\{\varphi(x)\}^{k-1}\right)' y_n^{(k)}(x) \\
&= w(x)\{\varphi(x)\}^k y_n^{(k+1)}(x) \\
&\qquad + \left((w(x)\varphi(x))'\{\varphi(x)\}^{k-1} + (k-1)w(x)\{\varphi(x)\}^{k-1}\varphi'(x)\right) y_n^{(k)}(x) \\
&= w(x)\{\varphi(x)\}^k y_n^{(k+1)}(x) + \\
&\qquad \left(\psi(x) + (k-1)\varphi'(x)\right) w(x)\{\varphi(x)\}^{k-1} y_n^{(k)}(x).
\end{aligned} \tag{3.4.6}$$

Hence (3.4.5) is equivalent to

$$\begin{aligned}
&w(x)\{\varphi(x)\}^k y_n^{(k+1)}(x) + \left(\psi(x) + (k-1)\varphi'(x)\right) w(x)\{\varphi(x)\}^{k-1} y_n^{(k)}(x) \\
&= (\lambda_n - \lambda_{k-1})w(x)\{\varphi(x)\}^{k-1} y_n^{(k-1)}(x).
\end{aligned}$$

Multiplying by $\varphi(x)$ and differentiating, we obtain

$$\begin{aligned}
&\left(w(x)\{\varphi(x)\}^{k+1} y_n^{(k+1)}(x)\right)' + \left(\psi'(x) + (k-1)\varphi''(x)\right) w(x)\{\varphi(x)\}^k y_n^{(k)}(x) \\
&\qquad + \left(\psi(x) + (k-1)\varphi'(x)\right)\left(w(x)\{\varphi(x)\}^k y_n^{(k)}(x)\right)' \\
&= (\lambda_n - \lambda_{k-1})\left(w(x)\{\varphi(x)\}^k\right)' y_n^{(k-1)}(x) \\
&\qquad + (\lambda_n - \lambda_{k-1})w(x)\{\varphi(x)\}^k y_n^{(k)}(x).
\end{aligned} \tag{3.4.7}$$

Now we use the fact that $\psi'(x) = 2\varepsilon$ and $\varphi''(x) = 2e$, so that

$$\lambda_n - \lambda_{k-1} = \lambda_n - \lambda_k + \psi'(x) + (k-1)\varphi''(x)$$

to conclude that

$$\left(w(x)\{\varphi(x)\}^{k+1}y_n^{(k+1)}(x)\right)' + (\psi(x)+(k-1)\varphi'(x))\left(w(x)\{\varphi(x)\}^k y_n^{(k)}(x)\right)'$$
$$= (\lambda_n-\lambda_{k-1})\left(w(x)\{\varphi(x)\}^k\right)' y_n^{(k-1)}(x) + (\lambda_n-\lambda_k)w(x)\{\varphi(x)\}^k y_n^{(k)}(x).$$

Since we have

$$\left(w(x)\varphi(x)\{\varphi(x)\}^{k-1}\right)' = (\psi(x)+(k-1)\varphi'(x))\,w(x)\{\varphi(x)\}^{k-1},$$

we obtain

$$\left(w(x)\{\varphi(x)\}^{k+1}y_n^{(k+1)}(x)\right)' + (\psi(x)+(k-1)\varphi'(x))\left(w(x)\{\varphi(x)\}^k y_n^{(k)}(x)\right)'$$
$$= (\lambda_n-\lambda_{k-1})(\psi(x)+(k-1)\varphi'(x))\,w(x)\{\varphi(x)\}^{k-1}y_n^{(k-1)}(x)$$
$$+ (\lambda_n-\lambda_k)w(x)\{\varphi(x)\}^k y_n^{(k)}(x).$$

Hence, if (3.4.5) holds for k, we have

$$\left(w(x)\{\varphi(x)\}^{k+1}y_n^{(k+1)}(x)\right)' = (\lambda_n-\lambda_k)w(x)\{\varphi(x)\}^k y_n^{(k)}(x),$$

which equals (3.4.5) with k replaced by $k+1$. This proves that (3.4.5) holds for all $k=1,2,3,\ldots$.

If we assume that the polynomials $\{y_n\}_{n=0}^{\infty}$ are monic, id est $y_n^{(n)}(x)=n!$, and if the regularity condition (2.3.3) holds, then we have by using (3.4.5)

$$\begin{aligned}
(w(x)\{\varphi(x)\}^n)^{(n)} &= \frac{1}{n!}\left(w(x)\{\varphi(x)\}^n y_n^{(n)}(x)\right)^{(n)}\\
&= \frac{1}{n!}(\lambda_n-\lambda_{n-1})\left(w(x)\{\varphi(x)\}^{n-1}y_n^{(n-1)}(x)\right)^{(n-1)}\\
&= \frac{1}{n!}(\lambda_n-\lambda_{n-1})(\lambda_n-\lambda_{n-2})\left(w(x)\{\varphi(x)\}^{n-2}y_n^{(n-2)}(x)\right)^{(n-2)}\\
&\;\;\vdots\\
&= \frac{1}{n!}\left\{\prod_{k=1}^{n}(\lambda_n-\lambda_{n-k})\right\}w(x)y_n(x), \quad n=1,2,3,\ldots.
\end{aligned}$$

Hence, together with $y_0(x)=1$, we have the Rodrigues formula

$$y_n(x) = \frac{K_n}{w(x)}D^n\left(w(x)\{\varphi(x)\}^n\right) = \frac{K_n}{w(x)}\frac{d^n}{dx^n}\left(w(x)\{\varphi(x)\}^n\right) \tag{3.4.8}$$

for $n=1,2,3,\ldots$ with

$$K_n = \frac{n!}{\prod_{k=1}^{n}(\lambda_n - \lambda_{n-k})} = \frac{1}{\prod_{k=1}^{n}(e(2n-k-1)+2\varepsilon)}, \quad n = 1,2,3,\ldots. \tag{3.4.9}$$

Now we use a similar procedure for the q-operator equation (2.2.7), which can be written as

$$\varphi(x)\left(\mathscr{A}_{q,\omega}^2 y_n\right)(x) + \psi(x)\left(\mathscr{A}_{q,\omega} y_n\right)(x) = \lambda_n y_n(qx+\omega), \quad n = 0,1,2,\ldots \tag{3.4.10}$$

where

$$\varphi(x) = ex^2 + 2fx + g, \quad \psi(x) = 2\varepsilon x + \gamma, \quad \lambda_n = \frac{[n]}{q^n}\left(e[n-1]+2\varepsilon\right).$$

In self-adjoint form this reads

$$\left(\mathscr{A}_{q,\omega}\left(w\left(\mathscr{S}_{q,\omega}^{-1}\varphi\right)\mathscr{A}_{q,\omega} y_n\right)\right)(x) = \lambda_n\left(\mathscr{S}_{q,\omega} w\right)(x)\left(\mathscr{S}_{q,\omega} y_n\right)(x) \tag{3.4.11}$$

where w satisfies the Pearson operator equation

$$\left(\mathscr{A}_{q,\omega}\left(w\left(\mathscr{S}_{q,\omega}^{-1}\varphi\right)\right)\right)(x) = \left(\mathscr{S}_{q,\omega} w\right)(x)\psi(x). \tag{3.4.12}$$

Since $\lambda_0 = 0$, (3.4.11) can be written as

$$\left(\mathscr{A}_{q,\omega}\left(w\left(\mathscr{S}_{q,\omega}^{-1}\varphi\right)\mathscr{A}_{q,\omega} y_n\right)\right)(x) = (\lambda_n - \lambda_0)\left(\mathscr{S}_{q,\omega} w\right)(x)\left(\mathscr{S}_{q,\omega} y_n\right)(x). \tag{3.4.13}$$

Applying the operator $\mathscr{S}_{q,\omega}^{-1}$ on both sides of (3.4.13), we obtain by using (2.5.3)

$$\left(\mathscr{A}_{q,\omega}\left(\left(\mathscr{S}_{q,\omega}^{-1} w\right)\left(\mathscr{S}_{q,\omega}^{-2}\varphi\right)\mathscr{S}_{q,\omega}^{-1}\left(\mathscr{A}_{q,\omega} y_n\right)\right)\right)(x) = \frac{\lambda_n - \lambda_0}{q} w(x) y_n(x). \tag{3.4.14}$$

This formula can be generalized to

$$\begin{aligned}&\left(\mathscr{A}_{q,\omega}\left(\left(\mathscr{S}_{q,\omega}^{-k} w\right)\left\{\prod_{i=1}^{k}\left(\mathscr{S}_{q,\omega}^{-i-1}\varphi\right)\right\}\mathscr{S}_{q,\omega}^{-k}\left(\mathscr{A}_{q,\omega}^{k} y_n\right)\right)\right)(x)\\ &= \frac{\lambda_n - \lambda_{k-1}}{q}\left(\left(\mathscr{S}_{q,\omega}^{-k+1} w\right)\left\{\prod_{i=1}^{k-1}\left(\mathscr{S}_{q,\omega}^{-i-1}\varphi\right)\right\}\mathscr{S}_{q,\omega}^{-k+1}\left(\mathscr{A}_{q,\omega}^{k-1} y_n\right)\right)(x)\end{aligned} \tag{3.4.15}$$

for $k = 1,2,3,\ldots$. To prove this we use induction on k. For $k = 1$ (3.4.15) equals (3.4.14) since the empty operator product $\prod_{i=1}^{0}$ at the right-hand side must be interpreted as the identity operator. For simplicity we leave out the argument x in the sequel. Then we obtain, by using the product rule (3.2.2) twice, for the left-hand side of (3.4.15)

$$\begin{aligned}
&\mathscr{A}_{q,\omega}\left(\left(\mathscr{S}_{q,\omega}^{-k}w\right)\left\{\prod_{i=1}^{k}(\mathscr{S}_{q,\omega}^{-i-1}\varphi)\right\}\mathscr{S}_{q,\omega}^{-k}\left(\mathscr{A}_{q,\omega}^{k}y_n\right)\right)\\
&=\left(\mathscr{S}_{q,\omega}^{-k}w\right)\left\{\prod_{i=1}^{k}(\mathscr{S}_{q,\omega}^{-i-1}\varphi)\right\}\left(\mathscr{A}_{q,\omega}\left(\mathscr{S}_{q,\omega}^{-k}\mathscr{A}_{q,\omega}^{k}y_n\right)\right)\\
&\quad+\mathscr{S}_{q,\omega}^{-k+1}\left(\mathscr{A}_{q,\omega}^{k}y_n\right)\left(\mathscr{A}_{q,\omega}\left(\mathscr{S}_{q,\omega}^{-k}w\right)\prod_{i=1}^{k}(\mathscr{S}_{q,\omega}^{-i-1}\varphi)\right)\\
&=\left(\mathscr{S}_{q,\omega}^{-k}w\right)\left\{\prod_{i=1}^{k}(\mathscr{S}_{q,\omega}^{-i-1}\varphi)\right\}\left(\mathscr{A}_{q,\omega}\left(\mathscr{S}_{q,\omega}^{-k}\mathscr{A}_{q,\omega}^{k}y_n\right)\right)\\
&\quad+\mathscr{S}_{q,\omega}^{-k+1}\left(\mathscr{A}_{q,\omega}^{k}y_n\right)\left\{\prod_{i=1}^{k-1}(\mathscr{S}_{q,\omega}^{-i-1}\varphi)\right\}\mathscr{A}_{q,\omega}\left(\left(\mathscr{S}_{q,\omega}^{-k}w\right)\left(\mathscr{S}_{q,\omega}^{-k-1}\varphi\right)\right)\\
&\quad+\mathscr{S}_{q,\omega}^{-k+1}\left(\mathscr{A}_{q,\omega}^{k}y_n\right)\left(\mathscr{S}_{q,\omega}^{-k+1}w\right)\left(\mathscr{S}_{q,\omega}^{-k}\varphi\right)\\
&\qquad\times\mathscr{A}_{q,\omega}\left(\prod_{i=1}^{k-1}(\mathscr{S}_{q,\omega}^{-i-1}\varphi)\right).
\end{aligned} \tag{3.4.16}$$

Now we use (2.5.3) and the Pearson operator equation (3.4.12) to obtain

$$\begin{aligned}
\mathscr{A}_{q,\omega}\left(\left(\mathscr{S}_{q,\omega}^{-k}w\right)\left(\mathscr{S}_{q,\omega}^{-k-1}\varphi\right)\right)&=\mathscr{A}_{q,\omega}\left(\mathscr{S}_{q,\omega}^{-k}\left(w\left(\mathscr{S}_{q,\omega}^{-1}\varphi\right)\right)\right)\\
&=q^{-k}\mathscr{S}_{q,\omega}^{-k}\left(\mathscr{A}_{q,\omega}\left(w\left(\mathscr{S}_{q,\omega}^{-1}\varphi\right)\right)\right)\\
&=q^{-k}\mathscr{S}_{q,\omega}^{-k}\left(\left(\mathscr{S}_{q,\omega}w\right)\psi\right)\\
&=q^{-k}\left(\mathscr{S}_{q,\omega}^{-k+1}w\right)\left(\mathscr{S}_{q,\omega}^{-k}\psi\right).
\end{aligned} \tag{3.4.17}$$

Further we use the following extension of the product rule (3.2.2)

$$\begin{aligned}
\mathscr{A}_{q,\omega}\left(\prod_{i=1}^{N}f_i\right)&=\left(\prod_{i=1}^{N-1}f_i\right)(\mathscr{A}_{q,\omega}f_N)+\left(\prod_{i=1}^{N-2}f_i\right)(\mathscr{A}_{q,\omega}f_{N-1})(\mathscr{S}_{q,\omega}f_N)\\
&\quad+\ldots+(\mathscr{A}_{q,\omega}f_1)\prod_{i=2}^{N}(\mathscr{S}_{q,\omega}f_i)
\end{aligned}$$

to find that

$$\begin{aligned}\mathscr{A}_{q,\omega}\left(\prod_{i=1}^{k-1}\left(\mathscr{S}_{q,\omega}^{-i-1}\varphi\right)\right) &= \left\{\prod_{i=1}^{k-2}\left(\mathscr{S}_{q,\omega}^{-i-1}\varphi\right)\right\}\mathscr{A}_{q,\omega}\left(\mathscr{S}_{q,\omega}^{-k}\varphi\right)\\ &\quad+\left\{\prod_{i=1}^{k-2}\left(\mathscr{S}_{q,\omega}^{-i-1}\varphi\right)\right\}\mathscr{A}_{q,\omega}\left(\mathscr{S}_{q,\omega}^{-k+1}\varphi\right)\\ &\quad+\ldots+\left\{\prod_{i=1}^{k-2}\left(\mathscr{S}_{q,\omega}^{-i-1}\varphi\right)\right\}\mathscr{A}_{q,\omega}\left(\mathscr{S}_{q,\omega}^{-2}\varphi\right)\\ &= \left\{\prod_{i=1}^{k-2}\left(\mathscr{S}_{q,\omega}^{-i-1}\varphi\right)\right\}\sum_{j=1}^{k-1}\mathscr{A}_{q,\omega}\left(\mathscr{S}_{q,\omega}^{-k+j-1}\varphi\right). \quad (3.4.18)\end{aligned}$$

Now we use (3.4.16), (3.4.17) and (3.4.18) to obtain

$$\begin{aligned}&\mathscr{A}_{q,\omega}\left(\left(\mathscr{S}_{q,\omega}^{-k}w\right)\left\{\prod_{i=1}^{k}\left(\mathscr{S}_{q,\omega}^{-i-1}\varphi\right)\right\}\mathscr{S}_{q,\omega}^{-k}\left(\mathscr{A}_{q,\omega}^{k}y_n\right)\right)\\ &= \left(\mathscr{S}_{q,\omega}^{-k}w\right)\left\{\prod_{i=1}^{k}\left(\mathscr{S}_{q,\omega}^{-i-1}\varphi\right)\right\}\left(\mathscr{A}_{q,\omega}\left(\mathscr{S}_{q,\omega}^{-k}\mathscr{A}_{q,\omega}^{k}y_n\right)\right)\\ &\quad+\left(\mathscr{S}_{q,\omega}^{-k+1}w\right)\left\{\prod_{i=1}^{k-1}\left(\mathscr{S}_{q,\omega}^{-i-1}\varphi\right)\right\}\mathscr{S}_{q,\omega}^{-k+1}\left(\mathscr{A}_{q,\omega}^{k}y_n\right)\\ &\quad\times\left(q^{-k}\mathscr{S}_{q,\omega}^{-k}\psi+\sum_{j=1}^{k-1}\mathscr{A}_{q,\omega}\left(\mathscr{S}_{q,\omega}^{-k+j-1}\varphi\right)\right). \qquad (3.4.19)\end{aligned}$$

Hence, combining (3.4.15) and (3.4.19), we have

$$\begin{aligned}&\left(\mathscr{S}_{q,\omega}^{-k}w\right)\left\{\prod_{i=1}^{k}\left(\mathscr{S}_{q,\omega}^{-i-1}\varphi\right)\right\}\left(\mathscr{A}_{q,\omega}\left(\mathscr{S}_{q,\omega}^{-k}\mathscr{A}_{q,\omega}^{k}y_n\right)\right)\\ &\quad+\left(\mathscr{S}_{q,\omega}^{-k+1}w\right)\left\{\prod_{i=1}^{k-1}\left(\mathscr{S}_{q,\omega}^{-i-1}\varphi\right)\right\}\mathscr{S}_{q,\omega}^{-k+1}\left(\mathscr{A}_{q,\omega}^{k}y_n\right)\\ &\quad\times\left(q^{-k}\mathscr{S}_{q,\omega}^{-k}\psi+\sum_{j=1}^{k-1}\mathscr{A}_{q,\omega}\left(\mathscr{S}_{q,\omega}^{-k+j-1}\varphi\right)\right)\\ &= \frac{\lambda_n-\lambda_{k-1}}{q}\left(\mathscr{S}_{q,\omega}^{-k+1}w\right)\left\{\prod_{i=1}^{k-1}\left(\mathscr{S}_{q,\omega}^{-i-1}\varphi\right)\right\}\mathscr{S}_{q,\omega}^{-k+1}\left(\mathscr{A}_{q,\omega}^{k-1}y_n\right). \quad (3.4.20)\end{aligned}$$

Now we apply $\mathscr{S}_{q,\omega}^{-1}$ on both sides of (3.4.20), multiply by $\mathscr{S}_{q,\omega}^{-2}\varphi$ and apply $\mathscr{A}_{q,\omega}$ on both sides, then we obtain by using (2.5.3)

$$\mathscr{A}_{q,\omega}\left(\left(\mathscr{S}_{q,\omega}^{-k-1}w\right)\left\{\prod_{i=1}^{k+1}(\mathscr{S}_{q,\omega}^{-i-1}\varphi)\right\}q^{-k}\mathscr{S}_{q,\omega}^{-k-1}\left(\mathscr{A}_{q,\omega}^{k+1}y_n\right)\right)$$

$$+\mathscr{A}_{q,\omega}\left(\left(\mathscr{S}_{q,\omega}^{-k}w\right)\left\{\prod_{i=1}^{k}(\mathscr{S}_{q,\omega}^{-i-1}\varphi)\right\}\mathscr{S}_{q,\omega}^{-k}\left(\mathscr{A}_{q,\omega}^{k}y_n\right)\right.$$

$$\left.\times\left(q^{-k}\mathscr{S}_{q,\omega}^{-k-1}\psi+\sum_{j=1}^{k-1}q^{-k+j-1}\mathscr{S}_{q,\omega}^{-k+j-2}(\mathscr{A}_{q,\omega}\varphi)\right)\right)$$

$$=\frac{\lambda_n-\lambda_{k-1}}{q}\mathscr{A}_{q,\omega}\left(\left(\mathscr{S}_{q,\omega}^{-k}w\right)\left\{\prod_{i=1}^{k}(\mathscr{S}_{q,\omega}^{-i-1}\varphi)\right\}\mathscr{S}_{q,\omega}^{-k}\left(\mathscr{A}_{q,\omega}^{k-1}y_n\right)\right). \quad (3.4.21)$$

Hence

$$q^{-k}\mathscr{A}_{q,\omega}\left(\left(\mathscr{S}_{q,\omega}^{-k-1}w\right)\left\{\prod_{i=1}^{k+1}(\mathscr{S}_{q,\omega}^{-i-1}\varphi)\right\}\mathscr{S}_{q,\omega}^{-k-1}\left(\mathscr{A}_{q,\omega}^{k+1}y_n\right)\right)$$

$$+\left(\mathscr{S}_{q,\omega}^{-k}w\right)\left\{\prod_{i=1}^{k}(\mathscr{S}_{q,\omega}^{-i-1}\varphi)\right\}\mathscr{S}_{q,\omega}^{-k}\left(\mathscr{A}_{q,\omega}^{k}y_n\right)$$

$$\times\mathscr{A}_{q,\omega}\left(q^{-k}\mathscr{S}_{q,\omega}^{-k-1}\psi+\sum_{j=1}^{k-1}q^{-k+j-1}\mathscr{S}_{q,\omega}^{-k+j-2}(\mathscr{A}_{q,\omega}\varphi)\right)$$

$$+\left(q^{-k}\mathscr{S}_{q,\omega}^{-k}\psi+\sum_{j=1}^{k-1}q^{-k+j-1}\mathscr{S}_{q,\omega}^{-k+j-1}(\mathscr{A}_{q,\omega}\varphi)\right)$$

$$\times\mathscr{A}_{q,\omega}\left(\left(\mathscr{S}_{q,\omega}^{-k}w\right)\left\{\prod_{i=1}^{k}(\mathscr{S}_{q,\omega}^{-i-1}\varphi)\right\}\mathscr{S}_{q,\omega}^{-k}\left(\mathscr{A}_{q,\omega}^{k}y_n\right)\right)$$

$$=\frac{\lambda_n-\lambda_{k-1}}{q}q^{-k}\left(\mathscr{S}_{q,\omega}^{-k}w\right)\left\{\prod_{i=1}^{k}(\mathscr{S}_{q,\omega}^{-i-1}\varphi)\right\}\mathscr{S}_{q,\omega}^{-k}\left(\mathscr{A}_{q,\omega}^{k}y_n\right)$$

$$+\frac{\lambda_n-\lambda_{k-1}}{q}\mathscr{S}_{q,\omega}^{-k+1}\left(\mathscr{A}_{q,\omega}^{k-1}y_n\right)$$

$$\times\mathscr{A}_{q,\omega}\left(\left(\mathscr{S}_{q,\omega}^{-k}w\right)\prod_{i=1}^{k}(\mathscr{S}_{q,\omega}^{-i-1}\varphi)\right). \quad (3.4.22)$$

Now we use the fact that

$$\lambda_n-\lambda_{k-1}=\lambda_n-\lambda_k+q^{-k}\left(e[k-1](1+q^{k-1})+2\varepsilon\right)$$

to conclude that

$$\begin{aligned}
&\frac{\lambda_n-\lambda_{k-1}}{q}q^{-k}\left(\mathscr{S}_{q,\omega}^{-k}w\right)\left\{\prod_{i=1}^{k}\left(\mathscr{S}_{q,\omega}^{-i-1}\varphi\right)\right\}\mathscr{S}_{q,\omega}^{-k}\left(\mathscr{A}_{q,\omega}^{k}y_n\right)\\
&=q^{-k-1}(\lambda_n-\lambda_k)\left(\mathscr{S}_{q,\omega}^{-k}w\right)\left\{\prod_{i=1}^{k}\left(\mathscr{S}_{q,\omega}^{-i-1}\varphi\right)\right\}\mathscr{S}_{q,\omega}^{-k}\left(\mathscr{A}_{q,\omega}^{k}y_n\right)\\
&\quad+q^{-2k-1}\left(\mathscr{S}_{q,\omega}^{-k}w\right)\left\{\prod_{i=1}^{k}\left(\mathscr{S}_{q,\omega}^{-i-1}\varphi\right)\right\}\\
&\qquad\times\mathscr{S}_{q,\omega}^{-k}\left(\mathscr{A}_{q,\omega}^{k}y_n\right)\left(e[k-1](1+q^{k-1})+2\varepsilon\right).
\end{aligned}\tag{3.4.23}$$

Now we use (3.4.17) and (3.4.18) to find that

$$\begin{aligned}
&\mathscr{A}_{q,\omega}\left(\left(\mathscr{S}_{q,\omega}^{-k}w\right)\prod_{i=1}^{k}\left(\mathscr{S}_{q,\omega}^{-i-1}\varphi\right)\right)\\
&=\mathscr{A}_{q,\omega}\left(\left(\mathscr{S}_{q,\omega}^{-k}w\right)\left(\mathscr{S}_{q,\omega}^{-k-1}\varphi\right)\prod_{i=1}^{k-1}\left(\mathscr{S}_{q,\omega}^{-i-1}\varphi\right)\right)\\
&=\left\{\prod_{i=1}^{k-1}\left(\mathscr{S}_{q,\omega}^{-i-1}\varphi\right)\right\}\mathscr{A}_{q,\omega}\left(\left(\mathscr{S}_{q,\omega}^{-k}w\right)\left(\mathscr{S}_{q,\omega}^{-k-1}\varphi\right)\right)\\
&\quad+\left(\mathscr{S}_{q,\omega}^{-k+1}w\right)\left(\mathscr{S}_{q,\omega}^{-k}\varphi\right)\mathscr{A}_{q,\omega}\left(\prod_{i=1}^{k-1}\left(\mathscr{S}_{q,\omega}^{-i-1}\varphi\right)\right)\\
&=\left\{\prod_{i=1}^{k-1}\left(\mathscr{S}_{q,\omega}^{-i-1}\varphi\right)\right\}\left(\mathscr{S}_{q,\omega}^{-k+1}w\right)q^{-k}\left(\mathscr{S}_{q,\omega}^{-k}\psi\right)\\
&\quad+\left(\mathscr{S}_{q,\omega}^{-k+1}w\right)\left\{\prod_{i=1}^{k-1}\left(\mathscr{S}_{q,\omega}^{-i-1}\varphi\right)\right\}\sum_{j=1}^{k-1}\mathscr{A}_{q,\omega}\left(\mathscr{S}_{q,\omega}^{-k+j-1}\varphi\right)\\
&=\left\{\prod_{i=1}^{k-1}\left(\mathscr{S}_{q,\omega}^{-i-1}\varphi\right)\right\}\left(\mathscr{S}_{q,\omega}^{-k+1}w\right)\\
&\quad\times\left(q^{-k}\mathscr{S}_{q,\omega}^{-k}\psi+\sum_{j=1}^{k-1}q^{-k+j-1}\mathscr{S}_{q,\omega}^{-k+j-1}\left(\mathscr{A}_{q,\omega}\varphi\right)\right).
\end{aligned}\tag{3.4.24}$$

Since $\psi(x)=2\varepsilon x+\gamma$, we have

$$\mathscr{A}_{q,\omega}\left(q^{-k}\mathscr{S}_{q,\omega}^{-k-1}\psi\right)=q^{-2k-1}\mathscr{S}_{q,\omega}^{-k-1}\left(\mathscr{A}_{q,\omega}\psi\right)=q^{-2k-1}2\varepsilon$$

and since $\varphi(x)=ex^2+2fx+g$, we have

$$\left(\mathscr{A}_{q,\omega}\varphi\right)(x)=e(1+q)x+e\omega+2f\quad\text{and}\quad\left(\mathscr{A}_{q,\omega}^2\varphi\right)(x)=e(1+q)$$

and therefore

$$
\begin{aligned}
&\mathscr{A}_{q,\omega}\left(\sum_{j=1}^{k-1} q^{-k+j-1}\mathscr{S}_{q,\omega}^{-k+j-2}\left(\mathscr{A}_{q,\omega}\varphi\right)\right)\\
&=q^{-2k-1}\left(\sum_{j=1}^{k-1} q^{2j-2}\mathscr{S}_{q,\omega}^{-k+j-2}\left(\mathscr{A}_{q,\omega}^{2}\varphi\right)\right)\\
&=q^{-2k-1}e(1+q)\sum_{j=1}^{k-1} q^{2j-2}=q^{-2k-1}e\frac{1-q^{2k-2}}{1-q}\\
&=q^{-2k-1}e[k-1](1+q^{k-1}).
\end{aligned}
\tag{3.4.25}
$$

If (3.4.15) holds for k, by combining (3.4.22), (3.4.23), (3.4.24) and (3.4.25), we obtain (3.4.15) with k replaced by $k+1$. This proves that (3.4.15) holds for all $k=1,2,3,\ldots$.

If we assume that the polynomials $\{y_n\}_{n=0}^{\infty}$ are monic, id est $\left(\mathscr{A}_{q,\omega}^{n}y_n\right)(x)=[n]!$, and the regularity condition (2.3.3) holds, then we have by using (3.4.15)

$$
\begin{aligned}
&\mathscr{A}_{q,\omega}^{n}\left(\left(\mathscr{S}_{q,\omega}^{-n}w\right)\prod_{i=1}^{n}\left(\mathscr{S}_{q,\omega}^{-i-1}\varphi\right)\right)\\
&=\frac{1}{[n]!}\mathscr{A}_{q,\omega}^{n}\left(\left(\mathscr{S}_{q,\omega}^{-n}w\right)\left\{\prod_{i=1}^{n}\left(\mathscr{S}_{q,\omega}^{-i-1}\varphi\right)\right\}\left(\mathscr{S}_{q,\omega}^{-n}\left(\mathscr{A}_{q,\omega}^{n}y_n\right)\right)\right)\\
&=\frac{\lambda_n-\lambda_{n-1}}{q\,[n]!}\mathscr{A}_{q,\omega}^{n-1}\left(\left(\mathscr{S}_{q,\omega}^{-n+1}w\right)\left\{\prod_{i=1}^{n-1}\left(\mathscr{S}_{q,\omega}^{-i-1}\varphi\right)\right\}\left(\mathscr{S}_{q,\omega}^{-n+1}\left(\mathscr{A}_{q,\omega}^{n-1}y_n\right)\right)\right)\\
&=\frac{(\lambda_n-\lambda_{n-1})(\lambda_n-\lambda_{n-2})}{q^2\,[n]!}\\
&\quad\times\mathscr{A}_{q,\omega}^{n-2}\left(\left(\mathscr{S}_{q,\omega}^{-n+2}w\right)\left\{\prod_{i=1}^{n-2}\left(\mathscr{S}_{q,\omega}^{-i-1}\varphi\right)\right\}\left(\mathscr{S}_{q,\omega}^{-n+2}\left(\mathscr{A}_{q,\omega}^{n-2}y_n\right)\right)\right)\\
&\ \vdots\\
&=\frac{\prod_{k=1}^{n}(\lambda_n-\lambda_{n-k})}{q^n\,[n]!}w\,y_n=q^{-n}\left\{\prod_{k=1}^{n}\frac{\lambda_n-\lambda_{n-k}}{[k]}\right\}w\,y_n,\quad n=1,2,3,\ldots.
\end{aligned}
$$

Hence, together with $y_0(x)=1$, we have the Rodrigues formula

$$
y_n(x)=\frac{K_n}{w(x)}\left[\mathscr{A}_{q,\omega}^{n}\left\{\left(\mathscr{S}_{q,\omega}^{-n}w\right)\prod_{k=1}^{n}\left(\mathscr{S}_{q,\omega}^{-k-1}\varphi\right)\right\}\right](x)
\tag{3.4.26}
$$

for $(q,\omega)\neq(1,0)$ and $n=1,2,3,\ldots$, where by using (2.3.2)

$$K_n = q^n \prod_{k=1}^{n} \frac{[k]}{\lambda_n - \lambda_{n-k}} = \frac{q^{n(n+1)}}{\prod_{k=1}^{n} (e[2n-k-1] + 2\varepsilon)}, \quad n = 1, 2, 3, \ldots. \tag{3.4.27}$$

In the case of the difference operator Δ (i.e. $q = 1$ and $\omega = 1$), this reads

$$y_n(x) = \frac{K_n}{w(x)} \Delta^n \left(w(x-n) \prod_{k=1}^{n} \varphi(x-k-1) \right), \quad n = 1, 2, 3, \ldots, \tag{3.4.28}$$

where K_n is given by (3.4.9).

3.5 Duality

In this section we consider the concept of dual polynomial systems. The following definition of dual polynomial systems is due to D.A. Leonard (see [371]):

Definition 3.1. Let $N \in \{1, 2, 3, \ldots\}$ or $N \to \infty$ and let $\{\kappa_n\}_{n=0}^{N}$ and $\{\lambda_n\}_{n=0}^{N}$ denote two (finite or infinite) sequences of complex numbers with

$$\kappa_m \neq \kappa_n \quad \text{and} \quad \lambda_m \neq \lambda_n, \quad m \neq n.$$

Then two (finite or infinite) polynomial systems $\{y_n\}_{n=0}^{N}$ and $\{z_n\}_{n=0}^{N}$ in $\mathscr{P}$ with $\text{degree}[y_n] = n$ and $\text{degree}[z_n] = n$ for $n = 0, 1, 2, \ldots, N$ are called dual polynomial systems with respect to the sequences $\{\kappa_n\}_{n=0}^{N}$ and $\{\lambda_n\}_{n=0}^{N}$ if

$$y_n(\kappa_m) = z_m(\lambda_n), \quad m, n = 0, 1, 2, \ldots, N. \tag{3.5.1}$$

The numbers $\kappa_n \in \mathbb{C}$ and $\lambda_n \in \mathbb{C}$ are called eigenvalues.

Remark. If $y_n = z_n$ for $n = 0, 1, 2, \ldots, N$ in the above definition, then the (finite or infinite) polynomial system $\{y_n\}_{n=0}^{N}$ is called self-dual.

Now we want to construct a dual polynomial system $\{z_n\}_{n=0}^{N}$ to a system of polynomial solutions $\{y_n\}_{n=0}^{N}$ of the eigenvalue problem (2.2.1). We will prove:

Theorem 3.6. *Let $N \in \{1, 2, 3, \ldots\}$ or $N \to \infty$, $q \in \mathbb{R} \setminus \{-1, 0\}$ and $\omega \in \mathbb{R}$ with $(q, \omega) \neq (1, 0)$. Define the (finite or infinite) sequence $\{\kappa_n\}_{n=0}^{N}$ by*

$$\kappa_n := q^n x_0 + [n]\omega, \quad x_0 \in \mathbb{C} \setminus \left\{ \frac{\omega}{1-q} \right\}, \quad n = 0, 1, 2, \ldots, N.$$

Let $\{y_n\}_{n=0}^{N}$ denote a (finite or infinite) system of polynomial solutions of the eigenvalue problem (2.2.12) and let C and D be functions defined by (2.2.13),

$$(D(x_0) =)\; qC(x_0) - \frac{2\varepsilon(x_0 - \omega) + \gamma q}{qx_0 + \omega - x_0} = 0$$

and

$$C(\kappa_n) \neq 0, \quad n = 0, 1, 2, \ldots, N-1.$$

If the regularity condition (2.3.3) holds, then there exists a (finite or infinite) polynomial system $\{z_n\}_{n=0}^N$ *in* $\mathscr{P}$ *such that* $\{y_n\}_{n=0}^N$ *and* $\{z_n\}_{n=0}^N$ *are dual polynomial systems with respect to the sequence* $\{\lambda_n\}_{n=0}^N$ *of eigenvalues of (2.2.12) and the sequence* $\{\kappa_n\}_{n=0}^N$*, id est we have (3.5.1). The polynomials* $\{z_n\}_{n=0}^N$ *satisfy the three-term recurrence relation*

$$C(\kappa_n)z_{n+1}(x) - \{C(\kappa_n) + D(\kappa_n)\}\, z_n(x) + D(\kappa_n)z_{n-1}(x) = xz_n(x) \tag{3.5.2}$$

for $n = 0, 1, 2, \ldots, N-1$ *with the convention that* $z_{-1}(x) := 0$.

Proof. Since degree$[y_0] = 0$, we have $y_0(x) = c \in \mathbb{C} \setminus \{0\}$. Hence from (3.5.1) we have $z_0(\lambda_0) = y_0(\kappa_0) = c$. Since degree$[z_0] = 0$, we conclude that $z_0(x) = c \neq 0$. Since $C(\kappa_n) \neq 0$ for $n = 0, 1, 2, \ldots, N-1$, the polynomial system $\{z_n\}_{n=0}^N$ is uniquely determined by the three-term recurrence relation (3.5.2).

If we substitute $x = \kappa_m$ in the symmetric form (2.2.12) of the operator equation, we find by using (3.5.1)

$$C(\kappa_m)y_n(\kappa_{m+1}) - \{C(\kappa_m) + D(\kappa_m)\}\, y_n(\kappa_m) + D(\kappa_m)y_n(\kappa_{m-1}) = \lambda_n y_n(\kappa_m).$$

This equation can be seen as a three-term recurrence relation for $\{y_n(\kappa_m)\}_{m=0}^N$. The coefficients of this recurrence relation coincide with those of (3.5.2). So we may conclude that $z_m(\lambda_n) = y_n(\kappa_m)$ if the initial conditions $z_0(\lambda_n) = y_n(\kappa_0)$ coincide for $n = 0, 1, 2, \ldots, N$. This can be achieved if and only if $y_n(\kappa_0) \neq 0$ for $n = 0, 1, 2, \ldots, N$.

To prove the last assumption, we first remark that $y_0(\kappa_0) = c \neq 0$. Next we assume that $y_1(\kappa_0) = 0$. Then we have

$$C(\kappa_0)y_1(\kappa_1) - C(\kappa_0)y_1(\kappa_0) = \lambda_1 y_1(\kappa_0) = 0.$$

Hence $y_1(\kappa_0) = y_1(\kappa_1) = 0$, which implies that $y_1(x) = 0$ since $C(\kappa_0) \neq 0$, $\kappa_0 \neq \kappa_1$ and degree$[y_1] = 1$. Hence $y_1(\kappa_0) \neq 0$. Similarly for each $n \in \{2, 3, 4, \ldots\}$ the assumption that $y_n(\kappa_0) = 0$ would imply that

$$y_n(\kappa_1) = y_n(\kappa_2) = y_n(\kappa_3) = \cdots = y_n(\kappa_n) = 0$$

and therefore $y_n(x) = 0$ since $\kappa_m \neq \kappa_n$ if $m \neq n$ and degree$[y_n] = n$. Hence $y_n(\kappa_0) \neq 0$ for $n = 0, 1, 2, \ldots, N$. □

Sometimes it is useful to have a bounded sequence $\{\kappa_n\}_{n=0}^N$ as we will see later on. In the preceding theorem this is not the case for $|q| > 1$. Therefore we will restate this theorem for a different sequence $\{\kappa_n\}_{n=0}^N$:

Theorem 3.7. *Let* $N \in \{1, 2, 3, \ldots\}$ *or* $N \to \infty$ *and let* $\{\kappa_n\}_{n=0}^N$ *be a finite or infinite sequence defined by*

$$\kappa_n = \frac{1}{q^n}(x_0 - [n]\omega), \quad x_0 \in \mathbb{C} \setminus \left\{\frac{\omega}{1-q}\right\}, \quad n = 0, 1, 2, \ldots, N.$$

Assume that the hypotheses of the preceding theorem hold with

$$(C(x_0)=)\ \frac{e(x_0-\omega)^2+2fq(x_0-\omega)+gq^2}{q(qx_0+\omega-x_0)^2}=0$$

and

$$D(\kappa_n)\neq 0,\quad n=0,1,2,\ldots,N-1.$$

Then there exists a (finite or infinite) sequence of dual polynomials $\{z_n\}_{n=0}^N$ *satisfying the three-term recurrence relation*

$$D(\kappa_n)z_{n+1}(x)-\{D(\kappa_n)+C(\kappa_n)\}z_n(x)+C(\kappa_n)z_{n-1}(x)=xz_n(x)$$

for $n=0,1,2,\ldots,N-1$ *with the convention that* $z_{-1}(x)=0$.

In some cases a theorem by G.K. Eagleson (see [189] in a generalized form) establishes a correlation between the orthogonality of two dual polynomial systems. For a given sequence $\{x_n\}_{n=0}^{\infty}$ of real numbers and a sequence $\{w_n\}_{n=0}^{\infty}$ of weights with $w_n>0$ for $n=0,1,2,\ldots$ we define the Hilbert space

$$L^2(\{x_n\},\{w_n\})=\left\{f:\mathbb{R}\to\mathbb{C}\,\middle|\,\sum_{n=0}^{\infty}w_n|f(x_n)|^2<\infty\right\}$$

with the scalar product

$$(f,g)=\sum_{n=0}^{\infty}w_nf(x_n)\overline{g}(x_n),$$

where $\overline{g}$ denotes the complex conjugate of g.

Now we will prove

Theorem 3.8. *Let* $N\in\{1,2,3,\ldots\}$ *or* $N\to\infty$ *and let* $\{x_\nu\}_{\nu=0}^N$ *be a finite or infinite sequence of real numbers. Let* $\{y_n\}_{n=0}^N$ *be a finite or infinite sequence of polynomials with degree*$[y_n]=n$ *satisfying the orthogonality relation*

$$\sum_{\nu=0}^{N}w_\nu y_m(x_\nu)y_n(x_\nu)=\sigma_n\,\delta_{mn},\quad w_\nu>0,\quad \sigma_n>0,\quad m,n=0,1,2,\ldots,N.$$

For $N\to\infty$ *let* $\left\{\frac{y_n}{\sqrt{\sigma_n}}\right\}_{n=0}^{\infty}$ *be a complete orthonormal set in* $L^2(\{x_n\},\{w_n\})$. *Then the dual orthogonality relation*

$$\sum_{n=0}^{N}\frac{1}{\sigma_n}y_n(x_\mu)y_n(x_\nu)=\frac{\delta_{\mu\nu}}{w_\nu},\quad \mu,\nu=0,1,2,\ldots,N \tag{3.5.3}$$

holds.

Proof. We prove the theorem for $N\to\infty$. Consider the function

$$u_\nu(x) = \begin{cases} 1 \text{ for } x = x_\nu, \\ \\ 0 \text{ for } x \neq x_\nu, \end{cases} \qquad x \in \mathbb{R}. \tag{3.5.4}$$

Because of the completeness of the orthonormal set $\left\{\frac{y_n}{\sqrt{\sigma_n}}\right\}_{n=0}^{\infty}$ we have the representation

$$u_\nu(x) = \sum_{n=0}^{\infty} \left(u_\nu, \frac{y_n}{\sqrt{\sigma_n}}\right) \frac{y_n(x)}{\sqrt{\sigma_n}},$$

where the Fourier coefficients can be computed as

$$\left(u_\nu, \frac{y_n}{\sqrt{\sigma_n}}\right) = \sum_{i=0}^{\infty} w_i u_\nu(x_i) \frac{y_n(x_i)}{\sqrt{\sigma_n}} = \frac{w_\nu y_n(x_\nu)}{\sqrt{\sigma_n}}.$$

By using Parseval's identity, we obtain

$$(u_\nu, u_\mu) = \sum_{n=0}^{\infty} \left(u_\nu, \frac{y_n}{\sqrt{\sigma_n}}\right)\left(u_\mu, \frac{y_n}{\sqrt{\sigma_n}}\right) = \sum_{n=0}^{\infty} \frac{w_\nu w_\mu y_n(x_\nu) y_n(x_\mu)}{\sigma_n}.$$

On the other hand we have

$$(u_\nu, u_\mu) = \sum_{i=0}^{\infty} w_i u_\nu(x_i) u_\mu(x_i) = w_\mu \delta_{\mu\nu}.$$

Comparing both results for (u_ν, u_μ) leads to (3.5.3). □

In terms of duality, the last result can be restated as follows.

Corollary 3.9. *Let $N \in \{1,2,3,\ldots\}$ or $N \to \infty$ and let the finite or infinite polynomial systems $\{y_n\}_{n=0}^{N}$ and $\{z_n\}_{n=0}^{N}$ be dual with respect to the finite or infinite sequences $\{\kappa_n\}_{n=0}^{N}$ and $\{\lambda_n\}_{n=0}^{N}$ and let the orthogonality relation (3.1.1) hold for $x_\nu = \kappa_\nu$. Then we also have the dual orthogonality relation*

$$\sum_{n=0}^{N} \frac{z_\mu(\lambda_n) z_\nu(\lambda_n)}{\sigma_n} = \frac{\delta_{\mu\nu}}{w_\nu}, \quad \mu, \nu = 0, 1, 2, \ldots, N.$$

In the case that $N \to \infty$, the completeness of $\left\{\frac{y_n}{\sqrt{\sigma_n}}\right\}_{n=0}^{\infty}$ in $L^2(\{x_n\},\{w_n\})$ is necessary.

The completeness of the orthonormal set $\left\{\frac{y_n}{\sqrt{\sigma_n}}\right\}_{n=0}^{\infty}$ in the infinite case (i.e. $N \to \infty$) can easily be established for a bounded sequence $\{x_n\}_{n=0}^{\infty}$:

Theorem 3.10. *Let $\{x_n\}_{n=0}^{\infty}$ be a bounded sequence.*
Then the orthonormal set $\left\{\frac{y_n}{\sqrt{\sigma_n}}\right\}_{n=0}^{\infty}$ is complete in $L^2(\{x_n\},\{w_n\})$.

Proof. It suffices to show that the functions u_ν defined by (3.5.4) can be approximated in the sense of the norm $||\cdot||$ of $L^2(\{x_n\},\{w_n\})$ by polynomials. Since $\{x_n\}_{n=0}^{\infty}$

is a bounded sequence, it is contained in some compact interval $[a,b]$ on the real line. First we show that the function u_ν can be approximated by continuous functions on $[a,b]$. For a given number $\varepsilon > 0$ we choose $n_0 \in \{1,2,3,\ldots\}$ such that

$$\sum_{\mu=n_0}^{\infty} w_\mu < \varepsilon$$

and we choose a function $f \in C[a,b]$ with $|f(x_\mu)| \leq 1$ for all $\mu \in \{0,1,2,\ldots\}$ and

$$f(x_\mu) = \begin{cases} 1 \text{ for } \mu = \nu, \\ \\ 0 \text{ for } \mu \neq \nu,\ \mu < n_0. \end{cases}$$

Then we have

$$||f - u_\nu|| = \sum_{\mu=0}^{\infty} w_\mu |f(x_\mu) - u_\nu(x_\mu)|^2 = \sum_{\mu=n_0}^{\infty} w_\mu |f(x_\mu) - u_\nu(x_\mu)|^2 \leq \sum_{\mu=n_0}^{\infty} w_\mu < \varepsilon.$$

Now it remains to prove that functions in $C[a,b]$ can be approximated by polynomials. According to the Weierstrass approximation theorem, for a given number $\varepsilon > 0$ and a function $f \in C[a,b]$, there exists a polynomial $p \in \mathscr{P}$ such that

$$\sup_{x\in[a,b]} |f(x) - p(x)| < \varepsilon.$$

Hence

$$||f - p|| = \sum_{\mu=0}^{\infty} w_\mu |f(x_\mu) - p(x_\mu)|^2 < \varepsilon \sum_{\mu=0}^{\infty} w_\mu.$$

□

Part I

Classical Orthogonal Polynomials

Chapter 4
Orthogonal Polynomial Solutions of Differential Equations

Continuous Classical Orthogonal Polynomials

4.1 Polynomial Solutions of Differential Equations

In the case of the ordinary differentiation operator $D = \mathscr{A}_{1,0}$, we have to deal with (cf. (2.2.9)):

$$(ex^2+2fx+g)y_n''(x)+(2\varepsilon x+\gamma)y_n'(x)=n(e(n-1)+2\varepsilon)y_n(x) \tag{4.1.1}$$

with $e,f,g,\varepsilon,\gamma \in \mathbb{C}$ and $n=0,1,2,\ldots$. We look for monic polynomial solutions of the form

$$y_n(x)=\sum_{k=0}^{n} a_{n,k}\frac{(x+c)^k}{k!}, \quad a_{n,n}=n!, \quad n=0,1,2,\ldots \tag{4.1.2}$$

where the coefficients satisfy the three-term recurrence relation

$$(n-k)\left(e(n+k-1)+2\varepsilon\right)a_{n,k}+\left(2(ec-f)k+2\varepsilon c-\gamma\right)a_{n,k+1}$$
$$-(ec^2-2fc+g)a_{n,k+2}=0, \quad a_{n,n+1}=0, \quad k=n-1,n-2,n-3,\ldots,0.$$

If c can be chosen such that $ec^2-2fc+g=0$, this leads to the two-term recurrence relation (cf. (2.4.9))

$$(n-k)\left(e(n+k-1)+2\varepsilon\right)a_{n,k}+\left(2(ec-f)k+2\varepsilon c-\gamma\right)a_{n,k+1}=0 \tag{4.1.3}$$

for $k=n-1,n-2,n-3,\ldots,0$. If $ec^2-2fc+g=0$ has no solution for c, id est $e=f=0$ and $g\neq 0$, then we find from (2.4.4) with $c=\gamma/2\varepsilon$ the two-term recurrence relation (cf. (2.4.10))

$$2\varepsilon(n-k)a_{n,k}-ga_{n,k+2}=0, \quad a_{n,n+1}=0, \quad k=n-1,n-2,n-3,\ldots,0. \tag{4.1.4}$$

R. Koekoek et al., *Hypergeometric Orthogonal Polynomials and Their q-Analogues*, Springer Monographs in Mathematics,
DOI 10.1007/978-3-642-05014-5_4,

In section 2.6 we found that the monic polynomial solutions $\{y_n\}_{n=0}^{\infty}$ satisfy the three-term recurrence relation

$$y_{n+1}(x) = (x - c_n)y_n(x) - d_n y_{n-1}(x), \quad n = 1,2,3,\ldots, \tag{4.1.5}$$

with initial values $y_0(x) = 1$ and $y_1(x) = x - c_0$, where

$$c_n = -\frac{2fn(e(n-1)+2\varepsilon) - \gamma(e-\varepsilon)}{2(e(n-1)+\varepsilon)(en+\varepsilon)}, \quad n = 0,1,2,\ldots \tag{4.1.6}$$

and

$$\begin{aligned} d_n = & \frac{n(e(n-2)+2\varepsilon)}{4(e(2n-3)+2\varepsilon)(e(n-1)+\varepsilon)^2(e(2n-1)+2\varepsilon)} \\ & \times \Big(\{2f(n-1)+\gamma\}\{2f(e(n-1)+2\varepsilon) - e\gamma\} \\ & - 4g(e(n-1)+\varepsilon)^2 \Big), \quad n = 1,2,3,\ldots. \end{aligned} \tag{4.1.7}$$

4.2 Classification of the Positive-Definite Orthogonal Polynomial Solutions

In this chapter we classify all positive-definite orthogonal polynomial solutions of the second-order differential equation (4.1.1). These solutions are called *continuous classical orthogonal polynomials*. Such a classification was first given in 1929 by S. Bochner in [106].

By Favard's theorem (theorem 3.1), a polynomial $y_n(x)$ of degree n satisfying both the differential equation (4.1.1) and the three-term recurrence relation (4.1.5) is orthogonal with respect to a positive-definite linear functional only if $c_n \in \mathbb{R}$ for all $n = 0,1,2,\ldots$ and $d_n > 0$ for all $n = 1,2,3,\ldots$.

We will consider three different cases depending on the form of $\varphi(x) = ex^2 + 2fx + g$:

Case I. Degree$[\varphi] = 0$: $e = f = 0$ and we may choose $g = 1$. Then we have

$$c_n = -\frac{\gamma}{2\varepsilon}, \quad n = 0,1,2,\ldots \quad \text{and} \quad d_n = -\frac{n}{2\varepsilon}, \quad n = 1,2,3,\ldots.$$

Hence positive-definite orthogonality occurs for $\varepsilon < 0$ and $\gamma \in \mathbb{R}$.

Case II. Degree$[\varphi] = 1$: $e = 0$ and we may choose $2f = 1$. Then we have

$$c_n = -\frac{2n+\gamma}{2\varepsilon}, \quad n = 0,1,2,\ldots \quad \text{and} \quad d_n = \frac{n(n-1+\gamma-2g\varepsilon)}{4\varepsilon^2}, \quad n = 1,2,3,\ldots.$$

Hence we have positive-definite orthogonality for $2g\varepsilon < \gamma$ and $g, \varepsilon, \gamma \in \mathbb{R}$.

Case III. Degree$[\varphi] = 2$: we may choose $e = 1$. Then we have

$$c_n = -\frac{2fn(n-1+2\varepsilon) - \gamma(1-\varepsilon)}{2(n-1+\varepsilon)(n+\varepsilon)}, \quad n = 0,1,2,\ldots$$

and

$$d_n = \frac{n(n-2+2\varepsilon)}{4(2n-3+2\varepsilon)(n-1+\varepsilon)^2(2n-1+2\varepsilon)} D_n, \quad n = 1,2,3,\ldots,$$

where

$$D_n = \{2f(n-1)+\gamma\}\{2f(n-1+2\varepsilon)-\gamma\} - 4g(n-1+\varepsilon)^2, \quad n = 1,2,3,\ldots.$$

Positive-definite orthogonality requires that $f, g, \varepsilon, \gamma \in \mathbb{R}$.

For $\varepsilon > 0$ we have

$$\frac{n(n-2+2\varepsilon)}{4(2n-3+2\varepsilon)(n-1+\varepsilon)^2(2n-1+2\varepsilon)} > 0, \quad n = 1,2,3,\ldots,$$

where the case $2\varepsilon = 1$ must be understood by continuity. Now we have

$$\begin{aligned} D_n &= \{2f(n-1)+\gamma\}\{2f(n-1+2\varepsilon)-\gamma\} - 4g(n-1+\varepsilon)^2 \\ &= \{2f(n-1)+\gamma\}\{2f(n-1)-\gamma\} + 4f\varepsilon\{2f(n-1)+\gamma\} - 4g(n-1+\varepsilon)^2 \\ &= 4(f^2-g)(n-1+\varepsilon)^2 - (\gamma-2f\varepsilon)^2, \quad n = 1,2,3,\ldots. \end{aligned}$$

Hence for $\varepsilon > 0$ the positivity of $d_n > 0$ for all $n = 1,2,3,\ldots$ requires that $f^2 - g > 0$, which implies that the polynomial $\varphi(x) = x^2 + 2fx + g$ has two different real zeros.

For $\varepsilon < 0$ we write

$$2\varepsilon = -2N - t \quad \text{with} \quad N \in \{1,2,3,\ldots\} \quad \text{and} \quad -1 < t \leq 1.$$

Note that this notation does not cover the (trivial) range $-1 \leq 2\varepsilon < 0$. However, we will be able to find finite orthogonal polynomial systems consisting of $N+1$ polynomials. Now we have

$$\frac{n(n-2+2\varepsilon)}{4(2n-3+2\varepsilon)(n-1+\varepsilon)^2(2n-1+2\varepsilon)} < 0, \quad n = 1,2,3,\ldots,N$$

which does not longer hold for $n = N+1$. So in this case we have

$$d_n > 0 \quad \text{for} \quad n = 1,2,3,\ldots,N \quad \Longleftrightarrow \quad D_n < 0 \quad \text{for} \quad n = 1,2,3,\ldots,N.$$

In this case the polynomial $\varphi(x) = x^2 + 2fx + g$ might have two different real zeros if $f^2 - g > 0$, two equal real zeros if $f^2 - g = 0$ or two nonreal (complex conjugate) zeros if $f^2 - g < 0$.

Concluding, positive-definite orthogonality occurs for

Case IIIa. $\varepsilon > 0$ and $4(f^2 - g)\varepsilon^2 > (\gamma - 2f\varepsilon)^2$. This implies that $f^2 > g$, which means that the polynomial φ has two different real zeros.

Moreover, we also find three finite orthogonal polynomial systems in the following three different cases:

Case IIIb. The polynomial φ has two different real zeros: $f^2 > g$ and $4(f^2 - g)\varepsilon^2 < (\gamma - 2f\varepsilon)^2$.

Case IIIc. The polynomial φ has two equal real zeros: $f^2 = g$ and $\gamma \neq 2f\varepsilon$.

Case IIId. The polynomial φ has two non-real (complex conjugate) zeros: $f^2 < g$.

As indicated in section 3.2, the orthogonality relations can be obtained in each case as follows. The differential equation (4.1.1) can be written in the self-adjoint form

$$\left(w\varphi y_n'\right)'(x) = \lambda_n w(x) y_n(x), \quad n = 0, 1, 2, \ldots \tag{4.2.1}$$

if w satisfies the Pearson differential equation (cf. (3.2.13))

$$(w\varphi)'(x) = w(x)\psi(x), \tag{4.2.2}$$

where $\varphi(x) = ex^2 + 2fx + g$, $\psi(x) = 2\varepsilon x + \gamma$ and $\lambda_n = n\left(e(n-1) + 2\varepsilon\right)$. If we multiply (4.2.1) by $y_m(x)$ and subtract from the resulting equation the same equation with m and n exchanged, then integration by parts over an interval (a,b) with $a, b \in \mathbb{R}$ and $a < b$ leads to (cf. (3.2.14))

$$\begin{aligned}(\lambda_n - \lambda_m) & \int_a^b w(x) y_m(x) y_n(x)\, dx \\ & = \left[w(x)\varphi(x)\left(y_m(x)y_n'(x) - y_n(x)y_m'(x)\right)\right]_a^b\end{aligned} \tag{4.2.3}$$

for $m, n \in \{0, 1, 2, \ldots\}$. If the regularity condition (2.3.3) holds, we have $\lambda_m \neq \lambda_n$ for $m \neq n$. This leads to an orthogonality relation on the interval (a,b) with respect to the weight function $w(x)$ if the boundary conditions

$$w(a)\varphi(a) = 0 \quad \text{and} \quad w(b)\varphi(b) = 0 \tag{4.2.4}$$

hold. Here we have $a < b$ with $a, b \in \mathbb{R}$ or possibly $a \to -\infty$ and/or $b \to \infty$. In the latter cases we have orthogonality on (a, ∞), $(-\infty, b)$ or $(-\infty, \infty)$ and the corresponding improper integrals of the first kind exist if the moments

$$M_n := \int_a^b w(x) x^n\, dx, \quad n = 0, 1, 2, \ldots,$$

where $a = -\infty$ and/or $b = \infty$, are all finite.

In the case of finite polynomial systems $\{y_n(x)\}_{n=0}^N$ consisting of $N+1$ polynomials with $N \in \{1,2,3,\ldots\}$ we must have that

$$M_n := \int_a^b w(x)x^n\,dx, \quad n = 0,1,2,\ldots,2N$$

are finite.

For the computation of the squared norm (cf. (3.1.4)) we must have $\Lambda[y_0^2] = \Lambda[1] = 1$. Therefore we define

$$d_0 := \int_a^b w(x)\,dx \tag{4.2.5}$$

which leads to the squared norm

$$\sigma_n := \int_a^b w(x)\{y_n(x)\}^2\,dx = \prod_{k=0}^{n} d_k, \quad n = 0,1,2,\ldots,N \tag{4.2.6}$$

with $N \in \{1,2,3,\ldots\}$ or $N \to \infty$.

Finally, we remark that the possible positive-definite orthogonal polynomials satisfying both the differential equation (4.1.1) and the three-term recurrence relation (4.1.5) are uniquely determined by Favard's theorem. However, the orthogonality relation can be obtained in several different ways.

In the next section we will derive a weight function, a second-order differential equation, a three-term recurrence relation, a hypergeometric representation, a Rodrigues formula and an orthogonality relation for the positive-definite orthogonal polynomials in all six cases. More details and properties will be given in chapter 9.

4.3 Properties of the Positive-Definite Orthogonal Polynomial Solutions

Case I. We have $e = f = 0$, $g = 1$ and $\varepsilon < 0$. In this case we have $\varphi(x) = 1$ and $\psi(x) = 2\varepsilon x + \gamma$. Hence by using the Pearson differential equation (4.2.2), we obtain

$$\frac{w'(x)}{w(x)} = 2\varepsilon x + \gamma$$

which leads to a positive-definite weight function of the form

$$w(x) = \mathrm{e}^{\varepsilon x^2 + \gamma x}$$

for the **Hermite** polynomials. The boundary conditions (4.2.4) lead to the interval of orthogonality $(-\infty, \infty)$. The weight function for the Hermite polynomials is connected with the normal distribution in stochastics.

The differential equation (4.1.1) reads

$$y_n''(x)+(2\varepsilon x+\gamma)y_n'(x)=2\varepsilon n y_n(x),\quad n=0,1,2,\ldots$$

and the three-term recurrence relation (4.1.5) can be written as

$$y_{n+1}(x)=\left(x+\frac{\gamma}{2\varepsilon}\right)y_n(x)+\frac{n}{2\varepsilon}y_{n-1}(x),\quad n=1,2,3,\ldots$$

with $y_0(x)=1$ and $y_1(x)=x+\gamma/2\varepsilon$.

For the coefficients of the representation (4.1.2) we use (4.1.4) with $c=\gamma/2\varepsilon$ to obtain

$$2\varepsilon(n-k)a_{n,k}=a_{n,k+2},\quad k=n-1,n-2,n-3,\ldots,0,\quad a_{n,n+1}=0,\quad a_{n,n}=n!.$$

Hence we have

$$a_{n,n-2k+1}=0\quad\text{and}\quad a_{n,n-2k}=\frac{n!}{(4\varepsilon)^k k!},\quad k=0,1,2,\ldots,\lfloor n/2\rfloor$$

which implies by using (4.1.2)

$$\begin{aligned}
y_n(x)&=\sum_{k=0}^{\lfloor n/2\rfloor}\frac{a_{n,n-2k}}{(n-2k)!}\left(x+\frac{\gamma}{2\varepsilon}\right)^{n-2k}=n!\sum_{k=0}^{\lfloor n/2\rfloor}\frac{1}{(4\varepsilon)^k k!(n-2k)!}\left(x+\frac{\gamma}{2\varepsilon}\right)^{n-2k}\\
&=\sum_{k=0}^{n}\frac{(-n)_{2k}}{(4\varepsilon)^k k!}\left(x+\frac{\gamma}{2\varepsilon}\right)^{n-2k}=\sum_{k=0}^{n}\frac{(-n/2)_k(-(n-1)/2)_k}{\varepsilon^k k!}\left(x+\frac{\gamma}{2\varepsilon}\right)^{n-2k}\\
&=\left(x+\frac{\gamma}{2\varepsilon}\right)^n {}_2F_0\left(\begin{matrix}-n/2,-(n-1)/2\\ -\end{matrix};\frac{1}{\varepsilon\left(x+\frac{\gamma}{2\varepsilon}\right)^2}\right),\quad n=0,1,2,\ldots,
\end{aligned}$$

where $\lfloor n/2\rfloor$ denotes the largest integer smaller than or equal to $n/2$.

The Rodrigues formula (3.4.8) equals

$$y_n(x)=\frac{\mathrm{e}^{-\varepsilon x^2-\gamma x}}{(2\varepsilon)^n}D^n\left[\mathrm{e}^{\varepsilon x^2+\gamma x}\right],\quad n=0,1,2,\ldots.$$

By using (1.2.5) we find that

$$d_0:=\int_{-\infty}^{\infty}w(x)\,dx=\int_{-\infty}^{\infty}\mathrm{e}^{\varepsilon x^2+\gamma x}\,dx=\sqrt{\frac{\pi}{-\varepsilon}}\,\mathrm{e}^{-\gamma^2/4\varepsilon}>0.$$

Together with (3.1.4), this leads to the orthogonality relation

$$\int_{-\infty}^{\infty}\mathrm{e}^{\varepsilon x^2+\gamma x}y_m(x)y_n(x)\,dx=\sqrt{\frac{\pi}{-\varepsilon}}\frac{n!}{(-2\varepsilon)^n}\mathrm{e}^{-\gamma^2/4\varepsilon}\,\delta_{mn},\quad m,n=0,1,2,\ldots.$$

Case II. We have $e = 0$, $2f = 1$ and $2g\varepsilon < \gamma$. By writing $g = -a$, we have $\varphi(x) = x-a$ and $\psi(x) = 2\varepsilon x+\gamma$. Hence by using the Pearson differential equation (4.2.2), we obtain

$$\frac{w'(x)}{w(x)} = \frac{2\varepsilon x+\gamma-1}{x-a} = 2\varepsilon + \frac{\alpha}{x-a}, \quad x \neq a$$

with $\gamma = \alpha+1-2a\varepsilon$, which leads to a positive-definite weight function of the form

$$w(x) = (x-a)^{\alpha} \mathrm{e}^{2\varepsilon x}, \quad a < x$$

for the **Laguerre** polynomials. The boundary conditions (4.2.4) lead to the interval of orthogonality (a,∞) and the conditions $\alpha+1>0$ and $\varepsilon<0$. Note that we have

$$\alpha+1>0 \quad \Longleftrightarrow \quad \gamma+2a\varepsilon>0 \quad \Longleftrightarrow \quad \gamma-2g\varepsilon>0,$$

which is equivalent to $2g\varepsilon < \gamma$. The weight function for the Laguerre polynomials is connected to the gamma distribution in stochastics.

The differential equation (4.1.1) reads

$$(x-a)y_n''(x) + \{2\varepsilon(x-a)+\alpha+1\}\, y_n'(x) = 2\varepsilon n y_n(x), \quad n = 0,1,2,\ldots$$

and the three-term recurrence relation (4.1.5) can be written as

$$y_{n+1}(x) = \left(x + \frac{2n-2a\varepsilon+\alpha+1}{2\varepsilon}\right) y_n(x) - \frac{n(n+\alpha)}{4\varepsilon^2} y_{n-1}(x), \quad n = 1,2,3,\ldots$$

with $y_0(x) = 1$ and $y_1(x) = x-a+(\alpha+1)/2\varepsilon$.

For the coefficients of the representation (4.1.2) we use (4.1.3) with $c = -a$ to obtain

$$2\varepsilon(n-k)a_{n,k} = (k+\alpha+1)a_{n,k+1}, \quad k = n-1, n-2, n-3, \ldots, 0, \quad a_{n,n} = n!.$$

Hence we have

$$a_{n,k} = \frac{(k+\alpha+1)_{n-k}}{(2\varepsilon)^{n-k}(n-k)!} n! = \frac{(\alpha+1)_n}{(2\varepsilon)^n} \frac{(-n)_k}{(\alpha+1)_k} (-2\varepsilon)^k, \quad k = 0,1,2,\ldots,n$$

which leads to

$$\begin{aligned} y_n(x) = \sum_{k=0}^{n} a_{n,k} \frac{(x-a)^k}{k!} &= \frac{(\alpha+1)_n}{(2\varepsilon)^n} \sum_{k=0}^{n} \frac{(-n)_k}{(\alpha+1)_k} \frac{(2\varepsilon)^k (a-x)^k}{k!} \\ &= \frac{(\alpha+1)_n}{(2\varepsilon)^n} \, {}_1F_1\left(\begin{matrix} -n \\ \alpha+1 \end{matrix}; 2\varepsilon(a-x)\right), \quad n = 0,1,2,\ldots. \end{aligned}$$

The Rodrigues formula (3.4.8) equals

$$y_n(x) = \frac{(x-a)^{-\alpha}\mathrm{e}^{-2\varepsilon x}}{(2\varepsilon)^n} D^n \left[(x-a)^{n+\alpha} \mathrm{e}^{2\varepsilon x}\right], \quad n = 0,1,2,\ldots.$$

By using (1.2.1) we find that

$$d_0 := \int_a^\infty w(x)\,dx = \int_a^\infty (x-a)^\alpha e^{2\varepsilon x}\,dx = \frac{\Gamma(\alpha+1)e^{2a\varepsilon}}{(-2\varepsilon)^{\alpha+1}} > 0.$$

Together with (3.1.4), this leads to the orthogonality relation

$$\int_a^\infty (x-a)^\alpha e^{2\varepsilon x} y_m(x) y_n(x)\,dx = \frac{\Gamma(n+\alpha+1)e^{2a\varepsilon}n!}{(-2\varepsilon)^{2n+\alpha+1}}\,\delta_{mn}, \quad m,n=0,1,2,\ldots.$$

Case IIIa. We have $e=1$, $f^2>g$, $\varepsilon>0$ and $4(f^2-g)\varepsilon^2 > (\gamma-2f\varepsilon)^2$. In this case we may write $\varphi(x) = x^2+2fx+g = (x-a)(x-b)$ with $2f=-a-b$ and $g=ab$. Then we obtain by using the Pearson differential equation (4.2.2)

$$\frac{w'(x)}{w(x)} = \frac{2(\varepsilon-1)x+\gamma+a+b}{(x-a)(x-b)} - \frac{\alpha}{x-a} - \frac{\beta}{b-x}, \quad x\neq a, \quad x\neq b, \quad a<b$$

with $2\varepsilon = \alpha+\beta+2$ and $\gamma = -a(\beta+1)-b(\alpha+1)$, which leads to a positive-definite weight function of the form

$$w(x) = (x-a)^\alpha (b-x)^\beta, \quad a<x<b$$

for the **Jacobi** polynomials. Note that we have

$$\begin{aligned}\frac{4(f^2-g)\varepsilon^2-(\gamma-2f\varepsilon)^2}{(b-a)^2} > 0 \quad &\Longleftrightarrow \quad \frac{-4g\varepsilon^2-\gamma^2+4f\varepsilon\gamma}{(b-a)^2} > 0\\ &\Longleftrightarrow \quad (\alpha+1)(\beta+1) > 0.\end{aligned}$$

The boundary conditions (4.2.4) lead to the interval of orthogonality (a,b) and the conditions $\alpha+1>0$ and $\beta+1>0$. The weight function for the Jacobi polynomials is connected to the beta distribution in stochastics.

The differential equation (4.1.1) reads

$$\begin{aligned}&(x-a)(x-b)y_n''(x) + \{(\alpha+1)(x-b)+(\beta+1)(x-a)\}y_n'(x)\\ &= n(n+\alpha+\beta+1)y_n(x)\end{aligned}$$

for $n=0,1,2,\ldots$ and the three-term recurrence relation (4.1.5) can be written as

$$\begin{aligned}&y_{n+1}(x)\\ &= \left(x - \frac{2n(n+\alpha+\beta+1)(a+b)+\{a(\beta+1)+b(\alpha+1)\}(\alpha+\beta)}{(2n+\alpha+\beta)(2n+\alpha+\beta+2)}\right)y_n(x)\\ &\quad - \frac{n(n+\alpha)(n+\beta)(n+\alpha+\beta)(b-a)^2}{(2n+\alpha+\beta-1)(2n+\alpha+\beta)^2(2n+\alpha+\beta+1)}y_{n-1}(x)\end{aligned}$$

for $n=1,2,3,\ldots$ with $y_0(x)=1$ and $y_1(x) = x-(a(\beta+1)+b(\alpha+1))/(\alpha+\beta+2)$.

For the coefficients of the representation (4.1.2) we use (4.1.3) to find

$$\begin{aligned}&(n-k)(n+k+\alpha+\beta+1)a_{n,k}\\&=-\{(2k+\alpha+\beta+2)c+(a+b)(k+1)+a\beta+b\alpha\}\,a_{n,k+1}\end{aligned}$$

for $k=n-1,n-2,n-3,\ldots,0$, where $(c+a)(c+b)=0$. So if we choose $c=-a$, we have

$$(n-k)(n+k+\alpha+\beta+1)a_{n,k}=(a-b)(k+\alpha+1)a_{n,k+1}$$

for $k=n-1,n-2,n-3,\ldots,0$ with $a_{n,n}=n!$, and if we choose $c=-b$, we obtain

$$(n-k)(n+k+\alpha+\beta+1)a_{n,k}=(b-a)(k+\beta+1)a_{n,k+1}$$

for $k=n-1,n-2,n-3,\ldots,0$ with $a_{n,n}=n!$. Hence for $c=-a$ we obtain

$$\begin{aligned}a_{n,k}&=\frac{(a-b)^{n-k}(k+\alpha+1)_{n-k}n!}{(n-k)!\,(n+k+\alpha+\beta+1)_{n-k}}\\&=\frac{(a-b)^n(\alpha+1)_n}{(n+\alpha+\beta+1)_n}\frac{(-n)_k(n+\alpha+\beta+1)_k}{(\alpha+1)_k}\left(\frac{1}{b-a}\right)^k,\end{aligned}$$

which leads to

$$\sum_{k=0}^{n}a_{n,k}\frac{(x-a)^k}{k!}=\frac{(a-b)^n(\alpha+1)_n}{(n+\alpha+\beta+1)_n}\,{}_2F_1\left(\begin{matrix}-n,n+\alpha+\beta+1\\\alpha+1\end{matrix};\frac{x-a}{b-a}\right)$$

for $n=0,1,2,\ldots$, and for $c=-b$ we obtain

$$\begin{aligned}a_{n,k}&=\frac{(b-a)^{n-k}(k+\beta+1)_{n-k}n!}{(n-k)!\,(n+k+\alpha+\beta+1)_{n-k}}\\&=\frac{(b-a)^n(\beta+1)_n}{(n+\alpha+\beta+1)_n}\frac{(-n)_k(n+\alpha+\beta+1)_k}{(\beta+1)_k}\left(\frac{1}{a-b}\right)^k,\end{aligned}$$

which leads to

$$\sum_{k=0}^{n}a_{n,k}\frac{(x-b)^k}{k!}=\frac{(b-a)^n(\beta+1)_n}{(n+\alpha+\beta+1)_n}\,{}_2F_1\left(\begin{matrix}-n,n+\alpha+\beta+1\\\beta+1\end{matrix};\frac{x-b}{a-b}\right)$$

for $n=0,1,2,\ldots$.

The Rodrigues formula (3.4.8) equals

$$y_n(x)=(-1)^n\frac{(x-a)^{-\alpha}(b-x)^{-\beta}}{(n+\alpha+\beta+1)_n}D^n\left[(x-a)^{n+\alpha}(b-x)^{n+\beta}\right],\quad n=0,1,2,\ldots.$$

By using (1.2.10) we find that

$$d_0 := \int_a^b w(x)\,dx = \int_a^b (x-a)^\alpha (b-x)^\beta\,dx = \frac{\Gamma(\alpha+1)\Gamma(\beta+1)}{\Gamma(\alpha+\beta+2)}(b-a)^{\alpha+\beta+1} > 0.$$

Together with (3.1.4), this leads to the orthogonality relation

$$\begin{aligned}&\int_a^b (x-a)^\alpha (b-x)^\beta y_m(x) y_n(x)\,dx\\&= \frac{\Gamma(n+\alpha+\beta+1)\Gamma(n+\alpha+1)\Gamma(n+\beta+1)n!}{\Gamma(2n+\alpha+\beta+1)\Gamma(2n+\alpha+\beta+2)}(b-a)^{2n+\alpha+\beta+1}\,\delta_{mn}\end{aligned}$$

for $m,n = 0,1,2,\ldots$.

Case IIIb. We have $e=1$, $f^2>g$, $4(f^2-g)\varepsilon^2 < (\gamma-2f\varepsilon)^2$ and $2\varepsilon = -2N-t$ with $N \in \{1,2,3,\ldots\}$ and $-1<t\leq 1$. Here we also have $\varphi(x) = x^2+2fx+g = (x-a)(x-b)$ with $2f=-a-b$ and $g=ab$. Then we find by using the Pearson differential equation (4.2.2)

$$\frac{w'(x)}{w(x)} = \frac{2(\varepsilon-1)x+\gamma+a+b}{(x-a)(x-b)} = \frac{\alpha}{x-a}+\frac{\beta}{x-b}, \quad x\neq a, \quad x\neq b, \quad a<b$$

with $2\varepsilon = \alpha+\beta+2$ and $\gamma = -a(\beta+1)-b(\alpha+1)$, which leads to a positive-definite weight function of the form

$$w(x) = (x-a)^\alpha (x-b)^\beta, \quad a<b<x$$

for the **Jacobi** polynomials. Now we have

$$\begin{aligned}\frac{4(f^2-g)\varepsilon^2-(\gamma-2f\varepsilon)^2}{(b-a)^2}<0 \quad &\Longleftrightarrow \quad \frac{-4g\varepsilon^2-\gamma^2+4f\varepsilon\gamma}{(b-a)^2}<0\\ &\Longleftrightarrow \quad (\alpha+1)(\beta+1)<0.\end{aligned}$$

The boundary conditions (4.2.4) lead to the interval of orthogonality (b,∞) and the conditions $\beta+1>0$ and $\alpha+\beta+1<-2N$.

The differential equation, the three-term recurrence relation and the hypergeometric representation are the same as in the previous case.

The Rodrigues formula (3.4.8) can now be written as

$$y_n(x) = \frac{(x-a)^{-\alpha}(x-b)^{-\beta}}{(n+\alpha+\beta+1)_n} D^n\left[(x-a)^{n+\alpha}(x-b)^{n+\beta}\right], \quad n=0,1,2,\ldots,N.$$

By using (1.2.10) we find that

$$\begin{aligned}d_0 := \int_b^\infty w(x)\,dx &= \int_b^\infty (x-a)^\alpha (x-b)^\beta\,dx\\ &= \frac{\Gamma(\beta+1)\Gamma(-\alpha-\beta-1)}{\Gamma(-\alpha)}(b-a)^{\alpha+\beta+1} > 0.\end{aligned}$$

Together with (3.1.4), this leads to the orthogonality relation

$$\int_b^\infty (x-a)^\alpha (x-b)^\beta y_m(x) y_n(x)\,dx$$
$$= \frac{\Gamma(n+\beta+1)\Gamma(-2n-\alpha-\beta-1)\Gamma(-2n-\alpha-\beta)n!}{\Gamma(-n-\alpha-\beta)\Gamma(-n-\alpha)}(b-a)^{2n+\alpha+\beta+1}\,\delta_{mn}$$

for $m,n=0,1,2,\ldots,N$.

Case IIIc. We have $e=1$, $f^2=g$, $\gamma\neq 2f\varepsilon$ and $2\varepsilon=-2N-t$ with $N\in\{1,2,3,\ldots\}$ and $-1<t\leq 1$. In this case we have $\varphi(x)=x^2+2fx+g=(x-a)^2$ with $f=-a$. Then we obtain by using the Pearson differential equation (4.2.2)

$$\frac{w'(x)}{w(x)}=\frac{2(\varepsilon-1)x+\gamma+2a}{(x-a)^2}=\frac{\alpha}{x-a}+\frac{\beta}{(x-a)^2},\quad x\neq a$$

with $2\varepsilon=\alpha+2$ and $\gamma=\beta-a(\alpha+2)$, which leads to a positive-definite weight function of the form

$$w(x)=(x-a)^\alpha \mathrm{e}^{-\frac{\beta}{x-a}},\quad x>a$$

for the **Bessel** polynomials. The boundary conditions (4.2.4) lead to the interval of orthogonality (a,∞) and the conditions $\alpha+1<-2N$ and $\beta>0$.

The differential equation (4.1.1) reads

$$(x-a)^2y_n''(x)+((\alpha+2)(x-a)+\beta)\,y_n'(x)=n(n+\alpha+1)y_n(x),\quad n=0,1,2,\ldots,N$$

and the three-term recurrence relation (4.1.5) can be written as

$$y_{n+1}(x)=\left(x-\frac{4n(n+\alpha+1)a+\{a(\alpha+2)-\beta\}\alpha}{(2n+\alpha)(2n+\alpha+2)}\right)y_n(x)$$
$$+\frac{n(n+\alpha)\beta^2}{(2n+\alpha-1)(2n+\alpha)^2(2n+\alpha+1)}y_{n-1}(x)$$

for $n=1,2,3,\ldots,N-1$ with $y_0(x)=1$ and $y_1(x)=x-(a(\alpha+2)-\beta)/(\alpha+2)$.

For the coefficients of the representation (4.1.2) we use (4.1.3) to find

$$(n-k)(n+k+\alpha+1)a_{n,k}=\beta a_{n,k+1},\quad k=n-1,n-2,n-3,\ldots,0,\quad a_{n,n}=n!$$

where $c=-a$. Hence we obtain

$$a_{n,k}=\frac{\beta^{n-k}n!}{(n-k)!\,(n+k+\alpha+1)_{n-k}}=\frac{(-1)^k(-n)_k(n+\alpha+1)_k\beta^{n-k}}{(n+\alpha+1)_n},$$

which leads to

$$\sum_{k=0}^n a_{n,k}\frac{(x-a)^k}{k!}=\frac{\beta^n}{(n+\alpha+1)_n}\,{}_2F_0\left(\begin{matrix}-n,n+\alpha+1\\-\end{matrix};-\frac{x-a}{\beta}\right)$$

for $n = 0, 1, 2, \ldots, N$.

The Rodrigues formula (3.4.8) equals

$$y_n(x) = \frac{(x-a)^{-\alpha} e^{\frac{\beta}{x-a}}}{(n+\alpha+1)_n} D^n \left[(x-a)^{2n+\alpha} e^{-\frac{\beta}{x-a}} \right], \quad n = 0, 1, 2, \ldots, N.$$

By using (1.2.1) we find that

$$d_0 := \int_a^\infty w(x)\, dx = \int_a^\infty (x-a)^\alpha e^{-\frac{\beta}{x-a}}\, dx = \beta^{\alpha+1} \Gamma(-\alpha-1) > 0.$$

Together with (3.1.4), this leads to the orthogonality relation

$$\int_a^\infty (x-a)^\alpha e^{-\frac{\beta}{x-a}} y_m(x) y_n(x)\, dx = \frac{\beta^{2n+\alpha+1} \Gamma(-2n-\alpha-1) \Gamma(-2n-\alpha) n!}{\Gamma(-n-\alpha)} \delta_{mn}$$

for $m, n = 0, 1, 2, \ldots, N$.

Case IIId. We have $e = 1$, $f^2 < g$ and $2\varepsilon = -2N - t$ with $N \in \{1, 2, 3, \ldots\}$ and $-1 < t \leq 1$. In this case we have $\varphi(x) = x^2 + 2fx + g = (x+f)^2 + g - f^2 = (x+f+i\zeta)(x+f-i\zeta)$ with $\zeta = \sqrt{g - f^2}$. Then we obtain by using the Pearson differential equation (4.2.2) for $x + f \neq \pm i\zeta$

$$\begin{aligned}\frac{w'(x)}{w(x)} &= \frac{2(\varepsilon-1)(x+f) + \gamma - 2f\varepsilon}{(x+f)^2 + \zeta^2} \\ &= \frac{2(\varepsilon-1)(x+f)}{(x+f)^2+\zeta^2} + \frac{\gamma - 2f\varepsilon}{2\zeta\{\zeta + i(x+f)\}} + \frac{\gamma - 2f\varepsilon}{2\zeta\{\zeta - i(x+f)\}},\end{aligned}$$

which leads to a positive-definite weight function of the form

$$\begin{aligned} w(x) &= \left\{ \zeta^2 + (x+f)^2 \right\}^{\varepsilon-1} e^{\{(\gamma-2f\varepsilon)/\zeta\} \arctan \frac{x+f}{\zeta}} \\ &= \{\zeta + i(x+f)\}^{\varepsilon-1-i\nu} \{\zeta - i(x+f)\}^{\varepsilon-1+i\nu} \end{aligned}$$

with $\nu = (\gamma - 2f\varepsilon)/2\zeta$ for the **pseudo Jacobi** polynomials. The boundary conditions (4.2.4) lead to the interval of orthogonality $(-\infty, \infty)$. The weight function for the pseudo Jacobi polynomials is connected to the student's t-distribution in stochastics.

The differential equation (4.1.1) reads

$$\left\{ (x+f)^2 + \zeta^2 \right\} y_n''(x) + (2\varepsilon x + \gamma) y_n'(x) = n(n-1+2\varepsilon) y_n(x), \quad n = 0, 1, 2, \ldots, N$$

and the three-term recurrence relation (4.1.5) can be written as

$$\begin{aligned} y_{n+1}(x) &= \left(x + \frac{2fn(n-1+2\varepsilon) - \gamma(1-\varepsilon)}{2(n-1+\varepsilon)(n+\varepsilon)} \right) y_n(x) \\ &\quad + \frac{n(n-2+2\varepsilon)\left(4(n-1+\varepsilon)^2\zeta^2 + (\gamma - 2f\varepsilon)^2\right)}{4(2n-3+2\varepsilon)(n-1+\varepsilon)^2(2n-1+2\varepsilon)} y_{n-1}(x) \end{aligned}$$

for $n = 1,2,3,\dots,N-1$ with $y_0(x) = 1$ and $y_1(x) = x + \gamma/2\varepsilon$.

For the coefficients of the representation (4.1.2) we use (4.1.3) to find

$$(n-k)(n+k-1+2\varepsilon)a_{n,k} = -(2(c-f)k + 2\varepsilon c - \gamma)\, a_{n,k+1}$$

for $k = n-1, n-2, n-3, \dots, 0$, where $c^2 - 2fc + g = 0 \iff (c-f)^2 + \zeta^2 = 0$. If we choose $c = f + i\zeta$, we have

$$(n-k)(n+k-1+2\varepsilon)a_{n,k} = (\gamma - 2f\varepsilon - 2i\zeta(k+\varepsilon))\, a_{n,k+1}$$

for $k = n-1, n-2, n-3, \dots, 0$ and if we choose $c = f - i\zeta$, we have

$$(n-k)(n+k-1+2\varepsilon)a_{n,k} = (\gamma - 2f\varepsilon + 2i\zeta(k+\varepsilon))\, a_{n,k+1}$$

for $k = n-1, n-2, n-3, \dots, 0$. Hence for $c = f + i\zeta$ we obtain

$$\begin{aligned} a_{n,k} &= \frac{(-2i\zeta)^{n-k}(k+\varepsilon+i\nu)_{n-k}}{(n-k)!(n+k-1+2\varepsilon)_{n-k}} n! \\ &= \frac{(-2i\zeta)^n(\varepsilon+i\nu)_n}{(n-1+2\varepsilon)_n} \frac{(-n)_k(n-1+2\varepsilon)_k}{(\varepsilon+i\nu)_k} \frac{1}{(2i\zeta)^k}, \end{aligned}$$

which leads to

$$\sum_{k=0}^{n} a_{n,k} \frac{(x+f+i\zeta)^k}{k!} = \frac{(-2i\zeta)^n(\varepsilon+i\nu)_n}{(n-1+2\varepsilon)_n} \, {}_2F_1\left(\begin{matrix} -n, n-1+2\varepsilon \\ \varepsilon+i\nu \end{matrix} ; \frac{x+f+i\zeta}{2i\zeta}\right)$$

for $n = 0,1,2,\dots,N$ and for $c = f - i\zeta$ we obtain

$$\begin{aligned} a_{n,k} &= \frac{(2i\zeta)^{n-k}(k+\varepsilon-i\nu)_{n-k}}{(n-k)!(n+k-1+2\varepsilon)_{n-k}} n! \\ &= \frac{(2i\zeta)^n(\varepsilon-i\nu)_n}{(n-1+2\varepsilon)_n} \frac{(-n)_k(n-1+2\varepsilon)_k}{(\varepsilon-i\nu)_k} \frac{1}{(-2i\zeta)^k}, \end{aligned}$$

which leads to

$$\sum_{k=0}^{n} a_{n,k} \frac{(x+f-i\zeta)^k}{k!} = \frac{(2i\zeta)^n(\varepsilon-i\nu)_n}{(n-1+2\varepsilon)_n} \, {}_2F_1\left(\begin{matrix} -n, n-1+2\varepsilon \\ \varepsilon-i\nu \end{matrix} ; \frac{x+f-i\zeta}{-2i\zeta}\right)$$

for $n = 0,1,2,\dots,N$.

The Rodrigues formula (3.4.8) equals

$$y_n(x) = \frac{\left\{(x+f)^2+\zeta^2\right\}^{1-\varepsilon} e^{-2\nu \arctan \frac{x+f}{\zeta}}}{(n-1+2\varepsilon)_n} D^n \left[\left\{(x+f)^2+\zeta^2\right\}^{n+\varepsilon-1} e^{2\nu \arctan \frac{x+f}{\zeta}}\right]$$

for $n = 0,1,2,\dots,N$.

By using the Cauchy integral (1.2.12), we obtain

$$d_0 := \int_{-\infty}^{\infty} w(x)\,dx = \int_{-\infty}^{\infty} \{\zeta + i(x+f)\}^{\varepsilon-1-i\nu} \{\zeta - i(x+f)\}^{\varepsilon-1+i\nu}\,dx$$
$$= \frac{2\pi\,\Gamma(1-2\varepsilon)(2\zeta)^{2\varepsilon-1}}{|\Gamma(1-\varepsilon+i\nu)|^2} > 0.$$

Together with (3.1.4), this leads to the orthogonality relation

$$\int_{-\infty}^{\infty} \{(x+f)^2+\zeta^2\}^{\varepsilon-1} \mathrm{e}^{2\nu \arctan \frac{x+f}{\zeta}} y_m(x) y_n(x)\,dx$$
$$= \frac{2\pi\,\Gamma(1-2n-2\varepsilon)\Gamma(2-2n-2\varepsilon)(2\zeta)^{2n+2\varepsilon-1} n!}{\Gamma(2-n-2\varepsilon)\,|\Gamma(1-n-\varepsilon+i\nu)|^2}\,\delta_{mn}, \quad m,n = 0,1,2,\ldots,N.$$

In this chapter we have proved:

Theorem 4.1. *The positive-definite orthogonal polynomial solutions $y_n(x)$ of the differential equation (4.1.1)*

$$(ex^2+2fx+g)y_n''(x)+(2\varepsilon x+\gamma)y_n'(x)=n(e(n-1)+2\varepsilon)y_n(x)$$

for $n=0,1,2,\ldots$ *consist of three infinite systems*

Case I. *Hermite polynomials with orthogonality on* $(-\infty,\infty)$ *with respect to* $w(x)=e^{\varepsilon x^2+\gamma x}$; $e=f=0$, $g=1$ *and* $\varepsilon<0$

Case II. *Laguerre polynomials with orthogonality on* (a,∞) *with respect to* $w(x)=(x-a)^{\alpha}e^{2\varepsilon x}$; $e=0$, $2f=1$, $\varepsilon<0$ *and* $\alpha>-1$

Case IIIa. *Jacobi polynomials with orthogonality on* (a,b) *with* $a<b$ *with respect to* $w(x)=(x-a)^{\alpha}(b-x)^{\beta}$; $e=1$, $f^2>g$, $\varepsilon>0$, $\alpha>-1$ *and* $\beta>-1$

and three finite systems of $N+1$ *polynomials*

Case IIIb. *Jacobi polynomials with orthogonality on* (b,∞) *with respect to* $w(x)=(x-a)^{\alpha}(x-b)^{\beta}$; $a<b$, $e=1$, $f^2>g$, $2\varepsilon=-2N-t$ *with* $N\in\{1,2,3,\ldots\}$ *and* $-1<t\leq 1$, $\alpha<-2N$ *and* $\beta>-1$

Case IIIc. *Bessel polynomials with orthogonality on* (a,∞) *with respect to* $w(x)=(x-a)^{\alpha}e^{-\beta/(x-a)}$; $e=1$, $f^2=g$, $2\varepsilon=-2N-t$ *with* $N\in\{1,2,3,\ldots\}$ *and* $-1<t\leq 1$, $\alpha<-2N-1$ *and* $\beta>0$

Case IIId. *Pseudo Jacobi polynomials with orthogonality on* $(-\infty,\infty)$ *with respect to* $w(x)=\left\{\zeta^2+(x+f)^2\right\}^{\varepsilon-1}e^{\{(\gamma-2f\varepsilon)/\zeta\}\arctan\frac{x+f}{\zeta}}$; $e=1$, $f^2<g$, $2\varepsilon=-2N-t$ *with* $N\in\{1,2,3,\ldots\}$ *and* $-1<t\leq 1$, $\zeta=\sqrt{g-f^2}>0$.

The finite cases of the Jacobi, Bessel and pseudo Jacobi polynomials were discovered in 1929 by V. Romanovski in [463].

These six systems constitute the class of continuous classical orthogonal polynomials.

Chapter 5
Orthogonal Polynomial Solutions of Real Difference Equations

Discrete Classical Orthogonal Polynomials I

5.1 Polynomial Solutions of Real Difference Equations

In the case of the difference operator $\Delta = \mathscr{A}_{1,1}$, we have to deal with (cf. (2.2.8)):

$$(ex^2+2fx+g)\left(\Delta^2 y_n\right)(x)+(2\varepsilon x+\gamma)\left(\Delta y_n\right)(x)=n(e(n-1)+2\varepsilon)y_n(x+1) \quad (5.1.1)$$

for $n=0,1,2,\ldots$ with $e,f,g,\varepsilon,\gamma \in \mathbb{R}$. This difference equation can also be written in the form (cf. (2.2.16))

$$\begin{aligned}&\left(e(x-1)^2+2f(x-1)+g\right)\left(\Delta\left(\nabla y_n\right)\right)(x)\\&\qquad+\left(2\varepsilon(x-1)+\gamma\right)\left(\nabla y_n\right)(x)=n(e(n-1)+2\varepsilon)y_n(x)\end{aligned}$$

or in the form (cf. (2.2.12))

$$C(x)y_n(x+1)-\{C(x)+D(x)\}\,y_n(x)+D(x)y_n(x-1)=n(e(n-1)+2\varepsilon)y_n(x)$$

for $n=0,1,2,\ldots$, where (cf. (2.2.13))

$$C(x)=e(x-1)^2+2f(x-1)+g \quad \text{and} \quad D(x)=C(x)-2\varepsilon(x-1)-\gamma. \quad (5.1.2)$$

We look for monic polynomial solutions of the form (cf. (2.4.14))

$$y_n(x)=\sum_{k=0}^{n} a_{n,k}\binom{x+c}{k}, \quad a_{n,n}=n!, \quad n=0,1,2,\ldots, \quad (5.1.3)$$

where the coefficients satisfy the two-term recurrence relation (cf. (2.4.15))

R. Koekoek et al., *Hypergeometric Orthogonal Polynomials and Their q-Analogues*, Springer Monographs in Mathematics,
DOI 10.1007/978-3-642-05014-5_5, © Springer-Verlag Berlin Heidelberg 2010

$$(n-k)\,(e(n+k-1)+2\varepsilon)\,a_{n,k} \\ -\left(e(k-1-c)^2+2f(k-1-c)+g\right)a_{n,k+1}=0 \qquad (5.1.4)$$

for $k=n-1,n-2,n-3,\ldots,0$ provided that c satisfies

$$e(c+1)^2-2f(c+1)+g=-2\varepsilon(c+1)+\gamma. \qquad (5.1.5)$$

In cases where c cannot be obtained from this equation, we use the representation (cf. (2.4.16))

$$y_n(x)=\sum_{k=0}^{n} b_{n,k}\binom{x+c+k-2}{k}, \quad b_{n,n}=n!, \quad n=0,1,2,\ldots, \qquad (5.1.6)$$

where the coefficients satisfy the two-term recurrence relation (cf. (2.4.17))

$$(n-k)\,(e(n+k-1)+2\varepsilon)\,b_{n,k} \\ +\left(ek^2+2(ec-f+\varepsilon)k+2\varepsilon c-\gamma\right)b_{n,k+1}=0 \qquad (5.1.7)$$

for $k=n-1,n-2,n-3,\ldots,0$, provided that c satisfies $ec^2-2fc+g=0$.

In section 2.6 we found that the monic polynomial solutions $\{y_n\}_{n=0}^{\infty}$ satisfy the three-term recurrence relation

$$y_{n+1}(x)=(x-c_n)y_n(x)-d_n y_{n-1}(x), \quad n=1,2,3,\ldots, \qquad (5.1.8)$$

with initial values $y_0(x)=1$ and $y_1(x)=x-c_0$, where (cf. (2.6.15))

$$c_n=\frac{n\,(e(n-1)+2\varepsilon)\,(2(e-f)+\varepsilon)+(e-\varepsilon)(\gamma-2\varepsilon)}{2\,(e(n-1)+\varepsilon)\,(en+\varepsilon)}, \quad n=0,1,2,\ldots$$

and (cf. (2.6.16))

$$d_n=-\frac{n\,(e(n-2)+2\varepsilon)}{4\,(e(2n-3)+2\varepsilon)\,(e(n-1)+\varepsilon)^2\,(e(2n-1)+2\varepsilon)} \\ \times\Big\{e(n-1)^2\,(e(n-1)+2\varepsilon)^2 \\ +2(n-1)\,(e(n-1)+2\varepsilon)\,(2eg+2f(\varepsilon-f)-e\gamma) \\ +4\varepsilon(g\varepsilon-f\gamma)+e\gamma^2\Big\}, \quad n=1,2,3,\ldots.$$

In order to rewrite the difference equation (5.1.1) in self-adjoint form, we need the product rule (3.2.2) with $q=1$ and $\omega=1$

$$\Delta\left(f_1(x)f_2(x)\right)=f_1(x+1)\Delta f_2(x)+f_2(x)\Delta f_1(x). \qquad (5.1.9)$$

By using (5.1.9) the difference equation (5.1.1) multiplied by $w(x+1)$ can be written in the self-adjoint form (cf. (3.2.5))

$$\Delta\left(w(x)\varphi(x-1)\Delta y_n(x)\right)=\lambda_n w(x+1)y_n(x+1),$$

provided that $w(x)$ satisfies the Pearson operator equation (cf. (3.2.6))

$$\Delta\left(w(x)\varphi(x-1)\right)=w(x+1)\psi(x),$$

where (cf. (2.2.2)) $\varphi(x)=ex^2+2fx+g$ and $\psi(x)=2\varepsilon x+\gamma$. This equation can be rewritten as

$$w(x+1)\left(\varphi(x)-\psi(x)\right)=w(x)\varphi(x-1),$$

which leads, by using (5.1.2), to the Pearson difference equation

$$\frac{w(x)}{w(x+1)}=\frac{\varphi(x)-\psi(x)}{\varphi(x-1)}=\frac{D(x+1)}{C(x)}, \tag{5.1.10}$$

where $C(x)$ and $D(x)$ are given by (5.1.2).

The orthogonality relation (3.2.18) reads

$$\sum_{\nu=0}^{N} w(A+\nu)y_m(A+\nu)y_n(A+\nu)=0, \quad m\neq n, \quad m,n\in\{0,1,2,\ldots,N\} \tag{5.1.11}$$

with boundary conditions

$$w(A-1)\varphi(A-2)=0 \quad \text{and} \quad w(A+N)\varphi(A+N-1)=0 \tag{5.1.12}$$

for $N\in\{1,2,3,\ldots\}$ or $N\to\infty$. For $N\to\infty$ we must have that the moments

$$\sum_{x=0}^{\infty} w(x)x^n, \quad n=0,1,2,\ldots$$

are all finite since we must be able to compute the norms of all polynomials.

5.2 Classification of the Positive-Definite Orthogonal Polynomial Solutions

Again we use Favard's theorem (theorem 3.1) to conclude from (5.1.8) that we have positive-definite orthogonality if $c_n\in\mathbb{R}$ for all $n=0,1,2,\ldots$ and $d_n>0$ for all $n=1,2,3,\ldots$.

Again we consider three different cases depending on the form of $\varphi(x)=ex^2+2fx+g$:

Case I. Degree$[\varphi]=0$: $e=f=0$ and we may choose $g=1$. Then we have

$$c_n=\frac{2(n+1)\varepsilon-\gamma}{2\varepsilon}, \quad n=0,1,2,\ldots \quad \text{and} \quad d_n=-\frac{n}{2\varepsilon}, \quad n=1,2,3,\ldots.$$

Hence positive-definite orthogonality occurs for $2\varepsilon < 0$.

Case II. Degree$[\varphi] = 1$: $e = 0$ and we may choose $2f = 1$. Then we have

$$c_n = \frac{2n(\varepsilon-1)+2\varepsilon-\gamma}{2\varepsilon}, \quad n = 0,1,2,\ldots$$

and

$$d_n = -\frac{n\left((n-1)(2\varepsilon-1)+2g\varepsilon-\gamma\right)}{4\varepsilon^2}, \quad n = 1,2,3,\ldots.$$

Hence we have positive-definite orthogonality for

Case IIa. $2\varepsilon \leq 1$ and $2g\varepsilon - \gamma < 0$. Note that the regularity condition (2.3.3) requires that $2\varepsilon \neq 0$.

For $2\varepsilon > 1$ we only have a finite orthogonal polynomial system with $N+1$ polynomials. Note that

$$\begin{aligned}
d_1 &= -\frac{2g\varepsilon-\gamma}{4\varepsilon^2}\\
d_2 &= -\frac{2(2\varepsilon-1)}{4\varepsilon^2}\left(1+\frac{2g\varepsilon-\gamma}{2\varepsilon-1}\right)\\
&\vdots\\
d_N &= -\frac{N(2\varepsilon-1)}{4\varepsilon^2}\left(N-1+\frac{2g\varepsilon-\gamma}{2\varepsilon-1}\right)\\
d_{N+1} &= -\frac{(N+1)(2\varepsilon-1)}{4\varepsilon^2}\left(N+\frac{2g\varepsilon-\gamma}{2\varepsilon-1}\right).
\end{aligned}$$

Hence for $2\varepsilon > 1$ and $-N \leq \dfrac{2g\varepsilon-\gamma}{2\varepsilon-1} < -N+1$ we have $d_1, d_2, d_3, \ldots, d_N$ positive and $d_{N+1} \leq 0$.

Case IIb. $2\varepsilon > 1$ and $-N \leq \dfrac{2g\varepsilon-\gamma}{2\varepsilon-1} < -N+1$.

Case III. Degree$[\varphi] = 2$: we may choose $e = 1$. Then we have

$$c_n = \frac{n(n-1+2\varepsilon)\left(2(1-f)+\varepsilon\right)+(1-\varepsilon)(\gamma-2\varepsilon)}{2(n-1+\varepsilon)(n+\varepsilon)}, \quad n = 0,1,2,\ldots$$

and

$$d_n = -\frac{n(n-2+2\varepsilon)}{4(2n-3+2\varepsilon)(n-1+\varepsilon)^2(2n-1+2\varepsilon)}$$
$$\times\Big\{(n-1)^2(n-1+2\varepsilon)^2$$
$$+2(n-1)(n-1+2\varepsilon)(2g+2f(\varepsilon-f)-\gamma)$$
$$+4\varepsilon(g\varepsilon-f\gamma)+\gamma^2\Big\},\quad n=1,2,3,\ldots.$$

The latter formula can also be rewritten in the form

$$d_n = -\frac{n(n-2+2\varepsilon)}{4(2n-3+2\varepsilon)(n-1+\varepsilon)^2(2n-1+2\varepsilon)}D_n,\quad n=1,2,3,\ldots,$$

where

$$D_n = \left\{(n-1+\varepsilon)^2-\delta^2-\eta^2\right\}^2-4\delta^2\eta^2,\quad n=1,2,3,\ldots \tag{5.2.1}$$

with

$$\delta^2=(f-\varepsilon)^2-g+\gamma \quad\text{and}\quad \eta^2=f^2-g.$$

Hence δ and η are either real or pure imaginary. The role of δ and η can be explained as follows:

$$(c+1)^2-2f(c+1)+g=-2\varepsilon(c+1)+\gamma \iff (c+1-f+\varepsilon)^2=\delta^2$$

and

$$\varphi(x)=x^2+2fx+g=(x+f)^2-\eta^2.$$

For $\varepsilon>0$ we have

$$-\frac{n(n-2+2\varepsilon)}{4(2n-3+2\varepsilon)(n-1+\varepsilon)^2(2n-1+2\varepsilon)}<0,\quad n=1,2,3,\ldots.$$

From (5.2.1) it follows that

$$D_n=(n-1+\varepsilon)^4-2(\delta^2+\eta^2)(n-1+\varepsilon)^2+(\delta^2-\eta^2)^2,\quad n=1,2,3,\ldots. \tag{5.2.2}$$

Hence for $d_n>0$ for $n=1,2,3,\ldots$ we must have $D_n<0$ for $n=1,2,3,\ldots$, which implies, by using (5.2.1) and (5.2.2), that both δ and η must be real. In that case we have for $n=1,2,3,\ldots$

$$D_n=\left\{(n-1+\varepsilon)^2-(\delta+\eta)^2\right\}\left\{(n-1+\varepsilon)^2-(\delta-\eta)^2\right\}, \tag{5.2.3}$$

with

$$\delta=\sqrt{(f-\varepsilon)^2-g+\gamma}\quad\text{and}\quad \eta=\sqrt{f^2-g}.$$

Finally, we remark that D_n can also be written in the form

$$D_n = (n-1+\varepsilon+\delta+\eta)(n-1+\varepsilon+\delta-\eta) \\ \times (n-1+\varepsilon-\delta+\eta)(n-1+\varepsilon-\delta-\eta) \tag{5.2.4}$$

for $n = 1,2,3,\ldots$.

Now we conclude that we have positive-definite orthogonality for

Case IIIa. $\varepsilon > 0$ and $(\delta-\eta)^2 < (n-1+\varepsilon)^2 < (\delta+\eta)^2$. Note that this cannot be true for all $n = 1,2,3,\ldots$. However, for $|\delta-\eta| < \varepsilon$ and $N-1 < \delta+\eta-\varepsilon$ $(\leq N)$ we have $d_n > 0$ for $n = 1,2,3,\ldots,N$.

Further we also find finite orthogonal polynomial systems in the following cases. First we define

$$2\varepsilon = -2N - t \quad \text{with} \quad N \in \{1,2,3,\ldots\} \quad \text{and} \quad -1 < t \leq 1.$$

Then we have

$$-\frac{n(n-2+2\varepsilon)}{4(2n-3+2\varepsilon)(n-1+\varepsilon)^2(2n-1+2\varepsilon)} > 0, \quad n = 1,2,3,\ldots,N$$

and for $n = N+1$ this is not longer true. So in that case, for $d_n > 0$ for $n = 1,2,3,\ldots,N$ we must have $D_n > 0$ for $n = 1,2,3,\ldots,N$. This implies that we have positive-definite orthogonality in the following two finite cases:

Case IIIb. $\delta^2 > 0$ and $\eta^2 > 0$ (the polynomial φ has two different real zeros). In that case we must have $(\delta+\eta)^2 < (n-1+\varepsilon)^2$ or $(n-1+\varepsilon)^2 < (\delta-\eta)^2$. Hence

$$\delta+\eta < |n-1+\varepsilon| \quad \text{or} \quad |n-1+\varepsilon| < |\delta-\eta|.$$

Since $n-1+\varepsilon = n-1-N-\dfrac{t}{2}$, this leads to [1]

$$\left(\frac{|t|}{2} \leq |\delta-\eta| \leq\right) \delta+\eta < 1+\frac{t}{2} \quad \text{or} \quad N+\frac{t}{2} < |\delta-\eta|.$$

Case IIIc. $\delta^2 \leq 0$ and/or $\eta^2 \leq 0$ (the polynomial φ has two equal real zeros or two nonreal, id est complex conjugate, zeros). In that case we must have δ pure imaginary (including zero) and/or η pure imaginary (including zero) and[2]

$$\varepsilon \pm \delta \pm \eta \neq 0, -1, -2, \ldots, -N+1.$$

Now we have proved:

[1] The extra condition $|\delta-\eta| \geq |t|/2$ implies that $d_n > 0$ does no longer hold for $n = N+1$.

[2] This condition prevents that $d_n = 0$ for $n \in \{1,2,3,\ldots,N\}$ in view of (5.2.4).

Theorem 5.1. *For the polynomial solutions y_n of the difference equation (5.1.1)*

$$(ex^2+2fx+g)\left(\Delta^2 y_n\right)(x)+(2\varepsilon x+\gamma)\left(\Delta y_n\right)(x)=n(e(n-1)+2\varepsilon)y_n(x+1)$$

with $e,f,g,\varepsilon,\gamma \in \mathbb{R}$ and $n=0,1,2,\ldots$ we only have positive-definite orthogonality in the following cases:

Case I. *$e=f=0$, $g=1$ and $2\varepsilon<0$. An infinite system of orthogonal polynomials.*

Case IIa. *$e=0$, $2f=1$, $2\varepsilon\leq 1$ and $2g\varepsilon-\gamma<0$. An infinite system of orthogonal polynomials.*

Case IIb. *$e=0$, $2f=1$, $2\varepsilon>1$ and $-N\leq \dfrac{2g\varepsilon-\gamma}{2\varepsilon-1}<-N+1$. A finite system of $N+1$ orthogonal polynomials.*

Case IIIa. *$e=1$, $\varepsilon>0$, $|\delta-\eta|<\varepsilon$ and $N-1<\delta+\eta-\varepsilon$ $(\leq N)$. A finite system of $N+1$ orthogonal polynomials.*

Case IIIb. *$e=1$, $2\varepsilon=-2N-t$ with $N\in\{1,2,3,\ldots\}$ and $-1<t\leq 1$. A finite system of $N+1$ orthogonal polynomials if*

$$\left(\frac{|t|}{2}\leq|\delta-\eta|\leq\right)\ \delta+\eta<1+\frac{t}{2}\quad \text{or}\quad N+\frac{t}{2}<|\delta-\eta|.$$

Case IIIc. *$e=1$, $2\varepsilon=-2N-t$ with $N\in\{1,2,3,\ldots\}$ and $-1<t\leq 1$, $\delta^2\leq 0$ and/or $\eta^2\leq 0$ and $\varepsilon\pm\delta\pm\eta\neq 0,-1,-2,\ldots,-N+1$. A finite system of $N+1$ orthogonal polynomials.*

5.3 Properties of the Positive-Definite Orthogonal Polynomial Solutions

Case I. We have $e=f=0$, $g=1$ and $2\varepsilon<0$. Then the Pearson difference equation (5.1.10) reads

$$\frac{w(x)}{w(x+1)}=-2\varepsilon x-\gamma+1.$$

In order to get $A=0$ (and $N\to\infty$) in the boundary conditions (5.1.12), we set $\gamma=2\varepsilon+1$, which leads to a positive-definite weight function of the form

$$w(x)=\frac{1}{(-2\varepsilon)^x\Gamma(x+1)},\qquad x=0,1,2,\ldots,\qquad 2\varepsilon<0$$

for the **Charlier** polynomials. This weight function for the Charlier polynomials is connected with the Poisson distribution in stochastics.

The difference equation (5.1.1) reads

$$\left(\Delta^2 y_n\right)(x) + (2\varepsilon(x+1)+1)\left(\Delta y_n\right)(x) = 2n\varepsilon y_n(x+1), \quad n = 0,1,2,\ldots$$

or

$$y_n(x+1) + (2\varepsilon x - 1)y_n(x) - 2\varepsilon x y_n(x-1) = 2n\varepsilon y_n(x), \quad n = 0,1,2,\ldots$$

and the three-term recurrence relation (5.1.8) can be written as

$$y_{n+1}(x) = \left(x - n + \frac{1}{2\varepsilon}\right) y_n(x) + \frac{n}{2\varepsilon} y_{n-1}(x), \quad n = 1,2,3,\ldots$$

with $y_0(x) = 1$ and $y_1(x) = x + 1/2\varepsilon$.

For the coefficients of the representation (5.1.3) we obtain from (5.1.4)

$$2\varepsilon(n-k)a_{n,k} = a_{n,k+1}, \quad k = n-1, n-2, n-3, \ldots 0$$

if c satisfies (5.1.5), which means that $1 = -2\varepsilon(c+1) + \gamma$. Since $a_{n,n} = n!$, this leads to

$$a_{n,k} = \frac{n!}{(2\varepsilon)^{n-k}(n-k)!} = \frac{1}{(2\varepsilon)^n}(-1)^k(-n)_k(2\varepsilon)^k, \quad k = 0,1,2,\ldots,n.$$

Since $\gamma = 2\varepsilon + 1$, we have $c = 0$. Hence by using (5.1.3), we obtain for $2\varepsilon < 0$

$$y_n(x) = \frac{1}{(2\varepsilon)^n}\sum_{k=0}^{n}(-1)^k(-n)_k(2\varepsilon)^k\binom{x}{k} = \frac{1}{(2\varepsilon)^n}\,{}_2F_0\left(\begin{matrix}-n, -x \\ -\end{matrix}; 2\varepsilon\right)$$

for $n = 0,1,2,\ldots$.

The Rodrigues formula (3.4.28) reads

$$y_n(x) = \frac{(-1)^n\Gamma(x+1)}{(-2\varepsilon)^{n-x}}\Delta^n\left(\frac{1}{(-2\varepsilon)^{x-n}\Gamma(x-n+1)}\right), \quad n = 0,1,2,\ldots.$$

Now we have

$$d_0 := \sum_{x=0}^{\infty} w(x) = \sum_{x=0}^{\infty}\frac{(-2\varepsilon)^{-x}}{\Gamma(x+1)} = \sum_{x=0}^{\infty}\frac{(-2\varepsilon)^{-x}}{x!} = \exp\left(-\frac{1}{2\varepsilon}\right) > 0.$$

Then using (5.1.11) and (3.1.4) the orthogonality relation can be written as

$$\sum_{x=0}^{\infty}\frac{y_m(x)y_n(x)}{(-2\varepsilon)^x x!} = \frac{\exp\left(-\frac{1}{2\varepsilon}\right)n!}{(-2\varepsilon)^n}\delta_{mn}, \quad m,n = 0,1,2,\ldots.$$

Case IIa. We have $e = 0$, $2f = 1$, $2\varepsilon \leq 1$, $2\varepsilon \neq 0$ and $2g\varepsilon < \gamma$. Now we have to distinguish between the cases that $2\varepsilon < 0$, $0 < 2\varepsilon < 1$ and $2\varepsilon = 1$.

Case IIa1. For $2\varepsilon < 0$ the two-term recurrence relation (5.1.4) can be used. In that case we have for (5.1.5)

$$-(c+1)+g=-2\varepsilon(c+1)+\gamma \quad \Longleftrightarrow \quad c+1=\frac{g-\gamma}{1-2\varepsilon}.$$

The Pearson difference equation (5.1.10) reads

$$\frac{w(x)}{w(x+1)}=\frac{(1-2\varepsilon)x+g-\gamma}{x+g-1}.$$

In order to get $A=0$ (and $N\to\infty$) in the boundary conditions (5.1.12), we set $g=\gamma-2\varepsilon+1$. This implies that $c=0$. With $\gamma>2\varepsilon$ this leads to a positive-definite weight function of the form

$$w(x)=\frac{\Gamma(x+\gamma-2\varepsilon)}{(1-2\varepsilon)^x\Gamma(x+1)}, \quad x=0,1,2,\dots$$

for the **Meixner** polynomials. This weight function for the Meixner polynomials is connected with the negative binomial distribution in stochastics.

The difference equation (5.1.1) reads

$$(x+\gamma-2\varepsilon+1)\left(\Delta^2 y_n\right)(x)+(2\varepsilon x+\gamma)\left(\Delta y_n\right)(x)=2n\varepsilon y_n(x+1), \quad n=0,1,2,\dots$$

or

$$\begin{aligned}(x+\gamma-2\varepsilon)y_n(x+1)-(2(1-\varepsilon)x+\gamma-2\varepsilon)\,y_n(x)&\\ +(1-2\varepsilon)xy_n(x-1)=2n\varepsilon y_n(x),&\quad n=0,1,2,\dots\end{aligned}$$

and the three-term recurrence relation (5.1.8) can be written as

$$y_{n+1}(x)=\left(x-\frac{2n(\varepsilon-1)+2\varepsilon-\gamma}{2\varepsilon}\right)y_n(x)-\frac{n(1-2\varepsilon)(n-1+\gamma-2\varepsilon)}{4\varepsilon^2}y_{n-1}(x)$$

for $n=1,2,3,\dots$ with $y_0(x)=1$ and $y_1(x)=x-1+\gamma/2\varepsilon$.

For the coefficients of the representation (5.1.3) we obtain from (5.1.4)

$$2\varepsilon(n-k)a_{n,k}=(k+\gamma-2\varepsilon)a_{n,k+1}, \quad k=n-1,n-2,n-3,\dots,0.$$

Since $a_{n,n}=n!$, this leads to

$$a_{n,k}=\frac{(k+\gamma-2\varepsilon)_{n-k}n!}{(2\varepsilon)^{n-k}(n-k)!}=\frac{(\gamma-2\varepsilon)_n}{(2\varepsilon)^n}\frac{(-n)_k}{(\gamma-2\varepsilon)_k}(-2\varepsilon)^k, \quad k=0,1,2,\dots,n.$$

Hence

$$y_n(x)=\sum_{k=0}^{n}a_{n,k}\binom{x}{k}=\frac{(\gamma-2\varepsilon)_n}{(2\varepsilon)^n}\,{}_2F_1\left(\begin{matrix}-n,-x\\ \gamma-2\varepsilon\end{matrix};2\varepsilon\right), \quad n=0,1,2,\dots, \quad \gamma>2\varepsilon.$$

The Rodrigues formula (3.4.28) reads

$$y_n(x) = \frac{(1-2\varepsilon)^x\Gamma(x+1)}{(2\varepsilon)^n\Gamma(x+\gamma-2\varepsilon)}\Delta^n\left(\frac{\Gamma(x+\gamma-2\varepsilon)}{(1-2\varepsilon)^{x-n}\Gamma(x-n+1)}\right), \quad n=0,1,2,\ldots.$$

Now we have

$$d_0 := \sum_{x=0}^{\infty}\frac{\Gamma(x+\gamma-2\varepsilon)}{(1-2\varepsilon)^x x!} = \Gamma(\gamma-2\varepsilon)\cdot {}_1F_0\left(\begin{matrix}\gamma-2\varepsilon\\ -\end{matrix};\frac{1}{1-2\varepsilon}\right)$$
$$= \Gamma(\gamma-2\varepsilon)\left(\frac{1-2\varepsilon}{-2\varepsilon}\right)^{\gamma-2\varepsilon} > 0.$$

Then using (5.1.11) and (3.1.4) the orthogonality relation can be written as

$$\sum_{x=0}^{\infty}\frac{\Gamma(x+\gamma-2\varepsilon)}{(1-2\varepsilon)^x x!}y_m(x)y_n(x) = \frac{(1-2\varepsilon)^{n+\gamma-2\varepsilon}\Gamma(n+\gamma-2\varepsilon)n!}{(-2\varepsilon)^{2n+\gamma-2\varepsilon}}\delta_{mn}$$

for $2\varepsilon<0$, $\gamma>2\varepsilon$ and $m,n=0,1,2,\ldots$.

Case IIa2. For $0<2\varepsilon<1$ the two-term recurrence relation (5.1.4) is still valid with

$$c+1 = \frac{g-\gamma}{1-2\varepsilon}.$$

The Pearson difference equation (5.1.10) reads

$$\frac{w(x)}{w(x+1)} = \frac{(1-2\varepsilon)x+g-\gamma}{x+g-1}.$$

In order to satisfy the boundary conditions (5.1.12), we have to take $A\to-\infty$. Then we may choose $A+N=0$, which implies that $g=1$. With $\gamma>2\varepsilon$ this leads to a positive-definite weight function of the form

$$w(x) = \frac{\Gamma(r-x)}{(1-2\varepsilon)^x\Gamma(1-x)}, \quad r=\frac{\gamma-2\varepsilon}{1-2\varepsilon}, \quad x=0,-1,-2,\ldots$$

for the **Meixner** polynomials. Note that this implies that $c=-r$.

The difference equation (5.1.1) reads

$$(x+1)\left(\Delta^2 y_n\right)(x)+(2\varepsilon x+\gamma)\left(\Delta y_n\right)(x) = 2n\varepsilon y_n(x+1), \quad n=0,1,2,\ldots$$

or

$$x y_n(x+1)-\left(2(1-\varepsilon)x+2\varepsilon-\gamma\right)y_n(x)$$
$$+\left((1-2\varepsilon)x+2\varepsilon-\gamma\right)y_n(x-1) = 2n\varepsilon y_n(x), \quad n=0,1,2,\ldots$$

and the three-term recurrence relation (5.1.8) can be written as

$$y_{n+1}(x) = \left(x-\frac{2n(\varepsilon-1)+2\varepsilon-\gamma}{2\varepsilon}\right)y_n(x)-\frac{n((n-1)(1-2\varepsilon)-2\varepsilon+\gamma)}{4\varepsilon^2}y_{n-1}(x)$$

for $n = 1, 2, 3, \ldots$ with $y_0(x) = 1$ and $y_1(x) = x - 1 + \gamma/2\varepsilon$.

For the coefficients of the representation (5.1.3) we obtain from (5.1.4)

$$2\varepsilon(n-k)a_{n,k} = (k+r)a_{n,k+1}, \quad k = n-1, n-2, n-3, \ldots, 0.$$

Since $a_{n,n} = n!$, this leads to

$$a_{n,k} = \frac{(k+r)_{n-k} n!}{(2\varepsilon)^{n-k}(n-k)!} = \frac{(r)_n}{(2\varepsilon)^n} \frac{(-n)_k}{(r)_k} (-2\varepsilon)^k, \quad k = 0, 1, 2, \ldots, n.$$

Hence by using $c = -r$, we obtain for $r > 0$

$$y_n(x) = \sum_{k=0}^{n} a_{n,k} \binom{x-r}{k} = \frac{(r)_n}{(2\varepsilon)^n} \, {}_2F_1\left(\begin{matrix} -n, r-x \\ r \end{matrix}; 2\varepsilon\right), \quad n = 0, 1, 2, \ldots.$$

The Rodrigues formula (3.4.28) reads

$$y_n(x) = \frac{(1-2\varepsilon)^x \Gamma(1-x)}{(-2\varepsilon)^n \Gamma(r-x)} \Delta^n \left(\frac{\Gamma(r-x+n)}{(1-2\varepsilon)^{x-n} \Gamma(1-x)} \right), \quad n = 0, 1, 2, \ldots.$$

Now we have

$$\begin{aligned} d_0 := \sum_{x=-\infty}^{0} w(x) &= \sum_{x=0}^{\infty} \frac{(1-2\varepsilon)^x \Gamma(x+r)}{\Gamma(x+1)} \\ &= \Gamma(r) \cdot {}_1F_0\left(\begin{matrix} r \\ - \end{matrix}; 1-2\varepsilon\right) = \frac{\Gamma(r)}{(2\varepsilon)^r} > 0. \end{aligned}$$

Then using (5.1.11) and (3.1.4) the orthogonality relation can be written as

$$\sum_{x=0}^{\infty} \frac{(1-2\varepsilon)^x \Gamma(x+r)}{x!} y_m(-x) y_n(-x) = \frac{(1-2\varepsilon)^n \Gamma(n+r) n!}{(2\varepsilon)^{2n+r}} \delta_{mn}$$

for $0 < 2\varepsilon < 1$, $\gamma > 2\varepsilon$ and $m, n = 0, 1, 2, \ldots$.

The connection between these Meixner polynomials and those found in the previous case can be described as follows. If we define

$$M_n^{(1)}(x) = \frac{(r)_n}{(2\varepsilon)^n} \, {}_2F_1\left(\begin{matrix} -n, -x \\ r \end{matrix}; 2\varepsilon\right), \quad n = 0, 1, 2, \ldots$$

and

$$M_n^{(2)}(x) = \frac{(r)_n}{(2\varepsilon)^n} \, {}_2F_1\left(\begin{matrix} -n, r-x \\ r \end{matrix}; 2\varepsilon\right), \quad n = 0, 1, 2, \ldots,$$

then the orthogonality relations for $M_n^{(1)}(x)$ and for $M_n^{(2)}(-x)$ are equal. Moreover we have by using (1.7.2)

$$(-1)^n M_n^{(2)}(-x) = \frac{(-1)^n (r)_n}{(2\varepsilon)^n} {}_2F_1\left(\begin{matrix} -n, r+x \\ r \end{matrix}; 2\varepsilon\right)$$
$$= \frac{(r)_n}{(2\varepsilon^*)^n} {}_2F_1\left(\begin{matrix} -n, -x \\ r \end{matrix}; 2\varepsilon^*\right) = M_n^{(1)}(x),$$

where

$$2\varepsilon^* = \frac{2\varepsilon}{2\varepsilon - 1} \quad \text{(note that this implies } 0 < 2\varepsilon^* < 1 \quad \Longleftrightarrow \quad 2\varepsilon < 0).$$

Case IIa3. For $2\varepsilon = 1$ the two-term recurrence relation (5.1.4) does not hold. In that case we use the representation (5.1.6) with $c = g$. The Pearson difference equation (5.1.10) reads

$$\frac{w(x)}{w(x+1)} = \frac{g-\gamma}{x+g-1}.$$

In order to satisfy the boundary conditions (5.1.12), we have to take $A \to -\infty$. Then we may choose $A + N = 0$, which implies that $g = 1$. With $\gamma > 1$ this leads to a positive-definite weight function of the form

$$w(x) = \frac{1}{(\gamma-1)^x \Gamma(1-x)}, \quad x = 0, -1, -2, \ldots$$

for the **Charlier** polynomials.

The difference equation (5.1.1) reads

$$(x+1)\left(\Delta^2 y_n\right)(x) + (x+\gamma)\left(\Delta y_n\right)(x) = n y_n(x+1), \quad n = 0, 1, 2, \ldots$$

or

$$x y_n(x+1) - (x+1-\gamma) y_n(x) + (1-\gamma) y_n(x-1) = n y_n(x), \quad n = 0, 1, 2, \ldots$$

and the three-term recurrence relation (5.1.8) can be written as

$$y_{n+1}(x) = (x+n-1+\gamma) y_n(x) + n(1-\gamma) y_{n-1}(x), \quad n = 1, 2, 3, \ldots$$

with $y_0(x) = 1$ and $y_1(x) = x - 1 + \gamma$.

For the coefficients of the representation (5.1.6) we obtain from (5.1.7)

$$(n-k) b_{n,k} = (\gamma-1) b_{n,k+1}, \quad k = n-1, n-2, n-3, \ldots, 0.$$

Since $b_{n,n} = n!$, this leads to

$$b_{n,k} = \frac{(\gamma-1)^{n-k} n!}{(n-k)!} = (\gamma-1)^n (-n)_k \left(\frac{1}{1-\gamma}\right)^k, \quad k = 0, 1, 2, \ldots, n.$$

Hence by using (5.1.6) with $c = 1$, we obtain for $n = 0, 1, 2, \ldots$

$$\begin{aligned} y_n(x) &= (\gamma-1)^n \sum_{k=0}^{n} (-n)_k \left(\frac{1}{1-\gamma}\right)^k \binom{x+k-1}{k} \\ &= (\gamma-1)^n \, {}_2F_0\left(\begin{matrix} -n, x \\ - \end{matrix}; \frac{1}{1-\gamma}\right). \end{aligned}$$

The Rodrigues formula (3.4.28) reads

$$y_n(x) = (-1)^n(\gamma-1)^x \Gamma(1-x)\, \Delta^n \left(\frac{(\gamma-1)^{n-x}}{\Gamma(1-x)}\right), \quad n = 0,1,2,\ldots.$$

Now we have

$$d_0 := \sum_{x=-\infty}^{0} w(x) = \sum_{x=0}^{\infty} \frac{(\gamma-1)^x}{x!} = \mathrm{e}^{\gamma-1} > 0.$$

Then using (5.1.11) and (3.1.4) the orthogonality relation can be written as

$$\sum_{x=0}^{\infty} \frac{(\gamma-1)^x}{x!} y_m(-x) y_n(-x) = \mathrm{e}^{\gamma-1} (\gamma-1)^n n!\, \delta_{mn}, \quad m,n = 0,1,2,\ldots.$$

The connection between these Charlier polynomials and those obtained in Case I can be described as follows. If we define

$$C_n^{(1)}(x) = \frac{1}{(2\varepsilon)^n} \, {}_2F_0\left(\begin{matrix} -n, -x \\ - \end{matrix}; 2\varepsilon\right), \quad n = 0,1,2,\ldots$$

and

$$C_n^{(2)}(x) = (\gamma-1)^n \, {}_2F_0\left(\begin{matrix} -n, x \\ - \end{matrix}; \frac{1}{1-\gamma}\right), \quad n = 0,1,2,\ldots,$$

then we simply have $(-1)^n C_n^{(1)}(-x) = C_n^{(2)}(x)$, where

$$1-\gamma = \frac{1}{2\varepsilon} \quad \text{(note that this implies } \gamma > 1 \quad \Longleftrightarrow \quad 2\varepsilon < 0).$$

Case IIb. We have $e = 0$, $2f = 1$, $2\varepsilon > 1$ and $-N \leq \dfrac{2g\varepsilon - \gamma}{2\varepsilon - 1} < -N+1$. Again the two-term recurrence relation (5.1.4) can be used with (cf. (5.1.5))

$$c+1 = \frac{g-\gamma}{1-2\varepsilon}.$$

Again we set $g = \gamma - 2\varepsilon + 1$ in order to get $A = 0$ in the boundary conditions (5.1.12). This implies that $c = 0$ and $-N \leq \gamma - 2\varepsilon < -N+1$. The Pearson difference equation (5.1.10) now reads

$$\frac{w(x)}{w(x+1)} = \frac{(1-2\varepsilon)(x+1)}{x+\gamma-2\varepsilon}.$$

Now we have to distinguish between two different cases.

Case IIb1. For $\gamma - 2\varepsilon = -N$ this leads to a positive-definite weight function of the form

$$w(x) = \frac{1}{(2\varepsilon-1)^x \Gamma(x+1)\Gamma(N+1-x)}, \qquad x = 0,1,2,\ldots,N$$

for the **Krawtchouk** polynomials. This weight function for the Krawtchouk polynomials is connected with the binomial distribution in stochastics.

The difference equation (5.1.1) reads

$$(x-N+1)\left(\Delta^2 y_n\right)(x) + (2\varepsilon(x+1)-N)\left(\Delta y_n\right)(x) = 2n\varepsilon y_n(x+1)$$

for $n = 0,1,2,\ldots,N$, or

$$(x-N)y_n(x+1) - (2(1-\varepsilon)x-N)\,y_n(x) + (1-2\varepsilon)xy_n(x-1) = 2n\varepsilon y_n(x)$$

for $n = 0,1,2,\ldots,N$, and the three-term recurrence relation (5.1.8) can be written as

$$y_{n+1}(x) = \left(x - \frac{2n(\varepsilon-1)+N}{2\varepsilon}\right) y_n(x) - \frac{n(1-2\varepsilon)(n-1-N)}{4\varepsilon^2} y_{n-1}(x)$$

for $n = 1,2,3,\ldots,N-1$ with $y_0(x) = 1$ and $y_1(x) = x - N/2\varepsilon$.

For the coefficients of the representation (5.1.3) we obtain from (5.1.4)

$$2\varepsilon(n-k)a_{n,k} = (k-N)a_{n,k+1}, \qquad k = n-1, n-2, n-3, \ldots, 0.$$

Since $a_{n,n} = n!$, this leads to

$$a_{n,k} = \frac{(k-N)_{n-k}\, n!}{(2\varepsilon)^{n-k}(n-k)!} = \frac{(-N)_n}{(2\varepsilon)^n} \frac{(-n)_k}{(-N)_k} (-2\varepsilon)^k, \qquad k = 0,1,2,\ldots,n.$$

Hence

$$y_n(x) = \sum_{k=0}^{n} a_{n,k} \binom{x}{k} = \frac{(-N)_n}{(2\varepsilon)^n}\, {}_2F_1\left(\begin{matrix} -n, -x \\ -N \end{matrix}; 2\varepsilon\right), \qquad n = 0,1,2,\ldots,N.$$

The Rodrigues formula (3.4.28) reads

$$y_n(x) = \frac{(2\varepsilon-1)^x \Gamma(x+1)\Gamma(N+1-x)}{(-2\varepsilon)^n} \times \Delta^n \left(\frac{1}{(2\varepsilon-1)^{x-n}\Gamma(x-n+1)\Gamma(N+1-x)} \right), \qquad n = 0,1,2,\ldots,N.$$

Now we have

$$d_0 := \sum_{x=0}^{N} \frac{1}{(2\varepsilon-1)^x x!\,(N-x)!} = \frac{(2\varepsilon)^N}{(2\varepsilon-1)^N N!} > 0.$$

Then using (5.1.11) and (3.1.4) the orthogonality relation can be written as

$$\sum_{x=0}^{N} \frac{1}{(2\varepsilon-1)^x x!\,(N-x)!} y_m(x) y_n(x) = \frac{(2\varepsilon)^{N-2n} n!}{(2\varepsilon-1)^{N-n}(N-n)!}\,\delta_{mn}$$

for $2\varepsilon > 1$, $\gamma - 2\varepsilon = -N$ and $m,n = 0,1,2,\ldots,N$.

Case IIb2. In the case that $-N < \gamma - 2\varepsilon < -N+1$ we were not able to find a solution for the Pearson difference equation such that the boundary conditions (5.1.12) can be fulfilled. However, the concept of orthogonality in section 3.2 can be generalized to an integration on the (possibly deformed) imaginary axis in the complex plane. Then we must have (cf. (3.2.14))

$$(\lambda_n - \lambda_m) \int_{-i\infty}^{i\infty} w(x) y_m(x) y_n(x)\, dx = \int_{-i\infty}^{i\infty} \{s_{n,m}(x) - s_{n,m}(x-1)\}\, dx \qquad (5.3.1)$$

where

$$s_{n,m}(x) = w(x)\varphi(x)\{y_n(x+1)y_m(x) - y_m(x+1)y_n(x)\}.$$

If the regularity condition (2.3.3) holds, id est $en + 2\varepsilon \neq 0$ for $n = 0,1,2,\ldots$ then this leads to an orthogonality relation if the right-hand side of (5.3.1) cancels. Hereby all moments of the form

$$\int_{-i\infty}^{i\infty} w(x) x^n\, dx, \quad n = 0,1,2,\ldots$$

should exist (this is usually guaranteed by the asymptotic behaviour of the gamma functions involved). By using Cauchy's integral theorem it can be shown that the right-hand side of (5.3.1) cancels. See [375].

With $g = \gamma - 2\varepsilon + 1$ we obtain a positive-definite weight function of the form

$$w(x) = (-1)^N \frac{\Gamma(x+\gamma-2\varepsilon)\Gamma(-x)}{(2\varepsilon-1)^x}$$

for the **Krawtchouk** polynomials.

Then the Rodrigues formula (3.4.28) reads

$$y_n(x) = \frac{(2\varepsilon-1)^x}{(2\varepsilon)^n \Gamma(x+\gamma-2\varepsilon)\Gamma(-x)} \Delta^n \left(\frac{\Gamma(x+\gamma-2\varepsilon)\Gamma(n-x)}{(2\varepsilon-1)^{x-n}} \right), \quad n = 0,1,2,\ldots.$$

Now we may use Barnes' integral representation (1.6.2) to find that

$$\frac{1}{2\pi i} \int_{-i\infty}^{i\infty} \frac{\Gamma(x+\gamma-2\varepsilon)\Gamma(-x)}{(2\varepsilon-1)^x}\, dx$$
$$= \Gamma(\gamma-2\varepsilon)\, {}_1F_0\left({\gamma-2\varepsilon \atop -} ; \frac{1}{1-2\varepsilon} \right) = \Gamma(\gamma-2\varepsilon) \left(\frac{2\varepsilon-1}{2\varepsilon} \right)^{\gamma-2\varepsilon}.$$

The latter integrand has poles in $\{0,1,2,\ldots\}$ and $\{2\varepsilon-\gamma,2\varepsilon-\gamma-1,2\varepsilon-\gamma-2,\ldots\}$. Since $N-1<2\varepsilon-\gamma<N$, these poles are all separated and the path can be taken as in (1.6.2) and we have

$$\begin{aligned} d_0 &:= \frac{(-1)^N}{2\pi i}\int_{-i\infty}^{i\infty}\frac{\Gamma(x+\gamma-2\varepsilon)\Gamma(-x)}{(2\varepsilon-1)^x}\,dx \\ &= (-1)^N\Gamma(\gamma-2\varepsilon)\left(\frac{2\varepsilon-1}{2\varepsilon}\right)^{\gamma-2\varepsilon} \\ &= \frac{(-1)^N\Gamma(\gamma-2\varepsilon+N)}{(\gamma-2\varepsilon)_N}\left(\frac{2\varepsilon-1}{2\varepsilon}\right)^{\gamma-2\varepsilon} > 0. \end{aligned}$$

If (5.1.11) and (3.1.4) are used, this leads to the orthogonality relation

$$\begin{aligned} &\frac{(-1)^N}{2\pi i}\int_{-i\infty}^{i\infty}\frac{\Gamma(x+\gamma-2\varepsilon)\Gamma(-x)}{(2\varepsilon-1)^x}y_m(x)y_n(x)\,dx \\ &= \frac{n!\,(-1)^{N+n}\Gamma(\gamma-2\varepsilon+N)(\gamma-2\varepsilon)_n}{(\gamma-2\varepsilon)_N}\left(\frac{2\varepsilon-1}{2\varepsilon}\right)^{n+\gamma-2\varepsilon}\delta_{mn} \end{aligned}$$

for $m,n=0,1,2,\ldots,N$.

Case III. We have $e=1$, $\delta=\sqrt{(f-\varepsilon)^2-g+\gamma}$ and $\eta=\sqrt{f^2-g}$. Then we find that

$$\begin{aligned} \varphi(x) = x^2+2fx+g &= (x+f)^2-(f^2-g) = (x+f)^2-\eta^2 \\ &= (x+f+\eta)(x+f-\eta) \end{aligned}$$

and

$$\begin{aligned} \psi(x) = 2\varepsilon x+\gamma &= 2\varepsilon x+(f-\varepsilon)^2-g+\gamma-(f^2-g)+2\varepsilon f-\varepsilon^2 \\ &= 2\varepsilon(x+f)+\delta^2-\eta^2-\varepsilon^2 = 2\varepsilon(x+f-\eta)+\delta^2-(\eta-\varepsilon)^2 \\ &= 2\varepsilon(x+f-\eta)+(\delta+\eta-\varepsilon)(\delta-\eta+\varepsilon). \end{aligned}$$

This implies that the difference equation (5.1.1) can be written as

$$\begin{aligned} &(x+f+\eta)(x+f-\eta)\left(\Delta^2 y_n\right)(x) \\ &\qquad + \{2\varepsilon(x+f-\eta)+(\delta+\eta-\varepsilon)(\delta-\eta+\varepsilon)\}\,\Delta y_n(x) \\ &= n(n-1+2\varepsilon)y_n(x+1), \quad n=0,1,2,\ldots. \end{aligned}$$

or equivalently

$$C(x)y_n(x+1)-\{C(x)+D(x)\}\,y_n(x)+D(x)y_n(x-1) = n(n-1+2\varepsilon)y_n(x)$$

for $n=0,1,2,\ldots$, where

$$C(x) = \varphi(x-1) = (x-1+f+\eta)(x-1+f-\eta)$$

and

$$\begin{aligned} D(x) &= C(x) - \psi(x-1) \\ &= (x-1+f+\eta)(x-1+f-\eta) - 2\varepsilon(x-1+f) - \delta^2 + \eta^2 + \varepsilon^2 \\ &= (x-1+f)^2 - 2\varepsilon(x-1+f) + \varepsilon^2 - \delta^2 = (x-1+f-\varepsilon)^2 - \delta^2 \\ &= (x-1+f-\varepsilon+\delta)(x-1+f-\varepsilon-\delta). \end{aligned}$$

For the hypergeometric representation we use (5.1.3) and (5.1.4) provided that (cf. (5.1.5))

$$\begin{aligned} &(c+1)^2 - 2f(c+1) + g = -2\varepsilon(c+1) + \gamma \\ &\iff \quad (c+1)^2 - 2(f-\varepsilon)(c+1) + g - \gamma = 0. \end{aligned}$$

This implies that

$$c+1 = f - \varepsilon \pm \sqrt{(f-\varepsilon)^2 - g + \gamma} = f - \varepsilon \pm \delta.$$

In that case we have

$$\begin{aligned} (k-1-c)^2 + 2f(k-1-c) + g &= (k-1-c)^2 + 2f(k-1-c) + g + (f^2-g) - \eta^2 \\ &= (k-1-c+f)^2 - \eta^2 \\ &= (k-1-c+f+\eta)(k-1-c+f-\eta) \\ &= (k+\varepsilon\mp\delta+\eta)(k+\varepsilon\mp\delta-\eta). \end{aligned}$$

Therefore, the two-term recurrence relation (5.1.4) reads

$$(n-k)(n+k-1+2\varepsilon)a_{n,k} = (k+\varepsilon\mp\delta+\eta)(k+\varepsilon\mp\delta-\eta)a_{n,k+1}$$

for $k = n-1, n-2, n-3, \ldots, 0$. By using $a_{n,n} = n!$, we obtain

$$\begin{aligned} a_{n,k} &= \frac{(k+\varepsilon\mp\delta+\eta)_{n-k}(k+\varepsilon\mp\delta-\eta)_{n-k}n!}{(n+k-1+2\varepsilon)_{n-k}(n-k)!} \\ &= \frac{(\varepsilon\mp\delta+\eta)_n(\varepsilon\mp\delta-\eta)_n}{(n-1+2\varepsilon)_n} \frac{(-n)_k(n-1+2\varepsilon)_k}{(\varepsilon\mp\delta+\eta)_k(\varepsilon\mp\delta-\eta)_k}(-1)^k \end{aligned}$$

for $k = 0, 1, 2, \ldots, n$, which implies that

$$\begin{aligned} y_n(x) &= \sum_{k=0}^{n} a_{n,k}\binom{x+c}{k} = \sum_{k=0}^{n} a_{n,k}(-1)^k \frac{(-x-c)_k}{k!} \\ &= \frac{(\varepsilon\mp\delta+\eta)_n(\varepsilon\mp\delta-\eta)_n}{(n-1+2\varepsilon)_n}\, {}_3F_2\left(\begin{matrix} -n, n-1+2\varepsilon, \varepsilon\mp\delta+1-f-x \\ \varepsilon\mp\delta+\eta, \varepsilon\mp\delta-\eta \end{matrix}; 1\right) \end{aligned}$$

for $n = 0, 1, 2, \ldots$. These polynomials are called **Hahn** polynomials.

In order to obtain an orthogonality relation, we use the Pearson difference equation (5.1.10), which can be written as

$$\frac{w(x)}{w(x+1)} = \frac{D(x+1)}{C(x)} = \frac{(x+f-\varepsilon+\delta)(x+f-\varepsilon-\delta)}{(x-1+f+\eta)(x-1+f-\eta)} \tag{5.3.2}$$

$$= \frac{(x+f-\varepsilon+\delta)(\varepsilon+\delta-f-x)}{(x-1+f+\eta)(1+\eta-f-x)} \tag{5.3.3}$$

$$= \frac{(\varepsilon-\delta-f-x)(\varepsilon+\delta-f-x)}{(x-1+f+\eta)(x-1+f-\eta)} \tag{5.3.4}$$

$$= \frac{(x+f-\varepsilon+\delta)(x+f-\varepsilon-\delta)}{(1-\eta-f-x)(1+\eta-f-x)} \tag{5.3.5}$$

$$= \frac{(\varepsilon-\delta-f-x)(\varepsilon+\delta-f-x)}{(1-\eta-f-x)(1+\eta-f-x)} \tag{5.3.6}$$

First we consider positive-definite orthogonality for the Hahn polynomials in the case that $\varepsilon > 0$.

Case IIIa. We have $\varepsilon > 0$, $|\delta-\eta| < \varepsilon$ and $N-1 < \delta+\eta-\varepsilon$ $(\leq N)$. In this case we use (5.3.3) to obtain a solution of the form

$$w(x) = \frac{\Gamma(x-1+f+\eta)\Gamma(1+\varepsilon+\delta-f-x)}{\Gamma(x+f-\varepsilon+\delta)\Gamma(2+\eta-f-x)}$$

for the Pearson difference equation. Now we have to distinguish between two different cases.

Case IIIa1. We have $\varepsilon > 0$ and $|\delta-\eta| < \varepsilon$. In order to get $A = 0$ in the boundary conditions (5.1.12), we set $f-\varepsilon+\delta = 1$ and $1-f+\eta = N$. Note that this implies that $\delta+\eta-\varepsilon = N$.

If we make the substitutions

$$\varepsilon+\delta-\eta = \alpha+1 \quad \text{and} \quad \varepsilon-\delta+\eta = \beta+1,$$

then we have $2\varepsilon = \alpha+\beta+2$, $2\delta = \alpha+N+1$, $2\eta = \beta+N+1$ and we obtain that $\varphi(x) = (x+\beta+2)(x-N+1)$ and $\psi(x) = (\alpha+\beta+2)(x-N+1)+(\alpha+1)N = (\alpha+\beta+2)(x+1)-(\beta+1)N$. We also have $C(x) = (x+\beta+1)(x-N)$ and $D(x) = x(x-\alpha-N-1)$. This implies that the difference equation can be written as

$$(x+\beta+2)(x-N+1)\left(\Delta^2 y_n\right)(x)+\{(\alpha+\beta+2)(x+1)-(\beta+1)N\}\Delta y_n(x) \\ = n(n+\alpha+\beta+1)y_n(x+1), \quad n = 0,1,2,\ldots$$

or equivalently

$$(x+\beta+1)(x-N)y_n(x+1)-\{(x+\beta+1)(x-N)+x(x-\alpha-N-1)\}y_n(x) \\ +x(x-\alpha-N-1)y_n(x-1)=n(n+\alpha+\beta+1)y_n(x), \quad n=0,1,2,\ldots.$$

Further we have

$$y_n(x)=\frac{(\beta+1)_n(-N)_n}{(n+\alpha+\beta+1)_n}\,{}_3F_2\left(\begin{matrix}-n,n+\alpha+\beta+1,-x\\ \beta+1,-N\end{matrix};1\right), \quad n=0,1,2,\ldots$$

and the weight function reads

$$w(x)=\frac{\Gamma(x+\beta+1)\Gamma(\alpha+N+1-x)}{\Gamma(x+1)\Gamma(N+1-x)}, \quad x=0,1,2,\ldots,N.$$

For positivity we must have $\alpha+1>0$ and $\beta+1>0$. This weight function for the **Hahn** polynomials is connected with the hypergeometric or Pólya distribution in stochastics. The Rodrigues formula (3.4.28) can be written as

$$y_n(x)=\frac{(-1)^n\Gamma(x+1)\Gamma(N+1-x)}{(n+\alpha+\beta+1)_n\Gamma(x+\beta+1)\Gamma(\alpha+N+1-x)} \\ \times\Delta^n\left(\frac{\Gamma(x+\beta+1)\Gamma(\alpha+N+n+1-x)}{\Gamma(N+1-x)\Gamma(x-n+1)}\right), \quad n=0,1,2,\ldots.$$

Now we have

$$\begin{aligned} d_0:=\sum_{x=0}^{N}w(x)&=\sum_{x=0}^{N}\frac{\Gamma(x+\beta+1)\Gamma(\alpha+N+1-x)}{x!\,(N-x)!}\\ &=\frac{\Gamma(\beta+1)\Gamma(\alpha+N+1)}{\Gamma(N+1)}\,{}_2F_1\left(\begin{matrix}-N,\beta+1\\ -\alpha-N\end{matrix};1\right)\\ &=\frac{\Gamma(\beta+1)\Gamma(\alpha+N+1)}{\Gamma(N+1)}\,\frac{(\alpha+\beta+2)_N}{(\alpha+1)_N}\\ &=\frac{\Gamma(\alpha+1)\Gamma(\beta+1)\Gamma(\alpha+\beta+N+2)}{\Gamma(\alpha+\beta+2)N!}>0. \end{aligned}$$

Then using (5.1.11) and (3.1.4) the orthogonality relation can be written as

$$\sum_{x=0}^{N}\frac{\Gamma(x+\beta+1)\Gamma(\alpha+N+1-x)}{x!\,(N-x)!}y_m(x)y_n(x) \\ =\frac{\Gamma(n+\alpha+1)\Gamma(n+\beta+1)\Gamma(n+\alpha+\beta+1)\Gamma(n+\alpha+\beta+N+2)n!}{\Gamma(2n+\alpha+\beta+1)\Gamma(2n+\alpha+\beta+2)(N-n)!}\delta_{mn}$$

for $m,n=0,1,2,\ldots,N$.

Case IIIa2. We have $\varepsilon>0$, $|\delta-\eta|<\varepsilon$ and $N-1<\delta+\eta-\varepsilon<N$. In this case we were not able to find a solution for the Pearson difference equation such that the boundary conditions (5.1.12) can be fulfilled. However we might proceed as on

page 109. By using (5.3.4) we obtain

$$w(x) = \Gamma(x-1+f+\eta)\Gamma(x-1+f-\eta)\Gamma(1+\varepsilon-\delta-f-x)\Gamma(1+\varepsilon+\delta-f-x)$$

as a possible solution for the Pearson difference equation. Here we also set $f-\varepsilon+\delta=1$ for simplicity.

Then the Rodrigues formula (3.4.28) reads

$$y_n(x) = \frac{\Delta^n\left(\Gamma(x+\varepsilon-\delta+\eta)\Gamma(x+\varepsilon-\delta-\eta)\Gamma(n-x)\Gamma(n+2\delta-x)\right)}{(n-1+2\varepsilon)_n\Gamma(x+\varepsilon-\delta+\eta)\Gamma(x+\varepsilon-\delta-\eta)\Gamma(-x)\Gamma(2\delta-x)}$$

for $n=0,1,2,\ldots,N$.

Then we may use the Mellin-Barnes integral (1.6.3) again to find that

$$\begin{aligned}\frac{1}{2\pi i}\int_{-i\infty}^{i\infty} w(x)dx &= \frac{1}{2\pi i}\int_{-i\infty}^{i\infty}\Gamma(x+\varepsilon-\delta+\eta)\Gamma(x+\varepsilon-\delta-\eta)\Gamma(-x)\Gamma(2\delta-x)\,dx\\ &= \frac{\Gamma(\varepsilon+\delta+\eta)\Gamma(\varepsilon+\delta-\eta)\Gamma(\varepsilon-\delta+\eta)\Gamma(\varepsilon-\delta-\eta)}{\Gamma(2\varepsilon)}.\end{aligned}$$

Note that the increasing poles are $\{0,1,2,\ldots\}$ and $\{2\delta,2\delta+1,2\delta+2,\ldots\}$ and the decreasing poles are $\{\delta\pm\eta-\varepsilon,\delta\pm\eta-\varepsilon-1,\delta\pm\eta-\varepsilon-2,\ldots\}$. Hence for $|\delta-\eta|<\varepsilon$ and $N-1<\delta+\eta-\varepsilon<N$ these increasing and decreasing poles stay separated. Since $-N<\varepsilon-\delta-\eta<-N+1$, we have

$$(-1)^N\Gamma(\varepsilon-\delta-\eta) = \frac{\Gamma(\varepsilon-\delta-\eta+N)}{(-1)^N(\varepsilon-\delta-\eta)_N} > 0.$$

This implies that

$$\begin{aligned}d_0 &:= \frac{(-1)^N}{2\pi i}\int_{-i\infty}^{i\infty}\Gamma(x+\varepsilon-\delta+\eta)\Gamma(x+\varepsilon-\delta-\eta)\Gamma(-x)\Gamma(2\delta-x)\,dx\\ &= \frac{(-1)^N\Gamma(\varepsilon+\delta+\eta)\Gamma(\varepsilon+\delta-\eta)\Gamma(\varepsilon-\delta+\eta)\Gamma(\varepsilon-\delta-\eta)}{\Gamma(2\varepsilon)} > 0.\end{aligned}$$

If (5.1.11) and (3.1.4) are used, this leads to the orthogonality relation (cf. page 109)

$$\begin{aligned}&\frac{(-1)^N}{2\pi i}\int_{-i\infty}^{i\infty}\Gamma(x+\varepsilon-\delta+\eta)\Gamma(x+\varepsilon-\delta-\eta)\Gamma(-x)\Gamma(2\delta-x)y_m(x)y_n(x)\,dx\\ &= \frac{n!\,(-1)^{N+n}\Gamma(n+2\varepsilon-1)}{\Gamma(2n+2\varepsilon-1)\Gamma(2n+2\varepsilon)}\\ &\qquad\times\Gamma(n+\varepsilon+\delta+\eta)\Gamma(n+\varepsilon+\delta-\eta)\Gamma(n+\varepsilon-\delta+\eta)\Gamma(n+\varepsilon-\delta-\eta)\,\delta_{mn}\end{aligned}$$

for $m,n=0,1,2,\ldots,N$.

Now we consider positive-definite orthogonality for the Hahn polynomials in the case that $\varepsilon<0$.

Case IIIb. We have $2\varepsilon = -2N - t$ with $N \in \{1,2,3,\ldots\}$ and $-1 < t \leq 1$ and

$$\left(\frac{|t|}{2} \leq |\delta - \eta| \leq\right) \delta + \eta < 1 + \frac{t}{2} \quad \text{or} \quad N + \frac{t}{2} < |\delta - \eta|.$$

Now we have to distinguish between three different cases.

Case IIIb1. First we consider the case that

$$\left(\frac{|t|}{2} \leq |\delta - \eta| \leq\right) \delta + \eta < 1 + \frac{t}{2} \quad \text{and} \quad \varepsilon \pm \delta \pm \eta = -N.$$

In this case we use (5.3.5) to obtain a solution of the form

$$w(x) = \frac{1}{\Gamma(x+f-\varepsilon+\delta)\Gamma(x+f-\varepsilon-\delta)\Gamma(2-\eta-f-x)\Gamma(2+\eta-f-x)}$$

for the Pearson difference equation. In order to get $A = 0$ in the boundary conditions (5.1.12), we set

$$\begin{cases} f - \varepsilon + \delta = 1 \quad \text{and} \quad 1 - \eta - f = N \quad \Longrightarrow \quad \delta - \eta - \varepsilon = N \\ \text{and we define} \quad \varepsilon + \delta + \eta = \alpha + 1 \quad \text{and} \quad \varepsilon - \delta - \eta = \beta + 1 \end{cases}$$

or

$$\begin{cases} f - \varepsilon + \delta = 1 \quad \text{and} \quad 1 + \eta - f = N \quad \Longrightarrow \quad \delta + \eta - \varepsilon = N \\ \text{and we define} \quad \varepsilon + \delta - \eta = \alpha + 1 \quad \text{and} \quad \varepsilon - \delta + \eta = \beta + 1 \end{cases}$$

or

$$\begin{cases} f - \varepsilon - \delta = 1 \quad \text{and} \quad 1 - \eta - f = N \quad \Longrightarrow \quad -\delta - \eta - \varepsilon = N \\ \text{and we define} \quad \varepsilon - \delta + \eta = \alpha + 1 \quad \text{and} \quad \varepsilon + \delta - \eta = \beta + 1 \end{cases}$$

or

$$\begin{cases} f - \varepsilon - \delta = 1 \quad \text{and} \quad 1 + \eta - f = N \quad \Longrightarrow \quad -\delta + \eta - \varepsilon = N \\ \text{and we define} \quad \varepsilon - \delta - \eta = \alpha + 1 \quad \text{and} \quad \varepsilon + \delta + \eta = \beta + 1. \end{cases}$$

In all cases this implies that $2\varepsilon = \alpha + \beta + 2$. Further we have $2\delta = \alpha + N + 1$ (first and second case) or $-2\delta = \alpha + N + 1$ (third and fourth case) and $2\eta = \beta + N + 1$ (second and fourth case) or $-2\eta = \beta + N + 1$ (first and third case). In all cases we find the same difference equation and hypergeometric representation as in Case IIIa. The weight function now reads

$$w(x) = \frac{1}{\Gamma(x+1)\Gamma(N+1-x)\Gamma(x-N-\alpha)\Gamma(-\beta-x)}, \quad x = 0,1,2,\ldots,N.$$

For positivity we must have $\alpha < -N$ and $\beta < -N$. The Rodrigues formula (3.4.28) can be written as

$$y_n(x) = \frac{\Gamma(x+1)\Gamma(N+1-x)\Gamma(x-N-\alpha)\Gamma(-\beta-x)}{(n+\alpha+\beta+1)_n} \times \Delta^n \left(\frac{1}{\Gamma(N+1-x)\Gamma(x-n+1)\Gamma(x-n-N-\alpha)\Gamma(-\beta-x)}\right)$$

for $n = 0,1,2,\ldots$. By using Dougall's bilateral sum (1.5.9), we find that

$$\begin{aligned}\sum_{x=-\infty}^{\infty} w(x) &= \sum_{x=-\infty}^{\infty} \frac{1}{\Gamma(x+f-\varepsilon+\delta)\Gamma(x+f-\varepsilon-\delta)\Gamma(2-\eta-f-x)\Gamma(2+\eta-f-x)} \\ &= \frac{\Gamma(1-2\varepsilon)}{\Gamma(1-\varepsilon+\delta-\eta)\Gamma(1-\varepsilon-\delta-\eta)\Gamma(1-\varepsilon+\delta+\eta)\Gamma(1-\varepsilon-\delta+\eta)},\end{aligned}$$

provided that $2f-2+1 < 2f-2\varepsilon$ or equivalently $2\varepsilon < 1$. For $f-\varepsilon\pm\delta = 1$ and $1\pm\eta-f = N$ this sum reduces to a finite sum over $\{0,1,2,\ldots,N\}$, id est

$$\begin{aligned} d_0 &:= \sum_{x=0}^{N} \frac{1}{x!\,(N-x)!\,\Gamma(x-N-\alpha)\Gamma(-\beta-x)} \\ &= \frac{\Gamma(-\alpha-\beta-1)}{N!\,\Gamma(-N-\alpha-\beta-1)\Gamma(-\alpha)\Gamma(-\beta)} > 0.\end{aligned}$$

Hence with (5.1.11) and (3.1.4), the orthogonality relation can be written as

$$\begin{aligned}&\sum_{x=0}^{N} \frac{1}{x!\,(N-x)!\,\Gamma(x-N-\alpha)\Gamma(-\beta-x)} y_m(x) y_n(x) \\ &= \frac{\Gamma(-2n-\alpha-\beta)\Gamma(-2n-\alpha-\beta-1)n!}{\Gamma(-n-\alpha-\beta)\Gamma(-n-\alpha)\Gamma(-n-\beta)\Gamma(-n-N-\alpha-\beta-1)(N-n)!}\delta_{mn}\end{aligned}$$

for $m,n = 0,1,2,\ldots,N$.

Case IIIb2. Now we consider the case that

$$\left(\frac{|t|}{2} \leq |\delta-\eta| \leq\right) \delta+\eta < 1+\frac{t}{2} \quad \text{and} \quad \varepsilon\pm\delta\pm\eta \neq -N.$$

In this case we use (5.3.5) to obtain a solution of the form

$$w(x) = \frac{1}{\Gamma(x+f-\varepsilon+\delta)\Gamma(x+f-\varepsilon-\delta)\Gamma(2-\eta-f-x)\Gamma(2+\eta-f-x)}$$

for the Pearson difference equation. For simplicity we now take $f-\varepsilon-\delta=1$. Then we have

$$w(x)=\frac{1}{\Gamma(x+1)\Gamma(x+2\delta+1)\Gamma(1-\varepsilon-\delta-\eta-x)\Gamma(1-\varepsilon-\delta+\eta-x)}$$

and the Rodrigues formula (3.4.28) reads

$$\begin{aligned} y_n(x)=&\frac{\Gamma(x+1)\Gamma(x+2\delta+1)\Gamma(1-\varepsilon-\delta-\eta-x)\Gamma(1-\varepsilon-\delta+\eta-x)}{(n-1+2\varepsilon)_n}\\ &\times\Delta^n\Bigg(\frac{1}{\Gamma(x-n+1)\Gamma(x-n+2\delta+1)}\\ &\qquad\times\frac{1}{\Gamma(1-\varepsilon-\delta-\eta-x)\Gamma(1-\varepsilon-\delta+\eta-x)}\Bigg)\end{aligned}$$

for $n=0,1,2,\ldots,N$. By using Dougall's bilateral sum (1.5.8), we obtain

$$\begin{aligned} d_0:=&\sum_{x=-\infty}^{\infty}w(x)\\ =&\sum_{x=0}^{\infty}\frac{1}{\Gamma(x+1)\Gamma(x+2\delta+1)\Gamma(1-\varepsilon-\delta-\eta-x)\Gamma(1-\varepsilon-\delta+\eta-x)}\\ =&\frac{\Gamma(1-2\varepsilon)}{\Gamma(1-\varepsilon+\delta+\eta)\Gamma(1-\varepsilon+\delta-\eta)\Gamma(1-\varepsilon-\delta+\eta)\Gamma(1-\varepsilon-\delta-\eta)}>0,\end{aligned}$$

and if (5.1.11) and (3.1.4) are used, this leads to the orthogonality relation

$$\begin{aligned} &\sum_{x=0}^{\infty}\frac{1}{x!\,\Gamma(x+2\delta+1)\Gamma(1-\varepsilon-\delta-\eta-x)\Gamma(1-\varepsilon-\delta+\eta-x)}y_m(x)y_n(x)\\ &=\frac{\Gamma(1-2\varepsilon-2n)\Gamma(2-2\varepsilon-2n)n!}{\Gamma(2-2\varepsilon-n)\Gamma(1-\varepsilon+\delta+\eta-n)\Gamma(1-\varepsilon-\delta-\eta-n)}\\ &\quad\times\frac{1}{\Gamma(1-\varepsilon+\delta-\eta-n)\Gamma(1-\varepsilon-\delta+\eta-n)}\delta_{mn},\quad m,n=0,1,2,\ldots,N.\end{aligned}$$

Case IIIb3. Further we consider the case that

$$|\delta-\eta|>N+\frac{t}{2}.$$

For $\delta>\eta>0$ this implies that $\delta-\eta>N+t/2$. In that case we use (5.3.2) to obtain

$$w(x)=\frac{\Gamma(x-1+f+\eta)\Gamma(x-1+f-\eta)}{\Gamma(x+f-\varepsilon+\delta)\Gamma(x+f-\varepsilon-\delta)}$$

as a possible solution for the Pearson difference equation. Again we take $f-\varepsilon-\delta=1$ for simplicity. Then the Rodrigues formula (3.4.28) reads

$$y_n(x) = \frac{\Gamma(x+1)\Gamma(x+2\delta+1)}{(n-1+2\varepsilon)_n\Gamma(x+\varepsilon+\delta+\eta)\Gamma(x+\varepsilon+\delta-\eta)}$$
$$\times \Delta^n\left(\frac{\Gamma(x+\varepsilon+\delta+\eta)\Gamma(x+\varepsilon+\delta-\eta)}{\Gamma(x-n+1)\Gamma(x-n+2\delta+1)}\right), \quad n=0,1,2,\ldots,N.$$

By using Gauss's summation formula (1.5.3), we obtain

$$d_0 := \sum_{x=-\infty}^{\infty} w(x) = \sum_{x=0}^{\infty} \frac{\Gamma(x+\varepsilon+\delta+\eta)\Gamma(x+\varepsilon+\delta-\eta)}{\Gamma(x+1)\Gamma(x+2\delta+1)}$$
$$= \frac{\Gamma(1-2\varepsilon)\Gamma(\varepsilon+\delta+\eta)\Gamma(\varepsilon+\delta-\eta)}{\Gamma(1-\varepsilon+\delta+\eta)\Gamma(1-\varepsilon+\delta-\eta)} > 0,$$

which, if (5.1.11) and (3.1.4) are used, leads to the orthogonality relation

$$\sum_{x=0}^{\infty} \frac{\Gamma(x+\varepsilon+\delta+\eta)\Gamma(x+\varepsilon+\delta-\eta)}{x!\,\Gamma(x+2\delta+1)} y_m(x)y_n(x)$$
$$= \frac{\Gamma(1-2\varepsilon-2n)\Gamma(2-2\varepsilon-2n)\Gamma(n+\varepsilon+\delta+\eta)\Gamma(n+\varepsilon+\delta-\eta)n!}{\Gamma(2-2\varepsilon-n)\Gamma(1-\varepsilon+\delta+\eta-n)\Gamma(1-\varepsilon+\delta-\eta-n)}\delta_{mn}$$

for $m,n=0,1,2,\ldots,N$.

For $\eta > \delta > 0$ this implies that $\eta-\delta > N+t/2$. In that case we use (5.3.6) to obtain

$$w(x) = \frac{\Gamma(1+\varepsilon-\delta-f-x)\Gamma(1+\varepsilon+\delta-f-x)}{\Gamma(2+\eta-f-x)\Gamma(2-\eta-f-x)}$$

as a possible solution for the Pearson difference equation. For simplicity we now take $f+\eta=1$. Then the Rodrigues formula (3.4.28) reads

$$y_n(x) = \frac{\Gamma(1-x)\Gamma(2\eta+1-x)}{(n-1+2\varepsilon)_n\Gamma(\varepsilon+\delta+\eta-x)\Gamma(\varepsilon-\delta+\eta-x)}$$
$$\times \Delta^n\left(\frac{\Gamma(n+\varepsilon+\delta+\eta-x)\Gamma(n+\varepsilon-\delta+\eta-x)}{\Gamma(1-x)\Gamma(2\eta+1-x)}\right)$$

for $n=0,1,2,\ldots,N$. By using Gauss's summation formula (1.5.3), we obtain

$$d_0 := \sum_{x=-\infty}^{\infty} w(x) = \sum_{x=-\infty}^{0} w(x) = \sum_{x=0}^{\infty} \frac{\Gamma(x+\varepsilon+\delta+\eta)\Gamma(x+\varepsilon-\delta+\eta)}{\Gamma(x+1)\Gamma(x+2\eta+1)}$$
$$= \frac{\Gamma(1-2\varepsilon)\Gamma(\varepsilon+\delta+\eta)\Gamma(\varepsilon-\delta+\eta)}{\Gamma(1-\varepsilon+\delta+\eta)\Gamma(1-\varepsilon-\delta+\eta)} > 0,$$

which, if (5.1.11) and (3.1.4) are used, leads to the orthogonality relation

$$\sum_{x=0}^{\infty} \frac{\Gamma(x+\varepsilon+\delta+\eta)\Gamma(x+\varepsilon-\delta+\eta)}{x!\,\Gamma(x+2\eta+1)} y_m(-x)y_n(-x)$$
$$= \frac{\Gamma(1-2\varepsilon-2n)\Gamma(2-2\varepsilon-2n)\Gamma(n+\varepsilon+\delta+\eta)\Gamma(n+\varepsilon-\delta+\eta)n!}{\Gamma(2-2\varepsilon-n)\Gamma(1-\varepsilon+\delta+\eta-n)\Gamma(1-\varepsilon-\delta+\eta-n)} \delta_{mn}$$

for $m,n = 0,1,2,\ldots,N$.

Case IIIc. We have $2\varepsilon = -2N-t$ with $N \in \{1,2,3,\ldots\}$ and $-1 < t \leq 1$ and δ pure imaginary (including zero) and/or η pure imaginary (including zero) and

$$\varepsilon \pm \delta \pm \eta \neq 0,-1,-2,\ldots,-N+1.$$

In this case we use (5.3.5) to obtain a solution of the form

$$w(x) = \frac{1}{\Gamma(x+f-\varepsilon+\delta)\Gamma(x+f-\varepsilon-\delta)\Gamma(2-\eta-f-x)\Gamma(2+\eta-f-x)}$$

for the Pearson difference equation.

In this case we may write $\delta = \delta_1 + i\delta_2$ and $\eta = \eta_1 + i\eta_2$ with $\delta_1\delta_2 = 0 = \eta_1\eta_2$ and $\delta_1, \delta_2, \eta_1, \eta_2 \in [0,\infty)$.

Then the Rodrigues formula (3.4.28) reads

$$y_n(x) = \frac{\Gamma(x+f-\varepsilon+\delta)\Gamma(x+f-\varepsilon-\delta)\Gamma(2-\eta-f-x)\Gamma(2+\eta-f-x)}{(n-1+2\varepsilon)_n}$$
$$\times \Delta^n\left(\frac{1}{\Gamma(x-n+f-\varepsilon+\delta)\Gamma(x-n+f-\varepsilon-\delta)} \times \frac{1}{\Gamma(2+\eta-f-x)\Gamma(2-\eta-f-x)}\right)$$

for $n = 0,1,2,\ldots,N$. By using Dougall's bilateral sum (1.5.9), we obtain

$$d_0 := \sum_{x=-\infty}^{\infty} w(x)$$
$$= \frac{\Gamma(1-2\varepsilon)}{\Gamma(1-\varepsilon+\delta+\eta)\Gamma(1-\varepsilon+\delta-\eta)\Gamma(1-\varepsilon-\delta+\eta)\Gamma(1-\varepsilon-\delta-\eta)} > 0,$$

which, if (5.1.11) with $A \to -\infty$ and $N \to \infty$ and (3.1.4) are used, leads to the orthogonality relation

$$\sum_{x=-\infty}^{\infty} \frac{1}{\Gamma(x+f-\varepsilon+\delta)\Gamma(x+f-\varepsilon-\delta)} \times \frac{1}{\Gamma(2-\eta-f-x)\Gamma(2+\eta-f-x)} y_m(x)y_n(x)$$
$$= \frac{\Gamma(1-2\varepsilon-2n)\Gamma(2-2\varepsilon-2n)n!}{\Gamma(2-2\varepsilon-n)\Gamma(1-\varepsilon+\delta+\eta-n)\Gamma(1-\varepsilon-\delta-\eta-n)} \times \frac{1}{\Gamma(1-\varepsilon+\delta-\eta-n)\Gamma(1-\varepsilon-\delta+\eta-n)} \delta_{mn}, \quad m,n = 0,1,2,\ldots,N.$$

In this chapter we have proved:

Theorem 5.2. *The positive-definite orthogonal polynomial solutions $y_n(x)$ of the difference equation (5.1.1)*

$$(ex^2+2fx+g)\left(\Delta^2 y_n\right)(x)+(2\varepsilon x+\gamma)\left(\Delta y_n\right)(x) = n(e(n-1)+2\varepsilon)y_n(x+1)$$

for $n = 0,1,2,\ldots$ with $e,f,g,\varepsilon,\gamma \in \mathbb{R}$ consist of four infinite systems

Case I. *Charlier polynomials with orthogonality on $\{0,1,2,\ldots\}$ with respect to*
$w(x) = 1/(-2\varepsilon)^x\Gamma(x+1)$ *;* $e = f = 0$*,* $g = 1$*,* $2\varepsilon < 0$*,* $\gamma = 2\varepsilon+1$

Case IIa1. *Meixner polynomials with orthogonality on $\{0,1,2,\ldots\}$ with respect to*
$w(x) = \Gamma(x+\gamma-2\varepsilon)/(1-2\varepsilon)^x\Gamma(x+1)$ *;* $e = 0$*,* $2f = 1$*,* $2\varepsilon < 0$*,* $g = \gamma-2\varepsilon+1$ *and* $\gamma > 2\varepsilon$

Case IIa2. *Meixner polynomials with orthogonality on $\{\ldots,-2,-1,0\}$ with respect to*
$w(x) = \Gamma(r-x)/(1-2\varepsilon)^x\Gamma(1-x)$ *with* $r = (\gamma-2\varepsilon)/(1-2\varepsilon)$ *;* $e = 0$*,* $2f = 1$*,* $0 < 2\varepsilon < 1$*,* $g = 1$ *and* $\gamma > 2\varepsilon$

Case IIa3. *Charlier polynomials with orthogonality on $\{\ldots,-2,-1,0\}$ with respect to*
$w(x) = 1/(\gamma-1)^x\Gamma(1-x)$ *;* $e = 0$*,* $2f = 1$*,* $2\varepsilon = 1$*,* $g = 1$ *and* $\gamma > 1$

and eight finite systems of $N+1$ polynomials

Case IIb1. *Krawtchouk polynomials with orthogonality on $\{0,1,2,\ldots,N\}$ with respect to* $w(x) = 1/(2\varepsilon-1)^x\Gamma(x+1)\Gamma(N+1-x)$ *;* $e = 0$*,* $2f = 1$*,* $2\varepsilon > 1$*,* $g = \gamma-2\varepsilon+1$ *and* $\gamma-2\varepsilon = -N$

Case IIb2. *Krawtchouk polynomials with orthogonality on $(-i\infty,i\infty)$ with respect to* $w(x) = (-1)^N\Gamma(x+\gamma-2\varepsilon)\Gamma(-x)/(2\varepsilon-1)^x$ *;* $e = 0$*,* $2f = 1$*,* $2\varepsilon > 1$*,* $g = \gamma-2\varepsilon+1$ *and* $-N < \gamma-2\varepsilon < -N+1$

Case IIIa1. *Hahn polynomials with orthogonality on* $\{0,1,2,\ldots,N\}$ *with respect to* $w(x)=\Gamma(x+\beta+1)\Gamma(\alpha+N+1-x)/\Gamma(x+1)\Gamma(N+1-x)$; $e=1$, $\varepsilon>0$, $\alpha>-1$, $\beta>-1$, $\varepsilon-\delta-\eta=-N$ ($\delta=\sqrt{(f-\varepsilon)^2-g+\gamma}$, $\eta=\sqrt{f^2-g}$ *and* $2\varepsilon=\alpha+\beta+2$)

Case IIIa2. *Hahn polynomials with orthogonality on* $(-i\infty,i\infty)$ *with respect to*
$w(x)=(-1)^N\Gamma(x+\varepsilon-\delta+\eta)\Gamma(x+\varepsilon-\delta-\eta)\Gamma(-x)\Gamma(-\beta-x)$; $e=1$, $\varepsilon>0$, $|\delta-\eta|<\varepsilon$ *and* $-N<\varepsilon-\delta-\eta<-N+1$

Case IIIb1. *Hahn polynomials with orthogonality on* $\{0,1,2,\ldots,N\}$ *with respect to* $w(x)=1/\Gamma(x+1)\Gamma(N+1-x)\Gamma(x-N-\alpha)\Gamma(-\beta-x)$; $e=1$, $2\varepsilon=-2N-t$ *with* $N\in\{1,2,3,\ldots\}$ *and* $-1<t\leq 1$, $\alpha<-N$, $\beta<-N$, $\delta+\eta<1+t/2$ *and* $\varepsilon\pm\delta\pm\eta=-N$

Case IIIb2. *Hahn polynomials with orthogonality on* $\{0,1,2,\ldots\}$ *with respect to*
$w(x)=1/\Gamma(x+1)\Gamma(x+2\delta+1)\Gamma(1-\varepsilon-\delta-\eta-x)\Gamma(1-\varepsilon-\delta+\eta-x)$; $e=1$, $2\varepsilon=-2N-t$ *with* $N\in\{1,2,3,\ldots\}$ *and* $-1<t\leq 1$, $\delta+\eta<1+t/2$ *and* $\varepsilon\pm\delta\pm\eta\neq -N$

Case IIIb3. *Hahn polynomials with orthogonality on* $\{\ldots,-2,-1,0\}$ *with respect to* $w(x)=\Gamma(x+\varepsilon+\delta+\eta)\Gamma(x+\varepsilon+|\delta-\eta|)/\Gamma(x+1)\Gamma(x+2\max(\delta,\eta)+1)$; $e=1$, $2\varepsilon=-2N-t$ *with* $N\in\{1,2,3,\ldots\}$ *and* $-1<t\leq 1$ *and* $|\delta-\eta|>N+t/2$

Case IIIc. *Hahn polynomials with orthogonality on* $\{\ldots,-1,0,1,\ldots\}$ *with respect to* $w(x)=1/\Gamma(x+f-\varepsilon+\delta)\Gamma(x+f-\varepsilon-\delta)\Gamma(2-\eta-f-x)\Gamma(2+\eta-f-x)$; $e=1$, $2\varepsilon=-2N-t$ *with* $N\in\{1,2,3,\ldots\}$ *and* $-1<t\leq 1$, $\varepsilon\pm\delta\pm\eta\neq 0,-1,-2,\ldots,-N+1$, $\delta=\delta_1+i\delta_2$, $\eta=\eta_1+i\eta_2$, $\delta_1\delta_2=0=\eta_1\eta_2$ *and* $\delta_1,\delta_2,\eta_1,\eta_2\in[0,\infty)$.

Chapter 6
Orthogonal Polynomial Solutions of Complex Difference Equations

Discrete Classical Orthogonal Polynomials II

6.1 Real Polynomial Solutions of Complex Difference Equations

We consider the difference equation (5.1.1) with complex coefficients, id est

$$(ez^2+2fz+g)\left(\Delta^2 y_n\right)(z)+(2\varepsilon z+\gamma)\left(\Delta y_n\right)(z)=n(e(n-1)+2\varepsilon)y_n(z+1) \quad (6.1.1)$$

for $n=0,1,2,\ldots$. This difference equation can also be written in the form

$$\begin{aligned}&\left(e(z-1)^2+2f(z-1)+g\right)\left(\Delta\left(\nabla y_n\right)\right)(z)+(2\varepsilon(z-1)+\gamma)\left(\nabla y_n\right)(z)\\&=n(e(n-1)+2\varepsilon)y_n(z)\end{aligned}$$

or in the form

$$C(z)y_n(z+1)-\{C(z)+D(z)\}\,y_n(z)+D(z)y_n(z-1)=n(e(n-1)+2\varepsilon)y_n(z)$$

for $n=0,1,2,\ldots$, where

$$C(z)=e(z-1)^2+2f(z-1)+g \quad \text{and} \quad D(z)=C(z)-2\varepsilon(z-1)-\gamma. \qquad (6.1.2)$$

We look for monic polynomial solutions of the form (cf. (2.4.14))

$$y_n(z)=\sum_{k=0}^{n}a_{n,k}\binom{z+c}{k}, \quad c\in\mathbb{C}, \quad a_{n,n}=n!, \quad n=0,1,2,\ldots, \qquad (6.1.3)$$

where the coefficients satisfy the two-term recurrence relation (cf. (2.4.15))

$$\begin{aligned}&(n-k)\left(e(n+k-1)+2\varepsilon\right)a_{n,k}\\&\quad-\left(e(k-1-c)^2+2f(k-1-c)+g\right)a_{n,k+1}=0 \qquad (6.1.4)\end{aligned}$$

R. Koekoek et al., *Hypergeometric Orthogonal Polynomials and Their q-Analogues*, Springer Monographs in Mathematics,
DOI 10.1007/978-3-642-05014-5_6, © Springer-Verlag Berlin Heidelberg 2010

for $k=n-1,n-2,n-3,\ldots,0$ provided that $c\in\mathbb{C}$ satisfies (cf. (5.1.5))

$$e(c+1)^2-2f(c+1)+g=-2\varepsilon(c+1)+\gamma.$$

In section 2.6 we found that the monic polynomial solutions $\{y_n\}_{n=0}^{\infty}$ satisfy the three-term recurrence relation

$$y_{n+1}(z)=(z-c_n)y_n(z)-d_ny_{n-1}(z),\quad n=1,2,3,\ldots, \tag{6.1.5}$$

with initial values $y_0(z)=1$ and $y_1(z)=z-c_0$, where (cf. (2.6.15))

$$c_n=\frac{n(e(n-1)+2\varepsilon)(2(e-f)+\varepsilon)+(e-\varepsilon)(\gamma-2\varepsilon)}{2(e(n-1)+\varepsilon)(en+\varepsilon)},\quad n=0,1,2,\ldots$$

and (cf. (2.6.16))

$$\begin{aligned}d_n=&-\frac{n(e(n-2)+2\varepsilon)}{4(e(2n-3)+2\varepsilon)(e(n-1)+\varepsilon)^2(e(2n-1)+2\varepsilon)}\\&\times\Big\{e(n-1)^2(e(n-1)+2\varepsilon)^2\\&\quad+2(n-1)(e(n-1)+2\varepsilon)(2eg+2f(\varepsilon-f)-e\gamma)\\&\quad+4\varepsilon(g\varepsilon-f\gamma)+e\gamma^2\Big\},\quad n=1,2,3,\ldots.\end{aligned}$$

In order to obtain real polynomial solutions, we set $z=a+ix$ with $a\in\mathbb{R}$ and $x\in\mathbb{R}$. Then we define

$$y_n(z)=y_n(a+ix)=i^n\widehat{y}_n(x).$$

Substitution into (6.1.5) leads to

$$\widehat{y}_{n+1}(x)=(x+i(c_n-a))\widehat{y}_n(x)+d_n\widehat{y}_{n-1}(x),\quad n=1,2,3,\ldots \tag{6.1.6}$$

with $\widehat{y}_0(x)=1$ and $\widehat{y}_1(x)=x+i(c_0-a)$. Since the coefficients have to be real, we conclude that c_n-a must be pure imaginary (including zero) for all $n=0,1,2,\ldots$ and d_n must be real for all $n=1,2,3,\ldots$. Again we consider three different cases as in section 5.2.

Case I. Degree$[\varphi]=0$: $e=f=0$ and we may choose $g=1$. Then we have

$$c_n=\frac{2(n+1)\varepsilon-\gamma}{2\varepsilon},\quad n=0,1,2,\ldots\quad\text{and}\quad d_n=-\frac{n}{2\varepsilon},\quad n=1,2,3,\ldots.$$

Since d_n has to be real, we must have $\varepsilon\in\mathbb{R}$. However, it is impossible to choose $\gamma=\gamma_1+i\gamma_2$ with $\gamma_1,\gamma_2\in\mathbb{R}$ such that c_n-a is pure imaginary for all $n=0,1,2,\ldots$.

Case II. Degree$[\varphi]=1$: $e=0$ and we may choose $2f=1$. Then we have

$$c_n = \frac{2n(\varepsilon-1)+2\varepsilon-\gamma}{2\varepsilon}, \quad n=0,1,2,\dots$$

and

$$d_n = -\frac{n((n-1)(2\varepsilon-1)+2g\varepsilon-\gamma)}{4\varepsilon^2}, \quad n=1,2,3,\dots.$$

If we set $g=g_1+ig_2$, $\varepsilon=\varepsilon_1+i\varepsilon_2$ and $\gamma=\gamma_1+i\gamma_2$ with $g_1,g_2,\varepsilon_1,\varepsilon_2,\gamma_1,\gamma_2\in\mathbb{R}$, we find that

$$\begin{aligned} c_n-a &= \frac{2n(\varepsilon_1-1+i\varepsilon_2)+2(1-a)(\varepsilon_1+i\varepsilon_2)-(\gamma_1+i\gamma_2)}{2(\varepsilon_1+i\varepsilon_2)} \\ &= \frac{2n(\varepsilon_1^2+\varepsilon_2^2)-2n(\varepsilon_1-i\varepsilon_2)+2(1-a)(\varepsilon_1^2+\varepsilon_2^2)-(\gamma_1+i\gamma_2)(\varepsilon_1-i\varepsilon_2)}{2(\varepsilon_1^2+\varepsilon_2^2)}. \end{aligned}$$

Since c_n-a must be pure imaginary for all $n=0,1,2,\dots$, we conclude that

$$\varepsilon_1^2+\varepsilon_2^2=\varepsilon_1 \quad \text{and} \quad 2(1-a)\varepsilon_1=\gamma_1\varepsilon_1+\gamma_2\varepsilon_2. \tag{6.1.7}$$

Further we have

$$\begin{aligned} d_n &= -\frac{n((n-1)(2\varepsilon_1-1+2i\varepsilon_2)+2(g_1+ig_2)(\varepsilon_1+i\varepsilon_2)-(\gamma_1+i\gamma_2))}{4(\varepsilon_1+i\varepsilon_2)^2} \\ &= -\frac{n\left(\varepsilon_1^2-\varepsilon_2^2-2i\varepsilon_1\varepsilon_2\right)}{4(\varepsilon_1^2+\varepsilon_2^2)^2} \\ &\qquad \times((n-1)(2\varepsilon_1-1+2i\varepsilon_2)+2(g_1+ig_2)(\varepsilon_1+i\varepsilon_2)-(\gamma_1+i\gamma_2)). \end{aligned}$$

Since d_n has to be real for all $n=1,2,3,\dots$, we conclude that the imaginary part must be zero. Hence by using (6.1.7), we find that

$$\begin{aligned} 0 &= 2(n-1)\varepsilon_2(\varepsilon_1^2-\varepsilon_2^2)-2(n-1)\varepsilon_1\varepsilon_2(2\varepsilon_1-1)+2(\varepsilon_1^2-\varepsilon_2^2)(g_1\varepsilon_2+g_2\varepsilon_1) \\ &\qquad -4\varepsilon_1\varepsilon_2(g_1\varepsilon_1-g_2\varepsilon_2)-\gamma_2(\varepsilon_1^2-\varepsilon_2^2)+2\varepsilon_1\varepsilon_2\gamma_1 \\ &= 2(n-1)\varepsilon_2\left(\varepsilon_1-\varepsilon_1^2-\varepsilon_2^2\right)-2g_1\varepsilon_2(\varepsilon_1^2+\varepsilon_2^2)+2g_2\varepsilon_1(\varepsilon_1^2+\varepsilon_2^2) \\ &\qquad -\gamma_2(\varepsilon_1^2+\varepsilon_2^2)+2\gamma_2\varepsilon_2^2+2\varepsilon_1\varepsilon_2\gamma_1 \\ &= (\varepsilon_1^2+\varepsilon_2^2)(2g_2\varepsilon_1-2g_1\varepsilon_2-\gamma_2)+2\varepsilon_2(\varepsilon_1\gamma_1+\varepsilon_2\gamma_2) \\ &= \varepsilon_1\{2(g_2\varepsilon_1-g_1\varepsilon_2)-\gamma_2+4(1-a)\varepsilon_2\}. \end{aligned}$$

Since $\varepsilon\neq 0$ in view of the regularity condition (2.3.3), we have $\varepsilon_1=\varepsilon_1^2+\varepsilon_2^2>0$ and therefore

$$\gamma_2=4(1-a)\varepsilon_2+2(g_2\varepsilon_1-g_1\varepsilon_2). \tag{6.1.8}$$

This implies that

$$i(c_n-a)=-\frac{2n\varepsilon_2+\gamma_1\varepsilon_2-\gamma_2\varepsilon_1}{2\varepsilon_1}, \quad n=0,1,2,\dots \tag{6.1.9}$$

and by using (6.1.7) and (6.1.8), we obtain

$$\begin{aligned}
4\varepsilon_1^2 d_n &= 4(\varepsilon_1^2+\varepsilon_2^2)^2 d_n \\
&= -n\left\{(n-1)(2\varepsilon_1-1)(\varepsilon_1^2-\varepsilon_2^2)+4(n-1)\varepsilon_1\varepsilon_2^2+2(g_1\varepsilon_1-g_2\varepsilon_2)(\varepsilon_1^2-\varepsilon_2^2)\right. \\
&\qquad \left.+4\varepsilon_1\varepsilon_2(g_1\varepsilon_2+g_2\varepsilon_1)-\gamma_1(\varepsilon_1^2-\varepsilon_2^2)-2\gamma_2\varepsilon_1\varepsilon_2\right\} \\
&= -n\left\{2(n-1)\varepsilon_1(\varepsilon_1^2+\varepsilon_2^2)-(n-1)(\varepsilon_1^2-\varepsilon_2^2)+2g_1\varepsilon_1(\varepsilon_1^2+\varepsilon_2^2)\right. \\
&\qquad \left.+2g_2\varepsilon_2(\varepsilon_1^2+\varepsilon_2^2)-\gamma_1(\varepsilon_1^2+\varepsilon_2^2-2\varepsilon_2^2)-2\gamma_2\varepsilon_1\varepsilon_2\right\} \\
&= -n\{(n-1)\varepsilon_1+\varepsilon_1(2g_1\varepsilon_1+2g_2\varepsilon_2-\gamma_1)+2\varepsilon_2(\gamma_1\varepsilon_2-\gamma_2\varepsilon_1)\} \\
&= -n\{(n-1)\varepsilon_1+2g_1\varepsilon_1+2\varepsilon_2(g_2\varepsilon_1-g_1\varepsilon_2)+\gamma_1\varepsilon_1-2\varepsilon_1(\gamma_1\varepsilon_1+\gamma_2\varepsilon_2)\} \\
&= -n\left\{(n-1)\varepsilon_1+2g_1\varepsilon_1+(1-2\varepsilon_1)(\gamma_1\varepsilon_1+\gamma_2\varepsilon_2)-4(1-a)\varepsilon_2^2\right\} \\
&= -n\left\{(n-1)\varepsilon_1+2g_1\varepsilon_1+2(1-a)\varepsilon_1-4(1-a)(\varepsilon_1^2+\varepsilon_2^2)\right\} \\
&= -n\varepsilon_1\{n-1+2g_1-2(1-a)\}.
\end{aligned}$$

Since $\varepsilon_1>0$, this implies that

$$d_n = -\frac{n(n+2g_1+2a-3)}{4\varepsilon_1}, \quad n=1,2,3,\ldots. \tag{6.1.10}$$

Case III. Degree$[\varphi]=2$: we may choose $e=1$. Then we have

$$c_n = \frac{n(n-1+2\varepsilon)(2(1-f)+\varepsilon)+(1-\varepsilon)(\gamma-2\varepsilon)}{2(n-1+\varepsilon)(n+\varepsilon)}, \quad n=0,1,2,\ldots$$

and

$$\begin{aligned}
d_n = &-\frac{n(n-2+2\varepsilon)}{4(2n-3+2\varepsilon)(n-1+\varepsilon)^2(2n-1+2\varepsilon)} \\
&\times\Big\{(n-1)^2(n-1+2\varepsilon)^2 \\
&\qquad +2(n-1)(n-1+2\varepsilon)(2g+2f(\varepsilon-f)-\gamma) \\
&\qquad\qquad +4\varepsilon(g\varepsilon-f\gamma)+\gamma^2\Big\}, \quad n=1,2,3,\ldots.
\end{aligned}$$

The latter formula can also be written in the form

$$d_n = -\frac{n(n-2+2\varepsilon)}{4(2n-3+2\varepsilon)(n-1+\varepsilon)^2(2n-1+2\varepsilon)}D_n, \quad n=1,2,3,\ldots, \tag{6.1.11}$$

where

$$D_n = \left\{(n-1+\varepsilon)^2-\delta^2-\eta^2\right\}^2-4\delta^2\eta^2, \quad n=1,2,3,\ldots$$

with

$$\delta^2=(f-\varepsilon)^2-g+\gamma \quad \text{and} \quad \eta^2=f^2-g. \tag{6.1.12}$$

Note that we have

$$c_n = \frac{(n-1+\varepsilon)(n+\varepsilon)\,(2(1-f)+\varepsilon)+(1-\varepsilon)(\gamma-2f\varepsilon+\varepsilon^2)}{2(n-1+\varepsilon)(n+\varepsilon)}, \qquad n=0,1,2,\ldots,$$

which implies that

$$c_n - a = \frac{(n-1+\varepsilon)(n+\varepsilon)\,(2(1-f-a)+\varepsilon)+(1-\varepsilon)(\gamma-2f\varepsilon+\varepsilon^2)}{2(n-1+\varepsilon)(n+\varepsilon)}$$

for $n=0,1,2,\ldots$. Since ε occurs in both $c_n - a$ and d_n in combination with n, we conclude that $c_n - a$ can only be pure imaginary for all $n=0,1,2,\ldots$ and d_n can only be real for all $n=1,2,3,\ldots$ if ε is real. If we now set $f=f_1+if_2$, $g=g_1+ig_2$ and $\gamma=\gamma_1+i\gamma_2$ with $f_1,f_2,g_1,g_2,\varepsilon,\gamma_1,\gamma_2\in\mathbb{R}$, we obtain

$$\begin{aligned} c_n - a = {} & \frac{(n-1+\varepsilon)(n+\varepsilon)\,(2(1-f_1-a)+\varepsilon)+(1-\varepsilon)(\gamma_1-2f_1\varepsilon+\varepsilon^2)}{2(n-1+\varepsilon)(n+\varepsilon)} \\ & -i\,\frac{2(n-1+\varepsilon)(n+\varepsilon)f_2-(1-\varepsilon)(\gamma_2-2f_2\varepsilon)}{2(n-1+\varepsilon)(n+\varepsilon)}. \end{aligned}$$

Since $c_n - a$ must be pure imaginary for all $n=0,1,2,\ldots$, we conclude that

$$\varepsilon = 2f_1+2a-2 \quad \text{and} \quad (1-\varepsilon)(\gamma_1-2f_1\varepsilon+\varepsilon^2)=0. \tag{6.1.13}$$

Then we have

$$i(c_n-a) = f_2 - \frac{(1-\varepsilon)(\gamma_2-2f_2\varepsilon)}{2(n-1+\varepsilon)(n+\varepsilon)}, \qquad n=0,1,2,\ldots.$$

As before, we may write

$$\begin{aligned} D_n &= \left\{(n-1+\varepsilon)^2-(\delta+\eta)^2\right\}\left\{(n-1+\varepsilon)^2-(\delta-\eta)^2\right\} && (6.1.14)\\ &= (n-1+\varepsilon)^4-2(\delta^2+\eta^2)(n-1+\varepsilon)^2+(\delta^2-\eta^2)^2 && (6.1.15) \end{aligned}$$

for $n=1,2,3,\ldots$. If we set $\delta=\delta_1+i\delta_2$ and $\eta=\eta_1+i\eta_2$ with $\delta_1,\delta_2,\eta_1,\eta_2\in\mathbb{R}$, we find that

$$(\delta+\eta)^2 = (\delta_1+\eta_1)^2-(\delta_2+\eta_2)^2+2i(\delta_1+\eta_1)(\delta_2+\eta_2)$$

and

$$(\delta-\eta)^2 = (\delta_1-\eta_1)^2-(\delta_2-\eta_2)^2+2i(\delta_1-\eta_1)(\delta_2-\eta_2).$$

Since d_n must be real for all $n=1,2,3,\ldots$ this implies, by using (6.1.11) and (6.1.14), that

$$(\delta_1+\eta_1)(\delta_2+\eta_2)=0 \quad \text{and} \quad (\delta_1-\eta_1)(\delta_2-\eta_2)=0.$$

Note that both δ and η are not uniquely determined in view of (6.1.12). Without loss of generality we may choose $\delta_1+i\delta_2=\eta_1-i\eta_2$ with $\delta_1=\eta_1\geq 0$. Hence δ and η are complex conjugates, id est $\delta=\overline{\eta}$. Then we have

$$(\delta+\eta)^2=4\delta_1^2 \quad \text{and} \quad (\delta-\eta)^2=-4\delta_2^2.$$

Now we use (6.1.12) again to find that

$$\begin{aligned}\delta^2+\eta^2 &= (f-\varepsilon)^2-g+\gamma+f^2-g=2f^2-2f\varepsilon+\varepsilon^2-2g+\gamma\\ &= 2(f_1+if_2)^2-2(f_1+if_2)\varepsilon+\varepsilon^2-2(g_1+ig_2)+\gamma_1+i\gamma_2\\ &= 2f_1^2-2f_2^2-2f_1\varepsilon+\varepsilon^2-2g_1+\gamma_1+i\{4f_1f_2-2f_2\varepsilon-2g_2+\gamma_2\}\end{aligned}$$

and

$$\begin{aligned}\delta^2-\eta^2 &= (f-\varepsilon)^2-g+\gamma-f^2+g=\gamma-2f\varepsilon+\varepsilon^2\\ &= \gamma_1+i\gamma_2-2(f_1+if_2)\varepsilon+\varepsilon^2=\gamma_1-2f_1\varepsilon+\varepsilon^2+i(\gamma_2-2f_2\varepsilon).\end{aligned}$$

Hence we have

$$(\delta^2-\eta^2)^2=(\gamma_1-2f_1\varepsilon+\varepsilon^2)^2-(\gamma_2-2f_2\varepsilon)^2+2i(\gamma_1-2f_1\varepsilon+\varepsilon^2)(\gamma_2-2f_2\varepsilon).$$

Since d_n must be real for all $n=1,2,3,\ldots$, this implies that, by using (6.1.11) and (6.1.15)

$$\gamma_2=2f_2\varepsilon-4f_1f_2+2g_2 \quad \text{and} \quad (\gamma_1-2f_1\varepsilon+\varepsilon^2)(\gamma_2-2f_2\varepsilon)=0.$$

For $\gamma_2=2f_2\varepsilon$ we obtain

$$(\delta^2-\eta^2)^2=(\gamma_1-2f_1\varepsilon+\varepsilon^2)^2\geq 0.$$

On the other hand we also have

$$\delta^2-\eta^2=(\delta_1+i\delta_2)^2-(\delta_1-i\delta_2)^2=4i\delta_1\delta_2 \quad \Longrightarrow \quad (\delta^2-\eta^2)^2=-16\delta_1^2\delta_2^2\leq 0.$$

Hence d_n can only be real for all $n=1,2,3,\ldots$ for

$$\gamma_1=2f_1\varepsilon-\varepsilon^2 \quad \text{and} \quad \gamma_2=2f_2\varepsilon-4f_1f_2+2g_2. \tag{6.1.16}$$

Combining (6.1.13) and (6.1.16), we conclude, since c_n-a must be pure imaginary for all $n=0,1,2,\ldots$ and d_n must be real for all $n=1,2,3,\ldots$, that we must have

$$\begin{aligned}\varepsilon=2f_1+2a-2, \quad \gamma_1&=2f_1\varepsilon-\varepsilon^2=2\varepsilon(1-a)\\ \text{and} \quad \gamma_2&=2f_2\varepsilon-4f_1f_2+2g_2.\end{aligned} \tag{6.1.17}$$

Note that we have

$$4(\delta_1^2-\delta_2^2)=(\delta+\eta)^2+(\delta-\eta)^2=2(\delta^2+\eta^2)=4f_1^2-4f_2^2-4g_1$$

and

$$-16\delta_1^2\delta_2^2=(\delta^2-\eta^2)^2=-(\gamma_2-2f_2\varepsilon)^2=-(2g_2-4f_1f_2)^2,$$

which implies that

$$\delta_1^2-\delta_2^2=f_1^2-f_2^2-g_1 \quad \text{and} \quad 4\delta_1^2\delta_2^2=(g_2-2f_1f_2)^2.$$

Finally, we write the difference equation (6.1.1) in a different form. Note that we have

$$\Delta y_n(z)=y_n(z+1)-y_n(z)=y_n(a+ix+1)-y_n(a+ix)=i^n\left(\widehat{y}_n(x-i)-\widehat{y}_n(x)\right)$$

and

$$\left(\Delta^2 y_n\right)(z)=y_n(z+2)-2y_n(z+1)+y_n(z)=i^n\left(\widehat{y}_n(x-2i)-2\widehat{y}_n(x-i)+\widehat{y}_n(x)\right).$$

The difference equation (6.1.1) can be written as

$$\varphi(z)\left(\Delta^2 y_n\right)(z)+\psi(z)\Delta y_n(z)=\lambda_n y_n(z+1), \quad n=0,1,2,\ldots$$

which is equivalent to

$$\begin{aligned}\varphi(a+ix)&\left(\widehat{y}_n(x-2i)-2\widehat{y}_n(x-i)+\widehat{y}_n(x)\right)\\&+\psi(a+ix)\left(\widehat{y}_n(x-i)-\widehat{y}_n(x)\right)=\lambda_n\widehat{y}_n(x-i)\end{aligned}$$

for $n=0,1,2,\ldots$. By shifting x to $x+i$ the latter formula can also be written in the form

$$\widehat{C}(x)\widehat{y}(x-i)-\left\{\widehat{C}(x)+\widehat{D}(x)\right\}\widehat{y}_n(x)+\widehat{D}(x)\widehat{y}_n(x+i)=\lambda_n\widehat{y}_n(x) \tag{6.1.18}$$

for $n=0,1,2,\ldots$, where

$$\widehat{C}(x)=\varphi(a+ix-1) \quad \text{and} \quad \widehat{D}(x)=\widehat{C}(x)-\psi(a+ix-1). \tag{6.1.19}$$

Note that the difference equation (6.1.1) can be written in the selfadjoint form

$$\Delta\left(w(z)\varphi(z+1)\Delta y_n(z)\right)=\lambda_n w(z+1)y_n(z+1),$$

where

$$\varphi(z+2)=ez^2+2fz+g, \quad \psi(z+1)=2\varepsilon z+\gamma \quad \text{and} \quad \lambda_n=n\left(e(n-1)+2\varepsilon\right),$$

provided that $w(z)$ satisfies the Pearson difference equation

$$\Delta\left(w(z)\varphi(z+1)\right)=w(z+1)\psi(z+1).$$

This implies that the Pearson difference equation (cf. (5.1.10)) can now be written in the form

$$\frac{w(z)}{w(z+1)}=\frac{\varphi(z+2)-\psi(z+1)}{\varphi(z+1)}=\frac{D(z+1)}{C(z)}, \tag{6.1.20}$$

where $C(z)$ and $D(z)$ are given by (6.1.2).

6.2 Classification of the Real Positive-Definite Orthogonal Polynomial Solutions

Again we use Favard's theorem (theorem 3.1) to conclude from (6.1.6) that we have positive-definite orthogonality if $i(c_n - a) \in \mathbb{R}$ for all $n = 0, 1, 2, \ldots$ and $d_n < 0$ for all $n = 1, 2, 3, \ldots$.

Again we consider three different cases depending on the form of $\varphi(x) = ex^2 + 2fx + g$:

Case I. Degree$[\varphi] = 0$: $e = f = 0$ and we may choose $g = 1$. In this case we do not have real polynomial solutions and therefore we cannot have orthogonal polynomial solutions.

Case II. Degree$[\varphi] = 1$: $e = 0$ and we may choose $2f = 1$. Further we have (6.1.7), (6.1.8) and

$$d_n = -\frac{n(n+2g_1+2a-3)}{4\varepsilon_1}, \quad n = 1, 2, 3, \ldots.$$

Hence positive-definite orthogonality occurs for $\varepsilon_1 = \varepsilon_1^2 + \varepsilon_2^2 > 0$ and $g_1 > 1 - a$.

Case III. Degree$[\varphi] = 2$: we may choose $e = 1$. Further we have (6.1.17) and

$$d_n = -\frac{n(n-2+2\varepsilon)}{4(2n-3+2\varepsilon)(n-1+\varepsilon)^2(2n-1+2\varepsilon)} D_n, \quad n = 1, 2, 3, \ldots$$

where

$$D_n = \left\{(n-1+\varepsilon)^2 - 4\delta_1^2\right\}\left\{(n-1+\varepsilon)^2 + 4\delta_2^2\right\}, \quad n = 1, 2, 3, \ldots$$

and

$$\delta_1^2 - \delta_2^2 = f_1^2 - f_2^2 - g_1 \quad \text{and} \quad 4\delta_1^2\delta_2^2 = (g_2 - 2f_1f_2)^2.$$

For $\varepsilon > 0$ we have

$$-\frac{n(n-2+2\varepsilon)}{4(2n-3+2\varepsilon)(n-1+\varepsilon)^2(2n-1+2\varepsilon)} < 0, \quad n = 1, 2, 3, \ldots.$$

This implies that positive-definite orthogonality occurs for

Case IIIa. $\varepsilon > 0$ and $2\delta_1 < \varepsilon$.

We also have a finite orthogonal polynomial system in this case. First we define

$$2\varepsilon = -2N - t \quad \text{with} \quad N \in \{1, 2, 3, \ldots\} \quad \text{and} \quad -1 < t \leq 1.$$

Then we have

$$-\frac{n(n-2+2\varepsilon)}{4(2n-3+2\varepsilon)(n-1+\varepsilon)^2(2n-1+2\varepsilon)} > 0, \quad n = 1,2,3,\ldots,N$$

and for $n = N+1$ this is not longer true. So in that case we have

$$d_n < 0 \quad \text{for} \quad n = 1,2,3,\ldots,N \quad \Longleftrightarrow \quad D_n < 0 \quad \text{for} \quad n = 1,2,3,\ldots,N.$$

This implies that we have positive-definite orthogonality for

Case IIIb. $2\delta_1 > -\varepsilon$.

6.3 Properties of the Positive-Definite Orthogonal Polynomial Solutions

Case I. We have $e = f = 0$ and $g = 1$. In this case we do not have real orthogonal polynomial solutions.

Case II. We have $e = 0$, $2f = 1$, $\varepsilon_1 > 0$ and $g_1 > 1-a$. From (6.1.7) and (6.1.8) it follows that

$$\begin{aligned}\gamma_1\varepsilon_1 &= 2(1-a)\varepsilon_1 - \gamma_2\varepsilon_2\\ &= 2(1-a)\varepsilon_1 - 4(1-a)\varepsilon_2^2 - 2\varepsilon_2(g_2\varepsilon_1 - g_1\varepsilon_2)\\ &= 2(1-a)\varepsilon_1 - 4(1-a)(\varepsilon_1-\varepsilon_1^2) - 2g_2\varepsilon_1\varepsilon_2 + 2g_1(\varepsilon_1-\varepsilon_1^2)\\ &= 2\varepsilon_1\left\{g_1(1-\varepsilon_1) - g_2\varepsilon_2 - (1-a)(1-2\varepsilon_1)\right\},\end{aligned}$$

which implies that

$$\gamma_1 = 2\left\{g_1(1-\varepsilon_1) - g_2\varepsilon_2 - (1-a)(1-2\varepsilon_1)\right\}. \tag{6.3.1}$$

The Pearson difference equation (6.1.20) reads

$$\frac{w(z)}{w(z+1)} = \frac{(1-2\varepsilon)z+g-\gamma}{z+g-1},$$

where

$$z = a+ix, \quad g = g_1+ig_2, \quad \varepsilon = \varepsilon_1+i\varepsilon_2 \quad \text{and} \quad \gamma = \gamma_1+i\gamma_2$$

with $a, x, g_1, g_2, \varepsilon_1, \varepsilon_2, \gamma_1, \gamma_2 \in \mathbb{R}$. By using (6.1.8) and (6.3.1), we obtain

$$\begin{aligned}g-\gamma &= g_1+ig_2-\gamma_1-i\gamma_2\\ &= -g_1+2(g_1\varepsilon_1+g_2\varepsilon_2)+2(1-a)(1-2\varepsilon_1)\\ &\qquad +i\{g_2-4(1-a)\varepsilon_2-2(g_2\varepsilon_1-g_1\varepsilon_2)\}\\ &= (2\varepsilon_1-1+2i\varepsilon_2)\{g_1-2(1-a)-ig_2\}.\end{aligned} \tag{6.3.2}$$

Hence we have

$$\frac{w(a+ix)}{w(a+ix+1)} = \frac{(2\varepsilon_1 - 1 + 2i\varepsilon_2)\{g_1 + a - 2 - i(x+g_2)\}}{g_1 + a - 1 + i(x+g_2)}.$$

Since $\varepsilon_1 = \varepsilon_1^2 + \varepsilon_2^2$, we have $(2\varepsilon_1 - 1)^2 + 4\varepsilon_2^2 = 1$. Hence we may define $\varphi \in \mathbb{R}$ such that

$$e^{i\varphi} := 2\varepsilon_1 - 1 + 2i\varepsilon_2 \quad \Longrightarrow \quad \cos\varphi = 2\varepsilon_1 - 1 \quad \text{and} \quad \sin\varphi = 2\varepsilon_2.$$

Then we obtain a positive-definite weight function of the form

$$w(a+ix) = |\Gamma(g_1 + a - 1 + i(x+g_2))|^2 \, e^{\varphi x} \tag{6.3.3}$$

for the **Meixner-Pollaczek** polynomials.

For the difference equation (6.1.18) we obtain

$$\widehat{C}(x)\widehat{y}_n(x-i) - \left\{\widehat{C}(x) + \widehat{D}(x)\right\}\widehat{y}_n(x) + \widehat{D}(x)\widehat{y}_n(x+i) = 2n(\varepsilon_1 + i\varepsilon_2)\widehat{y}_n(x-i),$$

where, by using (6.1.19), we get

$$\widehat{C}(x) = \varphi(a+ix-1) = a + ix - 1 + g = g_1 + a - 1 + i(x+g_2)$$

and

$$\begin{aligned}\widehat{D}(x) = \widehat{C}(x) - \psi(a+ix-1) &= a + ix - 1 + g - 2\varepsilon(a+ix-1) - \gamma \\ &= (1-2\varepsilon)(a+ix-1) + g - \gamma.\end{aligned}$$

Hence by using (6.3.2), we obtain

$$\widehat{D}(x) = (2\varepsilon_1 - 1 + 2i\varepsilon_2)(g_1 + a - 1 - i(x+g_2)).$$

By using (6.1.9) and (6.1.10), the three-term recurrence relation (6.1.6) reads

$$\widehat{y}_{n+1}(x) = \left(x - \frac{2n\varepsilon_2 + \gamma_1\varepsilon_2 - \gamma_2\varepsilon_1}{2\varepsilon_1}\right)\widehat{y}_n(x) - \frac{n(n+2g_1+2a-3)}{4\varepsilon_1}\widehat{y}_{n-1}(x)$$

for $n = 1, 2, 3, \ldots$ with $\widehat{y}_0(x) = 1$ and $\widehat{y}_1(x) = x - (\gamma_1\varepsilon_2 - \gamma_2\varepsilon_1)/2\varepsilon_1$.

For the coefficients of the representation (6.1.3) we obtain from (6.1.4)

$$2\varepsilon(n-k)a_{n,k} = (k-1-c+g)a_{n,k+1}, \quad k = n-1, n-2, n-3, \ldots, 0,$$

where $(1-2\varepsilon)(c+1) = g - \gamma$. Hence we have

$$-(2\varepsilon_1 - 1 + 2i\varepsilon_2)(c+1) = (2\varepsilon_1 - 1 + 2i\varepsilon_2)\{g_1 - 2(1-a) - ig_2\},$$

which implies that $c = 1 - 2a - g_1 + ig_2$ and $c + 1 - g = 2(1 - a - g_1)$, provided that $2\varepsilon \neq 1$. However, $2\varepsilon = 1$ (i.e. $2\varepsilon_1 = 1$ and $\varepsilon_2 = 0$) is impossible since $\varepsilon_1 = \varepsilon_1^2 + \varepsilon_2^2$.

Hence we have

$$2\varepsilon(n-k)a_{n,k} = (k+2g_1+2a-2)a_{n,k+1}, \quad k = n-1, n-2, n-3, \ldots, 0.$$

Since $a_{n,n} = n!$, this leads to

$$a_{n,k} = \frac{(k+2g_1+2a-2)_{n-k}\,n!}{(2\varepsilon)^{n-k}(n-k)!} = \frac{(2g_1+2a-2)_n}{(2\varepsilon)^n}\frac{(-n)_k(-2\varepsilon)^k}{(2g_1+2a-2)_k}$$

for $k = 0, 1, 2, \ldots, n$. Hence by using (6.1.3), we obtain

$$\begin{aligned} y_n(a+ix) &= \sum_{k=0}^{n} a_{n,k}\binom{a+ix+1-2a-g_1+ig_2}{k} \\ &= \frac{(2g_1+2a-2)_n}{(2\varepsilon)^n}\,{}_2F_1\left(\begin{matrix}-n, g_1+a-1-i(x+g_2)\\ 2g_1+2a-2\end{matrix}; 2\varepsilon\right) \end{aligned}$$

for $n = 0, 1, 2, \ldots$.

The Rodrigues formula (3.4.28) for $n = 0, 1, 2, \ldots$ reads

$$\begin{aligned} y_n(a+ix) &= \frac{1}{(1+\mathrm{e}^{i\varphi})^n \mathrm{e}^{\varphi x}\,|\Gamma(g_1+a-1+i(x+g_2))|^2} \\ &\quad \times \Delta^n\left(\Gamma(g_1+a-1+i(x+g_2))\Gamma(g_1+a+n-1-i(x+g_2))\mathrm{e}^{\varphi(x+in)}\right). \end{aligned}$$

Now we use (1.6.8) to find

$$\begin{aligned} d_0 &:= \frac{1}{2\pi}\int_{-\infty}^{\infty} |\Gamma(g_1+a-1+i(x+g_2))|^2\,\mathrm{e}^{\varphi x}\,dx \\ &= \Gamma(2g_1+2a-2)\mathrm{e}^{-g_2\varphi}(4\varepsilon_1)^{1-a-g_1} > 0. \end{aligned}$$

Then using (5.1.11) and (3.1.4) the orthogonality relation can be written as

$$\begin{aligned} &\frac{1}{2\pi}\int_{-\infty}^{\infty} |\Gamma(g_1+a-1+i(x+g_2))|^2\,\mathrm{e}^{\varphi x}\widehat{y}_m(x)\widehat{y}_n(x)\,dx \\ &= \Gamma(2g_1+2a+n-2)\mathrm{e}^{-g_2\varphi}(4\varepsilon_1)^{1-a-g_1-n}n!\,\delta_{mn}, \quad m,n = 0, 1, 2, \ldots. \end{aligned}$$

Case III. We have $e = 1$ and (6.1.17). Note that

$$\varphi(x) = x^2 + 2fx + g = (x+f)^2 + g - f^2 \quad \text{and} \quad \psi(x) = 2\varepsilon x + \gamma,$$

which, if (6.1.17) is used, leads to

$$\begin{aligned}\varphi(a+ix-1) &= \{a+f_1-1+i(x+f_2)\}^2+g_1+ig_2-(f_1+if_2)^2\\ &= -(x+f_2)^2+(a+f_1-1)^2+g_1-f_1^2+f_2^2\\ &\qquad +i\{2(a+f_1-1)(x+f_2)+g_2-2f_1f_2\}\\ &= -(x+f_2)^2+\frac{\varepsilon^2}{4}+g_1-f_1^2+f_2^2+i\{\varepsilon(x+f_2)+g_2-2f_1f_2\}\end{aligned}$$

and

$$\begin{aligned}\psi(a+ix-1) &= 2\varepsilon(a+ix-1)+\gamma_1+i\gamma_2=\gamma_1-2\varepsilon(1-a)+i(2\varepsilon x+\gamma_2)\\ &= 2i\{\varepsilon(x+f_2)+g_2-2f_1f_2\}.\end{aligned}$$

This implies that the difference equation (6.1.18) can be written as

$$\widehat{C}(x)\widehat{y}(x-i)-\left\{\widehat{C}(x)+\widehat{D}(x)\right\}\widehat{y}_n(x)+\widehat{D}(x)\widehat{y}_n(x+i)=\lambda_n\widehat{y}_n(x),\quad n=0,1,2,\ldots,$$

where, by using (6.1.17), we get

$$\begin{aligned}\widehat{C}(x) &= \varphi(a+ix-1)\\ &= -(x+f_2)^2+\frac{\varepsilon^2}{4}+g_1-f_1^2+f_2^2+i\{\varepsilon(x+f_2)+g_2-2f_1f_2\}\end{aligned}$$

and

$$\begin{aligned}\widehat{D}(x) &= \widehat{C}(x)-\psi(a+ix-1)\\ &= -(x+f_2)^2+\frac{\varepsilon^2}{4}+g_1-f_1^2+f_2^2-i\{\varepsilon(x+f_2)+g_2-2f_1f_2\}.\end{aligned}$$

Since

$$i(c_n-a)=f_2-\frac{(1-\varepsilon)(\gamma_2-2f_2\varepsilon)}{2(n-1+\varepsilon)(n+\varepsilon)}=f_2-\frac{(1-\varepsilon)(g_2-2f_1f_2)}{(n-1+\varepsilon)(n+\varepsilon)},\quad n=0,1,2,\ldots$$

and

$$d_n=-\frac{n(n-2+2\varepsilon)}{4(2n-3+2\varepsilon)(n-1+\varepsilon)^2(2n-1+2\varepsilon)}D_n,\quad n=1,2,3,\ldots$$

where

$$D_n=(n-1+\varepsilon)^4+4(g_1-f_1^2+f_2^2)(n-1+\varepsilon)^2-4(g_2-2f_1f_2)^2,\quad n=1,2,3,\ldots$$

the three-term recurrence relation (6.1.6) can be written as

$$\begin{aligned}\widehat{y}_{n+1}(x) = &\left(x+f_2-\frac{(1-\varepsilon)(g_2-2f_1f_2)}{(n-1+\varepsilon)(n+\varepsilon)}\right)\widehat{y}_n(x)\\ &-\frac{n(n-2+2\varepsilon)}{4(2n-3+2\varepsilon)(n-1+\varepsilon)^2(2n-1+2\varepsilon)}\\ &\times\left\{(n-1+\varepsilon)^4+4(g_1-f_1^2+f_2^2)(n-1+\varepsilon)^2\right.\\ &\left.-4(g_2-2f_1f_2)^2\right\}\widehat{y}_{n-1}(x)\end{aligned}$$

for $n=1,2,3,\ldots$ with $\widehat{y}_0(x)=1$ and $\widehat{y}_1(x)=x+f_2+(g_2-2f_1f_2)/\varepsilon$.

For the hypergeometric representation we use (6.1.3) and (6.1.4), provided that

$$\begin{aligned}&(c+1)^2-2f(c+1)+g=-2\varepsilon(c+1)+\gamma\\ &\iff\quad (c+1)^2-2(f-\varepsilon)(c+1)+g-\gamma=0.\end{aligned}$$

Since $g-\gamma=(f-\varepsilon)^2-\delta^2$, this implies that $(c+1-f+\varepsilon)^2=\delta^2$. In that case we have

$$\begin{aligned}(k-1-c)^2+2f(k-1-c)+g &= (k-1-c)^2+2f(k-1-c)+g+(f^2-g)-\eta^2\\ &=(k-1-c+f)^2-\eta^2\\ &=(k-1-c+f+\eta)(k-1-c+f-\eta)\\ &=(k+\varepsilon\mp\delta+\eta)(k+\varepsilon\mp\delta-\eta).\end{aligned}$$

Therefore, the two-term recurrence relation (6.1.4) reads

$$(n-k)(n+k-1+2\varepsilon)a_{n,k}=(k+\varepsilon\mp\delta+\eta)(k+\varepsilon\mp\delta-\eta)a_{n,k+1}$$

for $k=n-1,n-2,n-3,\ldots,0$. By using $a_{n,n}=n!$, we obtain

$$\begin{aligned}a_{n,k} &= \frac{(k+\varepsilon\mp\delta+\eta)_{n-k}(k+\varepsilon\mp\delta-\eta)_{n-k}n!}{(n+k-1+2\varepsilon)_{n-k}(n-k)!}\\ &=\frac{(\varepsilon\mp\delta+\eta)_n(\varepsilon\mp\delta-\eta)_n}{(n-1+2\varepsilon)_n}\frac{(-n)_k(n-1+2\varepsilon)_k}{(\varepsilon\mp\delta+\eta)_k(\varepsilon\mp\delta-\eta)_k}(-1)^k\end{aligned}$$

for $k=0,1,2,\ldots,n$, which implies that

$$\begin{aligned}y_n(a+ix) &= \sum_{k=0}^{n}a_{n,k}\binom{a+ix+c}{k}=\sum_{k=0}^{n}a_{n,k}(-1)^k\frac{(-a-ix-c)_k}{k!}\\ &=\frac{(\varepsilon\mp\delta+\eta)_n(\varepsilon\mp\delta-\eta)_n}{(n-1+2\varepsilon)_n}\\ &\quad\times{}_3F_2\left(\begin{array}{c}-n,n-1+2\varepsilon,\varepsilon\mp\delta+1-f-a-ix\\ \varepsilon\mp\delta+\eta,\varepsilon\mp\delta-\eta\end{array};1\right)\end{aligned}$$

for $n=0,1,2,\ldots$. Since $\delta=\delta_1+i\delta_2=\eta_1-i\eta_2=\overline{\eta}$, we have

$$\pm(\delta+\eta)=\pm(\delta+\overline{\delta})=\pm 2\delta_1 \quad \text{and} \quad \pm(\delta-\eta)=\pm(\delta-\overline{\delta})=\pm 2i\delta_2.$$

Hence by using (6.1.17), we now obtain

$$\widehat{y}_n(x)=\frac{(\varepsilon\mp 2\delta_1)_n(\varepsilon\mp 2i\delta_2)_n}{(n-1+2\varepsilon)_n\, i^n}\,{}_3F_2\left(\begin{matrix}-n,n-1+2\varepsilon,\varepsilon/2\mp\delta_1-i(x+f_2\pm\delta_2)\\ \varepsilon\mp 2\delta_1,\varepsilon\mp 2i\delta_2\end{matrix};1\right)$$

for $n=0,1,2,\ldots$. These polynomials are called **continuous Hahn** polynomials.
The Pearson difference equation (6.1.20) reads

$$\frac{w(z)}{w(z+1)}=\frac{z^2+2(f-\varepsilon)z+g-\gamma}{(z-1)^2+2f(z-1)+g}=\frac{(z+f-\varepsilon+\delta)(z+f-\varepsilon-\delta)}{(z-1+f+\eta)(z-1+f-\eta)}.$$

Case IIIa. $\varepsilon>0$ and $2\delta_1<\varepsilon$. In this case we use

$$w(z)=\Gamma(z-1+f+\eta)\Gamma(z-1+f-\eta)\Gamma(1+\varepsilon-\delta-f-z)\Gamma(1+\varepsilon+\delta-f-z)$$

as a possible solution for the Pearson difference equation.
First we set $z=a+ix$ and use $\eta=\overline{\delta}$ and (6.1.17) to obtain

$$\begin{aligned}w(a+ix)&=\Gamma(a+ix-1+f+\overline{\delta})\Gamma(a+ix-1+f-\overline{\delta})\\&\qquad\times\Gamma(1+\varepsilon-\delta-f-a-ix)\Gamma(1+\varepsilon+\delta-f-a-ix)\\&=\Gamma(\varepsilon/2+\delta_1+i(x+f_2-\delta_2))\Gamma(\varepsilon/2-\delta_1+i(x+f_2+\delta_2))\\&\qquad\times\Gamma(\varepsilon/2-\delta_1-i(x+f_2+\delta_2))\Gamma(\varepsilon/2+\delta_1-i(x+f_2-\delta_2))\\&=|\Gamma(\varepsilon/2+\delta_1+i(x+f_2-\delta_2))|^2\,|\Gamma(\varepsilon/2-\delta_1+i(x+f_2+\delta_2))|^2.\end{aligned}$$

This leads to the Rodrigues formula

$$\begin{aligned}\widehat{y}_n(x)&=\frac{1}{(n-1+2\varepsilon)_n}\\&\qquad\times\frac{1}{|\Gamma(\varepsilon/2+\delta_1+i(x+f_2-\delta_2))|^2\,|\Gamma(\varepsilon/2-\delta_1+i(x+f_2+\delta_2))|^2}\\&\qquad\times\Delta^n\left(\Gamma(\varepsilon/2+\delta_1+i(x+f_2-\delta_2))\Gamma(\varepsilon/2-\delta_1+i(x+f_2+\delta_2))\right.\\&\qquad\quad\times\Gamma(n+\varepsilon/2+\delta_1-i(x+f_2+\delta_2))\\&\qquad\qquad\left.\times\Gamma(n+\varepsilon/2-\delta_1-i(x+f_2-\delta_2))\right)\end{aligned}$$

for $n=0,1,2,\ldots$. Further we obtain by using the Mellin-Barnes integral (1.6.3) that

$$\begin{aligned}\frac{1}{2\pi}\int_{-\infty}^{\infty} w(a+ix)\,dx &= \frac{1}{2\pi i}\int_{-i\infty}^{i\infty} w(a+s)\,ds\\ &= \frac{\Gamma(\varepsilon-\delta+\eta)\Gamma(\varepsilon+\delta+\eta)\Gamma(\varepsilon-\delta-\eta)\Gamma(\varepsilon+\delta-\eta)}{\Gamma(2\varepsilon)}\\ &= \frac{\Gamma(\varepsilon+2\delta_1)\Gamma(\varepsilon-2\delta_1)\Gamma(\varepsilon+2i\delta_2)\Gamma(\varepsilon-2i\delta_2)}{\Gamma(2\varepsilon)},\end{aligned}$$

which implies that

$$d_0 := \frac{1}{2\pi}\int_{-\infty}^{\infty} w(a+ix)\,dx = \frac{\Gamma(\varepsilon+2\delta_1)\Gamma(\varepsilon-2\delta_1)\,|\Gamma(\varepsilon+2i\delta_2)|^2}{\Gamma(2\varepsilon)} > 0,$$

since $0 < 2\delta_1 < \varepsilon$. This leads to the orthogonality relation

$$\begin{aligned}&\frac{1}{2\pi}\int_{-\infty}^{\infty} |\Gamma(\varepsilon/2+\delta_1+i(x+f_2-\delta_2))|^2\\ &\qquad\times|\Gamma(\varepsilon/2-\delta_1+i(x+f_2+\delta_2))|^2\,\widehat{y}_m(x)\widehat{y}_n(x)\,dx\\ &= \frac{\Gamma(n-1+2\varepsilon)\Gamma(n+\varepsilon+2\delta_1)\Gamma(n+\varepsilon-2\delta_1)\,|\Gamma(n+\varepsilon+2i\delta_2)|^2\,n!}{\Gamma(2n-1+2\varepsilon)\Gamma(2n+2\varepsilon)}\,\delta_{mn}\end{aligned}$$

for $m,n = 0,1,2,\ldots$.

Case IIIb. We have $2\varepsilon = -2N-t$ with $N \in \{1,2,3,\ldots\}$ and $-1 < t \leq 1$ and $2\delta_1 > -\varepsilon$. In this case we use

$$w(z) = \frac{\Gamma(z-1+f+\eta)\Gamma(1+\varepsilon+\delta-f-z)}{\Gamma(z+f-\varepsilon+\delta)\Gamma(2+\eta-f-z)}$$

as a possible solution for the Pearson difference equation.

As before, we set $z = a+ix$ and use $\eta = \overline{\delta}$ and (6.1.17) to obtain

$$\begin{aligned}w(a+ix) &= \frac{\Gamma(a+ix-1+f+\overline{\delta})\Gamma(1+\varepsilon+\delta-f-a-ix)}{\Gamma(a+ix+f-\varepsilon+\delta)\Gamma(2+\overline{\delta}-f-a-ix)}\\ &= \frac{\Gamma(\varepsilon/2+\delta_1+i(x+f_2-\delta_2))\Gamma(\varepsilon/2+\delta_1-i(x+f_2-\delta_2))}{\Gamma(1-\varepsilon/2+\delta_1+i(x+f_2+\delta_2))\Gamma(1-\varepsilon/2+\delta_1-i(x+f_2+\delta_2))}\\ &= \left|\frac{\Gamma(\varepsilon/2+\delta_1+i(x+f_2-\delta_2))}{\Gamma(1-\varepsilon/2+\delta_1+i(x+f_2+\delta_2))}\right|^2.\end{aligned}$$

This leads to the Rodrigues formula

$$\widehat{y}_n(x) = \frac{(-1)^n}{(n-1+2\varepsilon)_n}\left|\frac{\Gamma(1-\varepsilon/2+\delta_1+i(x+f_2+\delta_2))}{\Gamma(\varepsilon/2+\delta_1+i(x+f_2-\delta_2))}\right|^2$$
$$\times\,\Delta^n\left(\frac{\Gamma(\varepsilon/2+\delta_1+i(x+f_2-\delta_2))}{\Gamma(1-\varepsilon/2+\delta_1-n+i(x+f_2-\delta_2))}\times\frac{\Gamma(n+\varepsilon/2+\delta_1-i(x+f_2-\delta_2))}{\Gamma(1-\varepsilon/2+\delta_1-i(x+f_2-\delta_2))}\right)$$

for $n=0,1,2,\ldots,N$. Now, by using (1.6.9), we obtain

$$\begin{aligned}\frac{1}{2\pi}\int_{-\infty}^{\infty} w(a+ix)\,dx &= \frac{1}{2\pi i}\int_{-i\infty}^{i\infty} w(a+s)\,ds\\
&= \frac{\Gamma(\varepsilon+\delta+\eta)\Gamma(1-2\varepsilon)}{\Gamma(1-\varepsilon+\delta+\eta)\Gamma(1-\varepsilon+\delta-\eta)\Gamma(1-\varepsilon-\delta+\eta)}\\
&= \frac{\Gamma(\varepsilon+2\delta_1)\Gamma(1-2\varepsilon)}{\Gamma(1-\varepsilon+2\delta_1)\Gamma(1-\varepsilon+2i\delta_2)\Gamma(1-\varepsilon-2i\delta_2)},\end{aligned}$$

which implies that

$$d_0 := \frac{1}{2\pi}\int_{-\infty}^{\infty} w(a+ix)\,dx = \frac{\Gamma(\varepsilon+2\delta_1)\Gamma(1-2\varepsilon)}{\Gamma(1-\varepsilon+2\delta_1)\,|\Gamma(1-\varepsilon+2i\delta_2)|^2} > 0,$$

since $0 < -\varepsilon < 2\delta_1$. This leads to the orthogonality relation

$$\begin{aligned}&\frac{1}{2\pi}\int_{-\infty}^{\infty}\left|\frac{\Gamma(\varepsilon/2+\delta_1+i(x+f_2-\delta_2))}{\Gamma(1-\varepsilon/2+\delta_1+i(x+f_2+\delta_2))}\right|^2 \widehat{y}_m(x)\widehat{y}_n(x)\,dx\\
&= \frac{\Gamma(2-2\varepsilon-2n)\Gamma(1-2\varepsilon-2n)\Gamma(n+\varepsilon+2\delta_1)n!}{\Gamma(2-2\varepsilon-n)\Gamma(1-\varepsilon+2\delta_1-n)\,|\Gamma(1-\varepsilon-n+2i\delta_2)|^2}\,\delta_{mn}\end{aligned}$$

for $m,n=0,1,2,\ldots,N$.

In this chapter we have proved:

Theorem 6.1. *The positive-definite orthogonal polynomial solutions $y_n(x)$ of the difference equation (5.1.1)*

$$(ex^2+2fx+g)\left(\Delta^2 y_n\right)(x)+(2\varepsilon x+\gamma)\left(\Delta y_n\right)(x)=n(e(n-1)+2\varepsilon)y_n(x+1)$$

for $n=0,1,2,\ldots$ with $e,f,g,\varepsilon,\gamma\in\mathbb{C}$ consist of two infinite systems

Case II. *Meixner-Pollaczek polynomials with orthogonality on $(-\infty,\infty)$ with respect to*
$w(x)=|\Gamma(g_1+a-1+i(x+g_2)|^2 e^{\varphi(x)}$; $e=0,\ 2f=1,\ \varepsilon_1>0,\ g_1>1-a$ *and*
$2\varepsilon_1-1+2i\varepsilon_2=e^{i\varphi}$

Case IIIa. *Continuous Hahn polynomials with orthogonality on $(-\infty,\infty)$ with respect to*
$w(x)=|\Gamma(\varepsilon/2+\delta_1+i(x+f_2-\delta_2))\Gamma(\varepsilon/2-\delta_1+i(x+f_2+\delta_2))|^2$; $e=1$, $\varepsilon>0$ *and* $2\delta_1<\varepsilon$

and one finite system

Case IIIb. *Continuous Hahn polynomials with orthogonality on $(-\infty,\infty)$ with respect to*
$w(x)=|\Gamma(\varepsilon/2+\delta_1+i(x+f_2-\delta_2))/\Gamma(1-\varepsilon/2+\delta_1+i(x+f_2+\delta_2))|^2$; $e=1$, $2\varepsilon=-2N-t$ *with* $N\in\{1,2,3,\ldots\}$ *and* $-1<t\leq 1$ *and* $2\delta_1>-\varepsilon$.

Chapter 7
Orthogonal Polynomial Solutions in $x(x+u)$ of Real Difference Equations

Discrete Classical Orthogonal Polynomials III

7.1 Motivation for Polynomials in $x(x+u)$ Through Duality

In the definition 3.1 we introduced the sequences $\{\kappa_n\}_{n=0}^N$ with $\kappa_m \neq \kappa_n$ for $m \neq n$ and $\{\lambda_n\}_{n=0}^N$ with $\lambda_m \neq \lambda_n$ for $m \neq n$ of eigenvalues and the sequences $\{y_n\}_{n=0}^N$ with degree$[y_n] = n$ and $\{z_n\}_{n=0}^N$ with degree$[z_n] = n$ of polynomials where $N \in \{1,2,3,\ldots\}$ or $N \to \infty$. The polynomials $y_n(x)$ and $z_n(x)$ are called dual polynomials with respect to the sequences of eigenvalues $\{\kappa_n\}_{n=0}^N$ and $\{\lambda_n\}_{n=0}^N$ when $y_n(\kappa_m) = z_m(\lambda_n)$ for all $m,n = 0,1,2,\ldots,N$.

In section 5.3 we obtained the difference equation

$$C(x)y_n(x+1) - \{C(x)+D(x)\}\,y_n(x) + D(x)y_n(x-1) = \lambda_n y_n(x) \tag{7.1.1}$$

for $n = 0,1,2,\ldots,N-1$ where $\lambda_n = n(n-1+2\varepsilon)$,

$$C(x) = (x-1+f+\eta)(x-1+f-\eta)$$
$$\text{and} \quad D(x) = (x-1+f-\varepsilon+\delta)(x-1+f-\varepsilon-\delta)$$

with

$$\delta = \sqrt{(f-\varepsilon)^2 - g + \gamma} \quad \text{and} \quad \eta = \sqrt{f^2 - g}$$

for the Hahn polynomials. If the regularity condition (2.3.3) holds, all eigenvalues are different. This implies because of theorem 3.6 that there exists a sequence of dual polynomials. In this case we have $\lambda_n = n(n-1+2\varepsilon)$ and $\kappa_m = m$ since $q = \omega = 1$ and $x_0 = 0$. If the conditions

$$(D(0) =)\, 1 - 2f + g + 2\varepsilon - \gamma = 0 \quad \text{and} \quad (C(n) =)\, (n-1)^2 + 2f(n-1) + g \neq 0$$

R. Koekoek et al., *Hypergeometric Orthogonal Polynomials and Their q-Analogues*, Springer Monographs in Mathematics,
DOI 10.1007/978-3-642-05014-5_7, © Springer-Verlag Berlin Heidelberg 2010

hold, the dual polynomials $\{z_m\}_{m=0}^{N}$ satisfy the three-term recurrence relation

$$C(m)z_{m+1}(x) - \{C(m)+D(m)\}\, z_m(x) + D(m)z_{m-1}(x) = xz_m(x) \tag{7.1.2}$$

for $m = 0,1,2,\ldots,N-1$ with the convention that $z_{-1}(x) := 0$. If we set $x = \kappa_m = m$ into (7.1.1), we find that

$$C(m)y_n(m+1) - \{C(m)+D(m)\}\, y_n(m) + D(m)y_n(m-1) = \lambda_n y_n(m) \tag{7.1.3}$$

for $n = 0,1,2,\ldots,N-1$. Since $y_n(\kappa_m) = y_n(m) = z_m(\lambda_n)$ for all $m,n = 0,1,2,\ldots,N$, this implies that for the Hahn polynomials there exist dual polynomials with argument $\lambda_n = n(n-1+2\varepsilon)$. This motivates the study of orthogonal polynomials in $x(x+u)$ with x a real variable and $u \in \mathbb{R}$ a constant.

7.2 Difference Equations Having Real Polynomial Solutions with Argument $x(x+u)$

In order to build a theory, we must first classify all second-order difference equations with real coefficients which have real polynomial solutions $\{y_n(x(x+u))\}_{n=0}^{N}$ with degree$[y_n] = n$ and $N \in \{1,2,3,\ldots\}$ or $N \to \infty$. Therefore we consider the difference equation

$$\begin{aligned}&\varphi(x+2)\Delta^2 y_n(x(x+u)) + \psi(x+1)\Delta y_n(x(x+u))\\ &= \lambda_n \rho(x+1)y_n((x+1)(x+1+u))\end{aligned} \tag{7.2.1}$$

for $n = 0,1,2,\ldots$, where $\Delta y_n(x(x+u)) = y_n((x+1)(x+1+u)) - y_n(x(x+u))$ and $\Delta^2 y_n(x(x+u)) = \Delta\left(\Delta y_n(x(x+u))\right)$. By replacing x by $x-1$, we can write this in the symmetric form

$$\begin{aligned}&C(x)y_n((x+1)(x+1+u)) - \{C(x)+D(x)\}\, y_n(x(x+u))\\ &\quad + D(x)y_n((x-1)(x-1+u)) = \lambda_n \rho(x)y_n(x(x+u))\end{aligned} \tag{7.2.2}$$

with

$$C(x) = \varphi(x+1) \quad \text{and} \quad D(x) = \varphi(x+1) - \psi(x).$$

Now we look for eigenvalues λ_n and coefficients $C(x)$, $D(x)$ and $\rho(x)$ so that for each eigenvalue λ_n there exists exactly one polynomial solution $y_n(x(x+u))$ with degree$[y_n] = n$ in $x(x+u)$ up to a constant factor. Since $(x(x+u))^n$ can be expressed as a linear combination of $(x-k+1)_k(x+u)_k$ for $k = 0,1,2,\ldots,n$, we set

$$y_n(x(x+u)) = \sum_{k=0}^{n} a_{n,k}\frac{(x-k+1)_k(x+u)_k}{k!}, \quad a_{n,n} \neq 0, \quad n = 0,1,2,\ldots. \tag{7.2.3}$$

Then we have

$$\Delta y_n(x(x+u)) = (2x+1+u)\sum_{k=1}^{n} a_{n,k}\frac{(x-k+2)_{k-1}(x+1+u)_{k-1}}{(k-1)!}.$$

This leads to the *first simplification*

$$C(x) = (2x-1+u)C^*(x), \quad D(x) = (2x+1+u)D^*(x)$$
$$\text{and} \quad \rho(x) = (2x-1+u)(2x+1+u)\rho^*(x). \tag{7.2.4}$$

For the moment we assume that $2x \neq -u \pm 1$. Then we have

$$C^*(x)\sum_{k=1}^{n} a_{n,k}\frac{(x-k+2)_{k-1}(x+1+u)_{k-1}}{(k-1)!}$$
$$-D^*(x)\sum_{k=1}^{n} a_{n,k}\frac{(x-k+1)_{k-1}(x+u)_{k-1}}{(k-1)!} = \lambda_n\rho^*(x)y_n(x(x+u)).$$

For a *second simplification* we note that

$$(x-k+2)_{k-1}(x+1+u)_{k-1} - (x-k+1)_{k-1}(x+u)_{k-1}$$
$$= (k-1)(2x+u)(x-k+2)_{k-2}(x+1+u)_{k-2}$$

for $k = 2,3,4,\ldots$. Now we define

$$C^*(x) - D^*(x) = (2x+u)B(x) \quad \text{and} \quad \rho^*(x) = (2x+u)r(x) \tag{7.2.5}$$

with the assumption that $2x \neq -u$. Without loss of generality we may choose $r(x) = 1$ so that we have

$$B(x)\sum_{k=1}^{n} a_{n,k}\frac{(x-k+1)_{k-1}(x+u)_{k-1}}{(k-1)!} + C^*(x)\sum_{k=2}^{n} a_{n,k}\frac{(x-k+2)_{k-2}(x+1+u)_{k-2}}{(k-2)!}$$
$$= \lambda_n\sum_{k=0}^{n} a_{n,k}\frac{(x-k+1)_k(x+u)_k}{k!} \tag{7.2.6}$$

for $n = 2,3,4,\ldots$. For $n = 0$ we have $y_0(x(x+u)) = a_{0,0}\,(\neq 0)$ which, if (7.2.1) is used, leads to $\lambda_0 = 0$ (except for the trivial situation that $r(x) = 0$). For $n = 1$ we have $y_1(x(x+u)) = a_{1,0} + a_{1,1}x(x+u)$ with $a_{1,1} \neq 0$, which leads to

$$B(x)a_{1,1} = \lambda_1\left\{a_{1,0} + a_{1,1}x(x+u)\right\}.$$

Since all eigenvalues must be different, we conclude that $\lambda_1 \neq 0\,(=\lambda_0)$. Hence

$$B(x) = v + wx(x+u) \quad \text{with} \quad v,w \in \mathbb{R}, \quad w \neq 0. \tag{7.2.7}$$

This form can be used as a *third simplification*. For the first term on the left-hand side of (7.2.6) we obtain

$$\sum_{k=0}^{n} wka_{n,k}\frac{(x-k+1)_k(x+u)_k}{k!}+\sum_{k=1}^{n}\left\{v+w(k-1)^2\right\}a_{n,k}\frac{(x-k+2)_{k-1}(x+u)_{k-1}}{(k-1)!}$$
$$-\sum_{k=2}^{n}(v+wx^2)(x+u)a_{n,k}\frac{(x-k+2)_{k-2}(x+1+u)_{k-2}}{(k-2)!}. \tag{7.2.8}$$

This implies that $C^*(x)$ must be of the form

$$\begin{aligned} C^*(x) &= (v+wx^2)(x+u)+\sigma x(x+u)+\tau x(x-1)(x+u)(x+1+u) \\ &= (x+u)\left\{v+\sigma x+wx^2+\tau x(x-1)(x+1+u)\right\}, \quad \sigma,\tau\in\mathbb{R}. \end{aligned} \tag{7.2.9}$$

This leads to the following theorem.

Theorem 7.1. *The difference equation (7.2.1) only has real polynomial solutions $y_n(x(x+u))$ with degree$[y_n]=n$ in $x(x+u)$ for $n=0,1,2,\ldots$ if the coefficients $C(x)$, $D(x)$ and $\rho(x)$ have the form*

$$C(x)=\varphi(x+1)=(x+u)(2x-1+u)\left\{v+\sigma x+wx^2+\tau x(x-1)(x+1+u)\right\},$$

$$\begin{aligned} D(x) &= \varphi(x+1)-\psi(x) \\ &= x(2x+1+u)\left\{-v+\sigma(x+u)-w(x+u)^2+\tau(x-1)(x+u)(x+1+u)\right\} \end{aligned}$$

and

$$\rho(x)=(2x-1+u)(2x+u)(2x+1+u)$$

with $u,v,w,\sigma,\tau\in\mathbb{R}$ and $w\neq 0$.

Note that the assumption that $2x\notin\{-u-1,-u,-u+1\}$ can be dropped.

7.3 The Hypergeometric Representation

In order to find the hypergeometric representation of the polynomials in the form (7.2.3), we use (7.2.7), (7.2.9) and (7.2.8) to obtain from (7.2.6)

$$\begin{aligned} &\sum_{k=0}^{n}\{w+(k-1)\tau\}a_{n,k}\frac{(x-k+1)_k(x+u)_k}{k!}k \\ &\qquad+\sum_{k=1}^{n}\{v+(k-1)\left(\sigma+(k-1)w+(k-2)(u+k)\tau\right)\} \\ &\qquad\qquad\times a_{n,k}\frac{(x-k+2)_{k-1}(x+u)_{k-1}}{(k-1)!} \\ &=\lambda_n\sum_{k=0}^{n}a_{n,k}\frac{(x-k+1)_k(x+u)_k}{k!}, \quad n=1,2,3,\ldots. \end{aligned}$$

By comparing the coefficients of $(x-k+1)_k(x+u)_k$ on both sides, we find that

$$k\{w+(k-1)\tau\}a_{n,k}+\{v+k(\sigma+kw+(k-1)(u+k+1)\tau)\}a_{n,k+1}=\lambda_n a_{n,k},$$

which holds for $k=0,1,2,\ldots,n$ if we define $a_{n,n+1}:=0$. This leads to the eigenvalues

$$\lambda_n=n((n-1)\tau+w),\quad n=0,1,2,\ldots \tag{7.3.1}$$

and the two-term recurrence relation

$$\begin{aligned}(n-k)&\{(n+k-1)\tau+w\}a_{n,k}\\&=\{v+k(\sigma+kw+(k-1)(u+k+1)\tau)\}a_{n,k+1}\end{aligned} \tag{7.3.2}$$

for $k=n-1,n-2,n-3,\ldots,0$. Hence the coefficients $\{a_{n,k}\}_{k=0}^{n}$ in (7.2.3) are uniquely determined in terms of $a_{n,n}\neq 0$ if

$$(n+k-1)\tau+w\neq 0 \tag{7.3.3}$$

for $k=n-1,n-2,n-3,\ldots,0$ and $n\in\{1,2,3,\ldots\}$. This condition holds if the eigenvalues in (7.3.1) are all different. In the sequel we will always assume that this holds.

Since $w\neq 0$, we only have to consider two different cases.

Case I. $\tau=0$ and $w=1$. Then we write

$$C^*(x)=(x+u)(x^2+\sigma x+v)=(x+u)(x+x_1)(x+x_2),\quad \sigma=x_1+x_2,\quad v=x_1x_2.$$

Then we use (7.2.5) and (7.2.7) to show that

$$\begin{aligned}D^*(x)&=C^*(x)-(2x+u)B(x)\\&=(x+u)(x+x_1)(x+x_2)-(2x+u)\left(x_1x_2+x(x+u)\right)\\&=-x(x+u-x_1)(x+u-x_2).\end{aligned}$$

Hence, if (7.2.4), (7.2.5) and (7.3.1) are used, the difference equation (7.2.1) reads

$$\begin{aligned}&(x+1+u)(2x+1+u)(x+1+x_1)(x+1+x_2)\Delta^2 y_n(x(x+u))\\&\quad+\{(x+1+u)(2x+1+u)(x+1+x_1)(x+1+x_2)\\&\qquad+(x+1)(2x+3+u)\\&\qquad\quad\times(x+1+u-x_1)(x+1+u-x_2)\}\,\Delta y_n(x(x+u))\\&=n(2x+1+u)(2x+2+u)(2x+3+u)y_n((x+1)(x+1+u))\end{aligned} \tag{7.3.4}$$

for $n=0,1,2,\ldots$, and the symmetric form (7.2.2) reads

$$\begin{aligned}
&(x+u)(2x-1+u)(x+x_1)(x+x_2)y_n((x+1)(x+1+u))\\
&\quad-\{(x+u)(2x-1+u)(x+x_1)(x+x_2)\\
&\qquad-x(2x+1+u)(x+u-x_1)(x+u-x_2)\}\,y_n(x(x+u))\\
&\quad-x(2x+1+u)(x+u-x_1)(x+u-x_2)y_n((x-1)(x-1+u))\\
&=n(2x-1+u)(2x+u)(2x+1+u)y_n(x(x+u)),\quad n=0,1,2,\ldots. \qquad (7.3.5)
\end{aligned}$$

For the two-term recurrence relation (7.3.2) we obtain

$$(n-k)a_{n,k}=(k+x_1)(k+x_2)a_{n,k+1},\quad k=n-1,n-2,n-3,\ldots,0.$$

This implies that for the monic polynomials, id est $a_{n,n}=n!$, we have

$$a_{n,k}=\frac{(k+x_1)_{n-k}(k+x_2)_{n-k}}{(n-k)!}n!=\frac{(x_1)_n(x_2)_n}{(x_1)_k(x_2)_k}(-1)^k(-n)_k,\quad k=0,1,2,\ldots,n.$$

By using (7.2.3) and $(-1)^k(x-k+1)_k=(-x)_k$, we obtain the representation

$$\begin{aligned}
y_n(x(x+u))&=(x_1)_n(x_2)_n\sum_{k=0}^{n}\frac{(-n)_k(-x)_k(x+u)_k}{(x_1)_k(x_2)_k k!}\\
&=(x_1)_n(x_2)_n\,{}_3F_2\left(\begin{matrix}-n,-x,x+u\\ x_1,x_2\end{matrix};1\right),\quad n=0,1,2,\ldots \qquad (7.3.6)
\end{aligned}$$

for the **dual Hahn** polynomials. This is explained by the observation that the (not normalized) Hahn polynomials can be written as (cf. section 5.3)

$$y_n^*(x)={}_3F_2\left(\begin{matrix}-n,n-1+2\varepsilon,\varepsilon-\delta+1-f-x\\ \varepsilon-\delta+\eta,\varepsilon-\delta-\eta\end{matrix};1\right),\quad n=0,1,2,\ldots.$$

If we now set $f-\varepsilon+\delta=1$, $\varepsilon-\delta+\eta=x_1$, $\varepsilon-\delta-\eta=x_2$ and $u=2\varepsilon-1$, then we obtain

$$y_n^*(x)={}_3F_2\left(\begin{matrix}-n,n+u,-x\\ x_1,x_2\end{matrix};1\right),\quad n=0,1,2,\ldots.$$

For the (not normalized) polynomials in (7.3.6) we have

$$z_n^*(x(x+u))={}_3F_2\left(\begin{matrix}-n,-x,x+u\\ x_1,x_2\end{matrix};1\right),\quad n=0,1,2,\ldots.$$

Apparently we have $y_n^*(x)=z_x^*(n(n+u))$ for $n,x=0,1,2,\ldots$; $\{y_n^*(x)\}$ and $\{z_n^*(x(x+u))\}$ are dual with respect to the sequences of eigenvalues $\{\kappa_n\}$ with $\kappa_n=n$ and $\{\lambda_n\}$ with $\lambda_n=n(n+u)$. Note that we also have $y_n(\kappa_0)=z_0(\lambda_n)$.

Case II. $\tau=1$. Then we write

$$\begin{aligned}
C^*(x)&=(x+u)\left\{x^3+(u+w)x^2+(\sigma-u-1)x+v\right\}\\
&=(x+u)(x+x_1)(x+x_2)(x+x_3)
\end{aligned}$$

where

$$w = x_1 + x_2 + x_3 - u, \quad \sigma = x_1x_2 + x_1x_3 + x_2x_3 + u + 1 \quad \text{and} \quad v = x_1x_2x_3.$$

Then we also obtain by using (7.2.5) and (7.2.7)

$$\begin{aligned} D^*(x) &= C^*(x) - (2x+u)B(x) \\ &= (x+u)(x+x_1)(x+x_2)(x+x_3) \\ &\qquad - (2x+u)\left(x_1x_2x_3 + x(x+u)(x_1+x_2+x_3-u)\right) \\ &= x(x+u-x_1)(x+u-x_2)(x+u-x_3). \end{aligned}$$

Hence, if (7.2.4), (7.2.5) and (7.3.1) are used, the difference equation (7.2.1) reads

$$\begin{aligned} &(x+1+u)(2x+1+u)(x+1+x_1)(x+1+x_2)(x+1+x_3)\Delta^2 y_n(x(x+u)) \\ &\quad + \{(x+1+u)(2x+1+u)(x+1+x_1)(x+1+x_2)(x+1+x_3) \\ &\qquad -(x+1)(2x+3+u)(x+1+u-x_1) \\ &\qquad\quad \times (x+1+u-x_2)(x+1+u-x_3)\}\,\Delta y_n(x(x+u)) \\ &= n(n-1+x_1+x_2+x_3-u) \\ &\quad \times (2x+1+u)(2x+2+u)(2x+3+u)y_n((x+1)(x+1+u)) \end{aligned} \tag{7.3.7}$$

and the symmetric form (7.2.2) reads

$$\begin{aligned} &(x+u)(2x-1+u)(x+x_1)(x+x_2)(x+x_3)y_n((x+1)(x+1+u)) \\ &\quad - \{(x+u)(2x-1+u)(x+x_1)(x+x_2)(x+x_3) \\ &\qquad + x(2x+1+u)(x+u-x_1)(x+u-x_2)(x+u-x_3)\}\,y_n(x(x+u)) \\ &\quad + x(2x+1+u)(x+u-x_1)(x+u-x_2)(x+u-x_3)y_n((x-1)(x-1+u)) \\ &= n(n-1+x_1+x_2+x_3-u)(2x-1+u)(2x+u)(2x+1+u)y_n(x(x+u)) \end{aligned} \tag{7.3.8}$$

for $n = 0, 1, 2, \ldots$. For the two-term recurrence relation (7.3.2) we obtain

$$(n-k)(n+k-1+x_1+x_2+x_3-u)a_{n,k} = (k+x_1)(k+x_2)(k+x_3)a_{n,k+1}$$

for $k = n-1, n-2, n-3, \ldots, 0$. This implies that for the monic polynomials, id est $a_{n,n} = n!$, we have

$$\begin{aligned} a_{n,k} &= \frac{(k+x_1)_{n-k}(k+x_2)_{n-k}(k+x_3)_{n-k}}{(n+k-1+x_1+x_2+x_3-u)_{n-k}(n-k)!}n! \\ &= \frac{(x_1)_n(x_2)_n(x_3)_n}{(n-1+x_1+x_2+x_3-u)_n}\,\frac{(n-1+x_1+x_2+x_3-u)_k}{(x_1)_k(x_2)_k(x_3)_k}(-1)^k(-n)_k \end{aligned}$$

for $k = 0, 1, 2, \ldots, n$. If (7.2.3) and $(-1)^k(x-k+1)_k = (-x)_k$ are used, this leads for $n = 0, 1, 2, \ldots$ to the representation

$$\begin{aligned} y_n(x(x+u)) &= \frac{(x_1)_n(x_2)_n(x_3)_n}{(n-1+x_1+x_2+x_3-u)_n} \\ &\quad \times \sum_{k=0}^{n} \frac{(-n)_k(n-1+x_1+x_2+x_3-u)_k(-x)_k(x+u)_k}{(x_1)_k(x_2)_k(x_3)_k k!} \\ &= \frac{(x_1)_n(x_2)_n(x_3)_n}{(n-1+x_1+x_2+x_3-u)_n} \\ &\quad \times {}_4F_3\left(\begin{matrix} -n, n-1+x_1+x_2+x_3-u, -x, x+u \\ x_1, x_2, x_3 \end{matrix}; 1\right) \end{aligned} \tag{7.3.9}$$

for the **Racah** polynomials. These polynomials have a certain symmetry in n and x which can be seen as follows. The (not normalized) Racah polynomials given by (7.3.9) can be written as

$$y_n^*(x(x+u)) = {}_4F_3\left(\begin{matrix} -n, n-1+x_1+x_2+x_3-u, -x, x+u \\ x_1, x_2, x_3 \end{matrix}; 1\right) \tag{7.3.10}$$

for $n = 0, 1, 2, \ldots$. If we replace u by $-1+x_1+x_2+x_3-u$, then we have

$$\begin{aligned} & z_n^*(x(x-1+x_1+x_2+x_3-u)) \\ &= {}_4F_3\left(\begin{matrix} -n, n+u, -x, x-1+x_1+x_2+x_3-u \\ x_1, x_2, x_3 \end{matrix}; 1\right) \end{aligned} \tag{7.3.11}$$

for $n = 0, 1, 2, \ldots$. Now we have $y_n^*(x(x+u)) = z_x^*(n(n-1+x_1+x_2+x_3-u))$ for $n, x = 0, 1, 2, \ldots$ and $\{y_n^*(x(x+u))\}$ and $\{z_n^*(x(x-1+x_1+x_2+x_3-u))\}$ are dual polynomial systems with respect to the sequences of eigenvalues $\{\kappa_n\}$ with $\kappa_n = n(n+u)$ and $\{\lambda_n\}$ with $\lambda_n = n(n-1+x_1+x_2+x_3-u)$. Note that we also have $y_n^*(\kappa_0) = z_0^*(\lambda_n)$. The polynomials $z_n^*(x(x-1+x_1+x_2+x_3-u))$ can be considered as dual Racah polynomials. This fact will be used in the next section.

7.4 The Three-Term Recurrence Relation

In order to obtain the three-term recurrence relation, we use the concept of duality. Again we consider the two different cases.

Case I. For the dual Hahn polynomials we start with the difference equation (7.1.1) for the Hahn polynomials. The coefficients of this difference equation give us the coefficients of the three-term recurrence relation for the dual Hahn polynomials. We have

$$C(x) = (x-1+f+\eta)(x-1+f-\eta) = (x+x_1)(x+x_2)$$

and

$$D(x) = (x-1+f-\varepsilon+\delta)(x-1+f-\varepsilon-\delta) = x(x-1+x_1+x_2-u)$$

with

$$f-\varepsilon+\delta=1,\quad x_1=f-1+\eta,\quad x_2=f-1-\eta\quad\text{and}\quad u=2\varepsilon-1.$$

This implies that the three-term recurrence relation for the dual Hahn polynomials $z_n(\lambda_x)$ can be written as

$$\begin{aligned}(n+x_1)(n+x_2)z_{n+1}(\lambda_x)-\{(n+x_1)(n+x_2)+n(n-1+x_1+x_2-u)\}\,z_n(\lambda_x)\\ +n(n-1+x_1+x_2-u)z_{n-1}(\lambda_x)=\lambda_x z_n(\lambda_x),\quad n=1,2,3,\ldots\end{aligned}$$

with $z_0(\lambda_x)=1$, $z_1(\lambda_x)=1+\lambda_x/x_1x_2$ $(x_1x_2\neq 0)$ and $\lambda_x=x(x+u)$. For the monic dual Hahn polynomials $y_n(x(x+u))$ this can be written as

$$\begin{aligned}y_{n+1}(x(x+u))=\{&x(x+u)+(n+x_1)(n+x_2)\\ &+n(n-1+x_1+x_2-u)\}\,y_n(x(x+u))\\ &-n(n-1+x_1)(n-1+x_2)\\ &\times(n-1+x_1+x_2-u)y_{n-1}(x(x+u))\end{aligned}\tag{7.4.1}$$

for $n=1,2,3,\ldots$ with $y_0(x(x+u))=1$ and $y_1(x(x+u))=x(x+u)+x_1x_2$.

Case II. For the Racah polynomials we start with the difference equation (cf. (7.2.2)) for the (not normalized) Racah polynomials $y_n^*(x(x+u))$ given by (7.3.10):

$$C(x)y_n^*(\kappa_{x+1})-\{C(x)+D(x)\}\,y_n^*(\kappa_x)+D(x)y_n^*(\kappa_{x-1})=\lambda_n\rho(x)y_n^*(\kappa_x).$$

Now we have $y_n^*(\kappa_x)=z_x^*(\lambda_n)$ with $\kappa_x=x(x+u)$ and $\lambda_n=n(n-1+x_1+x_2+x_3-u)$, which implies that the difference equation for the dual Racah polynomials $z_x^*(\lambda_n)$ given by (7.3.11) can be written as

$$C(x)z_{x+1}^*(\lambda_n)-\{C(x)+D(x)\}\,z_x^*(\lambda_n)+D(x)z_{x-1}^*(\lambda_n)=\lambda_n\rho(x)z_x^*(\lambda_n).$$

For the coefficients we now have

$$C(x)=(x+u)(2x-1+u)(x+x_1)(x+x_2)(x+x_3)$$

and

$$D(x)=x(2x+1+u)(x+u-x_1)(x+u-x_2)(x+u-x_3)$$

which leads to the three-term recurrence relation

$$\begin{aligned}&(n+u)(2n-1+u)(n+x_1)(n+x_2)(n+x_3)z_{n+1}^*(\lambda_x)\\ &\quad-\{(n+u)(2n-1+u)(n+x_1)(n+x_2)(n+x_3)\\ &\qquad+n(2n+1+u)(n+u-x_1)(n+u-x_2)(n+u-x_3)\}\,z_n^*(\lambda_x)\\ &\quad+n(2n+1+u)(n+u-x_1)(n+u-x_2)(n+u-x_3)z_{n-1}^*(\lambda_x)\\ &=\lambda_x(2n-1+u)(2n+u)(2n+1+u)z_n^*(\lambda_x),\quad n=1,2,3,\ldots\end{aligned}$$

with $z_0^*(\lambda_x)=1$ and $z_1^*(\lambda_x)=1+(u+1)\lambda_x/x_1x_2x_3$ $(x_1x_2x_3\neq 0)$.

When u is replaced by $-1+x_1+x_2+x_3-u(=-1+w)$, the polynomials $z_n^*(\lambda_x)$ given by (7.3.11) change into the polynomials $y_n^*(\kappa_x)$ given by (7.3.10). This implies that the three-term recurrence relation for the polynomials $y_n^*(\kappa_x)$ can be written as

$$\begin{aligned}
&(n-1+w)(2n-2+w)(n+x_1)(n+x_2)(n+x_3)y_{n+1}^*(\kappa_x)\\
&\quad-\{(n-1+w)(2n-2+w)(n+x_1)(n+x_2)(n+x_3)\\
&\qquad+n(2n+w)(n-1+w-x_1)\\
&\qquad\quad\times(n-1+w-x_2)(n-1+w-x_3)\}\,y_n^*(\kappa_x)\\
&\quad+n(2n+w)(n-1+w-x_1)(n-1+w-x_2)(n-1+w-x_3)y_{n-1}^*(\kappa_x)\\
&=\kappa_x(2n-2+w)(2n-1+w)(2n+w)y_n^*(\kappa_x),\quad n=1,2,3,\dots
\end{aligned}$$

with $y_0^*(\kappa_x)=1$, $y_1^*(\kappa_x)=1+w\kappa_x/x_1x_2x_3$ $(x_1x_2x_3\neq 0)$ and $\kappa_x=x(x+u)$. The connection with the monic Racah polynomials $y_n(\kappa_x)$ is given by

$$y_n^*(\kappa_x)=\frac{(n-1+w)_n}{(x_1)_n(x_2)_n(x_3)_n}y_n(\kappa_x),\quad n=0,1,2,\dots.$$

Hence the three-term recurrence relation for the monic Racah polynomials $y_n(\kappa_x)$ can be written in the form

$$y_{n+1}(\kappa_x)=\left(\kappa_x+c_n^{(1)}+c_n^{(2)}\right)y_n(\kappa_x)-c_{n-1}^{(1)}c_n^{(2)}y_{n-1}(\kappa_x),\quad n=1,2,3,\dots \tag{7.4.2}$$

with $y_0(\kappa_x)=1$ and $y_1(x)=\kappa_x+x_1x_2x_3/w$, where $\kappa_x=x(x+u)$ and $w=x_1+x_2+x_3-u$ with $w\neq 0,-1,-2,\dots$,

$$c_n^{(1)}=\frac{(n-1+w)(n+x_1)(n+x_2)(n+x_3)}{(2n-1+w)(2n+w)},\quad n=0,1,2,\dots \tag{7.4.3}$$

and

$$c_n^{(2)}=\frac{n(n-1+w-x_1)(n-1+w-x_2)(n-1+w-x_3)}{(2n-2+w)(2n-1+w)},\quad n=1,2,3,\dots. \tag{7.4.4}$$

7.5 Classification of the Positive-Definite Orthogonal Polynomial Solutions

Favard's theorem (theorem 3.1) can be extended for monic polynomials $y_n(x(x+u))$ of degree $n\in\{0,1,2,\dots\}$ with $u\in\mathbb{R}$. The polynomials given by

$$y_{n+1}(x(x+u))=\{x(x+u)-c_n\}y_n(x(x+u))-d_ny_{n-1}(x(x+u)) \tag{7.5.1}$$

for $n=1,2,3,\dots$ with $c_n,d_n\in\mathbb{R}$, $y_0(x(x+u))=1$ and $y_1(x(x+u))=x(x+u)-c_0$ $(c_0\in\mathbb{R})$ are orthogonal with respect to a positive-definite linear functional Λ, id est

$$\Lambda\left[y_m(x(x+u))y_n(x(x+u))\right] = \left(\prod_{k=0}^{n} d_k\right)\delta_{mn}, \quad m,n=0,1,2,\ldots, \tag{7.5.2}$$

where $\Lambda[y_0(x(x+u))] = d_0 \in \mathbb{R}$ and $\Lambda[y_n(x(x+u))] = 0$ for $n = 1,2,3,\ldots$ iff $d_0, d_1, d_2, \ldots, d_n$ are positive. The proof is similar to the proof of theorem 3.1.

Case I. For the monic dual Hahn polynomials given by (7.3.6) we have the three-term recurrence relation (7.4.1) with

$$c_n = -(n+x_1)(n+x_2) - n(n-1+x_1+x_2-u), \quad n=0,1,2,\ldots$$

and

$$d_n = n(n-1+x_1)(n-1+x_2)(n-1+x_1+x_2-u), \quad n=1,2,3,\ldots, \quad u\in\mathbb{R}.$$

Since $u \in \mathbb{R}$, we have $c_n \in \mathbb{R}$ for all $n = 0,1,2,\ldots$ if $x_1x_2 \in \mathbb{R}$ and $x_1 + x_2 \in \mathbb{R}$. This implies that $x_1, x_2 \in \mathbb{R}$ or x_1 and x_2 are complex conjugates. Positive-definite orthogonality occurs in six infinite and seven finite cases given by table 7.1, where x_1 and x_2 can be interchanged because of the symmetry.

case	conditions	positive
1	$x_{1,2} > 0,\ x_1+x_2-u > 0$	$d_1, d_2, d_3, \ldots$
2	$x_{1,2} = \alpha \pm i\beta\ (\alpha,\beta \in \mathbb{R},\ \beta \neq 0),\ 2\alpha - u > 0$	$d_1, d_2, d_3, \ldots$
3	$x_1 > 0,\ -N < x_2 < -N+1,\ -N < x_1+x_2-u < -N+1$	$d_1, d_2, d_3, \ldots$
4	$x_1 > 0,\ x_2 \leq -N,\ -N < x_1+x_2-u < -N+1$	$d_1, d_2, d_3, \ldots, d_N$
5	$x_1 > 0,\ -N < x_2 < -N+1,\ x_1+x_2-u \leq -N$	$d_1, d_2, d_3, \ldots, d_N$
6	$x_1 > 0,\ x_2 \leq -N,\ x_1+x_2-u \leq -N$	
	$-N-j < x_2 \leq -N-j+1,\ j \in \{1,2,3,\ldots\}$ $-N-k < x_1+x_2-u \leq -N-k+1,\ k \in \{1,2,3,\ldots\}$	
a	$j=k$ with $x_2 \neq -N-j+1$ and $x_1+x_2-u \neq -N-j+1$	$d_1, d_2, d_3, \ldots$
b	$j=k$ with $x_2 = -N-j+1$ or $x_1+x_2-u = -N-j+1$	$d_1, d_2, d_3, \ldots, d_{N+j-1}$
c	$j \neq k$	$d_1, d_2, d_3, \ldots, d_{N+\min(j,k)-1}$
7	$-N < x_{1,2} < -N+1,\ x_1+x_2-u > 0$	$d_1, d_2, d_3, \ldots$
8	$-N < x_1 < -N+1,\ x_2 \leq -N,\ x_1+x_2-u > 0$	$d_1, d_2, d_3, \ldots, d_N$
9	$x_{1,2} \leq -N,\ x_1+x_2-u > 0$	
	$-N-j < x_1 \leq -N-j+1,\ j \in \{1,2,3,\ldots\}$ $-N-k < x_2 \leq -N-k+1,\ k \in \{1,2,3,\ldots\}$	
a	$j=k$ with $x_1 \neq -N-j+1$ and $x_2 \neq -N-j+1$	$d_1, d_2, d_3, \ldots$
b	$j=k$ with $x_1 = -N-j+1$ or $x_2 = -N-j+1$	$d_1, d_2, d_3, \ldots, d_{N+j-1}$
c	$j \neq k$	$d_1, d_2, d_3, \ldots, d_{N+\min(j,k)-1}$

Table 7.1 Case I, $N \in \{1,2,3,\ldots\}$.

Case II. For the monic Racah polynomials given by (7.3.9) we have the three-term recurrence relation (7.4.2) with

$$c_0 = -c_0^{(1)}, \quad c_n = -c_n^{(1)} - c_n^{(2)}, \quad n = 1,2,3,\ldots$$

and

$$d_n = c_{n-1}^{(1)} c_n^{(2)} = \frac{n(n-2+w)}{(2n-3+w)(2n-2+w)^2(2n-1+w)} D_n, \quad n = 1,2,3,\ldots$$

and

$$\begin{aligned} D_n = {} & (n-1+x_1)(n-1+x_2)(n-1+x_3) \\ & \times (n-1+w-x_1)(n-1+w-x_2)(n-1+w-x_3) \end{aligned} \tag{7.5.3}$$

for $n = 1,2,3,\ldots$, where $w \in \mathbb{R}$. Since $w \in \mathbb{R}$, we have $c_n \in \mathbb{R}$ for all $n = 0,1,2,\ldots$ if $x_1x_2x_3 \in \mathbb{R}$, $x_1+x_2+x_3 \in \mathbb{R}$ and $x_1x_2+x_1x_3+x_2x_3 \in \mathbb{R}$. This implies that $x_1, x_2, x_3 \in \mathbb{R}$ or one is real and the other two are complex conjugates. Note the similarity with case III in section 4.2 with 2ε replaced by w. As in section 4.2 we conclude that for $w > 0$ we have

$$\frac{n(n-2+w)}{(2n-3+w)(2n-2+w)^2(2n-1+w)} > 0, \quad n = 1,2,3,\ldots$$

and for $w = -2N - t$ with $N \in \{1,2,3,\ldots\}$ and $-1 < t \leq 1$ we have

$$\frac{n(n-2+w)}{(2n-3+w)(2n-2+w)^2(2n-1+w)} < 0, \quad n = 1,2,3,\ldots,N$$

and this does no longer hold for $n = N+1$. In this case the range $-1 \leq w < 0$ must also be considered since this might also lead to positive-definite orthogonality. For $w = -1$ this is impossible since d_1 is not defined in that case. For $-1 < w < 0$ we have

$$\frac{n(n-2+w)}{(2n-3+w)(2n-2+w)^2(2n-1+w)} > 0, \quad n = 1,3,4,\ldots$$

and for $n = 2$

$$\frac{2w}{(w+1)(w+2)^2(w+3)} < 0.$$

Case IIa. $w > 0$. Then we must have $D_n > 0$ for $n = 1,2,3,\ldots$.
For $w > 0$ and $x_i < 0$ for $i = 1,2,3$ we have $w - x_i > 0$ for $i = 1,2,3$. However, $w > 0$ and $w - x_i < 0$ for $i = 1,2,3$ implies that $x_i > 0$ for $i = 1,2,3$. Therefore, D_1 can only be positive if all six factors in (7.5.3) with $n = 1$ are positive or four are positive and two are negative. This observation leads to positive-definite orthogonality in eight infinite and ten finite cases given by table 7.2.

case		conditions	positive
1		$0 < x_{1,2,3} < w$	$d_1, d_2, d_3, \ldots$
2		$0 < x_1 < w,\ x_{2,3} = \alpha \pm i\beta\ (\alpha, \beta \in \mathbb{R},\ \beta \neq 0)$	$d_1, d_2, d_3, \ldots$
3		$0 < x_1 < w,\ -N < x_{2,3} < -N+1$	$d_1, d_2, d_3, \ldots$
4		$0 < x_1 < w,\ x_2 \leq -N,\ -N < x_3 < -N+1$	$d_1, d_2, d_3, \ldots, d_N$
5		$0 < x_1 < w,\ x_{2,3} \leq -N$	
		$-N-j < x_2 \leq -N-j+1,\ j \in \{1,2,3,\ldots\}$ $-N-k < x_3 \leq -N-k+1,\ k \in \{1,2,3,\ldots\}$	
	a	$j = k$ with $x_2 \neq -N-j+1$ and $x_3 \neq -N-j+1$	$d_1, d_2, d_3, \ldots$
	b	$j = k$ with $x_2 = -N-j+1$ or $x_3 = -N-j+1$	$d_1, d_2, d_3, \ldots, d_{N+j-1}$
	c	$j \neq k$	$d_1, d_2, d_3, \ldots, d_{N+\min(j,k)-1}$
6		$0 < x_1 < w,\ -N < w - x_{2,3} < -N+1$	$d_1, d_2, d_3, \ldots$
7		$0 < x_1 < w,\ -N < w - x_2 < -N+1,\ w - x_3 \leq -N$	$d_1, d_2, d_3, \ldots, d_N$
8		$0 < x_1 < w,\ w - x_{2,3} \leq -N$	
		$-N-j < w - x_2 \leq -N-j+1,\ j \in \{1,2,3,\ldots\}$ $-N-k < w - x_3 \leq -N-k+1,\ k \in \{1,2,3,\ldots\}$	
	a	$j = k$ with $w - x_2 \neq -N-j+1$ and $w - x_3 \neq -N-j+1$	$d_1, d_2, d_3, \ldots$
	b	$j = k$ with $w - x_2 = -N-j+1$ or $w - x_3 = -N-j+1$	$d_1, d_2, d_3, \ldots, d_{N+j-1}$
	c	$j \neq k$	$d_1, d_2, d_3, \ldots, d_{N+\min(j,k)-1}$
9		$0 < x_1 < w,\ -N < x_2 < -N+1,\ -N < w - x_3 < -N+1$	$d_1, d_2, d_3, \ldots$
10		$0 < x_1 < w,\ x_2 \leq -N,\ -N < w - x_3 < -N+1$	$d_1, d_2, d_3, \ldots, d_N$
11		$0 < x_1 < w,\ -N < x_2 < -N+1,\ w - x_3 \leq -N$	$d_1, d_2, d_3, \ldots, d_N$
12		$0 < x_1 < w,\ x_2 \leq -N,\ w - x_3 \leq -N$	
		$-N-j < x_2 \leq -N-j+1,\ j \in \{1,2,3,\ldots\}$ $-N-k < w - x_3 \leq -N-k+1,\ k \in \{1,2,3,\ldots\}$	
	a	$j = k$ with $x_2 \neq -N-j+1$ and $w - x_3 \neq -N-j+1$	$d_1, d_2, d_3, \ldots$
	b	$j = k$ with $x_2 = -N-j+1$ or $w - x_3 = -N-j+1$	$d_1, d_2, d_3, \ldots, d_{N+j-1}$
	c	$j \neq k$	$d_1, d_2, d_3, \ldots, d_{N+\min(j,k)-1}$

Table 7.2 Case IIa, $N \in \{1,2,3,\ldots\}$.

Case IIb. $-1 < w < 0$. In this case we must have $D_2 < 0$ and $D_n > 0$ for $n = 1,3,4,\ldots$.
For $w < 0$ and $x_i > 0$ for $i = 1,2,3$ we have $w - x_i < 0$ for $i = 1,2,3$. However, $w < 0$ and $w - x_i > 0$ for $i = 1,2,3$ implies that $x_i < 0$ for $i = 1,2,3$. Therefore, D_1 can

case	conditions	positive
1	$w < x_1 < 0,\ -2 < x_2 < -1,\ -1 < x_3 < w$	$d_1, d_2, d_3, \ldots$
2	$w < x_1 < 0,\ -2 < x_2 < -1,\ 0 < x_3 < w+1$	$d_1, d_2, d_3, \ldots$
3	$w < x_1 < 0,\ w+1 < x_2 < w+2,\ -1 < x_3 < w$	$d_1, d_2, d_3, \ldots$
4	$w < x_1 < 0,\ w+1 < x_2 < w+2,\ 0 < x_3 < w+1$	$d_1, d_2, d_3, \ldots$

Table 7.3 Case IIb.

only be positive if all six factors in (7.5.3) with $n=1$ are negative or four factors are negative and two are positive.
For $w>-1$ and $x_i<0$ for $i=1,2,3$ we have $w-x_i>-1$ for $i=1,2,3$. However, $w>-1$ and $w-x_i<0$ for $i=1,2,3$ implies that $x_i>-1$ for $i=1,2,3$. We conclude that if all six factors in (7.5.3) with $n=1$ are negative, then all six factors for $n=2$ are positive. Therefore, D_2 can only be negative in the case that four factors in (7.5.3) with $n=2$ are negative and two are positive. Indeed, hereby only one negative pair $(x_i, w-x_i)$ with $i\in\{1,2,3\}$ may occur. This observation leads to positive-definite orthogonality in four infinite cases given by table 7.3.

Case IIc. $w=-2N-t$ with $N\in\{1,2,3,\ldots\}$ and $-1<t\le 1$. Then we must have $D_n<0$ for $n=1,2,3,\ldots$.

case	conditions	positive
1	$x_{1,2,3}>0$	$d_1,d_2,d_3,\ldots,d_N$
2	$x_{1,2}>0,\ x_3<-2N-t$	$d_1,d_2,d_3,\ldots,d_N$
3	$x_1>0,\ x_{2,3}=\alpha\pm i\beta\ (\alpha,\beta\in\mathbb{R},\ \beta\neq 0)$	$d_1,d_2,d_3,\ldots,d_N$
4	$x_1>0,\ x_{2,3}<-2N-t$	$d_1,d_2,d_3,\ldots,d_N$
5	$x_1>0,\ -N-t-1<x_{2,3}<-N+1$	$d_1,d_2,d_3,\ldots,d_N$
6	$x_1>0,\ -2N-t<x_2\le -N-t-1,$ $-N-t-1<x_3<-N+1$	
	$-N-t-j-1<x_2\le -N-t-j,\ j\in\{1,2,3,\ldots,N-1\}$	$d_1,d_2,d_3,\ldots,d_{N-j}$
7	$x_1>0,\ -2N-t<x_{2,3}\le -N-t-1$	
	$-N-t-j-1<x_2\le -N-t-j,\ j\in\{1,2,3,\ldots,N-1\}$ $-N-t-k-1<x_3\le -N-t-k,\ k\in\{1,2,3,\ldots,N-1\}$	
a	$j=k$ with $x_2\neq -N-t-j$ and $x_3\neq -N-t-j$	$d_1,d_2,d_3,\ldots,d_N$
b	$j=k$ with $x_2=-N-t-j$ or $x_3=-N-t-j$	$d_1,d_2,d_3,\ldots,d_{N-j}$
c	$j\neq k$	$d_1,d_2,d_3,\ldots,d_{N-\max(j,k)}$
8	$x_1>0,\ -N+1\le x_{2,3}<0$	
	$-N+j\le x_2<-N+j+1,\ j\in\{1,2,3,\ldots,N-1\}$ $-N+k\le x_3<-N+k+1,\ k\in\{1,2,3,\ldots,N-1\}$	
a	$j=k$ with $x_2\neq -N+j$ and $x_3\neq -N+j$	$d_1,d_2,d_3,\ldots,d_N$
b	$j=k$ with $x_2=-N+j$ or $x_3=-N+j$	$d_1,d_2,d_3,\ldots,d_{N-j}$
c	$j\neq k$	$d_1,d_2,d_3,\ldots,d_{N-\max(j,k)}$
9	$x_1>0,\ -N+1\le x_2<0,\ -N-t-1<x_3<-N+1$	
	$-N+j\le x_2<-N+j+1,\ j\in\{1,2,3,\ldots,N-1\}$	$d_1,d_2,d_3,\ldots,d_{N-j}$
10	$x_1>0,\ -N+1\le x_2<0,\ -2N-t<x_3\le -N-t-1$	
	$-N+j\le x_2<-N+j+1,\ j\in\{1,2,3,\ldots,N-1\}$ $-N-t-k-1<x_3\le -N-t-k,\ k\in\{1,2,3,\ldots,N-1\}$	
a	$j=k$ with $x_2\neq -N+j$ and $x_3\neq -N-t-j$	$d_1,d_2,d_3,\ldots,d_N$
b	$j=k$ with $x_2=-N+j$ or $x_3=-N-t-j$	$d_1,d_2,d_3,\ldots,d_{N-j}$
c	$j\neq k$	$d_1,d_2,d_3,\ldots,d_{N-\max(j,k)}$

Table 7.4 Case IIc1, $N\in\{1,2,3,\ldots\}$.

For $w < -1$ and $x_i > 0$ for $i = 1,2,3$ we have $w - x_i < -1$ for $i = 1,2,3$. However, $w < -1$ and $w - x_i > 0$ for $i = 1,2,3$ implies that $x_i < -1$ for $i = 1,2,3$. Therefore, D_1 can only be negative if three factors in (7.5.3) with $n = 1$ are negative and three are positive or one factor is positive and five are negative. This observation leads to positive-definite orthogonality in four infinite and forty finite cases given by table 7.4, table 7.5, table 7.6 and table 7.7.

In table 7.4 we list all cases with $x_1 > 0$.

case	conditions	positive
11	$x_{1,2,3} < -2N-t$	$d_1, d_2, d_3, \ldots, d_N$
12	$x_1 < -2N-t,\ x_{2,3} = \alpha \pm i\beta\ (\alpha, \beta \in \mathbb{R},\ \beta \neq 0)$	$d_1, d_2, d_3, \ldots, d_N$
13	$x_1 < -2N-t,\ -N-t-1 < x_{2,3} < -N+1$	$d_1, d_2, d_3, \ldots, d_N$
14	$x_1 < -2N-t,\ -2N-t < x_2 \leq -N-t-1,$ $-N-t-1 < x_3 < -N+1$	
	$-N-t-j-1 < x_2 \leq -N-t-j,\ j \in \{1,2,3,\ldots,N-1\}$	$d_1, d_2, d_3, \ldots, d_{N-j}$
15	$x_1 < -2N-t,\ -2N-t < x_{2,3} \leq -N-t-1$	
	$-N-t-j-1 < x_2 \leq -N-t-j,\ j \in \{1,2,3,\ldots,N-1\}$ $-N-t-k-1 < x_3 \leq -N-t-k,\ k \in \{1,2,3,\ldots,N-1\}$	
a	$j = k$ with $x_2 \neq -N-t-j$ and $x_3 \neq -N-t-j$	$d_1, d_2, d_3, \ldots, d_N$
b	$j = k$ with $x_2 = -N-t-j$ or $x_3 = -N-t-j$	$d_1, d_2, d_3, \ldots, d_{N-j}$
c	$j \neq k$	$d_1, d_2, d_3, \ldots, d_{N-\max(j,k)}$
16	$x_1 < -2N-t,\ -N+1 \leq x_{2,3} < 0$	
	$-N+j \leq x_2 < -N+j+1,\ j \in \{1,2,3,\ldots,N-1\}$ $-N+k \leq x_3 < -N+k+1,\ k \in \{1,2,3,\ldots,N-1\}$	
a	$j = k$ with $x_2 \neq -N+j$ and $x_3 \neq -N+j$	$d_1, d_2, d_3, \ldots, d_N$
b	$j = k$ with $x_2 = -N+j$ or $x_3 = -N+j$	$d_1, d_2, d_3, \ldots, d_{N-j}$
c	$j \neq k$	$d_1, d_2, d_3, \ldots, d_{N-\max(j,k)}$
17	$x_1 < -2N-t,\ -N+1 \leq x_2 < 0,\ -N-t-1 < x_3 < -N+1$	
	$-N+j \leq x_2 < -N+j+1,\ j \in \{1,2,3,\ldots,N-1\}$	$d_1, d_2, d_3, \ldots, d_{N-j}$
18	$x_1 < -2N-t,\ -N+1 \leq x_2 < 0,\ -2N-t < x_3 \leq -N-t-1$	
	$-N+j \leq x_2 < -N+j+1,\ j \in \{1,2,3,\ldots,N-1\}$ $-N-t-k-1 < x_3 \leq -N-t-k,\ k \in \{1,2,3,\ldots,N-1\}$	
a	$j = k$ with $x_2 \neq -N+j$ and $x_3 \neq -N-t-j$	$d_1, d_2, d_3, \ldots, d_N$
b	$j = k$ with $x_2 = -N+j$ or $x_3 = -N-t-j$	$d_1, d_2, d_3, \ldots, d_{N-j}$
c	$j \neq k$	$d_1, d_2, d_3, \ldots, d_{N-\max(j,k)}$

Table 7.5 Case IIc2, $N \in \{1,2,3,\ldots\}$.

In all cases, the condition $x_1 > 0$ can be replaced by the condition $x_1 < -2N-t$. This leads to the extra cases listed in table 7.5. Even more cases can be obtained by interchanging the role of x_1 and $w - x_1$ in table 7.5. If we distinguish between $-1 < t < 0$ and $0 < t < 1$ we have more cases. For $-1 < t < 0$ we have the cases listed in table 7.6 and for $0 < t < 1$ we have the cases listed in table 7.7.

case	conditions	positive
19	$-t < x_1 < -t+1,\ -N-t-1 < x_2 < -N,\ -N < x_3 < -N-t$	$d_1, d_2, d_3, \ldots$
20	$-t < x_1 < -t+1,\ -N-t < x_2 < -N+1,\ -N < x_3 < -N-t$	$d_1, d_2, d_3, \ldots$
21	$0 < x_1 \leq -t,\ -N-t-1 < x_2 < -N,\ -N < x_3 < -N-t$	$d_1, d_2, d_3, \ldots, d_{2N}$
22	$0 < x_1 \leq -t,\ -N-t < x_2 < -N+1,\ -N < x_3 < -N-t$	$d_1, d_2, d_3, \ldots, d_{2N}$
23	$x_1 \geq -t+1,\ -N-t-1 < x_2 < -N,\ -N < x_3 < -N-t$	$d_1, d_2, d_3, \ldots, d_{2N+1}$
24	$x_1 \geq -t+1,\ -N-t < x_2 < -N+1,\ -N < x_3 < -N-t$	$d_1, d_2, d_3, \ldots, d_{2N+1}$

Table 7.6 Case IIc3, $N \in \{1,2,3,\ldots\}$ and $-1 < t < 0$.

case	conditions	positive
25	$-t+1 < x_1 < -t+2,\ -N-1 < x_2 < -N-t,\ -N-t < x_3 < -N$	$d_1, d_2, d_3, \ldots$
26	$-t+1 < x_1 < -t+2,\ -N < x_2 < -N-t+1,\ -N-t < x_3 < -N$	$d_1, d_2, d_3, \ldots$
27	$0 < x_1 \leq -t+1,\ -N-1 < x_2 < -N-t,\ -N-t < x_3 < -N$	$d_1, d_2, d_3, \ldots, d_{2N+1}$
28	$0 < x_1 \leq -t+1,\ -N < x_2 < -N-t+1,\ -N-t < x_3 < -N$	$d_1, d_2, d_3, \ldots, d_{2N+1}$
29	$x_1 \geq -t+2,\ -N-1 < x_2 < -N-t,\ -N-t < x_3 < -N$	$d_1, d_2, d_3, \ldots, d_{2N+2}$
30	$x_1 \geq -t+2,\ -N < x_2 < -N-t+1,\ -N-t < x_3 < -N$	$d_1, d_2, d_3, \ldots, d_{2N+2}$
31	$x_1 > 0,\ -N-t+1 \leq x_2 < -N+1,\ -N-t < x_3 < -N$	$d_1, d_2, d_3, \ldots, d_{N+1}$
32	$x_1 > 0,\ -N-t-1 < x_2 \leq -N-1,\ -N-t < x_3 < -N$	$d_1, d_2, d_3, \ldots, d_{N+1}$

Table 7.7 Case IIc4, $N \in \{1,2,3,\ldots\}$ and $0 < t < 1$.

7.6 The Self-Adjoint Difference Equation

The orthogonality relations can be obtained in a similar way as in section 5.3. Therefore we write the difference equation (7.3.4) for the dual Hahn polynomials and the difference equation (7.3.7) for the Racah polynomials in the form (cf. (7.2.1))

$$\begin{aligned}&\varphi(x+2)\Delta^2 y_n(x(x+u)) + \psi(x+1)\Delta y_n(x(x+u))\\&= \lambda_n \rho(x+1) y_n((x+1)(x+1+u)), \quad n = 0,1,2,\ldots.\end{aligned}$$

If we multiply by $w(x+1)$, this can also be written in the self-adjoint form

$$\begin{aligned}&\Delta\left\{w(x)\varphi(x+1)\Delta y_n(x(x+u))\right\}\\&= \lambda_n \rho(x+1) w(x+1) y_n((x+1)(x+1+u)), \quad n = 0,1,2,\ldots, \end{aligned} \tag{7.6.1}$$

where $w(x)$ satisfies the Pearson difference equation

$$\Delta\left\{w(x)\varphi(x+1)\right\} = w(x+1)\psi(x+1). \tag{7.6.2}$$

As before, we multiply the difference equation (7.6.1) by $y_m((x+1)(x+1+u))$, subtract the result with m and n interchanged, and finally replace x by $x-1$. Then we obtain

$$(\lambda_n - \lambda_m)\rho(x)w(x)y_m(x(x+u))y_n(x(x+u)) = s_{n,m}(x) - s_{n,m}(x-1) \tag{7.6.3}$$

with

$$\begin{aligned} s_{n,m}(x) = w(x)\varphi(x+1)\{&y_n((x+1)(x+1+u))y_m(x(x+u)) \\ &- y_m((x+1)(x+1+u))y_n(x(x+u))\} \end{aligned} \tag{7.6.4}$$

for $m,n \in \{0,1,2,\ldots\}$. If the eigenvalues given by (7.3.1) are distinct, id est if (7.3.3) holds, then (7.6.3) and (7.6.4) lead to several different kinds of orthogonality relations with respect to the weight function $w^*(x) := \rho(x)w(x)$ similar to the case of chapter 5.

For the dual Hahn polynomials given by (7.3.6) and the Racah polynomials given by (7.3.9) we obtain an orthogonality relation of the form (cf. (5.1.11))

$$\sum_{x=A}^{A+N} w^*(x)y_m(x(x+u))y_n(x(x+u)) = \sigma_n\,\delta_{mn}, \quad m,n = 0,1,2,\ldots,N \tag{7.6.5}$$

with boundary conditions (cf. (5.1.12))

$$w^*(A-1)\varphi(A) = 0 \quad \text{and} \quad w^*(A+N)\varphi(A+N+1) = 0. \tag{7.6.6}$$

Compare with theorem 3.3, where we have $\varphi(x+1)$ in (7.6.1) instead of $\widehat{\varphi}(x-1)$ in (3.2.5). By using the theorem of Favard (cf. theorem 3.1, (3.1.4), (4.2.5) and (4.2.6)), we obtain

$$\sigma_n = \prod_{k=0}^{n} d_k, \quad n = 0,1,2,\ldots,N \quad \text{with} \quad d_0 = \sum_{x=A}^{A+N} w^*(x) \tag{7.6.7}$$

and $d_1, d_2, d_3, \ldots, d_N$ given by the three-term recurrence relation (7.4.1) or (7.4.2). If the second boundary condition in (7.6.6) cannot be satisfied for $N \in \{1,2,3,\ldots\}$, then it might be possible to use the condition

$$\lim_{x\to\infty} w^*(x)x^k = 0, \quad k = 0,1,2,\ldots \tag{7.6.8}$$

instead. In that case all moments of the form

$$\sum_{x=A}^{\infty} w^*(x)x^k, \quad k = 0,1,2,\ldots \tag{7.6.9}$$

should exist. As a third possibility the sum in (7.6.5) can be replaced by an improper integral over the (possibly deformed) imaginary axis (cf. page 109):

$$\begin{aligned} &\frac{1}{2\pi i}\int_{-i\infty}^{i\infty} w^*(x)y_m(x(x+u))y_n(x(x+u))\,dx \\ &= \sigma_n\,\delta_{mn}, \quad m,n = 0,1,2,\ldots,N. \end{aligned} \tag{7.6.10}$$

Also in this case all moments of the form

$$\frac{1}{2\pi i}\int_{-i\infty}^{i\infty} w^*(x)x^k\,dx, \quad k=0,1,2,\ldots \tag{7.6.11}$$

should exist. In the latter two cases we may also find infinite orthogonal polynomial systems.

For positive-definite orthogonality we must have $d_0 > 0$.

7.7 Orthogonality Relations for Dual Hahn Polynomials

In the case of the dual Hahn polynomials given by (7.3.6) we obtain by using (7.6.2) and (7.3.4) for the Pearson equation

$$\begin{aligned}\frac{w^*(x)}{w^*(x+1)} &= \frac{\rho(x)\{\varphi(x+2)-\psi(x+1)\}}{\rho(x+1)\varphi(x+1)}\\ &= -\frac{(2x+u)(x+1)(x+1+u-x_1)(x+1+u-x_2)}{(2x+2+u)(x+u)(x+x_1)(x+x_2)}.\end{aligned} \tag{7.7.1}$$

It will turn out that the seven finite and six infinite cases of table 7.1 for the dual Hahn polynomials can be treated (with only a few exceptions) by using two weight functions $w_1^*(x)$ and $w_2^*(x)$.

The Finite Cases

Note that (7.7.1) can be written in the form

$$\frac{w^*(x)}{w^*(x+1)} = \frac{(2x+u)(x+1)(x+1+u-x_1)(x+1+u-x_2)}{(2x+2+u)(x+u)(x+x_1)(-x-x_2)},$$

which leads (up to a factor of period 1 in x) to the solution

$$w_1^*(x) = \frac{(2x+u)\Gamma(x+u)\Gamma(x+x_1)}{\Gamma(x+1)\Gamma(x+1+u-x_1)\Gamma(x+1+u-x_2)\Gamma(1-x_2-x)}.$$

By using this weight function for the dual Hahn polynomials given by (7.3.6) the boundary conditions (7.6.6) can be fulfilled if we set $A=0$ and $x_2=-N$. By using (1.5.7), we obtain

$$
\begin{aligned}
d_0^{(1)} &= \sum_{x=0}^{N} w_1^*(x) \\
&= \sum_{x=0}^{N} \frac{2\Gamma(x+u/2+1)\Gamma(x+u)\Gamma(x+x_1)}{x!\,(N-x)!\,\Gamma(x+u/2)\Gamma(x+1+u-x_1)\Gamma(x+1+u+N)} \\
&= \frac{\Gamma(1+u)\Gamma(x_1)}{N!\,\Gamma(1+u-x_1)\Gamma(1+u+N)} \sum_{x=0}^{N} \frac{(-1)^x(u/2+1)_x(u)_x(x_1)_x(-N)_x}{x!\,(u/2)_x(1+u-x_1)_x(1+u+N)_x} \\
&= \frac{\Gamma(1+u)\Gamma(x_1)}{N!\,\Gamma(1+u-x_1)\Gamma(1+u+N)}\, {}_4F_3\left(\begin{matrix} u/2+1, u, x_1, -N \\ u/2, 1+u-x_1, 1+u+N \end{matrix}; -1\right) \\
&= \frac{\Gamma(1+u)\Gamma(x_1)}{N!\,\Gamma(1+u-x_1)\Gamma(1+u+N)} \cdot \frac{\Gamma(1+u-x_1)\Gamma(1+u+N)}{\Gamma(1+u)\Gamma(1+u-x_1+N)} \\
&= \frac{\Gamma(x_1)}{N!\,\Gamma(1+u-x_1+N)}.
\end{aligned}
$$

Further we obtain by using (7.4.1) and (7.6.7)

$$\sigma_n^{(1)} = \frac{\Gamma(n+x_1)n!}{\Gamma(1+u-x_1+N-n)(N-n)!}, \quad n = 0, 1, 2, \ldots, N.$$

This leads to the orthogonality relation

$$\sum_{x=0}^{N} w_1^*(x) y_m(x(x+u)) y_n(x(x+u)) = \sigma_n^{(1)} \delta_{mn}, \quad m, n = 0, 1, 2, \ldots, N \tag{7.7.2}$$

for the dual Hahn polynomials $y_n(x(x+u))$ given by (7.3.6).

In cases where it is not possible to satisfy the second boundary condition in (7.6.6) because $x_2 \neq -N$, one can try to use infinite sums instead. In that case we have

$$\sigma_n^{(1)} = \frac{\Gamma(n+x_1)n!}{\Gamma(1+u-x_1-x_2-n)\Gamma(1-x_2-n)}, \quad n = 0, 1, 2, \ldots, N$$

and the orthogonality relation

$$\sum_{x=0}^{\infty} w_1^*(x) y_m(x(x+u)) y_n(x(x+u)) = \sigma_n^{(1)} \delta_{mn}, \quad m, n = 0, 1, 2, \ldots, N. \tag{7.7.3}$$

For the convergence of this sum we need the existence of the moments

$$\sum_{x=0}^{\infty} w_1^*(x) x^k, \quad k = 0, 1, 2, \ldots, 4N.$$

By using the weight function $w_1^*(x)$ for the dual Hahn polynomials, we can treat the seven finite cases of table 7.1 with only one exception (see the remark below). We only need the positivity of

$$d_0^{(1)} = \frac{\Gamma(x_1)}{\Gamma(1-x_2)\Gamma(1+u-x_1-x_2)}.$$

Case I. We study the sign of $d_0^{(1)}$ for the seven finite cases of table 7.1.

In case 4 we have $x_1 > 0$, $x_2 \leq -N$ and $-N < x_1+x_2-u < -N+1$. This implies that $\Gamma(x_1) > 0$, $\Gamma(1-x_2) > 0$ and $\Gamma(1+u-x_1-x_2) > 0$. Hence we have $d_0^{(1)} > 0$.

In case 5 we have $x_1 > 0$, $-N < x_2 < -N+1$ and $x_1+x_2-u \leq -N$. This implies that $\Gamma(x_1) > 0$, $\Gamma(1-x_2) > 0$ and $\Gamma(1+u-x_1-x_2) > 0$. Hence we have $d_0^{(1)} > 0$. In this case $x_2 = -N$ is not possible.

In case 6 we have $x_1 > 0$, $x_2 \leq -N$ and $x_1+x_2-u \leq -N$. This implies that $\Gamma(x_1) > 0$, $\Gamma(1-x_2) > 0$ and $\Gamma(1+u-x_1-x_2) > 0$. Hence we have $d_0^{(1)} > 0$.

In case 8 we have $-N < x_1 < -N+1$, $x_2 \leq -N$ and $x_1+x_2-u > 0$. Then the sign of $\Gamma(x_1)$ equals $(-1)^N$ and $\Gamma(1-x_2) > 0$. Further we have $1+u-x_1-x_2 < 1$, hence if $-N-j < 1+u-x_1-x_2 < -N-j+1$ for $j \in \{-N,-N+1,-N+2,\ldots\}$, then the sign of $\Gamma(1+u-x_1-x_2)$ equals $(-1)^{N+j}$. For the positivity of $d_0^{(1)}$ we need a factor $(-1)^j$.

In case 9 we have $x_1 \leq -N$, $x_2 \leq -N$ and $x_1+x_2-u > 0$. Then we have $\Gamma(1-x_2) > 0$. If $-N-j < x_1 < -N-j+1$ and $-N-k < 1+u-x_1-x_2 < -N-k+1$ for $j,k \in \{1,2,3,\ldots\}$, then the sign of $\Gamma(x_1)$ equals $(-1)^{N+j}$ and the sign of $\Gamma(1+u-x_1-x_2)$ equals $(-1)^{N+k}$. For the positivity of $d_0^{(1)}$ we need a factor $(-1)^{j+k}$.

Remark. The cases where $x_1 \in \{-N,-N-1,-N-2,\ldots\}$ cannot be treated by using the weight function $w_1^*(x)$. A separate treatment of these cases is left out here.

The Infinite Cases

In order to deal with the six infinite cases of table 7.1 for the dual Hahn polynomials, we write instead of (7.7.1)

$$\frac{w^*(x)}{w^*(x+1)} = \frac{(x+u/2)(x+1/2+u/2)(-x-1)(-x-1-u+x_1)(-x-1-u+x_2)}{(-x-1-u/2)(-x-1/2-u/2)(x+u)(x+x_1)(x+x_2)}$$

which leads (up to a factor of period 1 in x) to the solution

$$w_2^*(x) = \frac{\Gamma(x+u)\Gamma(x+x_1)\Gamma(x+x_2)\Gamma(-x)\Gamma(x_1-u-x)\Gamma(x_2-u-x)}{\Gamma(x+u/2)\Gamma(x+1/2+u/2)\Gamma(-u/2-x)\Gamma(1/2-u/2-x)}.$$

For this weight function it is impossible to satisfy the boundary conditions (7.6.6). However, the concept of orthogonality of section 3.2 can be generalized to integration over the (possibly deformed) imaginary axis (cf. page 109). In that case we need the existence of the required moments, which is in general taken care of by the asymptotic properties of the gamma functions involved.

By using (1.6.10), we obtain

$$\begin{aligned} d_0^{(2)} &= \frac{1}{2\pi i}\int_{-i\infty}^{i\infty} w_2^*(x)\,dx \\ &= \frac{1}{2\pi i}\int_{-i\infty}^{i\infty} \frac{\Gamma(x+u)\Gamma(x+x_1)\Gamma(x+x_2)\Gamma(-x)\Gamma(x_1-u-x)\Gamma(x_2-u-x)}{\Gamma(x+u/2)\Gamma(x+1/2+u/2)\Gamma(-u/2-x)\Gamma(1/2-u/2-x)}\,dx \\ &= \frac{1}{2\pi}\Gamma(x_1)\Gamma(x_2)\Gamma(x_1+x_2-u). \end{aligned} \tag{7.7.4}$$

Further we obtain by using (7.4.1) and (7.6.7)

$$\sigma_n^{(2)} = \frac{1}{2\pi}\Gamma(n+x_1)\Gamma(n+x_2)\Gamma(n+x_1+x_2-u)n!, \quad n=0,1,2,\ldots,$$

which leads to the orthogonality relation

$$\begin{aligned} &\frac{1}{2\pi i}\int_{-i\infty}^{i\infty} w_2^*(x)y_m(x(x+u))y_n(x(x+u))\,dx \\ &= \frac{1}{2\pi}\Gamma(n+x_1)\Gamma(n+x_2)\Gamma(n+x_1+x_2-u)n!\,\delta_{mn}, \quad m,n=0,1,2,\ldots \end{aligned} \tag{7.7.5}$$

By using the weight function $w_2^*(x)$ for the dual Hahn polynomials, we can treat the six infinite cases of table 7.1 with only one exception (see the remark below). We only need the positivity of

$$d_0^{(2)} = \frac{1}{2\pi}\Gamma(x_1)\Gamma(x_2)\Gamma(x_1+x_2-u).$$

Case I. We study the sign of $d_0^{(2)}$ for the six infinite cases of table 7.1.

In case 1 we have $x_1>0$, $x_2>0$ and $x_1+x_2-u>0$. This implies that $\Gamma(x_1)>0$, $\Gamma(x_2)>0$ and $\Gamma(x_1+x_2-u)>0$. Hence we have $d_0^{(2)}>0$.

In case 2 we have $x_{1,2}=\alpha\pm i\beta$ with $\alpha,\beta\in\mathbb{R}$ and $\beta\neq 0$ and $2\alpha-u>0$. This implies that $\Gamma(\alpha+i\beta)\Gamma(\alpha-i\beta)=|\Gamma(\alpha+i\beta)|^2>0$ and $\Gamma(x_1+x_2-u)=\Gamma(2\alpha-u)>0$. Hence we have $d_0^{(2)}>0$.

In case 3 we have $x_1>0$, $-N<x_2<-N+1$ and $-N<x_1+x_2-u<-N+1$. This implies that $\Gamma(x_1)>0$ and the sign of both $\Gamma(x_2)$ and $\Gamma(x_1+x_2-u)$ equals $(-1)^N$. Hence we have $d_0^{(2)}>0$.

In case 6 we have $x_1>0$, $x_2\leq -N$ and $x_1+x_2-u\leq -N$. This implies that $\Gamma(x_1)>0$ and if $-N-j<x_2<-N-j+1$ and $-N-j<x_1+x_2-u<-N-j+1$ for $j\in\{-N+1,-N+2,-N+3,\ldots\}$ the sign of both $\Gamma(x_2)$ and $\Gamma(x_1+x_2-u)$ equals $(-1)^{N+j}$. Hence we have $d_0^{(2)}>0$.

In case 7 we have $-N < x_1 < -N+1$, $-N < x_2 < -N+1$ and $x_1 + x_2 - u > 0$. This implies that $\Gamma(x_1 + x_2 - u) > 0$ and the sign of both $\Gamma(x_1)$ and $\Gamma(x_2)$ equals $(-1)^N$. Hence we have $d_0^{(2)} > 0$.

In case 9 we have $x_1 \leq -N$, $x_2 \leq -N$ and $x_1 + x_2 - u > 0$. This implies that $\Gamma(x_1 + x_2 - u) > 0$ and if $-N - j < x_{1,2} < -N - j + 1$ for $j \in \{-N+1, -N+2, -N+3, \ldots\}$ then the sign of both $\Gamma(x_1)$ and $\Gamma(x_2)$ equals $(-1)^{N+j}$. Hence we have $d_0^{(2)} > 0$.

Remark. The cases where $x_{1,2} \in \{-N, -N-1, -N-2, \ldots\}$ cannot be treated by using the weight function $w_2^*(x)$. A separate treatment of these cases is left out.

7.8 Orthogonality Relations for Racah Polynomials

In the case of the Racah polynomials given by (7.3.9), we obtain by using (7.6.2) and (7.3.7) for the Pearson equation

$$\begin{aligned}\frac{w^*(x)}{w^*(x+1)} &= \frac{\rho(x)\{\varphi(x+2) - \psi(x+1)\}}{\rho(x+1)\varphi(x+1)} \\ &= \frac{(2x+u)(x+1)}{(2x+2+u)} \\ &\quad \times \frac{(x+1+u-x_1)(x+1+u-x_2)(x+1+u-x_3)}{(x+u)(x+x_1)(x+x_2)(x+x_3)}. \qquad (7.8.1)\end{aligned}$$

It will turn out that the finite and infinite cases of table 7.2 up to table 7.7 for the Racah polynomials can be treated (with only a few exceptions) by using two weight functions $w_3^*(x)$ and $w_4^*(x)$.

The Finite Cases

Note that (7.8.1) can be written in the form

$$\frac{w^*(x)}{w^*(x+1)} = \frac{(2x+u)(x+1)(x+1+u-x_1)(x+1+u-x_2)(x+1+u-x_3)}{(2x+2+u)(x+u)(x+x_1)(-x-x_2)(-x-x_3)},$$

which leads (up to a factor of period 1 in x) to the solution

$$\begin{aligned}w_3^*(x) &= \frac{2\Gamma(x+u/2+1)\Gamma(x+u)}{\Gamma(x+u/2)\Gamma(x+1+u-x_1)\Gamma(x+1+u-x_2)\Gamma(x+1+u-x_3)} \\ &\quad \times \frac{\Gamma(x+x_1)}{\Gamma(x+1)\Gamma(1-x_2-x)\Gamma(1-x_3-x)}.\end{aligned}$$

If this weight function for the Racah polynomials given by (7.3.9) is used, the boundary conditions (7.6.6) can be fulfilled if we set $A=0$ and $x_2=-N$ (or $x_3=-N$). By using (1.5.6), we obtain

$$\begin{aligned}
d_0^{(3)} &= \sum_{x=0}^{N} w_3^*(x)\\
&= \sum_{x=0}^{N} \frac{2\Gamma(x+u/2+1)\Gamma(x+u)}{\Gamma(x+u/2)\Gamma(x+1+u-x_1)\Gamma(x+1+u+N)\Gamma(x+1+u-x_3)}\\
&\quad\times \frac{\Gamma(x+x_1)}{x!\,(N-x)!\,\Gamma(1-x_3-x)}\\
&= \frac{2\Gamma(u/2+1)\Gamma(u)\Gamma(x_1)}{N!\,\Gamma(u/2)\Gamma(1+u-x_1)\Gamma(1+u+N)\Gamma(1+u-x_3)\Gamma(1-x_3)}\\
&\quad\times \sum_{x=0}^{N} \frac{(-1)^{2x}(u/2+1)_x(u)_x(x_1)_x(-N)_x(x_3)_x}{x!\,(u/2)_x(1+u-x_1)_x(1+u+N)_x(1+u-x_3)_x}\\
&= \frac{\Gamma(1+u)\Gamma(x_1)}{N!\,\Gamma(1+u-x_1)\Gamma(1+u+N)\Gamma(1+u-x_3)\Gamma(1-x_3)}\\
&\quad\times {}_5F_4\left(\begin{matrix}1+u/2,u,x_1,-N,x_3\\ u/2,1+u-x_1,1+u+N,1+u-x_3\end{matrix};1\right)\\
&= \frac{\Gamma(1+u)\Gamma(x_1)}{N!\,\Gamma(1+u-x_1)\Gamma(1+u+N)\Gamma(1+u-x_3)\Gamma(1-x_3)}\\
&\quad\times \frac{\Gamma(1+u-x_1)\Gamma(1+u+N)\Gamma(1+u-x_3)\Gamma(1+u-x_1-x_3+N)}{\Gamma(1+u)\Gamma(1+u-x_1+N)\Gamma(1+u-x_1-x_3)\Gamma(1+u-x_3+N)}\\
&= \frac{\Gamma(x_1)\Gamma(1+u-x_1-x_3+N)}{N!\,\Gamma(1+u-x_1+N)\Gamma(1+u-x_3+N)\Gamma(1+u-x_1-x_3)\Gamma(1-x_3)}.
\end{aligned}$$

Further we get by using (7.4.2), (7.6.7) and $w=x_1-N+x_3-u$

$$\begin{aligned}
\sigma_n^{(3)} &= \frac{\Gamma(w)\Gamma(1-w)\Gamma(n+w-1)}{\Gamma(2n+w)\Gamma(2n+w-1)\Gamma(1-w+x_1-n)\Gamma(1-w-N-n)}\\
&\quad\times \frac{\Gamma(n+x_1)(-1)^n n!}{\Gamma(1-w+x_3-n)\Gamma(1-x_3-n)(N-n)!}, \qquad n=0,1,2,\dots,N.
\end{aligned}$$

This leads to the orthogonality relation

$$\sum_{x=0}^{N} w_3^*(x)y_m(x(x+u))y_n(x(x+u)) = \sigma_n^{(3)}\,\delta_{mn}, \quad m,n=0,1,2,\dots,N \tag{7.8.2}$$

for the Racah polynomials $y_n(x(x+u))$ given by (7.3.9).

In cases where it is not possible to satisfy the second boundary condition in (7.6.6) because $x_2\neq -N$, one can try to use infinite sums instead. In that case we have

$$\sigma_n^{(3)} = \frac{\Gamma(w)\Gamma(1-w)\Gamma(n+w-1)}{\Gamma(2n+w)\Gamma(2n+w-1)\Gamma(1-w+x_1-n)\Gamma(1-w+x_2-n)} \times \frac{\Gamma(n+x_1)(-1)^n n!}{\Gamma(1-w+x_3-n)\Gamma(1-x_2-n)\Gamma(1-x_3-n)}, \quad n=0,1,2,\ldots,N$$

with $w = x_1 + x_2 + x_3 - u$ and the orthogonality relation

$$\sum_{x=0}^{\infty} w_3^*(x) y_m(x(x+u)) y_n(x(x+u)) = \sigma_n^{(3)} \delta_{mn}, \quad m,n=0,1,2,\ldots,N. \tag{7.8.3}$$

For the convergence of this sum we need the existence of the moments

$$\sum_{x=0}^{\infty} w_3^*(x) x^k, \quad k=0,1,2,\ldots,4N.$$

By using this weight function $w_3^*(x)$ for the Racah polynomials, we can treat almost all finite cases of table 7.2 and table 7.4 up to table 7.7. We only need the positivity of

$$d_0^{(3)} = \frac{\Gamma(x_1)\Gamma(1-w)}{\Gamma(1-x_2)\Gamma(1-x_3)\Gamma(1-w+x_1)\Gamma(1-w+x_2)\Gamma(1-w+x_3)}.$$

Case IIa. In table 7.2 we have $w > 0$. In all cases we have $0 < x_1 < w$ which implies that $\Gamma(x_1) > 0$, $\Gamma(w) > 0$ and $\Gamma(w - x_1) > 0$.

Further we obtain by using (1.2.3)

$$d_0^{(3)} = \frac{\Gamma(x_1)\Gamma(w-x_1)\Gamma(w-x_2)\Gamma(w-x_3)}{\Gamma(w)\Gamma(1-x_2)\Gamma(1-x_3)} \times \frac{\sin\pi(w-x_1)\sin\pi(w-x_2)\sin\pi(w-x_3)}{\pi^2 \sin\pi w}.$$

In case 4 and case 5 we have $x_2 < 0$ and $x_3 < 0$, which implies that we also have $\Gamma(1-x_2) > 0$, $\Gamma(1-x_3) > 0$, $\Gamma(w-x_2) > 0$ and $\Gamma(w-x_3) > 0$. In these cases we need a factor with the same sign as the quotient of sines above. Note that we sometimes need a special treatment for the cases where $w, w-x_1, w-x_2, w-x_3 \in \{0, \pm 1, \pm 2, \ldots\}$. This will be left out here.

By using (1.2.3), we may also write

$$d_0^{(3)} = \frac{\Gamma(x_1)\Gamma(x_2)\Gamma(x_3)\Gamma(w-x_1)}{\Gamma(w)\Gamma(1-w+x_2)\Gamma(1-w+x_3)} \cdot \frac{\sin\pi(w-x_1)\sin\pi x_2 \sin\pi x_3}{\pi^2 \sin\pi w}.$$

In case 7 and case 8 we have $w - x_2 < 0$ and $w - x_3 < 0$, which implies that $\Gamma(x_2) > 0$, $\Gamma(x_3) > 0$, $\Gamma(1-w+x_2) > 0$ and $\Gamma(1-w+x_3) > 0$. In these cases we need a factor with the same sign as the quotient of sines above. Note that we sometimes need a special treatment for the cases where $w, w-x_1, x_2, x_3 \in \{0, \pm 1, \pm 2, \ldots\}$. This will be left out here.

Finally, we also obtain by using (1.2.3)

$$d_0^{(3)} = \frac{\Gamma(x_1)\Gamma(x_3)\Gamma(w-x_1)\Gamma(w-x_2)}{\Gamma(w)\Gamma(1-x_2)\Gamma(1-w+x_3)} \cdot \frac{\sin\pi(w-x_1)\sin\pi(w-x_2)\sin\pi x_3}{\pi^2\sin\pi w}.$$

In case 10, case 11 and case 12 we have $x_2 < 0$ and $w - x_3 < 0$, which implies that $\Gamma(1-x_2) > 0$, $\Gamma(x_3) > 0$, $\Gamma(w-x_2) > 0$ and $\Gamma(1-w+x_3) > 0$. In these cases we need a factor with the same sign as the quotient of sines above. Note that we sometimes need a special treatment for the cases where $w, x_3, w-x_1, w-x_2 \in \{0, \pm 1, \pm 2, \ldots\}$. This will be left out here.

Remark. In the cases 7, 8 and 11 it is impossible to take $x_2 = -N$.

Case IIb. In table 7.3 we have no finite cases.

Case IIc1. In table 7.4 we have $w = -2N - t < 0$ with $N \in \{1, 2, 3, \ldots\}$ and $-1 < t \leq 1$, which implies that $\Gamma(1-w) > 0$. Further we have $x_1 > 0$ and therefore $\Gamma(x_1) > 0$ and $\Gamma(1-w+x_1) > 0$.

Again we obtain by using (1.2.3)

$$d_0^{(3)} = \frac{\Gamma(x_1)\Gamma(x_2)\Gamma(x_3)\Gamma(1-w)}{\Gamma(1-w+x_1)\Gamma(1-w+x_2)\Gamma(1-w+x_3)} \cdot \frac{\sin\pi x_2 \sin\pi x_3}{\pi^2}.$$

In case 1 we also have $x_2 > 0$ and $x_3 > 0$, which implies that $\Gamma(x_2) > 0$, $\Gamma(x_3) > 0$, $\Gamma(1-w+x_2) > 0$ and $\Gamma(1-w+x_3) > 0$. In this case we need a factor with the same sign as the product of sines above. Note that we sometimes need a special treatment for the cases where $x_2, x_3 \in \{0, \pm 1, \pm 2, \ldots\}$. This will be left out here.

By using (1.2.3), we may also write

$$d_0^{(3)} = \frac{\Gamma(x_1)\Gamma(x_2)\Gamma(w-x_3)\Gamma(1-w)}{\Gamma(1-x_3)\Gamma(1-w+x_1)\Gamma(1-w+x_2)} \cdot \frac{\sin\pi x_2 \sin\pi(w-x_3)}{\pi^2}.$$

In case 2 we have $x_2 > 0$ and $x_3 < w < 0$, which implies that $\Gamma(x_2) > 0$, $\Gamma(1-x_3) > 0$, $\Gamma(1-w+x_2) > 0$ and $\Gamma(w-x_3) > 0$. In this case we need a factor with the same sign as the product of sines above. Note that we sometimes need a special treatment for the cases where $x_2, w-x_3 \in \{0, \pm 1, \pm 2, \ldots\}$. This will be left out here.

Again we use (1.2.3) to write

$$d_0^{(3)} = \frac{\Gamma(x_1)\Gamma(w-x_2)\Gamma(w-x_3)\Gamma(1-w)}{\Gamma(1-x_2)\Gamma(1-x_3)\Gamma(1-w+x_1)} \cdot \frac{\sin\pi(w-x_2)\sin\pi(w-x_3)}{\pi^2}.$$

In case 4 we have $x_2 < w < 0$ and $x_3 < w < 0$, which implies that $\Gamma(1-x_2) > 0$, $\Gamma(1-x_3) > 0$, $\Gamma(w-x_2) > 0$ and $\Gamma(w-x_3) > 0$. In this case we need a factor with the same sign as the product of sines above. Note that we sometimes need a special treatment for the cases where $w-x_2, w-x_3 \in \{0, \pm 1, \pm 2, \ldots\}$. This will be left out here.

In case 3 we have

$$\Gamma(1-x_2)\Gamma(1-x_3)=\Gamma(1-\alpha-i\beta)\Gamma(1-\alpha+i\beta)=|\Gamma(1-\alpha+i\beta)|^2>0$$

and

$$\begin{aligned}\Gamma(1-w+x_2)\Gamma(1-w+x_3)&=\Gamma(1-w+\alpha+i\beta)\Gamma(1-w+\alpha-i\beta)\\&=|\Gamma(1-w+\alpha+i\beta)|^2>0.\end{aligned}$$

Hence in case 3 and also in case 5 through case 10 we simply have

$$d_0^{(3)}=\frac{\Gamma(x_1)\Gamma(1-w)}{\Gamma(1-x_2)\Gamma(1-x_3)\Gamma(1-w+x_1)\Gamma(1-w+x_2)\Gamma(1-w+x_3)}>0.$$

Remark. Only in case 5 it is possible to take $x_2=-N$.

Case IIc2. In table 7.5 we have $w=-2N-t<0$ with $N\in\{1,2,3,\ldots\}$ and $-1<t\leq 1$. Further we have $x_1<w<0$ and therefore $\Gamma(1-x_1)>0$, $\Gamma(1-w)>0$ and $\Gamma(w-x_1)>0$.

Again we obtain by using (1.2.3)

$$\begin{aligned}d_0^{(3)}&=\frac{\Gamma(w-x_1)\Gamma(w-x_2)\Gamma(w-x_3)\Gamma(1-w)}{\Gamma(1-x_1)\Gamma(1-x_2)\Gamma(1-x_3)}\\&\quad\times\frac{\sin\pi(w-x_1)\sin\pi(w-x_2)\sin\pi(w-x_3)}{\pi^2\sin\pi x_1}.\end{aligned}$$

In case 11 we also have $x_2<w<0$ and $x_3<w<0$, which implies that $\Gamma(1-x_2)>0$, $\Gamma(1-x_3)>0$, $\Gamma(w-x_2)>0$ and $\Gamma(w-x_3)>0$. In this case we need a factor with the same sign as the quotient of sines above. Note that we sometimes need a special treatment for the cases where $x_1,w-x_1,w-x_2,w-x_3\in\{0,\pm1,\pm2,\ldots\}$. This will be left out here.

For the other cases we use

$$d_0^{(3)}=\frac{\Gamma(w-x_1)\Gamma(1-w)}{\Gamma(1-x_1)\Gamma(1-x_2)\Gamma(1-x_3)\Gamma(1-w+x_2)\Gamma(1-w+x_3)}\cdot\frac{\sin\pi(w-x_1)}{\sin\pi x_1}.$$

In case 12 we have $x_{2,3}=\alpha\pm i\beta$, which implies, as before, that both $\Gamma(1-x_2)\Gamma(1-x_3)>0$ and $\Gamma(1-w+x_2)\Gamma(1-w+x_3)>0$. In case 13 through case 18 we have $w<x_2<0$ and $w<x_3<0$, which implies that $\Gamma(1-x_2)>0$, $\Gamma(1-x_3)>0$, $\Gamma(1-w+x_2)>0$ and $\Gamma(1-w+x_3)>0$. Hence in case 12 through case 18 we need a factor with the same sign as the quotient of sines above. Note that sometimes we need a special treatment for the cases where $x_1,w-x_1\in\{0,\pm1,\pm2,\ldots\}$. This will be left out here.

Remark. Only in case 13 it is possible to take $x_2=-N$.

Case IIc3. In table 7.6 we have $w = -2N - t < 0$ with $N \in \{1,2,3,\ldots\}$ and $-1 < t < 0$. In all finite cases we have $\Gamma(x_1) > 0$, $\Gamma(1-w) > 0$, $\Gamma(1-x_2) > 0$, $\Gamma(1-x_3) > 0$, $\Gamma(1-w+x_1) > 0$, $\Gamma(1-w+x_2) > 0$ and $\Gamma(1-w+x_3) > 0$, which implies that

$$d_0^{(3)} = \frac{\Gamma(x_1)\Gamma(1-w)}{\Gamma(1-x_2)\Gamma(1-x_3)\Gamma(1-w+x_1)\Gamma(1-w+x_2)\Gamma(1-w+x_3)} > 0.$$

Remark. In these cases it is impossible to take $x_2 = -N$.

Case IIc4. In table 7.7 we have $w = -2N - t < 0$ with $N \in \{1,2,3,\ldots\}$ and $0 < t < 1$. In all finite cases we have $\Gamma(x_1) > 0$, $\Gamma(1-w) > 0$, $\Gamma(1-x_2) > 0$, $\Gamma(1-x_3) > 0$, $\Gamma(1-w+x_1) > 0$, $\Gamma(1-w+x_2) > 0$ and $\Gamma(1-w+x_3) > 0$ which implies that

$$d_0^{(3)} = \frac{\Gamma(x_1)\Gamma(1-w)}{\Gamma(1-x_2)\Gamma(1-x_3)\Gamma(1-w+x_1)\Gamma(1-w+x_2)\Gamma(1-w+x_3)} > 0.$$

Remark. In these cases it is impossible to take $x_2 = -N$.

The Infinite Cases

In order to deal with the infinite cases of table 7.2 through table 7.7 for the Racah polynomials, we write instead of (7.8.1)

$$\begin{aligned}\frac{w^*(x)}{w^*(x+1)} &= \frac{(x+u/2)(x+1/2+u/2)}{(x+u)(x+x_1)(x+x_2)(x+x_3)} \\ &\quad\times \frac{(-x-1)(-x-1-u+x_1)(-x-1-u+x_2)(-x-1-u+x_3)}{(-x-1-u/2)(-x-1/2-u/2)}\end{aligned}$$

which leads (up to a factor of period 1 in x) to the solution

$$\begin{aligned}w_4^*(x) &= \frac{\Gamma(x+u)\Gamma(x+x_1)\Gamma(x+x_2)\Gamma(x+x_3)}{\Gamma(x+u/2)\Gamma(x+1/2+u/2)} \\ &\quad\times \frac{\Gamma(-x)\Gamma(x_1-u-x)\Gamma(x_2-u-x)\Gamma(x_3-u-x)}{\Gamma(-u/2-x)\Gamma(1/2-u/2-x)}.\end{aligned}$$

For this weight function it is impossible to satisfy the boundary conditions (7.6.6). However, the concept of orthogonality of section 3.2 can be generalized to integration over the (possibly deformed) imaginary axis (cf. page 109). In that case we need the existence of the required moments, which is in general taken care of by the asymptotic properties of the gamma functions involved.

By using (1.6.6), we have

$$\begin{aligned} d_0^{(4)} &= \frac{1}{2\pi i}\int_{-i\infty}^{i\infty} w_4^*(x)\,dx \\ &= \frac{1}{2\pi i}\int_{-i\infty}^{i\infty} \frac{\Gamma(x+u)\Gamma(x_1-u-x)\Gamma(x_2-u-x)\Gamma(x_3-u-x)}{\Gamma(x+u/2)\Gamma(x+1/2+u/2)\Gamma(-u/2-x)\Gamma(1/2-u/2-x)} \\ &\qquad \times \Gamma(x+x_1)\Gamma(x+x_2)\Gamma(x+x_3)\Gamma(-x)\,dx \\ &= \frac{1}{2\pi}\frac{\Gamma(x_1)\Gamma(x_2)\Gamma(x_3)\Gamma(w-x_1)\Gamma(w-x_2)\Gamma(w-x_3)}{\Gamma(w)}, \end{aligned} \tag{7.8.4}$$

where $w = x_1 + x_2 + x_3 - u$. Further we obtain by using (7.4.2) and (7.6.7)

$$\sigma_n^{(4)} = \frac{\Gamma(n+x_1)\Gamma(n+x_2)\Gamma(n+x_3)\Gamma(n+w-1)}{2\pi\Gamma(2n+w)\Gamma(2n-1+w)} \times \Gamma(n+w-x_1)\Gamma(n+w-x_2)\Gamma(n+w-x_3)n!$$

for $n = 0, 1, 2, \ldots$, which leads to the orthogonality relation

$$\frac{1}{2\pi i}\int_{-i\infty}^{i\infty} w_4^*(x)y_m(x(x+u))y_n(x(x+u))\,dx = \sigma_n^{(4)}\,\delta_{mn}, \quad m,n = 0,1,2,\ldots. \tag{7.8.5}$$

By using the weight function $w_4^*(x)$ for the Racah polynomials, we can treat almost all infinite cases of table 7.2 through table 7.7. We only need the positivity of

$$d_0^{(4)} = \frac{1}{2\pi}\frac{\Gamma(x_1)\Gamma(x_2)\Gamma(x_3)\Gamma(w-x_1)\Gamma(w-x_2)\Gamma(w-x_3)}{\Gamma(w)}.$$

Case IIa. In table 7.2 we have $w > 0$. In all cases we have $0 < x_1 < w$ and therefore $\Gamma(x_1) > 0$, $\Gamma(w) > 0$ and $\Gamma(w - x_1) > 0$.

In case 1 we also have $0 < x_2 < w$ and $0 < x_3 < w$, which implies that $\Gamma(x_2) > 0$, $\Gamma(x_3) > 0$, $\Gamma(w-x_2) > 0$ and $\Gamma(w-x_3) > 0$. Hence we have $d_0^{(4)} > 0$.

In case 2 we have $x_{2,3} = \alpha \pm i\beta$, which implies that both

$$\Gamma(x_2)\Gamma(x_3) = \Gamma(\alpha+i\beta)\Gamma(\alpha-i\beta) = |\Gamma(\alpha+i\beta)|^2 > 0$$

and

$$\Gamma(w-x_2)\Gamma(w-x_3) = \Gamma(w-\alpha-i\beta)\Gamma(w-\alpha+i\beta) = |\Gamma(w-\alpha+i\beta)|^2 > 0.$$

Hence we have $d_0^{(4)} > 0$.

In case 3 and case 5 we have $x_2 < 0$ and $x_3 < 0$, which implies that $\Gamma(w-x_2) > 0$ and $\Gamma(w-x_3) > 0$. If $-N-j < x_{2,3} < -N-j+1$ for $j \in \{0,1,2,\ldots\}$ then the sign of both $\Gamma(x_2)$ and $\Gamma(x_3)$ equals $(-1)^{N+j}$. Hence we have $d_0^{(4)} > 0$.

In case 6 and case 8 we have $w-x_2<0$ and $w-x_3<0$, which implies that $\Gamma(x_2)>0$ and $\Gamma(x_3)>0$. If $-N-j<w-x_{2,3}<-N-j+1$ for $j\in\{0,1,2,\ldots\}$ then the sign of both $\Gamma(w-x_2)$ and $\Gamma(w-x_3)$ equals $(-1)^{N+j}$. Hence we have $d_0^{(4)}>0$.

In case 9 and case 12 we have $x_2<0$ and $w-x_3<0$, which implies that $\Gamma(w-x_2)>0$ and $\Gamma(x_3)>0$. If $-N-j<x_2<-N-j+1$ and $-N-j<w-x_3<-N-j+1$ for $j\in\{0,1,2,\ldots\}$ then the sign of both $\Gamma(x_2)$ and $\Gamma(w-x_3)$ equals $(-1)^{N+j}$. Hence we have $d_0^{(4)}>0$.

Case IIb. In table 7.3 we have $-1<w<0$. In all cases we have $w<x_1<0$ and therefore $\Gamma(x_1)<0$, $\Gamma(w)<0$ and $\Gamma(w-x_1)<0$.

In case 1 we have $-2<x_2<-1$ and $-1<x_3<w$, which implies that $\Gamma(x_2)>0$, $\Gamma(x_3)<0$, $\Gamma(w-x_2)>0$ and $\Gamma(w-x_3)>0$. Hence we have $d_0^{(4)}>0$.

In case 2 we have $-2<x_2<-1$ and $0<x_3<w+1$, which implies that $\Gamma(x_2)>0$, $\Gamma(x_3)>0$, $\Gamma(w-x_2)>0$ and $\Gamma(w-x_3)<0$. Hence we have $d_0^{(4)}>0$.

In case 3 we have $w+1<x_2<w+2$ and $-1<x_3<w$, which implies that $\Gamma(x_2)>0$, $\Gamma(x_3)<0$, $\Gamma(w-x_2)>0$ and $\Gamma(w-x_3)>0$. Hence we have $d_0^{(4)}>0$.

In case 4 we have $w+1<x_2<w+2$ and $0<x_3<w+1$, which implies that $\Gamma(x_2)>0$, $\Gamma(x_3)>0$, $\Gamma(w-x_2)>0$ and $\Gamma(w-x_3)<0$. Hence we have $d_0^{(4)}>0$.

Case IIc1. In table 7.4 we have no infinite cases.

Case IIc2. In table 7.5 we have no infinite cases.

Case IIc3. In table 7.6 we have $w=-2N-t<0$ with $N\in\{1,2,3,\ldots\}$ and $-1<t<0$. In both cases we have $\Gamma(w)>0$ and $\Gamma(x_1)\Gamma(x_2)\Gamma(x_3)\Gamma(w-x_1)\Gamma(w-x_2)\Gamma(w-x_3)>0$, which implies that $d_0^{(4)}>0$.

Case IIc4. In table 7.7 we have $w=-2N-t<0$ with $N\in\{1,2,3,\ldots\}$ and $0<t<1$. In both cases we have $\Gamma(w)<0$ and $\Gamma(x_1)\Gamma(x_2)\Gamma(x_3)\Gamma(w-x_1)\Gamma(w-x_2)\Gamma(w-x_3)<0$ which implies that $d_0^{(4)}>0$.

In this chapter we have proved:

Theorem 7.2. *The positive-definite orthogonal polynomial solutions* $y(x(x+u))$ *with* $x \in \mathbb{R}$ *and* $u \in \mathbb{R}$ *of the difference equation (7.2.1)*

$$\varphi(x+2)\Delta^2 y_n(x(x+u)) + \psi(x+1)\Delta y_n(x(x+u))$$
$$= \lambda_n \rho(x+1) y_n((x+1)(x+1+u)), \quad n = 0, 1, 2, \ldots,$$

consist of the polynomial solutions of the (real) difference equations (7.3.4) and (7.3.7). In fact this leads to both infinite and finite systems of

- *dual Hahn polynomials and*
- *Racah polynomials.*

Chapter 8
Orthogonal Polynomial Solutions in $z(z+u)$ of Complex Difference Equations

Discrete Classical Orthogonal Polynomials IV

8.1 Real Polynomial Solutions of Complex Difference Equations

As in chapter 6, the difference equations (7.3.4) for the dual Hahn polynomials and (7.3.7) for the Racah polynomials can also be considered in the case that the coefficients are complex. In that case we look for polynomial solutions $y_n(z(z+u))$ with $z \in \mathbb{C}$ and $u \in \mathbb{R}$. The three-term recurrence relations (7.4.1) and (7.4.2) still hold with x replaced by z. Again we set $z = a + ix$ with $a \in \mathbb{R}$ and $x \in \mathbb{R}$. Then we have

$$z(z+u) = (a+ix)(a+ix+u) = a(a+u) + i(2a+u)x - x^2.$$

For $u = -2a$ the imaginary part cancels and we have $z(z+u) = -a^2 - x^2$. Now we define

$$y_n(z(z+u)) = y_n(-a^2 - x^2) = (-1)^n \widehat{y}_n(x^2), \quad n = 0,1,2,\ldots. \tag{8.1.1}$$

Case I. For the (monic) dual Hahn polynomials given by (7.3.6) this leads to the (monic) **continuous dual Hahn** polynomials

$$\widehat{y}_n(x^2) = (-1)^n (x_1)_n (x_2)_n \, {}_3F_2\left(\begin{array}{c} -n, -a-ix, -a+ix \\ x_1, x_2 \end{array}; 1\right) \tag{8.1.2}$$

for $n = 0,1,2,\ldots$. These continuous dual Hahn polynomials satisfy the three-term recurrence relation

$$\begin{aligned}\widehat{y}_{n+1}(x^2) &= \left\{x^2 + a^2 - (n+x_1)(n+x_2) - n(n-1+2a+x_1+x_2)\right\} \widehat{y}_n(x^2) \\ &\quad - n(n-1+x_1)(n-1+x_2)(n-1+2a+x_1+x_2)\widehat{y}_{n-1}(x^2)\end{aligned} \tag{8.1.3}$$

R. Koekoek et al., *Hypergeometric Orthogonal Polynomials and Their q-Analogues*, Springer Monographs in Mathematics,
DOI 10.1007/978-3-642-05014-5_8, © Springer-Verlag Berlin Heidelberg 2010

for $n=1,2,3,\ldots$ with $\widehat{y}_0(x^2)=1$ and $\widehat{y}_1(x^2)=x^2+a^2-x_1x_2$.

Since the coefficients in $\widehat{y}_n(x^2)$ can only be real for x_1 and x_2 both real or complex conjugates, the positive-definite orthogonality can be obtained in a similar way as in the real case.

Case II. For the (monic) Racah polynomials given by (7.3.9), this leads to the (monic) **Wilson** polynomials

$$\widehat{y}_n(x^2)=(-1)^n\frac{(x_1)_n(x_2)_n(x_3)_n}{(n-1+w)_n}\,{}_4F_3\left(\begin{matrix}-n,n-1+w,-a-ix,-a+ix\\ x_1,x_2,x_3\end{matrix};1\right) \quad (8.1.4)$$

for $n=0,1,2,\ldots$, where $w=2a+x_1+x_2+x_3$. These Wilson polynomials satisfy the three-term recurrence relation

$$\widehat{y}_{n+1}(x^2)=\left\{x^2+a^2-c_n^{(1)}-c_n^{(2)}\right\}\widehat{y}_n(x^2)-c_{n-1}^{(1)}c_n^{(2)}\widehat{y}_{n-1}(x^2) \quad (8.1.5)$$

for $n=1,2,3,\ldots$ with $\widehat{y}_0(x^2)=1$, $\widehat{y}_1(x^2)=x^2+a^2-x_1x_2x_3/w$ and $c_n^{(1)}$ and $c_n^{(2)}$ given by (7.4.3) and (7.4.4) respectively. Note that

$$\widehat{y}_2(x^2)=(x^2+a^2)^2-\left\{\frac{2(x_1+1)(x_2+1)(x_3+1)}{w+2}-2a-1\right\}(x^2+a^2)$$
$$+\frac{x_1x_2x_3(x_1+1)(x_2+1)(x_3+1)}{(w+1)(w+2)}.$$

Since $a\in\mathbb{R}$, the coefficients of $\widehat{y}_n(x^2)$ can only be real if

$$\frac{x_1x_2x_3}{w}\in\mathbb{R},\quad \frac{(x_1+1)(x_2+1)(x_3+1)}{w+2}\in\mathbb{R}\quad\text{and}\quad\frac{x_1x_2x_3}{w+1}\in\mathbb{R}.$$

For $x_1x_2x_3\neq 0$ this implies that $w\in\mathbb{R}$ and therefore $x_1x_2x_3\in\mathbb{R}$ and $(x_1+1)(x_2+1)(x_3+1)\in\mathbb{R}$. Moreover, $d_1\in\mathbb{R}$ leads to $(w-x_1)(w-x_2)(w-x_3)\in\mathbb{R}$ and therefore $x_1+x_2+x_3\in\mathbb{R}$ and $x_1x_2+x_1x_3+x_2x_3\in\mathbb{R}$. Note that $x_i=0$ for some $i\in\{1,2,3\}$ or $x_j=w$ for some $j\in\{1,2,3\}$ leads to $d_1=0$. This implies that there are only two possibilities, id est $x_1,x_2,x_3\in\mathbb{R}$ or one is real and the other two are complex conjugates. In either case both $c_n^{(1)}$ given by (7.4.3) and $c_n^{(2)}$ given by (7.4.4) are real for all $n=1,2,3,\ldots$. This shows that in contrast with chapter 6, the coefficients in the difference equation for $\widehat{y}_n(x^2)$ are real except for the powers of $a+ix$ involved. This implies that the positive-definite orthogonality can be obtained in a similar way as in the real case.

8.2 Orthogonality Relations for Continuous Dual Hahn Polynomials

For the dual Hahn polynomials given by (7.3.6) and the Racah polynomials given by (7.3.9) we obtained orthogonality relations of the form (7.6.5) with boundary conditions (7.6.6).

We use the transformation (8.1.1). If we set $a+ix=t$, then we obtain $x=i(a-t)$ and therefore $x^2=-(a-t)^2$. Hence, for the continuous dual Hahn polynomials given by (8.1.2) and the Wilson polynomials given by (8.1.4) we obtain an orthogonality relation of the form

$$\sum_{t=A}^{A+N} w^*(t)\widehat{y}_m(-(a-t)^2)\widehat{y}_n(-(a-t)^2)=\sigma_n\,\delta_{mn},\quad m,n=0,1,2,\ldots,N \tag{8.2.1}$$

with the boundary conditions

$$w^*(A-1)\varphi(A)=0 \quad\text{and}\quad w^*(A+N)\varphi(A+N+1)=0. \tag{8.2.2}$$

Then we have

$$\sigma_n=\prod_{k=0}^{n} d_k,\quad n=0,1,2,\ldots,N \quad\text{with}\quad d_0=\sum_{t=A}^{A+N} w^*(t) \tag{8.2.3}$$

and $d_1,d_2,d_3,\ldots,d_N$ are given by the three-term recurrence relation (8.1.3) or (8.1.5). Also in this case it is possible to use an infinite sum or an improper integral over the (possibly deformed) imaginary axis instead of the finite sum in (8.2.1).

For positive-definite orthogonality we must have $d_0>0$.

For the continuous dual Hahn polynomials $\widehat{y}_n(x^2)$ given by (8.1.2), the Pearson equation (7.7.1) still holds.

The Finite Cases

In order to deal with the seven finite cases in table 7.1, we write instead of (7.7.1)

$$\frac{w^*(x)}{w^*(x+1)}=\frac{(-x-u/2)(-x-1/2-u/2)(x+1)(-x-1-u+x_1)(x+1+u-x_2)}{(x+1+u/2)(x+1/2+u/2)(-x-u)(x+x_1)(-x-x_2)}$$

which leads (up to a factor of period 1 in x) to the solution

$$w_5^*(x)=\Gamma(x+1+u/2)\Gamma(x+1/2+u/2)\\ \times\frac{\Gamma(1-u/2-x)\Gamma(1/2-u/2-x)\Gamma(x_1-u-x)\Gamma(x+x_1)}{\Gamma(x+1)\Gamma(x+1+u-x_2)\Gamma(1-u-x)\Gamma(1-x_2-x)}.$$

By using Legendre's duplication formula (1.2.6), we have

$$\Gamma(x+1+u/2)\Gamma(x+1/2+u/2)\Gamma(1-u/2-x)\Gamma(1/2-u/2-x)$$
$$=\pi\Gamma(2x+1+u)\Gamma(1-u-2x)$$

for $2x+u\neq\pm1,\pm2,\pm3,\ldots$, which implies that

$$w_5^*(x)=\frac{\pi\Gamma(2x+1+u)\Gamma(1-u-2x)\Gamma(x_1-u-x)\Gamma(x+x_1)}{\Gamma(x+1)\Gamma(x+1+u-x_2)\Gamma(1-u-x)\Gamma(1-x_2-x)}.$$

If we now replace x by $a+ix$ and u by $-2a$, we obtain

$$w_5^*(a+ix)=\pi\left|\frac{\Gamma(1+2ix)\Gamma(x_1+a+ix)}{\Gamma(1+a+ix)\Gamma(1-a-x_2+ix)}\right|^2.$$

In this case the boundary conditions (7.6.6) can be satisfied by taking $A=0$ and $x_2=-N$. By using (1.5.7) and (1.2.6) we obtain

$$\begin{aligned}
d_0^{(5)} &= \sum_{x=0}^{N} w_5^*(x)\\
&= \sum_{x=0}^{N}\Gamma(x+1+u/2)\Gamma(x+1/2+u/2)\\
&\qquad\times\frac{\Gamma(1-u/2-x)\Gamma(1/2-u/2-x)\Gamma(x_1-u-x)\Gamma(x+x_1)}{x!\,\Gamma(x+1+u+N)\Gamma(1-u-x)(N-x)!}\\
&= \frac{\Gamma(1+u/2)\Gamma(1/2+u/2)\Gamma(1-u/2)\Gamma(1/2-u/2)\Gamma(x_1)\Gamma(x_1-u)}{N!\,\Gamma(1-u)\Gamma(1+u+N)}\\
&\qquad\times\sum_{x=0}^{N}\frac{(-1)^x(-N)_x(1+u/2)_x(u)_x(x_1)_x}{x!\,(u/2)_x(1+u-x_1)_x(1+u+N)_x}\\
&= \frac{\Gamma(1+u/2)\Gamma(1/2+u/2)\Gamma(1-u/2)\Gamma(1/2-u/2)\Gamma(x_1)\Gamma(x_1-u)}{N!\,\Gamma(1-u)\Gamma(1+u+N)}\\
&\qquad\times{}_4F_3\left(\begin{matrix}-N,1+u/2,u,x_1\\ u/2,1+u-x_1,1+u+N\end{matrix};-1\right)\\
&= \frac{\Gamma(1+u/2)\Gamma(1/2+u/2)\Gamma(1-u/2)\Gamma(1/2-u/2)\Gamma(x_1)\Gamma(x_1-u)}{N!\,\Gamma(1-u)\Gamma(1+u+N)}\\
&\qquad\times\frac{\Gamma(1+u-x_1)\Gamma(1+u+N)}{\Gamma(1+u)\Gamma(1+u-x_1+N)}\\
&= \frac{\pi\Gamma(x_1)\Gamma(x_1-u)\Gamma(1+u-x_1)}{N!\,\Gamma(1+u-x_1+N)}.
\end{aligned}$$

Further we obtain by using (8.1.3) and (7.6.7)

$$\sigma_n^{(5)}=\frac{\pi\Gamma(n+x_1)\Gamma(x_1-u)\Gamma(1+u-x_1)n!}{\Gamma(1+u-x_1+N-n)(N-n)!},\qquad n=0,1,2,\ldots,N,$$

which leads to the orthogonality relation (cf. (8.2.1))

$$\begin{aligned}\sum_{t=0}^{N} &\frac{\Gamma(1-2a+2t)\Gamma(1+2a-2t)\Gamma(x_1+2a-t)\Gamma(x_1+t)}{\Gamma(1+t)\Gamma(1-2a+N+t)\Gamma(1+2a-t)\Gamma(1+N-t)} \\ &\qquad\qquad\times \widehat{y}_m(-(a-t)^2)\widehat{y}_n(-(a-t)^2) \\ &= \frac{\Gamma(n+x_1)\Gamma(2a+x_1)\Gamma(1-2a-x_1)n!}{\Gamma(1-2a-x_1+N-n)(N-n)!}\,\delta_{mn}, \quad m,n=0,1,2,\ldots,N. \quad (8.2.4)\end{aligned}$$

If the boundary conditions (8.2.2) cannot be satisfied, one may use infinite sums. In that case we have

$$\sigma_n^{(5)} = \frac{\pi\Gamma(n+x_1)\Gamma(2a+x_1)\Gamma(1-2a-x_1)n!}{\Gamma(1-2a-x_1-x_2-n)\Gamma(1-x_2-n)}, \quad n=0,1,2,\ldots,N,$$

which leads to the orthogonality relation (cf. (8.2.1))

$$\begin{aligned}\sum_{t=0}^{\infty} &\frac{\Gamma(1-2a+2t)\Gamma(1+2a-2t)\Gamma(x_1+2a-t)\Gamma(x_1+t)}{\Gamma(1+t)\Gamma(1-2a-x_2+t)\Gamma(1+2a-t)\Gamma(1-x_2-t)} \\ &\qquad\qquad\times \widehat{y}_m(-(a-t)^2)\widehat{y}_n(-(a-t)^2) \\ &= \frac{\Gamma(n+x_1)\Gamma(2a+x_1)\Gamma(1-2a-x_1)n!}{\Gamma(1-2a-x_1-x_2-n)\Gamma(1-x_2-n)}\,\delta_{mn}, \quad m,n=0,1,2,\ldots,N. \quad (8.2.5)\end{aligned}$$

Again we need the existence of the moments involved in order to have convergent sums.

Table 7.1 also holds for the continuous dual Hahn polynomials $\widehat{y}_n(x^2)$ given by (8.1.2). The seven finite cases can be treated by using the weight function $w_5^*(x)$ for the continuous dual Hahn polynomials with only one exception, as in the preceding section. The positivity of $d_0^{(5)}$ is dealt with in the same way as before. The factor $\Gamma(2a+x_1)\Gamma(1-2a-x_1)$ with $2a+x_1 \neq 0,\pm1,\pm2,\ldots$ determines the sign which is necessary for positivity.

The Infinite Cases

For the six infinite cases of table 7.1 we may use the weight function $w_2^*(x)$ for the dual Hahn polynomials again. By using Legendre's duplication formula (1.2.6), we have

$$\begin{aligned}&\Gamma(x+u/2)\Gamma(x+1/2+u/2)\Gamma(-u/2-x)\Gamma(1/2-u/2-x) \\ &\quad = 4\pi\Gamma(2x+u)\Gamma(-u-2x)\end{aligned}$$

for $2x+u \neq 0,\pm1,\pm2,\ldots$, which implies that

$$w_2^*(x) = \frac{\Gamma(x+u)\Gamma(-x)\Gamma(x+x_1)\Gamma(x_1-u-x)\Gamma(x+x_2)\Gamma(x_2-u-x)}{4\pi\Gamma(2x+u)\Gamma(-u-2x)}.$$

If we now replace x by $a+ix$ and u by $-2a$, we obtain

$$w_2^*(a+ix) = \frac{1}{4\pi}\left|\frac{\Gamma(-a+ix)\Gamma(x_1+a+ix)\Gamma(x_2+a+ix)}{\Gamma(2ix)}\right|^2.$$

In this case the boundary conditions (8.2.2) cannot be satisfied. However, as before, the concept of orthogonality can be extended to integration over the (possibly deformed) imaginary axis (cf. page 109). Then we need again the existence of the required moments which is in general taken care of by the asymptotic properties of the gamma functions involved. By using $z=a+\xi$ with $u=-2a$, we have

$$\begin{aligned} w_2^*(z) &= \frac{\Gamma(z-2a)\Gamma(-z)\Gamma(z+x_1)\Gamma(x_1+2a-z)\Gamma(z+x_2)\Gamma(x_2+2a-z)}{\Gamma(z-a)\Gamma(z-a+1/2)\Gamma(a-z)\Gamma(1/2+a-z)} \\ &= \frac{\Gamma(\xi-a)\Gamma(-a-\xi)}{\Gamma(\xi)\Gamma(-\xi)} \\ &\qquad \times \frac{\Gamma(\xi+a+x_1)\Gamma(x_1+a-\xi)\Gamma(\xi+a+x_2)\Gamma(x_2+a-\xi)}{\Gamma(\xi+1/2)\Gamma(1/2-\xi)}. \end{aligned}$$

Analogous to (7.7.4) we obtain by using (1.6.10)

$$\frac{1}{2\pi i}\int_{-i\infty}^{i\infty} w_2^*(a+\xi)\,d\xi = \frac{1}{2\pi}\Gamma(x_1)\Gamma(x_2)\Gamma(x_1+x_2+2a)$$

which, if $\xi = ix$ is used, leads to

$$\left(d_0^{(2)} =\right)\int_{-\infty}^{\infty} w_2^*(a+ix)\,dx = \Gamma(x_1)\Gamma(x_2)\Gamma(x_1+x_2+2a).$$

Then by using (8.1.3), we obtain

$$\sigma_n^{(2)} = \Gamma(n+x_1)\Gamma(n+x_2)\Gamma(n+x_1+x_2+2a)n!, \quad n=0,1,2,\ldots$$

and therefore the orthogonality relation

$$\begin{aligned} &\frac{1}{4\pi}\int_{-\infty}^{\infty}\left|\frac{\Gamma(-a+ix)\Gamma(x_1+a+ix)\Gamma(x_2+a+ix)}{\Gamma(2ix)}\right|^2 \widehat{y}_m(x^2)\widehat{y}_n(x^2)\,dx \\ &= \Gamma(n+x_1)\Gamma(n+x_2)\Gamma(n+x_1+x_2+2a)n!\,\delta_{mn}, \quad m,n=0,1,2,\ldots. \qquad (8.2.6) \end{aligned}$$

The six infinite cases of table 7.1 can be treated by using this weight function $w_2^*(x)$ with only one exception. The positivity of $d_0^{(2)}$ is treated as before.

8.3 Orthogonality Relations for Wilson Polynomials

For the Wilson polynomials $\widehat{y}_n(x^2)$ given by (8.1.4), the Pearson equation (7.8.1) still holds.

The Finite Cases

In order to deal with the finite cases in table 7.2 through table 7.7, we write instead of (7.8.1)

$$\frac{w^*(x)}{w^*(x+1)} = \frac{(-x-u/2)(-x-1/2-u/2)(x+1)(-x-1-u+x_1)}{(x+1+u/2)(x+1/2+u/2)(-x-u)(x+x_1)} \times \frac{(x+1+u-x_2)(x+1+u-x_3)}{(-x-x_2)(-x-x_3)}$$

which leads (up to a factor of period 1 in x) to the solution

$$w_6^*(x) = \frac{\Gamma(x+1+u/2)\Gamma(x+1/2+u/2)\Gamma(1-u/2-x)\Gamma(1/2-u/2-x)}{\Gamma(x+1)\Gamma(1-u-x)} \times \frac{\Gamma(x_1-u-x)\Gamma(x+x_1)}{\Gamma(1-x_2-x)\Gamma(1-x_3-x)\Gamma(x+1+u-x_2)\Gamma(x+1+u-x_3)}.$$

By using Legendre's duplication formula (1.2.6) we obtain

$$\Gamma(x+1+u/2)\Gamma(x+1/2+u/2)\Gamma(1-u/2-x)\Gamma(1/2-u/2-x) = \pi\Gamma(2x+1+u)\Gamma(1-u-2x)$$

for $2x+u \neq \pm 1, \pm 2, \pm 3, \ldots$, which implies that

$$w_6^*(x) = \frac{\pi\Gamma(2x+1+u)\Gamma(1-u-2x)}{\Gamma(x+1)\Gamma(1-u-x)} \times \frac{\Gamma(x_1-u-x)\Gamma(x+x_1)}{\Gamma(1-x_2-x)\Gamma(1-x_3-x)\Gamma(x+1+u-x_2)\Gamma(x+1+u-x_3)}.$$

If we now replace x by $a+ix$ and u by $-2a$ then we obtain

$$w_6^*(a+ix) = \pi \left| \frac{\Gamma(1+2ix)\Gamma(x_1+a+ix)}{\Gamma(1+a+ix)\Gamma(1-a-x_2+ix)\Gamma(1-a-x_3+ix)} \right|^2.$$

In this case the boundary conditions (7.6.6) can be satisfied by taking $A=0$ and $x_2=-N$. By using (1.5.6) and (1.2.6) we obtain

$$
\begin{aligned}
d_0^{(6)} &= \sum_{x=0}^{N} w_6^*(x) \\
&= \sum_{x=0}^{N} \frac{\Gamma(x+1+u/2)\Gamma(x+1/2+u/2)\Gamma(1-u/2-x)\Gamma(1/2-u/2-x)}{\Gamma(x+1)\Gamma(1-u-x)} \\
&\quad \times \frac{\Gamma(x_1-u-x)\Gamma(x+x_1)}{\Gamma(1+N-x)\Gamma(1-x_3-x)\Gamma(x+1+u+N)\Gamma(x+1+u-x_3)} \\
&= \frac{\Gamma(1+u/2)\Gamma(1/2+u/2)\Gamma(1-u/2)\Gamma(1/2-u/2)\Gamma(x_1)\Gamma(x_1-u)}{N!\Gamma(1-u)\Gamma(1+u+N)\Gamma(1-x_3)\Gamma(1+u-x_3)} \\
&\quad \times \sum_{x=0}^{N} \frac{(-1)^{2x}(-N)_x(1+u/2)_x(u)_x(x_1)_x(x_3)_x}{x!\,(u/2)_x(1+u-x_1)_x(1+u+N)_x(1+u-x_3)_x} \\
&= \frac{\Gamma(1+u/2)\Gamma(1/2+u/2)\Gamma(1-u/2)\Gamma(1/2-u/2)\Gamma(x_1)\Gamma(x_1-u)}{N!\Gamma(1-u)\Gamma(1+u+N)\Gamma(1-x_3)\Gamma(1+u-x_3)} \\
&\quad \times {}_5F_4\left(\begin{matrix} 1+u/2, u, x_1, -N, x_3 \\ u/2, 1+u-x_1, 1+u+N, 1+u-x_3 \end{matrix}; 1\right) \\
&= \frac{\Gamma(1+u/2)\Gamma(1/2+u/2)\Gamma(1-u/2)\Gamma(1/2-u/2)\Gamma(x_1)\Gamma(x_1-u)}{N!\Gamma(1-u)\Gamma(1+u+N)\Gamma(1-x_3)\Gamma(1+u-x_3)} \\
&\quad \times \frac{\Gamma(1+u-x_1)\Gamma(1+u+N)\Gamma(1+u-x_3)\Gamma(1+u-x_1-x_3+N)}{\Gamma(1+u)\Gamma(1+u-x_1+N)\Gamma(1+u-x_1-x_3)\Gamma(1+u-x_3+N)} \\
&= \frac{\pi\Gamma(x_1)\Gamma(x_1-u)\Gamma(1+u-x_1)\Gamma(1-w)}{N!\Gamma(1-x_3)\Gamma(1+x_1-w)\Gamma(1+x_3-w)\Gamma(1-w-N)},
\end{aligned}
$$

where $w = x_1 - N + x_3 - u$. Further we get by using (8.1.5) and (7.6.7)

$$
\begin{aligned}
\sigma_n^{(6)} &= \frac{\pi\Gamma(w)\Gamma(1-w)\Gamma(w-1+n)\Gamma(x_1-u)\Gamma(1+u-x_1)}{\Gamma(2n+w)\Gamma(2n-1+w)\Gamma(1-x_3-n)} \\
&\quad \times \frac{\Gamma(n+x_1)(-1)^n n!}{\Gamma(1+x_1-w-n)\Gamma(1-N-w-n)\Gamma(1+x_3-w-n)(N-n)!}
\end{aligned}
$$

for $n = 0,1,2,\ldots,N$, which leads to the orthogonality relation (cf. (8.2.1))

$$
\begin{aligned}
&\sum_{t=0}^{N} \frac{\Gamma(1-2a+2t)\Gamma(1+2a-2t)\Gamma(x_1+2a-t)\Gamma(x_1+t)}{\Gamma(1+t)\Gamma(1+2a-t)\Gamma(1+N-t)\Gamma(1-x_3-t)} \\
&\quad \times \frac{1}{\Gamma(1-2a+N+t)\Gamma(1-2a-x_3+t)} \widehat{y}_m(-(a-t)^2)\widehat{y}_n(-(a-t)^2) \\
&= \frac{1}{\pi}\sigma_n^{(6)}\,\delta_{mn}, \quad m,n = 0,1,2,\ldots,N. \qquad (8.3.1)
\end{aligned}
$$

If the boundary conditions (8.2.2) cannot be satisfied, one may use infinite sums. In that case we have $w = x_1 + x_2 + x_3 - u$ and

$$\sigma_n^{(6)} = \frac{\pi\,\Gamma(w)\Gamma(1-w)\Gamma(w-1+n)\Gamma(x_1-u)\Gamma(1+u-x_1)}{\Gamma(2n+w)\Gamma(2n-1+w)\Gamma(1-x_2-n)\Gamma(1-x_3-n)} \times \frac{\Gamma(n+x_1)(-1)^n n!}{\Gamma(1+x_1-w-n)\Gamma(1+x_2-w-n)\Gamma(1+x_3-w-n)}$$

for $n=0,1,2,\ldots,N$, which leads to the orthogonality relation (cf. (8.2.1))

$$\sum_{t=0}^{\infty} \frac{\Gamma(1-2a+2t)\Gamma(1+2a-2t)\Gamma(x_1+2a-t)\Gamma(x_1+t)}{\Gamma(1+t)\Gamma(1+2a-t)\Gamma(1-x_2-t)\Gamma(1-x_3-t)} \times \frac{1}{\Gamma(1-2a-x_2+t)\Gamma(1-2a-x_3+t)}\widehat{y}_m(-(a-t)^2)\widehat{y}_n(-(a-t)^2)$$
$$= \frac{1}{\pi}\sigma_n^{(6)}\,\delta_{mn}, \quad m,n=0,1,2,\ldots,N. \tag{8.3.2}$$

Again we need the existence of the moments involved in order to have convergent sums.

Table 7.2 through table 7.7 also hold for the Wilson polynomials $\widehat{y}_n(x^2)$ given by (8.1.4). Almost all finite cases can be treated by using the weight function $w_6^*(x)$ for the Wilson polynomials. The positivity of $d_0^{(6)}$ is dealt with in the same way as in the preceding section. The factor $\Gamma(2a+x_1)\Gamma(1-2a-x_1)$ with $2a+x_1 \neq 0,\pm 1,\pm 2,\ldots$ determines the sign which is necessary for positivity.

The Infinite Cases

The infinite cases of table 7.2 through table 7.7 can be treated by using the weight function $w_4^*(x)$ for the Racah polynomials again. By using Legendre's duplication formula (1.2.6), we obtain

$$\Gamma(x+u/2)\Gamma(x+1/2+u/2)\Gamma(-u/2-x)\Gamma(1/2-u/2-x)$$
$$= 4\pi\,\Gamma(2x+u)\Gamma(-u-2x)$$

for $2x+u \neq 0,\pm 1,\pm 2,\ldots$, which implies that

$$w_4^*(x) = \frac{\Gamma(x+u)\Gamma(-x)\Gamma(x+x_1)\Gamma(x_1-u-x)}{4\pi\,\Gamma(2x+u)\Gamma(-u-2x)} \times \Gamma(x+x_2)\Gamma(x_2-u-x)\Gamma(x+x_3)\Gamma(x_3-u-x)$$

If we now replace x by $a+ix$ and u by $-2a$, then we obtain

$$w_4^*(a+ix) = \frac{1}{4\pi}\left|\frac{\Gamma(-a+ix)\Gamma(x_1+a+ix)\Gamma(x_2+a+ix)\Gamma(x_3+a+ix)}{\Gamma(2ix)}\right|^2.$$

In this case the boundary conditions (8.2.2) cannot be satisfied. However, as before, the concept of orthogonality can be extended to integration over the (possibly deformed) imaginary axis (cf. page 109). Then we need again the existence of the required moments which is in general taken care of by the asymptotic properties of the gamma functions involved. By using $z=a+\xi$ with $u=-2a$, we obtain

$$\begin{aligned} w_4^*(z) &= \frac{\Gamma(z-2a)\Gamma(-z)\Gamma(z+x_1)\Gamma(x_1+2a-z)}{\Gamma(z-a)\Gamma(z-a+1/2)\Gamma(a-z)\Gamma(1/2+a-z)} \\ &\qquad \times \Gamma(z+x_2)\Gamma(x_2+2a-z)\Gamma(z+x_3)\Gamma(x_3+2a-z) \\ &= \frac{\Gamma(\xi-a)\Gamma(-a-\xi)\Gamma(\xi+a+x_1)\Gamma(x_1+a-\xi)}{\Gamma(\xi)\Gamma(\xi+1/2)\Gamma(-\xi)\Gamma(1/2-\xi)} \\ &\qquad \times \Gamma(\xi+a+x_2)\Gamma(x_2+a-\xi)\Gamma(\xi+a+x_3)\Gamma(x_3+a-\xi). \end{aligned}$$

Analogous to (7.7.4) we obtain by using (1.6.6)

$$\frac{1}{2\pi i}\int_{-i\infty}^{i\infty} w_4^*(a+\xi)\,d\xi = \frac{\Gamma(x_1)\Gamma(x_2)\Gamma(x_3)\Gamma(w-x_1)\Gamma(w-x_2)\Gamma(w-x_3)}{2\pi\,\Gamma(w)}$$

which, if $\xi=ix$ is used, leads to

$$\left(d_0^{(4)}=\right)\int_{-\infty}^{\infty} w_4^*(a+ix)\,dx = \frac{\Gamma(x_1)\Gamma(x_2)\Gamma(x_3)\Gamma(w-x_1)\Gamma(w-x_2)\Gamma(w-x_3)}{\Gamma(w)}.$$

Then we obtain by using (8.1.5)

$$\begin{aligned} \sigma_n^{(4)} &= \frac{\Gamma(n-1+w)\Gamma(n+x_1)\Gamma(n+x_2)\Gamma(n+x_3)}{\Gamma(2n+w)\Gamma(2n-1+w)} \\ &\qquad \times \Gamma(n+w-x_1)\Gamma(n+w-x_2)\Gamma(n+w-x_3)n! \end{aligned}$$

for $n=0,1,2,\ldots$ and therefore the orthogonality relation

$$\begin{aligned} &\frac{1}{4\pi}\int_{-\infty}^{\infty}\left|\frac{\Gamma(-a+ix)\Gamma(x_1+a+ix)\Gamma(x_2+a+ix)\Gamma(x_3+a+ix)}{\Gamma(2ix)}\right|^2 \widehat{y}_m(x^2)\widehat{y}_n(x^2)\,dx \\ &= \sigma_n^{(4)}\,\delta_{mn}, \quad m,n=0,1,2,\ldots. \end{aligned} \tag{8.3.3}$$

Almost all infinite cases of table 7.2 through table 7.7 can be treated by using this weight function $w_4^*(x)$. The positivity of $d_0^{(4)}$ is treated as before.

In this chapter we have proved:

Theorem 8.1. *The positive-definite orthogonal polynomial solutions $y(z(z+u))$ with $z \in \mathbb{C}$ and $u \in \mathbb{R}$ of the difference equation (7.2.1)*

$$\varphi(z+2)\Delta^2 y_n(z(z+u)) + \psi(z+1)\Delta y_n(z(z+u))$$
$$= \lambda_n \rho(z+1) y_n((z+1)(z+1+u)), \quad n = 0, 1, 2, \ldots,$$

consist of the polynomial solutions of the (real) difference equations (7.3.4) and (7.3.7). In fact this leads to both infinite and finite systems of

- *continuous dual Hahn polynomials and*
- *Wilson polynomials.*

ASKEY SCHEME
OF
HYPERGEOMETRIC
ORTHOGONAL POLYNOMIALS

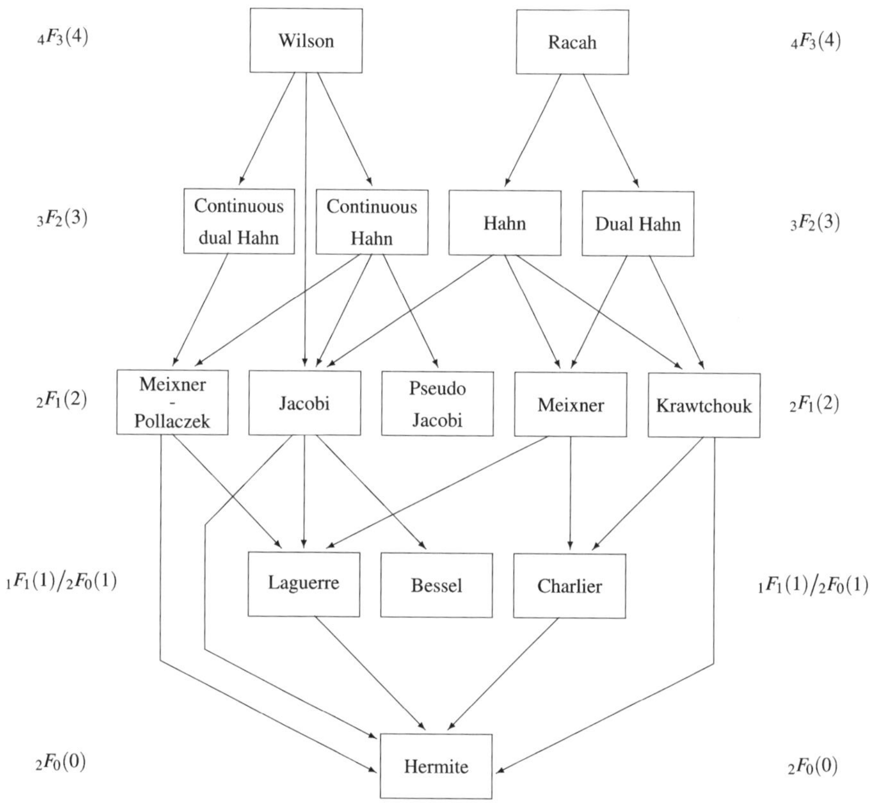

Chapter 9
Hypergeometric Orthogonal Polynomials

In this chapter we deal with all families of hypergeometric orthogonal polynomials appearing in the Askey scheme on page 183. For each family of orthogonal polynomials we state the most important properties such as a representation as a hypergeometric function, orthogonality relation(s), the three-term recurrence relation, the second-order differential or difference equation, the forward shift (or degree lowering) and backward shift (or degree raising) operator, a Rodrigues-type formula and some generating functions. In each case we use the notation which seems to be most common in the literature. Moreover, in each case we mention the connection between various families by stating the appropriate limit relations. See also [500] for an algebraic approach of this Askey scheme and [498] for a view from asymptotic analysis. For notations the reader is referred to chapter 1.

9.1 Wilson

Hypergeometric Representation

$$\frac{W_n(x^2;a,b,c,d)}{(a+b)_n(a+c)_n(a+d)_n} = {}_4F_3\left(\begin{matrix}-n,n+a+b+c+d-1,a+ix,a-ix\\ a+b,a+c,a+d\end{matrix};1\right). \tag{9.1.1}$$

R. Koekoek et al., *Hypergeometric Orthogonal Polynomials and Their q-Analogues*, Springer Monographs in Mathematics,
DOI 10.1007/978-3-642-05014-5_9, © Springer-Verlag Berlin Heidelberg 2010

Orthogonality Relation

If $Re(a,b,c,d) > 0$ and non-real parameters occur in conjugate pairs, then

$$\frac{1}{2\pi}\int_0^\infty \left|\frac{\Gamma(a+ix)\Gamma(b+ix)\Gamma(c+ix)\Gamma(d+ix)}{\Gamma(2ix)}\right|^2 \times W_m(x^2;a,b,c,d)W_n(x^2;a,b,c,d)\,dx = \frac{\Gamma(n+a+b)\cdots\Gamma(n+c+d)}{\Gamma(2n+a+b+c+d)}(n+a+b+c+d-1)_n n!\,\delta_{mn}, \quad (9.1.2)$$

where

$$\Gamma(n+a+b)\cdots\Gamma(n+c+d) = \Gamma(n+a+b)\Gamma(n+a+c)\Gamma(n+a+d)\Gamma(n+b+c)\Gamma(n+b+d)\Gamma(n+c+d).$$

If $a<0$ and $a+b$, $a+c$, $a+d$ are positive or a pair of complex conjugates occur with positive real parts, then

$$\frac{1}{2\pi}\int_0^\infty \left|\frac{\Gamma(a+ix)\Gamma(b+ix)\Gamma(c+ix)\Gamma(d+ix)}{\Gamma(2ix)}\right|^2 \times W_m(x^2;a,b,c,d)W_n(x^2;a,b,c,d)\,dx + \frac{\Gamma(a+b)\Gamma(a+c)\Gamma(a+d)\Gamma(b-a)\Gamma(c-a)\Gamma(d-a)}{\Gamma(-2a)} \times \sum_{\substack{k=0,1,2\ldots \\ a+k<0}} \frac{(2a)_k(a+1)_k(a+b)_k(a+c)_k(a+d)_k}{(a)_k(a-b+1)_k(a-c+1)_k(a-d+1)_k k!} \times W_m(-(a+k)^2;a,b,c,d)W_n(-(a+k)^2;a,b,c,d) = \frac{\Gamma(n+a+b)\cdots\Gamma(n+c+d)}{\Gamma(2n+a+b+c+d)}(n+a+b+c+d-1)_n n!\,\delta_{mn}. \quad (9.1.3)$$

Recurrence Relation

$$-\left(a^2+x^2\right)\tilde{W}_n(x^2) = A_n\tilde{W}_{n+1}(x^2) - (A_n+C_n)\tilde{W}_n(x^2) + C_n\tilde{W}_{n-1}(x^2), \quad (9.1.4)$$

where

$$\tilde{W}_n(x^2) := \tilde{W}_n(x^2;a,b,c,d) = \frac{W_n(x^2;a,b,c,d)}{(a+b)_n(a+c)_n(a+d)_n}$$

and

$$\begin{cases} A_n = \dfrac{(n+a+b+c+d-1)(n+a+b)(n+a+c)(n+a+d)}{(2n+a+b+c+d-1)(2n+a+b+c+d)} \\[2ex] C_n = \dfrac{n(n+b+c-1)(n+b+d-1)(n+c+d-1)}{(2n+a+b+c+d-2)(2n+a+b+c+d-1)}. \end{cases}$$

Normalized Recurrence Relation

$$xp_n(x) = p_{n+1}(x) + (A_n + C_n - a^2)p_n(x) + A_{n-1}C_n p_{n-1}(x), \tag{9.1.5}$$

where

$$W_n(x^2;a,b,c,d) = (-1)^n(n+a+b+c+d-1)_n p_n(x^2).$$

Difference Equation

$$\begin{aligned} &n(n+a+b+c+d-1)y(x) \\ &= B(x)y(x+i) - [B(x)+D(x)]\,y(x) + D(x)y(x-i), \end{aligned} \tag{9.1.6}$$

where

$$y(x) = W_n(x^2;a,b,c,d)$$

and

$$\begin{cases} B(x) = \dfrac{(a-ix)(b-ix)(c-ix)(d-ix)}{2ix(2ix-1)} \\[2ex] D(x) = \dfrac{(a+ix)(b+ix)(c+ix)(d+ix)}{2ix(2ix+1)}. \end{cases}$$

Forward Shift Operator

$$\begin{aligned} &W_n((x+\tfrac{1}{2}i)^2;a,b,c,d) - W_n((x-\tfrac{1}{2}i)^2;a,b,c,d) \\ &= -2inx(n+a+b+c+d-1)W_{n-1}(x^2;a+\tfrac{1}{2},b+\tfrac{1}{2},c+\tfrac{1}{2},d+\tfrac{1}{2}) \end{aligned} \tag{9.1.7}$$

or equivalently

$$\begin{aligned} &\frac{\delta W_n(x^2;a,b,c,d)}{\delta x^2} \\ &= -n(n+a+b+c+d-1)W_{n-1}(x^2;a+\tfrac{1}{2},b+\tfrac{1}{2},c+\tfrac{1}{2},d+\tfrac{1}{2}). \end{aligned} \tag{9.1.8}$$

Backward Shift Operator

$$\begin{aligned}&(a-\tfrac{1}{2}-ix)(b-\tfrac{1}{2}-ix)(c-\tfrac{1}{2}-ix)(d-\tfrac{1}{2}-ix)W_n((x+\tfrac{1}{2}i)^2;a,b,c,d)\\&\qquad-(a-\tfrac{1}{2}+ix)(b-\tfrac{1}{2}+ix)(c-\tfrac{1}{2}+ix)(d-\tfrac{1}{2}+ix)W_n((x-\tfrac{1}{2}i)^2;a,b,c,d)\\&=-2ixW_{n+1}(x^2;a-\tfrac{1}{2},b-\tfrac{1}{2},c-\tfrac{1}{2},d-\tfrac{1}{2})\end{aligned} \tag{9.1.9}$$

or equivalently

$$\begin{aligned}&\frac{\delta\left[\omega(x;a,b,c,d)W_n(x^2;a,b,c,d)\right]}{\delta x^2}\\&=\omega(x;a-\tfrac{1}{2},b-\tfrac{1}{2},c-\tfrac{1}{2},d-\tfrac{1}{2})W_{n+1}(x^2;a-\tfrac{1}{2},b-\tfrac{1}{2},c-\tfrac{1}{2},d-\tfrac{1}{2}),\end{aligned} \tag{9.1.10}$$

where

$$\omega(x;a,b,c,d):=\frac{1}{2ix}\left|\frac{\Gamma(a+ix)\Gamma(b+ix)\Gamma(c+ix)\Gamma(d+ix)}{\Gamma(2ix)}\right|^2.$$

Rodrigues-Type Formula

$$\begin{aligned}&\omega(x;a,b,c,d)W_n(x^2;a,b,c,d)\\&=\left(\frac{\delta}{\delta x^2}\right)^n\left[\omega(x;a+\tfrac{1}{2}n,b+\tfrac{1}{2}n,c+\tfrac{1}{2}n,d+\tfrac{1}{2}n)\right].\end{aligned} \tag{9.1.11}$$

Generating Functions

$${}_2F_1\left(\begin{matrix}a+ix,b+ix\\a+b\end{matrix};t\right){}_2F_1\left(\begin{matrix}c-ix,d-ix\\c+d\end{matrix};t\right)=\sum_{n=0}^{\infty}\frac{W_n(x^2;a,b,c,d)t^n}{(a+b)_n(c+d)_n n!}. \tag{9.1.12}$$

$${}_2F_1\left(\begin{matrix}a+ix,c+ix\\a+c\end{matrix};t\right){}_2F_1\left(\begin{matrix}b-ix,d-ix\\b+d\end{matrix};t\right)=\sum_{n=0}^{\infty}\frac{W_n(x^2;a,b,c,d)t^n}{(a+c)_n(b+d)_n n!}. \tag{9.1.13}$$

$${}_2F_1\left(\begin{matrix}a+ix,d+ix\\a+d\end{matrix};t\right){}_2F_1\left(\begin{matrix}b-ix,c-ix\\b+c\end{matrix};t\right)=\sum_{n=0}^{\infty}\frac{W_n(x^2;a,b,c,d)t^n}{(a+d)_n(b+c)_n n!}. \tag{9.1.14}$$

$$(1-t)^{1-a-b-c-d}$$
$$\times {}_4F_3\left(\begin{array}{c}\frac{1}{2}(a+b+c+d-1),\frac{1}{2}(a+b+c+d),a+ix,a-ix\\ a+b,a+c,a+d\end{array};-\frac{4t}{(1-t)^2}\right)$$
$$=\sum_{n=0}^{\infty}\frac{(a+b+c+d-1)_n}{(a+b)_n(a+c)_n(a+d)_n n!}W_n(x^2;a,b,c,d)t^n. \tag{9.1.15}$$

Limit Relations

Wilson → Continuous Dual Hahn

The continuous dual Hahn polynomials given by (9.3.1) can be found from the Wilson polynomials by dividing by $(a+d)_n$ and letting $d\to\infty$:

$$\lim_{d\to\infty}\frac{W_n(x^2;a,b,c,d)}{(a+d)_n}=S_n(x^2;a,b,c). \tag{9.1.16}$$

Wilson → Continuous Hahn

The continuous Hahn polynomials given by (9.4.1) are obtained from the Wilson polynomials by the substitutions $a\to a-it$, $b\to b-it$, $c\to c+it$, $d\to d+it$ and $x\to x+t$ and the limit $t\to\infty$ in the following way:

$$\lim_{t\to\infty}\frac{W_n((x+t)^2;a-it,b-it,c+it,d+it)}{(-2t)^n n!}=p_n(x;a,b,c,d). \tag{9.1.17}$$

Wilson → Jacobi

The Jacobi polynomials given by (9.8.1) can be found from the Wilson polynomials by substituting $a=b=\frac{1}{2}(\alpha+1)$, $c=\frac{1}{2}(\beta+1)+it$, $d=\frac{1}{2}(\beta+1)-it$ and $x\to t\sqrt{\frac{1}{2}(1-x)}$ in the definition (9.1.1) of the Wilson polynomials and taking the limit $t\to\infty$. In fact we have

$$\lim_{t\to\infty}\frac{W_n(\frac{1}{2}(1-x)t^2;\frac{1}{2}(\alpha+1),\frac{1}{2}(\alpha+1),\frac{1}{2}(\beta+1)+it,\frac{1}{2}(\beta+1)-it)}{t^{2n}n!}$$
$$=P_n^{(\alpha,\beta)}(x). \tag{9.1.18}$$

Remarks

Note that for $k<n$ we have

$$\frac{(a+b)_n(a+c)_n(a+d)_n}{(a+b)_k(a+c)_k(a+d)_k}=(a+b+k)_{n-k}(a+c+k)_{n-k}(a+d+k)_{n-k},$$

which implies that the Wilson polynomials defined by (9.1.1) can also be seen as polynomials in the parameters a, b, c and d.
If we set

$$a=\tfrac{1}{2}(\gamma+\delta+1),\quad b=\tfrac{1}{2}(2\alpha-\gamma-\delta+1),$$
$$c=\tfrac{1}{2}(2\beta-\gamma+\delta+1),\quad d=\tfrac{1}{2}(\gamma-\delta+1)$$

and

$$ix\rightarrow x+\tfrac{1}{2}(\gamma+\delta+1)$$

in

$$\tilde{W}_n(x^2;a,b,c,d)=\frac{W_n(x^2;a,b,c,d)}{(a+b)_n(a+c)_n(a+d)_n},$$

given by (9.1.1) and take

$$\alpha+1=-N\quad\text{or}\quad\beta+\delta+1=-N\quad\text{or}\quad\gamma+1=-N$$

with N a nonnegative integer, we obtain the Racah polynomials given by (9.2.1).

References

[51], [71], [72], [80], [277], [278], [281], [282], [339], [340], [387], [401], [407], [410], [503], [512], [513].

9.2 Racah

Hypergeometric Representation

$$R_n(\lambda(x);\alpha,\beta,\gamma,\delta)$$
$$={}_4F_3\left(\begin{matrix}-n,n+\alpha+\beta+1,-x,x+\gamma+\delta+1\\ \alpha+1,\beta+\delta+1,\gamma+1\end{matrix};1\right),\quad n=0,1,2,\ldots,N,\ (9.2.1)$$

where

$$\lambda(x)=x(x+\gamma+\delta+1)$$

and

$$\alpha+1=-N \quad \text{or} \quad \beta+\delta+1=-N \quad \text{or} \quad \gamma+1=-N$$

with N a nonnegative integer.

Orthogonality Relation

$$\sum_{x=0}^{N} \frac{(\alpha+1)_x(\beta+\delta+1)_x(\gamma+1)_x(\gamma+\delta+1)_x((\gamma+\delta+3)/2)_x}{(-\alpha+\gamma+\delta+1)_x(-\beta+\gamma+1)_x((\gamma+\delta+1)/2)_x(\delta+1)_x x!}$$
$$\times R_m(\lambda(x))R_n(\lambda(x))$$
$$= M\frac{(n+\alpha+\beta+1)_n(\alpha+\beta-\gamma+1)_n(\alpha-\delta+1)_n(\beta+1)_n n!}{(\alpha+\beta+2)_{2n}(\alpha+1)_n(\beta+\delta+1)_n(\gamma+1)_n}\,\delta_{mn}, \tag{9.2.2}$$

where

$$R_n(\lambda(x)) := R_n(\lambda(x);\alpha,\beta,\gamma,\delta)$$

and

$$M = \begin{cases} \dfrac{(-\beta)_N(\gamma+\delta+2)_N}{(-\beta+\gamma+1)_N(\delta+1)_N} & \text{if} \quad \alpha+1=-N \\[2ex] \dfrac{(-\alpha+\delta)_N(\gamma+\delta+2)_N}{(-\alpha+\gamma+\delta+1)_N(\delta+1)_N} & \text{if} \quad \beta+\delta+1=-N \\[2ex] \dfrac{(\alpha+\beta+2)_N(-\delta)_N}{(\alpha-\delta+1)_N(\beta+1)_N} & \text{if} \quad \gamma+1=-N. \end{cases}$$

Recurrence Relation

$$\lambda(x)R_n(\lambda(x)) = A_nR_{n+1}(\lambda(x)) - (A_n+C_n)R_n(\lambda(x)) + C_nR_{n-1}(\lambda(x)), \tag{9.2.3}$$

where

$$R_n(\lambda(x)) := R_n(\lambda(x);\alpha,\beta,\gamma,\delta)$$

and

$$\begin{cases} A_n = \dfrac{(n+\alpha+1)(n+\alpha+\beta+1)(n+\beta+\delta+1)(n+\gamma+1)}{(2n+\alpha+\beta+1)(2n+\alpha+\beta+2)} \\[2ex] C_n = \dfrac{n(n+\alpha+\beta-\gamma)(n+\alpha-\delta)(n+\beta)}{(2n+\alpha+\beta)(2n+\alpha+\beta+1)}, \end{cases}$$

hence

$$A_n = \begin{cases} \dfrac{(n+\beta-N)(n+\beta+\delta+1)(n+\gamma+1)(n-N)}{(2n+\beta-N)(2n+\beta-N+1)} & \text{if} \quad \alpha+1=-N \\ \dfrac{(n+\alpha+1)(n+\alpha+\beta+1)(n+\gamma+1)(n-N)}{(2n+\alpha+\beta+1)(2n+\alpha+\beta+2)} & \text{if} \quad \beta+\delta+1=-N \\ \dfrac{(n+\alpha+1)(n+\alpha+\beta+1)(n+\beta+\delta+1)(n-N)}{(2n+\alpha+\beta+1)(2n+\alpha+\beta+2)} & \text{if} \quad \gamma+1=-N \end{cases}$$

and

$$C_n = \begin{cases} \dfrac{n(n+\beta)(n+\beta-\gamma-N-1)(n-\delta-N-1)}{(2n+\beta-N-1)(2n+\beta-N)} & \text{if} \quad \alpha+1=-N \\ \dfrac{n(n+\alpha+\beta+N+1)(n+\alpha+\beta-\gamma)(n+\beta)}{(2n+\alpha+\beta)(2n+\alpha+\beta+1)} & \text{if} \quad \beta+\delta+1=-N \\ \dfrac{n(n+\alpha+\beta+N+1)(n+\alpha-\delta)(n+\beta)}{(2n+\alpha+\beta)(2n+\alpha+\beta+1)} & \text{if} \quad \gamma+1=-N. \end{cases}$$

Normalized Recurrence Relation

$$xp_n(x) = p_{n+1}(x) - (A_n + C_n)p_n(x) + A_{n-1}C_n p_{n-1}(x), \tag{9.2.4}$$

where

$$R_n(\lambda(x);\alpha,\beta,\gamma,\delta) = \frac{(n+\alpha+\beta+1)_n}{(\alpha+1)_n(\beta+\delta+1)_n(\gamma+1)_n} p_n(\lambda(x)).$$

Difference Equation

$$n(n+\alpha+\beta+1)y(x) = B(x)y(x+1) - [B(x)+D(x)]\,y(x) + D(x)y(x-1), \tag{9.2.5}$$

where

$$y(x) = R_n(\lambda(x);\alpha,\beta,\gamma,\delta)$$

and

$$\begin{cases} B(x) = \dfrac{(x+\alpha+1)(x+\beta+\delta+1)(x+\gamma+1)(x+\gamma+\delta+1)}{(2x+\gamma+\delta+1)(2x+\gamma+\delta+2)} \\ D(x) = \dfrac{x(x-\alpha+\gamma+\delta)(x-\beta+\gamma)(x+\delta)}{(2x+\gamma+\delta)(2x+\gamma+\delta+1)}. \end{cases}$$

Forward Shift Operator

$$\begin{aligned}
&R_n(\lambda(x+1);\alpha,\beta,\gamma,\delta)-R_n(\lambda(x);\alpha,\beta,\gamma,\delta)\\
&=\frac{n(n+\alpha+\beta+1)}{(\alpha+1)(\beta+\delta+1)(\gamma+1)}\\
&\qquad\times(2x+\gamma+\delta+2)R_{n-1}(\lambda(x);\alpha+1,\beta+1,\gamma+1,\delta)
\end{aligned}\tag{9.2.6}$$

or equivalently

$$\begin{aligned}
&\frac{\Delta R_n(\lambda(x);\alpha,\beta,\gamma,\delta)}{\Delta\lambda(x)}\\
&=\frac{n(n+\alpha+\beta+1)}{(\alpha+1)(\beta+\delta+1)(\gamma+1)}R_{n-1}(\lambda(x);\alpha+1,\beta+1,\gamma+1,\delta).
\end{aligned}\tag{9.2.7}$$

Backward Shift Operator

$$\begin{aligned}
&(x+\alpha)(x+\beta+\delta)(x+\gamma)(x+\gamma+\delta)R_n(\lambda(x);\alpha,\beta,\gamma,\delta)\\
&\qquad-x(x-\beta+\gamma)(x-\alpha+\gamma+\delta)(x+\delta)R_n(\lambda(x-1);\alpha,\beta,\gamma,\delta)\\
&=\alpha\gamma(\beta+\delta)(2x+\gamma+\delta)R_{n+1}(\lambda(x);\alpha-1,\beta-1,\gamma-1,\delta)
\end{aligned}\tag{9.2.8}$$

or equivalently

$$\begin{aligned}
&\frac{\nabla\left[\omega(x;\alpha,\beta,\gamma,\delta)R_n(\lambda(x);\alpha,\beta,\gamma,\delta)\right]}{\nabla\lambda(x)}\\
&=\frac{1}{\gamma+\delta}\omega(x;\alpha-1,\beta-1,\gamma-1,\delta)R_{n+1}(\lambda(x);\alpha-1,\beta-1,\gamma-1,\delta),
\end{aligned}\tag{9.2.9}$$

where

$$\omega(x;\alpha,\beta,\gamma,\delta)=\frac{(\alpha+1)_x(\beta+\delta+1)_x(\gamma+1)_x(\gamma+\delta+1)_x}{(-\alpha+\gamma+\delta+1)_x(-\beta+\gamma+1)_x(\delta+1)_x x!}.$$

Rodrigues-Type Formula

$$\begin{aligned}
&\omega(x;\alpha,\beta,\gamma,\delta)R_n(\lambda(x);\alpha,\beta,\gamma,\delta)\\
&=(\gamma+\delta+1)_n\left(\nabla_\lambda\right)^n\left[\omega(x;\alpha+n,\beta+n,\gamma+n,\delta)\right],
\end{aligned}\tag{9.2.10}$$

where

$$\nabla_\lambda := \frac{\nabla}{\nabla\lambda(x)}.$$

Generating Functions

For $x = 0,1,2,\ldots,N$ we have

$$\begin{aligned}
&{}_2F_1\left(\begin{array}{c}-x,-x+\alpha-\gamma-\delta\\ \alpha+1\end{array};t\right)\,{}_2F_1\left(\begin{array}{c}x+\beta+\delta+1,x+\gamma+1\\ \beta+1\end{array};t\right)\\
&=\sum_{n=0}^{N}\frac{(\beta+\delta+1)_n(\gamma+1)_n}{(\beta+1)_n n!}R_n(\lambda(x);\alpha,\beta,\gamma,\delta)t^n,\\
&\qquad\text{if}\quad \beta+\delta+1=-N \quad\text{or}\quad \gamma+1=-N.
\end{aligned} \tag{9.2.11}$$

$$\begin{aligned}
&{}_2F_1\left(\begin{array}{c}-x,-x+\beta-\gamma\\ \beta+\delta+1\end{array};t\right)\,{}_2F_1\left(\begin{array}{c}x+\alpha+1,x+\gamma+1\\ \alpha-\delta+1\end{array};t\right)\\
&=\sum_{n=0}^{N}\frac{(\alpha+1)_n(\gamma+1)_n}{(\alpha-\delta+1)_n n!}R_n(\lambda(x);\alpha,\beta,\gamma,\delta)t^n,\\
&\qquad\text{if}\quad \alpha+1=-N \quad\text{or}\quad \gamma+1=-N.
\end{aligned} \tag{9.2.12}$$

$$\begin{aligned}
&{}_2F_1\left(\begin{array}{c}-x,-x-\delta\\ \gamma+1\end{array};t\right)\,{}_2F_1\left(\begin{array}{c}x+\alpha+1,x+\beta+\delta+1\\ \alpha+\beta-\gamma+1\end{array};t\right)\\
&=\sum_{n=0}^{N}\frac{(\alpha+1)_n(\beta+\delta+1)_n}{(\alpha+\beta-\gamma+1)_n n!}R_n(\lambda(x);\alpha,\beta,\gamma,\delta)t^n,\\
&\qquad\text{if}\quad \alpha+1=-N \quad\text{or}\quad \beta+\delta+1=-N.
\end{aligned} \tag{9.2.13}$$

$$\begin{aligned}
&\Bigg[(1-t)^{-\alpha-\beta-1}\\
&\quad\times{}_4F_3\left(\begin{array}{c}\frac{1}{2}(\alpha+\beta+1),\frac{1}{2}(\alpha+\beta+2),-x,x+\gamma+\delta+1\\ \alpha+1,\beta+\delta+1,\gamma+1\end{array};-\frac{4t}{(1-t)^2}\right)\Bigg]_N\\
&=\sum_{n=0}^{N}\frac{(\alpha+\beta+1)_n}{n!}R_n(\lambda(x);\alpha,\beta,\gamma,\delta)t^n.
\end{aligned} \tag{9.2.14}$$

Limit Relations

Racah → Hahn

The Hahn polynomials given by (9.5.1) can be obtained from the Racah polynomials by taking $\gamma+1=-N$ and letting $\delta\to\infty$:

$$\lim_{\delta\to\infty} R_n(\lambda(x);\alpha,\beta,-N-1,\delta)=Q_n(x;\alpha,\beta,N). \tag{9.2.15}$$

The Hahn polynomials given by (9.5.1) can also be obtained from the Racah polynomials by taking $\delta=-\beta-N-1$ and letting $\gamma\to\infty$:

$$\lim_{\gamma\to\infty} R_n(\lambda(x);\alpha,\beta,\gamma,-\beta-N-1)=Q_n(x;\alpha,\beta,N). \tag{9.2.16}$$

Another way to do this is to take $\alpha+1=-N$ and $\beta\to\beta+\gamma+N+1$ and then take the limit $\delta\to\infty$. In that case we obtain the Hahn polynomials given by (9.5.1) in the following way:

$$\lim_{\delta\to\infty} R_n(\lambda(x);-N-1,\beta+\gamma+N+1,\gamma,\delta)=Q_n(x;\gamma,\beta,N). \tag{9.2.17}$$

Racah → Dual Hahn

The dual Hahn polynomials given by (9.6.1) are obtained from the Racah polynomials if we take $\alpha+1=-N$ and let $\beta\to\infty$:

$$\lim_{\beta\to\infty} R_n(\lambda(x);-N-1,\beta,\gamma,\delta)=R_n(\lambda(x);\gamma,\delta,N). \tag{9.2.18}$$

The dual Hahn polynomials given by (9.6.1) are also obtained from the Racah polynomials if we take $\beta=-\delta-N-1$ and let $\alpha\to\infty$:

$$\lim_{\alpha\to\infty} R_n(\lambda(x);\alpha,-\delta-N-1,\gamma,\delta)=R_n(\lambda(x);\gamma,\delta,N). \tag{9.2.19}$$

Finally, the dual Hahn polynomials given by (9.6.1) are also obtained from the Racah polynomials if we take $\gamma+1=-N$ and $\delta\to\alpha+\delta+N+1$ and take the limit $\beta\to\infty$:

$$\lim_{\beta\to\infty} R_n(\lambda(x);\alpha,\beta,-N-1,\alpha+\delta+N+1)=R_n(\lambda(x);\alpha,\delta,N). \tag{9.2.20}$$

Remark

If we set $\alpha = a+b-1$, $\beta = c+d-1$, $\gamma = a+d-1$, $\delta = a-d$ and $x \to -a+ix$ in the definition (9.2.1) of the Racah polynomials we obtain the Wilson polynomials given by (9.1.1):

$$\begin{aligned} &R_n(\lambda(-a+ix);a+b-1,c+d-1,a+d-1,a-d) \\ &= \tilde{W}_n(x^2;a,b,c,d) = \frac{W_n(x^2;a,b,c,d)}{(a+b)_n(a+c)_n(a+d)_n}. \end{aligned}$$

References

[51], [70], [72], [80], [82], [187], [340], [376], [381], [387], [416], [417], [426], [437], [439], [512].

9.3 Continuous Dual Hahn

Hypergeometric Representation

$$\frac{S_n(x^2;a,b,c)}{(a+b)_n(a+c)_n} = {}_3F_2\left(\begin{matrix} -n, a+ix, a-ix \\ a+b, a+c \end{matrix} ; 1\right). \tag{9.3.1}$$

Orthogonality Relation

If a, b and c are positive except possibly for a pair of complex conjugates with positive real parts, then

$$\begin{aligned} &\frac{1}{2\pi}\int_0^\infty \left|\frac{\Gamma(a+ix)\Gamma(b+ix)\Gamma(c+ix)}{\Gamma(2ix)}\right|^2 S_m(x^2;a,b,c)S_n(x^2;a,b,c)\,dx \\ &= \Gamma(n+a+b)\Gamma(n+a+c)\Gamma(n+b+c)n!\,\delta_{mn}. \end{aligned} \tag{9.3.2}$$

If $a<0$ and $a+b$, $a+c$ are positive or a pair of complex conjugates with positive real parts, then

$$\frac{1}{2\pi}\int_0^{\infty}\left|\frac{\Gamma(a+ix)\Gamma(b+ix)\Gamma(c+ix)}{\Gamma(2ix)}\right|^2 S_m(x^2;a,b,c)S_n(x^2;a,b,c)\,dx$$
$$+\frac{\Gamma(a+b)\Gamma(a+c)\Gamma(b-a)\Gamma(c-a)}{\Gamma(-2a)}$$
$$\times\sum_{\substack{k=0,1,2\ldots\\ a+k<0}}\frac{(2a)_k(a+1)_k(a+b)_k(a+c)_k}{(a)_k(a-b+1)_k(a-c+1)_k k!}(-1)^k$$
$$\times S_m(-(a+k)^2;a,b,c)S_n(-(a+k)^2;a,b,c)$$
$$=\Gamma(n+a+b)\Gamma(n+a+c)\Gamma(n+b+c)n!\,\delta_{mn}. \tag{9.3.3}$$

Recurrence Relation

$$-\left(a^2+x^2\right)\tilde{S}_n(x^2)=A_n\tilde{S}_{n+1}(x^2)-(A_n+C_n)\tilde{S}_n(x^2)+C_n\tilde{S}_{n-1}(x^2), \tag{9.3.4}$$

where

$$\tilde{S}_n(x^2):=\tilde{S}_n(x^2;a,b,c)=\frac{S_n(x^2;a,b,c)}{(a+b)_n(a+c)_n}$$

and

$$\begin{cases} A_n=(n+a+b)(n+a+c)\\ C_n=n(n+b+c-1).\end{cases}$$

Normalized Recurrence Relation

$$xp_n(x)=p_{n+1}(x)+(A_n+C_n-a^2)p_n(x)+A_{n-1}C_np_{n-1}(x), \tag{9.3.5}$$

where

$$S_n(x^2;a,b,c)=(-1)^n p_n(x^2).$$

Difference Equation

$$ny(x)=B(x)y(x+i)-[B(x)+D(x)]y(x)+D(x)y(x-i), \tag{9.3.6}$$

where

$$y(x)=S_n(x^2;a,b,c)$$

and

$$\begin{cases} B(x) = \dfrac{(a-ix)(b-ix)(c-ix)}{2ix(2ix-1)} \\ \\ D(x) = \dfrac{(a+ix)(b+ix)(c+ix)}{2ix(2ix+1)}. \end{cases}$$

Forward Shift Operator

$$\begin{aligned} &S_n((x+\tfrac{1}{2}i)^2;a,b,c) - S_n((x-\tfrac{1}{2}i)^2;a,b,c) \\ &= -2inxS_{n-1}(x^2;a+\tfrac{1}{2},b+\tfrac{1}{2},c+\tfrac{1}{2}) \end{aligned} \tag{9.3.7}$$

or equivalently

$$\frac{\delta S_n(x^2;a,b,c)}{\delta x^2} = -nS_{n-1}(x^2;a+\tfrac{1}{2},b+\tfrac{1}{2},c+\tfrac{1}{2}). \tag{9.3.8}$$

Backward Shift Operator

$$\begin{aligned} &(a-\tfrac{1}{2}-ix)(b-\tfrac{1}{2}-ix)(c-\tfrac{1}{2}-ix)S_n((x+\tfrac{1}{2}i)^2;a,b,c) \\ &\qquad -(a-\tfrac{1}{2}+ix)(b-\tfrac{1}{2}+ix)(c-\tfrac{1}{2}+ix)S_n((x-\tfrac{1}{2}i)^2;a,b,c) \\ &= -2ixS_{n+1}(x^2;a-\tfrac{1}{2},b-\tfrac{1}{2},c-\tfrac{1}{2}) \end{aligned} \tag{9.3.9}$$

or equivalently

$$\begin{aligned} &\frac{\delta\left[\omega(x;a,b,c)S_n(x^2;a,b,c)\right]}{\delta x^2} \\ &= \omega(x;a-\tfrac{1}{2},b-\tfrac{1}{2},c-\tfrac{1}{2})S_{n+1}(x^2;a-\tfrac{1}{2},b-\tfrac{1}{2},c-\tfrac{1}{2}), \end{aligned} \tag{9.3.10}$$

where

$$\omega(x;a,b,c) = \frac{1}{2ix}\left|\frac{\Gamma(a+ix)\Gamma(b+ix)\Gamma(c+ix)}{\Gamma(2ix)}\right|^2.$$

Rodrigues-Type Formula

$$\omega(x;a,b,c)S_n(x^2;a,b,c) = \left(\frac{\delta}{\delta x^2}\right)^n \left[\omega(x;a+\tfrac{1}{2}n,b+\tfrac{1}{2}n,c+\tfrac{1}{2}n)\right]. \tag{9.3.11}$$

Generating Functions

$$(1-t)^{-c+ix}\,{}_2F_1\left(\begin{array}{c} a+ix,b+ix \\ a+b \end{array};t\right)=\sum_{n=0}^{\infty}\frac{S_n(x^2;a,b,c)}{(a+b)_n n!}t^n. \tag{9.3.12}$$

$$(1-t)^{-b+ix}\,{}_2F_1\left(\begin{array}{c} a+ix,c+ix \\ a+c \end{array};t\right)=\sum_{n=0}^{\infty}\frac{S_n(x^2;a,b,c)}{(a+c)_n n!}t^n. \tag{9.3.13}$$

$$(1-t)^{-a+ix}\,{}_2F_1\left(\begin{array}{c} b+ix,c+ix \\ b+c \end{array};t\right)=\sum_{n=0}^{\infty}\frac{S_n(x^2;a,b,c)}{(b+c)_n n!}t^n. \tag{9.3.14}$$

$$\mathrm{e}^t\,{}_2F_2\left(\begin{array}{c} a+ix,a-ix \\ a+b,a+c \end{array};-t\right)=\sum_{n=0}^{\infty}\frac{S_n(x^2;a,b,c)}{(a+b)_n(a+c)_n n!}t^n. \tag{9.3.15}$$

$$(1-t)^{-\gamma}\,{}_3F_2\left(\begin{array}{c} \gamma,a+ix,a-ix \\ a+b,a+c \end{array};\frac{t}{t-1}\right)$$
$$=\sum_{n=0}^{\infty}\frac{(\gamma)_n S_n(x^2;a,b,c)}{(a+b)_n(a+c)_n n!}t^n,\quad \gamma \text{ arbitrary}. \tag{9.3.16}$$

Limit Relations

Wilson → Continuous Dual Hahn

The continuous dual Hahn polynomials can be found from the Wilson polynomials given by (9.1.1) by dividing by $(a+d)_n$ and letting $d\to\infty$:

$$\lim_{d\to\infty}\frac{W_n(x^2;a,b,c,d)}{(a+d)_n}=S_n(x^2;a,b,c).$$

Continuous Dual Hahn → Meixner-Pollaczek

The Meixner-Pollaczek polynomials given by (9.7.1) can be obtained from the continuous dual Hahn polynomials by the substitutions $x\to x-t$, $a=\lambda+it$, $b=\lambda-it$ and $c=t\cot\phi$ and the limit $t\to\infty$:

$$\lim_{t\to\infty}\frac{S_n((x-t)^2;\lambda+it,\lambda-it,t\cot\phi)}{t^n n!}=\frac{P_n^{(\lambda)}(x;\phi)}{(\sin\phi)^n}. \tag{9.3.17}$$

Remark

Since we have for $k < n$

$$\frac{(a+b)_n(a+c)_n}{(a+b)_k(a+c)_k} = (a+b+k)_{n-k}(a+c+k)_{n-k},$$

the continuous dual Hahn polynomials defined by (9.3.1) can also be seen as polynomials in the parameters a, b and c.

References

[72], [280], [339], [340], [378], [380], [381], [388], [389], [410], [415], [503].

9.4 Continuous Hahn

Hypergeometric Representation

$$\begin{aligned} &p_n(x;a,b,c,d) \\ &= i^n \frac{(a+c)_n(a+d)_n}{n!} \, {}_3F_2\left(\begin{matrix} -n, n+a+b+c+d-1, a+ix \\ a+c, a+d \end{matrix}; 1\right). \end{aligned} \tag{9.4.1}$$

Orthogonality Relation

If $Re(a,b,c,d) > 0$, $c = \bar{a}$ and $d = \bar{b}$, then

$$\begin{aligned} &\frac{1}{2\pi}\int_{-\infty}^{\infty} \Gamma(a+ix)\Gamma(b+ix)\Gamma(c-ix)\Gamma(d-ix)p_m(x;a,b,c,d)p_n(x;a,b,c,d)\,dx \\ &= \frac{\Gamma(n+a+c)\Gamma(n+a+d)\Gamma(n+b+c)\Gamma(n+b+d)}{(2n+a+b+c+d-1)\Gamma(n+a+b+c+d-1)n!}\,\delta_{mn}. \end{aligned} \tag{9.4.2}$$

Recurrence Relation

$$(a+ix)\tilde{p}_n(x) = A_n\tilde{p}_{n+1}(x) - (A_n + C_n)\,\tilde{p}_n(x) + C_n\tilde{p}_{n-1}(x), \tag{9.4.3}$$

where

$$\tilde{p}_n(x) := \tilde{p}_n(x;a,b,c,d) = \frac{n!}{i^n(a+c)_n(a+d)_n} p_n(x;a,b,c,d)$$

and

$$\begin{cases} A_n = -\dfrac{(n+a+b+c+d-1)(n+a+c)(n+a+d)}{(2n+a+b+c+d-1)(2n+a+b+c+d)} \\ \\ C_n = \dfrac{n(n+b+c-1)(n+b+d-1)}{(2n+a+b+c+d-2)(2n+a+b+c+d-1)}. \end{cases}$$

Normalized Recurrence Relation

$$xp_n(x) = p_{n+1}(x) + i(A_n + C_n + a)p_n(x) - A_{n-1}C_n p_{n-1}(x), \tag{9.4.4}$$

where

$$p_n(x;a,b,c,d) = \frac{(n+a+b+c+d-1)_n}{n!} p_n(x).$$

Difference Equation

$$\begin{aligned} &n(n+a+b+c+d-1)y(x) \\ &= B(x)y(x+i) - [B(x)+D(x)]\, y(x) + D(x)y(x-i), \end{aligned} \tag{9.4.5}$$

where

$$y(x) = p_n(x;a,b,c,d)$$

and

$$\begin{cases} B(x) = (c-ix)(d-ix) \\ \\ D(x) = (a+ix)(b+ix). \end{cases}$$

Forward Shift Operator

$$\begin{aligned} &p_n(x+\tfrac{1}{2}i;a,b,c,d) - p_n(x-\tfrac{1}{2}i;a,b,c,d) \\ &= i(n+a+b+c+d-1)p_{n-1}(x;a+\tfrac{1}{2},b+\tfrac{1}{2},c+\tfrac{1}{2},d+\tfrac{1}{2}) \end{aligned} \tag{9.4.6}$$

or equivalently

$$\frac{\delta p_n(x;a,b,c,d)}{\delta x} = (n+a+b+c+d-1)p_{n-1}(x;a+\tfrac{1}{2},b+\tfrac{1}{2},c+\tfrac{1}{2},d+\tfrac{1}{2}). \tag{9.4.7}$$

Backward Shift Operator

$$\begin{aligned}
&(c-\tfrac{1}{2}-ix)(d-\tfrac{1}{2}-ix)p_n(x+\tfrac{1}{2}i;a,b,c,d)\\
&\qquad -(a-\tfrac{1}{2}+ix)(b-\tfrac{1}{2}+ix)p_n(x-\tfrac{1}{2}i;a,b,c,d)\\
&=\frac{n+1}{i}p_{n+1}(x;a-\tfrac{1}{2},b-\tfrac{1}{2},c-\tfrac{1}{2},d-\tfrac{1}{2})
\end{aligned} \tag{9.4.8}$$

or equivalently

$$\begin{aligned}
&\frac{\delta\left[\omega(x;a,b,c,d)p_n(x;a,b,c,d)\right]}{\delta x}\\
&=-(n+1)\omega(x;a-\tfrac{1}{2},b-\tfrac{1}{2},c-\tfrac{1}{2},d-\tfrac{1}{2})\\
&\qquad\times p_{n+1}(x;a-\tfrac{1}{2},b-\tfrac{1}{2},c-\tfrac{1}{2},d-\tfrac{1}{2}),
\end{aligned} \tag{9.4.9}$$

where

$$\omega(x;a,b,c,d)=\Gamma(a+ix)\Gamma(b+ix)\Gamma(c-ix)\Gamma(d-ix).$$

Rodrigues-Type Formula

$$\begin{aligned}
&\omega(x;a,b,c,d)p_n(x;a,b,c,d)\\
&=\frac{(-1)^n}{n!}\left(\frac{\delta}{\delta x}\right)^n\left[\omega(x;a+\tfrac{1}{2}n,b+\tfrac{1}{2}n,c+\tfrac{1}{2}n,d+\tfrac{1}{2}n)\right].
\end{aligned} \tag{9.4.10}$$

Generating Functions

$${}_1F_1\left(\begin{matrix}a+ix\\a+c\end{matrix};-it\right){}_1F_1\left(\begin{matrix}d-ix\\b+d\end{matrix};it\right)=\sum_{n=0}^{\infty}\frac{p_n(x;a,b,c,d)}{(a+c)_n(b+d)_n}t^n. \tag{9.4.11}$$

$${}_1F_1\left(\begin{matrix}a+ix\\a+d\end{matrix};-it\right){}_1F_1\left(\begin{matrix}c-ix\\b+c\end{matrix};it\right)=\sum_{n=0}^{\infty}\frac{p_n(x;a,b,c,d)}{(a+d)_n(b+c)_n}t^n. \tag{9.4.12}$$

$$\begin{aligned}
&(1-t)^{1-a-b-c-d}\,{}_3F_2\left(\begin{matrix}\tfrac{1}{2}(a+b+c+d-1),\tfrac{1}{2}(a+b+c+d),a+ix\\a+c,a+d\end{matrix};-\frac{4t}{(1-t)^2}\right)\\
&=\sum_{n=0}^{\infty}\frac{(a+b+c+d-1)_n}{(a+c)_n(a+d)_ni^n}p_n(x;a,b,c,d)t^n.
\end{aligned} \tag{9.4.13}$$

Limit Relations

Wilson → Continuous Hahn

The continuous Hahn polynomials are obtained from the Wilson polynomials given by (9.1.1) by the substitution $a \to a - it$, $b \to b - it$, $c \to c + it$, $d \to d + it$ and $x \to x + t$ and the limit $t \to \infty$ in the following way:

$$\lim_{t\to\infty} \frac{W_n((x+t)^2; a-it, b-it, c+it, d+it)}{(-2t)^n n!} = p_n(x; a, b, c, d).$$

Continuous Hahn → Meixner-Pollaczek

The Meixner-Pollaczek polynomials given by (9.7.1) can be obtained from the continuous Hahn polynomials by setting $x \to x + t$, $a = \lambda - it$, $c = \lambda + it$ and $b = d = t\tan\phi$ and taking the limit $t \to \infty$:

$$\lim_{t\to\infty} \frac{p_n(x+t; \lambda - it, t\tan\phi, \lambda + it, t\tan\phi)}{t^n} = \frac{P_n^{(\lambda)}(x;\phi)}{(\cos\phi)^n}. \tag{9.4.14}$$

Continuous Hahn → Jacobi

The Jacobi polynomials given by (9.8.1) follow from the continuous Hahn polynomials by the substitution $x \to \frac{1}{2}xt$, $a = \frac{1}{2}(\alpha + 1 - it)$, $b = \frac{1}{2}(\beta + 1 + it)$, $c = \frac{1}{2}(\alpha + 1 + it)$ and $d = \frac{1}{2}(\beta + 1 - it)$, division by t^n and the limit $t \to \infty$:

$$\lim_{t\to\infty} \frac{p_n(\frac{1}{2}xt; \frac{1}{2}(\alpha+1-it), \frac{1}{2}(\beta+1+it), \frac{1}{2}(\alpha+1+it), \frac{1}{2}(\beta+1-it))}{t^n} = P_n^{(\alpha,\beta)}(x). \tag{9.4.15}$$

Continuous Hahn → Pseudo Jacobi

The pseudo Jacobi polynomials given by (9.9.1) follow from the continuous Hahn polynomials by the substitution $x \to xt$, $a = \frac{1}{2}(-N + i\nu - 2t)$, $b = \frac{1}{2}(-N - i\nu + 2t)$, $c = \frac{1}{2}(-N - i\nu - 2t)$ and $d = \frac{1}{2}(-N + i\nu + 2t)$, division by t^n and the limit $t \to \infty$:

$$\lim_{t\to\infty} \frac{p_n(xt; \frac{1}{2}(-N+i\nu-2t), \frac{1}{2}(-N-i\nu+2t), \frac{1}{2}(-N+i\nu-2t), \frac{1}{2}(-N-i\nu+2t))}{t^n} = \frac{(n-2N-1)_n}{n!} P_n(x;\nu,N). \tag{9.4.16}$$

Remark

Since we have for $k < n$

$$\frac{(a+b)_n(a+c)_n}{(a+b)_k(a+c)_k} = (a+b+k)_{n-k}(a+c+k)_{n-k},$$

the continuous Hahn polynomials defined by (9.4.1) can also be seen as polynomials in the parameters a, b and c.

References

[48], [51], [80], [81], [89], [258], [327], [340], [378], [381], [383].

9.5 Hahn

Hypergeometric Representation

$$Q_n(x;\alpha,\beta,N) = {}_3F_2\left(\begin{matrix} -n, n+\alpha+\beta+1, -x \\ \alpha+1, -N \end{matrix}; 1\right), \quad n = 0,1,2,\ldots,N. \tag{9.5.1}$$

Orthogonality Relation

For $\alpha > -1$ and $\beta > -1$, or for $\alpha < -N$ and $\beta < -N$, we have

$$\sum_{x=0}^{N} \binom{\alpha+x}{x}\binom{\beta+N-x}{N-x} Q_m(x;\alpha,\beta,N)Q_n(x;\alpha,\beta,N)$$
$$= \frac{(-1)^n(n+\alpha+\beta+1)_{N+1}(\beta+1)_n n!}{(2n+\alpha+\beta+1)(\alpha+1)_n(-N)_n N!}\,\delta_{mn}. \tag{9.5.2}$$

Recurrence Relation

$$-xQ_n(x) = A_nQ_{n+1}(x) - (A_n+C_n)Q_n(x) + C_nQ_{n-1}(x), \tag{9.5.3}$$

where

$$Q_n(x) := Q_n(x;\alpha,\beta,N)$$

and

$$\begin{cases} A_n = \dfrac{(n+\alpha+\beta+1)(n+\alpha+1)(N-n)}{(2n+\alpha+\beta+1)(2n+\alpha+\beta+2)} \\ C_n = \dfrac{n(n+\alpha+\beta+N+1)(n+\beta)}{(2n+\alpha+\beta)(2n+\alpha+\beta+1)}. \end{cases}$$

Normalized Recurrence Relation

$$xp_n(x) = p_{n+1}(x) + (A_n + C_n)\, p_n(x) + A_{n-1}C_n p_{n-1}(x), \tag{9.5.4}$$

where

$$Q_n(x;\alpha,\beta,N) = \frac{(n+\alpha+\beta+1)_n}{(\alpha+1)_n(-N)_n} p_n(x).$$

Difference Equation

$$n(n+\alpha+\beta+1)y(x) = B(x)y(x+1) - [B(x)+D(x)]\, y(x) + D(x)y(x-1), \tag{9.5.5}$$

where

$$y(x) = Q_n(x;\alpha,\beta,N)$$

and

$$\begin{cases} B(x) = (x+\alpha+1)(x-N) \\ D(x) = x(x-\beta-N-1). \end{cases}$$

Forward Shift Operator

$$\begin{aligned} &Q_n(x+1;\alpha,\beta,N) - Q_n(x;\alpha,\beta,N) \\ &= -\frac{n(n+\alpha+\beta+1)}{(\alpha+1)N} Q_{n-1}(x;\alpha+1,\beta+1,N-1) \end{aligned} \tag{9.5.6}$$

or equivalently

$$\Delta Q_n(x;\alpha,\beta,N) = -\frac{n(n+\alpha+\beta+1)}{(\alpha+1)N} Q_{n-1}(x;\alpha+1,\beta+1,N-1). \tag{9.5.7}$$

Backward Shift Operator

$$(x+\alpha)(N+1-x)Q_n(x;\alpha,\beta,N)-x(\beta+N+1-x)Q_n(x-1;\alpha,\beta,N)$$
$$=\alpha(N+1)Q_{n+1}(x;\alpha-1,\beta-1,N+1) \tag{9.5.8}$$

or equivalently

$$\nabla\left[\omega(x;\alpha,\beta,N)Q_n(x;\alpha,\beta,N)\right]$$
$$=\frac{N+1}{\beta}\omega(x;\alpha-1,\beta-1,N+1)Q_{n+1}(x;\alpha-1,\beta-1,N+1), \tag{9.5.9}$$

where

$$\omega(x;\alpha,\beta,N)=\binom{\alpha+x}{x}\binom{\beta+N-x}{N-x}.$$

Rodrigues-Type Formula

$$\omega(x;\alpha,\beta,N)Q_n(x;\alpha,\beta,N)$$
$$=\frac{(-1)^n(\beta+1)_n}{(-N)_n}\nabla^n\left[\omega(x;\alpha+n,\beta+n,N-n)\right]. \tag{9.5.10}$$

Generating Functions

For $x=0,1,2,\ldots,N$ we have

$${}_1F_1\left(\begin{array}{c}-x\\ \alpha+1\end{array};-t\right)\,{}_1F_1\left(\begin{array}{c}x-N\\ \beta+1\end{array};t\right)=\sum_{n=0}^{N}\frac{(-N)_n}{(\beta+1)_n n!}Q_n(x;\alpha,\beta,N)t^n. \tag{9.5.11}$$

$${}_2F_0\left(\begin{array}{c}-x,-x+\beta+N+1\\ -\end{array};-t\right)\,{}_2F_0\left(\begin{array}{c}x-N,x+\alpha+1\\ -\end{array};t\right)$$
$$=\sum_{n=0}^{N}\frac{(-N)_n(\alpha+1)_n}{n!}Q_n(x;\alpha,\beta,N)t^n. \tag{9.5.12}$$

$$\left[(1-t)^{-\alpha-\beta-1}\,{}_3F_2\left(\begin{array}{c}\frac{1}{2}(\alpha+\beta+1),\frac{1}{2}(\alpha+\beta+2),-x\\ \alpha+1,-N\end{array};-\frac{4t}{(1-t)^2}\right)\right]_N$$
$$=\sum_{n=0}^{N}\frac{(\alpha+\beta+1)_n}{n!}Q_n(x;\alpha,\beta,N)t^n. \qquad (9.5.13)$$

Limit Relations

Racah → Hahn

If we take $\gamma+1=-N$ and let $\delta\to\infty$ in the definition (9.2.1) of the Racah polynomials, we obtain the Hahn polynomials. Hence

$$\lim_{\delta\to\infty}R_n(\lambda(x);\alpha,\beta,-N-1,\delta)=Q_n(x;\alpha,\beta,N).$$

And if we take $\delta=-\beta-N-1$ and let $\gamma\to\infty$ in the definition (9.2.1) of the Racah polynomials, we also obtain the Hahn polynomials:

$$\lim_{\gamma\to\infty}R_n(\lambda(x);\alpha,\beta,\gamma,-\beta-N-1)=Q_n(x;\alpha,\beta,N).$$

Another way to do this is to take $\alpha+1=-N$ and $\beta\to\beta+\gamma+N+1$ in the definition (9.2.1) of the Racah polynomials and then take the limit $\delta\to\infty$. In that case we obtain the Hahn polynomials in the following way:

$$\lim_{\delta\to\infty}R_n(\lambda(x);-N-1,\beta+\gamma+N+1,\gamma,\delta)=Q_n(x;\gamma,\beta,N).$$

Hahn → Jacobi

To find the Jacobi polynomials given by (9.8.1) from the Hahn polynomials we take $x\to Nx$ and let $N\to\infty$. In fact we have

$$\lim_{N\to\infty}Q_n(Nx;\alpha,\beta,N)=\frac{P_n^{(\alpha,\beta)}(1-2x)}{P_n^{(\alpha,\beta)}(1)}. \qquad (9.5.14)$$

Hahn → Meixner

The Meixner polynomials given by (9.10.1) can be obtained from the Hahn polynomials by taking $\alpha=b-1$, $\beta=N(1-c)c^{-1}$ and letting $N\to\infty$:

$$\lim_{N\to\infty}Q_n(x;b-1,N(1-c)c^{-1},N)=M_n(x;b,c). \qquad (9.5.15)$$

Hahn → Krawtchouk

The Krawtchouk polynomials given by (9.11.1) are obtained from the Hahn polynomials if we take $\alpha = pt$ and $\beta = (1-p)t$ and let $t \to \infty$:

$$\lim_{t\to\infty} Q_n(x;pt,(1-p)t,N) = K_n(x;p,N). \tag{9.5.16}$$

Remark

If we interchange the role of x and n in (9.5.1) we obtain the dual Hahn polynomials given by (9.6.1).
Since

$$Q_n(x;\alpha,\beta,N) = R_x(\lambda(n);\alpha,\beta,N)$$

we obtain the dual orthogonality relation for the Hahn polynomials from the orthogonality relation (9.6.2) of the dual Hahn polynomials:

$$\sum_{n=0}^{N} \frac{(2n+\alpha+\beta+1)(\alpha+1)_n(-N)_nN!}{(-1)^n(n+\alpha+\beta+1)_{N+1}(\beta+1)_n n!} Q_n(x;\alpha,\beta,N)Q_n(y;\alpha,\beta,N)$$
$$= \frac{\delta_{xy}}{\binom{\alpha+x}{x}\binom{\beta+N-x}{N-x}}, \quad x,y \in \{0,1,2,\ldots,N\}.$$

References

[16], [34], [39], [46], [51], [54], [59], [72], [80], [82], [146], [152], [155], [164], [184], [185], [226], [228], [265], [268], [277], [308], [337], [340], [363], [364], [367], [372], [374], [375], [377], [381], [391], [415], [416], [417], [432], [434], [435], [440], [465], [488], [489], [512], [515], [521].

9.6 Dual Hahn

Hypergeometric Representation

$$R_n(\lambda(x);\gamma,\delta,N) = {}_3F_2\left(\begin{matrix} -n,-x,x+\gamma+\delta+1 \\ \gamma+1,-N \end{matrix};1\right), \quad n=0,1,2,\ldots,N, \tag{9.6.1}$$

where

$$\lambda(x) = x(x+\gamma+\delta+1).$$

Orthogonality Relation

For $\gamma > -1$ and $\delta > -1$, or for $\gamma < -N$ and $\delta < -N$, we have

$$\sum_{x=0}^{N} \frac{(2x+\gamma+\delta+1)(\gamma+1)_x(-N)_xN!}{(-1)^x(x+\gamma+\delta+1)_{N+1}(\delta+1)_xx!} R_m(\lambda(x);\gamma,\delta,N)R_n(\lambda(x);\gamma,\delta,N)$$
$$= \frac{\delta_{mn}}{\binom{\gamma+n}{n}\binom{\delta+N-n}{N-n}}. \tag{9.6.2}$$

Recurrence Relation

$$\lambda(x)R_n(\lambda(x)) = A_nR_{n+1}(\lambda(x)) - (A_n+C_n)R_n(\lambda(x)) + C_nR_{n-1}(\lambda(x)), \tag{9.6.3}$$

where

$$R_n(\lambda(x)) := R_n(\lambda(x);\gamma,\delta,N)$$

and

$$\begin{cases} A_n = (n+\gamma+1)(n-N) \\ C_n = n(n-\delta-N-1). \end{cases}$$

Normalized Recurrence Relation

$$xp_n(x) = p_{n+1}(x) - (A_n+C_n)p_n(x) + A_{n-1}C_np_{n-1}(x), \tag{9.6.4}$$

where

$$R_n(\lambda(x);\gamma,\delta,N) = \frac{1}{(\gamma+1)_n(-N)_n}p_n(\lambda(x)).$$

Difference Equation

$$-ny(x) = B(x)y(x+1) - [B(x)+D(x)]y(x) + D(x)y(x-1), \tag{9.6.5}$$

where

$$y(x) = R_n(\lambda(x);\gamma,\delta,N)$$

and

$$\begin{cases} B(x) = \dfrac{(x+\gamma+1)(x+\gamma+\delta+1)(N-x)}{(2x+\gamma+\delta+1)(2x+\gamma+\delta+2)} \\ D(x) = \dfrac{x(x+\gamma+\delta+N+1)(x+\delta)}{(2x+\gamma+\delta)(2x+\gamma+\delta+1)}. \end{cases}$$

Forward Shift Operator

$$\begin{aligned} &R_n(\lambda(x+1);\gamma,\delta,N) - R_n(\lambda(x);\gamma,\delta,N) \\ &= -\frac{n(2x+\gamma+\delta+2)}{(\gamma+1)N} R_{n-1}(\lambda(x);\gamma+1,\delta,N-1) \end{aligned} \tag{9.6.6}$$

or equivalently

$$\frac{\Delta R_n(\lambda(x);\gamma,\delta,N)}{\Delta\lambda(x)} = -\frac{n}{(\gamma+1)N} R_{n-1}(\lambda(x);\gamma+1,\delta,N-1). \tag{9.6.7}$$

Backward Shift Operator

$$\begin{aligned} &(x+\gamma)(x+\gamma+\delta)(N+1-x)R_n(\lambda(x);\gamma,\delta,N) \\ &\qquad -x(x+\gamma+\delta+N+1)(x+\delta)R_n(\lambda(x-1);\gamma,\delta,N) \\ &= \gamma(N+1)(2x+\gamma+\delta)R_{n+1}(\lambda(x);\gamma-1,\delta,N+1) \end{aligned} \tag{9.6.8}$$

or equivalently

$$\begin{aligned} &\frac{\nabla\left[\omega(x;\gamma,\delta,N)R_n(\lambda(x);\gamma,\delta,N)\right]}{\nabla\lambda(x)} \\ &= \frac{1}{\gamma+\delta}\omega(x;\gamma-1,\delta,N+1)R_{n+1}(\lambda(x);\gamma-1,\delta,N+1), \end{aligned} \tag{9.6.9}$$

where

$$\omega(x;\gamma,\delta,N) = \frac{(-1)^x(\gamma+1)_x(\gamma+\delta+1)_x(-N)_x}{(\gamma+\delta+N+2)_x(\delta+1)_x x!}.$$

Rodrigues-Type Formula

$$\omega(x;\gamma,\delta,N)R_n(\lambda(x);\gamma,\delta,N)=(\gamma+\delta+1)_n\left(\nabla_\lambda\right)^n\left[\omega(x;\gamma+n,\delta,N-n)\right], \quad (9.6.10)$$

where

$$\nabla_\lambda:=\frac{\nabla}{\nabla\lambda(x)}.$$

Generating Functions

For $x=0,1,2,\ldots,N$ we have

$$(1-t)^{N-x}\,{}_2F_1\left(\begin{array}{c}-x,-x-\delta\\ \gamma+1\end{array};t\right)=\sum_{n=0}^{N}\frac{(-N)_n}{n!}R_n(\lambda(x);\gamma,\delta,N)t^n. \quad (9.6.11)$$

$$(1-t)^x\,{}_2F_1\left(\begin{array}{c}x-N,x+\gamma+1\\ -\delta-N\end{array};t\right)$$
$$=\sum_{n=0}^{N}\frac{(\gamma+1)_n(-N)_n}{(-\delta-N)_n n!}R_n(\lambda(x);\gamma,\delta,N)t^n. \quad (9.6.12)$$

$$\left[\mathrm{e}^t\,{}_2F_2\left(\begin{array}{c}-x,x+\gamma+\delta+1\\ \gamma+1,-N\end{array};-t\right)\right]_N=\sum_{n=0}^{N}\frac{R_n(\lambda(x);\gamma,\delta,N)}{n!}t^n. \quad (9.6.13)$$

$$\left[(1-t)^{-\varepsilon}\,{}_3F_2\left(\begin{array}{c}\varepsilon,-x,x+\gamma+\delta+1\\ \gamma+1,-N\end{array};\frac{t}{t-1}\right)\right]_N$$
$$=\sum_{n=0}^{N}\frac{(\varepsilon)_n}{n!}R_n(\lambda(x);\gamma,\delta,N)t^n, \quad \varepsilon \text{ arbitrary}. \quad (9.6.14)$$

Limit Relations

Racah → Dual Hahn

If we take $\alpha+1=-N$ and let $\beta\to\infty$ in the definition (9.2.1) of the Racah polynomials, then we obtain the dual Hahn polynomials:

$$\lim_{\beta\to\infty}R_n(\lambda(x);-N-1,\beta,\gamma,\delta)=R_n(\lambda(x);\gamma,\delta,N).$$

And if we take $\beta = -\delta - N - 1$ and let $\alpha \to \infty$ in the definition (9.2.1) of the Racah polynomials, then we also obtain the dual Hahn polynomials:

$$\lim_{\alpha\to\infty} R_n(\lambda(x);\alpha,-\delta-N-1,\gamma,\delta) = R_n(\lambda(x);\gamma,\delta,N).$$

Finally, if we take $\gamma+1=-N$ and $\delta \to \alpha+\delta+N+1$ in the definition (9.2.1) of the Racah polynomials and take the limit $\beta \to \infty$ we find the dual Hahn polynomials in the following way:

$$\lim_{\beta\to\infty} R_n(\lambda(x);\alpha,\beta,-N-1,\alpha+\delta+N+1) = R_n(\lambda(x);\alpha,\delta,N).$$

Dual Hahn → Meixner

The Meixner polynomials given by (9.10.1) are obtained from the dual Hahn polynomials if we take $\gamma = \beta - 1$ and $\delta = N(1-c)c^{-1}$ and let $N \to \infty$:

$$\lim_{N\to\infty} R_n(\lambda(x);\beta-1,N(1-c)c^{-1},N = M_n(x;\beta,c). \tag{9.6.15}$$

Dual Hahn → Krawtchouk

The Krawtchouk polynomials given by (9.11.1) can be obtained from the dual Hahn polynomials by setting $\gamma = pt$, $\delta = (1-p)t$ and letting $t \to \infty$:

$$\lim_{t\to\infty} R_n(\lambda(x);pt,(1-p)t,N) = K_n(x;p,N). \tag{9.6.16}$$

Remark

If we interchange the role of x and n in the definition (9.6.1) of the dual Hahn polynomials we obtain the Hahn polynomials given by (9.5.1).
Since

$$R_n(\lambda(x);\gamma,\delta,N) = Q_x(n;\gamma,\delta,N)$$

we obtain the dual orthogonality relation for the dual Hahn polynomials from the orthogonality relation (9.5.2) for the Hahn polynomials:

$$\sum_{n=0}^{N} \binom{\gamma+n}{n}\binom{\delta+N-n}{N-n} R_n(\lambda(x);\gamma,\delta,N)R_n(\lambda(y);\gamma,\delta,N)$$
$$= \frac{(-1)^x(x+\gamma+\delta+1)_{N+1}(\delta+1)_x x!}{(2x+\gamma+\delta+1)(\gamma+1)_x(-N)_x N!}\delta_{xy}, \quad x,y \in \{0,1,2,\ldots,N\}.$$

References

[54], [72], [80], [82], [277], [308], [337], [340], [376], [377], [380], [381], [416], [417], [439], [488], [512].

9.7 Meixner-Pollaczek

Hypergeometric Representation

$$P_n^{(\lambda)}(x;\phi) = \frac{(2\lambda)_n}{n!} e^{in\phi}\, {}_2F_1\left(\begin{array}{c} -n, \lambda + ix \\ 2\lambda \end{array}; 1 - e^{-2i\phi}\right). \tag{9.7.1}$$

Orthogonality Relation

$$\frac{1}{2\pi}\int_{-\infty}^{\infty} e^{(2\phi-\pi)x} \left|\Gamma(\lambda+ix)\right|^2 P_m^{(\lambda)}(x;\phi) P_n^{(\lambda)}(x;\phi)\,dx$$
$$= \frac{\Gamma(n+2\lambda)}{(2\sin\phi)^{2\lambda} n!}\delta_{mn}, \quad \lambda > 0 \quad \text{and} \quad 0 < \phi < \pi. \tag{9.7.2}$$

Recurrence Relation

$$(n+1)P_{n+1}^{(\lambda)}(x;\phi) - 2\left[x\sin\phi + (n+\lambda)\cos\phi\right] P_n^{(\lambda)}(x;\phi)$$
$$+ (n+2\lambda-1)P_{n-1}^{(\lambda)}(x;\phi) = 0. \tag{9.7.3}$$

Normalized Recurrence Relation

$$xp_n(x) = p_{n+1}(x) - \left(\frac{n+\lambda}{\tan\phi}\right) p_n(x) + \frac{n(n+2\lambda-1)}{4\sin^2\phi} p_{n-1}(x), \tag{9.7.4}$$

where

$$P_n^{(\lambda)}(x;\phi) = \frac{(2\sin\phi)^n}{n!} p_n(x).$$

Difference Equation

$$\begin{aligned} &\mathrm{e}^{i\phi}(\lambda-ix)y(x+i)+2i\left[x\cos\phi-(n+\lambda)\sin\phi\right]y(x)\\ &\quad-\mathrm{e}^{-i\phi}(\lambda+ix)y(x-i)=0,\quad y(x)=P_n^{(\lambda)}(x;\phi). \end{aligned} \tag{9.7.5}$$

Forward Shift Operator

$$P_n^{(\lambda)}(x+\tfrac{1}{2}i;\phi)-P_n^{(\lambda)}(x-\tfrac{1}{2}i;\phi)=(\mathrm{e}^{i\phi}-\mathrm{e}^{-i\phi})P_{n-1}^{(\lambda+\frac{1}{2})}(x;\phi) \tag{9.7.6}$$

or equivalently

$$\frac{\delta P_n^{(\lambda)}(x;\phi)}{\delta x}=2\sin\phi\, P_{n-1}^{(\lambda+\frac{1}{2})}(x;\phi). \tag{9.7.7}$$

Backward Shift Operator

$$\begin{aligned} &\mathrm{e}^{i\phi}(\lambda-\tfrac{1}{2}-ix)P_n^{(\lambda)}(x+\tfrac{1}{2}i;\phi)+\mathrm{e}^{-i\phi}(\lambda-\tfrac{1}{2}+ix)P_n^{(\lambda)}(x-\tfrac{1}{2}i;\phi)\\ &=(n+1)P_{n+1}^{(\lambda-\frac{1}{2})}(x;\phi) \end{aligned} \tag{9.7.8}$$

or equivalently

$$\frac{\delta\left[\omega(x;\lambda,\phi)P_n^{(\lambda)}(x;\phi)\right]}{\delta x}=-(n+1)\omega(x;\lambda-\tfrac{1}{2},\phi)P_{n+1}^{(\lambda-\frac{1}{2})}(x;\phi), \tag{9.7.9}$$

where

$$\omega(x;\lambda,\phi)=\Gamma(\lambda+ix)\Gamma(\lambda-ix)\mathrm{e}^{(2\phi-\pi)x}.$$

Rodrigues-Type Formula

$$\omega(x;\lambda,\phi)P_n^{(\lambda)}(x;\phi)=\frac{(-1)^n}{n!}\left(\frac{\delta}{\delta x}\right)^n\left[\omega(x;\lambda+\tfrac{1}{2}n,\phi)\right]. \tag{9.7.10}$$

Generating Functions

$$(1-e^{i\phi}t)^{-\lambda+ix}(1-e^{-i\phi}t)^{-\lambda-ix}=\sum_{n=0}^{\infty}P_n^{(\lambda)}(x;\phi)t^n. \tag{9.7.11}$$

$$e^t\,{}_1F_1\left(\begin{matrix}\lambda+ix\\2\lambda\end{matrix};(e^{-2i\phi}-1)t\right)=\sum_{n=0}^{\infty}\frac{P_n^{(\lambda)}(x;\phi)}{(2\lambda)_n e^{in\phi}}t^n. \tag{9.7.12}$$

$$(1-t)^{-\gamma}{}_2F_1\left(\begin{matrix}\gamma,\lambda+ix\\2\lambda\end{matrix};\frac{(1-e^{-2i\phi})t}{t-1}\right)$$
$$=\sum_{n=0}^{\infty}\frac{(\gamma)_n}{(2\lambda)_n}\frac{P_n^{(\lambda)}(x;\phi)}{e^{in\phi}}t^n,\quad \gamma \text{ arbitrary}. \tag{9.7.13}$$

Limit Relations

Continuous Dual Hahn → Meixner-Pollaczek

The Meixner-Pollaczek polynomials can be obtained from the continuous dual Hahn polynomials given by (9.3.1) by the substitutions $x\to x-t$, $a=\lambda+it$, $b=\lambda-it$ and $c=t\cot\phi$ and the limit $t\to\infty$:

$$\lim_{t\to\infty}\frac{S_n((x-t)^2;\lambda+it,\lambda-it,t\cot\phi)}{t^n n!}=\frac{P_n^{(\lambda)}(x;\phi)}{(\sin\phi)^n}.$$

Continuous Hahn → Meixner-Pollaczek

By setting $x\to x+t$, $a=\lambda-it$, $c=\lambda+it$ and $b=d=t\tan\phi$ in the definition (9.4.1) of the continuous Hahn polynomials and taking the limit $t\to\infty$ we obtain the Meixner-Pollaczek polynomials:

$$\lim_{t\to\infty}\frac{p_n(x+t;\lambda-it,t\tan\phi,\lambda+it,t\tan\phi)}{t^n n!}=\frac{P_n^{(\lambda)}(x;\phi)}{(\cos\phi)^n}.$$

Meixner-Pollaczek → Laguerre

The Laguerre polynomials given by (9.12.1) can be obtained from the Meixner-Pollaczek polynomials by the substitution $\lambda = \frac{1}{2}(\alpha+1)$, $x \to -\frac{1}{2}\phi^{-1}x$ and the limit $\phi \to 0$:

$$\lim_{\phi\to 0} P_n^{(\frac{1}{2}\alpha+\frac{1}{2})}(-\tfrac{1}{2}\phi^{-1}x;\phi) = L_n^{(\alpha)}(x). \tag{9.7.14}$$

Meixner-Pollaczek → Hermite

The Hermite polynomials given by (9.15.1) are obtained from the Meixner-Pollaczek polynomials if we substitute $x \to (\sin\phi)^{-1}(x\sqrt{\lambda} - \lambda\cos\phi)$ and then let $\lambda \to \infty$:

$$\lim_{\lambda\to\infty} \lambda^{-\frac{1}{2}n} P_n^{(\lambda)}((\sin\phi)^{-1}(x\sqrt{\lambda} - \lambda\cos\phi);\phi) = \frac{H_n(x)}{n!}. \tag{9.7.15}$$

Remark

Since we have for $k < n$

$$\frac{(2\lambda)_n}{(2\lambda)_k} = (2\lambda+k)_{n-k},$$

the Meixner-Pollaczek polynomials defined by (9.7.1) can also be seen as polynomials in the parameter λ.

References

[16], [20], [34], [36], [37], [51], [72], [80], [82], [135], [138], [146], [270], [273], [277], [283], [295], [317], [340], [342], [363], [364], [381], [392], [406], [416], [434], [512], [517].

9.8 Jacobi

Hypergeometric Representation

$$P_n^{(\alpha,\beta)}(x) = \frac{(\alpha+1)_n}{n!}\, {}_2F_1\left(\begin{matrix} -n, n+\alpha+\beta+1 \\ \alpha+1 \end{matrix}; \frac{1-x}{2}\right). \tag{9.8.1}$$

Orthogonality Relation

For $\alpha > -1$ and $\beta > -1$ we have

$$\int_{-1}^{1} (1-x)^{\alpha}(1+x)^{\beta} P_m^{(\alpha,\beta)}(x) P_n^{(\alpha,\beta)}(x)\,dx$$
$$= \frac{2^{\alpha+\beta+1}}{2n+\alpha+\beta+1} \frac{\Gamma(n+\alpha+1)\Gamma(n+\beta+1)}{\Gamma(n+\alpha+\beta+1)n!}\,\delta_{mn}. \tag{9.8.2}$$

For $\alpha+\beta < -2N-1$, $\beta > -1$ and $m,n \in \{0,1,2,\ldots,N\}$ we also have

$$\int_{1}^{\infty} (x+1)^{\alpha}(x-1)^{\beta} P_m^{(\alpha,\beta)}(-x) P_n^{(\alpha,\beta)}(-x)\,dx$$
$$= -\frac{2^{\alpha+\beta+1}}{2n+\alpha+\beta+1} \frac{\Gamma(-n-\alpha-\beta)\Gamma(n+\alpha+\beta+1)}{\Gamma(-n-\alpha)n!}\,\delta_{mn}. \tag{9.8.3}$$

Recurrence Relation

$$xP_n^{(\alpha,\beta)}(x) = \frac{2(n+1)(n+\alpha+\beta+1)}{(2n+\alpha+\beta+1)(2n+\alpha+\beta+2)} P_{n+1}^{(\alpha,\beta)}(x)$$
$$+ \frac{\beta^2-\alpha^2}{(2n+\alpha+\beta)(2n+\alpha+\beta+2)} P_n^{(\alpha,\beta)}(x)$$
$$+ \frac{2(n+\alpha)(n+\beta)}{(2n+\alpha+\beta)(2n+\alpha+\beta+1)} P_{n-1}^{(\alpha,\beta)}(x). \tag{9.8.4}$$

Normalized Recurrence Relation

$$xp_n(x) = p_{n+1}(x) + \frac{\beta^2-\alpha^2}{(2n+\alpha+\beta)(2n+\alpha+\beta+2)} p_n(x)$$
$$+ \frac{4n(n+\alpha)(n+\beta)(n+\alpha+\beta)}{(2n+\alpha+\beta-1)(2n+\alpha+\beta)^2(2n+\alpha+\beta+1)} p_{n-1}(x), \tag{9.8.5}$$

where

$$P_n^{(\alpha,\beta)}(x) = \frac{(n+\alpha+\beta+1)_n}{2^n n!} p_n(x).$$

Differential Equation

$$(1-x^2)y''(x)+[\beta-\alpha-(\alpha+\beta+2)x]y'(x) \\ +n(n+\alpha+\beta+1)y(x)=0, \quad y(x)=P_n^{(\alpha,\beta)}(x). \tag{9.8.6}$$

Forward Shift Operator

$$\frac{d}{dx}P_n^{(\alpha,\beta)}(x)=\frac{n+\alpha+\beta+1}{2}P_{n-1}^{(\alpha+1,\beta+1)}(x). \tag{9.8.7}$$

Backward Shift Operator

$$(1-x^2)\frac{d}{dx}P_n^{(\alpha,\beta)}(x)+[(\beta-\alpha)-(\alpha+\beta)x]P_n^{(\alpha,\beta)}(x) \\ =-2(n+1)P_{n+1}^{(\alpha-1,\beta-1)}(x) \tag{9.8.8}$$

or equivalently

$$\frac{d}{dx}\left[(1-x)^{\alpha}(1+x)^{\beta}P_n^{(\alpha,\beta)}(x)\right] \\ =-2(n+1)(1-x)^{\alpha-1}(1+x)^{\beta-1}P_{n+1}^{(\alpha-1,\beta-1)}(x). \tag{9.8.9}$$

Rodrigues-Type Formula

$$(1-x)^{\alpha}(1+x)^{\beta}P_n^{(\alpha,\beta)}(x)=\frac{(-1)^n}{2^n n!}\left(\frac{d}{dx}\right)^n\left[(1-x)^{n+\alpha}(1+x)^{n+\beta}\right]. \tag{9.8.10}$$

Generating Functions

$$\frac{2^{\alpha+\beta}}{R(1+R-t)^{\alpha}(1+R+t)^{\beta}}=\sum_{n=0}^{\infty}P_n^{(\alpha,\beta)}(x)t^n, \quad R=\sqrt{1-2xt+t^2}. \tag{9.8.11}$$

$$
{}_0F_1\left(\begin{array}{c} - \\ \alpha+1 \end{array}; \frac{(x-1)t}{2}\right) {}_0F_1\left(\begin{array}{c} - \\ \beta+1 \end{array}; \frac{(x+1)t}{2}\right)
= \sum_{n=0}^{\infty} \frac{P_n^{(\alpha,\beta)}(x)}{(\alpha+1)_n(\beta+1)_n} t^n. \tag{9.8.12}
$$

$$
(1-t)^{-\alpha-\beta-1} {}_2F_1\left(\begin{array}{c} \frac{1}{2}(\alpha+\beta+1), \frac{1}{2}(\alpha+\beta+2) \\ \alpha+1 \end{array}; \frac{2(x-1)t}{(1-t)^2}\right)
= \sum_{n=0}^{\infty} \frac{(\alpha+\beta+1)_n}{(\alpha+1)_n} P_n^{(\alpha,\beta)}(x) t^n. \tag{9.8.13}
$$

$$
(1+t)^{-\alpha-\beta-1} {}_2F_1\left(\begin{array}{c} \frac{1}{2}(\alpha+\beta+1), \frac{1}{2}(\alpha+\beta+2) \\ \beta+1 \end{array}; \frac{2(x+1)t}{(1+t)^2}\right)
= \sum_{n=0}^{\infty} \frac{(\alpha+\beta+1)_n}{(\beta+1)_n} P_n^{(\alpha,\beta)}(x) t^n. \tag{9.8.14}
$$

$$
{}_2F_1\left(\begin{array}{c} \gamma, \alpha+\beta+1-\gamma \\ \alpha+1 \end{array}; \frac{1-R-t}{2}\right) {}_2F_1\left(\begin{array}{c} \gamma, \alpha+\beta+1-\gamma \\ \beta+1 \end{array}; \frac{1-R+t}{2}\right)
= \sum_{n=0}^{\infty} \frac{(\gamma)_n(\alpha+\beta+1-\gamma)_n}{(\alpha+1)_n(\beta+1)_n} P_n^{(\alpha,\beta)}(x) t^n, \quad R = \sqrt{1-2xt+t^2} \tag{9.8.15}
$$

with γ arbitrary.

Limit Relations

Wilson → Jacobi

The Jacobi polynomials can be found from the Wilson polynomials given by (9.1.1) by substituting $a = b = \frac{1}{2}(\alpha+1)$, $c = \frac{1}{2}(\beta+1)+it$, $d = \frac{1}{2}(\beta+1)-it$ and $x \rightarrow t\sqrt{\frac{1}{2}(1-x)}$ in the definition (9.1.1) of the Wilson polynomials and taking the limit $t \rightarrow \infty$. In fact we have

$$
\lim_{t\rightarrow\infty} \frac{W_n(\frac{1}{2}(1-x)t^2; \frac{1}{2}(\alpha+1), \frac{1}{2}(\alpha+1), \frac{1}{2}(\beta+1)+it, \frac{1}{2}(\beta+1)-it)}{t^{2n}n!} = P_n^{(\alpha,\beta)}(x).
$$

Continuous Hahn → Jacobi

The Jacobi polynomials follow from the continuous Hahn polynomials given by (9.4.1) by using the substitution $x \to \frac{1}{2}xt$, $a = \frac{1}{2}(\alpha+1-it)$, $b = \frac{1}{2}(\beta+1+it)$, $c = \frac{1}{2}(\alpha+1+it)$ and $d = \frac{1}{2}(\beta+1-it)$ in (9.4.1), division by t^n and the limit $t \to \infty$:

$$\lim_{t\to\infty} \frac{p_n(\frac{1}{2}xt;\frac{1}{2}(\alpha+1-it),\frac{1}{2}(\beta+1+it),\frac{1}{2}(\alpha+1+it),\frac{1}{2}(\beta+1-it))}{t^n} = P_n^{(\alpha,\beta)}(x).$$

Hahn → Jacobi

To find the Jacobi polynomials from the Hahn polynomials given by (9.5.1) we take $x \to Nx$ in (9.5.1) and let $N \to \infty$. In fact we have

$$\lim_{N\to\infty} Q_n(Nx;\alpha,\beta,N) = \frac{P_n^{(\alpha,\beta)}(1-2x)}{P_n^{(\alpha,\beta)}(1)}.$$

Jacobi → Laguerre

The Laguerre polynomials given by (9.12.1) can be obtained from the Jacobi polynomials by setting $x \to 1-2\beta^{-1}x$ and then the limit $\beta \to \infty$:

$$\lim_{\beta\to\infty} P_n^{(\alpha,\beta)}(1-2\beta^{-1}x) = L_n^{(\alpha)}(x). \tag{9.8.16}$$

Jacobi → Bessel

The Bessel polynomials given by (9.13.1) are obtained from the Jacobi polynomials if we take $\beta = a-\alpha$ and let $\alpha \to -\infty$:

$$\lim_{\alpha\to-\infty} \frac{P_n^{(\alpha,a-\alpha)}(1+\alpha x)}{P_n^{(\alpha,a-\alpha)}(1)} = y_n(x;a). \tag{9.8.17}$$

Jacobi → Hermite

The Hermite polynomials given by (9.15.1) follow from the Jacobi polynomials by taking $\beta = \alpha$ and letting $\alpha \to \infty$ in the following way:

$$\lim_{\alpha\to\infty} \alpha^{-\frac{1}{2}n} P_n^{(\alpha,\alpha)}(\alpha^{-\frac{1}{2}}x) = \frac{H_n(x)}{2^n n!}. \tag{9.8.18}$$

Remarks

The definition (9.8.1) of the Jacobi polynomials can also be written as:

$$P_n^{(\alpha,\beta)}(x) = \frac{1}{n!}\sum_{k=0}^{n}\frac{(-n)_k}{k!}(n+\alpha+\beta+1)_k(\alpha+k+1)_{n-k}\left(\frac{1-x}{2}\right)^k.$$

In this way the Jacobi polynomials can also be seen as polynomials in the parameters α and β. Therefore they can be defined for all α and β. Then we have the following connection with the Meixner polynomials given by (9.10.1):

$$\frac{(\beta)_n}{n!}M_n(x;\beta,c) = P_n^{(\beta-1,-n-\beta-x)}((2-c)c^{-1}).$$

The Jacobi polynomials are related to the pseudo Jacobi polynomials defined by (9.9.1) in the following way:

$$P_n(x;\nu,N) = \frac{(-2i)^n n!}{(n-2N-1)_n}P_n^{(-N-1+i\nu,-N-1-i\nu)}(ix).$$

The Jacobi polynomials are also related to the Gegenbauer (or ultraspherical) polynomials given by (9.8.19) by the quadratic transformations:

$$C_{2n}^{(\lambda)}(x) = \frac{(\lambda)_n}{(\frac{1}{2})_n}P_n^{(\lambda-\frac{1}{2},-\frac{1}{2})}(2x^2-1)$$

and

$$C_{2n+1}^{(\lambda)}(x) = \frac{(\lambda)_{n+1}}{(\frac{1}{2})_{n+1}}xP_n^{(\lambda-\frac{1}{2},\frac{1}{2})}(2x^2-1).$$

References

[2], [4], [9], [11], [15], [19], [34], [35], [41], [42], [43], [44], [45], [46], [47], [51], [54], [55], [56], [57], [58], [69], [72], [91], [109], [117], [130], [132], [137], [146], [149], [152], [155], [165], [166], [167], [171], [172], [173], [175], [178], [187], [193], [196], [198], [202], [216], [219], [220], [221], [222], [223], [224], [225], [226], [227], [228], [229], [239], [241], [242], [243], [244], [247], [251], [253], [262], [264], [266], [267], [268], [274], [277], [283], [287], [314], [316], [317], [327], [330], [331], [332], [334], [335], [336], [339], [340], [357], [358], [363], [364], [367], [381], [382], [393], [394], [397], [399], [400], [403], [405], [408], [412], [416], [417], [424], [428], [430], [431], [433], [438], [450], [456], [464], [470], [477], [479], [480], [483], [485], [489], [493], [496], [505], [516], [518], [519], [521], [522].

Special Cases

9.8.1 Gegenbauer / Ultraspherical

Hypergeometric Representation

The Gegenbauer (or ultraspherical) polynomials are Jacobi polynomials with $\alpha = \beta = \lambda - \frac{1}{2}$ and another normalization:

$$\begin{aligned} C_n^{(\lambda)}(x) &= \frac{(2\lambda)_n}{(\lambda+\frac{1}{2})_n} P_n^{(\lambda-\frac{1}{2},\lambda-\frac{1}{2})}(x) \\ &= \frac{(2\lambda)_n}{n!} \, {}_2F_1\left(\begin{matrix} -n, n+2\lambda \\ \lambda+\frac{1}{2} \end{matrix}; \frac{1-x}{2}\right), \quad \lambda \neq 0. \end{aligned} \tag{9.8.19}$$

Orthogonality Relation

$$\begin{aligned} &\int_{-1}^{1} (1-x^2)^{\lambda-\frac{1}{2}} C_m^{(\lambda)}(x) C_n^{(\lambda)}(x)\, dx \\ &= \frac{\pi \Gamma(n+2\lambda) 2^{1-2\lambda}}{\{\Gamma(\lambda)\}^2 (n+\lambda) n!} \delta_{mn}, \quad \lambda > -\frac{1}{2} \quad \text{and} \quad \lambda \neq 0. \end{aligned} \tag{9.8.20}$$

Recurrence Relation

$$2(n+\lambda) x C_n^{(\lambda)}(x) = (n+1) C_{n+1}^{(\lambda)}(x) + (n+2\lambda-1) C_{n-1}^{(\lambda)}(x). \tag{9.8.21}$$

Normalized Recurrence Relation

$$x p_n(x) = p_{n+1}(x) + \frac{n(n+2\lambda-1)}{4(n+\lambda-1)(n+\lambda)} p_{n-1}(x), \tag{9.8.22}$$

where

$$C_n^{(\lambda)}(x) = \frac{2^n (\lambda)_n}{n!} p_n(x).$$

Differential Equation

$$(1-x^2)y''(x)-(2\lambda+1)xy'(x)+n(n+2\lambda)y(x)=0, \quad y(x)=C_n^{(\lambda)}(x). \tag{9.8.23}$$

Forward Shift Operator

$$\frac{d}{dx}C_n^{(\lambda)}(x)=2\lambda C_{n-1}^{(\lambda+1)}(x). \tag{9.8.24}$$

Backward Shift Operator

$$(1-x^2)\frac{d}{dx}C_n^{(\lambda)}(x)+(1-2\lambda)xC_n^{(\lambda)}(x)=-\frac{(n+1)(2\lambda+n-1)}{2(\lambda-1)}C_{n+1}^{(\lambda-1)}(x) \tag{9.8.25}$$

or equivalently

$$\begin{aligned}&\frac{d}{dx}\left[(1-x^2)^{\lambda-\frac{1}{2}}C_n^{(\lambda)}(x)\right]\\&=-\frac{(n+1)(2\lambda+n-1)}{2(\lambda-1)}(1-x^2)^{\lambda-\frac{3}{2}}C_{n+1}^{(\lambda-1)}(x).\end{aligned} \tag{9.8.26}$$

Rodrigues-Type Formula

$$(1-x^2)^{\lambda-\frac{1}{2}}C_n^{(\lambda)}(x)=\frac{(2\lambda)_n(-1)^n}{(\lambda+\frac{1}{2})_n 2^n n!}\left(\frac{d}{dx}\right)^n\left[(1-x^2)^{\lambda+n-\frac{1}{2}}\right]. \tag{9.8.27}$$

Generating Functions

$$(1-2xt+t^2)^{-\lambda}=\sum_{n=0}^{\infty}C_n^{(\lambda)}(x)t^n. \tag{9.8.28}$$

$$R^{-1}\left(\frac{1+R-xt}{2}\right)^{\frac{1}{2}-\lambda}=\sum_{n=0}^{\infty}\frac{(\lambda+\frac{1}{2})_n}{(2\lambda)_n}C_n^{(\lambda)}(x)t^n, \quad R=\sqrt{1-2xt+t^2}. \tag{9.8.29}$$

$$ {}_0F_1\left(\begin{array}{c} - \\ \lambda+\frac{1}{2} \end{array}; \frac{(x-1)t}{2}\right) {}_0F_1\left(\begin{array}{c} - \\ \lambda+\frac{1}{2} \end{array}; \frac{(x+1)t}{2}\right) = \sum_{n=0}^{\infty} \frac{C_n^{(\lambda)}(x)}{(2\lambda)_n(\lambda+\frac{1}{2})_n} t^n. \quad (9.8.30) $$

$$ \mathrm{e}^{xt}\,{}_0F_1\left(\begin{array}{c} - \\ \lambda+\frac{1}{2} \end{array}; \frac{(x^2-1)t^2}{4}\right) = \sum_{n=0}^{\infty} \frac{C_n^{(\lambda)}(x)}{(2\lambda)_n} t^n. \quad (9.8.31) $$

$$ {}_2F_1\left(\begin{array}{c} \gamma, 2\lambda-\gamma \\ \lambda+\frac{1}{2} \end{array}; \frac{1-R-t}{2}\right) {}_2F_1\left(\begin{array}{c} \gamma, 2\lambda-\gamma \\ \lambda+\frac{1}{2} \end{array}; \frac{1-R+t}{2}\right) $$
$$ = \sum_{n=0}^{\infty} \frac{(\gamma)_n(2\lambda-\gamma)_n}{(2\lambda)_n(\lambda+\frac{1}{2})_n} C_n^{(\lambda)}(x) t^n, \quad R = \sqrt{1-2xt+t^2}, \quad \gamma \text{ arbitrary}. \quad (9.8.32) $$

$$ (1-xt)^{-\gamma}\,{}_2F_1\left(\begin{array}{c} \frac{1}{2}\gamma, \frac{1}{2}\gamma+\frac{1}{2} \\ \lambda+\frac{1}{2} \end{array}; \frac{(x^2-1)t^2}{(1-xt)^2}\right) $$
$$ = \sum_{n=0}^{\infty} \frac{(\gamma)_n}{(2\lambda)_n} C_n^{(\lambda)}(x) t^n, \quad \gamma \text{ arbitrary}. \quad (9.8.33) $$

Limit Relation

Gegenbauer / Ultraspherical → Hermite

The Hermite polynomials given by (9.15.1) follow from the Gegenbauer (or ultraspherical) polynomials by taking $\lambda = \alpha + \frac{1}{2}$ and letting $\alpha \to \infty$ in the following way:

$$ \lim_{\alpha\to\infty} \alpha^{-\frac{1}{2}n} C_n^{(\alpha+\frac{1}{2})}(\alpha^{-\frac{1}{2}}x) = \frac{H_n(x)}{n!}. \quad (9.8.34) $$

Remarks

The case $\lambda = 0$ needs another normalization. In that case we have the Chebyshev polynomials of the first kind described in the next subsection.
The Gegenbauer (or ultraspherical) polynomials are related to the Jacobi polynomials given by (9.8.1) by the quadratic transformations:

$$ C_{2n}^{(\lambda)}(x) = \frac{(\lambda)_n}{(\frac{1}{2})_n} P_n^{(\lambda-\frac{1}{2},-\frac{1}{2})}(2x^2-1) $$

and

$$C_{2n+1}^{(\lambda)}(x) = \frac{(\lambda)_{n+1}}{(\frac{1}{2})_{n+1}} x P_n^{(\lambda-\frac{1}{2},\frac{1}{2})}(2x^2-1).$$

References

[2], [5], [35], [38], [40], [45], [46], [51], [55], [66], [99], [104], [107], [109], [110], [117], [119], [120], [121], [123], [124], [130], [146], [154], [156], [163], [167], [168], [169], [170], [174], [183], [190], [191], [194], [198], [200], [219], [225], [231], [253], [267], [277], [317], [340], [366], [390], [398], [403], [413], [416], [417], [450], [456], [458], [467], [479], [493], [496], [508], [522].

9.8.2 Chebyshev

Hypergeometric Representation

The Chebyshev polynomials of the first kind can be obtained from the Jacobi polynomials by taking $\alpha = \beta = -\frac{1}{2}$:

$$T_n(x) = \frac{P_n^{(-\frac{1}{2},-\frac{1}{2})}(x)}{P_n^{(-\frac{1}{2},-\frac{1}{2})}(1)} = {}_2F_1\left(\begin{array}{c}-n, n\\ \frac{1}{2}\end{array}; \frac{1-x}{2}\right) \tag{9.8.35}$$

and the Chebyshev polynomials of the second kind can be obtained from the Jacobi polynomials by taking $\alpha = \beta = \frac{1}{2}$:

$$U_n(x) = (n+1)\frac{P_n^{(\frac{1}{2},\frac{1}{2})}(x)}{P_n^{(\frac{1}{2},\frac{1}{2})}(1)} = (n+1)\,{}_2F_1\left(\begin{array}{c}-n, n+2\\ \frac{3}{2}\end{array}; \frac{1-x}{2}\right). \tag{9.8.36}$$

Orthogonality Relation

$$\int_{-1}^{1}(1-x^2)^{-\frac{1}{2}}T_m(x)T_n(x)\,dx = \begin{cases} \dfrac{\pi}{2}\,\delta_{mn}, & n \neq 0 \\ \pi\,\delta_{mn}, & n = 0. \end{cases} \tag{9.8.37}$$

$$\int_{-1}^{1}(1-x^2)^{\frac{1}{2}}U_m(x)U_n(x)\,dx = \frac{\pi}{2}\,\delta_{mn}. \tag{9.8.38}$$

Recurrence Relations

$$2xT_n(x) = T_{n+1}(x) + T_{n-1}(x), \quad T_0(x) = 1 \quad \text{and} \quad T_1(x) = x. \tag{9.8.39}$$

$$2xU_n(x) = U_{n+1}(x) + U_{n-1}(x), \quad U_0(x) = 1 \quad \text{and} \quad U_1(x) = 2x. \tag{9.8.40}$$

Normalized Recurrence Relations

$$xp_n(x) = p_{n+1}(x) + \frac{1}{4}p_{n-1}(x), \tag{9.8.41}$$

where

$$T_1(x) = p_1(x) = x \quad \text{and} \quad T_n(x) = 2^n p_n(x), \quad n \neq 1.$$

$$xp_n(x) = p_{n+1}(x) + \frac{1}{4}p_{n-1}(x), \tag{9.8.42}$$

where

$$U_n(x) = 2^n p_n(x).$$

Differential Equations

$$(1-x^2)y''(x) - xy'(x) + n^2 y(x) = 0, \quad y(x) = T_n(x). \tag{9.8.43}$$

$$(1-x^2)y''(x) - 3xy'(x) + n(n+2)y(x) = 0, \quad y(x) = U_n(x). \tag{9.8.44}$$

Forward Shift Operator

$$\frac{d}{dx}T_n(x) = nU_{n-1}(x). \tag{9.8.45}$$

Backward Shift Operator

$$(1-x^2)\frac{d}{dx}U_n(x) - xU_n(x) = -(n+1)T_{n+1}(x) \tag{9.8.46}$$

or equivalently

$$\frac{d}{dx}\left[\left(1-x^2\right)^{\frac{1}{2}}U_n(x)\right]=-(n+1)\left(1-x^2\right)^{-\frac{1}{2}}T_{n+1}(x). \tag{9.8.47}$$

Rodrigues-Type Formulas

$$(1-x^2)^{-\frac{1}{2}}T_n(x)=\frac{(-1)^n}{(\frac{1}{2})_n 2^n}\left(\frac{d}{dx}\right)^n\left[(1-x^2)^{n-\frac{1}{2}}\right]. \tag{9.8.48}$$

$$(1-x^2)^{\frac{1}{2}}U_n(x)=\frac{(n+1)(-1)^n}{(\frac{3}{2})_n 2^n}\left(\frac{d}{dx}\right)^n\left[(1-x^2)^{n+\frac{1}{2}}\right]. \tag{9.8.49}$$

Generating Functions

$$\frac{1-xt}{1-2xt+t^2}=\sum_{n=0}^{\infty}T_n(x)t^n. \tag{9.8.50}$$

$$R^{-1}\sqrt{\frac{1}{2}(1+R-xt)}=\sum_{n=0}^{\infty}\frac{\left(\frac{1}{2}\right)_n}{n!}T_n(x)t^n,\quad R=\sqrt{1-2xt+t^2}. \tag{9.8.51}$$

$${}_0F_1\left(\begin{array}{c}-\\ \frac{1}{2}\end{array};\frac{(x-1)t}{2}\right){}_0F_1\left(\begin{array}{c}-\\ \frac{1}{2}\end{array};\frac{(x+1)t}{2}\right)=\sum_{n=0}^{\infty}\frac{T_n(x)}{\left(\frac{1}{2}\right)_n n!}t^n. \tag{9.8.52}$$

$$\mathrm{e}^{xt}\,{}_0F_1\left(\begin{array}{c}-\\ \frac{1}{2}\end{array};\frac{(x^2-1)t^2}{4}\right)=\sum_{n=0}^{\infty}\frac{T_n(x)}{n!}t^n. \tag{9.8.53}$$

$$\begin{aligned}&{}_2F_1\left(\begin{array}{c}\gamma,-\gamma\\ \frac{1}{2}\end{array};\frac{1-R-t}{2}\right){}_2F_1\left(\begin{array}{c}\gamma,-\gamma\\ \frac{1}{2}\end{array};\frac{1-R+t}{2}\right)\\ &=\sum_{n=0}^{\infty}\frac{(\gamma)_n(-\gamma)_n}{\left(\frac{1}{2}\right)_n n!}T_n(x)t^n,\quad R=\sqrt{1-2xt+t^2},\quad \gamma\text{ arbitrary.}\end{aligned} \tag{9.8.54}$$

$$\begin{aligned}&(1-xt)^{-\gamma}{}_2F_1\left(\begin{array}{c}\frac{1}{2}\gamma,\frac{1}{2}\gamma+\frac{1}{2}\\ \frac{1}{2}\end{array};\frac{(x^2-1)t^2}{(1-xt)^2}\right)\\ &=\sum_{n=0}^{\infty}\frac{(\gamma)_n}{n!}T_n(x)t^n,\quad \gamma\text{ arbitrary.}\end{aligned} \tag{9.8.55}$$

$$\frac{1}{1-2xt+t^2}=\sum_{n=0}^{\infty}U_n(x)t^n. \tag{9.8.56}$$

$$\frac{1}{R\sqrt{\frac{1}{2}(1+R-xt)}} = \sum_{n=0}^{\infty} \frac{\left(\frac{3}{2}\right)_n}{(n+1)!} U_n(x) t^n, \quad R = \sqrt{1-2xt+t^2}. \tag{9.8.57}$$

$${}_0F_1\left(\begin{array}{c} - \\ \frac{3}{2} \end{array}; \frac{(x-1)t}{2}\right) {}_0F_1\left(\begin{array}{c} - \\ \frac{3}{2} \end{array}; \frac{(x+1)t}{2}\right) = \sum_{n=0}^{\infty} \frac{U_n(x)}{\left(\frac{3}{2}\right)_n (n+1)!} t^n. \tag{9.8.58}$$

$$\mathrm{e}^{xt}\, {}_0F_1\left(\begin{array}{c} - \\ \frac{3}{2} \end{array}; \frac{(x^2-1)t^2}{4}\right) = \sum_{n=0}^{\infty} \frac{U_n(x)}{(n+1)!} t^n. \tag{9.8.59}$$

$$\begin{aligned} &{}_2F_1\left(\begin{array}{c} \gamma, 2-\gamma \\ \frac{3}{2} \end{array}; \frac{1-R-t}{2}\right) {}_2F_1\left(\begin{array}{c} \gamma, 2-\gamma \\ \frac{3}{2} \end{array}; \frac{1-R+t}{2}\right) \\ &= \sum_{n=0}^{\infty} \frac{(\gamma)_n (2-\gamma)_n}{\left(\frac{3}{2}\right)_n (n+1)!} U_n(x) t^n, \quad R = \sqrt{1-2xt+t^2}, \quad \gamma \text{ arbitrary}. \end{aligned} \tag{9.8.60}$$

$$\begin{aligned} &(1-xt)^{-\gamma}\, {}_2F_1\left(\begin{array}{c} \frac{1}{2}\gamma, \frac{1}{2}\gamma+\frac{1}{2} \\ \frac{3}{2} \end{array}; \frac{(x^2-1)t^2}{(1-xt)^2}\right) \\ &= \sum_{n=0}^{\infty} \frac{(\gamma)_n}{(n+1)!} U_n(x) t^n, \quad \gamma \text{ arbitrary}. \end{aligned} \tag{9.8.61}$$

Remarks

The Chebyshev polynomials can also be written as:

$$T_n(x) = \cos(n\theta), \quad x = \cos\theta$$

and

$$U_n(x) = \frac{\sin(n+1)\theta}{\sin\theta}, \quad x = \cos\theta.$$

Further we have

$$U_n(x) = C_n^{(1)}(x)$$

where $C_n^{(\lambda)}(x)$ denotes the Gegenbauer (or ultraspherical) polynomial given by (9.8.19) in the preceding subsection.

References

[2], [55], [60], [61], [94], [146], [156], [168], [198], [253], [264], [277], [317], [399], [403], [416], [417], [456], [457], [459], [466], [493], [496], [514], [522], [525], [526].

9.8.3 Legendre / Spherical

Hypergeometric Representation

The Legendre (or spherical) polynomials are Jacobi polynomials with $\alpha = \beta = 0$:

$$P_n(x) = P_n^{(0,0)}(x) = {}_2F_1\left(\begin{matrix}-n, n+1\\ 1\end{matrix}; \frac{1-x}{2}\right). \tag{9.8.62}$$

Orthogonality Relation

$$\int_{-1}^{1} P_m(x)P_n(x)\,dx = \frac{2}{2n+1}\,\delta_{mn}. \tag{9.8.63}$$

Recurrence Relation

$$(2n+1)xP_n(x) = (n+1)P_{n+1}(x) + nP_{n-1}(x). \tag{9.8.64}$$

Normalized Recurrence Relation

$$xp_n(x) = p_{n+1}(x) + \frac{n^2}{(2n-1)(2n+1)}p_{n-1}(x), \tag{9.8.65}$$

where

$$P_n(x) = \binom{2n}{n}\frac{1}{2^n}p_n(x).$$

Differential Equation

$$(1-x^2)y''(x) - 2xy'(x) + n(n+1)y(x) = 0, \quad y(x) = P_n(x). \tag{9.8.66}$$

Rodrigues-Type Formula

$$P_n(x) = \frac{(-1)^n}{2^n n!}\left(\frac{d}{dx}\right)^n\left[(1-x^2)^n\right]. \tag{9.8.67}$$

Generating Functions

$$\frac{1}{\sqrt{1-2xt+t^2}} = \sum_{n=0}^{\infty} P_n(x)t^n. \tag{9.8.68}$$

$${}_0F_1\left(\begin{array}{c} - \\ 1 \end{array}; \frac{(x-1)t}{2}\right) {}_0F_1\left(\begin{array}{c} - \\ 1 \end{array}; \frac{(x+1)t}{2}\right) = \sum_{n=0}^{\infty} \frac{P_n(x)}{(n!)^2} t^n. \tag{9.8.69}$$

$$\mathrm{e}^{xt}\,{}_0F_1\left(\begin{array}{c} - \\ 1 \end{array}; \frac{(x^2-1)t^2}{4}\right) = \sum_{n=0}^{\infty} \frac{P_n(x)}{n!} t^n. \tag{9.8.70}$$

$$\begin{aligned} &{}_2F_1\left(\begin{array}{c} \gamma, 1-\gamma \\ 1 \end{array}; \frac{1-R-t}{2}\right) {}_2F_1\left(\begin{array}{c} \gamma, 1-\gamma \\ 1 \end{array}; \frac{1-R+t}{2}\right) \\ &= \sum_{n=0}^{\infty} \frac{(\gamma)_n(1-\gamma)_n}{(n!)^2} P_n(x)t^n, \quad R = \sqrt{1-2xt+t^2}, \quad \gamma \text{ arbitrary}. \end{aligned} \tag{9.8.71}$$

$$\begin{aligned} &(1-xt)^{-\gamma}\,{}_2F_1\left(\begin{array}{c} \frac{1}{2}\gamma, \frac{1}{2}\gamma+\frac{1}{2} \\ 1 \end{array}; \frac{(x^2-1)t^2}{(1-xt)^2}\right) \\ &= \sum_{n=0}^{\infty} \frac{(\gamma)_n}{n!} P_n(x)t^n, \quad \gamma \text{ arbitrary}. \end{aligned} \tag{9.8.72}$$

References

[2], [6], [16], [103], [109], [127], [146], [156], [158], [168], [195], [198], [253], [403], [416], [417], [424], [456], [493], [496], [522].

9.9 Pseudo Jacobi

Hypergeometric Representation

$$P_n(x;\nu,N) = \frac{(-2i)^n(-N+i\nu)_n}{(n-2N-1)_n}\,{}_2F_1\left(\begin{matrix}-n,n-2N-1\\ -N+i\nu\end{matrix};\frac{1-ix}{2}\right) \tag{9.9.1}$$
$$= (x+i)^n\,{}_2F_1\left(\begin{matrix}-n,N+1-n-i\nu\\ 2N+2-2n\end{matrix};\frac{2}{1-ix}\right), \quad n=0,1,2,\ldots,N.$$

Orthogonality Relation

$$\frac{1}{2\pi}\int_{-\infty}^{\infty}(1+x^2)^{-N-1}\mathrm{e}^{2\nu\arctan x}P_m(x;\nu,N)P_n(x;\nu,N)\,dx$$
$$= \frac{\Gamma(2N+1-2n)\Gamma(2N+2-2n)2^{2n-2N-1}n!}{\Gamma(2N+2-n)\,|\Gamma(N+1-n+i\nu)|^2}\,\delta_{mn}. \tag{9.9.2}$$

Recurrence Relation

$$xP_n(x;\nu,N) = P_{n+1}(x;\nu,N) + \frac{(N+1)\nu}{(n-N-1)(n-N)}P_n(x;\nu,N)$$
$$-\frac{n(n-2N-2)}{(2n-2N-3)(n-N-1)^2(2n-2N-1)}$$
$$\times(n-N-1-i\nu)(n-N-1+i\nu)P_{n-1}(x;\nu,N). \tag{9.9.3}$$

Normalized Recurrence Relation

$$xp_n(x) = p_{n+1}(x) + \frac{(N+1)\nu}{(n-N-1)(n-N)}p_n(x)$$
$$-\frac{n(n-2N-2)(n-N-1-i\nu)(n-N-1+i\nu)}{(2n-2N-3)(n-N-1)^2(2n-2N-1)}p_{n-1}(x), \tag{9.9.4}$$

where

$$P_n(x;\nu,N) = p_n(x).$$

Differential Equation

$$(1+x^2)y''(x)+2(\nu-Nx)y'(x)-n(n-2N-1)y(x)=0, \tag{9.9.5}$$

where

$$y(x)=P_n(x;\nu,N).$$

Forward Shift Operator

$$\frac{d}{dx}P_n(x;\nu,N)=nP_{n-1}(x;\nu,N-1). \tag{9.9.6}$$

Backward Shift Operator

$$\begin{aligned}&(1+x^2)\frac{d}{dx}P_n(x;\nu,N)+2[\nu-(N+1)x]P_n(x;\nu,N)\\&=(n-2N-2)P_{n+1}(x;\nu,N+1)\end{aligned} \tag{9.9.7}$$

or equivalently

$$\begin{aligned}&\frac{d}{dx}\left[(1+x^2)^{-N-1}\mathrm{e}^{2\nu\arctan x}P_n(x;\nu,N)\right]\\&=(n-2N-2)(1+x^2)^{-N-2}\mathrm{e}^{2\nu\arctan x}P_{n+1}(x;\nu,N+1).\end{aligned} \tag{9.9.8}$$

Rodrigues-Type Formula

$$P_n(x;\nu,N)=\frac{(1+x^2)^{N+1}\mathrm{e}^{-2\nu\arctan x}}{(n-2N-1)_n}\left(\frac{d}{dx}\right)^n\left[(1+x^2)^{n-N-1}\mathrm{e}^{2\nu\arctan x}\right]. \tag{9.9.9}$$

Generating Function

$$\begin{aligned}&\left[{}_0F_1\left(\begin{matrix}-\\-N+i\nu\end{matrix};(x+i)t\right){}_0F_1\left(\begin{matrix}-\\-N-i\nu\end{matrix};(x-i)t\right)\right]_N\\&=\sum_{n=0}^{N}\frac{(n-2N-1)_n}{(-N+i\nu)_n(-N-i\nu)_n n!}P_n(x;\nu,N)t^n.\end{aligned} \tag{9.9.10}$$

Limit Relation

Continuous Hahn → Pseudo Jacobi

The pseudo Jacobi polynomials follow from the continuous Hahn polynomials given by (9.4.1) by the substitutions $x \to xt$, $a = \frac{1}{2}(-N+i\nu-2t)$, $b = \frac{1}{2}(-N-i\nu+2t)$, $c = \frac{1}{2}(-N-i\nu-2t)$ and $d = \frac{1}{2}(-N+i\nu+2t)$, division by t^n and the limit $t \to \infty$:

$$\lim_{t\to\infty} \frac{p_n(xt; \frac{1}{2}(-N+i\nu-2t), \frac{1}{2}(-N-i\nu+2t), \frac{1}{2}(-N+i\nu-2t), \frac{1}{2}(-N-i\nu+2t))}{t^n}$$
$$= \frac{(n-2N-1)_n}{n!} P_n(x;\nu,N).$$

Remarks

Since we have for $k < n$

$$\frac{(-N+i\nu)_n}{(-N+i\nu)_k} = (-N+i\nu+k)_{n-k},$$

the pseudo Jacobi polynomials given by (9.9.1) can also be seen as polynomials in the parameter ν.
The weight function for the pseudo Jacobi polynomials can be written as

$$(1+x^2)^{-N-1}\mathrm{e}^{2\nu\arctan x} = (1+ix)^{-N-1-i\nu}(1-ix)^{-N-1+i\nu}.$$

The pseudo Jacobi polynomials are related to the Jacobi polynomials defined by (9.8.1) in the following way:

$$P_n(x;\nu,N) = \frac{(-2i)^n n!}{(n-2N-1)_n} P_n^{(-N-1+i\nu,-N-1-i\nu)}(ix).$$

If we set $x \to \nu x$ in the definition (9.9.1) of the pseudo Jacobi polynomials and take the limit $\nu \to \infty$ we obtain a special case of the Bessel polynomials given by (9.13.1) in the following way:

$$\lim_{\nu\to\infty} \frac{P_n(\nu x;\nu,N)}{\nu^n} = \frac{2^n}{(n-2N-1)_n} y_n(x;-2N-2).$$

References

[50], [108], [382].

9.10 Meixner

Hypergeometric Representation

$$M_n(x;\beta,c) = {}_2F_1\left(\begin{matrix}-n,-x\\ \beta\end{matrix};1-\frac{1}{c}\right). \tag{9.10.1}$$

Orthogonality Relation

$$\sum_{x=0}^{\infty}\frac{(\beta)_x}{x!}c^x M_m(x;\beta,c)M_n(x;\beta,c)$$
$$= \frac{c^{-n}n!}{(\beta)_n(1-c)^{\beta}}\delta_{mn}, \quad \beta>0 \quad \text{and} \quad 0<c<1. \tag{9.10.2}$$

Recurrence Relation

$$(c-1)xM_n(x;\beta,c) = c(n+\beta)M_{n+1}(x;\beta,c)$$
$$-\left[n+(n+\beta)c\right]M_n(x;\beta,c)+nM_{n-1}(x;\beta,c). \tag{9.10.3}$$

Normalized Recurrence Relation

$$xp_n(x) = p_{n+1}(x)+\frac{n+(n+\beta)c}{1-c}p_n(x)+\frac{n(n+\beta-1)c}{(1-c)^2}p_{n-1}(x), \tag{9.10.4}$$

where

$$M_n(x;\beta,c) = \frac{1}{(\beta)_n}\left(\frac{c-1}{c}\right)^n p_n(x).$$

Difference Equation

$$n(c-1)y(x) = c(x+\beta)y(x+1)-\left[x+(x+\beta)c\right]y(x)+xy(x-1), \tag{9.10.5}$$

where

$$y(x) = M_n(x;\beta,c).$$

Forward Shift Operator

$$M_n(x+1;\beta,c)-M_n(x;\beta,c)=\frac{n}{\beta}\left(\frac{c-1}{c}\right)M_{n-1}(x;\beta+1,c) \qquad (9.10.6)$$

or equivalently

$$\Delta M_n(x;\beta,c)=\frac{n}{\beta}\left(\frac{c-1}{c}\right)M_{n-1}(x;\beta+1,c). \qquad (9.10.7)$$

Backward Shift Operator

$$c(\beta+x-1)M_n(x;\beta,c)-xM_n(x-1;\beta,c)=c(\beta-1)M_{n+1}(x;\beta-1,c) \qquad (9.10.8)$$

or equivalently

$$\nabla\left[\frac{(\beta)_x c^x}{x!}M_n(x;\beta,c)\right]=\frac{(\beta-1)_x c^x}{x!}M_{n+1}(x;\beta-1,c). \qquad (9.10.9)$$

Rodrigues-Type Formula

$$\frac{(\beta)_x c^x}{x!}M_n(x;\beta,c)=\nabla^n\left[\frac{(\beta+n)_x c^x}{x!}\right]. \qquad (9.10.10)$$

Generating Functions

$$\left(1-\frac{t}{c}\right)^x(1-t)^{-x-\beta}=\sum_{n=0}^{\infty}\frac{(\beta)_n}{n!}M_n(x;\beta,c)t^n. \qquad (9.10.11)$$

$$\mathrm{e}^t\,{}_1F_1\left(\begin{matrix}-x\\ \beta\end{matrix};\left(\frac{1-c}{c}\right)t\right)=\sum_{n=0}^{\infty}\frac{M_n(x;\beta,c)}{n!}t^n. \qquad (9.10.12)$$

$$(1-t)^{-\gamma}\,{}_2F_1\left(\begin{matrix}\gamma,-x\\ \beta\end{matrix};\frac{(1-c)t}{c(1-t)}\right)=\sum_{n=0}^{\infty}\frac{(\gamma)_n}{n!}M_n(x;\beta,c)t^n, \quad \gamma \text{ arbitrary}. \qquad (9.10.13)$$

Limit Relations

Hahn → Meixner

If we take $\alpha = b-1$, $\beta = N(1-c)c^{-1}$ in the definition (9.5.1) of the Hahn polynomials and let $N \to \infty$ we find the Meixner polynomials:

$$\lim_{N\to\infty} Q_n(x;b-1,N(1-c)c^{-1},N) = M_n(x;b,c).$$

Dual Hahn → Meixner

To obtain the Meixner polynomials from the dual Hahn polynomials we have to take $\gamma = \beta - 1$ and $\delta = N(1-c)c^{-1}$ in the definition (9.6.1) of the dual Hahn polynomials and let $N \to \infty$:

$$\lim_{N\to\infty} R_n(\lambda(x);\beta-1,N(1-c)c^{-1},N) = M_n(x;\beta,c).$$

Meixner → Laguerre

The Laguerre polynomials given by (9.12.1) are obtained from the Meixner polynomials if we take $\beta = \alpha + 1$ and $x \to (1-c)^{-1}x$ and let $c \to 1$:

$$\lim_{c\to 1} M_n((1-c)^{-1}x;\alpha+1,c) = \frac{L_n^{(\alpha)}(x)}{L_n^{(\alpha)}(0)}. \tag{9.10.14}$$

Meixner → Charlier

The Charlier polynomials given by (9.14.1) are obtained from the Meixner polynomials if we take $c = (a+\beta)^{-1}a$ and let $\beta \to \infty$:

$$\lim_{\beta\to\infty} M_n(x;\beta,(a+\beta)^{-1}a) = C_n(x;a). \tag{9.10.15}$$

Remarks

The Meixner polynomials are related to the Jacobi polynomials given by (9.8.1) in the following way:

$$\frac{(\beta)_n}{n!} M_n(x;\beta,c) = P_n^{(\beta-1,-n-\beta-x)}((2-c)c^{-1}).$$

The Meixner polynomials are also related to the Krawtchouk polynomials given by (9.11.1) in the following way:

$$K_n(x;p,N) = M_n(x;-N,(p-1)^{-1}p).$$

References

[7], [11], [16], [20], [22], [29], [34], [39], [46], [51], [54], [59], [61], [72], [80], [82], [96], [97], [125], [146], [155], [198], [215], [217], [218], [226], [228], [265], [277], [279], [283], [289], [295], [303], [307], [317], [340], [363], [364], [375], [377], [381], [391], [406], [416], [417], [434], [503], [506], [521], [523].

9.11 Krawtchouk

Hypergeometric Representation

$$K_n(x;p,N) = {}_2F_1\left(\begin{matrix} -n, -x \\ -N \end{matrix}; \frac{1}{p}\right), \quad n = 0,1,2,\ldots,N. \tag{9.11.1}$$

Orthogonality Relation

$$\sum_{x=0}^{N} \binom{N}{x} p^x (1-p)^{N-x} K_m(x;p,N) K_n(x;p,N)$$
$$= \frac{(-1)^n n!}{(-N)_n} \left(\frac{1-p}{p}\right)^n \delta_{mn}, \quad 0 < p < 1. \tag{9.11.2}$$

Recurrence Relation

$$\begin{aligned} -xK_n(x;p,N) = {} & p(N-n)K_{n+1}(x;p,N) \\ & - \left[p(N-n) + n(1-p)\right] K_n(x;p,N) \\ & + n(1-p)K_{n-1}(x;p,N). \end{aligned} \tag{9.11.3}$$

Normalized Recurrence Relation

$$\begin{aligned} xp_n(x) = {} & p_{n+1}(x) + [p(N-n)+n(1-p)]\,p_n(x) \\ & + np(1-p)(N+1-n)p_{n-1}(x), \end{aligned} \tag{9.11.4}$$

where

$$K_n(x;p,N) = \frac{1}{(-N)_n p^n} p_n(x).$$

Difference Equation

$$\begin{aligned} -ny(x) = {} & p(N-x)y(x+1) \\ & - [p(N-x)+x(1-p)]\,y(x) + x(1-p)y(x-1), \end{aligned} \tag{9.11.5}$$

where

$$y(x) = K_n(x;p,N).$$

Forward Shift Operator

$$K_n(x+1;p,N) - K_n(x;p,N) = -\frac{n}{Np} K_{n-1}(x;p,N-1) \tag{9.11.6}$$

or equivalently

$$\Delta K_n(x;p,N) = -\frac{n}{Np} K_{n-1}(x;p,N-1). \tag{9.11.7}$$

Backward Shift Operator

$$\begin{aligned} & (N+1-x)K_n(x;p,N) - x\left(\frac{1-p}{p}\right) K_n(x-1;p,N) \\ & = (N+1)K_{n+1}(x;p,N+1) \end{aligned} \tag{9.11.8}$$

or equivalently

$$\nabla\left[\binom{N}{x}\left(\frac{p}{1-p}\right)^x K_n(x;p,N)\right] = \binom{N+1}{x}\left(\frac{p}{1-p}\right)^x K_{n+1}(x;p,N+1). \tag{9.11.9}$$

Rodrigues-Type Formula

$$\binom{N}{x}\left(\frac{p}{1-p}\right)^x K_n(x;p,N) = \nabla^n\left[\binom{N-n}{x}\left(\frac{p}{1-p}\right)^x\right]. \tag{9.11.10}$$

Generating Functions

For $x = 0,1,2,\ldots,N$ we have

$$\left(1-\frac{(1-p)}{p}t\right)^x (1+t)^{N-x} = \sum_{n=0}^{N}\binom{N}{n}K_n(x;p,N)t^n. \tag{9.11.11}$$

$$\left[\mathrm{e}^t\,{}_1F_1\left(\begin{matrix}-x\\-N\end{matrix};-\frac{t}{p}\right)\right]_N = \sum_{n=0}^{N}\frac{K_n(x;p,N)}{n!}t^n. \tag{9.11.12}$$

$$\left[(1-t)^{-\gamma}\,{}_2F_1\left(\begin{matrix}\gamma,-x\\-N\end{matrix};\frac{t}{p(t-1)}\right)\right]_N = \sum_{n=0}^{N}\frac{(\gamma)_n}{n!}K_n(x;p,N)t^n, \quad \gamma \text{ arbitrary}. \tag{9.11.13}$$

Limit Relations

Hahn → Krawtchouk

If we take $\alpha = pt$ and $\beta = (1-p)t$ in the definition (9.5.1) of the Hahn polynomials and let $t \to \infty$ we obtain the Krawtchouk polynomials:

$$\lim_{t\to\infty} Q_n(x;pt,(1-p)t,N) = K_n(x;p,N).$$

Dual Hahn → Krawtchouk

The Krawtchouk polynomials follow from the dual Hahn polynomials given by (9.6.1) if we set $\gamma = pt$, $\delta = (1-p)t$ and let $t \to \infty$:

$$\lim_{t\to\infty} R_n(\lambda(x);pt,(1-p)t,N) = K_n(x;p,N).$$

Krawtchouk → Charlier

The Charlier polynomials given by (9.14.1) can be found from the Krawtchouk polynomials by taking $p = N^{-1}a$ and letting $N \to \infty$:

$$\lim_{N\to\infty} K_n(x;N^{-1}a,N) = C_n(x;a). \tag{9.11.14}$$

Krawtchouk → Hermite

The Hermite polynomials given by (9.15.1) follow from the Krawtchouk polynomials by setting $x \to pN + x\sqrt{2p(1-p)N}$ and then letting $N \to \infty$:

$$\lim_{N\to\infty} \sqrt{\binom{N}{n}} K_n(pN + x\sqrt{2p(1-p)N};p,N) = \frac{(-1)^n H_n(x)}{\sqrt{2^n n! \left(\frac{p}{1-p}\right)^n}}. \tag{9.11.15}$$

Remarks

The Krawtchouk polynomials are self-dual, which means that

$$K_n(x;p,N) = K_x(n;p,N), \quad n,x \in \{0,1,2,\ldots,N\}.$$

By using this relation we easily obtain the so-called dual orthogonality relation from the orthogonality relation (9.11.2):

$$\sum_{n=0}^{N} \binom{N}{n} p^n (1-p)^{N-n} K_n(x;p,N) K_n(y;p,N) = \frac{\left(\frac{1-p}{p}\right)^x}{\binom{N}{x}} \delta_{xy},$$

where $0 < p < 1$ and $x,y \in \{0,1,2,\ldots,N\}$.
The Krawtchouk polynomials are related to the Meixner polynomials given by (9.10.1) in the following way:

$$K_n(x;p,N) = M_n(x;-N,(p-1)^{-1}p).$$

References

[16], [34], [39], [46], [51], [59], [72], [80], [125], [141], [146], [164], [182], [184], [187], [188], [198], [203], [226], [228], [265], [277], [307], [338], [340], [363],

[364], [372], [375], [377], [381], [391], [416], [417], [429], [434], [436], [488], [489], [493], [521], [523].

9.12 Laguerre

Hypergeometric Representation

$$L_n^{(\alpha)}(x) = \frac{(\alpha+1)_n}{n!} \, {}_1F_1\left(\begin{matrix} -n \\ \alpha+1 \end{matrix}; x\right). \tag{9.12.1}$$

Orthogonality Relation

$$\int_0^\infty e^{-x} x^\alpha L_m^{(\alpha)}(x) L_n^{(\alpha)}(x)\,dx = \frac{\Gamma(n+\alpha+1)}{n!}\,\delta_{mn}, \quad \alpha > -1. \tag{9.12.2}$$

Recurrence Relation

$$(n+1)L_{n+1}^{(\alpha)}(x) - (2n+\alpha+1-x)L_n^{(\alpha)}(x) + (n+\alpha)L_{n-1}^{(\alpha)}(x) = 0. \tag{9.12.3}$$

Normalized Recurrence Relation

$$xp_n(x) = p_{n+1}(x) + (2n+\alpha+1)p_n(x) + n(n+\alpha)p_{n-1}(x), \tag{9.12.4}$$

where

$$L_n^{(\alpha)}(x) = \frac{(-1)^n}{n!} p_n(x).$$

Differential Equation

$$xy''(x) + (\alpha+1-x)y'(x) + ny(x) = 0, \quad y(x) = L_n^{(\alpha)}(x). \tag{9.12.5}$$

Forward Shift Operator

$$\frac{d}{dx}L_n^{(\alpha)}(x) = -L_{n-1}^{(\alpha+1)}(x). \tag{9.12.6}$$

Backward Shift Operator

$$x\frac{d}{dx}L_n^{(\alpha)}(x) + (\alpha - x)L_n^{(\alpha)}(x) = (n+1)L_{n+1}^{(\alpha-1)}(x) \tag{9.12.7}$$

or equivalently

$$\frac{d}{dx}\left[\mathrm{e}^{-x}x^{\alpha}L_n^{(\alpha)}(x)\right] = (n+1)\mathrm{e}^{-x}x^{\alpha-1}L_{n+1}^{(\alpha-1)}(x). \tag{9.12.8}$$

Rodrigues-Type Formula

$$\mathrm{e}^{-x}x^{\alpha}L_n^{(\alpha)}(x) = \frac{1}{n!}\left(\frac{d}{dx}\right)^n\left[\mathrm{e}^{-x}x^{n+\alpha}\right]. \tag{9.12.9}$$

Generating Functions

$$(1-t)^{-\alpha-1}\exp\left(\frac{xt}{t-1}\right) = \sum_{n=0}^{\infty} L_n^{(\alpha)}(x)t^n. \tag{9.12.10}$$

$$\mathrm{e}^t\,{}_0F_1\left(\genfrac{}{}{0pt}{}{-}{\alpha+1};-xt\right) = \sum_{n=0}^{\infty}\frac{L_n^{(\alpha)}(x)}{(\alpha+1)_n}t^n. \tag{9.12.11}$$

$$(1-t)^{-\gamma}\,{}_1F_1\left(\genfrac{}{}{0pt}{}{\gamma}{\alpha+1};\frac{xt}{t-1}\right) = \sum_{n=0}^{\infty}\frac{(\gamma)_n}{(\alpha+1)_n}L_n^{(\alpha)}(x)t^n, \quad \gamma \text{ arbitrary}. \tag{9.12.12}$$

Limit Relations

Meixner-Pollaczek → Laguerre

The Laguerre polynomials can be obtained from the Meixner-Pollaczek polynomials given by (9.7.1) by the substitution $\lambda = \frac{1}{2}(\alpha+1)$, $x \to -\frac{1}{2}\phi^{-1}x$ and the limit $\phi \to 0$:

$$\lim_{\phi\to 0} P_n^{(\frac{1}{2}\alpha+\frac{1}{2})}(-\tfrac{1}{2}\phi^{-1}x;\phi) = L_n^{(\alpha)}(x).$$

Jacobi → Laguerre

The Laguerre polynomials are obtained from the Jacobi polynomials given by (9.8.1) if we set $x \to 1-2\beta^{-1}x$ and then take the limit $\beta \to \infty$:

$$\lim_{\beta\to\infty} P_n^{(\alpha,\beta)}(1-2\beta^{-1}x) = L_n^{(\alpha)}(x).$$

Meixner → Laguerre

If we take $\beta = \alpha+1$ and $x \to (1-c)^{-1}x$ in the definition (9.10.1) of the Meixner polynomials and let $c \to 1$ we obtain the Laguerre polynomials:

$$\lim_{c\to 1} M_n((1-c)^{-1}x;\alpha+1,c) = \frac{L_n^{(\alpha)}(x)}{L_n^{(\alpha)}(0)}.$$

Laguerre → Hermite

The Hermite polynomials given by (9.15.1) can be obtained from the Laguerre polynomials by taking the limit $\alpha \to \infty$ in the following way:

$$\lim_{\alpha\to\infty}\left(\frac{2}{\alpha}\right)^{\frac{1}{2}n} L_n^{(\alpha)}((2\alpha)^{\frac{1}{2}}x+\alpha) = \frac{(-1)^n}{n!}H_n(x). \tag{9.12.13}$$

Remarks

The definition (9.12.1) of the Laguerre polynomials can also be written as:

$$L_n^{(\alpha)}(x) = \frac{1}{n!}\sum_{k=0}^{n}\frac{(-n)_k}{k!}(\alpha+k+1)_{n-k}x^k.$$

In this way the Laguerre polynomials can also be seen as polynomials in the parameter α. Therefore they can be defined for all α.
The Laguerre polynomials are related to the Bessel polynomials given by (9.13.1) in the following way:

$$L_n^{(\alpha)}(x) = \frac{(-x)^n}{n!}y_n(2x^{-1};-2n-\alpha-1).$$

The Laguerre polynomials are related to the Charlier polynomials given by (9.14.1) in the following way:

$$\frac{(-a)^n}{n!}C_n(x;a) = L_n^{(x-n)}(a).$$

The Laguerre polynomials and the Hermite polynomials given by (9.15.1) are also connected by the following quadratic transformations:

$$H_{2n}(x) = (-1)^n n! 2^{2n} L_n^{(-\frac{1}{2})}(x^2)$$

and

$$H_{2n+1}(x) = (-1)^n n! 2^{2n+1} x L_n^{(\frac{1}{2})}(x^2).$$

In combinatorics the Laguerre polynomials with $\alpha = 0$ are often called Rook polynomials.

References

[1], [2], [4], [7], [10], [11], [15], [16], [19], [20], [34], [35], [41], [46], [51], [54], [58], [59], [61], [65], [72], [93], [95], [109], [111], [112], [117], [123], [124], [128], [129], [130], [131], [133], [136], [137], [144], [146], [153], [155], [158], [165], [166], [170], [176], [192], [197], [198], [201], [202], [227], [229], [240], [245], [246], [248], [249], [251], [253], [263], [267], [268], [277], [279], [283], [287], [289], [295], [297], [300], [307], [313], [315], [317], [333], [335], [339], [340], [354], [356], [359], [360], [363], [364], [365], [368], [369], [370], [381], [382], [390], [397], [403], [406], [416], [417], [424], [425], [427], [456], [466], [475], [476], [477], [478], [479], [493], [496], [501], [502], [506].

9.13 Bessel

Hypergeometric Representation

$$\begin{aligned} y_n(x;a) &= {}_2F_0\left(\begin{matrix} -n, n+a+1 \\ - \end{matrix}; -\frac{x}{2}\right) \qquad (9.13.1) \\ &= (n+a+1)_n \left(\frac{x}{2}\right)^n {}_1F_1\left(\begin{matrix} -n \\ -2n-a \end{matrix}; \frac{2}{x}\right), \quad n = 0, 1, 2, \ldots, N. \end{aligned}$$

Orthogonality Relation

$$\int_0^\infty x^a e^{-\frac{2}{x}} y_m(x;a) y_n(x;a)\,dx$$
$$= -\frac{2^{a+1}}{2n+a+1}\Gamma(-n-a)n!\,\delta_{mn}, \quad a < -2N-1. \tag{9.13.2}$$

Recurrence Relation

$$2(n+a+1)(2n+a)y_{n+1}(x;a)$$
$$= (2n+a+1)\,[2a+(2n+a)(2n+a+2)x]\,y_n(x;a)$$
$$+2n(2n+a+2)y_{n-1}(x;a). \tag{9.13.3}$$

Normalized Recurrence Relation

$$xp_n(x) = p_{n+1}(x) - \frac{2a}{(2n+a)(2n+a+2)}p_n(x)$$
$$-\frac{4n(n+a)}{(2n+a-1)(2n+a)^2(2n+a+1)}p_{n-1}(x), \tag{9.13.4}$$

where

$$y_n(x;a) = \frac{(n+a+1)_n}{2^n}p_n(x).$$

Differential Equation

$$x^2y''(x) + [(a+2)x+2]y'(x) - n(n+a+1)y(x) = 0, \quad y(x) = y_n(x;a). \tag{9.13.5}$$

Forward Shift Operator

$$\frac{d}{dx}y_n(x;a) = \frac{n(n+a+1)}{2}y_{n-1}(x;a+2). \tag{9.13.6}$$

Backward Shift Operator

$$x^2\frac{d}{dx}y_n(x;a)+(ax+2)y_n(x;a)=2y_{n+1}(x;a-2) \tag{9.13.7}$$

or equivalently

$$\frac{d}{dx}\left[x^a\mathrm{e}^{-\frac{2}{x}}y_n(x;a)\right]=2x^{a-2}\mathrm{e}^{-\frac{2}{x}}y_{n+1}(x;a-2). \tag{9.13.8}$$

Rodrigues-Type Formula

$$y_n(x;a)=2^{-n}x^{-a}\mathrm{e}^{\frac{2}{x}}D^n\left(x^{2n+a}\mathrm{e}^{-\frac{2}{x}}\right). \tag{9.13.9}$$

Generating Function

$$(1-2xt)^{-\frac{1}{2}}\left(\frac{2}{1+\sqrt{1-2xt}}\right)^a\exp\left(\frac{2t}{1+\sqrt{1-2xt}}\right)=\sum_{n=0}^{\infty}y_n(x;a)\frac{t^n}{n!}. \tag{9.13.10}$$

Limit Relation

Jacobi → Bessel

If we take $\beta=a-\alpha$ in the definition (9.8.1) of the Jacobi polynomials and let $\alpha\rightarrow-\infty$ we find the Bessel polynomials:

$$\lim_{\alpha\rightarrow-\infty}\frac{P_n^{(\alpha,a-\alpha)}(1+\alpha x)}{P_n^{(\alpha,a-\alpha)}(1)}=y_n(x;a).$$

Remarks

The following notations are also used for the Bessel polynomials:

$$y_n(x;a,b)=y_n(2b^{-1}x;a) \quad \text{and} \quad \theta_n(x;a,b)=x^ny_n(x^{-1};a,b).$$

However, the Bessel polynomials essentially depend on only one parameter.
The Bessel polynomials are related to the Laguerre polynomials given by (9.12.1) in the following way:

$$L_n^{(\alpha)}(x) = \frac{(-x)^n}{n!} y_n(2x^{-1}; -2n-\alpha-1).$$

The special case $a = -2N - 2$ of the Bessel polynomials can be obtained from the pseudo Jacobi polynomials by setting $x \to \nu x$ in the definition (9.9.1) of the pseudo Jacobi polynomials and taking the limit $\nu \to \infty$ in the following way:

$$\lim_{\nu\to\infty} \frac{P_n(\nu x; \nu, N)}{\nu^n} = \frac{2^n}{(n-2N-1)_n} y_n(x; -2N-2).$$

References

[32], [102], [126], [159], [179], [181], [255], [277], [352], [384], [417].

9.14 Charlier

Hypergeometric Representation

$$C_n(x;a) = {}_2F_0\left(\begin{matrix} -n, -x \\ - \end{matrix}; -\frac{1}{a}\right). \tag{9.14.1}$$

Orthogonality Relation

$$\sum_{x=0}^{\infty} \frac{a^x}{x!} C_m(x;a) C_n(x;a) = a^{-n} \mathrm{e}^a n!\, \delta_{mn}, \quad a > 0. \tag{9.14.2}$$

Recurrence Relation

$$-xC_n(x;a) = aC_{n+1}(x;a) - (n+a)C_n(x;a) + nC_{n-1}(x;a). \tag{9.14.3}$$

Normalized Recurrence Relation

$$xp_n(x) = p_{n+1}(x) + (n+a)p_n(x) + nap_{n-1}(x), \tag{9.14.4}$$

where

$$C_n(x;a) = \left(-\frac{1}{a}\right)^n p_n(x).$$

Difference Equation

$$-ny(x) = ay(x+1) - (x+a)y(x) + xy(x-1), \quad y(x) = C_n(x;a). \tag{9.14.5}$$

Forward Shift Operator

$$C_n(x+1;a) - C_n(x;a) = -\frac{n}{a}C_{n-1}(x;a) \tag{9.14.6}$$

or equivalently

$$\Delta C_n(x;a) = -\frac{n}{a}C_{n-1}(x;a). \tag{9.14.7}$$

Backward Shift Operator

$$C_n(x;a) - \frac{x}{a}C_n(x-1;a) = C_{n+1}(x;a) \tag{9.14.8}$$

or equivalently

$$\nabla\left[\frac{a^x}{x!}C_n(x;a)\right] = \frac{a^x}{x!}C_{n+1}(x;a). \tag{9.14.9}$$

Rodrigues-Type Formula

$$\frac{a^x}{x!}C_n(x;a) = \nabla^n\left[\frac{a^x}{x!}\right]. \tag{9.14.10}$$

Generating Function

$$\mathrm{e}^t\left(1-\frac{t}{a}\right)^x = \sum_{n=0}^{\infty}\frac{C_n(x;a)}{n!}t^n. \tag{9.14.11}$$

Limit Relations

Meixner $\rightarrow$ Charlier

If we take $c = (a+\beta)^{-1}a$ in the definition (9.10.1) of the Meixner polynomials and let $\beta \rightarrow \infty$ we find the Charlier polynomials:

$$\lim_{\beta\rightarrow\infty} M_n(x;\beta,(a+\beta)^{-1}a) = C_n(x;a).$$

Krawtchouk $\rightarrow$ Charlier

The Charlier polynomials can be found from the Krawtchouk polynomials given by (9.11.1) by taking $p = N^{-1}a$ and letting $N \rightarrow \infty$:

$$\lim_{N\rightarrow\infty} K_n(x;N^{-1}a,N) = C_n(x;a).$$

Charlier $\rightarrow$ Hermite

The Hermite polynomials given by (9.15.1) are obtained from the Charlier polynomials if we set $x \rightarrow (2a)^{1/2}x+a$ and let $a \rightarrow \infty$. In fact we have

$$\lim_{a\rightarrow\infty} (2a)^{\frac{1}{2}n} C_n((2a)^{\frac{1}{2}}x+a;a) = (-1)^n H_n(x). \tag{9.14.12}$$

Remark

The Charlier polynomials are related to the Laguerre polynomials given by (9.12.1) in the following way:

$$\frac{(-a)^n}{n!} C_n(x;a) = L_n^{(x-n)}(a).$$

References

[7], [11], [16], [20], [22], [34], [39], [46], [59], [72], [80], [96], [98], [146], [147], [184], [198], [226], [228], [252], [265], [279], [317], [340], [353], [363], [364], [365], [372], [375], [377], [381], [391], [397], [406], [416], [417], [492], [493], [506], [521], [523].

9.15 Hermite

Hypergeometric Representation

$$H_n(x) = (2x)^n \, {}_2F_0 \left(\begin{matrix} -n/2, -(n-1)/2 \\ - \end{matrix} ; -\frac{1}{x^2} \right). \tag{9.15.1}$$

Orthogonality Relation

$$\frac{1}{\sqrt{\pi}} \int_{-\infty}^{\infty} e^{-x^2} H_m(x) H_n(x) \, dx = 2^n n! \, \delta_{mn}. \tag{9.15.2}$$

Recurrence Relation

$$H_{n+1}(x) - 2xH_n(x) + 2nH_{n-1}(x) = 0. \tag{9.15.3}$$

Normalized Recurrence Relation

$$xp_n(x) = p_{n+1}(x) + \frac{n}{2} p_{n-1}(x), \tag{9.15.4}$$

where

$$H_n(x) = 2^n p_n(x).$$

Differential Equation

$$y''(x) - 2xy'(x) + 2ny(x) = 0, \quad y(x) = H_n(x). \tag{9.15.5}$$

Forward Shift Operator

$$\frac{d}{dx} H_n(x) = 2nH_{n-1}(x). \tag{9.15.6}$$

Backward Shift Operator

$$\frac{d}{dx}H_n(x) - 2xH_n(x) = -H_{n+1}(x) \tag{9.15.7}$$

or equivalently

$$\frac{d}{dx}\left[\mathrm{e}^{-x^2}H_n(x)\right] = -\mathrm{e}^{-x^2}H_{n+1}(x). \tag{9.15.8}$$

Rodrigues-Type Formula

$$\mathrm{e}^{-x^2}H_n(x) = (-1)^n\left(\frac{d}{dx}\right)^n\left[\mathrm{e}^{-x^2}\right]. \tag{9.15.9}$$

Generating Functions

$$\exp\left(2xt - t^2\right) = \sum_{n=0}^{\infty}\frac{H_n(x)}{n!}t^n. \tag{9.15.10}$$

$$\begin{cases} \mathrm{e}^t\cos(2x\sqrt{t}) = \displaystyle\sum_{n=0}^{\infty}\frac{(-1)^n}{(2n)!}H_{2n}(x)t^n \\ \dfrac{\mathrm{e}^t}{\sqrt{t}}\sin(2x\sqrt{t}) = \displaystyle\sum_{n=0}^{\infty}\frac{(-1)^n}{(2n+1)!}H_{2n+1}(x)t^n. \end{cases} \tag{9.15.11}$$

$$\begin{cases} \mathrm{e}^{-t^2}\cosh(2xt) = \displaystyle\sum_{n=0}^{\infty}\frac{H_{2n}(x)}{(2n)!}t^{2n} \\ \mathrm{e}^{-t^2}\sinh(2xt) = \displaystyle\sum_{n=0}^{\infty}\frac{H_{2n+1}(x)}{(2n+1)!}t^{2n+1}. \end{cases} \tag{9.15.12}$$

$$\begin{cases} (1+t^2)^{-\gamma}\,{}_1F_1\left(\begin{matrix}\gamma \\ \frac{1}{2}\end{matrix}; \dfrac{x^2t^2}{1+t^2}\right) = \displaystyle\sum_{n=0}^{\infty}\frac{(\gamma)_n}{(2n)!}H_{2n}(x)t^{2n} \\ \dfrac{xt}{\sqrt{1+t^2}}\,{}_1F_1\left(\begin{matrix}\gamma+\frac{1}{2} \\ \frac{3}{2}\end{matrix}; \dfrac{x^2t^2}{1+t^2}\right) = \displaystyle\sum_{n=0}^{\infty}\frac{(\gamma+\frac{1}{2})_n}{(2n+1)!}H_{2n+1}(x)t^{2n+1} \end{cases} \tag{9.15.13}$$

with γ arbitrary.

$$\frac{1+2xt+4t^2}{(1+4t^2)^{\frac{3}{2}}}\exp\left(\frac{4x^2t^2}{1+4t^2}\right) = \sum_{n=0}^{\infty}\frac{H_n(x)}{\lfloor n/2\rfloor!}t^n, \tag{9.15.14}$$

where $\lfloor n/2\rfloor$ denotes the largest integer smaller than or equal to $n/2$.

Limit Relations

Meixner-Pollaczek → Hermite

If we take $x \to (\sin\phi)^{-1}(x\sqrt{\lambda} - \lambda\cos\phi)$ in the definition (9.7.1) of the Meixner-Pollaczek polynomials and then let $\lambda \to \infty$ we obtain the Hermite polynomials:

$$\lim_{\lambda\to\infty} \lambda^{-\frac{1}{2}n} P_n^{(\lambda)}((\sin\phi)^{-1}(x\sqrt{\lambda} - \lambda\cos\phi);\phi) = \frac{H_n(x)}{n!}.$$

Jacobi → Hermite

The Hermite polynomials follow from the Jacobi polynomials given by (9.8.1) by taking $\beta = \alpha$ and letting $\alpha \to \infty$ in the following way:

$$\lim_{\alpha\to\infty} \alpha^{-\frac{1}{2}n} P_n^{(\alpha,\alpha)}(\alpha^{-\frac{1}{2}}x) = \frac{H_n(x)}{2^n n!}.$$

Gegenbauer / Ultraspherical → Hermite

The Hermite polynomials follow from the Gegenbauer (or ultraspherical) polynomials given by (9.8.19) by taking $\lambda = \alpha + \frac{1}{2}$ and letting $\alpha \to \infty$ in the following way:

$$\lim_{\alpha\to\infty} \alpha^{-\frac{1}{2}n} C_n^{(\alpha+\frac{1}{2})}(\alpha^{-\frac{1}{2}}x) = \frac{H_n(x)}{n!}.$$

Krawtchouk → Hermite

The Hermite polynomials follow from the Krawtchouk polynomials given by (9.11.1) by setting $x \to pN + x\sqrt{2p(1-p)N}$ and then letting $N \to \infty$:

$$\lim_{N\to\infty} \sqrt{\binom{N}{n}} K_n(pN + x\sqrt{2p(1-p)N}; p, N) = \frac{(-1)^n H_n(x)}{\sqrt{2^n n! \left(\frac{p}{1-p}\right)^n}}.$$

Laguerre → Hermite

The Hermite polynomials can be obtained from the Laguerre polynomials given by (9.12.1) by taking the limit $\alpha \to \infty$ in the following way:

$$\lim_{\alpha\to\infty}\left(\frac{2}{\alpha}\right)^{\frac{1}{2}n}L_n^{(\alpha)}((2\alpha)^{\frac{1}{2}}x+\alpha)=\frac{(-1)^n}{n!}H_n(x).$$

Charlier → Hermite

If we set $x \to (2a)^{1/2}x + a$ in the definition (9.14.1) of the Charlier polynomials and let $a \to \infty$ we find the Hermite polynomials. In fact we have

$$\lim_{a\to\infty}(2a)^{\frac{1}{2}n}C_n((2a)^{\frac{1}{2}}x+a;a)=(-1)^nH_n(x).$$

Remarks

The Hermite polynomials can also be written as:

$$\frac{H_n(x)}{n!}=\sum_{k=0}^{\lfloor n/2\rfloor}\frac{(-1)^k(2x)^{n-2k}}{k!\,(n-2k)!},$$

where $\lfloor n/2\rfloor$ denotes the largest integer smaller than or equal to $n/2$.
The Hermite polynomials and the Laguerre polynomials given by (9.12.1) are also connected by the following quadratic transformations:

$$H_{2n}(x)=(-1)^n n!\,2^{2n}L_n^{(-\frac{1}{2})}(x^2)$$

and

$$H_{2n+1}(x)=(-1)^n n!\,2^{2n+1}xL_n^{(\frac{1}{2})}(x^2).$$

References

[2], [11], [16], [19], [20], [34], [35], [38], [41], [46], [51], [58], [72], [88], [99], [105], [109], [111], [112], [134], [136], [146], [153], [156], [165], [166], [177], [198], [202], [245], [246], [250], [251], [253], [267], [277], [295], [317], [340], [355], [361], [365], [381], [382], [390], [396], [403], [406], [416], [417], [424], [427], [456], [466], [479], [484], [493], [496], [497], [501], [506], [510], [520].

Part II

Classical q-Orthogonal Polynomials

Chapter 10
Orthogonal Polynomial Solutions of q-Difference Equations

Classical q-Orthogonal Polynomials I

10.1 Polynomial Solutions of q-Difference Equations

In the case of the q-derivative operator $\mathscr{D}_q := \mathscr{A}_{q,0}$, we have to deal with (cf. (2.2.7)):

$$\begin{aligned}(ex^2+2fx+g)\left(\mathscr{D}_q^2 y_n\right)(x)+(2\varepsilon x+\gamma)\left(\mathscr{D}_q y_n\right)(x)\\ =\frac{[n]}{q^n}(e[n-1]+2\varepsilon)y_n(qx),\end{aligned} \tag{10.1.1}$$

for $n=0,1,2,\ldots$ and $q\in\mathbb{R}\setminus\{-1,0,1\}$. In the symmetric form (cf. (2.2.12)) this reads

$$\begin{aligned}C(x)y_n(qx)-\{C(x)+D(x)\}\,y_n(x)+D(x)y_n(q^{-1}x)\\ =\frac{q^{-n}-1}{(q-1)^2}\left\{e(1-q^{n-1})+2\varepsilon(1-q)\right\}y_n(x)\end{aligned}$$

with (cf. (2.2.13))

$$C(x)=\frac{ex^2+2fqx+gq^2}{q(q-1)^2x^2}$$
$$\text{and}\quad D(x)=\frac{\{e+2\varepsilon(1-q)\}x^2+\{2f+\gamma(1-q)\}\,qx+gq^2}{(q-1)^2x^2}.$$

Hence

R. Koekoek et al., *Hypergeometric Orthogonal Polynomials and Their q-Analogues*,
Springer Monographs in Mathematics,
DOI 10.1007/978-3-642-05014-5_10,

$$\begin{aligned}
&\left(ex^2+2fqx+gq^2\right)y_n(qx)\\
&\quad-\left(\{e(1+q)+2\varepsilon q(1-q)\}x^2\right.\\
&\qquad\left.+\{2f(1+q)+\gamma q(1-q)\}qx+gq^2(1+q)\right)y_n(x)\\
&\quad+\left(\{e+2\varepsilon(1-q)\}x^2+\{2f+\gamma(1-q)\}qx+gq^2\right)qy_n(q^{-1}x)\\
&=(q^{-n}-1)\left\{e(1-q^{n-1})+2\varepsilon(1-q)\right\}qx^2y_n(x),\quad n=0,1,2,\ldots. \qquad (10.1.2)
\end{aligned}$$

The regularity condition (2.3.3) implies that $\varepsilon\neq 0$.

It will turn out to be convenient to introduce

$$\alpha:=e+2\varepsilon(1-q)\quad\text{and}\quad\beta:=2f+\gamma(1-q). \qquad (10.1.3)$$

Then (10.1.2) can be written as

$$\begin{aligned}
&\left(ex^2+2fqx+gq^2\right)y_n(qx)-\left\{(e+\alpha q)x^2+(2f+\beta q)qx+gq^2(1+q)\right\}y_n(x)\\
&\quad+\left(\alpha x^2+\beta qx+gq^2\right)qy_n(q^{-1}x)\\
&=(q^{-n}-1)\left(\alpha-eq^{n-1}\right)qx^2y_n(x),\quad n=0,1,2,\ldots \qquad (10.1.4)
\end{aligned}$$

with $e,f,g,\alpha,\beta\in\mathbb{C}$. Note that, in view of the homogeneity, one of the coefficients can be chosen arbitrarily. Furthermore, without loss of generality we may assume that $e\in\mathbb{R}$. In section 10.4 we will see that this implies that all coefficients e,f,g,α and β must be real.

10.2 The Basic Hypergeometric Representation

Since $\omega=0$, we have (cf. (2.4.2))

$$\begin{bmatrix}x;c\\0\end{bmatrix}:=1\quad\text{and}\quad\begin{bmatrix}x;c\\k\end{bmatrix}:=\prod_{i=1}^{k}\frac{x+cq^{i-1}}{[i]},\quad k=1,2,3,\ldots.$$

Now we have

$$\prod_{i=1}^{k}\frac{x+cq^{i-1}}{[i]}=(1-q)^kx^k\prod_{i=1}^{k}\frac{1+cx^{-1}q^{i-1}}{1-q^i}=\frac{(-cx^{-1};q)_k}{(q;q)_k}(1-q)^kx^k.$$

Therefore we try to find polynomial solutions of the form (cf. (2.4.3))

$$y_n(x)=\sum_{k=0}^{n}a_{n,k}\frac{(-cx^{-1};q)_k}{(q;q)_k}(1-q)^kx^k,\quad a_{n,n}\neq 0,\quad n=0,1,2,\ldots. \qquad (10.2.1)$$

Substitution of (10.2.1) into (10.1.4) leads to a three-term recurrence relation for the coefficients $\{a_{n,k}\}_{k=0}^{n}$ (cf. (2.4.4)). If c satisfies the relation (cf. (2.4.5))

$$\alpha c^2 - \beta qc + gq^2 = 0 \tag{10.2.2}$$

the recurrence relation reduces to the two-term recurrence relation (cf. (2.4.6))

$$[n-k]\left(\alpha - eq^{n+k-1}\right)a_{n,k} = -\left\{c(\alpha - eq^{2k}) - q(\beta - 2fq^k)\right\}q^{n-k-1}a_{n,k+1}$$

for $k = n-1, n-2, n-3, \ldots, 0$. For $c = 0$ (and therefore $g = 0$) this can be written as

$$[n-k]\left(\alpha - eq^{n+k-1}\right)a_{n,k} = \left(\beta - 2fq^k\right)q^{n-k}a_{n,k+1} \tag{10.2.3}$$

for $k = n-1, n-2, n-3, \ldots, 0$, and for $c \neq 0$ this can be written as

$$[n-k]\left(\alpha - eq^{n+k-1}\right)ca_{n,k} = \left(eq^{2k-2}c^2 - 2fq^{k-1}c + g\right)q^{n-k+1}a_{n,k+1} \tag{10.2.4}$$

for $k = n-1, n-2, n-3, \ldots, 0$. Now we have by using (10.1.3)

$$e[n] + 2\varepsilon = \frac{\alpha - eq^n}{1-q}, \quad n = 0, 1, 2, \ldots.$$

The regularity condition (2.3.3) holds if and only if all eigenvalues in (2.2.6) are different. In that case we have $\alpha - eq^{n+k-1} \neq 0$ which implies that the coefficients $\{a_{n,k}\}_{k=0}^n$ are determined uniquely in terms of $a_{n,n} \neq 0$.

Now we will distinguish between two different cases: $g = 0$ and $g \neq 0$ (cf. [385]).

For $g = 0$ the condition (10.2.2) can be satisfied for $c = 0$. Then the representation (10.2.1) reads

$$y_n(x) = \sum_{k=0}^n a_{n,k}\frac{(1-q)^k}{(q;q)_k}x^k = \sum_{k=0}^n a_{n,k}\frac{x^k}{[k]!}, \quad a_{n,n} \neq 0, \quad n = 0, 1, 2, \ldots.$$

Note that this is equivalent to the second approach (2.4.12). In that case the coefficients $\{a_{n,k}\}_{k=0}^n$ satisfy the two-term recurrence relation (10.2.3). Hence the coefficients $\{a_{n,k}\}_{k=0}^n$ are determined uniquely in terms of $a_{n,n} \neq 0$. In fact we have

$$a_{n,k} = \left(\prod_{i=1}^{n-k}\frac{q^i\left(\beta - 2fq^{n-i}\right)}{[i]\left(\alpha - eq^{2n-i-1}\right)}\right)a_{n,n}, \quad k = 0, 1, 2, \ldots, n-1.$$

Now we have

$$\prod_{i=1}^{n-k} q^i = q^{\binom{n-k+1}{2}} = q^{(n-k)(n-k+1)/2}$$

and by using (1.8.16)

$$\prod_{i=1}^{n-k}\frac{1}{[i]} = \frac{(1-q)^{n-k}}{(q;q)_{n-k}} = \frac{(q^{-n};q)_k}{(q;q)_n}(-1)^k(1-q)^{n-k}q^{-\binom{k}{2}+nk}. \tag{10.2.5}$$

In order to find monic polynomials we choose $a_{n,n} = [n]! = (q;q)_n/(1-q)^n$. Then we have from (10.2.1)

$$y_n(x) = q^{n(n+1)/2} \sum_{k=0}^{n} \frac{(q^{-n};q)_k}{(q;q)_k} \left(\prod_{i=1}^{n-k} \frac{\beta - 2fq^{n-i}}{\alpha - eq^{2n-i-1}} \right) (-x)^k \tag{10.2.6}$$

for $n = 0,1,2,\ldots$.

Case I. $g=0$, $\beta=0$ and $\alpha=0$. Since $\beta=0$ we have $f \neq 0$ and since $\alpha=0$ we have $e \neq 0$. Hence by using (10.2.6), this leads to the basic hypergeometric representation

$$\begin{aligned} y_n^{(I)}(x;q) &= \left(\frac{2f}{e}\right)^n q^{-n(n-3)/2} \sum_{k=0}^{n} \frac{(q^{-n};q)_k}{(q;q)_k} \left(\frac{-eq^{n-1}x}{2f}\right)^k \\ &= \left(\frac{2f}{e}\right)^n q^{-n(n-3)/2} {}_1\phi_0\left(\begin{matrix} q^{-n} \\ - \end{matrix}; q, -\frac{eq^{n-1}x}{2f}\right), \quad n=0,1,2,\ldots. \end{aligned}$$

The q-polynomials in this class have, besides q, in fact only one free parameter e/f or f/e. Later we will see that there are no *orthogonal* polynomial solutions in this case.

Case II. $g=0$, $\beta=0$ and $\alpha \neq 0$. Since $\beta=0$ we have $f \neq 0$. Hence by using (10.2.6), this leads to the basic hypergeometric representation

$$\begin{aligned} y_n^{(II)}(x;q) &= \left(-\frac{2fq^n}{\alpha}\right)^n \frac{1}{(e\alpha^{-1}q^{n-1};q)_n} \sum_{k=0}^{n} \frac{(q^{-n};q)_k (e\alpha^{-1}q^{n-1};q)_k}{(q;q)_k} q^{-\binom{k}{2}} \left(\frac{\alpha x}{2f}\right)^k \\ &= \left(-\frac{2fq^n}{\alpha}\right)^n \frac{1}{(e\alpha^{-1}q^{n-1};q)_n} {}_2\phi_0\left(\begin{matrix} q^{-n}, e\alpha^{-1}q^{n-1} \\ - \end{matrix}; q, -\frac{\alpha x}{2f}\right) \end{aligned}$$

for $n=0,1,2,\ldots$. The q-polynomials in this class have, besides q, two free parameters e/α and f/α. A special case of q-polynomials in this class is the family of **Stieltjes-Wigert** polynomials without free parameters.

Case III. $g=0$, $\beta \neq 0$ and $\alpha=0$. Since $\alpha=0$ we have $e \neq 0$. Hence by using (10.2.6), this leads to the basic hypergeometric representation

$$\begin{aligned} y_n^{(III)}(x;q) &= \left(-\frac{\beta}{e}\right)^n q^{-n(n-2)} (2f\beta^{-1};q)_n \sum_{k=0}^{n} \frac{(q^{-n};q)_k}{(2f\beta^{-1};q)_k (q;q)_k} q^{\binom{k}{2}} \left(\frac{eq^{n-1}x}{\beta}\right)^k \\ &= \left(-\frac{\beta}{e}\right)^n q^{-n(n-2)} (2f\beta^{-1};q)_n \, {}_1\phi_1\left(\begin{matrix} q^{-n} \\ 2f\beta^{-1} \end{matrix}; q, \frac{-eq^{n-1}x}{\beta}\right) \end{aligned}$$

for $n=0,1,2,\ldots$. The q-polynomials in this class have, besides q, two free parameters e/β and f/β. A special case of q-polynomials in this class is the family of **q-Laguerre** polynomials with only one free parameter.

Another special case of q-polynomials in this class is again the family of **Stieltjes-Wigert** polynomials without free parameters. Sometimes families of q-orthogonal polynomials belong to different classes for $0 < q < 1$ or $q > 1$. For instance, in the case of the Stieltjes-Wigert polynomials we have by using (1.8.7): if

$$S_n(x;q) = \frac{1}{(q;q)_n}\,{}_1\phi_1\left(\begin{matrix} q^{-n} \\ 0 \end{matrix}\;;q,-q^{n+1}x\right), \quad 0<q<1,$$

then

$$S_n(x;q^{-1}) = \frac{(-1)^n q^{n+\binom{n}{2}}}{(q;q)_n}\,{}_2\phi_0\left(\begin{matrix} q^{-n},0 \\ - \end{matrix}\;;q,-x\right), \quad q>1.$$

Case IV. $g=0$, $\beta \neq 0$ and $\alpha \neq 0$. By using (10.2.6), this leads to the basic hypergeometric representation

$$\begin{aligned} y_n^{(IV)}(x;q) &= \left(\frac{\beta}{\alpha}\right)^n q^{n(n+1)/2}\frac{(2f\beta^{-1};q)_n}{(e\alpha^{-1}q^{n-1};q)_n} \\ &\quad\times \sum_{k=0}^{n}\frac{(q^{-n};q)_k(e\alpha^{-1}q^{n-1};q)_k}{(2f\beta^{-1};q)_k(q;q)_k}\left(-\frac{\alpha x}{\beta}\right)^k \\ &= \left(\frac{\beta}{\alpha}\right)^n q^{n(n+1)/2}\frac{(2f\beta^{-1};q)_n}{(e\alpha^{-1}q^{n-1};q)_n}\,{}_2\phi_1\left(\begin{matrix} q^{-n},e\alpha^{-1}q^{n-1} \\ 2f\beta^{-1} \end{matrix}\;;q,-\frac{\alpha x}{\beta}\right) \end{aligned}$$

for $n=0,1,2,\ldots$. The q-polynomials in this class have, besides q, three free parameters e/α, f/β and β/α. Special cases of q-polynomials in this class are the **little q-Jacobi** polynomials with two free parameters and the **little q-Laguerre** and the **q-Bessel** polynomials with only one free parameter.

If $g \neq 0$ we cannot use (10.2.2) in the case that both $\alpha = 0$ and $\beta = 0$. In that case we use the second approach (2.4.12). Since $\omega = 0$, we have $v = c(1-q)$, and therefore (cf. (2.4.11))

$$\begin{aligned} \begin{bmatrix} x;c+[k-2]vq^{2-k} \\ k \end{bmatrix} &= \prod_{i=1}^{k}\frac{x+c+[k-i-1]vq^{i+1-k}}{[i]} \\ &= \prod_{i=1}^{k}\frac{x+c+c(1-q^{k-i-1})q^{i+1-k}}{[i]} \\ &= \prod_{i=1}^{k}\frac{x+cq^{i+1-k}}{[i]} = \frac{(-cx^{-1}q^{2-k};q)_k}{(q;q)_k}(1-q)^k x^k. \end{aligned}$$

Hence, the representation (2.4.12) reads

$$y_n(x) = \sum_{k=0}^{n} b_{n,k}\frac{(-cx^{-1}q^{2-k};q)_k}{(q;q)_k}(1-q)^k x^k, \quad b_{n,n}\neq 0, \quad n=0,1,2,\ldots.$$

Since $\alpha = 0$ we have $e \neq 0$, which implies that

$$ec^2 - 2fc + g = 0 \tag{10.2.7}$$

can be solved for $c \neq 0$. This implies that we have the two-term recurrence relation (cf. (2.4.13))

$$[n-k]\,(e[n+k-1]+2\varepsilon)\,b_{n,k} + \Big\{(e[k]+2\varepsilon)([k+1]q^{-k}-1)v \\ + (2[k](ec-f)+2\varepsilon c-\gamma)\Big\}\,q^{n-k}b_{n,k+1} = 0, \quad k = n-1, n-2, n-3, \ldots, 0.$$

Now we use (10.1.3) and the fact that $\alpha = 0$, $\beta = 0$ and $v = c(1-q)$ to obtain

$$eq^{n+k-1}(1-q^{n-k})b_{n,k} = (2f-ec)(1-q)q^n b_{n,k+1}, \quad k = n-1, n-2, n-3, \ldots, 0$$

or equivalently, by using (10.2.7),

$$ecq^{k-1}(1-q^{n-k})b_{n,k} = g(1-q)b_{n,k+1}, \quad k = n-1, n-2, n-3, \ldots, 0.$$

This implies that

$$b_{n,k} = \left(\frac{g}{ec}\right)^{n-k}\left(\prod_{i=1}^{n-k}\frac{1}{[i]q^{n-i-1}}\right)b_{n,n}, \quad k = 0, 1, 2, \ldots, n-1.$$

Now we have as before by using (1.8.16)

$$\prod_{i=1}^{n-k}\frac{1}{[i]q^{n-i-1}} = q^{-(n-1)(n-k)}\prod_{i=1}^{n-k}\frac{q^i}{[i]} = q^{-(n-1)(n-k)}\frac{(1-q)^{n-k}q^{(n-k)(n-k+1)/2}}{(q;q)_{n-k}} \\ = \frac{(q^{-n};q)_k}{(q;q)_n}(-1)^k(1-q)^{n-k}q^{-n(n-3)/2+(n-1)k}.$$

In order to find monic polynomials we choose $b_{n,n} = (q;q)_n/(1-q)^n$. Hence we have

$$b_{n,k} = \left(\frac{g}{ec}\right)^{n-k}\frac{(q^{-n};q)_k}{(1-q)^k}(-1)^k q^{-n(n-3)/2+(n-1)k}, \quad k = 0, 1, 2, \ldots, n.$$

By using (1.8.14) this leads to the representation

$$y_n(x) = \left(\frac{g}{ec}\right)^n q^{-n(n-3)/2} \\ \times \sum_{k=0}^{n}\frac{(q^{-n};q)_k(-c^{-1}xq^{-1};q)_k}{(q;q)_k}(-1)^k q^{-\binom{k}{2}}\left(\frac{ec^2q^n}{g}\right)^k \tag{10.2.8}$$

for $n = 0, 1, 2, \ldots$.

Case V. $g \neq 0$, $\alpha = 0$ and $\beta = 0$. Since $\alpha = 0$ we have $e \neq 0$. We also have $c \neq 0$ and by using (10.2.8) this leads to the basic hypergeometric representation

$$\begin{aligned} y_n^{(V)}(x;q) &= \left(\frac{g}{ec}\right)^n q^{-n(n-3)/2} \sum_{k=0}^{n} \frac{(q^{-n};q)_k(-c^{-1}xq^{-1};q)_k}{(q;q)_k} (-1)^k q^{-\binom{k}{2}} \left(\frac{ec^2q^n}{g}\right)^k \\ &= \left(\frac{g}{ec}\right)^n q^{-n(n-3)/2} {}_2\phi_0\left(\begin{matrix} q^{-n}, -c^{-1}xq^{-1} \\ - \end{matrix}; q, \frac{ec^2q^n}{g}\right), \quad n = 0,1,2,\ldots. \end{aligned}$$

The q-polynomials in this class have, besides q, two free parameters g/e (or e/g) and $c(\neq 0)$, a solution of (10.2.7). A special case of q-polynomials in this class is the family of **discrete q-Hermite II** polynomials with no free parameters.

If $g \neq 0$ and $(\alpha,\beta) \neq (0,0)$ we use the representation (10.2.1). In that case we also have $c \neq 0$ in view of (10.2.2). Then we have by using (10.2.4)

$$a_{n,k} = c^{k-n} \left(\prod_{i=1}^{n-k} \frac{eq^{2n-i-1}c^2 - 2fq^nc + gq^{i+1}}{[i]\,(\alpha - eq^{2n-i-1})}\right) a_{n,n}, \quad k = 0,1,2,\ldots,n-1.$$

Now we have

$$\begin{aligned} &\prod_{i=1}^{n-k} \left\{eq^{2n-i-1}c^2 - 2fq^nc + gq^{i+1}\right\} \\ &\quad = g^{n-k} q^{(n-k)(n-k+3)/2} \prod_{i=1}^{n-k} \left\{1 - \frac{2f}{g}q^{n-i-1}c + \frac{e}{g}q^{2n-2i-2}c^2\right\} \\ &\quad = g^{n-k} q^{(n-k)(n-k+3)/2} \prod_{i=1}^{n-k} \left\{\left(1 + c\gamma_1 q^{n-i-1}\right)\left(1 + c\gamma_2 q^{n-i-1}\right)\right\}, \end{aligned}$$

where $\gamma_1\gamma_2 = e/g$ and $\gamma_1 + \gamma_2 = -2f/g$. Hence we have

$$g\gamma_i^2 + 2f\gamma_i + e = 0, \quad i = 1,2.$$

Later we will see that e, f and g must be real , which implies that γ_1 and γ_2 are both real in the case that $f^2 \geq eg$, id est

$$\gamma_1 = \frac{-f - \sqrt{f^2 - eg}}{g} \quad \text{and} \quad \gamma_2 = \frac{-f + \sqrt{f^2 - eg}}{g},$$

and that they are complex conjugates in the case that $f^2 < eg$. As before we choose $a_{n,n} = [n]! = (q;q)_n/(1-q)^n$ in order to get monic polynomials. Then we have, by using (10.2.5), from (10.2.1)

$$y_n(x) = \left(\frac{g}{c}\right)^n q^{n(n+3)/2} \sum_{k=0}^{n} \frac{(q^{-n};q)_k(-cx^{-1};q)_k}{(q;q)_k} \times \left(\prod_{i=1}^{n-k} \frac{\left(1+c\gamma_1 q^{n-i-1}\right)\left(1+c\gamma_2 q^{n-i-1}\right)}{\alpha - eq^{2n-i-1}}\right)\left(-\frac{cx}{gq}\right)^k \tag{10.2.9}$$

for $n = 0,1,2,\ldots$.

Case VI. $g \neq 0$, $\alpha = 0$ and $\beta \neq 0$. Since $\alpha = 0$ we have $e \neq 0$. Note that in this case (10.2.2) implies that $c = gq/\beta$. Hence by using (10.2.9), this leads to the basic hypergeometric representation

$$\begin{aligned} y_n^{(VI)}(x;q) &= \left(-\frac{g}{ec}\right)^n q^{-n(n-3)} \left(-c\gamma_1 q^{-1}, -c\gamma_2 q^{-1};q\right)_n \\ &\quad \times \sum_{k=0}^{n} \frac{(q^{-n};q)_k(-cx^{-1};q)_k}{(-c\gamma_1 q^{-1}, -c\gamma_2 q^{-1};q)_k\,(q;q)_k} q^{\binom{k}{2}} \left(\frac{ecq^{n-2}x}{g}\right)^k \\ &= \left(-\frac{\beta}{eq}\right)^n q^{-n(n-3)} \left(-g\beta^{-1}\gamma_1, -g\beta^{-1}\gamma_2;q\right)_n \\ &\quad \times {}_2\phi_2\left(\begin{matrix} q^{-n}, -g\beta^{-1}qx^{-1} \\ -g\beta^{-1}\gamma_1, -g\beta^{-1}\gamma_2 \end{matrix}; q, -\frac{eq^{n-1}x}{\beta}\right) \end{aligned}$$

for $n = 0,1,2,\ldots$. The q-polynomials in this class have, besides q, three free parameters e/β, f/β and g/β. By using the fact that $\gamma_1\gamma_2 = e/g$ we may apply (1.13.5) to find that

$$y_n^{(VI)}(x;q) = \left(-\frac{\beta}{\gamma_1\gamma_2 g}\right)^n q^{-n(n-2)} \left(-g\beta^{-1}\gamma_1;q\right)_n \, {}_2\phi_1\left(\begin{matrix} q^{-n}, \gamma_1 q^{-1}x \\ -g\beta^{-1}\gamma_1 \end{matrix}; q, -\frac{g\gamma_2 q^n}{\beta}\right)$$

for $n = 0,1,2,\ldots$.

Case VII. $g \neq 0$ and $\alpha \neq 0$. By using (10.2.9), this leads to the basic hypergeometric representation

$$\begin{aligned} y_n^{(VII)}(x;q) &= \left(\frac{g}{c\alpha}\right)^n q^{n(n+3)/2} \frac{\left(-c\gamma_1 q^{-1}, -c\gamma_2 q^{-1};q\right)_n}{(e\alpha^{-1}q^{n-1};q)_n} \\ &\quad \times \sum_{k=0}^{n} \frac{(q^{-n};q)_k(-cx^{-1};q)_k(e\alpha^{-1}q^{n-1};q)_k}{(-c\gamma_1 q^{-1}, -c\gamma_2 q^{-1};q)_k\,(q;q)_k} \left(-\frac{c\alpha x}{gq}\right)^k \\ &= \left(\frac{g}{c\alpha}\right)^n q^{n(n+3)/2} \frac{\left(-c\gamma_1 q^{-1}, -c\gamma_2 q^{-1};q\right)_n}{(e\alpha^{-1}q^{n-1};q)_n} \\ &\quad \times {}_3\phi_2\left(\begin{matrix} q^{-n}, -cx^{-1}, e\alpha^{-1}q^{n-1} \\ -c\gamma_1 q^{-1}, -c\gamma_2 q^{-1} \end{matrix}; q, -\frac{c\alpha x}{gq}\right) \end{aligned}$$

for $n = 0, 1, 2, \ldots$, where c satisfies (10.2.2). The q-polynomials in this class have, besides q, four free parameters e/α, f/α, g/α and β/α. Special cases of q-polynomials in this class are the **big q-Jacobi** polynomials with three free parameters, the **big q-Laguerre** polynomials with two free parameters, the **Al-Salam-Carlitz I and II** polynomials with one free parameter and the **discrete q-Hermite I and II** polynomials with no free parameters.

We will see that we have:

Theorem 10.1. *All orthogonal polynomial solutions $y_n(x)$ of the q-difference equation (10.1.4)*

$$\begin{aligned}
&\left(ex^2+2fqx+gq^2\right)y_n(qx)\\
&\qquad -\left\{(e+\alpha q)x^2+(2f+\beta q)qx+gq^2(1+q)\right\}y_n(x)\\
&\qquad\qquad +\left(\alpha x^2+\beta qx+gq^2\right)qy_n(q^{-1}x)\\
&=\left(q^{-n}-1\right)\left(\alpha-eq^{n-1}\right)qx^2y_n(x),\quad n=0,1,2,\ldots,
\end{aligned}$$

where $\alpha = e + 2\varepsilon(1-q)$ and $\beta = 2f + \gamma(1-q)$, can be divided into six different cases:

Case II. *$g = 0$, $\beta = 0$ and $\alpha \neq 0$*

Case III. *$g = 0$, $\beta \neq 0$ and $\alpha = 0$*

Case IV. *$g = 0$, $\beta \neq 0$ and $\alpha \neq 0$*

Case V. *$g \neq 0$, $\alpha = 0$ and $\beta = 0$*

Case VI. *$g \neq 0$, $\alpha = 0$ and $\beta \neq 0$*

Case VII. *$g \neq 0$ and $\alpha \neq 0$.*

In chapters 4 and 5 we have indicated that weight functions for classical orthogonal polynomials are connected to certain probability distributions in stochastics. In the case of classical q-orthogonal polynomials weight functions are connected to certain (discrete) q-distributions in probability and statistics as well. Examples of q-distributions of this kind are the Euler, Heine and Kemp distributions, which appear in models of specific processes in physics, biology and mathematical economy. See for instance [309], [310] and [311].

10.3 The Three-Term Recurrence Relation

In section 2.6 we have seen that the monic polynomial solutions $\{y_n\}_{n=0}^{\infty}$ of (10.1.1) satisfy the three-term recurrence relation

$$y_{n+1}(x) = (x - c_n)y_n(x) - d_n y_{n-1}(x), \quad n = 1,2,3,\ldots, \tag{10.3.1}$$

with initial values $y_0(x) = 1$ and $y_1(x) = x - c_0$, where (cf. (2.6.11)) $c_0 = -\gamma q/2\varepsilon$,

$$c_n = -\frac{q^n\left\{(e[n-1]+2\varepsilon)\left(2f[n](1+q)+\gamma q\right) - e\gamma[n+1]q^{n-1}\right\}}{(e[2n-2]+2\varepsilon)(e[2n]+2\varepsilon)}, \quad n = 1,2,3,\ldots$$

and (cf. (2.6.12))

$$\begin{aligned} d_n = &\frac{q^{n+1}[n](e[n-2]+2\varepsilon)}{(e[2n-3]+2\varepsilon)(e[2n-2]+2\varepsilon)^2(e[2n-1]+2\varepsilon)} \\ &\times\left\{q^{n-1}(2f[n-1]+\gamma)\left(2f\{e[n-1]+2\varepsilon\} - q^{n-1}e\gamma\right)\right. \\ &\left. -g(e[2n-2]+2\varepsilon)^2\right\}, \quad n = 1,2,3,\ldots. \end{aligned}$$

By using (10.1.3) we obtain $c_0 = (2f-\beta)q/(\alpha - e)$,

$$\begin{aligned} c_n = &-\frac{2fq^n\left\{\alpha - \alpha(1+q)q^n + eq^{2n-1}\right\}}{(\alpha - eq^{2n-2})(\alpha - eq^{2n})} \\ &-\frac{\beta q^{n+1}\left\{\alpha - e(1+q)q^{n-2} + eq^{2n-1}\right\}}{(\alpha - eq^{2n-2})(\alpha - eq^{2n})}, \quad n = 1,2,3,\ldots \end{aligned} \tag{10.3.2}$$

and

$$\begin{aligned} d_n = &\frac{q^{n+1}(1-q^n)(\alpha - eq^{n-2})}{(\alpha - eq^{2n-3})(\alpha - eq^{2n-2})^2(\alpha - eq^{2n-1})} \\ &\times\left\{q^{n-1}(\beta - 2fq^{n-1})(2f\alpha - e\beta q^{n-1}) - g(\alpha - eq^{2n-2})^2\right\} \end{aligned} \tag{10.3.3}$$

for $n = 1,2,3,\ldots$. In the case that $g = 0$ we have $c_0 = A_0$, $c_n = A_n + C_n$ and $d_n = A_{n-1}C_n$ for $n = 1,2,3,\ldots$ where

$$A_{n-1} = -\frac{q^n(\alpha - eq^{n-2})(\beta - 2fq^{n-1})}{(\alpha - eq^{2n-3})(\alpha - eq^{2n-2})}, \quad n = 1,2,3,\ldots$$

and

$$C_n = -\frac{q^n(1-q^n)(2f\alpha - e\beta q^{n-1})}{(\alpha - eq^{2n-2})(\alpha - eq^{2n-1})}, \quad n = 1,2,3,\ldots.$$

10.4 Classification of the Positive-Definite Orthogonal Polynomial Solutions

Again we will use Favard's theorem (theorem 3.1) to conclude that there exist positive definite orthogonal polynomial solutions of (10.1.1) if and only if $c_n \in \mathbb{R}$ for all $n = 0, 1, 2, \ldots$ and $d_n > 0$ for all $n = 1, 2, 3, \ldots$. First of all we write instead of (10.3.2)

$$c_n = -\frac{q^n}{(\alpha - eq^{2n-2})(\alpha - eq^{2n})} \times \left\{(\alpha + eq^{2n-1})(2f + \beta q) - q^{n-1}(1+q)(2f\alpha q + e\beta)\right\},$$

which holds for $n = 0, 1, 2, \ldots$. Note that

$$c_0 = \frac{(2f - \beta)q}{\alpha - e} \quad \text{and} \quad c_1 = \frac{2f - \beta}{\alpha - e}\left\{1 - \frac{\alpha(1+q)(1-q^2)}{\alpha - eq^2}\right\} - \frac{\beta q(1-q^2)}{\alpha - eq^2}.$$

Further we write for (10.3.3)

$$d_n = q^{n+1}(1-q^n)D_n^{(1)}D_n^{(2)}, \quad n = 1, 2, 3, \ldots \tag{10.4.1}$$

with

$$D_n^{(1)} = \frac{\alpha - eq^{n-2}}{(\alpha - eq^{2n-3})(\alpha - eq^{2n-2})^2(\alpha - eq^{2n-1})}, \quad n = 1, 2, 3, \ldots \tag{10.4.2}$$

and

$$\begin{aligned} D_n^{(2)} &= q^{n-1}(\beta - 2fq^{n-1})(2f\alpha - e\beta q^{n-1}) - g(\alpha - eq^{2n-2})^2 \\ &= -\alpha(\beta - 2fq^{n-1})^2 + \beta(\alpha - eq^{2n-2})(\beta - 2fq^{n-1}) \\ &\quad - g(\alpha - eq^{2n-2})^2, \quad n = 1, 2, 3, \ldots. \end{aligned} \tag{10.4.3}$$

Note that

$$d_1 = -\frac{q^2(1-q)}{(\alpha - e)^2(\alpha - eq)}\left\{\alpha(\beta - 2f)^2 - \beta(\alpha - e)(\beta - 2f) + g(\alpha - e)^2\right\}.$$

For $\alpha \neq 0$ we may write

$$\begin{aligned} D_n^{(2)} &= -\alpha\left\{\beta - 2fq^{n-1} + \delta_1(\alpha - eq^{2n-2})\right\}\left\{\beta - 2fq^{n-1} + \delta_2(\alpha - eq^{2n-2})\right\} \\ &= -\alpha\left(2fq^{n-1} + \delta_1 eq^{2n-2} + \delta_2\alpha\right)\left(2fq^{n-1} + \delta_1\alpha + \delta_2 eq^{2n-2}\right) \end{aligned} \tag{10.4.4}$$

for $n = 1, 2, 3, \ldots$, where $\delta_1\delta_2 = g/\alpha$ and $\delta_1 + \delta_2 = -\beta/\alpha$. Hence we have

$$\alpha\delta_i^2+\beta\delta_i+g=0,\quad i=1,2.$$

Note that $\delta_1,\delta_2\in\mathbb{R}$ in the case that $\beta^2\geq 4g\alpha$ and that $\delta_1,\delta_2\in\mathbb{C}$ with $\delta_2=\overline{\delta_1}$ in the case that $\beta^2<4g\alpha$.

We also distinguish between the cases $\alpha=0$ and $\alpha\neq 0$ in the q-difference equation (10.1.4).

For $\alpha=0$ we have

$$\begin{aligned}&\left(ex^2+2fqx+gq^2\right)y_n(qx)-\left\{ex^2+(2f+\beta q)qx+gq^2(1+q)\right\}y_n(x)\\&\quad+\left(\beta qx+gq^2\right)qy_n(q^{-1}x)\\&=eq^n(1-q^{-n})x^2y_n(x),\quad n=0,1,2,\ldots.\end{aligned}\tag{10.4.5}$$

In view of the homogeneity, one of the coefficients e, f, g and β can be chosen arbitrarily. We remark that we have $e\neq 0$ in view of (10.1.3) and the fact that $\varepsilon\neq 0$. Further we have

$$D_n^{(1)}=-\frac{q^{-7n+6}}{e^3}\quad\text{and}\quad D_n^{(2)}=-eq^{2n-2}\left\{\beta(\beta-2fq^{n-1})+egq^{2n-2}\right\},$$

which implies that

$$d_n=q^{n+1}(1-q^n)D_n^{(1)}D_n^{(2)}=\frac{q^{n+1}(1-q^n)}{e^2q^{5n-4}}\left\{\beta(\beta-2fq^{n-1})+egq^{2n-2}\right\}$$

for $n=1,2,3,\ldots$. Note that

$$d_1=q(1-q)\left\{\frac{\beta}{e}\left(\frac{\beta}{e}-\frac{2f}{e}\right)+\frac{g}{e}\right\}.$$

Hence for $\alpha=0$ we conclude that $c_0\in\mathbb{R}$ implies that $(\beta-2f)/e\in\mathbb{R}$ and $c_1\in\mathbb{R}$ further implies that also $\beta/e\in\mathbb{R}$. Therefore we also have $2f/e\in\mathbb{R}$. Finally, we conclude that $d_1>0$ further implies that $g/e\in\mathbb{R}$. In view of the homogeneity of (10.4.5) we may assume that $e\in\mathbb{R}$ without loss of generality. This implies that in (10.4.5) all coefficients e, f, g and β are real.

For $\alpha\neq 0$ the q-difference equation (10.1.4) can be divided by α. Then we obtain

$$\begin{aligned}&\left(e'x^2+2f'qx+g'q^2\right)y_n(qx)-\left\{(e'+q)x^2+(2f'+\beta'q)qx+g'q^2(1+q)\right\}y_n(x)\\&\quad+\left(x^2+\beta'qx+g'q^2\right)qy_n(q^{-1}x)\\&=(q^{-n}-1)\left(1-e'q^{n-1}\right)qx^2y_n(x),\quad n=0,1,2,\ldots\end{aligned}\tag{10.4.6}$$

with

$$e'=\frac{e}{\alpha},\quad f'=\frac{f}{\alpha},\quad g'=\frac{g}{\alpha}\quad\text{and}\quad\beta'=\frac{\beta}{\alpha}.$$

Note that this q-difference equation is equivalent to (10.1.4) with $\alpha = 1$ and e, f, g and β replaced by e', f', g' and β', respectively. In the sequel we will avoid the notation e', f', g', β' by simply setting $\alpha = 1$ and using e, f, g, β instead of e', f', g', β'. We remark that through $\alpha = 1$ the arbitrary choice of one of the parameters in (10.1.4) has been made. Then we have

$$c_n = -\frac{q^n}{(1-eq^{2n-2})(1-eq^{2n})}\left\{(1+eq^{2n-1})(2f+\beta q) - q^{n-1}(1+q)(2fq+e\beta)\right\}$$

for $n = 0, 1, 2, \ldots$. Without loss of generality, we may assume that $e \in \mathbb{R}$. Note that

$$c_0 = \frac{(2f-\beta)q}{1-e} \quad \text{and} \quad c_1 = \frac{2f-\beta}{1-e}\left\{1 - \frac{(1+q)(1-q^2)}{1-eq^2}\right\} - \frac{\beta q(1-q^2)}{1-eq^2}.$$

Further we have

$$d_1 = -\frac{q^2(1-q)}{1-eq}\left\{\left(\frac{\beta-2f}{1-e}\right)^2 - \frac{\beta(\beta-2f)}{1-e} + g\right\}.$$

Hence $c_0 \in \mathbb{R}$ implies that $2f - \beta \in \mathbb{R}$ and $c_1 \in \mathbb{R}$ further implies that also $\beta \in \mathbb{R}$. Therefore we also have $f \in \mathbb{R}$. Finally, we conclude that $d_1 > 0$ further implies that $g \in \mathbb{R}$.

The sign of $q^{n+1}(1-q^n)$ for $n = 1, 2, 3, \ldots$ is given in table 10.1 and for $\alpha = 1$ the sign of $D_n^{(1)}$ for $n = 1, 2, 3, \ldots$ is given in table 10.2.

$\mathbf{q}$	$q < -1$	$-1 < q < 0$	$0 < q < 1$	$q > 1$
$\mathbf{q^{n+1}(1-q^n)}$	$+$	$(-1)^{n+1}$	$+$	$-$

Table 10.1 sign of $q^{n+1}(1-q^n)$, $n = 1, 2, 3, \ldots$

Case I. $g = 0$, $\beta = 0$ and $\alpha = 0$. In this case we have $d_n = 0$ for all $n = 1, 2, 3, \ldots$, which implies that there exists no positive-definite orthogonal polynomials.

Case II. $g = 0$, $\beta = 0$ and $\alpha \neq 0$. In this case we set $\alpha = 1$ and by using (10.4.3) we conclude that

$$D_n^{(2)} = -4f^2q^{2n-2}, \quad n = 1, 2, 3, \ldots.$$

This implies, once more, that we must have $f \neq 0$. Hence by using table 10.1 and table 10.2, we conclude that we have positive-definite orthogonality for an infinite system of polynomials only in one case:

Case IIa1. $q > 1$, $f \neq 0$ and $e \leq 0$.

q	**extra conditions**	$\mathbf{D_n^{(1)}}$	**for**
$q<-1$	$eq>1$	$(-1)^n$	$n=1,2,3,\dots$
	$0<eq<1$ with $q<eq^{2N}\leq q^{-1}$	$+$	$n=1,2,3,\dots,N$
	$e=0$	$+$	$n=1,2,3,\dots$
	$0<e<1$ with $eq^{2N}=1$	$+$	$n=1,2,3,\dots,N$
	$0<e<1$ with $1<eq^{2N}<q^2$	$+$	$n=1,2,3,\dots,2N+1$
	$e>1$	$(-1)^{n+1}$	$n=1,2,3,\dots$
$-1<q<0$	$eq>1$ with $q^{-1}\leq eq^{2N}<q$	$(-1)^n$	$n=1,2,3,\dots,N$
	$q<eq<1$	$+$	$n=1,2,3,\dots$
	$e>1$ with $eq^{2N}=1$	$(-1)^{n+1}$	$n=1,2,3,\dots,N$
	$e>1$ with $q^2<eq^{2N}<1$	$(-1)^{n+1}$	$n=1,2,3,\dots,2N+1$
$0<q<1$	$e<1$	$+$	$n=1,2,3,\dots$
	$e>1$ with $q<eq^{2N}\leq q^{-1}$	$-$	$n=1,2,3,\dots,N$
$q>1$	$e\leq 0$	$+$	$n=1,2,3,\dots$
	$0<e<1$ with $q^{-1}\leq eq^{2N}<q$	$+$	$n=1,2,3,\dots,N$
	$e>1$	$-$	$n=1,2,3,\dots$

Table 10.2 sign of $D_n^{(1)}$, $\alpha=1$ and $N\in\{1,2,3,\dots\}$

It is also possible to have positive-definite orthogonality for finite systems of $N+1$ polynomials with $N\in\{1,2,3,\dots\}$ in the following three cases:

Case IIb1. $-1<q<0$, $f\neq 0$ and $eq>1$ with $q^{-1}\leq eq^{2N}<q$.

Case IIb2. $0<q<1$, $f\neq 0$ and $e>1$ with $q<eq^{2N}\leq q^{-1}$.

Case IIb3. $q>1$, $f\neq 0$ and $0<e<1$ with $q^{-1}\leq eq^{2N}<q$.

Case III. $g=0$, $\beta\neq 0$ and $\alpha=0$. In this case we have $e\neq 0$ and

$$d_n=\frac{q^{n+1}(1-q^n)}{q^{5n-4}}\left(\frac{\beta}{e}\right)^2\left(1-\frac{2f}{\beta}q^{n-1}\right),\quad n=1,2,3,\dots.$$

By using table 10.1 we conclude that for $q<-1$ we must have

$$(-1)^n\left(1-\frac{2f}{\beta}q^{n-1}\right)>0.$$

For $n=1$ this reads

$$-\left(1-\frac{2f}{\beta}\right)>0 \quad \Longleftrightarrow \quad \frac{2f}{\beta}>1.$$

Since $q<-1$ this implies that

$$\frac{2f}{\beta}q<q<-1<1 \quad \Longrightarrow \quad 1-\frac{2f}{\beta}q>0.$$

Hence, $(-1)^n\left(1-\frac{2f}{\beta}q^{n-1}\right)>0$ also holds for $n=2$. Further we have $q^2>1$ which implies that

$$1<\frac{2f}{\beta}<\frac{2f}{\beta}q^2<\frac{2f}{\beta}q^4<\ldots$$

and

$$\ldots<\frac{2f}{\beta}q^5<\frac{2f}{\beta}q^3<\frac{2f}{\beta}q<q<-1<1.$$

Hence, if $(-1)^n\left(1-\frac{2f}{\beta}q^{n-1}\right)>0$ is true for $n=1$, then it holds for all $n=1,2,3,\ldots$.

For $-1<q<0$ we must have $1-\frac{2f}{\beta}q^{n-1}<0$. This cannot be true for both $n=1$ and $n=2$, since $\frac{2f}{\beta}>1$ and $\frac{2f}{\beta}q>1$ cannot hold simultaneously if $-1<q<0$.

For $0<q<1$ we must have $1-\frac{2f}{\beta}q^{n-1}>0$. For $n=1$ this reads

$$1-\frac{2f}{\beta}>0 \quad \Longleftrightarrow \quad \frac{2f}{\beta}<1.$$

In the case that $0\leq\frac{2f}{\beta}<1$ we then have

$$\ldots\leq\frac{2f}{\beta}q^2\leq\frac{2f}{\beta}q\leq\frac{2f}{\beta}<1$$

and in the case that $\frac{2f}{\beta}\leq 0$ we conclude that

$$\frac{2f}{\beta}\leq\frac{2f}{\beta}q\leq\frac{2f}{\beta}q^2\leq\ldots\leq 0<1.$$

Hence if $1-\frac{2f}{\beta}q^{n-1}>0$ is true for $n=1$, then it holds for all $n=1,2,3,\ldots$.

For $q > 1$ we must have $1 - \frac{2f}{\beta} q^{n-1} < 0$. For $n = 1$ this reads

$$1 - \frac{2f}{\beta} < 0 \quad \Longleftrightarrow \quad \frac{2f}{\beta} > 1,$$

which implies that

$$1 < \frac{2f}{\beta} < \frac{2f}{\beta} q < \frac{2f}{\beta} q^2 < \ldots.$$

Hence if $1 - \frac{2f}{\beta} q^{n-1} < 0$ is true for $n = 1$, then it holds for all $n = 1, 2, 3, \ldots$.

Hence, we conclude that we have positive-definite orthogonality in the following three infinite cases:

Case IIIa1. $q < -1$ and $\frac{2f}{\beta} > 1$.

Case IIIa2. $0 < q < 1$ and $\frac{2f}{\beta} < 1$.

Case IIIa3. $q > 1$ and $\frac{2f}{\beta} > 1$.

We also conclude that it is impossible to have positive-definite orthogonality for finite systems of $N + 1$ polynomials with $N \in \{1, 2, 3, \ldots\}$.

Case IV. $g = 0$, $\beta \neq 0$ and $\alpha \neq 0$. In this case we set $\alpha = 1$ and by using (10.4.3) we have

$$D_n^{(2)} = \beta^2 q^{n-1} \left(1 - \frac{2f}{\beta} q^{n-1}\right) \left(\frac{2f}{\beta} - e q^{n-1}\right), \quad n = 1, 2, 3, \ldots. \tag{10.4.7}$$

For $f = 0$ we must have $e \neq 0$ and

$$D_n^{(2)} = -e\beta^2 q^{2n-2}, \quad n = 1, 2, 3, \ldots.$$

Then we conclude by using (10.4.1), table 10.1 and table 10.2 that $d_n > 0$ for all $n = 1, 2, 3, \ldots$ is only possible in the case that $0 < q < 1$ and $e < 0$.

Now we assume that $f \neq 0$. In order to study the positivity of d_n we need the signs of the factors $1 - \frac{2f}{\beta} q^{n-1}$ and $\frac{2f}{\beta} - e q^{n-1}$. Therefore the sign of $A q^{n-1}$ for $n = 1, 2, 3, \ldots$ is given in table 10.3.

q	**A**	$\mathbf{A\,q^{n-1}}$
$q<-1$	$A>0$	$\ldots<Aq^5<Aq^3<Aq<0<A<Aq^2<Aq^4<\ldots$
	$A<0$	$\ldots<Aq^4<Aq^2<A<0<Aq<Aq^3<Aq^5<\ldots$
$-1<q<0$	$A>0$	$Aq<Aq^3<Aq^5<\ldots<0<\ldots<Aq^4<Aq^2<A$
	$A<0$	$A<Aq^2<Aq^4<\ldots<0<\ldots<Aq^5<Aq^3<Aq$
$0<q<1$	$A>0$	$0<\ldots<Aq^2<Aq<A$
	$A<0$	$A<Aq<Aq^2<\ldots<0$
$q>1$	$A>0$	$0<A<Aq<Aq^2<\ldots$
	$A<0$	$\ldots<Aq^2<Aq<A<0$

Table 10.3 sign of Aq^{n-1}, $n=1,2,3,\ldots$

By using (10.4.1), (10.4.2), (10.4.7), table 10.1 and table 10.2 we conclude that in order to have $d_n>0$ for all $n=1,2,3,\ldots$ we must have that the sign of $D_n^{(2)}$ for $n=1,2,3,\ldots$ should be as in table 10.4.

q	**extra conditions**	$\mathbf{D_n^{(2)}}$
$q<-1$	$eq>1$	$(-1)^n$
	$e=0$	$+$
	$e>1$	$(-1)^{n+1}$
$-1<q<0$	$q<eq<1$	$(-1)^{n+1}$
$0<q<1$	$e<1$	$+$
$q>1$	$e\le 0$	$-$
	$e>1$	$+$

Table 10.4 sign of $D_n^{(2)}$ needed for $d_n>0$ for all $n=1,2,3,\ldots$

In the case $q<-1$ and $e=0$ we have

$$d_n=q^{n+1}(1-q^n)\beta^2\left(1-\frac{2f}{\beta}q^{n-1}\right)\frac{2f}{\beta}q^{n-1},\quad n=1,2,3,\ldots.$$

For $\frac{2f}{\beta} > 0$ it follows from $d_1 > 0$ that $1 - \frac{2f}{\beta} > 0$ or equivalently $\frac{2f}{\beta} < 1$. Hence $0 < \frac{2f}{\beta} < 1$. However, $d_2 > 0$ then requires that $1 - \frac{2f}{\beta}q < 0$ or equivalently $\frac{2f}{\beta}q > 1$, which cannot be true.

For $\frac{2f}{\beta} < 0$ it follows from $d_1 > 0$ that $1 - \frac{2f}{\beta} < 0$ or equivalently $\frac{2f}{\beta} > 1$, which cannot be true.

By using (10.4.7) and table 10.4 we conclude that for all other cases the product

$$\left(1 - \frac{2f}{\beta}q^{n-1}\right)\left(\frac{2f}{\beta} - eq^{n-1}\right)$$

should be either positive or negative for all $n = 1, 2, 3, \ldots$.

Now we first consider the sign of the factor $1 - \frac{2f}{\beta}q^{n-1}$ for $n = 1, 2, 3, \ldots$ with the aid of table 10.3.

1. For $q < -1$ and $\frac{2f}{\beta} > 0$ we cannot have that $1 - \frac{2f}{\beta}q^{n-1} > 0$ for all $n = 1, 2, 3, \ldots$ since $q^{2n} \to \infty$ for $n \to \infty$.
2. For $q < -1$ and $\frac{2f}{\beta} < 0$ we cannot have that $1 - \frac{2f}{\beta}q^{n-1} > 0$ for all $n = 1, 2, 3, \ldots$ since $q^{2n+1} \to -\infty$ for $n \to \infty$.
3. For $q < -1$ and $\frac{2f}{\beta} > 0$ we cannot have that $1 - \frac{2f}{\beta}q^{n-1} < 0$ for all $n = 1, 2, 3, \ldots$ since $q^{2n+1} \to -\infty$ for $n \to \infty$.
4. For $q < -1$ and $\frac{2f}{\beta} < 0$ we cannot have that $1 - \frac{2f}{\beta}q^{n-1} < 0$ for all $n = 1, 2, 3, \ldots$ since $q^{2n} \to \infty$ for $n \to \infty$.
5. For $-1 < q < 0$ and $\frac{2f}{\beta} > 0$ it is possible to have $1 - \frac{2f}{\beta}q^{n-1} > 0$ for all $n = 1, 2, 3, \ldots$ provided that $\frac{2f}{\beta} < 1$ since $q^n \to 0$ for $n \to \infty$. Hence we need that $0 < \frac{2f}{\beta} < 1$ in this case.

6. For $-1 < q < 0$ and $\frac{2f}{\beta} < 0$ it is possible to have $1 - \frac{2f}{\beta}q^{n-1} > 0$ for all $n = 1,2,3,\ldots$ provided that $\frac{2f}{\beta}q < 1$ since $q^n \to 0$ for $n \to \infty$. Hence we need that $q^{-1} < \frac{2f}{\beta} < 0$ in this case.

7. For $-1 < q < 0$ and $\frac{2f}{\beta} > 0$ we cannot have that $1 - \frac{2f}{\beta}q^{n-1} < 0$ for all $n = 1,2,3,\ldots$ since $q^n \to 0$ for $n \to \infty$.

8. For $-1 < q < 0$ and $\frac{2f}{\beta} < 0$ we cannot have that $1 - \frac{2f}{\beta}q^{n-1} < 0$ for all $n = 1,2,3,\ldots$ since $q^n \to 0$ for $n \to \infty$.

9. For $0 < q < 1$ and $\frac{2f}{\beta} > 0$ it is possible to have $1 - \frac{2f}{\beta}q^{n-1} > 0$ for all $n = 1,2,3,\ldots$ provided that $\frac{2f}{\beta} < 1$ since $q^n \to 0$ for $n \to \infty$. Hence we need that $0 < \frac{2f}{\beta} < 1$ in this case.

10. For $0 < q < 1$ and $\frac{2f}{\beta} < 0$ it is possible to have $1 - \frac{2f}{\beta}q^{n-1} > 0$ for all $n = 1,2,3,\ldots$. In fact, for $\frac{2f}{\beta} < 0$ this is always true since $q^n \to 0$ for $n \to \infty$.

11. For $0 < q < 1$ and $\frac{2f}{\beta} > 0$ we cannot have that $1 - \frac{2f}{\beta}q^{n-1} < 0$ for all $n = 1,2,3,\ldots$ since $q^n \to 0$ for $n \to \infty$.

12. For $0 < q < 1$ and $\frac{2f}{\beta} < 0$ we cannot have that $1 - \frac{2f}{\beta}q^{n-1} < 0$ for all $n = 1,2,3,\ldots$ since $q^n \to 0$ for $n \to \infty$.

13. For $q > 1$ and $\frac{2f}{\beta} > 0$ we cannot have that $1 - \frac{2f}{\beta}q^{n-1} > 0$ for all $n = 1,2,3,\ldots$ since $q^n \to \infty$ for $n \to \infty$.

14. For $q > 1$ and $\frac{2f}{\beta} < 0$ it is possible to have $1 - \frac{2f}{\beta}q^{n-1} > 0$ for all $n = 1,2,3,\ldots$. In fact, for $\frac{2f}{\beta} < 0$ this is always true since $q^n \to \infty$ for $n \to \infty$.

15. For $q>1$ and $\frac{2f}{\beta}>0$ it is possible to have $1-\frac{2f}{\beta}q^{n-1}<0$ for all $n=1,2,3,\ldots$ provided that $\frac{2f}{\beta}>1$ since $q^n\to\infty$ for $n\to\infty$. Hence we need that $\frac{2f}{\beta}>1$ in this case.

16. For $q>1$ and $\frac{2f}{\beta}<0$ we cannot have that $1-\frac{2f}{\beta}q^{n-1}<0$ for all $n=1,2,3,\ldots$ since $q^n\to\infty$ for $n\to\infty$.

For the factor $\frac{2f}{\beta}-eq^{n-1}$ we only need to consider the six possible cases 5, 6, 9, 10, 14 and 15 above. Again we use table 10.3.

1. For $-1<q<0$, $0<\frac{2f}{\beta}<1$ and $e>0$ we must have $\frac{2f}{\beta}-eq^{n-1}>0$ for all $n=1,2,3,\ldots$ which is possible provided that $\frac{2f}{\beta}>e$. Hence we need that $0<e<\frac{2f}{\beta}<1$ in this case.

2. For $-1<q<0$, $0<\frac{2f}{\beta}<1$ and $e\leq 0$ we must have $\frac{2f}{\beta}-eq^{n-1}>0$ for all $n=1,2,3,\ldots$ which is possible provided that $\frac{2f}{\beta}>eq$. Hence we need that $0\leq eq<\frac{2f}{\beta}<1$ in this case.

3. For $-1<q<0$, $q^{-1}<\frac{2f}{\beta}<0$ and $e>0$ we must have $\frac{2f}{\beta}-eq^{n-1}>0$ for all $n=1,2,3,\ldots$ which is not possible since $q^n\to 0$ for $n\to\infty$.

4. For $-1<q<0$, $q^{-1}<\frac{2f}{\beta}<0$ and $e\leq 0$ we must have $\frac{2f}{\beta}-eq^{n-1}>0$ for all $n=1,2,3,\ldots$ which is not possible since $q^n\to 0$ for $n\to\infty$.

5. For $0<q<1$, $0<\frac{2f}{\beta}<1$ and $e>0$ we must have $\frac{2f}{\beta}-eq^{n-1}>0$ for all $n=1,2,3,\ldots$ which is possible provided that $\frac{2f}{\beta}>e$. Hence we need that $0<e<\frac{2f}{\beta}<1$ in this case.

6. For $0 < q < 1$, $0 < \dfrac{2f}{\beta} < 1$ and $e \leq 0$ we must have $\dfrac{2f}{\beta} - eq^{n-1} > 0$ for all $n = 1,2,3,\ldots$ which is possible. In fact, for $0 < \dfrac{2f}{\beta} < 1$ and $e \leq 0$ this is always true.

7. For $0 < q < 1$, $\dfrac{2f}{\beta} < 0$ and $e > 0$ we must have $\dfrac{2f}{\beta} - eq^{n-1} > 0$ for all $n = 1,2,3,\ldots$ which is not possible since $q^n \to 0$ for $n \to \infty$.

8. For $0 < q < 1$, $\dfrac{2f}{\beta} < 0$ and $e \leq 0$ we must have $\dfrac{2f}{\beta} - eq^{n-1} > 0$ for all $n = 1,2,3,\ldots$ which is not possible since $q^n \to 0$ for $n \to \infty$.

9. For $q > 1$, $\dfrac{2f}{\beta} < 0$ and $e > 0$ we must have $\dfrac{2f}{\beta} - eq^{n-1} > 0$ for all $n = 1,2,3,\ldots$ which is not possible since $q^n \to \infty$ for $n \to \infty$.

10. For $q > 1$, $\dfrac{2f}{\beta} < 0$ and $e \leq 0$ we must have $\dfrac{2f}{\beta} - eq^{n-1} < 0$ for all $n = 1,2,3,\ldots$ which is not possible since $q^n \to \infty$ for $n \to \infty$.

11. For $q > 1$, $\dfrac{2f}{\beta} > 1$ and $e > 0$ we must have $\dfrac{2f}{\beta} - eq^{n-1} < 0$ for all $n = 1,2,3,\ldots$ which is possible provided that $\dfrac{2f}{\beta} < e$. Hence we need that $1 < \dfrac{2f}{\beta} < e$ in this case.

12. For $q > 1$, $\dfrac{2f}{\beta} > 1$ and $e \leq 0$ we must have $\dfrac{2f}{\beta} - eq^{n-1} > 0$ for all $n = 1,2,3,\ldots$ which is possible. In fact, for $\dfrac{2f}{\beta} > 1$ and $e \leq 0$ this is always true.

Hence, we conclude that we have positive-definite orthogonality in the following seven infinite cases:

Case IVa1. $-1 < q < 0$ and $0 < e < \dfrac{2f}{\beta} < 1$.

Case IVa2. $-1 < q < 0$ and $0 \leq eq < \dfrac{2f}{\beta} < 1$.

Case IVa3. $0 < q < 1$ and $0 < e < \dfrac{2f}{\beta} < 1$.

Case IVa4. $0 < q < 1$, $e \leq 0$ and $0 < \dfrac{2f}{\beta} < 1$.

Case IVa5. $0 < q < 1$, $e < 0$ and $f = 0$.

Case IVa6. $q > 1$ and $1 < \dfrac{2f}{\beta} < e$.

Case IVa7. $q > 1$, $e \leq 0$ and $\dfrac{2f}{\beta} > 1$.

It is also possible to have positive-definite orthogonality for finite systems of $N+1$ polynomials with $N \in \{1,2,3,\ldots\}$.

By using (10.4.1), (10.4.2), (10.4.7), table 10.1 and table 10.2 we conclude that in order to have $d_n > 0$ we must have that the sign of $D_n^{(2)}$ should be as in table 10.5.

q	**extra conditions**	$\mathbf{D_n^{(2)}}$	**for**
$q < -1$	$0 < eq < 1$ with $q < eq^{2N} \leq q^{-1}$	$+$	$n = 1,2,3,\ldots,N$
	$0 < e < 1$ with $eq^{2N} = 1$	$+$	$n = 1,2,3,\ldots,N$
	$0 < e < 1$ with $1 < eq^{2N} < q^2$	$+$	$n = 1,2,3,\ldots,2N+1$
$-1 < q < 0$	$eq > 1$ with $q^{-1} \leq eq^{2N} < q$	$-$	$n = 1,2,3,\ldots,N$
	$e > 1$ with $eq^{2N} = 1$	$+$	$n = 1,2,3,\ldots,N$
	$e > 1$ with $q^2 < eq^{2N} < 1$	$+$	$n = 1,2,3,\ldots,2N+1$
$0 < q < 1$	$e > 1$ with $q < eq^{2N} \leq q^{-1}$	$-$	$n = 1,2,3,\ldots,N$
$q > 1$	$0 < e < 1$ with $q^{-1} \leq eq^{2N} < q$	$-$	$n = 1,2,3,\ldots,N$

Table 10.5 sign of $D_n^{(2)}$ needed for $d_n > 0$ with $N \in \{1,2,3,\ldots\}$

By using (10.4.7) and table 10.5 we conclude that the sign of the product

$$\left(1 - \frac{2f}{\beta}q^{n-1}\right)\left(\frac{2f}{\beta} - eq^{n-1}\right)$$

should alternate in the cases that $q < -1$ or $-1 < q < 0$ and should be negative in the cases that $0 < q < 1$ or $q > 1$.

First of all we note that for $q < -1$ and for $-1 < q < 0$ it is possible that the factor $1 - \dfrac{2f}{\beta}q^{n-1}$ keeps the same sign for $n = 1,2,3,\ldots,N$ for certain $N \in \{1,2,3,\ldots\}$

and that the factor $\frac{2f}{\beta} - eq^{n-1}$ alternates for these values of n (in the case that $f = 0$ for instance).

As before, for $f = 0$ we must have $e \neq 0$ and

$$D_n^{(2)} = -e\beta^2 q^{2n-2}, \quad n = 1,2,3,\ldots.$$

By using table 10.5 we conclude that we only have $d_n > 0$ for $n = 1,2,3,\ldots,N$ in the cases that $q < -1$ and $0 < eq < 1$ with $q < eq^{2N} \leq q^{-1}$, or $0 < q < 1$ and $e > 1$ with $q < eq^{2N} \leq q^{-1}$, or $q > 1$ and $0 < e < 1$ with $q^{-1} \leq eq^{2N} < q$.

Now we assume that $f \neq 0$ and we consider the sign of the factor $1 - \frac{2f}{\beta}q^{n-1}$ for $n = 1,2,3,\ldots,N$ or $n = 1,2,3,\ldots,2N+1$ with the aid of table 10.3.

1. For $q < -1$ and $\frac{2f}{\beta} > 0$ it is possible to have $1 - \frac{2f}{\beta}q^{n-1} > 0$ for all $n = 1,2,3,\ldots,N$ provided that $\frac{2f}{\beta}q^{N-2} < 1$ and N even.
2. For $q < -1$ and $\frac{2f}{\beta} < 0$ it is possible to have $1 - \frac{2f}{\beta}q^{n-1} > 0$ for all $n = 1,2,3,\ldots,N$ provided that $\frac{2f}{\beta}q^{N-2} < 1$ and N odd.
3. For $q < -1$ and $\frac{2f}{\beta} > 0$ we cannot have that $1 - \frac{2f}{\beta}q^{n-1} < 0$ for all $n = 1,2,3,\ldots,N$ with $N \geq 2$ since $1 - \frac{2f}{\beta}q < 0$ then implies that $\frac{2f}{\beta} < q^{-1} < 0$.
4. For $q < -1$ and $\frac{2f}{\beta} < 0$ we cannot have that $1 - \frac{2f}{\beta}q^{n-1} < 0$ for all $n = 1,2,3,\ldots,N$ since then we have $1 - \frac{2f}{\beta} > 1 > 0$.
5. For $-1 < q < 0$ and $\frac{2f}{\beta} > 0$ we have: if $1 - \frac{2f}{\beta}q^{n-1} > 0$ is true for $n = 1$, which implies that $0 < \frac{2f}{\beta} < 1$, then it holds for all $n = 1,2,3,\ldots$.

6. For $-1 < q < 0$ and $\frac{2f}{\beta} < 0$ we have: if $1 - \frac{2f}{\beta}q^{n-1} > 0$ is true for $n = 1$ and $n = 2$, which implies that $q^{-1} < \frac{2f}{\beta} < 0$, then it holds for all $n = 1,2,3,\ldots$.

7. For $-1 < q < 0$ and $\frac{2f}{\beta} > 0$ we cannot have that $1 - \frac{2f}{\beta}q^{n-1} < 0$ for all $n = 1,2,3,\ldots,N$ with $N \geq 2$ since $1 - \frac{2f}{\beta}q < 0$ then implies that $\frac{2f}{\beta} < q^{-1} < 0$.

8. For $-1 < q < 0$ and $\frac{2f}{\beta} < 0$ we cannot have that $1 - \frac{2f}{\beta}q^{n-1} < 0$ for all $n = 1,2,3,\ldots,N$ since then we have $1 - \frac{2f}{\beta} > 1 > 0$.

9. For $0 < q < 1$ and $\frac{2f}{\beta} > 0$ we have: if $1 - \frac{2f}{\beta}q^{n-1} > 0$ is true for $n = 1$, which implies that $0 < \frac{2f}{\beta} < 1$, then it holds for all $n = 1,2,3,\ldots$.

10. For $0 < q < 1$ and $\frac{2f}{\beta} < 0$ we have: $1 - \frac{2f}{\beta}q^{n-1} > 0$ is true for all $n = 1,2,3,\ldots$.

11. For $0 < q < 1$ and $\frac{2f}{\beta} > 0$ it is possible to have $1 - \frac{2f}{\beta}q^{n-1} < 0$ for all $n = 1,2,3,\ldots,N$ provided that $\frac{2f}{\beta}q^{N-1} > 1$.

12. For $0 < q < 1$ and $\frac{2f}{\beta} < 0$ we have: $1 - \frac{2f}{\beta}q^{n-1} < 0$ is false for all $n = 1,2,3,\ldots$.

13. For $q > 1$ and $\frac{2f}{\beta} > 0$ it is possible to have $1 - \frac{2f}{\beta}q^{n-1} > 0$ for all $n = 1,2,3,\ldots,N$ provided that $\frac{2f}{\beta}q^{N-1} < 1$.

14. For $q > 1$ and $\frac{2f}{\beta} < 0$ we have: $1 - \frac{2f}{\beta}q^{n-1} > 0$ is true for all $n = 1,2,3,\ldots$.

15. For $q > 1$ and $\frac{2f}{\beta} > 0$ we have: if $1 - \frac{2f}{\beta}q^{n-1} < 0$ is true for $n = 1$, which implies that $\frac{2f}{\beta} > 1$, then it holds for all $n = 1,2,3,\ldots$.

16. For $q>1$ and $\dfrac{2f}{\beta}<0$ we have: $1-\dfrac{2f}{\beta}q^{n-1}<0$ is false for all $n=1,2,3,\ldots$.

For the factor $\dfrac{2f}{\beta}-eq^{n-1}$ we only need to consider the ten possible cases 1, 2, 5, 6, 9, 10, 11, 13, 14 and 15 above. Again we use table 10.3.

1. For $q<-1$, $\dfrac{2f}{\beta}>0$ and $\dfrac{2f}{\beta}q^{N-2}<1$ with N even, we conclude that the sign of $\dfrac{2f}{\beta}-eq^{n-1}$ equals $(-1)^n$ for all $n=1,2,3,\ldots,N$ provided that $e>0$ and $\dfrac{2f}{\beta}<e$ and that it equals $(-1)^{n+1}$ for all $n=1,2,3,\ldots,N$ provided that $e<0$ and $\dfrac{2f}{\beta}<eq$.

2. For $q<-1$, $\dfrac{2f}{\beta}<0$ and $\dfrac{2f}{\beta}q^{N-2}<1$ with N odd, we conclude that the sign of $\dfrac{2f}{\beta}-eq^{n-1}$ equals $(-1)^n$ for all $n=1,2,3,\ldots,N$ provided that $e>0$ and $\dfrac{2f}{\beta}>eq$ and that it equals $(-1)^{n+1}$ for all $n=1,2,3,\ldots,N$ provided that $e<0$ and $\dfrac{2f}{\beta}>e$.

3. For $-1<q<0$ and $0<\dfrac{2f}{\beta}<1$, we conclude that the sign of $\dfrac{2f}{\beta}-eq^{n-1}$ equals $(-1)^n$ for $n=1$ provided that $e>\dfrac{2f}{\beta}>0$. Then we use table 10.3 to conclude that for $(-1)^n\left(\dfrac{2f}{\beta}-eq^{n-1}\right)>0$ for all $n=1,2,3,\ldots,N$ we must have

$$eq<eq^3<eq^5<\ldots<0<\ldots<\frac{2f}{\beta}<eq^{N-1}<\ldots<eq^4<eq^2<e$$

with N odd.

4. For $-1<q<0$ and $q^{-1}<\dfrac{2f}{\beta}<0$, we conclude that the sign of $\dfrac{2f}{\beta}-eq^{n-1}$ equals $(-1)^{n+1}$ for $n=1$ provided that $e<\dfrac{2f}{\beta}<0$. Then we use table 10.3 to conclude that for $(-1)^{n+1}\left(\dfrac{2f}{\beta}-eq^{n-1}\right)>0$ for all $n=1,2,3,\ldots,N$ we must have

$$e<eq^2<eq^4<\ldots<eq^{N-1}<\frac{2f}{\beta}<\ldots<0<\ldots<eq^5<eq^3<eq$$

with N even.

5. For $0<q<1$ and $0<\dfrac{2f}{\beta}<1$, we conclude that $\dfrac{2f}{\beta}-eq^{n-1}<0$ for all $n=1,2,3,\ldots,N$ provided that $\dfrac{2f}{\beta}<eq^{N-1}$.

6. For $0<q<1$ and $\dfrac{2f}{\beta}<0$, we conclude that $\dfrac{2f}{\beta}-eq^{n-1}<0$ for all $n=1,2,3,\ldots,N$ provided that $\dfrac{2f}{\beta}<e$. In fact, then we have that $\dfrac{2f}{\beta}-eq^{n-1}<0$ for all $n=1,2,3,\ldots$.

7. For $0<q<1$, $\dfrac{2f}{\beta}>0$ and $\dfrac{2f}{\beta}q^{N-1}>1$, we conclude that $\dfrac{2f}{\beta}-eq^{n-1}>0$ for all $n=1,2,3,\ldots,N$ provided that $\dfrac{2f}{\beta}>e$. In fact, then we have that $\dfrac{2f}{\beta}-eq^{n-1}>0$ for all $n=1,2,3,\ldots$.

8. For $q>1$, $\dfrac{2f}{\beta}>0$ and $\dfrac{2f}{\beta}q^{N-1}<1$, we conclude that $\dfrac{2f}{\beta}-eq^{n-1}<0$ for all $n=1,2,3,\ldots,N$ provided that $\dfrac{2f}{\beta}<e$. In fact, then we have that $\dfrac{2f}{\beta}-eq^{n-1}<0$ for all $n=1,2,3,\ldots$.

9. For $q>1$ and $\dfrac{2f}{\beta}<0$, we conclude that $\dfrac{2f}{\beta}-eq^{n-1}<0$ for $n=1$ provided that $\dfrac{2f}{\beta}<e$. If $e>0$, which implies that $\dfrac{2f}{\beta}<0<e$, then we have that $\dfrac{2f}{\beta}-eq^{n-1}<0$ for all $n=1,2,3,\ldots$. If $e<0$, which implies that $\dfrac{2f}{\beta}<e<0$, then we have that $\dfrac{2f}{\beta}-eq^{n-1}<0$ for all $n=1,2,3,\ldots,N$ provided that $\dfrac{2f}{\beta}<eq^{N-1}$.

10. For $q>1$ and $\dfrac{2f}{\beta}>1$, we conclude that $\dfrac{2f}{\beta}-eq^{n-1}>0$ for all $n=1,2,3,\ldots,N$ provided that $\dfrac{2f}{\beta}>eq^{N-1}$.

Hence, we conclude that we have positive-definite orthogonality in the following eight finite cases:

Case IVb1. $q<-1$, $0<\dfrac{2f}{\beta}<eq<1$ with $q<eq^{2N}\leq q^{-1}$ and $0<\dfrac{2f}{\beta}q^{N-2}<1$ with N even. In that case we have a finite system of $N+1$ polynomials.

Case IVb2. $q < -1$, $q^{-1} < e < \dfrac{2f}{\beta} < 0$ with $q < eq^{2N} \leq q^{-1}$ and $0 < \dfrac{2f}{\beta}q^{N-2} < 1$ with N odd. In that case we have a finite system of $N+1$ polynomials.

Case IVb3. $0 < q < 1$, $e > 1$ with $q < eq^{2N} \leq q^{-1}$, $0 < \dfrac{2f}{\beta} < 1$ and $\dfrac{2f}{\beta} < eq^{N-1}$. In that case we have a finite system of $N+1$ polynomials.

Case IVb4. $0 < q < 1$, $e > 1$ with $q < eq^{2N} \leq q^{-1}$ and $\dfrac{2f}{\beta} < 0$. In that case we have a finite system of $N+1$ polynomials.

Case IVb5. $0 < q < 1$, $e > 1$ with $q < eq^{2N} \leq q^{-1}$, $\dfrac{2f}{\beta} > e$ and $\dfrac{2f}{\beta}q^{N-1} > 1$. In that case we have a finite system of $N+1$ polynomials.

Case IVb6. $q > 1$, $0 < e < 1$ with $q^{-1} \leq eq^{2N} < q$, $0 < \dfrac{2f}{\beta} < e$ and $\dfrac{2f}{\beta}q^{N-1} < 1$. In that case we have a finite system of $N+1$ polynomials.

Case IVb7. $q > 1$, $0 < e < 1$ with $q^{-1} \leq eq^{2N} < q$ and $\dfrac{2f}{\beta} < 0$. In that case we have a finite system of $N+1$ polynomials.

Case IVb8. $q > 1$, $0 < e < 1$ with $q^{-1} \leq eq^{2N} < q$, $\dfrac{2f}{\beta} > 1$ and $\dfrac{2f}{\beta} > eq^{N-1}$. In that case we have a finite system of $N+1$ polynomials.

Case V. $g \neq 0$, $\alpha = 0$ and $\beta = 0$. In this case we have $e \neq 0$ and

$$d_n = \frac{g}{e}\,\frac{q^{n+1}(1-q^n)}{q^{3n-2}}, \quad n = 1,2,3,\ldots.$$

By using table 10.1 we conclude that we have positive-definite orthogonality in the following three cases:

Case Va1. $-1 < q < 0$ and $\dfrac{g}{e} < 0$.

Case Va2. $0 < q < 1$ and $\dfrac{g}{e} > 0$.

Case Va3. $q > 1$ and $\dfrac{g}{e} < 0$.

We also conclude that it is impossible to have positive-definite orthogonality for finite systems of $N+1$ polynomials with $N \in \{1,2,3,\ldots\}$.

Case VI. $g \neq 0$, $\alpha = 0$ and $\beta \neq 0$. In this case we have $e \neq 0$ and

$$\begin{aligned} d_n &= \frac{q^{n+1}(1-q^n)}{q^{5n-4}} \left\{ \frac{\beta}{e}\left(\frac{\beta}{e} - \frac{2f}{e}q^{n-1}\right) + \frac{g}{e}q^{2n-2} \right\} \\ &= \frac{q^{n+1}(1-q^n)}{q^{5n-4}} \left(\frac{\beta}{e}\right)^2 \left(1 - \frac{2f}{\beta}q^{n-1} + \frac{eg}{\beta^2}q^{2n-2}\right) \\ &= \frac{q^{n+1}(1-q^n)}{q^{5n-4}} \left(\frac{\beta}{e}\right)^2 \left(1 + \frac{\gamma_1 g}{\beta}q^{n-1}\right)\left(1 + \frac{\gamma_2 g}{\beta}q^{n-1}\right), \quad n = 1,2,3,\ldots, \end{aligned}$$

with $\gamma_1 + \gamma_2 = -2f/g$ and $\gamma_1\gamma_2 = e/g$. We remark that for $f^2 < eg$ we have

$$1 - \frac{2f}{\beta}q^{n-1} + \frac{eg}{\beta^2}q^{2n-2} = \left(1 - \frac{f}{\beta}q^{n-1}\right)^2 + \frac{eg - f^2}{\beta^2}q^{2n-2} > 0$$

for all $n = 1,2,3,\ldots$. In that case γ_1 and γ_2 are complex conjugates. In the case that $f^2 \geq eg$ they are real:

$$\gamma_1 = \frac{-f - \sqrt{f^2 - eg}}{g} \quad \text{and} \quad \gamma_2 = \frac{-f + \sqrt{f^2 - eg}}{g}.$$

By using table 10.1 we conclude that for $q < -1$ we must have

$$(-1)^n \left(1 - \frac{2f}{\beta}q^{n-1} + \frac{eg}{\beta^2}q^{2n-2}\right) > 0,$$

which cannot be true for all $n = 1,2,3,\ldots$, since $(-1)^n q^{2n-2} \to -\infty$ for odd $n \to \infty$ and $(-1)^n q^{2n-2} \to \infty$ for even $n \to \infty$. Moreover, the constant eg is either positive or negative.

For $-1 < q < 0$ we must have

$$1 - \frac{2f}{\beta}q^{n-1} + \frac{eg}{\beta^2}q^{2n-2} < 0,$$

which cannot be true for all $n = 1,2,3,\ldots$, since $q^{n-1} \to 0$ for $n \to \infty$.

For $0 < q < 1$ we must have

$$1 - \frac{2f}{\beta}q^{n-1} + \frac{eg}{\beta^2}q^{2n-2} > 0,$$

which is true for all $n = 1,2,3,\ldots$ in the case that $f^2 < eg$. In the case that $f^2 \geq eg$ we must have

$$1 - \frac{2f}{\beta}q^{n-1} + \frac{eg}{\beta^2}q^{2n-2} = \left(1 + \frac{\gamma_1 g}{\beta}q^{n-1}\right)\left(1 + \frac{\gamma_2 g}{\beta}q^{n-1}\right) > 0.$$

Since $q^n \to 0$ for $n \to \infty$ this can only be true for all $n = 1,2,3,\ldots$ if both factors are positive, which implies that we must have

$$\frac{\gamma_1 g}{\beta} > -1 \quad \text{and} \quad \frac{\gamma_2 g}{\beta} > -1.$$

For $q > 1$ we must have

$$1 - \frac{2f}{\beta}q^{n-1} + \frac{eg}{\beta^2}q^{2n-2} < 0,$$

which is false in the case $f^2 < eg$. In the case that $f^2 \geq eg$ we must have

$$1 - \frac{2f}{\beta}q^{n-1} + \frac{eg}{\beta^2}q^{2n-2} = \left(1 + \frac{\gamma_1 g}{\beta}q^{n-1}\right)\left(1 + \frac{\gamma_2 g}{\beta}q^{n-1}\right) < 0.$$

Since $q^n \to \infty$ for $n \to \infty$ this can only be true for all $n = 1,2,3,\ldots$ if $\gamma_1\gamma_2 = \dfrac{e}{g} < 0$. We conclude that we must have either

$$\frac{\gamma_1 g}{\beta} < -1 \quad \text{and} \quad \frac{\gamma_2 g}{\beta} > 0,$$

or

$$\frac{\gamma_1 g}{\beta} > 0 \quad \text{and} \quad \frac{\gamma_2 g}{\beta} < -1.$$

Hence, we conclude that we have positive-definite orthogonality in the following four infinite cases:

Case VIa1. $0 < q < 1$ and $f^2 < eg$.

Case VIa2. $0 < q < 1$, $f^2 \geq eg$, $\dfrac{\gamma_1 g}{\beta} > -1$ and $\dfrac{\gamma_2 g}{\beta} > -1$.

Case VIa3. $q > 1$, $\dfrac{\gamma_1 g}{\beta} < -1$ and $\dfrac{\gamma_2 g}{\beta} > 0$.

Case VIa4. $q > 1$, $\dfrac{\gamma_1 g}{\beta} > 0$ and $\dfrac{\gamma_2 g}{\beta} < -1$.

It is also possible to have positive-definite orthogonality for finite systems of $N+1$ polynomials with $N \in \{1,2,3,\ldots\}$.

For $q<-1$ we must have that

$$(-1)^n\left(1-\frac{2f}{\beta}q^{n-1}+\frac{eg}{\beta^2}q^{2n-2}\right)>0$$

is true for all $n \leq N$ and false for $n = N+1$.

For $-1<q<0$ we must have that

$$1-\frac{2f}{\beta}q^{n-1}+\frac{eg}{\beta^2}q^{2n-2}<0$$

is true for all $n \leq N$ and false for $n = N+1$.

We recall that both are impossible in the case that $f^2 < eg$. In the case that $f^2 \geq eg$ we have

$$1-\frac{2f}{\beta}q^{n-1}+\frac{eg}{\beta^2}q^{2n-2}=\left(1+\frac{\gamma_1 g}{\beta}q^{n-1}\right)\left(1+\frac{\gamma_2 g}{\beta}q^{n-1}\right)$$

with

$$\gamma_1=\frac{-f-\sqrt{f^2-eg}}{g} \quad \text{and} \quad \gamma_2=\frac{-f+\sqrt{f^2-eg}}{g}.$$

Without loss of generality we will assume that $\dfrac{\gamma_1 g}{\beta} \leq \dfrac{\gamma_2 g}{\beta}$ (since γ_1 and γ_2 are real).

For $q<-1$ and $\dfrac{\gamma_2 g}{\beta}<0$ we must have

$$-\left(1+\frac{\gamma_1 g}{\beta}\right)\left(1+\frac{\gamma_2 g}{\beta}\right)>0 \quad \Longrightarrow \quad \frac{\gamma_1 g}{\beta}<-1<\frac{\gamma_2 g}{\beta}<0.$$

Now we use table 10.3 to conclude that

$$\ldots<-\frac{\gamma_2 g}{\beta}q^5<-\frac{\gamma_2 g}{\beta}q^3<-\frac{\gamma_2 g}{\beta}q<0<-\frac{\gamma_2 g}{\beta}<-\frac{\gamma_2 g}{\beta}q^2<-\frac{\gamma_2 g}{\beta}q^4<\ldots,$$

which implies that

$$\ldots<1+\frac{\gamma_2 g}{\beta}q^4<1+\frac{\gamma_2 g}{\beta}q^2<1+\frac{\gamma_2 g}{\beta}<1$$

and

$$1<1+\frac{\gamma_2 g}{\beta}q<1+\frac{\gamma_2 g}{\beta}q^3<1+\frac{\gamma_2 g}{\beta}q^5<\ldots.$$

Hence we have that $(-1)^n\left(1+\frac{\gamma_2 g}{\beta}q^{n-1}\right)>0$ for all $n=1,2,3,\ldots$. Now we have $1+\frac{\gamma_1 g}{\beta}q^{n-1}>0$ for $n=1,2,3,\ldots,N$ and $1+\frac{\gamma_1 g}{\beta}q^N\leq 0$ provided that $\frac{\gamma_1 g}{\beta}q^N\leq -1<\frac{\gamma_1 g}{\beta}q^{N-2}$ and N even.

For $q<-1$ and $\frac{\gamma_2 g}{\beta}>0$ it is possible to have $1+\frac{\gamma_2 g}{\beta}q^{n-1}>0$ for $n=1,2,3,\ldots,N$ and $1+\frac{\gamma_2 g}{\beta}q^N\leq 0$ provided that $\frac{\gamma_2 g}{\beta}q^N\leq -1<\frac{\gamma_2 g}{\beta}q^{N-2}$ with N odd. Combined with $(-1)^n\left(1+\frac{\gamma_1 g}{\beta}q^{n-1}\right)>0$ for all $n=1,2,3,\ldots$ as above this leads to another finite case with $-1<\frac{\gamma_1 g}{\beta}<0<\frac{\gamma_2 g}{\beta}$.

For $-1<q<0$ and $\frac{\gamma_2 g}{\beta}<0$ we must have

$$\left(1+\frac{\gamma_1 g}{\beta}\right)\left(1+\frac{\gamma_2 g}{\beta}\right)<0 \quad\Longrightarrow\quad \frac{\gamma_1 g}{\beta}<-1<\frac{\gamma_2 g}{\beta}<0.$$

Now we use table 10.3 to conclude that

$$-\frac{\gamma_2 g}{\beta}q<-\frac{\gamma_2 g}{\beta}q^3<-\frac{\gamma_2 g}{\beta}q^5<\ldots<0<\ldots<-\frac{\gamma_2 g}{\beta}q^4<-\frac{\gamma_2 g}{\beta}q^2<-\frac{\gamma_2 g}{\beta}<1,$$

which implies that $1+\frac{\gamma_2 g}{\beta}q^{n-1}>0$ for all $n=1,2,3,\ldots$. However, then we cannot have that $1+\frac{\gamma_1 g}{\beta}q^{n-1}<0$ for $n=1,2,3,\ldots,N$.

For $-1<q<0$ and $\frac{\gamma_2 g}{\beta}>0$ we must have

$$\left(1+\frac{\gamma_1 g}{\beta}\right)\left(1+\frac{\gamma_2 g}{\beta}\right)<0 \quad\Longrightarrow\quad \frac{\gamma_1 g}{\beta}<-1<0<\frac{\gamma_2 g}{\beta}.$$

Now we use table 10.3 to conclude that

$$-\frac{\gamma_2 g}{\beta}<-\frac{\gamma_2 g}{\beta}q^2<-\frac{\gamma_2 g}{\beta}q^4<\ldots<0<\ldots<-\frac{\gamma_2 g}{\beta}q^5<-\frac{\gamma_2 g}{\beta}q^3<-\frac{\gamma_2 g}{\beta}q,$$

which implies that

$$1+\frac{\gamma_2 g}{\beta}q<1+\frac{\gamma_2 g}{\beta}q^3<1+\frac{\gamma_2 g}{\beta}q^5<\ldots<1$$

and

$$1 < \ldots < 1 + \frac{\gamma_2 g}{\beta} q^4 < 1 + \frac{\gamma_2 g}{\beta} q^2 < 1 + \frac{\gamma_2 g}{\beta}.$$

Furthermore we have

$$-\frac{\gamma_1 g}{\beta} q < -\frac{\gamma_1 g}{\beta} q^3 < -\frac{\gamma_1 g}{\beta} q^5 < \ldots < 0 < \ldots < -\frac{\gamma_1 g}{\beta} q^4 < -\frac{\gamma_1 g}{\beta} q^2 < -\frac{\gamma_1 g}{\beta},$$

which implies that

$$1 + \frac{\gamma_1 g}{\beta} < 1 + \frac{\gamma_1 g}{\beta} q^2 < 1 + \frac{\gamma_1 g}{\beta} q^4 < \ldots < 1$$

and

$$1 < \ldots < 1 + \frac{\gamma_1 g}{\beta} q^5 < 1 + \frac{\gamma_1 g}{\beta} q^3 < 1 + \frac{\gamma_1 g}{\beta} q.$$

Now we want that

$$\left(1 + \frac{\gamma_1 g}{\beta} q^{n-1}\right)\left(1 + \frac{\gamma_2 g}{\beta} q^{n-1}\right) = 1 - \frac{2f}{\beta} q^{n-1} + \frac{eg}{\beta^2} q^{2n-2} < 0, \quad n = 1,2,3,\ldots,N$$

and $1 - \frac{2f}{\beta} q^N + \frac{eg}{\beta^2} q^{2N} \geq 0$. This implies that we must have

$$1 + \frac{\gamma_1 g}{\beta} q^{N-2} < 0 \leq 1 + \frac{\gamma_1 g}{\beta} q^N \quad \text{with } N \text{ even}$$

or

$$1 + \frac{\gamma_2 g}{\beta} q^N \leq 0 < 1 + \frac{\gamma_2 g}{\beta} q^{N-2} \quad \text{with } N \text{ odd.}$$

Hence, we must have either $\frac{\gamma_1 g}{\beta} q^{N-2} < -1 \leq \frac{\gamma_1 g}{\beta} q^N$ with N even, or $\frac{\gamma_2 g}{\beta} q^N \leq -1 < \frac{\gamma_2 g}{\beta} q^{N-2}$ with N odd.

For $0 < q < 1$ we must have

$$\left(1 + \frac{\gamma_1 g}{\beta} q^{n-1}\right)\left(1 + \frac{\gamma_2 g}{\beta} q^{n-1}\right) > 0, \quad n = 1,2,3,\ldots,N$$

with both $\frac{\gamma_1 g}{\beta} < -1$ and $\frac{\gamma_2 g}{\beta} < -1$. With the assumption that $\frac{\gamma_1 g}{\beta} \leq \frac{\gamma_2 g}{\beta}$ we must have that

$$1 + \frac{\gamma_2 g}{\beta} q^{N-1} < 0 \leq 1 + \frac{\gamma_2 g}{\beta} q^N \quad \Longrightarrow \quad \frac{\gamma_2 g}{\beta} q^{N-1} < -1 \leq \frac{\gamma_2 g}{\beta} q^N.$$

For $q > 1$ we must have

$$\left(1+\frac{\gamma_1 g}{\beta}q^{n-1}\right)\left(1+\frac{\gamma_2 g}{\beta}q^{n-1}\right)<0, \quad n=1,2,3,\ldots,N$$

and $\left(1+\frac{\gamma_1 g}{\beta}q^{N}\right)\left(1+\frac{\gamma_2 g}{\beta}q^{N}\right)>0$. If $\frac{\gamma_1 g}{\beta}<-1$, then we have $1+\frac{\gamma_1 g}{\beta}q^{n-1}<0$ for all $n=1,2,3,\ldots$. Hence, then we must have that

$$-1<\frac{\gamma_2 g}{\beta}<0 \quad \text{with} \quad \frac{\gamma_2 g}{\beta}q^N \leq -1 < \frac{\gamma_2 g}{\beta}q^{N-1}.$$

If $\frac{\gamma_2 g}{\beta}>0$, then we have $1+\frac{\gamma_2 g}{\beta}q^{n-1}>0$ for all $n=1,2,3,\ldots$. Then it is impossible to have $1+\frac{\gamma_1 g}{\beta}q^{n-1}<0$ for $n=1,2,3,\ldots,N$ and $1+\frac{\gamma_1 g}{\beta}q^{N}>0$.

Hence, we conclude that we have positive-definite orthogonality in the following six finite cases:

Case VIb1. $q<-1$ and $\frac{\gamma_1 g}{\beta}<-1<\frac{\gamma_2 g}{\beta}<0$ with $\frac{\gamma_1 g}{\beta}q^N \leq -1 < \frac{\gamma_1 g}{\beta}q^{N-2}$ and N even.

Case VIb2. $q<-1$ and $-1<\frac{\gamma_1 g}{\beta}<0<\frac{\gamma_2 g}{\beta}$ with $\frac{\gamma_2 g}{\beta}q^N \leq -1 < \frac{\gamma_2 g}{\beta}q^{N-2}$ and N odd.

Case VIb3. $-1<q<0$, $\frac{\gamma_1 g}{\beta}<-1<0<\frac{\gamma_2 g}{\beta}$ with $\frac{\gamma_1 g}{\beta}q^{N-2}<-1\leq\frac{\gamma_1 g}{\beta}q^N$ and N even.

Case VIb4. $-1<q<0$, $\frac{\gamma_1 g}{\beta}<-1<0<\frac{\gamma_2 g}{\beta}$ with $\frac{\gamma_2 g}{\beta}q^N \leq -1 < \frac{\gamma_2 g}{\beta}q^{N-2}$ and N odd.

Case VIb5. $0<q<1$ and $\frac{\gamma_1 g}{\beta}\leq\frac{\gamma_2 g}{\beta}<-1$ with $\frac{\gamma_2 g}{\beta}q^{N-1}<-1\leq\frac{\gamma_2 g}{\beta}q^N$.

Case VIb6. $q>1$ and $\frac{\gamma_1 g}{\beta}<-1<\frac{\gamma_2 g}{\beta}<0$ with $\frac{\gamma_2 g}{\beta}q^N \leq -1 < \frac{\gamma_2 g}{\beta}q^{N-1}$.

Case VII. $g \neq 0$ and $\alpha \neq 0$. In this case we set $\alpha = 1$ and by using (10.4.4) we have

$$D_n^{(2)} = -\left(2fq^{n-1} + \delta_1 eq^{2n-2} + \delta_2\right)\left(2fq^{n-1} + \delta_1 + \delta_2 eq^{2n-2}\right), \quad n = 1,2,3,\ldots.$$

In the case that $\beta^2 < 4g$ we have $\delta_2 = \overline{\delta_1}$ which implies that

$$\begin{aligned} D_n^{(2)} &= -\left(2fq^{n-1} + \delta_1 eq^{2n-2} + \overline{\delta_1}\right)\left(2fq^{n-1} + \delta_1 + \overline{\delta_1} eq^{2n-2}\right) \\ &= -\left(2fq^{n-1} + \delta_1 eq^{2n-2} + \overline{\delta_1}\right)\overline{\left(2fq^{n-1} + \overline{\delta_1} + \delta_1 eq^{2n-2}\right)} \\ &= -\left|2fq^{n-1} + \delta_1 eq^{2n-2} + \overline{\delta_1}\right|^2 < 0, \quad n = 1,2,3,\ldots. \end{aligned}$$

By using $\gamma_1\gamma_2 = e/g$, $\gamma_1 + \gamma_2 = -2f/g$ and $\delta_1\delta_2 = g$, we obtain that

$$e = \gamma_1\gamma_2\delta_1\delta_2 \quad \text{and} \quad 2f = -(\gamma_1 + \gamma_2)\delta_1\delta_2,$$

which implies that

$$D_n^{(2)} = -\delta_1\delta_2\left(1 - \gamma_1\delta_1 q^{n-1}\right)\left(1 - \gamma_1\delta_2 q^{n-1}\right)\left(1 - \gamma_2\delta_1 q^{n-1}\right)\left(1 - \gamma_2\delta_2 q^{n-1}\right)$$

for $n = 1,2,3,\ldots$. In the case that $\beta^2 \geq 4g$ both δ_1 and δ_2 are real. Moreover, in the case that $f^2 < eg$ we have $\gamma_2 = \overline{\gamma_1}$ which implies that

$$\begin{aligned} D_n^{(2)} &= -\delta_1\delta_2\left(1 - \gamma_1\delta_1 q^{n-1}\right)\left(1 - \gamma_1\delta_2 q^{n-1}\right)\left(1 - \overline{\gamma_1}\delta_1 q^{n-1}\right)\left(1 - \overline{\gamma_1}\delta_2 q^{n-1}\right) \\ &= -\delta_1\delta_2\left(1 - \gamma_1\delta_1 q^{n-1}\right)\left(1 - \gamma_1\delta_2 q^{n-1}\right)\overline{\left(1 - \gamma_1\delta_1 q^{n-1}\right)}\,\overline{\left(1 - \gamma_1\delta_2 q^{n-1}\right)} \\ &= -\delta_1\delta_2\left|1 - \gamma_1\delta_1 q^{n-1}\right|^2\left|1 - \gamma_1\delta_2 q^{n-1}\right|^2, \quad n = 1,2,3,\ldots. \end{aligned}$$

This implies that $D_n^{(2)}$ has the same sign as $-\delta_1\delta_2 = -g$ in that case.

Finally, we consider the case that $\gamma_1, \gamma_2, \delta_1, \delta_2 \in \mathbb{R}$. Again we use table 10.1 and table 10.2 to conclude that for $q < -1$ we must have either $eq > 1$ and $(-1)^n D_n^{(2)} > 0$, or $e = 0$ and $D_n^{(2)} > 0$, or $e > 1$ and $(-1)^n D_n^{(2)} < 0$, in order to have $d_n > 0$ for all $n = 1,2,3,\ldots$. However, note that $(-1)^n D_n^{(2)} > 0$ or $(-1)^n D_n^{(2)} < 0$ cannot be true for all $n = 1,2,3,\ldots$. Hence we must have $e = 0$ and $D_n^{(2)} > 0$ or

$$\left(2fq^{n-1} + \delta_2\right)\left(2fq^{n-1} + \delta_1\right) < 0,$$

which is also impossible for all $n = 1,2,3,\ldots$.

For $-1 < q < 0$ we must have $q < eq < 1$ and $(-1)^{n+1} D_n^{(2)} > 0$. Also in this case $(-1)^{n+1} D_n^{(2)} > 0$ cannot be true for all $n = 1,2,3,\ldots$, since $D_n^{(2)} \to -\delta_1\delta_2 = -g$ for $n \to \infty$.

For $0 < q < 1$ we must have $e < 1$ and $D_n^{(2)} > 0$, which implies that

$$\left(2fq^{n-1} + \delta_1 eq^{2n-2} + \delta_2\right)\left(2fq^{n-1} + \delta_1 + \delta_2 eq^{2n-2}\right) < 0, \quad n = 1,2,3,\ldots.$$

For $n = 1$ this reads

$$(2f + \delta_1 e + \delta_2)(2f + \delta_1 + \delta_2 e) < 0,$$

which implies that

$$\delta_1 e + \delta_2 < -2f < \delta_1 + \delta_2 e \quad \text{or} \quad \delta_1 + \delta_2 e < -2f < \delta_1 e + \delta_2.$$

The first inequality implies that

$$(1-e)\delta_2 < (1-e)\delta_1 \quad \Longrightarrow \quad \delta_2 < \delta_1,$$

since $1 - e > 0$. This contradicts the fact that $\delta_1 < \delta_2$. So we must have

$$\delta_1 + \delta_2 e < -2f < \delta_1 e + \delta_2.$$

Further we must have that $g = \delta_1\delta_2 < 0$, which implies that $\delta_1 < 0 < \delta_2$. We conclude that it is possible to have orthogonality for an infinite system of polynomials in that case.

For $q > 1$ we must have either $e \leq 0$ and $D_n^{(2)} < 0$, or $e > 1$ and $D_n^{(2)} > 0$.

So, for $e \leq 0$ we must have

$$\left(2fq^{n-1} + \delta_1 eq^{2n-2} + \delta_2\right)\left(2fq^{n-1} + \delta_1 + \delta_2 eq^{2n-2}\right) > 0, \quad n = 1,2,3,\ldots.$$

For $n = 1$ this reads

$$(2f + \delta_1 e + \delta_2)(2f + \delta_1 + \delta_2 e) > 0,$$

which implies that both factors are either positive or negative. Further we must have that $g = \delta_1\delta_2 > 0$, which implies that either $\delta_1 \leq \delta_2 < 0$ or $0 < \delta_1 \leq \delta_2$. We conclude that it is possible to have orthogonality for an infinite system of polynomials in this case too.

For $e > 1$ we must have

$$\left(2fq^{n-1} + \delta_1 eq^{2n-2} + \delta_2\right)\left(2fq^{n-1} + \delta_1 + \delta_2 eq^{2n-2}\right) < 0, \quad n = 1,2,3,\ldots.$$

In this case we must have

$$\delta_1 e + \delta_2 < -2f < \delta_1 + \delta_2 e \quad \text{and} \quad g = \delta_1\delta_2 < 0.$$

Hence, we have $\delta_1 < 0 < \delta_2$, which implies that $\delta_1 e < 0$ and $\delta_2 e > 0$. Also in this case it is possible to have orthogonality for an infinite system of polynomials.

Hence, we conclude that we have positive-definite orthogonality in the following three infinite cases:

Case VIIa1. $0 < q < 1$, $e < 1$, $\delta_1 < 0 < \delta_2$ and

$$\left(2fq^{n-1} + \delta_1 eq^{2n-2} + \delta_2\right)\left(2fq^{n-1} + \delta_1 + \delta_2 eq^{2n-2}\right) < 0, \quad n = 1,2,3,\ldots.$$

Case VIIa2. $q > 1$, $e \leq 0$, $\delta_1\delta_2 > 0$ with $\delta_1, \delta_2 \in \mathbb{R}$ or $\delta_2 = \overline{\delta_1}$, and

$$\left(2fq^{n-1} + \delta_1 eq^{2n-2} + \delta_2\right)\left(2fq^{n-1} + \delta_1 + \delta_2 eq^{2n-2}\right) > 0, \quad n = 1,2,3,\ldots.$$

Case VIIa3. $q > 1$, $e > 1$, $\delta_1 < 0 < \delta_2$ and

$$\left(2fq^{n-1} + \delta_1 eq^{2n-2} + \delta_2\right)\left(2fq^{n-1} + \delta_1 + \delta_2 eq^{2n-2}\right) < 0, \quad n = 1,2,3,\ldots.$$

It is also possible to have positive-definite orthogonality for a finite system of $N+1$ or $2N+2$ polynomials with $N \in \{1,2,3,\ldots\}$. We only consider the cases where $D_n^{(1)}$ has opposite sign for $n = N$ and $n = N+1$ according to table 10.5. Skipping further details, we conclude that we have positive-definite orthogonality, at least in the following eight finite cases:

Case VIIb1. $q < -1$, $0 < eq < 1$ with $q < eq^{2N} \leq q^{-1}$ and $D_n^{(2)} > 0$ for $n = 1,2,3,\ldots,N$. In that case we have a finite system of $N+1$ polynomials.

Case VIIb2. $q < -1$, $0 < e < 1$ with $eq^{2N} = 1$ and $D_n^{(2)} > 0$ for $n = 1,2,3,\ldots,N$. In that case we have a finite system of $N+1$ polynomials.

Case VIIb3. $q < -1$, $0 < e < 1$ with $1 < eq^{2N} < q^2$ and $D_n^{(2)} > 0$ for $n = 1,2,3,\ldots,$ $2N+1$. In that case we have a finite system of $2N+2$ polynomials.

Case VIIb4. $-1 < q < 0$, $eq > 1$ with $q^{-1} \leq eq^{2N} < q$ and $D_n^{(2)} < 0$ for $n = 1,2,3,\ldots,N$. In that case we have a finite system of $N+1$ polynomials.

Case VIIb5. $-1 < q < 0$, $e > 1$ with $eq^{2N} = 1$ and $D_n^{(2)} > 0$ for $n = 1,2,3,\ldots,N$. In that case we have a finite system of $N+1$ polynomials.

Case VIIb6. $-1 < q < 0$, $e > 1$ with $q^2 < eq^{2N} < 1$ and $D_n^{(2)} > 0$ for $n = 1, 2, 3, \ldots, 2N+1$. In that case we have a finite system of $2N+2$ polynomials.

Case VIIb7. $0 < q < 1$, $e > 1$ with $q < eq^{2N} \leq q^{-1}$ and $D_n^{(2)} < 0$ for $n = 1, 2, 3, \ldots, N$. In that case we have a finite system of $N+1$ polynomials.

Case VIIb8. $q > 1$, $0 < e < 1$ with $q^{-1} \leq eq^{2N} < q$ and $D_n^{(2)} < 0$ for $n = 1, 2, 3, \ldots, N$. In that case we have a finite system of $N+1$ polynomials.

We remark that it is also possible to have positive-definite orthogonality for finite systems of $N+1$ polynomials with $N \in \{1, 2, 3, \ldots\}$ in cases where $D_n^{(2)}$ has opposite sign for $n = N$ and $n = N+1$.

10.5 Solutions of the q-Pearson Equation

We look for solutions of the q-Pearson equation (cf. (3.2.8))

$$w(x)C(x) = qw(qx)D(qx), \quad 0 < |q| < 1$$

with

$$C(x) = \frac{ex^2 + 2fqx + gq^2}{q(q-1)^2x^2}$$

and

$$D(x) = \frac{\{e + 2\varepsilon(1-q)\}x^2 + \{2f + \gamma(1-q)\}qx + gq^2}{(q-1)^2x^2} = \frac{\alpha x^2 + \beta qx + gq^2}{(q-1)^2x^2}.$$

Hence we have by using (10.1.3)

$$\begin{aligned} \frac{w(x)}{w(qx)} &= \frac{q^2\left(\{e + 2\varepsilon(1-q)\}x^2 + \{2f + \gamma(1-q)\}x + g\right)}{ex^2 + 2fqx + gq^2} \\ &= \frac{q^2(\alpha x^2 + \beta x + g)}{ex^2 + 2fqx + gq^2}. \end{aligned} \tag{10.5.1}$$

We consider two types of solutions:

A. *Continuous* solutions for $x \in \mathbb{R}$ in terms of (convergent) infinite products. For the convergence of these infinite products we refer to the book [471] by L.J. Slater.

B. *Discrete* solutions for $\{x_\nu\}_{\nu=0}^N$ with $N \to \infty$ or $\{x_\nu\}_{\nu=-\infty}^\infty$, id est $x_\nu = Aq^\nu$ with $A \in \mathbb{C}$ and $\nu = 0, \pm 1, \pm 2, \ldots$, in terms of finite products. Note that $x_{\nu+1} = qx_\nu$. Without loss of generality we can choose $A = 1$ in each case.

In order to find solutions of the q-Pearson equation (10.5.1), we distinguish between $0 < |q| < 1$ and $|q| > 1$.

For $0 < |q| < 1$ we make the following observations. For

$$w(x) = (-rx;q)_\infty = \prod_{k=0}^{\infty}(1+rxq^k), \quad r \in \mathbb{C} \tag{10.5.2}$$

we have

$$w(qx) = (-rqx;q)_\infty = \prod_{k=0}^{\infty}(1+rxq^{k+1}),$$

which implies that

$$\frac{w(x)}{w(qx)} = 1+rx. \tag{10.5.3}$$

For

$$w(x) = \left(-rx, -\frac{q}{rx};q\right)_\infty = \prod_{k=0}^{\infty}\left(1+rxq^k\right)\left(1+\frac{q^{k+1}}{rx}\right), \quad r \in \mathbb{C}, \quad r \neq 0 \tag{10.5.4}$$

we have

$$w(qx) = \prod_{k=0}^{\infty}\left(1+rxq^{k+1}\right)\left(1+\frac{q^k}{rx}\right),$$

which implies that

$$\frac{w(x)}{w(qx)} = \frac{1+rx}{1+\frac{1}{rx}} = rx. \tag{10.5.5}$$

Finally, for

$$w(x) = \left(-sx, -\frac{q}{tx};q\right)_\infty = \prod_{k=0}^{\infty}\left(1+sxq^k\right)\left(1+\frac{q^{k+1}}{tx}\right), \quad s,t \in \mathbb{C}, \quad t \neq 0 \tag{10.5.6}$$

we have

$$w(qx) = \prod_{k=0}^{\infty}\left(1+sxq^{k+1}\right)\left(1+\frac{q^k}{tx}\right),$$

which implies that

$$\frac{w(x)}{w(qx)} = \frac{1+sx}{1+\frac{1}{tx}} = \frac{1+sx}{1+tx}\cdot tx. \tag{10.5.7}$$

Combining the latter two results, we conclude that

$$w(x) = \frac{\left(-sx, -\frac{q}{tx};q\right)_\infty}{\left(-rx, -\frac{q}{rx};q\right)_\infty}, \quad r,s,t \in \mathbb{C}, \quad r \neq 0, \quad t \neq 0 \tag{10.5.8}$$

implies that

$$\frac{w(x)}{w(qx)} = \frac{1+sx}{1+tx} \cdot \frac{tx}{rx} = \frac{1+sx}{1+tx} \cdot \frac{t}{r}. \tag{10.5.9}$$

For $|q| > 1$ we rewrite the q-Pearson equation (10.5.1) in the following way. If we set $q = p^{-1}$, then we have

$$\begin{aligned} \frac{w(x)}{w(p^{-1}x)} &= \frac{p^{-2}\left(\{e+2\varepsilon(1-p^{-1})\}x^2 + \{2f+\gamma(1-p^{-1})\}x+g\right)}{ex^2+2fp^{-1}x+gp^{-2}} \\ &= \frac{p^{-2}(\alpha' x^2+\beta' x+g)}{ex^2+2fp^{-1}x+gp^{-2}} = \frac{\alpha' x^2+\beta' x+g}{ep^2x^2+2fpx+g}, \end{aligned} \tag{10.5.10}$$

where

$$\alpha' := e+2\varepsilon(1-p^{-1}) \quad \text{and} \quad \beta' := 2f+\gamma(1-p^{-1}). \tag{10.5.11}$$

For $|q| > 1$ this implies that $0 < |p| < 1$. Now we replace x by px to obtain

$$\frac{w(x)}{w(px)} = \frac{ep^4x^2+2fp^2x+g}{\alpha' p^2x^2+\beta' px+g}, \quad 0 < |p| < 1. \tag{10.5.12}$$

Note that if $\alpha = e+2\varepsilon(1-q)$ does not depend on q, for instance in the case that $\alpha = 0$ or $\alpha = 1$, then $\alpha' = \alpha$. The same holds for β.

Case II-A. $g = 0$, $\beta = 0 = \beta'$, $\alpha \neq 0$ respectively $\alpha' \neq 0$ and $f \neq 0$. From (10.5.1) we have

$$\frac{w(x)}{w(qx)} = \frac{\alpha q^2x^2}{ex^2+2fqx} = \frac{1}{1+\frac{ex}{2fq}} \cdot \frac{\alpha qx}{2f}$$

and by using (10.5.2) and (10.5.3) with $r = e/2fq$ and by using (10.5.4) and (10.5.5) with $r = \alpha q/2f$, we obtain the solution

$$w^{(II)}(x;q) = \frac{\left(-\frac{\alpha qx}{2f}, -\frac{2f}{\alpha x};q\right)_\infty}{\left(-\frac{ex}{2fq};q\right)_\infty}, \quad 0 < |q| < 1.$$

From (10.5.12) we have

$$\frac{w(x)}{w(px)} = \frac{ep^4x^2+2fp^2x}{\alpha' p^2x^2} = \left(1+\frac{ep^2x}{2f}\right) \cdot \frac{2f}{\alpha' x}$$

and by using (10.5.2) and (10.5.3) with $r = ep^2/2f$ and by using (10.5.4) and (10.5.5) with $r = \alpha'/2f$, we obtain the solution

$$w^{(II)}(x;p) = \frac{\left(-\frac{ep^2x}{2f};p\right)_\infty}{\left(-\frac{\alpha' x}{2f}, -\frac{2fp}{\alpha' x};p\right)_\infty}, \quad 0 < |p| < 1.$$

The special case $0 < p < 1$, $e = 0$ and $\alpha' = 2f$ leads to the weight function

$$w(x;p) = \frac{1}{\left(-x, -\frac{p}{x};p\right)_\infty}$$

for the **Stieltjes-Wigert** polynomials.

Case II-B. $g = 0$, $\beta = 0 = \beta'$, $\alpha \neq 0$ respectively $\alpha' \neq 0$ and $f \neq 0$. Then we have from (10.5.1) with $x_\nu = q^\nu$

$$\frac{w(x_\nu)}{w(x_{\nu+1})} = \frac{1}{1+\frac{eq^\nu}{2fq}} \cdot \frac{\alpha q}{2f} \cdot q^\nu$$

with possible solution

$$w^{(II)}(x_\nu;q) = \left(-\frac{e}{2fq};q\right)_\nu \left(\frac{2f}{\alpha q}\right)^\nu q^{-\binom{\nu}{2}}.$$

From (10.5.10) we have with $x_\nu = p^{-\nu}$

$$\frac{w(x_\nu)}{w(x_{\nu+1})} = \frac{1}{1+\frac{ep^{1-\nu}}{2f}} \cdot \frac{\alpha'}{2fp} \cdot p^{-\nu}$$

with possible solution

$$w^{(II)}(x_\nu;p) = \left(-\frac{ep^{2-\nu}}{2f};p\right)_\nu \left(\frac{2fp}{\alpha'}\right)^\nu p^{\binom{\nu}{2}}.$$

Case III-A. $g = 0$, $\beta \neq 0$ respectively $\beta' \neq 0$, $\alpha = 0 = \alpha'$ and $e \neq 0$. From (10.5.1) we have for $f \neq 0$

$$\frac{w(x)}{w(qx)} = \frac{\beta q^2 x}{ex^2 + 2fqx} = \frac{1}{1+\frac{ex}{2fq}} \cdot \frac{\beta q}{2f}$$

and by using (10.5.8) and (10.5.9) with $r = e/\beta q^2$, $s = 0$ and $t = e/2fq$, we obtain the solution

$$w^{(III)}(x;q) = \frac{\left(-\frac{2fq^2}{ex};q\right)_\infty}{\left(-\frac{ex}{\beta q^2}, -\frac{\beta q^3}{ex};q\right)_\infty}, \quad 0 < |q| < 1.$$

For $f=0$ we have

$$\frac{w(x)}{w(qx)}=\frac{\beta q^2}{ex}$$

and by using (10.5.4) and (10.5.5) with $r=e/\beta q^2$, we obtain the solution

$$w^{(III)}(x;q)\Big|_{f=0}=\frac{1}{\left(-\frac{ex}{\beta q^2},-\frac{\beta q^3}{ex};q\right)_\infty},\quad 0<|q|<1,$$

which implies that the solution above is also valid for $f=0$.

The special case $0<q<1$, $f=0$ and $e=\beta q^2$ leads to the weight function

$$w(x;q)=\frac{1}{\left(-x,-\frac{q}{x};q\right)_\infty}$$

for the **Stieltjes-Wigert** polynomials.

For $\dfrac{2f}{\beta}>0$ we may write $\dfrac{2f}{\beta}=q^\nu$ with $q>0$, $q\neq 1$ and $\nu=\dfrac{\ln\frac{2f}{\beta}}{\ln q}\in\mathbb{R}$. Then we have

$$\frac{w(x)}{w(qx)}=\frac{q^{1-\nu}}{1+\frac{ex}{2fq}}$$

with possible solution

$$w^{(III)}(x;q)=\frac{x^{\nu-1}}{\left(-\frac{ex}{2fq};q\right)_\infty},\quad 0<|q|<1.$$

The special case $e=2fq$ leads to the weight function

$$w(x;q)=\frac{x^{\nu-1}}{(-x;q)_\infty}$$

for the q-**Laguerre** polynomials.

From (10.5.12) we have for $f\neq 0$

$$\frac{w(x)}{w(px)}=\frac{ep^4x^2+2fp^2x}{\beta' px}=\left(1+\frac{ep^2x}{2f}\right)\cdot\frac{2fp}{\beta'}$$

and by using (10.5.8) and (10.5.9) with $r=ep^3/\beta'$, $s=0$ and $t=ep^2/2f$, we obtain the solution

$$w^{(III)}(x;p)=\frac{\left(-\frac{ep^3x}{\beta'},-\frac{\beta'}{ep^2x};p\right)_\infty}{\left(-\frac{2f}{epx};p\right)_\infty},\quad 0<|p|<1.$$

For $f=0$ we have

$$\frac{w(x)}{w(px)}=\frac{ep^4x^2}{\beta' px}=\frac{ep^3x}{\beta'}.$$

By using (10.5.4) and (10.5.5) with $r=ep^3/\beta'$, we obtain the solution

$$w^{(III)}(x;p)\Big|_{f=0}=\left(-\frac{ep^3x}{\beta'},-\frac{\beta'}{ep^2x};p\right)_\infty,\quad 0<|p|<1,$$

which implies that the solution above is also valid for $f=0$.

Case III-B. $g=0$, $\beta\neq 0$ respectively $\beta'\neq 0$, $\alpha=0=\alpha'$ and $e\neq 0$. For $f\neq 0$ we have from (10.5.1) with $x_\nu=q^\nu$

$$\frac{w(x_\nu)}{w(x_{\nu+1})}=\frac{1}{1+\frac{eq^\nu}{2fq}}\cdot\frac{\beta q}{2f}$$

with possible solution

$$w^{(III)}(x_\nu;q)=\left(-\frac{e}{2fq};q\right)_\nu\left(\frac{2f}{\beta q}\right)^\nu.$$

For $f=0$ we have

$$\frac{w(x_\nu)}{w(x_{\nu+1})}=\frac{\beta q^2}{e}\cdot q^{-\nu}$$

with possible solution

$$w^{(III)}(x_\nu;q)\Big|_{f=0}=\left(\frac{e}{\beta q^2}\right)^\nu q^{\binom{\nu}{2}}.$$

For $f\neq 0$ we have from (10.5.10) with $x_\nu=p^{-\nu}$

$$\frac{w(x_\nu)}{w(x_{\nu+1})}=\frac{1}{1+\frac{ep^{1-\nu}}{2f}}\cdot\frac{\beta'}{2fp}$$

with possible solution

$$w^{(III)}(x_\nu;p)=\left(-\frac{ep^{2-\nu}}{2f};p\right)_\nu\left(\frac{2fp}{\beta'}\right)^\nu.$$

For $f = 0$ we have

$$\frac{w(x_\nu)}{w(x_{\nu+1})} = \frac{\beta'}{ep^2} \cdot p^\nu$$

with possible solution

$$w^{(III)}(x_\nu; p)\Big|_{f=0} = \left(\frac{ep^2}{\beta'}\right)^\nu p^{-\binom{\nu}{2}}.$$

Case IV-A. $g = 0$, $\beta \neq 0$ respectively $\beta' \neq 0$ and $\alpha \neq 0$ respectively $\alpha' \neq 0$. From (10.5.1) we have for $f \neq 0$

$$\frac{w(x)}{w(qx)} = \frac{\alpha q^2 x^2 + \beta q^2 x}{ex^2 + 2fqx} = \frac{1 + \frac{\alpha x}{\beta}}{1 + \frac{ex}{2fq}} \cdot \frac{\beta q}{2f}$$

and by using (10.5.8) and (10.5.9) with $r = e/\beta q^2$, $s = \alpha/\beta$ and $t = e/2fq$, we obtain the solution

$$w^{(IV)}(x; q) = \frac{\left(-\frac{\alpha x}{\beta}, -\frac{2fq^2}{ex}; q\right)_\infty}{\left(-\frac{ex}{\beta q^2}, -\frac{\beta q^3}{ex}; q\right)_\infty}, \quad e \neq 0, \quad 0 < |q| < 1.$$

For $f = 0$ we have $e \neq 0$ and

$$\frac{w(x)}{w(qx)} = \frac{\alpha q^2 x^2 + \beta q^2 x}{ex^2} = \left(1 + \frac{\alpha x}{\beta}\right) \cdot \frac{\beta q^2}{ex}.$$

By using (10.5.2) and (10.5.3) with $r = \alpha/\beta$ and (10.5.4) and (10.5.5) with $r = e/\beta q^2$, we obtain the solution

$$w^{(IV)}(x; q)\Big|_{f=0} = \frac{\left(-\frac{\alpha x}{\beta}; q\right)_\infty}{\left(-\frac{ex}{\beta q^2}, -\frac{\beta q^3}{ex}; q\right)_\infty}, \quad 0 < |q| < 1,$$

which implies that the solution above is also valid for $f = 0$.

For $e = 0$ we have $f \neq 0$ and

$$\frac{w(x)}{w(qx)} = \left(1 + \frac{\alpha x}{\beta}\right) \cdot \frac{\beta q}{2f}$$

By using (10.5.8) and (10.5.9) with $r = \alpha q/2f$, $s = 0$ and $t = \alpha/\beta$, we obtain the solution

$$w^{(IV)}(x;q)\Big|_{e=0}=\frac{\left(-\frac{\alpha qx}{2f},-\frac{2f}{\alpha x};q\right)_\infty}{\left(-\frac{\beta q}{\alpha x};q\right)_\infty},\quad 0<|q|<1.$$

From (10.5.12) we have for $f\neq 0$

$$\frac{w(x)}{w(px)}=\frac{ep^4x^2+2fp^2x}{\alpha' p^2x^2+\beta' px}=\frac{1+\frac{ep^2x}{2f}}{1+\frac{\alpha' px}{\beta'}}\cdot\frac{2fp}{\beta'}$$

and by using (10.5.8) and (10.5.9) with $r=\alpha'/2f$, $s=ep^2/2f$ and $t=\alpha' p/\beta'$, we obtain the solution

$$w^{(IV)}(x;p)=\frac{\left(-\frac{ep^2x}{2f},-\frac{\beta'}{\alpha' x};p\right)_\infty}{\left(-\frac{\alpha' x}{2f},-\frac{2fp}{\alpha' x};p\right)_\infty},\quad 0<|p|<1.$$

For $f=0$ we have $e\neq 0$ and

$$\frac{w(x)}{w(px)}=\frac{ep^3x}{\alpha' px+\beta'}=\frac{1}{1+\frac{\alpha' px}{\beta'}}\cdot\frac{ep^3x}{\beta'}.$$

By using (10.5.2) and (10.5.3) with $r=\alpha' p/\beta'$ and by using (10.5.4) and (10.5.5) with $r=ep^3/\beta'$, we obtain the solution

$$w^{(IV)}(x;p)\Big|_{f=0}=\frac{\left(-\frac{ep^3x}{\beta'},-\frac{\beta'}{ep^2x};p\right)_\infty}{\left(-\frac{\alpha' px}{\beta'};p\right)_\infty},\quad 0<|p|<1.$$

Case IV-B. $g=0$, $\beta\neq 0$ respectively $\beta'\neq 0$ and $\alpha\neq 0$ respectively $\alpha'\neq 0$. For $f\neq 0$ we have from (10.5.1) with $x_\nu=q^\nu$

$$\frac{w(x_\nu)}{w(x_{\nu+1})}=\frac{1+\frac{\alpha q^\nu}{\beta}}{1+\frac{eq^\nu}{2fq}}\cdot\frac{\beta q}{2f}$$

with possible solution

$$w^{(IV)}(x_\nu;q)=\frac{\left(-\frac{e}{2fq};q\right)_\nu}{\left(-\frac{\alpha}{\beta};q\right)_\nu}\left(\frac{2f}{\beta q}\right)^\nu.$$

The special case $e = abq^2$, $2f = -a$, $\alpha = 1$ and $\beta = -q^{-1}$ leads to the weight function

$$w(x_\nu;q) = \frac{(bq;q)_\nu}{(q;q)_\nu} a^\nu$$

for the **little q-Jacobi** polynomials.

The special case $e = 0$, $2f = -a$, $\alpha = 1$ and $\beta = -q^{-1}$ leads to the weight function

$$w(x_\nu;q) = \frac{1}{(q;q)_\nu} a^\nu$$

for the **little q-Laguerre** polynomials.

For $f = 0$ we have $e \neq 0$ and

$$\frac{w(x_\nu)}{w(x_{\nu+1})} = \left(1 + \frac{\alpha q^\nu}{\beta}\right) \cdot \frac{\beta q^2}{e} \cdot q^{-\nu}$$

with possible solution

$$w^{(IV)}(x_\nu;q)\Big|_{f=0} = \frac{1}{\left(-\frac{\alpha}{\beta};q\right)_\nu} \left(\frac{e}{\beta q^2}\right)^\nu q^{\binom{\nu}{2}}.$$

The special case $e = -aq$, $\alpha = 1$ and $\beta = -q^{-1}$ leads to the weight function

$$w(x_\nu;q) = \frac{1}{(q;q)_\nu} a^\nu q^{\binom{\nu}{2}}$$

for the **q-Bessel** polynomials.

For $f \neq 0$ we have from (10.5.10) with $x_\nu = p^{-\nu}$

$$\frac{w(x_\nu)}{w(x_{\nu+1})} = \frac{1 + \frac{\alpha' p^{-\nu}}{\beta'}}{1 + \frac{ep^{1-\nu}}{2f}} \cdot \frac{\beta'}{2fp}$$

with possible solution

$$w^{(IV)}(x_\nu;p) = \frac{\left(-\frac{ep^{2-\nu}}{2f};p\right)_\nu}{\left(-\frac{\alpha' p^{1-\nu}}{\beta'};p\right)_\nu} \left(\frac{2fp}{\beta'}\right)^\nu.$$

For $f = 0$ we have $e \neq 0$ and

$$\frac{w(x_\nu)}{w(x_{\nu+1})} = \left(1+\frac{\alpha' p^{-\nu}}{\beta'}\right)\cdot\frac{\beta'}{ep^2}\cdot p^\nu$$

with possible solution

$$w^{(IV)}(x_\nu;p)\Big|_{f=0} = \frac{1}{\left(-\frac{\alpha' p^{1-\nu}}{\beta'};p\right)_\nu}\left(\frac{ep^2}{\beta'}\right)^\nu p^{-\binom{\nu}{2}}.$$

Case V-A. $g\neq 0$, $\alpha = 0 = \alpha'$ and $\beta = 0 = \beta'$. From (10.5.1) we have

$$\frac{w(x)}{w(qx)} = \frac{gq^2}{ex^2+2fqx+gq^2} = \frac{1}{\left(1+\frac{2fx}{gq}+\frac{ex^2}{gq^2}\right)} = \frac{1}{\left(1-\frac{\gamma_1 x}{q}\right)\left(1-\frac{\gamma_2 x}{q}\right)},$$

where $\gamma_1\gamma_2 = e/g$ and $\gamma_1+\gamma_2 = -2f/g$, which implies that for $f^2\geq eg$

$$\gamma_1 = \frac{-f-\sqrt{f^2-eg}}{g} \quad \text{and} \quad \gamma_2 = \frac{-f+\sqrt{f^2-eg}}{g}.$$

In the case that $f^2 < eg$ we have $\gamma_2 = \overline{\gamma_1}$. By using (10.5.2) and (10.5.3), once with $r = -\gamma_1/q$ and once with $r = -\gamma_2/q$, we obtain the solution

$$w^{(V)}(x;q) = \frac{1}{\left(\frac{\gamma_1 x}{q},\frac{\gamma_2 x}{q};q\right)_\infty}, \qquad 0<|q|<1.$$

The special case $e = q^2$, $f = 0$ and $g = 1$, which implies that $\gamma_1 = iq$ and $\gamma_2 = -iq$, leads to the weight function

$$w(x_\nu;q) = \frac{1}{(ix,-ix;q)_\infty} = \frac{1}{(x^2;q^2)_\infty}$$

for the **discrete q-Hermite II** polynomials. Note that γ_1 and γ_2 are non-real in this case.

From (10.5.12) we have

$$\frac{w(x)}{w(px)} = \frac{ep^4x^2+2fp^2x+g}{g} = 1+\frac{2fp^2x}{g}+\frac{ep^4x^2}{g} = \left(1-\gamma_1 p^2 x\right)\left(1-\gamma_2 p^2 x\right).$$

By using (10.5.2) and (10.5.3), once with $r = -\gamma_1 p^2$ and once with $r = -\gamma_2 p^2$, we obtain the solution

$$w^{(V)}(x;p) = \left(\gamma_1 p^2 x, \gamma_2 p^2 x; p\right)_\infty, \qquad 0<|p|<1.$$

Case V-B. $g \neq 0$, $\alpha = 0 = \alpha'$ and $\beta = 0 = \beta'$. Then we have from (10.5.1) with $x_\nu = q^\nu$

$$\frac{w(x_\nu)}{w(x_{\nu+1})} = \frac{1}{(1-\gamma_1 q^{\nu-1})(1-\gamma_2 q^{\nu-1})}$$

with possible solution

$$w^{(V)}(x_\nu;q) = \left(\frac{\gamma_1}{q}, \frac{\gamma_2}{q}; q\right)_\nu .$$

From (10.5.10) we have with $x_\nu = p^{-\nu}$

$$\frac{w(x_\nu)}{w(x_{\nu+1})} = \frac{1}{(1-\gamma_1 p^{1-\nu})(1-\gamma_2 p^{1-\nu})}$$

with possible solution

$$w^{(V)}(x_\nu;p) = \left(\gamma_1 p^{2-\nu}, \gamma_2 p^{2-\nu}; p\right)_\nu .$$

Case VI-A. $g \neq 0$, $\alpha = 0 = \alpha'$ and $\beta \neq 0$ respectively $\beta' \neq 0$. From (10.5.1) we have

$$\frac{w(x)}{w(qx)} = \frac{q^2(\beta x + g)}{ex^2 + 2fqx + gq^2} = \frac{gq^2\left(1+\frac{\beta x}{g}\right)}{gq^2\left(1+\frac{2fx}{gq}+\frac{ex^2}{gq^2}\right)} = \frac{1+\frac{\beta x}{g}}{\left(1-\frac{\gamma_1 x}{q}\right)\left(1-\frac{\gamma_2 x}{q}\right)},$$

where, as before, $\gamma_1\gamma_2 = e/g$ and $\gamma_1+\gamma_2 = -2f/g$, which implies that for $f^2 \geq eg$

$$\gamma_1 = \frac{-f-\sqrt{f^2-eg}}{g} \quad \text{and} \quad \gamma_2 = \frac{-f+\sqrt{f^2-eg}}{g}.$$

In the case that $f^2 < eg$ we have $\gamma_2 = \overline{\gamma_1}$. By using (10.5.2) and (10.5.3), once with $r = \beta/g$, once with $r = -\gamma_1/q$ and once with $r = -\gamma_2/q$, we obtain the solution

$$w^{(VI)}(x;q) = \frac{\left(-\frac{\beta x}{g}; q\right)_\infty}{\left(\frac{\gamma_1 x}{q}, \frac{\gamma_2 x}{q}; q\right)_\infty}, \quad 0 < |q| < 1.$$

From (10.5.12) we have

$$\frac{w(x)}{w(px)} = \frac{ep^4x^2 + 2fp^2x + g}{\beta' px + g} = \frac{1+\frac{2fp^2x}{g}+\frac{ep^4x^2}{g}}{1+\frac{\beta' px}{g}} = \frac{\left(1-\gamma_1 p^2 x\right)\left(1-\gamma_2 p^2 x\right)}{1+\frac{\beta' px}{g}}.$$

By using (10.5.2) and (10.5.3), once with $r = -\gamma_1 p^2$, once with $r = -\gamma_2 p^2$ and once with $r = \beta' p/g$, we obtain the solution

$$w^{(VI)}(x;p) = \frac{(\gamma_1 p^2 x, \gamma_2 p^2 x; p)_\infty}{\left(-\frac{\beta' p x}{g}; p\right)_\infty}, \quad 0 < |p| < 1.$$

Case VI-B. $g \neq 0$, $\alpha = 0 = \alpha'$ and $\beta \neq 0$ respectively $\beta' \neq 0$. Then we have from (10.5.1) with $x_\nu = q^\nu$

$$\frac{w(x_\nu)}{w(x_{\nu+1})} = \frac{1 + \frac{\beta q^\nu}{g}}{(1-\gamma_1 q^{\nu-1})(1-\gamma_2 q^{\nu-1})}$$

with possible solution

$$w^{(VI)}(x_\nu;q) = \frac{\left(\frac{\gamma_1}{q}, \frac{\gamma_2}{q}; q\right)_\nu}{\left(-\frac{\beta}{g}; q\right)_\nu}.$$

From (10.5.10) we have with $x_\nu = p^{-\nu}$

$$\frac{w(x_\nu)}{w(x_{\nu+1})} = \frac{1 + \frac{\beta' p^{-\nu}}{g}}{(1-\gamma_1 p^{1-\nu})(1-\gamma_2 p^{1-\nu})}$$

with possible solution

$$w^{(VI)}(x_\nu;p) = \frac{(\gamma_1 p^{2-\nu}, \gamma_2 p^{2-\nu}; p)_\nu}{\left(-\frac{\beta' p^{1-\nu}}{g}; p\right)_\nu}.$$

Case VII-A. $g \neq 0$ and $\alpha \neq 0$ respectively $\alpha' \neq 0$. From (10.5.1) we have

$$\frac{w(x)}{w(qx)} = \frac{q^2(\alpha x^2 + \beta x + g)}{e x^2 + 2 f q x + g q^2} = \frac{1 + \frac{\beta x}{g} + \frac{\alpha x^2}{g}}{1 + \frac{2 f x}{g q} + \frac{e x^2}{g q^2}} = \frac{\left(1 - \frac{x}{\delta_1}\right)\left(1 - \frac{x}{\delta_2}\right)}{\left(1 - \frac{\gamma_1 x}{q}\right)\left(1 - \frac{\gamma_2 x}{q}\right)},$$

where, as before, in the case that $f^2 \geq eg$

$$\gamma_1 = \frac{-f - \sqrt{f^2 - eg}}{g} \quad \text{and} \quad \gamma_2 = \frac{-f + \sqrt{f^2 - eg}}{g}$$

and $\delta_1 \delta_2 = g/\alpha$ and $\delta_1 + \delta_2 = -\beta/\alpha$, which implies that for $\beta^2 \geq 4g\alpha$

$$\delta_1 = \frac{-\beta - \sqrt{\beta^2 - 4g\alpha}}{2\alpha} \quad \text{and} \quad \delta_2 = \frac{-\beta + \sqrt{\beta^2 - 4g\alpha}}{2\alpha}.$$

In the case that $f^2 < eg$ we have $\gamma_2 = \overline{\gamma_1}$ and in the case that $\beta^2 < 4g\alpha$ we have $\delta_2 = \overline{\delta_1}$. By using (10.5.2) and (10.5.3), we obtain the solution

$$w^{(VII)}(x;q) = \frac{\left(\frac{x}{\delta_1}, \frac{x}{\delta_2}; q\right)_\infty}{\left(\frac{\gamma_1 x}{q}, \frac{\gamma_2 x}{q}; q\right)_\infty}, \quad 0 < |q| < 1.$$

The special case $e = abq^2$, $2f = -a(b+c)q$, $g = ac$, $\alpha = 1$ and $\beta = -(a+c)$ leads to $\gamma_1 = q$, $\gamma_2 = bq/c$, $\delta_1 = c$ and $\delta_2 = a$. Then we obtain the weight function

$$w(x;q) = \frac{\left(\frac{x}{a}, \frac{x}{c}; q\right)_\infty}{\left(x, \frac{bx}{c}; q\right)_\infty}$$

for the **big q-Jacobi** polynomials.

The special case $e = 0$, $2f = -abq$, $g = ab$, $\alpha = 1$ and $\beta = -(a+b)$ leads to $\gamma_1 = q$, $\gamma_2 = 0$, $\delta_1 = b$ and $\delta_2 = a$. Then we obtain the weight function

$$w(x;q) = \frac{\left(\frac{x}{a}, \frac{x}{b}; q\right)_\infty}{(x;q)_\infty}$$

for the **big q-Laguerre** polynomials.

The special case $e = f = 0$, $g = a/q^2$, $\alpha = 1$ and $\beta = -(a+1)/q$ leads to $\gamma_1 = \gamma_2 = 0$, $\delta_1 = a/q$ and $\delta_2 = 1/q$. Then we obtain the weight function

$$w(x;q) = \left(qx, \frac{qx}{a}; q\right)_\infty$$

for the **Al-Salam-Carlitz I** polynomials.

The special case $e = f = 0$, $g = -1/q^2$, $\alpha = 1$ and $\beta = 0$ leads to $\gamma_1 = \gamma_2 = 0$, $\delta_1 = -1/q$ and $\delta_2 = 1/q$. Then we obtain by using (1.8.24) the weight function

$$w(x;q) = (qx, -qx; q)_\infty = \left(q^2x^2; q^2\right)_\infty$$

for the **discrete q-Hermite I** polynomials.

From (10.5.12) we have

$$\frac{w(x)}{w(px)} = \frac{ep^4x^2 + 2fp^2x + g}{\alpha' p^2x^2 + \beta' px + g} = \frac{1 + \frac{2fp^2x}{g} + \frac{ep^4x^2}{g}}{1 + \frac{\beta' px}{g} + \frac{\alpha' p^2x^2}{g}} = \frac{(1-\gamma_1 p^2x)(1-\gamma_2 p^2x)}{\left(1 - \frac{px}{\delta_1'}\right)\left(1 - \frac{px}{\delta_2'}\right)},$$

where $\delta_1'\delta_2' = g/\alpha'$ and $\delta_1' + \delta_2' = -\beta'/\alpha'$, which implies that for $(\beta')^2 \geq 4g\alpha'$

$$\delta_1' = \frac{-\beta' - \sqrt{(\beta')^2 - 4g\alpha'}}{2\alpha'} \quad \text{and} \quad \delta_2' = \frac{-\beta' + \sqrt{(\beta')^2 - 4g\alpha'}}{2\alpha'}.$$

In the case that $(\beta')^2 < 4g\alpha'$ we have $\delta_2' = \overline{\delta_1'}$. By using (10.5.2) and (10.5.3), we obtain the solution

$$w^{(VII)}(x;p) = \frac{(\gamma_1 p^2 x, \gamma_2 p^2 x; p)_\infty}{\left(\frac{px}{\delta_1'}, \frac{px}{\delta_2'}; p\right)_\infty}, \quad 0 < |p| < 1.$$

The special case $e = f = 0$, $g = p^2$, $\alpha^* = 1$ and $\beta^* = 0$ leads to $\gamma_1 = \gamma_2 = 0$, $\delta_1' = -ip$ and $\delta_2' = ip$. Then we obtain by using (1.8.24) the weight function

$$w(x;p) = \frac{1}{(ix, -ix; p)_\infty} = \frac{1}{(x^2; p^2)_\infty}$$

for the **discrete q-Hermite II** polynomials. Note that δ_1' and δ_2' are non-real in this case.

Case VII-B. $g \neq 0$ and $\alpha \neq 0$. Then we have from (10.5.1) with $x_\nu = q^\nu$

$$\frac{w(x_\nu)}{w(x_{\nu+1})} = \frac{\left(1 - \frac{q^\nu}{\delta_1}\right)\left(1 - \frac{q^\nu}{\delta_2}\right)}{(1 - \gamma_1 q^{\nu-1})(1 - \gamma_2 q^{\nu-1})}$$

with possible solution

$$w^{(VII)}(x_\nu;q) = \frac{\left(\frac{\gamma_1}{q}, \frac{\gamma_2}{q}; q\right)_\nu}{\left(\frac{1}{\delta_1}, \frac{1}{\delta_2}; q\right)_\nu}.$$

From (10.5.10) we obtain with $x_\nu = p^{-\nu}$

$$\frac{w(x_\nu)}{w(x_{\nu+1})} = \frac{\left(1 - \frac{p^{-\nu}}{\delta_1'}\right)\left(1 - \frac{p^{-\nu}}{\delta_2'}\right)}{(1 - \gamma_1 p^{1-\nu})(1 - \gamma_2 p^{1-\nu})}$$

with possible solution

$$w^{(VII)}(x_\nu;p) = \frac{(\gamma_1 p^{2-\nu}, \gamma_2 p^{2-\nu}; p)_\nu}{\left(\frac{p^{1-\nu}}{\delta_1'}, \frac{p^{1-\nu}}{\delta_2'}; p\right)_\nu}.$$

The special case $e = f = 0$, $g = ap^2$, $\alpha' = 1$ and $\beta' = -(a+1)p$ leads to $\gamma_1 = \gamma_2 = 0$, $\delta_1' = ap$ and $\delta_2' = p$. Then we obtain by using (1.8.14) the weight function

$$w(x_\nu;p)=\frac{1}{\left(p^{-\nu},\frac{p^{-\nu}}{a};p\right)_\nu}=\frac{1}{(p,ap;p)_\nu}a^\nu p^{2\nu+2\binom{\nu}{2}}=\frac{a^\nu p^{\nu(\nu+1)}}{(p,ap;p)_\nu}$$

for the **Al-Salam-Carlitz II** polynomials.

10.6 Orthogonality Relations

In the preceding section we have obtained both continuous and discrete solutions of the q-Pearson equation (10.5.1) and the p^{-1}-Pearson equation (10.5.10) or (10.5.12). In this section we will derive orthogonality relations for several cases obtained in section 10.4. We will not give explicit orthogonality relations for each different case, but we will restrict to the most important cases (either continuous or discrete).

We remark that for each different case the appropriate boundary conditions (3.2.12) or (3.2.19) should be satisfied. Therefore we have to consider

$$w(q^{-1}x;q)\varphi(q^{-2}x)\quad\text{with}\quad\varphi(x)=ex^2+2fx+g$$

and $w(x;q)$ the involving weight function which satisfies the q-Pearson equation. This product should vanish at both ends of the interval of orthogonality.

Case II. In this case we have $g=0$, $\beta=0=\beta'$ and $\alpha\neq 0$. In section 10.4 we have seen that it is only possible to have positive-definite orthogonality for an infinite system of polynomials in the case that $q>1$, $f\neq 0$ and $e\leq 0$. Further we have seen that it is possible to have positive-definite orthogonality for a finite system of polynomials in three different cases. Here we will only treat the infinite case.

Case IIa1. $q>1$, $f\neq 0$ and $e\leq 0$. If we write $q=p^{-1}$, then we have

$$q>1\quad\Longleftrightarrow\quad 0<p<1.$$

Now we use the weight function

$$w^{(II)}(x;p)=\frac{\left(-\frac{ep^2x}{2f};p\right)_\infty}{\left(-\frac{\alpha'x}{2f},-\frac{2fp}{\alpha'x};p\right)_\infty},\quad 0<|p|<1.$$

Since $\alpha'\neq 0$ we set $\alpha'=1$. Then we have

$$w^{(II)}(px;p)\varphi(p^2x) = \frac{\left(-\frac{ep^3x}{2f};p\right)_\infty}{\left(-\frac{px}{2f},-\frac{2f}{x};p\right)_\infty}\cdot(ep^4x^2+2fp^2x)$$
$$= \frac{\left(-\frac{ep^2x}{2f};p\right)_\infty}{\left(-\frac{px}{2f},-\frac{2f}{x};p\right)_\infty}\cdot 2fp^2x,$$

which vanishes for $x \to 0$. It also vanishes for $x \to \infty$ in the case that $e = 0$. Then we have orthogonality on the interval $(0,\infty)$ and by using (1.12.3) we have for $f > 0$

$$d_0 := \int_0^\infty \frac{dx}{\left(-\frac{x}{2f},-\frac{2fp}{x};p\right)_\infty} = 2f\int_0^\infty \frac{dt}{\left(-t,-\frac{p}{t};p\right)_\infty} = -2f\ln p\,(p;p)_\infty > 0.$$

Further we find by using (10.4.1), (10.4.2) and (10.4.3) with $e = 0$ and $q = p^{-1}$

$$d_n = -4f^2q^{3n-1}(1-q^n) = 4f^2p^{-4n+1}(1-p^n), \quad n = 1,2,3,\ldots$$

which implies that

$$\prod_{k=1}^n d_k = (4f^2)^n p^{-n(2n+1)}(p;p)_n, \quad n = 1,2,3,\ldots.$$

This leads to the orthogonality relation

$$\int_0^\infty \frac{y_m^{(II)}(x;p)y_n^{(II)}(x;p)}{\left(-\frac{x}{2f},-\frac{2fp}{x};p\right)_\infty}\,dx = -2f\ln p\,(p;p)_\infty(4f^2)^n p^{-n(2n+1)}(p;p)_n\,\delta_{mn}$$

for $f > 0$ and $m,n = 0,1,2,\ldots$.

The special case $2f = 1$ leads to the orthogonality relation

$$\int_0^\infty \frac{y_m(x;p)y_n(x;p)}{\left(-x,-\frac{p}{x};p\right)_\infty}\,dx = -\ln p\,(p;p)_\infty p^{-n(2n+1)}(p;p)_n\,\delta_{mn}, \quad m,n = 0,1,2,\ldots$$

for the **Stieltjes-Wigert** polynomials.

Case III. In this case we have $g = 0$, $\beta \neq 0$ respectively $\beta' \neq 0$ and $\alpha = 0 = \alpha'$. In section 10.4 we have seen that it is possible to have positive-definite orthogonality for an infinite system of polynomials in three different cases and that it is impossible to have positive-definite orthogonality for a finite system of polynomials. Here we will only treat one case.

Case IIIa2. $0<q<1$ and $\dfrac{2f}{\beta}<1$. We use the weight function

$$w^{(III)}(x;q)=\frac{\left(-\frac{2fq^2}{ex};q\right)_\infty}{\left(-\frac{ex}{\beta q^2},-\frac{\beta q^3}{ex};q\right)_\infty},\quad 0<|q|<1.$$

Then we have

$$\begin{aligned}w^{(III)}(q^{-1}x;q)\varphi(q^{-2}x)&=\frac{\left(-\frac{2fq^3}{ex};q\right)_\infty}{\left(-\frac{ex}{\beta q^3},-\frac{\beta q^4}{ex};q\right)_\infty}\cdot(eq^{-4}x^2+2fq^{-2}x)\\&=\frac{\left(-\frac{2fq^2}{ex};q\right)_\infty}{\left(-\frac{ex}{\beta q^3},-\frac{\beta q^4}{ex};q\right)_\infty}\cdot eq^{-4}x^2,\end{aligned}$$

which vanishes for $x\to 0$ if $f=0$. Then it also vanishes for $x\to\infty$. Hence we have orthogonality on the interval $(0,\infty)$ if $f=0$ and by using (1.12.3) we obtain for $\dfrac{\beta}{e}>0$

$$\begin{aligned}d_0&:=\int_0^\infty\frac{1}{\left(-\frac{ex}{\beta q^2},-\frac{\beta q^3}{ex};q\right)_\infty}\,dx=\frac{\beta q^2}{e}\int_0^\infty\frac{1}{\left(-t,-\frac{q}{t};q\right)_\infty}\,dt\\&=-\frac{\beta q^2}{e}\ln q\,(q;q)_\infty>0.\end{aligned}$$

In this case we have for $f=0$

$$d_n=\frac{q^{n+1}(1-q^n)}{e^2q^{5n-4}}\beta(\beta-2fq^{n-1})=\frac{\beta^2}{e^2}q^{-4n+5}(1-q^n),\quad n=1,2,3,\ldots,$$

which implies that

$$\prod_{k=1}^n d_k=\left(\frac{\beta}{e}\right)^{2n}q^{-n(2n-3)}(q;q)_n,\quad n=1,2,3,\ldots.$$

This leads to the orthogonality relation

$$\begin{aligned}&\int_0^\infty\frac{1}{\left(-\frac{ex}{\beta q^2},-\frac{\beta q^3}{ex};q\right)_\infty}y_m^{(III)}(x;q)y_n^{(III)}(x;q)\,dx\\&=-\frac{\beta q^2}{e}\ln q\,(q;q)_\infty\left(\frac{\beta}{e}\right)^{2n}q^{-n(2n-3)}(q;q)_n\,\delta_{mn},\quad m,n=0,1,2,\ldots\end{aligned}$$

for $\dfrac{\beta}{e}>0$.

The special case $e=\beta q^2$ leads to the orthogonality relation

$$\int_0^\infty \frac{y_m(x;q)y_n(x;q)}{\left(-x,-\frac{q}{x};q\right)_\infty}\,dx=-\ln q\,(q;q)_\infty q^{-n(2n+1)}(q;q)_n\,\delta_{mn},\quad m,n=0,1,2,\ldots$$

for the **Stieltjes-Wigert** polynomials.

For $0<\dfrac{2f}{\beta}<1$ we may use the weight function

$$w^{(III)}(x;q)=\frac{x^{\nu-1}}{\left(-\frac{ex}{2fq};q\right)_\infty},\quad \nu=\frac{\ln\frac{2f}{\beta}}{\ln q}>0.$$

Then we have

$$w^{(III)}(q^{-1}x;q)\varphi(q^{-2}x)=\frac{q^{-\nu+1}x^{\nu-1}}{\left(-\frac{ex}{2fq^2};q\right)_\infty}\cdot\left(eq^{-4}x^2+2fq^{-2}x\right)=\frac{2fq^{-\nu-1}x^\nu}{\left(-\frac{ex}{2fq};q\right)_\infty},$$

which vanishes for $x=0$, since $\nu>0$, and also for $x\to\infty$. Hence we have orthogonality on the interval $(0,\infty)$ and by using (1.12.2) we obtain for $\dfrac{f}{e}>0$

$$d_0:=\int_0^\infty\frac{x^{\nu-1}}{\left(-\frac{ex}{2fq};q\right)_\infty}\,dx=\left(\frac{2fq}{e}\right)^\nu\int_0^\infty\frac{t^{\nu-1}}{(-t;q)_\infty}\,dt=\left(\frac{2fq}{e}\right)^\nu\frac{\pi}{\sin\pi\nu}\,\frac{(q^{1-\nu};q)_\infty}{(q;q)_\infty}.$$

In this case we have $\dfrac{2f}{\beta}=q^\nu$, which implies that

$$\prod_{k=1}^n d_k=\left(\frac{2f}{eq^\nu}\right)^{2n}q^{-n(2n-3)}(q;q)_n(q^\nu;q)_n,\quad n=1,2,3,\ldots.$$

This leads to the orthogonality relation

$$\int_0^\infty\frac{x^{\nu-1}}{\left(-\frac{ex}{2fq};q\right)_\infty}y_m^{(III)}(x;q)y_n^{(III)}(x;q)\,dx$$
$$=\left(\frac{2fq}{e}\right)^\nu\frac{\pi}{\sin\pi\nu}\,\frac{(q^{1-\nu};q)_\infty}{(q;q)_\infty}\left(\frac{2f}{eq^\nu}\right)^{2n}q^{-n(2n-3)}(q;q)_n(q^\nu;q)_n\,\delta_{mn}$$

for $\nu>0$, $\dfrac{f}{e}>0$ and $m,n=0,1,2,\ldots$.

The special case $e=2fq$ leads to the orthogonality relation (for $\nu>0$)

$$\int_0^\infty \frac{x^{\nu-1}}{(-x;q)_\infty} y_m(x;q) y_n(x;q)\,dx$$
$$= \frac{\pi}{\sin \pi\nu} \frac{(q^{1-\nu};q)_\infty}{(q;q)_\infty} q^{-n(2n+2\nu-1)} (q;q)_n (q^\nu;q)_n \,\delta_{mn}, \quad m,n = 0,1,2,\ldots$$

for the q-**Laguerre** polynomials.

Case IV. $g = 0$, $\beta \neq 0$ respectively $\beta' \neq 0$ and $\alpha \neq 0$ respectively $\alpha' \neq 0$. In section 10.4 we have seen that it is possible to have positive-definite orthogonality for an infinite system of polynomials in seven different cases and for a finite system of polynomials in eight different cases. Here we will only treat three infinite cases.

Case IVa3. $0 < q < 1$ and $0 < e < \dfrac{2f}{\beta} < 1$. We use the weight function

$$w^{(IV)}(x_\nu;q) = \frac{\left(-\frac{e}{2fq};q\right)_\nu}{\left(-\frac{\alpha}{\beta};q\right)_\nu} \left(\frac{2f}{\beta q}\right)^\nu.$$

Since $\alpha \neq 0$ we set $\alpha = 1$. Then we have by using $x_\nu = q^\nu$

$$w^{(IV)}(q^{-1}x_\nu;q)\varphi(q^{-2}x_\nu) = \frac{\left(-\frac{e}{2fq};q\right)_{\nu-1}}{\left(-\frac{1}{\beta};q\right)_{\nu-1}} \left(\frac{2f}{\beta q}\right)^{\nu-1} \cdot \left(eq^{2\nu-4} + 2fq^{\nu-2}\right)$$
$$= \frac{\left(-\frac{e}{2fq};q\right)_\nu}{\left(-\frac{1}{\beta};q\right)_{\nu-1}} \left(\frac{2f}{\beta}\right)^\nu \frac{\beta}{q},$$

which should vanish for $\nu = 0$. Since $\beta \neq 0$ this implies by using (1.8.6) that

$$\frac{1}{\left(-\frac{1}{\beta};q\right)_{-1}} = \left(-\frac{1}{\beta q};q\right)_1 = 1 + \frac{1}{\beta q} = 0 \quad \Longrightarrow \quad \beta = -q^{-1}.$$

Note that it also vanishes for $\nu \to \infty$ since $0 < \dfrac{2f}{\beta} < 1$. Then we have by using (1.11.1) for $x_\nu = q^\nu$, $0 < e < 1$ and $0 < -2fq < 1$

$$d_0 := \sum_{\nu=0}^\infty w^{(IV)}(x_\nu;q)x_\nu = \sum_{\nu=0}^\infty \frac{\left(-\frac{e}{2fq};q\right)_\nu}{(q;q)_\nu} (-2fq)^\nu = \frac{(e;q)_\infty}{(-2fq;q)_\infty} > 0.$$

Further we have by using (10.4.1), (10.4.2) and (10.4.3) with $\beta = -q^{-1}$

$$\begin{aligned} d_n &= q^{n+1}(1-q^n)\frac{1-eq^{n-2}}{(1-eq^{2n-3})(1-eq^{2n-2})^2(1-eq^{2n-1})} \\ &\quad\times q^{n-1}(\beta-2fq^{n-1})(2f-e\beta q^{n-1}) \\ &= -2fq^{2n-1}\frac{(1-q^n)(1-eq^{n-2})}{(1-eq^{2n-3})(1-eq^{2n-2})^2(1-eq^{2n-1})}(1+2fq^n)\left(1+\frac{e}{2f}q^{n-2}\right) \end{aligned}$$

for $n=1,2,3,\ldots$, which implies that

$$\prod_{k=1}^{n} d_k = (-2f)^n q^{n^2}\frac{(q,eq^{-1};q)_n}{(eq^{-1},e;q)_{2n}}\left(-2fq,-\frac{e}{2fq};q\right)_n, \quad n=1,2,3,\ldots.$$

This leads to the orthogonality relation

$$\begin{aligned} &\sum_{\nu=0}^{\infty}\frac{\left(-\frac{e}{2fq};q\right)_\nu}{(q;q)_\nu}(-2fq)^\nu y_m^{(IV)}(q^\nu;q)y_n^{(IV)}(q^\nu;q) \\ &= \frac{(e;q)_\infty}{(-2fq;q)_\infty}(-2f)^n q^{n^2}\frac{(q,eq^{-1};q)_n}{(eq^{-1},e;q)_{2n}}\left(-2fq,-\frac{e}{2fq};q\right)_n\delta_{mn} \end{aligned}$$

for $0<e<1$ and $0<-2fq<1$ and $m,n=0,1,2,\ldots$.

Case IVa4. $0<q<1$, $e\leq 0$ and $0<\dfrac{2f}{\beta}<1$. In this case we can use the same weight function as above leading to the same orthogonality relation for $e\leq 0$ and $0<-2fq<1$. This implies that this orthogonality relation holds for $e<1$ and $0<-2fq<1$.

Then the special case $e=abq^2$ and $2f=-a$ leads to the orthogonality relation (for $0<aq<1$ and $bq<1$)

$$\begin{aligned} &\sum_{\nu=0}^{\infty}\frac{(bq;q)_\nu}{(q;q)_\nu}(aq)^\nu y_m(q^\nu;q)y_n(q^\nu;q) \\ &= \frac{(abq^2;q)_\infty}{(aq;q)_\infty}a^n q^{n^2}\frac{(q,abq;q)_n}{(abq,abq^2;q)_{2n}}(aq,bq;q)_n\,\delta_{mn}, \quad m,n=0,1,2,\ldots \end{aligned}$$

for the **little q-Jacobi** polynomials.

The special case $e=0$ and $2f=-a$ now leads to the orthogonality relation (for $0<aq<1$)

$$\sum_{\nu=0}^{\infty}\frac{(aq)^\nu}{(q;q)_\nu}y_m(q^\nu;q)y_n(q^\nu;q)=\frac{a^n q^{n^2}}{(aq;q)_\infty}(q;q)_n(aq;q)_n\,\delta_{mn}, \quad m,n=0,1,2,\ldots$$

for the **little q-Laguerre** polynomials.

Case IVa5. $0 < q < 1$, $e < 0$ and $f = 0$. Now we use the weight function

$$w^{(IV)}(x_\nu;q)\Big|_{f=0} = \frac{1}{\left(-\frac{\alpha}{\beta};q\right)_\nu}\left(\frac{e}{\beta q^2}\right)^\nu q^{\binom{\nu}{2}}.$$

Since $\alpha \neq 0$ we set $\alpha = 1$. Then we have

$$w^{(IV)}(q^{-1}x_\nu;q)\Big|_{f=0}\varphi(q^{-2}x_\nu) = \frac{1}{\left(-\frac{1}{\beta};q\right)_{\nu-1}}\left(\frac{e}{\beta q^2}\right)^{\nu-1} q^{\binom{\nu-1}{2}}\cdot eq^{2\nu-4},$$

which vanishes for $\nu = 0$ if as before by using (1.8.6)

$$\frac{1}{\left(-\frac{1}{\beta};q\right)_{-1}} = \left(-\frac{1}{\beta q};q\right)_1 = 1+\frac{1}{\beta q} = 0 \quad\Longrightarrow\quad \beta = -q^{-1}.$$

Then we have by using (1.14.2) since $e < 0$

$$d_0 := \sum_{\nu=0}^{\infty} w^{(IV)}(x_\nu;q)\Big|_{f=0} x_\nu = \sum_{\nu=0}^{\infty}\frac{(-e)^\nu}{(q;q)_\nu}q^{\binom{\nu}{2}} = (e;q)_\infty > 0.$$

Further we have by using (10.4.1), (10.4.2) and (10.4.3) with $f = 0$ and $\beta = -q^{-1}$

$$\begin{aligned} d_n &= q^{n+1}(1-q^n)\frac{1-eq^{n-2}}{(1-eq^{2n-3})(1-eq^{2n-2})^2(1-eq^{2n-1})} \\ &\quad \times q^{n-1}(\beta - 2fq^{n-1})(2f - e\beta q^{n-1}) \\ &= -eq^{3n-3}\frac{(1-q^n)(1-eq^{n-2})}{(1-eq^{2n-3})(1-eq^{2n-2})^2(1-eq^{2n-1})}, \quad n = 0,1,2,\ldots, \end{aligned}$$

which implies that

$$\prod_{k=1}^{n} d_k = (-e)^n q^{3n(n-1)/2}\frac{(q,eq^{-1};q)_n}{(eq^{-1},e;q)_{2n}}, \quad n = 1,2,3,\ldots.$$

This leads to the orthogonality relation

$$\sum_{\nu=0}^{\infty}\frac{(-e)^\nu}{(q;q)_\nu}q^{\binom{\nu}{2}} y_m^{(IV)}(q^\nu;q)y_n^{(IV)}(q^\nu;q) = (e;q)_\infty(-e)^n q^{3n(n-1)/2}\frac{(q,eq^{-1};q)_n}{(eq^{-1},e;q)_{2n}}\,\delta_{mn}$$

for $e < 0$ and $m,n = 0,1,2,\ldots$.

The special case $e = -aq$ leads to the orthogonality relation (for $a > 0$)

$$\sum_{\nu=0}^{\infty}\frac{(aq)^{\nu}}{(q;q)_{\nu}}q^{\binom{\nu}{2}}y_m(q^{\nu};q)y_n(q^{\nu};q)$$
$$=(-aq;q)_{\infty}a^nq^{n(3n-1)/2}\frac{(q,-a;q)_n}{(-a,-aq;q)_{2n}}\,\delta_{mn},\quad m,n=0,1,2,\ldots$$

for the q-**Bessel** polynomials.

Case V. $g\neq 0$, $\alpha=0=\alpha'$ and $\beta=0=\beta'$. In section 10.4 we have seen that it is possible to have positive-definite orthogonality for an infinite system of polynomials in three different cases and that it is impossible to have positive-definite orthogonality for a finite system of polynomials. Here we will only treat one case.

Case Va2. $0<q<1$ and $\dfrac{g}{e}>0$. We use the weight function

$$w^{(V)}(x;q)=\frac{1}{\left(\frac{\gamma_1x}{q},\frac{\gamma_2x}{q};q\right)_{\infty}},\quad 0<|q|<1.$$

Then we have

$$w^{(V)}(q^{-1}x;q)\varphi(q^{-2}x)=\frac{eq^{-4}x^2+2fq^{-2}x+g}{\left(\frac{\gamma_1x}{q^2},\frac{\gamma_2x}{q^2};q\right)_{\infty}}=\frac{g\left(1-\frac{\gamma_1x}{q^2}\right)\left(1-\frac{\gamma_2x}{q^2}\right)}{\left(\frac{\gamma_1x}{q^2},\frac{\gamma_2x}{q^2};q\right)_{\infty}}$$
$$=\frac{g}{\left(\frac{\gamma_1x}{q},\frac{\gamma_2x}{q};q\right)_{\infty}},$$

which clearly vanishes for $x\to\pm\infty$ since $\gamma_1\gamma_2\neq 0$. Hence we have for $q^2<|\gamma_1|<q$, $q^2<|\gamma_2|<q$ and $\gamma_1\gamma_2>0$ (if $\gamma_1,\gamma_2\in\mathbb{R}$), or for $\gamma_2=\overline{\gamma_1}$ (if $\gamma_1,\gamma_2\notin\mathbb{R}$) by using (1.15.13)

$$d_0:=\int_{-\infty}^{\infty}\frac{1}{\left(\frac{\gamma_1x}{q},\frac{\gamma_2x}{q};q\right)_{\infty}}\,d_qx=(1-q)\frac{\left(q,-q,-1,-\frac{\gamma_1\gamma_2}{q^2},-\frac{q^3}{\gamma_1\gamma_2};q\right)_{\infty}}{\left(\frac{\gamma_1}{q},-\frac{\gamma_1}{q},\frac{q^2}{\gamma_1},-\frac{q^2}{\gamma_1},\frac{\gamma_2}{q},-\frac{\gamma_2}{q},\frac{q^2}{\gamma_2},-\frac{q^2}{\gamma_2};q\right)_{\infty}}>0.$$

Further we have by using $e=\gamma_1\gamma_2g$

$$d_n=\frac{1}{\gamma_1\gamma_2}q^{-2n+3}(1-q^n),\quad n=1,2,3,\ldots,$$

which implies that

$$\prod_{k=1}^{n}d_k=\left(\frac{1}{\gamma_1\gamma_2}\right)^nq^{-n(n-2)}(q;q)_n,\quad n=1,2,3,\ldots.$$

This leads to the orthogonality relation

$$\int_{-\infty}^{\infty} \frac{1}{\left(\frac{\gamma_1 x}{q}, \frac{\gamma_2 x}{q}; q\right)_\infty} y_m^{(V)}(x;q) y_n^{(V)}(x;q)\, d_q x$$
$$= (1-q) \frac{\left(q, -q, -1, -\frac{\gamma_1\gamma_2}{q^2}, -\frac{q^3}{\gamma_1\gamma_2}; q\right)_\infty}{\left(\frac{\gamma_1}{q}, -\frac{\gamma_1}{q}, \frac{q^2}{\gamma_1}, -\frac{q^2}{\gamma_1}, \frac{\gamma_2}{q}, -\frac{\gamma_2}{q}, \frac{q^2}{\gamma_2}, -\frac{q^2}{\gamma_2}; q\right)_\infty} \left(\frac{1}{\gamma_1\gamma_2}\right)^n q^{-n(n-2)} (q;q)_n\, \delta_{mn}$$

for $q^2 < |\gamma_1| < q$, $q^2 < |\gamma_2| < q$, $\gamma_1\gamma_2 > 0$ (if $\gamma_1, \gamma_2 \in \mathbb{R}$), or for $\gamma_2 = \overline{\gamma_1}$ (if $\gamma_1, \gamma_2 \notin \mathbb{R}$) and $m, n = 0, 1, 2, \ldots$.

The special case $e = q^2$, $f = 0$ and $g = 1$, which implies that $\gamma_1 = iq$ and $\gamma_2 = -iq$, leads to the orthogonality relation

$$\int_{-\infty}^{\infty} \frac{1}{(ix, -ix; q)_\infty} y_m(x;q) y_n(x;q)\, d_q x$$
$$= (1-q) \frac{(q, -q, -1, -1, -q; q)_\infty}{(i, -i, -iq, iq, -i, i, iq, -iq; q)_\infty} q^{-n^2} (q;q)_n\, \delta_{mn}, \quad m, n = 0, 1, 2, \ldots$$

for the **discrete q-Hermite II** polynomials.

Case VI. $g \neq 0$, $\alpha = 0 = \alpha'$ and $\beta \neq 0$ respectively $\beta' \neq 0$. In section 10.4 we have seen that it is possible to have positive-definite orthogonality for an infinite system of polynomials in four different cases and for a finite system of polynomials in six different cases. Here we will only treat one infinite and one finite case.

Case VIa1/2. $0 < q < 1$, $f^2 < eg$ or $f^2 \geq eg$ with $\dfrac{\gamma_1 g}{\beta} > -1$ and $\dfrac{\gamma_2 g}{\beta} > -1$. We use the weight function

$$w^{(VI)}(x;q) = \frac{\left(-\frac{\beta x}{g}; q\right)_\infty}{\left(\frac{\gamma_1 x}{q}, \frac{\gamma_2 x}{q}; q\right)_\infty}, \quad 0 < |q| < 1.$$

Then we have

$$w^{(VI)}(q^{-1}x;q)\varphi(q^{-2}x) = \frac{\left(-\frac{\beta x}{gq}; q\right)_\infty}{\left(\frac{\gamma_1 x}{q^2}, \frac{\gamma_2 x}{q^2}; q\right)_\infty} \cdot (eq^{-4}x^2 + 2fq^{-2}x + g)$$
$$= \frac{\left(-\frac{\beta x}{gq}; q\right)_\infty}{\left(\frac{\gamma_1 x}{q^2}, \frac{\gamma_2 x}{q^2}; q\right)_\infty} \cdot g\left(1 - \frac{\gamma_1 x}{q^2}\right)\left(1 - \frac{\gamma_2 x}{q^2}\right)$$
$$= \frac{\left(-\frac{\beta x}{gq}; q\right)_\infty}{\left(\frac{\gamma_1 x}{q}, \frac{\gamma_2 x}{q}; q\right)_\infty} \cdot g,$$

which vanishes for $x \to \infty$ since $\gamma_1\gamma_2 \neq 0$. Since $\beta \neq 0$ it also vanishes for $x = -gq/\beta$. Hence we have for $\dfrac{g}{\beta} > 0$, $q^2 < |\gamma_1| < q$, $q^2 < |\gamma_2| < q$ and $\gamma_1\gamma_2 > 0$ (if $\gamma_1\,\gamma_2 \in \mathbb{R}$), or for $\dfrac{g}{\beta} > 0$ and $\gamma_2 = \overline{\gamma_1}$ (if $\gamma_1, \gamma_2 \notin \mathbb{R}$) by using (1.15.12)

$$d_0 := \int_{-\frac{gq}{\beta}}^{\infty} \frac{\left(-\frac{\beta x}{g};q\right)_\infty}{\left(\frac{\gamma_1 x}{q},\frac{\gamma_2 x}{q};q\right)_\infty} d_qx = (1-q)\frac{\left(q,-\frac{gq}{\beta},-\frac{\beta}{g},-\frac{\gamma_1\gamma_2 g}{\beta q},-\frac{\beta q^2}{\gamma_1\gamma_2 g};q\right)_\infty}{\left(-\frac{\gamma_1 g}{\beta},-\frac{\gamma_2 g}{\beta},\frac{\gamma_1}{q},\frac{q^2}{\gamma_1},\frac{\gamma_2}{q},\frac{q^2}{\gamma_2};q\right)_\infty} > 0.$$

Further we have by using $e = \gamma_1\gamma_2 g$

$$d_n = \left(\frac{\beta}{\gamma_1\gamma_2 g}\right)^2 q^{-4n+5}(1-q^n)\left(1+\frac{\gamma_1 g q^{n-1}}{\beta}\right)\left(1+\frac{\gamma_2 g q^{n-1}}{\beta}\right), \quad n = 1,2,3,\ldots,$$

which implies that

$$\prod_{k=1}^{n} d_k = \left(\frac{\beta}{\gamma_1\gamma_2 g}\right)^{2n} q^{-n(2n-3)}\left(q,-\frac{\gamma_1 g}{\beta},-\frac{\gamma_2 g}{\beta};q\right)_n, \quad n = 1,2,3,\ldots.$$

This leads to the orthogonality relation

$$\begin{aligned}
&\int_{-\frac{gq}{\beta}}^{\infty} \frac{\left(-\frac{\beta x}{g};q\right)_\infty}{\left(\frac{\gamma_1 x}{q},\frac{\gamma_2 x}{q};q\right)_\infty} y_m^{(VI)}(x;q)y_n^{(VI)}(x;q)\,d_qx \\
&= (1-q)\frac{\left(q,-\frac{gq}{\beta},-\frac{\beta}{g},-\frac{\gamma_1\gamma_2 g}{\beta q},-\frac{\beta q^2}{\gamma_1\gamma_2 g};q\right)_\infty}{\left(-\frac{\gamma_1 g}{\beta},-\frac{\gamma_2 g}{\beta},\frac{\gamma_1}{q},\frac{q^2}{\gamma_1},\frac{\gamma_2}{q},\frac{q^2}{\gamma_2};q\right)_\infty} \\
&\quad \times \left(\frac{\beta}{\gamma_1\gamma_2 g}\right)^{2n} q^{-n(2n-3)}\left(q,-\frac{\gamma_1 g}{\beta},-\frac{\gamma_2 g}{\beta};q\right)_n \delta_{mn}, \quad m,n = 0,1,2,\ldots
\end{aligned}$$

for $\dfrac{g}{\beta} > 0$, $q^2 < |\gamma_1| < q$, $q^2 < |\gamma_2| < q$ and $\gamma_1\gamma_2 > 0$ (if $\gamma_1, \gamma_2 \in \mathbb{R}$), or for $\dfrac{g}{\beta} > 0$ and $\gamma_2 = \overline{\gamma_1}$ (if $\gamma_1, \gamma_2 \notin \mathbb{R}$).

Case VIb5. $0 < q < 1$ and $\dfrac{\gamma_1 g}{\beta} \leq \dfrac{\gamma_2 g}{\beta} < -1$ with $\dfrac{\gamma_2 g}{\beta}q^{N-1} < -1 \leq \dfrac{\gamma_2 g}{\beta}q^N$. We use the weight function

$$w^{(VI)}(x_\nu;q) = \frac{\left(\frac{\gamma_1}{q},\frac{\gamma_2}{q};q\right)_\nu}{\left(-\frac{\beta}{g};q\right)_\nu}.$$

Then we have

$$w^{(VI)}(q^{-1}x_\nu;q)\varphi(q^{-2}x_\nu) = \frac{\left(\frac{\gamma_1}{q},\frac{\gamma_2}{q};q\right)_{\nu-1}}{\left(-\frac{\beta}{g};q\right)_\nu}\cdot\left(eq^{2\nu-4}+2fq^{\nu-2}+g\right)$$
$$= \frac{\left(\frac{\gamma_1}{q},\frac{\gamma_2}{q};q\right)_{\nu-1}}{\left(-\frac{\beta}{g};q\right)_{\nu-1}}\cdot g\left(1-\gamma_1 q^{\nu-2}\right)\left(1-\gamma_2 q^{\nu-2}\right)$$
$$= \frac{\left(\frac{\gamma_1}{q},\frac{\gamma_2}{q};q\right)_{\nu}}{\left(-\frac{\beta}{g};q\right)_{\nu-1}}\cdot g,$$

which vanishes for $\nu = 0$ if we have by using (1.8.6)

$$\frac{1}{\left(-\frac{\beta}{g};q\right)_{-1}} = \left(-\frac{\beta}{gq};q\right)_1 = 1+\frac{\beta}{gq} = 0 \quad\Longrightarrow\quad \beta = -gq.$$

It also vanishes for $\nu = N+1$ if $\frac{\gamma_2}{q} = q^{-N}$. Hence we take $\gamma_2 = q^{-N+1}$ and obtain by using (1.11.5) with $x_\nu = q^\nu$ for $\gamma_1 > 0$

$$d_0 := \sum_{\nu=0}^{N} w^{(VI)}(x_\nu;q)x_\nu = \sum_{\nu=0}^{N}\frac{\left(\frac{\gamma_1}{q},q^{-N};q\right)_\nu}{(q;q)_\nu}q^\nu = \left(\frac{\gamma_1}{q}\right)^N > 0.$$

Further we have

$$d_n = \left(\frac{q^N}{\gamma_1}\right)^2 q^{-4n+5}(1-q^n)\left(1-\gamma_1 q^{n-2}\right)\left(1-q^{n-N-1}\right), \quad n = 1,2,3,\ldots,$$

which implies that

$$\prod_{k=1}^{n} d_k = \left(\frac{q^N}{\gamma_1}\right)^{2n} q^{-n(2n-3)}\left(q,\frac{\gamma_1}{q},q^{-N};q\right)_n, \quad n = 1,2,3,\ldots.$$

This leads to the orthogonality relation

$$\sum_{\nu=0}^{N}\frac{\left(\frac{\gamma_1}{q},q^{-N};q\right)_\nu}{(q;q)_\nu}q^\nu y_m^{(VI)}(q^\nu;q)y_n^{(VI)}(q^\nu;q)$$
$$= \left(\frac{\gamma_1}{q}\right)^N\left(\frac{q^N}{\gamma_1}\right)^{2n} q^{-n(2n-3)}\left(q,\frac{\gamma_1}{q},q^{-N};q\right)_n\delta_{mn}, \quad m,n = 0,1,2,\ldots,N$$

for $\gamma_1 > 0$.

Case VII. $g \neq 0$ and $\alpha \neq 0$ respectively $\alpha' \neq 0$. In section 10.4 we have seen that it is possible to have positive-definite orthogonality for an infinite system of polynomials in three different cases and for a finite system of polynomials in eight different cases. Here we will only treat two infinite cases.

Case VIIa1. $0 < q < 1$, $e < 1$, $\delta_1, \delta_2 \in \mathbb{R}$ and

$$\left(2fq^{n-1} + \delta_1 eq^{2n-2} + \delta_2\right)\left(2fq^{n-1} + \delta_1 + \delta_2 eq^{2n-2}\right) < 0, \quad n = 1,2,3,\ldots.$$

We use the weight function

$$w^{(VII)}(x;q) = \frac{\left(\frac{x}{\delta_1}, \frac{x}{\delta_2}; q\right)_\infty}{\left(\frac{\gamma_1 x}{q}, \frac{\gamma_2 x}{q}; q\right)_\infty}, \quad 0 < |q| < 1.$$

Then we have

$$\begin{aligned} w^{(VII)}(q^{-1}x;q)\varphi(q^{-2}x) &= \frac{\left(\frac{x}{\delta_1 q}, \frac{x}{\delta_2 q}; q\right)_\infty}{\left(\frac{\gamma_1 x}{q^2}, \frac{\gamma_2 x}{q^2}; q\right)_\infty} \cdot (eq^{-4}x^2 + 2fq^{-2}x + g) \\ &= \frac{\left(\frac{x}{\delta_1 q}, \frac{x}{\delta_2 q}; q\right)_\infty}{\left(\frac{\gamma_1 x}{q^2}, \frac{\gamma_2 x}{q^2}; q\right)_\infty} \cdot g\left(1 - \frac{\gamma_1 x}{q^2}\right)\left(1 - \frac{\gamma_2 x}{q^2}\right) \\ &= \frac{\left(\frac{x}{\delta_1 q}, \frac{x}{\delta_2 q}; q\right)_\infty}{\left(\frac{\gamma_1 x}{q}, \frac{\gamma_2 x}{q}; q\right)_\infty} \cdot g, \end{aligned}$$

which clearly vanishes for both $x = \delta_1 q$ and $x = \delta_2 q$. Hence we have for $\delta_1 < \delta_2$, $\dfrac{\delta_2 q}{\delta_1} < 1$, $\dfrac{\delta_1 q}{\delta_2} < 1$, $\gamma_1\delta_1 < 1$, $\gamma_1\delta_2 < 1$, $\gamma_2\delta_1 < 1$ and $\gamma_2\delta_2 < 1$ by using (1.15.11)

$$d_0 := \int_{\delta_1 q}^{\delta_2 q} \frac{\left(\frac{x}{\delta_1}, \frac{x}{\delta_2}; q\right)_\infty}{\left(\frac{\gamma_1 x}{q}, \frac{\gamma_2 x}{q}; q\right)_\infty} d_q x = (\delta_2 - \delta_1)q(1-q)\frac{\left(q, \frac{\delta_2 q}{\delta_1}, \frac{\delta_1 q}{\delta_2}, \gamma_1\gamma_2\delta_1\delta_2; q\right)_\infty}{(\gamma_1\delta_1, \gamma_1\delta_2, \gamma_2\delta_1, \gamma_2\delta_2; q)_\infty} > 0.$$

Further we have for $\delta_1\delta_2 = g < 0$ by using $e = \gamma_1\gamma_2\delta_1\delta_2$, (10.4.1), (10.4.2) and (10.4.3)

$$\begin{aligned} d_n &= -\delta_1\delta_2 q^{n+1}(1-q^n) \\ &\quad\times \frac{1 - \gamma_1\gamma_2\delta_1\delta_2 q^{n-2}}{(1-\gamma_1\gamma_2\delta_1\delta_2 q^{2n-3})(1-\gamma_1\gamma_2\delta_1\delta_2 q^{2n-2})^2(1-\gamma_1\gamma_2\delta_1\delta_2 q^{2n-1})} \\ &\quad\times \left(1-\gamma_1\delta_1 q^{n-1}\right)\left(1-\gamma_1\delta_2 q^{n-1}\right)\left(1-\gamma_2\delta_1 q^{n-1}\right)\left(1-\gamma_2\delta_2 q^{n-1}\right) \end{aligned}$$

for $n = 1,2,3,\ldots$, which implies that

$$\prod_{k=1}^{n} d_k = (-\delta_1\delta_2)^n q^{n(n+3)/2} \frac{(q,\gamma_1\gamma_2\delta_1\delta_2q^{-1};q)_n}{(\gamma_1\gamma_2\delta_1\delta_2q^{-1},\gamma_1\gamma_2\delta_1\delta_2;q)_{2n}} (\gamma_1\delta_1,\gamma_1\delta_2,\gamma_2\delta_1,\gamma_2\delta_2;q)_n$$

for $n = 1,2,3,\ldots$. This leads to the orthogonality relation

$$\int_{\delta_1 q}^{\delta_2 q} \frac{\left(\frac{x}{\delta_1},\frac{x}{\delta_2};q\right)_\infty}{\left(\frac{\gamma_1 x}{q},\frac{\gamma_2 x}{q};q\right)_\infty} y_m^{(VII)}(x;q) y_n^{(VII)}(x;q)\, d_qx$$
$$= (\delta_2-\delta_1)q(1-q)\frac{\left(q,\frac{\delta_2 q}{\delta_1},\frac{\delta_1 q}{\delta_2},\gamma_1\gamma_2\delta_1\delta_2;q\right)_\infty}{(\gamma_1\delta_1,\gamma_1\delta_2,\gamma_2\delta_1,\gamma_2\delta_2;q)_\infty}(-\delta_1\delta_2)^n q^{n(n+3)/2}$$
$$\times \frac{(q,\gamma_1\gamma_2\delta_1\delta_2q^{-1};q)_n}{(\gamma_1\gamma_2\delta_1\delta_2q^{-1},\gamma_1\gamma_2\delta_1\delta_2;q)_{2n}} (\gamma_1\delta_1,\gamma_1\delta_2,\gamma_2\delta_1,\gamma_2\delta_2;q)_n\,\delta_{mn}$$

for $\delta_1 < 0 < \delta_2$, $\dfrac{\delta_2 q}{\delta_1} < 1$, $\dfrac{\delta_1 q}{\delta_2} < 1$, $\gamma_1\delta_1 < 1$, $\gamma_1\delta_2 < 1$, $\gamma_2\delta_1 < 1$, $\gamma_2\delta_2 < 1$ and $m,n = 0,1,2,\ldots$.

The special case $\gamma_1 = q$, $\gamma_2 = bq/c$, $\delta_1 = c$ and $\delta_2 = a$ leads to the orthogonality relation (for $0 < aq < 1$, $0 \le bq < 1$ and $c < 0$)

$$\int_{cq}^{aq} \frac{\left(\frac{x}{a},\frac{x}{c};q\right)_\infty}{\left(x,\frac{bx}{c};q\right)_\infty} y_m(x;q)y_n(x;q)\,d_qx$$
$$= (a-c)q(1-q)\frac{(q,ac^{-1}q,a^{-1}cq,abq^2;q)_\infty}{(aq,bq,cq,abc^{-1}q;q)_\infty}(-ac)^n q^{n(n+3)/2}$$
$$\times \frac{(q,abq;q)_n}{(abq,abq^2;q)_{2n}}(aq,bq,cq,abc^{-1}q;q)_n\,\delta_{mn}, \quad m,n=0,1,2,\ldots$$

for the **big q-Jacobi** polynomials.

The special case $\gamma_1 = q$, $\gamma_2 = 0$, $\delta_1 = b$ and $\delta_2 = a$ leads to the orthogonality relation (for $0 < aq < 1$ and $b < 0$)

$$\int_{bq}^{aq} \frac{\left(\frac{x}{a},\frac{x}{b};q\right)_\infty}{(x;q)_\infty} y_m(x;q)y_n(x;q)\,d_qx$$
$$= (a-b)q(1-q)\frac{(q,ab^{-1}q,a^{-1}bq;q)_\infty}{(aq,bq;q)_\infty}$$
$$\times (-ab)^n q^{n(n+3)/2}(q;q)_n(aq,bq;q)_n\,\delta_{mn}, \quad m,n=0,1,2,\ldots$$

for the **big q-Laguerre** polynomials.

The special case $\gamma_1 = \gamma_2 = 0$, $\delta_1 = a/q$ and $\delta_2 = 1/q$ leads to the orthogonality relation (for $a < 0$)

$$\int_a^1 \left(qx, \frac{qx}{a};q\right)_\infty y_m(x;q)y_n(x;q)\,d_qx$$
$$= (1-a)(1-q)(q,aq,a^{-1}q;q)_\infty(-a)^n q^{\binom{n}{2}}(q;q)_n\,\delta_{mn}, \quad m,n=0,1,2,\ldots$$

for the **Al-Salam-Carlitz I** polynomials.

The special case $\gamma_1 = \gamma_2 = 0$, $\delta_1 = -1/q$ and $\delta_2 = 1/q$ leads to the orthogonality relation

$$\int_{-1}^1 (qx,-qx;q)_\infty y_m(x;q)y_n(x;q)\,d_qx$$
$$= 2(1-q)(q,-q,-q;q)_\infty q^{\binom{n}{2}}(q;q)_n\,\delta_{mn}, \quad m,n=0,1,2,\ldots$$

for the **discrete q-Hermite I** polynomials.

Case VIIa2. $q>1$, $e\le 0$, $\delta_1,\delta_2\in\mathbb{R}$ or $\delta_2=\overline{\delta_1}$ and

$$\left(2fq^{n-1}+\delta_1 eq^{2n-2}+\delta_2\right)\left(2fq^{n-1}+\delta_1+\delta_2 eq^{2n-2}\right)>0, \quad n=1,2,3,\ldots.$$

We write $q=p^{-1}$, which implies that $0<p<1$, and use the weight function

$$w^{(VII)}(x;p) = \frac{(\gamma_1 p^2x,\gamma_2 p^2x;p)_\infty}{\left(\frac{px}{\delta_1'},\frac{px}{\delta_2'};p\right)_\infty}, \quad 0<|p|<1.$$

Then we have

$$\begin{aligned} w^{(VII)}(px;p)\varphi(p^2x) &= \frac{(\gamma_1 p^3x,\gamma_2 p^3x;p)_\infty}{\left(\frac{p^2x}{\delta_1'},\frac{p^2x}{\delta_2'};p\right)_\infty}\cdot\left(ep^4x^2+2fp^2x+g\right) \\ &= \frac{(\gamma_1 p^3x,\gamma_2 p^3x;p)_\infty}{\left(\frac{p^2x}{\delta_1'},\frac{p^2x}{\delta_2'};p\right)_\infty}\cdot g\left(1-\gamma_1 p^2x\right)\left(1-\gamma_2 p^2x\right) \\ &= \frac{(\gamma_1 p^2x,\gamma_2 p^2x;p)_\infty}{\left(\frac{p^2x}{\delta_1'},\frac{p^2x}{\delta_2'};p\right)_\infty}\cdot g, \end{aligned}$$

which clearly vanishes for $x\to\pm\infty$ if $\gamma_1=\gamma_2=0$, which implies that $e=f=0$. Then we have for $p<|\delta_1'|<1$, $p<|\delta_2'|<1$ and $\delta_1'\delta_2'=g>0$ by using (1.15.13)

$$d_0 := \int_{-\infty}^{\infty}\frac{1}{\left(\frac{px}{\delta_1'},\frac{px}{\delta_2'};p\right)_\infty}\,d_px = (1-p)\frac{\left(p,-p,-1,-\frac{p^2}{\delta_1'\delta_2'},-\frac{\delta_1'\delta_2'}{p};p\right)_\infty}{\left(\frac{p}{\delta_1'},-\frac{p}{\delta_1'},\delta_1',-\delta_1',\frac{p}{\delta_2'},-\frac{p}{\delta_2'},\delta_2',-\delta_2';p\right)_\infty}>0.$$

Further we have for $\gamma_1=\gamma_2=0$ and $q=p^{-1}$

$$d_n = -\delta_1'\delta_2' q^{n+1}(1-q^n) = \delta_1'\delta_2' p^{-2n-1}(1-p^n), \quad n = 1,2,3,\ldots,$$

which implies that

$$\prod_{k=1}^{n} d_k = (\delta_1'\delta_2')^n p^{-n(n+2)}(p;p)_n, \quad n = 1,2,3,\ldots.$$

This leads to the orthogonality relation

$$\int_{-\infty}^{\infty} \frac{1}{\left(\frac{px}{\delta_1'}, \frac{px}{\delta_2'};p\right)_\infty} y_m^{(VII)}(x;p) y_n^{(VII)}(x;p)\, d_p x$$
$$= (1-p)\frac{\left(p,-p,-1,-\frac{p^2}{\delta_1'\delta_2'},-\frac{\delta_1'\delta_2'}{p};p\right)_\infty}{\left(\frac{p}{\delta_1'},-\frac{p}{\delta_1'},\delta_1',-\delta_1',\frac{p}{\delta_2'},-\frac{p}{\delta_2'},\delta_2',-\delta_2';p\right)_\infty}(\delta_1'\delta_2')^n p^{-n(n+2)}(p;p)_n\,\delta_{mn}$$

for $p < |\delta_1'| < 1$, $p < |\delta_2'| < 1$, $\delta_1'\delta_2' = g > 0$ and $m,n = 0,1,2,\ldots$.

The special case $\delta_1' = -ip$ and $\delta_2' = ip$, which implies that $\delta_1'\delta_2' = p^2 > 0$, leads to the orthogonality relation

$$\int_{-\infty}^{\infty} \frac{1}{(ix,-ix;p)_\infty} y_m(x;p) y_n(x;p)\, d_p x$$
$$= (1-p)\frac{(p,-p,-1,-1,-p;p)_\infty}{(i,-i,-ip,ip,-i,i,ip,-ip;p)_\infty} p^{-n^2}(p;p)_n\,\delta_{mn}, \quad m,n = 0,1,2,\ldots$$

for the **discrete q-Hermite II** polynomials.

Alternatively, we use the weight function

$$w^{(VII)}(x_\nu;p) = \frac{(\gamma_1 p^{2-\nu},\gamma_2 p^{2-\nu};p)_\nu}{\left(\frac{p^{1-\nu}}{\delta_1'},\frac{p^{1-\nu}}{\delta_2'};p\right)_\nu}.$$

For $\gamma_1 = \gamma_2 = 0$, which implies that $e = f = 0$, we have by using (1.8.14)

$$w^{(VII)}(x_\nu;p) = \frac{1}{\left(\frac{p^{1-\nu}}{\delta_1'},\frac{p^{1-\nu}}{\delta_2'};p\right)_\nu} = \frac{(\delta_1'\delta_2')^\nu p^{2\binom{\nu}{2}}}{(\delta_1',\delta_2';p)_\nu}.$$

Then we have

$$w^{(VII)}(px_\nu;p)\varphi(p^2x_\nu) = \frac{(\delta_1'\delta_2')^{\nu-1} p^{2\binom{\nu-1}{2}}}{(\delta_1',\delta_2';p)_{\nu-1}}\cdot g,$$

which vanishes for $\nu = 0$ if $\delta_1' = p$ or $\delta_2' = p$. Hence we have by using (1.11.8) with $\delta_2' = p$ and $x_\nu = p^{-\nu}$ for $0 < \delta_1' < 1$

$$d_0 := \sum_{\nu=0}^{\infty} w^{(VII)}(x_\nu;p)x_\nu = \sum_{\nu=0}^{\infty} \frac{(\delta_1')^\nu p^{2\binom{\nu}{2}}}{\left(\delta_1', p;p\right)_\nu} = \frac{1}{(\delta_1';p)_\infty} > 0.$$

Further we have for $\gamma_1 = \gamma_2 = 0$, $\delta_2' = p$, $q = p^{-1}$

$$d_n = -\delta_1'\delta_2'q^{n+1}(1-q^n) = \delta_1' p^{-2n}(1-p^n), \quad n = 1,2,3,\ldots,$$

which implies that

$$\prod_{k=1}^{n} d_k = (\delta_1')^n p^{-n(n+1)}(p;p)_n, \quad n = 1,2,3,\ldots.$$

This leads to the orthogonality relation

$$\sum_{\nu=0}^{\infty} \frac{(\delta_1')^\nu p^{2\binom{\nu}{2}}}{\left(\delta_1', p;p\right)_\nu} y_m^{(VII)}(p^{-\nu};p) y_n^{(VII)}(p^{-\nu};p) = \frac{1}{(\delta_1';p)_\infty}(\delta_1')^n p^{-n(n+1)}(p;p)_n\,\delta_{mn}$$

for $0 < \delta_1' < 1$ and $m,n = 0,1,2,\ldots$.

The special case $\delta_1' = ap$ leads to the orthogonality relation (for $0 < ap < 1$)

$$\sum_{\nu=0}^{\infty} \frac{(ap)^\nu p^{2\binom{\nu}{2}}}{(p,ap;p)_\nu} y_m(p^{-\nu};p) y_n(p^{-\nu};p) = \frac{1}{(ap;p)_\infty} a^n p^{-n^2}(p;p)_n\,\delta_{mn}$$

with $m,n = 0,1,2,\ldots$, for the **Al-Salam-Carlitz II** polynomials.

We remark that we cannot take $\delta_1' = -p$ in order to obtain an orthogonality relation for the **discrete q-Hermite II** polynomials, since then we have $\delta_1'\delta_2' = -p^2 < 0$.

Chapter 11
Orthogonal Polynomial Solutions in q^{-x} of q-Difference Equations

Classical q-Orthogonal Polynomials II

11.1 Polynomial Solutions in q^{-x} of q-Difference Equations

In the case that $\omega = 0$, $q > 0$ and $q \neq 1$ we might replace q by q^{-1} in (3.2.1) and then replace x by q^{-x}. Then we have

$$\left(\mathscr{D}_{1/q}f\right)(q^{-x}) = \frac{f(q^{-x-1}) - f(q^{-x})}{q^{-x-1} - q^{-x}} = \frac{f(q^{-x-1}) - f(q^{-x})}{q^{-x-1}(1-q)}, \quad q > 0, \quad q \neq 1.$$

In this case the eigenvalue problem reads (cf. (10.1.1))

$$\begin{aligned}&(eq^{-2x} + 2fq^{-x} + g)\left(\mathscr{D}_{1/q}^2 y_n\right)(q^{-x}) + (2\varepsilon q^{-x} + \gamma)\left(\mathscr{D}_{1/q} y_n\right)(q^{-x})\\&= \frac{q(q^n - 1)}{(q-1)^2}\left\{eq(1 - q^{-n+1}) - 2\varepsilon(1-q)\right\} y_n(q^{-x-1})\end{aligned} \qquad (11.1.1)$$

for $n = 0, 1, 2, \ldots$. This can also be written in the symmetric form

$$\begin{aligned}&A^*(x)y_n(q^{-x-1}) - \left\{A^*(x) + B^*(x)\right\} y_n(q^{-x}) + B^*(x)y_n(q^{-x+1})\\&= (q^n - 1)\left\{eq(1 - q^{-n+1}) - 2\varepsilon(1-q)\right\} y_n(q^{-x})\end{aligned} \qquad (11.1.2)$$

for $n = 0, 1, 2, \ldots$, with

$$A^*(x) = eq^2 + 2fq^{x+1} + gq^{2x}$$

and

$$B^*(x) = eq - 2\varepsilon(1-q) + \left\{2fq - \gamma(1-q)\right\} q^{x-1} + gq^{2x-1}.$$

The regularity condition (2.3.3) implies that $\varepsilon \neq 0$.

DOI 10.1007/978-3-642-05014-5_11,

It will turn out to be convenient to introduce (cf. (10.1.3))

$$\alpha^* := eq - 2\varepsilon(1-q) \quad \text{and} \quad \beta^* := 2fq - \gamma(1-q). \tag{11.1.3}$$

Then we have

$$A^*(x) = eq^2 + 2fq^{x+1} + gq^{2x} \quad \text{and} \quad B^*(x) = \alpha^* + \beta^* q^{x-1} + gq^{2x-1}, \tag{11.1.4}$$

where $e, f, g, \alpha^*, \beta^* \in \mathbb{C}$. In view of the homogeneity, one of the coefficients can be chosen arbitrarily. In this chapter, without loss of generality we may assume that $\alpha^* \in \mathbb{R}$. In section 11.6 we will see that this implies that all coefficients e, f, g, α^* and β^* must be real.

11.2 The Basic Hypergeometric Representation

We try to find solutions of the form

$$y_n(q^{-x}) = \sum_{k=0}^{n} a_{n,k} \frac{(-c^{-1}q^{-x};q)_k}{(q;q)_k} c^k (1-q)^k, \quad a_{n,n} \neq 0, \quad n = 0,1,2,\ldots \tag{11.2.1}$$

for (11.1.2) with $c \neq 0$. Substitution of (11.2.1) into (11.1.2) leads to a three-term recurrence relation (cf. (2.4.4)) for the coefficients $\{a_{n,k}\}_{k=0}^{n}$. If c satisfies the relation (cf. (2.4.5))

$$\alpha^* qc^2 - \beta^* c + g = 0 \tag{11.2.2}$$

the recurrence relation reduces to the two-term recurrence relation (cf. (2.4.7))

$$[n-k]\left(\alpha^* - eq^{-n-k+2}\right) c a_{n,k} = q^{-k}\left\{ec^2 q^{-k+1} - 2fc + gq^{k-1}\right\} a_{n,k+1} \tag{11.2.3}$$

for $k = n-1, n-2, n-3, \ldots, 0$ and $n = 1,2,3,\ldots$ for $a_{n,k}$. In this case the regularity condition (2.3.3) implies that

$$\alpha^* - eq^{-n+1} \neq 0, \quad n = 0,1,2,\ldots. \tag{11.2.4}$$

This implies that all coefficients $a_{n,k}$ are uniquely determined in terms of $a_{n,n} (\neq 0)$ provided that $c \neq 0$. In fact we have

$$a_{n,k} = \left(\prod_{i=1}^{n-k} \frac{ec^2 q^{-n+i+1} - 2fc + gq^{n-i-1}}{[i]\left(\alpha^* q^{n-i} - eq^{-n+2}\right) c}\right) a_{n,n}.$$

In order to have monic polynomials we choose

$$a_{n,n} = [n]!\, q^{-\binom{n}{2}} = \frac{(q;q)_n}{(1-q)^n} q^{-\binom{n}{2}}.$$

Hence, by using (10.2.5) we obtain

$$a_{n,k} = \left(\prod_{i=1}^{n-k} \frac{ec^2q^{-n+i+1} - 2fc + gq^{n-i-1}}{\alpha^* q^{n-i} - eq^{-n+2}}\right) \frac{(q^{-n};q)_k}{(1-q)^k c^{n-k}} (-1)^k q^{-\binom{n}{2}-\binom{k}{2}+nk}.$$

If $e = 0$ we have $\alpha^* = -2\varepsilon(1-q) \neq 0$ and

$$\prod_{i=1}^{n-k} \left(\alpha^* q^{n-i} - eq^{-n+2}\right) = (\alpha^*)^{n-k} q^{(n-k)(n+k-1)/2}.$$

If $e \neq 0$ we have

$$\begin{aligned}\prod_{i=1}^{n-k} \left(\alpha^* q^{n-i} - eq^{-n+2}\right) &= \left(-eq^{-n+2}\right)^{n-k} \left(\frac{\alpha^* q^{n-2}}{e} q^k; q\right)_{n-k} \\ &= \left(-eq^{-n+2}\right)^{n-k} \frac{\left(\frac{\alpha^*}{e} q^{n-2}; q\right)_n}{\left(\frac{\alpha^*}{e} q^{n-2}; q\right)_k}.\end{aligned}$$

Case I. If $e = f = 0$ and $g \neq 0$ we have $\alpha^* = -2\varepsilon(1-q) \neq 0$ and

$$\prod_{i=1}^{n-k} \left(ec^2q^{-n+i+1} - 2fc + gq^{n-i-1}\right) = g^{n-k} q^{(n-k)(n+k-3)/2}.$$

Hence for $e = f = 0$ and $g \neq 0$ the monic polynomials can be written as

$$\begin{aligned}y_n^{(I)}(q^{-x};q) &= \left(\frac{g}{\alpha^* cq}\right)^n q^{-\binom{n}{2}} \sum_{k=0}^{n} \frac{(q^{-n};q)_k(-c^{-1}q^{-x};q)_k}{(q;q)_k} q^{-\binom{k}{2}} \left(-\frac{\alpha^* c^2 q^{n+1}}{g}\right)^k \\ &= \left(\frac{g}{\alpha^* cq}\right)^n q^{-\binom{n}{2}} {}_2\phi_0\left(\begin{matrix} q^{-n}, -c^{-1}q^{-x} \\ - \end{matrix}; q, \frac{\alpha^* c^2 q^{n+1}}{g}\right)\end{aligned}$$

for $n = 0, 1, 2, \ldots$. The q-polynomials in this class have, besides q, one free parameter α^*/g $(g \neq 0)$ or g/α^* $(\alpha^* \neq 0)$. Special cases of q-polynomials in this class are the **Al-Salam-Carlitz I and II** polynomials with one free parameter and the **discrete q-Hermite I and II** polynomials with no free parameters.

Case II. If $e = 0$ and $f \neq 0$ we have $\alpha^* = -2\varepsilon(1-q) \neq 0$ and by using (1.8.12)

$$\begin{aligned}\prod_{i=1}^{n-k}\left(ec^2q^{-n+i+1}-2fc+gq^{n-i-1}\right) &= (-2fc)^{n-k}\prod_{i=1}^{n-k}\left(1-\frac{g}{2fc}q^{n-i-1}\right)\\ &= (-2fc)^{n-k}\left(\frac{g}{2fcq}q^k;q\right)_{n-k}\\ &= (-2fc)^{n-k}\frac{\left(\frac{g}{2fcq};q\right)_n}{\left(\frac{g}{2fcq};q\right)_k}.\end{aligned}$$

Hence for $e=0$ and $f\neq 0$ the monic polynomials can be written as

$$\begin{aligned}y_n^{(II)}(q^{-x};q) &= \left(-\frac{2f}{\alpha^*}\right)^n\left(\frac{g}{2fcq};q\right)_n q^{-2\binom{n}{2}}\\ &\quad\times\sum_{k=0}^{n}\frac{(q^{-n};q)_k(-c^{-1}q^{-x};q)_k}{\left(\frac{g}{2fcq};q\right)_k(q;q)_k}\left(\frac{\alpha^* cq^n}{2f}\right)^k\\ &= \left(-\frac{2f}{\alpha^* q^{n-1}}\right)^n\left(\frac{g}{2fcq};q\right)_n {}_2\phi_1\left(\begin{matrix}q^{-n},-c^{-1}q^{-x}\\ \frac{g}{2fcq}\end{matrix};q,\frac{\alpha^* cq^n}{2f}\right)\end{aligned}$$

for $n=0,1,2,\ldots$. The q-polynomials in this class have, besides q, two free parameters α^*/f and g/f ($f\neq 0$) or f/α^* and g/α^* ($\alpha^*\neq 0$). Special cases of q-polynomials in this class are the **q-Meixner** polynomials and the **quantum q-Krawtchouk** polynomials with two free parameters each and the **q-Charlier** polynomials with only one free parameter.

Case III. If $e\neq 0$ we have

$$\begin{aligned}&\prod_{i=1}^{n-k}\left(ec^2q^{-n+i+1}-2fc+gq^{n-i-1}\right)\\ &=\prod_{i=1}^{n-k}\left\{ec^2q^{-n+i+1}\left(1-\frac{2f}{ec}q^{n-i-1}+\frac{g}{ec^2}q^{2n-2i-2}\right)\right\}\\ &=\left(ec^2\right)^{n-k}q^{-(n-k)(n+k-3)/2}\prod_{i=1}^{n-k}\left\{\left(1+c^{-1}\xi_1q^{n-i-1}\right)\left(1+c^{-1}\xi_2q^{n-i-1}\right)\right\},\end{aligned}$$

where $\xi_1\xi_2=g/e$ and $\xi_1+\xi_2=-2f/e$. This implies that for $f^2\geq eg$

$$\xi_1=\frac{-f-\sqrt{f^2-eg}}{e}\quad\text{and}\quad\xi_2=\frac{-f+\sqrt{f^2-eg}}{e}.$$

Note that for $g\neq 0$ we have that $\xi_1=\gamma_2^{-1}$ and $\xi_2=\gamma_1^{-1}$ with $\gamma_1\gamma_2=e/g$ and $\gamma_1+\gamma_2=-2f/g$ and therefore for $f^2\geq eg$

$$\gamma_1 = \frac{-f-\sqrt{f^2-eg}}{g} \quad \text{and} \quad \gamma_2 = \frac{-f+\sqrt{f^2-eg}}{g}$$

as in the previous chapter. In the case that $f^2 < eg$ we have both $\gamma_2 = \overline{\gamma_1}$ and $\xi_2 = \overline{\xi_1}$.

Hence for $e \neq 0$ the monic polynomials can be written as

$$\begin{aligned} y_n^{(III)}(q^{-x};q) &= \frac{(-c)^n}{\left(\frac{\alpha^*}{e}q^{n-2};q\right)_n} \sum_{k=0}^{n} \left(-c^{-1}\xi_1 q^{k-1}, -c^{-1}\xi_2 q^{k-1};q\right)_{n-k} \\ &\quad \times \left(\frac{\alpha^*}{e}q^{n-2};q\right)_k \frac{(q^{-n};q)_k(-c^{-1}q^{-x};q)_k}{(q;q)_k} q^k \\ &= (-c)^n \frac{\left(-c^{-1}\xi_1 q^{-1}, -c^{-1}\xi_2 q^{-1};q\right)_n}{\left(\frac{\alpha^*}{e}q^{n-2};q\right)_n} \\ &\quad \times {}_3\phi_2\left(\begin{matrix} q^{-n}, -c^{-1}q^{-x}, \frac{\alpha^*}{e}q^{n-2} \\ -c^{-1}\xi_1 q^{-1}, -c^{-1}\xi_2 q^{-1} \end{matrix}; q, q\right) \end{aligned}$$

for $n = 0,1,2,\ldots$. The q-polynomials in this class have, besides q, three free parameters ξ_1, ξ_2 and α^*/e ($e \neq 0$). Special cases of q-polynomials in this class are the **q-Hahn** polynomials with three free parameters and the **q-Krawtchouk** and the **affine q-Krawtchouk** polynomials with two free parameters each.

Now we have:

Theorem 11.1. *All orthogonal polynomial solutions $y_n(q^{-x})$ of the q-difference equation (11.1.2)*

$$\begin{aligned} &\left(eq^2+2fq^{x+1}+gq^{2x}\right) y_n(q^{-x-1}) \\ &\quad -\left\{eq^2+2fq^{x+1}+gq^{2x}+\alpha^*+\beta^* q^{x-1}+gq^{2x-1}\right\} y_n(q^{-x}) \\ &\qquad +\left(\alpha^*+\beta^* q^{x-1}+gq^{2x-1}\right) y_n(q^{-x+1}) \\ &= (q^n-1)\left(\alpha^*-eq^{-n+2}\right) y_n(q^{-x}), \quad n=0,1,2,\ldots, \end{aligned}$$

where $\alpha^* := eq - 2\varepsilon(1-q)$ and $\beta^* := 2fq - \gamma(1-q)$, can be divided into three different cases:

Case I. *$e = f = 0$ and $g \neq 0$*

Case II. *$e = 0$ and $f \neq 0$*

Case III. *$e \neq 0$.*

11.3 The Three-Term Recurrence Relation

Analogous to section 10.3 we obtain for the monic polynomial solutions of (11.1.1) the three-term recurrence relation

$$y_{n+1}(q^{-x}) = (q^{-x} - c_n)y_n(q^{-x}) - d_n y_{n-1}(q^{-x}), \quad n = 1,2,3,\ldots \tag{11.3.1}$$

with initial values $y_0(q^{-x}) = 1$ and $y_1(q^{-x}) = q^{-x} - c_0$ where $c_0 = -\gamma/2\varepsilon q$. If we replace q by q^{-1} we find by using (10.1.3) and (11.1.3) that

$$\alpha = e + 2\varepsilon(1-q) \quad \longrightarrow \quad e + 2\varepsilon(1-q^{-1}) = (eq - 2\varepsilon(1-q))q^{-1} = \alpha^* q^{-1}$$

and

$$\beta = 2f + \gamma(1-q) \quad \longrightarrow \quad 2f + \gamma(1-q^{-1}) = (2fq - \gamma(1-q))q^{-1} = \beta^* q^{-1}.$$

Further we obtain from (10.3.2) and (10.3.3) that $c_0 = -(2fq - \beta^*)/(eq - \alpha^*)q$,

$$\begin{aligned} c_n &= -\frac{2f\left\{\alpha^* q^{-1} - \alpha^* q^{-1}(1+q^{-1})q^{-n} + eq^{-2n+1}\right\}}{q^n\left(\alpha^* q^{-1} - eq^{-2n+2}\right)\left(\alpha^* q^{-1} - eq^{-2n}\right)} \\ &\quad - \frac{\beta^* q^{-2}\left\{\alpha^* q^{-1} - e(1+q^{-1})q^{-n+2} + eq^{-2n+1}\right\}}{q^n\left(\alpha^* q^{-1} - eq^{-2n+2}\right)\left(\alpha^* q^{-1} - eq^{-2n}\right)} \\ &= -\frac{2fq\left\{\alpha^* q - \alpha^*(1+q)q^{-n} + eq^{-2n+3}\right\}}{q^{n+1}\left(\alpha^* - eq^{-2n+3}\right)\left(\alpha^* - eq^{-2n+1}\right)} \\ &\quad - \frac{\beta^*\left\{\alpha^* - e(1+q)q^{-n+2} + eq^{-2n+2}\right\}}{q^{n+1}\left(\alpha^* - eq^{-2n+3}\right)\left(\alpha^* - eq^{-2n+1}\right)} \end{aligned} \tag{11.3.2}$$

for $n = 1,2,3,\ldots$ and

$$\begin{aligned} d_n &= \frac{q^{-n-1}(1-q^{-n})\left(\alpha^* q^{-1} - eq^{-n+2}\right)}{\left(\alpha^* q^{-1} - eq^{-2n+3}\right)\left(\alpha^* q^{-1} - eq^{-2n+2}\right)^2\left(\alpha^* q^{-1} - eq^{-2n+1}\right)} \\ &\qquad \times \left\{q^{-n+1}(\beta^* q^{-1} - 2fq^{-n+1})(2f\alpha^* q^{-1} - e\beta^* q^{-1} q^{-n+1}) \right. \\ &\qquad \left. - g(\alpha^* q^{-1} - eq^{-2n+2})^2\right\} \\ &= \frac{q^{-n}(1-q^{-n})\left(\alpha^* - eq^{-n+3}\right)}{\left(\alpha^* - eq^{-2n+4}\right)\left(\alpha^* - eq^{-2n+3}\right)^2\left(\alpha^* - eq^{-2n+2}\right)} \\ &\qquad \times \left\{q^{-n+1}\left(\beta^* - 2fq^{-n+2}\right)\left(2f\alpha^* - e\beta^* q^{-n+1}\right) \right. \\ &\qquad \left. - g\left(\alpha^* - eq^{-2n+3}\right)^2\right\} \end{aligned} \tag{11.3.3}$$

for $n = 1,2,3,\ldots$.

11.4 Orthogonality and the Self-Adjoint Operator Equation

In this section we will use the following form of Hahn's q-operator (cf. (3.2.1))

$$\begin{aligned}\left(\mathscr{D}_{1/q}f\right)(q^{-x}) &:= \left(\mathscr{A}_{1/q,0}f\right)(q^{-x})\\ &= \frac{f(q^{-x-1})-f(q^{-x})}{q^{-x-1}-q^{-x}}, \quad q>0, \quad q\neq 1, \end{aligned} \tag{11.4.1}$$

where f is a complex-valued function in q^{-x} whose domain contains both q^{-x} and q^{-x-1} for each $x\in\mathbb{R}$. For two such functions f_1 and f_2, the product rule (3.2.2) reads

$$\begin{aligned}&\left(\mathscr{D}_{1/q}(f_1f_2)\right)(q^{-x})\\ &= \left(\mathscr{D}_{1/q}f_1\right)(q^{-x})f_2(q^{-x})+f_1(q^{-x-1})\left(\mathscr{D}_{1/q}f_2\right)(q^{-x}).\end{aligned} \tag{11.4.2}$$

Analogous to (3.2.3) and (3.2.4) we now define

$$\begin{aligned}&\left(\mathscr{S}_{1/q}f\right)(q^{-x}) = f(q^{-x-1})\\ &\text{and}\quad \widehat{f}(q^{-x}) = \left(\mathscr{S}_{1/q}^{-1}f\right)(q^{-x}) = f(q^{-x+1}).\end{aligned} \tag{11.4.3}$$

As in section 3.2

$$\begin{aligned}&\left(\mathscr{S}_{1/q}w\right)(q^{-x})\varphi(q^{-x})\left(\mathscr{D}_{1/q}^2y_n\right)(q^{-x})\\ &\quad+\left(\mathscr{S}_{1/q}w\right)(q^{-x})\psi(q^{-x})\left(\mathscr{D}_{1/q}y_n\right)(q^{-x}) = \lambda_n\left(\mathscr{S}_{1/q}w\right)(q^{-x})\left(\mathscr{S}_{1/q}y_n\right)(q^{-x})\end{aligned}$$

where $\lambda_n = q(q^n-1)\left\{eq(1-q^{-n+1})-2\varepsilon(1-q)\right\}/(q-1)^2$,

$$\varphi(q^{-x}) = eq^{-2x}+2fq^{-x}+g \quad\text{and}\quad \psi(q^{-x}) = 2\varepsilon q^{-x}+\gamma$$

leads to the self-adjoint form

$$\left(\mathscr{D}_{1/q}\left(w\widehat{\varphi}\mathscr{D}_{1/q}y_n\right)\right)(q^{-x}) = \lambda_n\left(\mathscr{S}_{1/q}w\right)(q^{-x})\left(\mathscr{S}_{1/q}y_n\right)(q^{-x}) \tag{11.4.4}$$

if the Pearson operator equation

$$\left(\mathscr{D}_{1/q}(w\widehat{\varphi})\right)(q^{-x}) = \left(\mathscr{S}_{1/q}w\right)(q^{-x})\psi(q^{-x}) \tag{11.4.5}$$

holds. By using the product rule (11.4.2), we find that

$$\begin{aligned}\left(\mathscr{D}_{1/q}(w\widehat{\varphi})\right)(q^{-x}) &= \widehat{\varphi}(q^{-x})\frac{w(q^{-x-1})-w(q^{-x})}{q^{-x-1}-q^{-x}}\\ &\qquad+w(q^{-x-1})\frac{\widehat{\varphi}(q^{-x-1})-\widehat{\varphi}(q^{-x})}{q^{-x-1}-q^{-x}}\\ &= \frac{w(q^{-x-1})\varphi(q^{-x})-w(q^{-x})\widehat{\varphi}(q^{-x})}{q^{-x-1}-q^{-x}}.\end{aligned}$$

Note that the right-hand side of (11.4.5) can be written in the form $w(q^{-x-1})\psi(q^{-x})$. Hence (11.4.5) becomes

$$w(q^{-x})\widehat{\varphi}(q^{-x}) = w(q^{-x-1})\left\{\varphi(q^{-x}) - \psi(q^{-x})(q^{-x-1} - q^{-x})\right\}. \tag{11.4.6}$$

Further we have from (2.2.13) with $\omega = 0$

$$C(x) = \frac{ex^2 + 2fqx + gq^2}{q(q-1)^2x^2} = \frac{q\left(ex^2/q^2 + 2fx/q + g\right)}{(q-1)^2x^2} = \frac{q\varphi(x/q)}{(q-1)^2x^2} = \frac{A(x)}{q(q-1)^2x^2},$$

where $A(x) = ex^2 + 2fqx + gq^2$. If we replace q by q^{-1} and then x by q^{-x}, we obtain that

$$\frac{q}{(1-q)^2}(eq^2 + 2fq^{x+1} + gq^{2x}) = \frac{q^{2x+1}}{(1-q)^2}\varphi(q^{-x+1}) = \frac{q}{(1-q)^2}A^*(x).$$

Moreover, we also find from (2.2.13) with $\omega = 0$ that

$$D(qx) = qC(qx) - \frac{\psi(x)}{(q-1)x} = \frac{1}{(q-1)^2x^2}\left\{\varphi(x) + (1-q)x\psi(x)\right\}.$$

If we replace q by q^{-1} and then x by q^{-x}, the latter becomes

$$\begin{aligned} &\frac{q^{2x+2}}{(1-q)^2}\left\{\varphi(q^{-x}) + (1-q^{-1})q^{-x}\psi(q^{-x})\right\} \\ &= \frac{q^{2x+2}}{(1-q)^2}\left\{\varphi(q^{-x}) - (1-q)q^{-x-1}\psi(q^{-x})\right\}. \end{aligned}$$

Note that

$$\begin{aligned} B^*(x) &= eq + 2fq^x + gq^{2x-1} - (1-q)(2\varepsilon + \gamma q^{x-1}) \\ &= q^{2x-1}\varphi(q^{-x+1}) - (1-q)q^{x-1}\psi(q^{-x+1}). \end{aligned}$$

Hence

$$qB^*(x+1) = q^{2x+2}\left\{\varphi(q^{-x}) - (1-q)q^{-x-1}\psi(q^{-x})\right\}.$$

This implies that

$$\frac{q^{2x+2}}{(1-q)^2}\left\{\varphi(q^{-x}) - (1-q)q^{-x-1}\psi(q^{-x})\right\} = \frac{q}{(1-q)^2}B^*(x+1).$$

Hence the Pearson equation (11.4.5) is equivalent to

$$qw(q^{-x})A^*(x) = w(q^{-x-1})B^*(x+1). \tag{11.4.7}$$

Now we multiply (11.4.4) by $\left(\mathscr{S}_{1/q}y_m\right)(q^{-x})$ and subtract from the resulting equation the same equation with m and n exchanged. Then we apply $\mathscr{S}_{1/q}^{-1}$ to the result and use the commutation relation (cf. (2.5.3))

$$\mathscr{S}_{1/q}^{-1}\mathscr{D}_{1/q} = q^{-1}\mathscr{D}_{1/q}\mathscr{S}_{1/q}^{-1} \tag{11.4.8}$$

to find that

$$\begin{aligned}
&(\lambda_n - \lambda_m)w(q^{-x})y_m(q^{-x})y_n(q^{-x}) \\
&= q^{-1}\left(\mathscr{D}_{1/q}\mathscr{S}_{1/q}^{-1}\left(w\widehat{\varphi}\mathscr{D}_{1/q}y_n\right)\right)(q^{-x})y_m(q^{-x}) \\
&\quad - q^{-1}\left(\mathscr{D}_{1/q}\mathscr{S}_{1/q}^{-1}\left(w\widehat{\varphi}\mathscr{D}_{1/q}y_m\right)\right)(q^{-x})y_n(q^{-x}).
\end{aligned}$$

As before this leads to two kinds of orthogonality.

A. Consider the interval (a,b) on the real line and the q-integration by parts formula (cf. (3.2.10))

$$\begin{aligned}
&\int_a^b \left(\mathscr{D}_{1/q}f_1\right)(q^{-x})f_2(q^{-x})\,d_qx \\
&= \Big[f_1(q^{-x})f_2(q^{-x})\Big]_a^b - \int_a^b \left(\mathscr{S}_{1/q}f_1\right)(q^{-x})\left(\mathscr{D}_{1/q}f_2\right)(q^{-x})\,d_qx
\end{aligned} \tag{11.4.9}$$

for arbitrary complex-valued functions f_1 and f_2 which are q-integrable on the interval (a,b). Then we have

$$\begin{aligned}
&(\lambda_n - \lambda_m)\int_a^b w(q^{-x})y_m(q^{-x})y_n(q^{-x})\,d_qx \\
&= q^{-1}\Big(\left(\mathscr{S}_{1/q}^{-1}(w\widehat{\varphi})\right)(q^{-x})\Big\{\left(\mathscr{S}_{1/q}^{-1}\left(\mathscr{D}_{1/q}y_n\right)\right)(q^{-x})y_m(q^{-x}) \\
&\quad - \left(\mathscr{S}_{1/q}^{-1}\left(\mathscr{D}_{1/q}y_m\right)\right)(q^{-x})y_n(q^{-x})\Big\}\Big)_a^b.
\end{aligned}$$

In view of the regularity condition (11.2.4), we have $\lambda_m \neq \lambda_n$ for $m \neq n$. So we have

Theorem 11.2. *Let $\{y_n\}_{n=0}^{\infty}$ denote the polynomial solutions of the eigenvalue problem (11.1.1), let the regularity condition (11.2.4) hold for $n = 0,1,2,\ldots$ and let w denote a complex valued function which is q-integrable on the interval (a,b) on the real line and which satisfies the Pearson operator equation (11.4.5). Then we have the orthogonality relation*

$$\int_a^b w(q^{-x})y_m(q^{-x})y_n(q^{-x})\,d_qx = 0, \quad m \neq n, \quad m,n \in \{0,1,2,\ldots\} \tag{11.4.10}$$

with weight function w if the boundary conditions

$$\left(\mathscr{S}_{1/q}^{-1}(w\widehat{\varphi})\right)(q^{-a}) = 0 \quad \text{and} \quad \left(\mathscr{S}_{1/q}^{-1}(w\widehat{\varphi})\right)(q^{-b}) = 0 \tag{11.4.11}$$

hold. Here a continuous extension of $w\widehat{\varphi}$ might be necessary.

If the necessary convergence conditions hold, the integral in (11.4.10) can also be taken over (a,∞), $(-\infty,b)$ or $(-\infty,\infty)$ with appropriate boundary conditions.

B. For $N \in \{1,2,3,\ldots\}$ we consider the set of points

$$x_\nu := Aq^{-\nu}, \quad A \in \mathbb{C}, \quad \nu = 0,1,2,\ldots,N+1. \tag{11.4.12}$$

Then we have the summation by parts formula

$$\begin{aligned}
&\sum_{\nu=0}^{N} \left(\mathscr{D}_{1/q} f_1\right)(x_\nu) f_2(x_\nu)\left(q^{-1}(1-q)x_\nu\right) \\
&= \Big[f_1(x_\nu) f_2(x_\nu)\Big]_{\nu=0}^{N+1} \\
&\qquad - \sum_{\nu=0}^{N} \left(\mathscr{S}_{1/q} f_1\right)(x_\nu)\left(\mathscr{D}_{1/q} f_2\right)(x_\nu)\left(q^{-1}(1-q)x_\nu\right)
\end{aligned} \tag{11.4.13}$$

for arbitrary complex-valued functions f_1 and f_2 in q^{-x} whose domain contains the set of points $\{x_\nu\}_{\nu=0}^{N}$. Hence we have

$$\begin{aligned}
&(\lambda_n - \lambda_m) \sum_{\nu=0}^{N} w(x_\nu) y_m(x_\nu) y_n(x_\nu)\left(q^{-1}(1-q)x_\nu\right) \\
&= q^{-1}\Bigg(\left(\mathscr{S}_{1/q}^{-1}(w\widehat{\varphi})\right)(x_\nu)\Big\{\left(\mathscr{S}_{1/q}^{-1}\left(\mathscr{D}_{1/q} y_n\right)\right)(x_\nu) y_m(x_\nu) \\
&\qquad - \left(\mathscr{S}_{1/q}^{-1}\left(\mathscr{D}_{1/q} y_m\right)\right)(x_\nu) y_n(x_\nu)\Big\}\Bigg)_{\nu=0}^{N+1}.
\end{aligned}$$

In view of the regularity condition (11.2.4), we have $\lambda_m \neq \lambda_n$ for $m \neq n$. So we have

Theorem 11.3. *Let $\{y_n\}_{n=0}^{\infty}$ denote the polynomial solutions of the eigenvalue problem (11.1.1), let the regularity condition (11.2.4) hold for $n = 0,1,2,\ldots$ and let w denote a function whose domain contains the set of points $\{x_\nu\}_{\nu=0}^{N+1}$ and that satisfies the Pearson operator equation (11.4.5). Then we have the orthogonality relation*

$$\sum_{\nu=0}^{N} w(x_\nu) y_m(x_\nu) y_n(x_\nu) x_\nu = 0, \quad m \neq n, \quad m,n \in \{0,1,2,\ldots,N\} \tag{11.4.14}$$

if the boundary conditions

$$\left(\mathscr{S}_{1/q}^{-1}(w\widehat{\varphi})\right)(x_0) = 0 \quad \text{and} \quad \left(\mathscr{S}_{1/q}^{-1}(w\widehat{\varphi})\right)(x_{N+1}) = 0 \tag{11.4.15}$$

hold. Here a continuous extension of $w\widehat{\varphi}$ might be necessary.

11.5 Rodrigues Formulas

We start with the self-adjoint operator equation (cf. (11.4.4))

$$\begin{aligned}&\left(\mathscr{D}_{1/q}\left(w\left(\mathscr{S}_{1/q}^{-1}\varphi\right)\left(\mathscr{D}_{1/q}y_n\right)\right)\right)(q^{-x})\\&=\lambda_n\left(\mathscr{S}_{1/q}w\right)(q^{-x})\left(\mathscr{S}_{1/q}y_n\right)(q^{-x}).\end{aligned}\tag{11.5.1}$$

We apply the operator $\mathscr{S}_{1/q}^{-1}$ on both sides of (11.5.1) and use (11.4.8) and the fact that $\lambda_0=0$ to obtain

$$\begin{aligned}&\left(\mathscr{D}_{1/q}\left(\left(\mathscr{S}_{1/q}^{-1}w\right)\left(\mathscr{S}_{1/q}^{-2}\varphi\right)\mathscr{S}_{1/q}^{-1}\left(\mathscr{D}_{1/q}y_n\right)\right)\right)(q^{-x})\\&=q(\lambda_n-\lambda_0)w(q^{-x})y_n(q^{-x}).\end{aligned}$$

This formula can be generalized to

$$\begin{aligned}&\left(\mathscr{D}_{1/q}\left(\left(\mathscr{S}_{1/q}^{-k}w\right)\left\{\prod_{i=1}^{k}\left(\mathscr{S}_{1/q}^{-i-1}\varphi\right)\right\}\mathscr{S}_{1/q}^{-k}\left(\mathscr{D}_{1/q}^{k}y_n\right)\right)\right)(q^{-x})\\&=q(\lambda_n-\lambda_{k-1})\\&\quad\times\left(\left(\mathscr{S}_{1/q}^{-k+1}w\right)\left\{\prod_{i=1}^{k-1}\left(\mathscr{S}_{1/q}^{-i-1}\varphi\right)\right\}\mathscr{S}_{1/q}^{-k+1}\left(\mathscr{D}_{1/q}^{k-1}y_n\right)\right)(q^{-x})\end{aligned}$$

for $k=1,2,3,\ldots$. This can be proved by using induction similar to the situation in section 3.4.

If we assume that the polynomials $\{y_n\}_{n=0}^{\infty}$ are monic, id est $\left(\mathscr{D}_{1/q}^{n}y_n\right)(q^{-x})=q^{-\binom{n}{2}}[n]!$, and the regularity condition (11.2.4) holds, then we have the Rodrigues formula

$$y_n(q^{-x})=\frac{K_n}{w(q^{-x})}\left[\mathscr{D}_{1/q}^{n}\left\{\left(\mathscr{S}_{1/q}^{-n}w\right)\prod_{k=1}^{n}\left(\mathscr{S}_{1/q}^{-k-1}\varphi\right)\right\}\right](q^{-x})\tag{11.5.2}$$

for $n=1,2,3,\ldots$, where

$$K_n=\frac{1}{q^{\binom{n+1}{2}}}\prod_{k=1}^{n}\frac{[k]}{\lambda_n-\lambda_{n-k}},\quad n=1,2,3,\ldots.\tag{11.5.3}$$

11.6 Classification of the Positive-Definite Orthogonal Polynomial Solutions

Again we will use Favard's theorem (theorem 3.1) to conclude that there exist positive definite orthogonal polynomial solutions iff $c_n \in \mathbb{R}$ for $n = 0, 1, 2, \ldots$ and $d_n > 0$ for all $n = 1, 2, 3, \ldots$. In section 11.3 we have obtained

$$c_n = -\frac{2fq\left\{\alpha^* q - \alpha^*(1+q)q^{-n} + eq^{-2n+3}\right\}}{q^{n+1}\left(\alpha^* - eq^{-2n+3}\right)\left(\alpha^* - eq^{-2n+1}\right)} - \frac{\beta^*\left\{\alpha^* - e(1+q)q^{-n+2} + eq^{-2n+2}\right\}}{q^{n+1}\left(\alpha^* - eq^{-2n+3}\right)\left(\alpha^* - eq^{-2n+1}\right)}$$

for $n = 0, 1, 2, \ldots$ and

$$d_n = \frac{q^{-n}\left(1-q^{-n}\right)\left(\alpha^* - eq^{-n+3}\right)}{\left(\alpha^* - eq^{-2n+4}\right)\left(\alpha^* - eq^{-2n+3}\right)^2\left(\alpha^* - eq^{-2n+2}\right)} \times \left\{q^{-n+1}\left(\beta^* - 2fq^{-n+2}\right)\left(2f\alpha^* - e\beta^* q^{-n+1}\right) - g\left(\alpha^* - eq^{-2n+3}\right)^2\right\}$$

for $n = 1, 2, 3, \ldots$. Now we write

$$d_n = q^{-n}\left(1-q^{-n}\right)D_n^{(1)}D_n^{(2)} = -q^{-2n}(1-q^n)D_n^{(1)}D_n^{(2)}, \quad n = 1, 2, 3, \ldots \tag{11.6.1}$$

with

$$D_n^{(1)} = \frac{\alpha^* - eq^{-n+3}}{\left(\alpha^* - eq^{-2n+4}\right)\left(\alpha^* - eq^{-2n+3}\right)^2\left(\alpha^* - eq^{-2n+2}\right)}, \quad n = 1, 2, 3, \ldots \tag{11.6.2}$$

and

$$D_n^{(2)} = q^{-n+1}\left(\beta^* - 2fq^{-n+2}\right)\left(2f\alpha^* - e\beta^* q^{-n+1}\right) - g\left(\alpha^* - eq^{-2n+3}\right)^2 \tag{11.6.3}$$

for $n = 1, 2, 3, \ldots$.

Case I. $e = f = 0$ and $g \neq 0$. Then we have $\alpha^* = -2\varepsilon(1-q) \neq 0$ and

$$c_n = -\frac{\beta^*}{\alpha^* q^{n+1}}, \quad n = 0, 1, 2, \ldots.$$

Further we have

$$d_n = q^{-2n}(1-q^n)\frac{g}{\alpha^*}, \quad n = 1, 2, 3, \ldots. \tag{11.6.4}$$

Note that $c_n \in \mathbb{R}$ for $n = 0, 1, 2, \ldots$ implies that we must have $\beta^*/\alpha^* \in \mathbb{R}$ and that positive-definite orthogonality for an infinite system of polynomials occurs for $g(1-q^n)/\alpha^* > 0$. Since $\alpha^* \in \mathbb{R}$ this implies that $\beta^* \in \mathbb{R}$ and $g \in \mathbb{R}$. Now we have:

Case Ia1. $0<q<1$, $e=f=0$, $g\neq 0$ and $\dfrac{g}{\alpha^*}>0$.

Case Ia2. $q>1$, $e=f=0$, $g\neq 0$ and $\dfrac{g}{\alpha^*}<0$.

In this case we have no finite systems of positive-definite orthogonal polynomials.

Case II. $e=0$ and $f\neq 0$. Then we have $\alpha^*=-2\varepsilon(1-q)\neq 0$ and

$$c_n=\frac{2f(1+q-q^{n+1})-\beta^*q^{n-1}}{\alpha^*q^{2n}},\quad n=0,1,2,\ldots$$

and for $n=1,2,3,\ldots$

$$\begin{aligned}d_n&=-\frac{q^{-2n}(1-q^n)}{(\alpha^*)^2}\left\{2fq^{-n+1}(\beta^*-2fq^{-n+2})-g\alpha^*\right\}\\&=\frac{4f^2}{(\alpha^*)^2}q^{-4n+3}(1-q^n)\left\{1-\frac{\beta^*}{2f}q^{n-2}+\frac{g\alpha^*}{4f^2}q^{2n-3}\right\}\\&=\left(\frac{2f}{\alpha^*}\right)^2q^{-4n+3}(1-q^n)\left(1-\frac{\eta_1}{2f}q^{n-1}\right)\left(1-\frac{\eta_2 g}{2f}q^{n-2}\right),\end{aligned}\tag{11.6.5}$$

where

$$\eta_1\eta_2=\alpha^*\quad\text{and}\quad\eta_1 q+\eta_2 g=\beta^*.$$

Note that we have

$$c_0=\frac{2f-\beta^*q^{-1}}{\alpha^*}=\frac{2fq-\beta^*}{\alpha^*q}$$

and

$$c_1=\frac{2f(1+q-q^2)-\beta^*}{\alpha^*q^2}=\frac{2fq-\beta^*}{\alpha^*q^2}+\frac{2f(1-q^2)}{\alpha^*q^2}.$$

It can be shown that

$$c_n=\frac{2fq-\beta^*}{\alpha^*q^{n+1}}+\frac{2f(1+q)(1-q^n)}{\alpha^*q^{2n}},\quad n=0,1,2,\ldots.$$

Now $c_0\in\mathbb{R}$ and $c_1\in\mathbb{R}$ implies that $2f/\alpha^*\in\mathbb{R}$ and also $\beta^*/\alpha^*\in\mathbb{R}$. Since $\alpha^*\in\mathbb{R}$ this implies that we also have $f\in\mathbb{R}$ and $\beta^*\in\mathbb{R}$. Then we must also have $g\in\mathbb{R}$ since $d_n\in\mathbb{R}$.

In the case that $(\beta^*)^2<4g\alpha^*q$ we have

$$1-\frac{\beta^*}{2f}q^{n-2}+\frac{g\alpha^*}{4f^2}q^{2n-3}=\left(1-\frac{\beta^*}{4f}q^{n-2}\right)^2+\frac{4g\alpha^*q-(\beta^*)^2}{16f^2}q^{2n-4}>0$$

for all $n = 1,2,3,\ldots$. This implies that $d_n > 0$ for all $n = 1,2,3,\ldots$ in the case that $0 < q < 1$ and $(\beta^*)^2 < 4g\alpha^* q$ and that $d_n < 0$ for all $n = 1,2,3,\ldots$ in the case that $q > 1$ and $(\beta^*)^2 < 4g\alpha^* q$.

In the case that $(\beta^*)^2 \geq 4g\alpha^* q$ we have

$$\eta_1 = \frac{\beta^* \pm \sqrt{(\beta^*)^2 - 4g\alpha^* q}}{2q} \quad \text{and} \quad \eta_2 = \frac{\beta^* \mp \sqrt{(\beta^*)^2 - 4g\alpha^* q}}{2g}.$$

Note that for $0 < q < 1$ we have $q^{-4n+3}(1-q^n) > 0$ for $n = 1,2,3,\ldots$ and for $q > 1$ we have $q^{-4n+3}(1-q^n) < 0$ for $n = 1,2,3,\ldots$.

For $0 < q < 1$ we must have

$$\left(1 - \frac{\eta_1}{2f} q^{n-1}\right)\left(1 - \frac{\eta_2 g}{2f} q^{n-2}\right) > 0, \quad n = 1,2,3,\ldots,$$

which implies that both

$$1 - \frac{\eta_1}{2f} q^{n-1} > 0 \quad \text{and} \quad 1 - \frac{\eta_2 g}{2f} q^{n-2} > 0, \quad n = 1,2,3,\ldots,$$

since $q^n \to 0$ for $n \to \infty$. For $n = 1$ this reads $\dfrac{\eta_1}{2f} < 1$ and $\dfrac{\eta_2 g}{2fq} < 1$. And, as before, this implies that

$$\left(1 - \frac{\eta_1}{2f} q^{n-1}\right)\left(1 - \frac{\eta_2 g}{2f} q^{n-2}\right) > 0$$

holds for all $n = 1,2,3,\ldots$.

For $q > 1$ we must have

$$\left(1 - \frac{\eta_1}{2f} q^{n-1}\right)\left(1 - \frac{\eta_2 g}{2f} q^{n-2}\right) < 0, \quad n = 1,2,3,\ldots,$$

which implies that either

$$1 - \frac{\eta_1}{2f} q^{n-1} < 0 \quad \text{and} \quad 1 - \frac{\eta_2 g}{2f} q^{n-2} > 0, \quad n = 1,2,3,\ldots,$$

or

$$1 - \frac{\eta_1}{2f} q^{n-1} > 0 \quad \text{and} \quad 1 - \frac{\eta_2 g}{2f} q^{n-2} < 0, \quad n = 1,2,3,\ldots.$$

This implies that we must either have $\dfrac{\eta_1}{2f} > 1$ and $\dfrac{\eta_2 g}{2fq} < 0$, or $\dfrac{\eta_1}{2f} < 0$ and $\dfrac{\eta_2 g}{2fq} > 1$.

Hence we have positive-definite orthogonality for an infinite system of polynomials in the following four cases:

Case IIa1. $0 < q < 1$, $e = 0$, $f \neq 0$, $(\beta^*)^2 < 4g\alpha^* q$.

Case IIa2. $0 < q < 1$, $e = 0$, $f \neq 0$, $(\beta^*)^2 \geq 4g\alpha^* q$, $\dfrac{\eta_1}{2f} < 1$ and $\dfrac{\eta_2 g}{2fq} < 1$.

Case IIa3. $q > 1$, $e = 0$, $f \neq 0$, $(\beta^*)^2 \geq 4g\alpha^* q$, $\dfrac{\eta_1}{2f} > 1$ and $\dfrac{\eta_2 g}{2f} < 0$.

Case IIa4. $q > 1$, $e = 0$, $f \neq 0$, $(\beta^*)^2 \geq 4g\alpha^* q$, $\dfrac{\eta_1}{2f} < 0$ and $\dfrac{\eta_2 g}{2fq} > 1$.

In an analogous way we can show that it is also possible to have positive-definite orthogonality for finite systems of $N+1$ polynomials with $N \in \{1,2,3,\ldots\}$ in the following three cases:

Case IIb1. $0 < q < 1$, $e = 0$, $f \neq 0$, $\dfrac{\eta_1}{2f} > 1$ with $\dfrac{\eta_1}{2f}q^{N_1} \leq 1 < \dfrac{\eta_1}{2f}q^{N_1-1}$, $\dfrac{\eta_2 g}{2fq} > 1$ with $\dfrac{\eta_2 g}{2f}q^{N_2-1} \leq 1 < \dfrac{\eta_2 g}{2f}q^{N_2-2}$ and $N = \min(N_1, N_2)$.

Case IIb2. $q > 1$, $e = 0$, $f \neq 0$, $0 < \dfrac{\eta_1}{2f} < 1$ with $\dfrac{\eta_1}{2f}q^{N-1} < 1 \leq \dfrac{\eta_1}{2f}q^N$ and $\dfrac{\eta_2 g}{2fq} > 1$.

Case IIb3. $q > 1$, $e = 0$, $f \neq 0$, $0 < \dfrac{\eta_2 g}{2fq} < 1$ with $\dfrac{\eta_2 g}{2f}q^{N-2} < 1 \leq \dfrac{\eta_2 g}{2f}q^{N-1}$ and $\dfrac{\eta_1}{2f} > 1$.

Case III. $e \neq 0$. Note that we have

$$c_0 = \frac{2fq-\beta^*}{q(\alpha^*-eq)}, \quad c_1 = \frac{2fq-\beta^*}{q^2(\alpha^*-eq^{-1})} + \frac{(1-q^2)(2f\alpha^*-e\beta^*)}{q^2(\alpha^*-eq)(\alpha^*-eq^{-1})}$$

and

$$c_2 = \frac{2fq-\beta^*}{q^3(\alpha^*-eq^{-3})} + \frac{(1+q)(1-q^2)(2fq\alpha^*-e\beta^*)}{q^5(\alpha^*-eq^{-1})(\alpha^*-eq^{-3})}.$$

In fact, in general we have

$$c_n = \frac{2fq-\beta^*}{q^{n+1}(\alpha^*-eq^{-2n+1})} + \frac{(1+q)(1-q^n)\left(2fq^{n-1}\alpha^*-e\beta^*\right)}{q^{3n-1}(\alpha^*-eq^{-2n+1})(\alpha^*-eq^{-2n+3})}$$

for $n = 0,1,2,\ldots$. Further we have

$$d_n = -q^{-2n}(1-q^n)D_n^{(1)}D_n^{(2)}, \quad n = 1,2,3,\ldots \tag{11.6.6}$$

with

$$\begin{aligned} D_n^{(1)} &= \frac{\alpha^* - eq^{-n+3}}{(\alpha^* - eq^{-2n+4})(\alpha^* - eq^{-2n+3})^2(\alpha^* - eq^{-2n+2})} \\ &= -\frac{q^{7n-9}}{e^3} \cdot \frac{1 - \frac{\alpha^*}{e}q^{n-3}}{\left(1 - \frac{\alpha^*}{e}q^{2n-4}\right)\left(1 - \frac{\alpha^*}{e}q^{2n-3}\right)^2\left(1 - \frac{\alpha^*}{e}q^{2n-2}\right)} \end{aligned} \tag{11.6.7}$$

for $n = 1,2,3,\ldots$, and

$$D_n^{(2)} = q^{-n+1}\left(\beta^* - 2fq^{-n+2}\right)\left(2f\alpha^* - e\beta^*q^{-n+1}\right) - g\left(\alpha^* - eq^{-2n+3}\right)^2 \tag{11.6.8}$$

for $n = 1,2,3,\ldots$. For $g = 0$ we have for $n = 1,2,3,\ldots$

$$\begin{aligned} D_n^{(2)} &= q^{-n+1}\left(\beta^* - 2fq^{-n+2}\right)\left(2f\alpha^* - e\beta^*q^{-n+1}\right) \\ &= -eq^{-2n+2}\left(\beta^* - 2fq^{-n+2}\right)\left(\beta^* - \frac{2f\alpha^*}{e}q^{n-1}\right) \end{aligned} \tag{11.6.9}$$

and for $g \neq 0$ we have for $n = 1,2,3,\ldots$

$$\begin{aligned} D_n^{(2)} &= q^{-n+1}\left(\beta^* - 2fq^{-n+2}\right)\left(2f\alpha^* - e\beta^*q^{-n+1}\right) - g\left(\alpha^* - eq^{-2n+3}\right)^2 \\ &= -eq^{-2n+2}\left\{\beta^* - 2fq^{-n+2} + \xi_1 q^{n-1}\left(\alpha^* - eq^{-2n+3}\right)\right\} \\ &\qquad \times \left\{\beta^* - 2fq^{-n+2} + \xi_2 q^{n-1}\left(\alpha^* - eq^{-2n+3}\right)\right\} \\ &= -eq^{-2n+2}\left\{\beta^* + \xi_2 eq^{-n+2} + \xi_1\alpha^* q^{n-1}\right\} \\ &\qquad \times \left\{\beta^* + \xi_1 eq^{-n+2} + \xi_2\alpha^* q^{n-1}\right\} \\ &= -eq^{-2n+2}\left(1 + \xi_1\eta_2 q^{n-2}\right)\left(\eta_1 q + \xi_2 eq^{-n+2}\right) \\ &\qquad \times \left(1 + \xi_2\eta_2 q^{n-2}\right)\left(\eta_1 q + \xi_1 eq^{-n+2}\right), \end{aligned} \tag{11.6.10}$$

where as before $\eta_1\eta_2 = \alpha^*$, $\eta_1 q + \eta_2 g = \beta^*$, $\xi_1\xi_2 = g/e$ and $\xi_1 + \xi_2 = -2f/e$, which implies that for $f^2 \geq eg$

$$\xi_1 = \frac{-f - \sqrt{f^2 - eg}}{e} \quad \text{and} \quad \xi_2 = \frac{-f + \sqrt{f^2 - eg}}{e}.$$

In the case that $f^2 < eg$ we have $\xi_1, \xi_2 \in \mathbb{C}$ with $\xi_2 = \overline{\xi_1}$. Finally, we remark that for $g \neq 0$ and $\alpha^* = 0$ we have

$$D_n^{(2)} = -eq^{-4n+6}\left[\left(\beta^* q^{n-2} - f\right)^2 + eg - f^2\right], \quad n = 1,2,3,\ldots.$$

Hence, for $g \neq 0$ and $\alpha^* = 0$ we obtain that

$$d_n = -\frac{q^{n-3}}{e^2}(1-q^n)\left[\left(\beta^* q^{n-2} - f\right)^2 + eg - f^2\right], \quad n = 1,2,3,\ldots.$$

This implies that in the case that $f^2 < eg$ we have $d_n < 0$ for all $n = 1,2,3,\ldots$ for $0 < q < 1$ and $d_n > 0$ for all $n = 1,2,3,\ldots$ for $q > 1$. In the case that $f^2 \geq eg$ we have $\xi_1, \xi_2 \in \mathbb{R}$ and

$$\begin{aligned} d_n &= -\xi_1 \xi_2 q^{n-3}(1-q^n)\left(1 + \frac{\beta^*}{\xi_1 e} q^{n-2}\right)\left(1 + \frac{\beta^*}{\xi_2 e} q^{n-2}\right) \\ &= -\frac{g}{e} q^{n-3}(1-q^n)\left(1 + \frac{\xi_1 \beta^*}{g} q^{n-2}\right)\left(1 + \frac{\xi_2 \beta^*}{g} q^{n-2}\right), \quad n = 1,2,3,\ldots. \end{aligned}$$

This eventually leads to some finite systems of orthogonal polynomials in the case that $0 < q < 1$, $g \neq 0$ and $\alpha^* = 0$.

Note that for $0 < q < 1$ we have $q^{-2n}(1-q^n) > 0$ for $n = 1,2,3,\ldots$ and for $q > 1$ we have $q^{-2n}(1-q^n) < 0$ for $n = 1,2,3,\ldots$.

Analogous to the situation in the previous chapter (cf. table 10.2 on page 270) the sign of $D_n^{(1)}$ is given in table 11.1.

q	extra conditions	$\mathbf{D}_n^{(1)}$	for
$0 < q < 1$	$e < 0$ and $\alpha^* > eq$	$+$	$n = 1,2,3,\ldots$
	$e > 0$ and $\alpha^* < eq$	$-$	$n = 1,2,3,\ldots$
	$\alpha^* < eq < 0$ with $e \leq \alpha^* q^{2N} < eq^2$	$-$	$n = 1,2,3,\ldots,N$
	$0 < eq < \alpha^*$ with $eq^2 < \alpha^* q^{2N} \leq e$	$+$	$n = 1,2,3,\ldots,N$
$q > 1$	$e < 0 \leq \alpha^*$	$+$	$n = 1,2,3,\ldots$
	$\alpha^* \leq 0 < e$	$-$	$n = 1,2,3,\ldots$
	$eq < \alpha^* < 0$ with $eq^2 < \alpha^* q^{2N} \leq e$	$+$	$n = 1,2,3,\ldots,N$
	$0 < \alpha^* < eq$ with $e \leq \alpha^* q^{2N} < eq^2$	$-$	$n = 1,2,3,\ldots,N$
	$\alpha^* < eq < 0$	$-$	$n = 1,2,3,\ldots$
	$0 < eq < \alpha^*$	$+$	$n = 1,2,3,\ldots$

Table 11.1 sign of $D_n^{(1)}$, $N \in \{1,2,3,\ldots\}$

Hence, for $0 < q < 1$, $e < 0$ and $\alpha^* > eq$ we must have $D_n^{(2)} < 0$. In the case that $g = 0$ this implies that we must have

$$\left(\beta^* - 2fq^{-n+2}\right)\left(\beta^* - \frac{2f\alpha^*}{e} q^{n-1}\right) < 0,$$

which is true for $\beta^* = 0$ provided that $\alpha^* > 0$, and for $\beta^* \neq 0$ if

$$\left(1 - \frac{2f}{\beta^*} q^{-n+2}\right) \left(1 - \frac{2f\alpha^*}{e\beta^*} q^{n-1}\right) < 0,$$

which is true for all $n = 1, 2, 3, \ldots$ if and only if

$$\frac{2fq}{\beta^*} > 1 \quad \text{and} \quad \frac{2f\alpha^*}{e\beta^*} < 1.$$

And for $0 < q < 1$, $e > 0$ and $\alpha^* < eq$ we must have $D_n^{(2)} > 0$. In the case that $g = 0$ this implies that we must have

$$(\beta^* - 2fq^{-n+2}) \left(\beta^* - \frac{2f\alpha^*}{e} q^{n-1}\right) < 0,$$

which is true for $\beta^* = 0$ provided that $\alpha^* < 0$, and for $\beta^* \neq 0$ if

$$\left(1 - \frac{2f}{\beta^*} q^{-n+2}\right) \left(1 - \frac{2f\alpha^*}{e\beta^*} q^{n-1}\right) < 0,$$

which is true for all $n = 1, 2, 3, \ldots$ if and only if

$$\frac{2fq}{\beta^*} > 1 \quad \text{and} \quad \frac{2f\alpha^*}{e\beta^*} < 1.$$

Similarly, for $q > 1$, $e < 0 \leq \alpha^*$ we must have $D_n^{(2)} > 0$. In the case that $g = 0$ this implies that we must have

$$(\beta^* - 2fq^{-n+2}) \left(\beta^* - \frac{2f\alpha^*}{e} q^{n-1}\right) > 0,$$

which is false for $\beta^* = 0$. For $\beta^* \neq 0$ we must have

$$\left(1 - \frac{2f}{\beta^*} q^{-n+2}\right) \left(1 - \frac{2f\alpha^*}{e\beta^*} q^{n-1}\right) > 0,$$

which is true for all $n = 1, 2, 3 \ldots$ if and only if

$$\frac{2fq}{\beta^*} < 1 \quad \text{and} \quad \frac{2f\alpha^*}{e\beta^*} \leq 0.$$

For $q > 1$, $0 < eq < \alpha^*$ we must have $D_n^{(2)} > 0$. In the case that $g = 0$ this implies that we must have

$$(\beta^* - 2fq^{-n+2}) \left(\beta^* - \frac{2f\alpha^*}{e} q^{n-1}\right) < 0,$$

which is false for $\beta^* = 0$. For $\beta^* \neq 0$ we must have

$$\left(1 - \frac{2f}{\beta^*} q^{-n+2}\right) \left(1 - \frac{2f\alpha^*}{e\beta^*} q^{n-1}\right) < 0,$$

which is true for all $n = 1, 2, 3 \ldots$ if and only if

$$\frac{2fq}{\beta^*} < 1 \quad \text{and} \quad \frac{2f\alpha^*}{e\beta^*} > 1.$$

And for $q > 1$, $\alpha^* < eq < 0$ we must have $D_n^{(2)} < 0$. In the case that $g = 0$ this implies that we must have

$$\left(\beta^* - 2fq^{-n+2}\right) \left(\beta^* - \frac{2f\alpha^*}{e} q^{n-1}\right) < 0,$$

which is false for $\beta^* = 0$. For $\beta^* \neq 0$ we must have

$$\left(1 - \frac{2f}{\beta^*} q^{-n+2}\right) \left(1 - \frac{2f\alpha^*}{e\beta^*} q^{n-1}\right) < 0,$$

which is true for all $n = 1, 2, 3 \ldots$ if and only if

$$\frac{2fq}{\beta^*} < 1 \quad \text{and} \quad \frac{2f\alpha^*}{e\beta^*} > 1.$$

For $q > 1$, $\alpha^* \leq 0 < e$ we must have $D_n^{(2)} < 0$. In the case that $g = 0$ this implies that we must have

$$\left(\beta^* - 2fq^{-n+2}\right) \left(\beta^* - \frac{2f\alpha^*}{e} q^{n-1}\right) > 0,$$

which is false for $\beta^* = 0$. For $\beta^* \neq 0$ we must have

$$\left(1 - \frac{2f}{\beta^*} q^{-n+2}\right) \left(1 - \frac{2f\alpha^*}{e\beta^*} q^{n-1}\right) > 0,$$

which is true for all $n = 1, 2, 3 \ldots$ if and only if

$$\frac{2fq}{\beta^*} < 1 \quad \text{and} \quad \frac{2f\alpha^*}{e\beta^*} \leq 0.$$

Finally, we remark that in the case that $0 < q < 1$ and $g \neq 0$ we have

$$D_n^{(2)} \sim -ge^2 q^{-4n+6} \quad \text{for} \quad n \to \infty,$$

which implies that $D_n^{(2)} > 0$ for all $n = 1,2,3,\ldots$ can only be true if $g < 0$ and that $D_n^{(2)} < 0$ for all $n = 1,2,3,\ldots$ can only be true if $g > 0$. Similarly, in the case that $q > 1$ and $g \neq 0$ we have

$$D_n^{(2)} \sim -g\,(\alpha^*)^2 \quad \text{for} \quad n \to \infty,$$

which implies that $D_n^{(2)} > 0$ for all $n = 1,2,3,\ldots$ can only be true if $g < 0$ and that $D_n^{(2)} < 0$ for all $n = 1,2,3,\ldots$ can only be true if $g > 0$.

Hence we conclude that we have positive-definite orthogonality for an infinite system of polynomials, at least in the following nine cases:

Case IIIa1. $0 < q < 1$, $g = 0$, $\beta^* = 0$ and $e < 0 < \alpha^*$.

Case IIIa2. $0 < q < 1$, $g = 0$, $\beta^* = 0$ and $\alpha^* < 0 < e$.

Case IIIa3. $0 < q < 1$, $g = 0$, $\beta^* \neq 0$, $e < 0$, $\alpha^* > eq$, $\dfrac{2fq}{\beta^*} > 1$ and $\dfrac{2f\alpha^*}{e\beta^*} < 1$.

Case IIIa4. $0 < q < 1$, $g = 0$, $\beta^* \neq 0$, $e > 0$, $\alpha^* < eq$, $\dfrac{2fq}{\beta^*} > 1$ and $\dfrac{2f\alpha^*}{e\beta^*} < 1$.

Case IIIa5. $q > 1$, $g = 0$, $\beta^* \neq 0$, $e < 0 \leq \alpha^*$, $\dfrac{2fq}{\beta^*} < 1$ and $\dfrac{2f\alpha^*}{e\beta^*} \leq 0$.

Case IIIa6. $q > 1$, $g = 0$, $\beta^* \neq 0$, $0 < eq < \alpha^*$, $\dfrac{2fq}{\beta^*} < 1$ and $\dfrac{2f\alpha^*}{e\beta^*} > 1$.

Case IIIa7. $q > 1$, $g = 0$, $\beta^* \neq 0$, $\alpha^* < eq < 0$, $\dfrac{2fq}{\beta^*} < 1$ and $\dfrac{2f\alpha^*}{e\beta^*} > 1$.

Case IIIa8. $q > 1$, $g = 0$, $\beta^* \neq 0$, $\alpha^* \leq 0 < e$, $\dfrac{2fq}{\beta^*} < 1$ and $\dfrac{2f\alpha^*}{e\beta^*} \leq 0$.

Case IIIa9. $q > 1$, $g \neq 0$, $\alpha^* = 0$ and $f^2 < eg$.

It is also possible to have positive-definite orthogonality for finite systems of polynomials. We only consider the cases where $g \neq 0$ and $\alpha^* = 0$, and, for $\alpha^* \neq 0$, the cases where $D_n^{(1)}$ has opposite sign for $n = N$ and $n = N+1$ according to table 11.1. Skipping the details, we conclude that we have positive-definite orthogonality, at least in the following finite cases:

Case IIIb1. $0 < q < 1$, $\dfrac{g}{e} < 0$, $\alpha^* = 0$, $\dfrac{\xi_1\beta^*}{gq} < -1$ with $\dfrac{\xi_1\beta^*}{g}q^{N_1-2} < -1 \leq \dfrac{\xi_1\beta^*}{g}q^{N_1-1}$, $\dfrac{\xi_2\beta^*}{gq} < -1$ with $\dfrac{\xi_2\beta^*}{g}q^{N_2-2} < -1 \leq \dfrac{\xi_2\beta^*}{g}q^{N_2-1}$ and $N = \min(N_1, N_2)$,

Case IIIb2. $0 < q < 1$, $\dfrac{g}{e} > 0$, $\alpha^* = 0$, $\dfrac{\xi_1\beta^*}{gq} < -1$ with $\dfrac{\xi_1\beta^*}{g}q^{N-2} < -1 \leq \dfrac{\xi_1\beta^*}{g}q^{N-1}$ and $\dfrac{\xi_2\beta^*}{gq} > -1$.

Case IIIb3. $0 < q < 1$, $\dfrac{g}{e} > 0$, $\alpha^* = 0$, $\dfrac{\xi_2\beta^*}{gq} < -1$ with $\dfrac{\xi_2\beta^*}{g}q^{N-2} < -1 \leq \dfrac{\xi_2\beta^*}{g}q^{N-1}$ and $\dfrac{\xi_1\beta^*}{gq} > -1$.

Case IIIb4. $0 < q < 1$, $\alpha^* < eq < 0$ with $e \leq \alpha^* q^{2N} < eq^2$ and $D_n^{(2)} > 0$.

Case IIIb5. $0 < q < 1$, $0 < eq < \alpha^*$ with $eq^2 < \alpha^* q^{2N} \leq e$ and $D_n^{(2)} < 0$.

Case IIIb6. $q > 1$, $eq < \alpha^* < 0$ with $eq^2 < \alpha^* q^{2N} \leq e$ and $D_n^{(2)} > 0$.

Case IIIb7. $q > 1$, $0 < \alpha^* < eq$ with $e \leq \alpha^* q^{2N} < eq^2$ and $D_n^{(2)} < 0$.

We remark that it is also possible to have positive-definite orthogonality for finite systems of $N+1$ polynomials with $N \in \{1, 2, 3, \ldots\}$ in cases where $D_n^{(2)}$ has opposite sign for $n = N$ and $n = N+1$. As an example we mention

Case IIIb8. $q > 1$, $g = 0$, $\beta^* \neq 0$, $\alpha^* \leq 0 < e$, $\dfrac{2fq}{\beta^*} < 1$ and $0 < \dfrac{2f\alpha^*}{e\beta^*} < 1$ with $\dfrac{2f\alpha^*}{e\beta^*}q^N \leq 1 < \dfrac{2f\alpha^*}{e\beta^*}q^{N-1}$.

11.7 Solutions of the q^{-1}-Pearson Equation

We look for solutions of the q^{-1}-Pearson equation (11.4.7)

$$qw(q^{-x})A^*(x) = w(q^{-x-1})B^*(x+1)$$

with

$$A^*(x) = eq^2 + 2fq^{x+1} + gq^{2x} \quad \text{and} \quad B^*(x+1) = \alpha^* + \beta^* q^x + gq^{2x+1}.$$

Hence we have

$$\frac{w(q^{-x})}{w(q^{-x-1})} = \frac{B^*(x+1)}{qA^*(x)} = \frac{\alpha^* + \beta^* q^x + gq^{2x+1}}{q(eq^2 + 2fq^{x+1} + gq^{2x})}.$$

For simplicity we set $y = q^{-x}$, which leads to

$$\frac{w(y)}{w(q^{-1}y)} = \frac{\alpha^* + \beta^* y^{-1} + gqy^{-2}}{q(eq^2 + 2fqy^{-1} + gy^{-2})} = \frac{\alpha^* y^2 + \beta^* y + gq}{q(eq^2 y^2 + 2fqy + g)}. \tag{11.7.1}$$

In order to find solutions of the q^{-1}-Pearson equation (11.7.1), we distinguish between $0 < q < 1$ and $q > 1$. For $q > 1$ we set $q = p^{-1}$, which implies that $0 < p < 1$. Then we have

$$\frac{w(p^x)}{w(p^{x+1})} = \frac{p\left(\alpha^* + \beta^* p^{-x} + gp^{-2x-1}\right)}{ep^{-2} + 2fp^{-x-1} + gp^{-2x}} = \frac{\alpha^* p^{2x+1} + \beta^* p^{x+1} + g}{ep^{2x-2} + 2fp^{x-1} + g}.$$

Now we set $z = p^x$ for simplicity, which leads to

$$\frac{w(z)}{w(pz)} = \frac{\alpha^* pz^2 + \beta^* pz + g}{ep^{-2}z^2 + 2fp^{-1}z + g}. \tag{11.7.2}$$

We consider two types of solutions:

A. *Continuous* solutions for $y \in \mathbb{R}$ in terms of (convergent) infinite products. For the convergence of these infinite products we refer to the book [471] by L.J. Slater.

B. *Discrete* solutions for $\{y_\nu\}_{\nu=0}^N$ with $N \to \infty$ or $\{y_\nu\}_{\nu=-\infty}^{\infty}$, id est $y_\nu = Aq^{-\nu}$ with $A \in \mathbb{C}$ and $\nu = 0, \pm 1, \pm 2, \ldots$, in terms of finite products. Note that $y_{\nu+1} = q^{-1} y_\nu$. Without loss of generality we can choose $A = 1$ in each case.

The solutions are obtained in a similar way as in the previous chapter. In this chapter some of the details (analogous to the previous chapter) are left for the reader.

Case I-A. $e = f = 0$ and $g \neq 0$. From (11.7.1) we have

$$\frac{w(y)}{w(q^{-1}y)}=\frac{\alpha^*y^2+\beta^*y+gq}{gq}=1+\frac{\beta^*y}{gq}+\frac{\alpha^*y^2}{gq}=\left(1+\frac{\eta_1y}{g}\right)\left(1+\frac{\eta_2y}{q}\right),$$

where $\eta_1\eta_2=\alpha^*$ and $\eta_1q+\eta_2g=\beta^*$ as before, which implies that

$$\eta_1=\frac{\beta^*\pm\sqrt{(\beta^*)^2-4g\alpha^*q}}{2q}\quad\text{and}\quad\eta_2=\frac{\beta^*\mp\sqrt{(\beta^*)^2-4g\alpha^*q}}{2g},$$

provided that $(\beta^*)^2\geq 4g\alpha^*q$. Then we easily find the solution

$$w^{(I)}(y;q^{-1})=\frac{1}{\left(-\frac{\eta_1qy}{g},-\eta_2y;q\right)_\infty}.$$

The special case $e=f=0$, $g=q$, $\alpha^*=1$ and $\beta^*=0$, which implies that $\eta_1=-i$ and $\eta_2=i$, leads by using (1.8.24) to the weight function

$$w(y;q)=\frac{1}{(iy,-iy;q)_\infty}=\frac{1}{(y^2;q^2)_\infty}$$

for the **discrete q-Hermite II** polynomials. Note that η_1 and η_2 are non-real in this case.

From (11.7.2) we have

$$\frac{w(z)}{w(pz)}=\frac{\alpha^*pz^2+\beta^*pz+g}{g}=1+\frac{\beta^*pz}{g}+\frac{\alpha^*pz^2}{g}=\left(1+\frac{\varepsilon_1z}{g}\right)(1+\varepsilon_2pz),$$

where $\varepsilon_1\varepsilon_2=\alpha^*$ and $\varepsilon_1p^{-1}+\varepsilon_2g=\beta^*$, which implies that

$$\varepsilon_1=\frac{\beta^*\pm\sqrt{(\beta^*)^2-4g\alpha^*p^{-1}}}{2p^{-1}}\quad\text{and}\quad\varepsilon_2=\frac{\beta^*\mp\sqrt{(\beta^*)^2-4g\alpha^*p^{-1}}}{2g},$$

provided that $(\beta^*)^2\geq 4g\alpha^*p^{-1}$. Then we easily find the solution

$$w^{(I)}(z;p)=\left(-\frac{\varepsilon_1z}{g},-\varepsilon_2pz;p\right)_\infty.$$

The special case $e=f=0$, $g=a/p$, $\alpha^*=1$ and $\beta^*=-(a+1)/p$, which implies that $\varepsilon_1=\varepsilon_2=-1$, leads to the weight function

$$w(z;p)=\left(\frac{pz}{a},pz;p\right)_\infty$$

for the **Al-Salam-Carlitz I** polynomials.

The special case $e = f = 0$, $g = -1/p$, $\alpha^* = 1$ and $\beta^* = 0$, which implies that $\varepsilon_1 = \varepsilon_2 = -1$, leads to the weight function

$$w(z;p) = (-pz, pz; p)_\infty$$

for the **discrete q-Hermite I** polynomials.

Note that for $q \to p^{-1}$ we have that $\eta_1 \to \varepsilon_1$ and $\eta_2 \to \varepsilon_2$.

Case I-B. $e = f = 0$ and $g \neq 0$. Note that $e = 0$ implies that $\eta_1\eta_2 = \alpha^* = -2\varepsilon(1-q) \neq 0$. Then we have from (11.7.1) with $y_\nu = q^{-\nu}$

$$\frac{w(y_\nu)}{w(y_{\nu+1})} = \frac{\alpha^* q^{-2\nu} + \beta^* q^{-\nu} + gq}{gq} = \left(1 + \frac{gq^\nu}{\eta_1}\right)\left(1 + \frac{q^{\nu+1}}{\eta_2}\right) \cdot \frac{\eta_1\eta_2}{gq} \cdot q^{-2\nu}$$

with possible solution

$$w^{(I)}(y_\nu; q^{-1}) = \frac{1}{\left(-\frac{g}{\eta_1}, -\frac{q}{\eta_2}; q\right)_\nu} \left(\frac{gq}{\eta_1\eta_2}\right)^\nu q^{2\binom{\nu}{2}}.$$

The special case $e = f = 0$, $g = aq$, $\alpha^* = 1$ and $\beta^* = -(a+1)q$, which implies that $\eta_1 = \eta_2 = -1$, leads to the weight function

$$w(y_\nu; q^{-1}) = \frac{1}{(aq, q; q)_\nu} a^\nu q^{\nu(\nu+1)}$$

for the **Al-Salam-Carlitz II** polynomials.

From (11.7.2) we have with $z_\nu = p^\nu$

$$\frac{w(z_\nu)}{w(z_{\nu+1})} = \frac{\alpha^* p^{2\nu+1} + \beta^* p^{\nu+1} + g}{g} = \left(1 + \frac{\varepsilon_1 p^\nu}{g}\right)\left(1 + \varepsilon_2 p^{\nu+1}\right)$$

with possible solution

$$w^{(I)}(z_\nu; p) = \frac{1}{\left(-\frac{\varepsilon_1}{g}, -\varepsilon_2 p; p\right)_\nu}.$$

Case II-A. $e = 0$ and $f \neq 0$. From (11.7.1) we have for $g \neq 0$

$$\frac{w(y)}{w(q^{-1}y)} = \frac{\alpha^* y^2 + \beta^* y + gq}{q(2fqy + g)} = \frac{\left(1 + \frac{\eta_1 y}{g}\right)\left(1 + \frac{\eta_2 y}{q}\right)}{1 + \frac{2fqy}{g}}$$

with possible solution

$$w^{(II)}(y;q^{-1}) = \frac{\left(-\frac{2fq^2y}{g};q\right)_\infty}{\left(-\frac{\eta_1 qy}{g}, -\eta_2 y;q\right)_\infty}.$$

For $g=0$ we use the fact that $\alpha^* = -2\varepsilon(1-q) \neq 0$ to obtain

$$\frac{w(y)}{w(q^{-1}y)} = \frac{\alpha^* y^2 + \beta^* y}{2fq^2 y} = \left(1 + \frac{\beta^*}{\alpha^* y}\right) \cdot \frac{\alpha^* y}{2fq^2}$$

with possible solution

$$w^{(II)}(y;q^{-1})\Big|_{g=0} = \frac{\left(-\frac{\beta^*}{\alpha^* y};q\right)_\infty}{\left(-\frac{\alpha^* y}{2fq}, -\frac{2fq^2}{\alpha^* y};q\right)_\infty}.$$

From (11.7.2) we have for $g \neq 0$

$$\frac{w(z)}{w(pz)} = \frac{\alpha^* pz^2 + \beta^* pz + g}{2fp^{-1}z + g} = \frac{\left(1 + \frac{\varepsilon_1 z}{g}\right)(1 + \varepsilon_2 pz)}{1 + \frac{2fz}{gp}}$$

with possible solution

$$w^{(II)}(z;p) = \frac{\left(-\frac{\varepsilon_1 z}{g}, -\varepsilon_2 pz;p\right)_\infty}{\left(-\frac{2fz}{gp};p\right)_\infty}.$$

For $g=0$ we use the fact that $\alpha^* = -2\varepsilon(1-q) \neq 0$ to obtain

$$\frac{w(z)}{w(pz)} = \frac{\alpha^* pz^2 + \beta^* pz}{2fp^{-1}z} = \left(1 + \frac{\beta^*}{\alpha^* z}\right) \cdot \frac{\alpha^* p^2 z}{2f}$$

with possible solution

$$w^{(II)}(z;p)\Big|_{g=0} = \frac{\left(-\frac{\alpha^* p^2 z}{2f}, -\frac{2f}{\alpha^* pz};p\right)_\infty}{\left(-\frac{\beta^* p}{\alpha^* z};p\right)_\infty}.$$

Case II-B. $e=0$ and $f \neq 0$. Note that $e=0$ implies that $\eta_1\eta_2 = \alpha^* = -2\varepsilon(1-q) \neq 0$. Then we have from (11.7.1) with $y_\nu = q^{-\nu}$

$$\frac{w(y_\nu)}{w(y_{\nu+1})} = \frac{\alpha^* q^{-2\nu} + \beta^* q^{-\nu} + gq}{2fq^{-\nu+2} + gq} = \frac{\left(1+\eta_1^{-1} g q^\nu\right)\left(1+\eta_2^{-1} q^{\nu+1}\right)}{1 + \frac{gq^{\nu-1}}{2f}} \cdot \frac{\eta_1 \eta_2}{2fq^2} \cdot q^{-\nu}$$

with possible solution

$$w^{(II)}(y_\nu; q^{-1}) = \frac{\left(-\frac{g}{2fq}; q\right)_\nu}{\left(-\eta_1^{-1} g, -\eta_2^{-1} q; q\right)_\nu} \left(\frac{2fq^2}{\eta_1 \eta_2}\right)^\nu q^{\binom{\nu}{2}}.$$

The special case $e = 0$, $2f = cq^{-1}$, $g = -bcq$, $\alpha^* = 1$ and $\beta^* = -q(1-bc)$, which implies that $\eta_1 = \eta_2 = -1$, leads to the weight function

$$w(y_\nu; q^{-1}) = \frac{(bq;q)_\nu}{(-bcq, q; q)_\nu} c^\nu q^{\binom{\nu+1}{2}}$$

for the q-**Meixner** polynomials.

The special case $e = 0$, $2f = -b^{-1}q^{-1}$, $g = b^{-1}q^{-N}$, $\alpha^* = 1$ and $\beta^* = -q - b^{-1}q^{-N}$, which also implies that $\eta_1 = \eta_2 = -1$, leads to the weight function

$$w(y_\nu; q^{-1}) = \frac{(q^{-N};q)_\nu}{(b^{-1}q^{-N}, q; q)_\nu} \left(-b^{-1}\right)^\nu q^{\binom{\nu+1}{2}}.$$

Then we use (1.8.17) to obtain the weight function

$$w(y_\nu; q^{-1}) = \frac{(q;q)_N}{(bq;q)_N} \frac{(bq;q)_{N-\nu}}{(q;q)_{N-\nu}} \frac{(-1)^\nu}{(q;q)_\nu} q^{\binom{\nu+1}{2}},$$

for the **quantum** q-**Krawtchouk** polynomials.

The special case $e = 0$, $2f = aq^{-1}$, $g = 0$, $\alpha^* = 1$ and $\beta^* = -q$, which again implies that $\eta_1 = \eta_2 = -1$, leads to the weight function

$$w(y_\nu; q^{-1}) = \frac{a^\nu}{(q;q)_\nu} q^{\binom{\nu+1}{2}}$$

for the q-**Charlier** polynomials.

From (11.7.2) we have with $z_\nu = p^\nu$ for $g \neq 0$

$$\frac{w(z_\nu)}{w(z_{\nu+1})} = \frac{\alpha^* p^{2\nu+1} + \beta^* p^{\nu+1} + g}{2fp^{\nu-1} + g} = \frac{\left(1 + \frac{\varepsilon_1 p^\nu}{g}\right)\left(1 + \varepsilon_2 p^{\nu+1}\right)}{1 + \frac{2fp^{\nu-1}}{g}}$$

with possible solution

$$w^{(II)}(z_\nu;p) = \frac{\left(-\frac{2f}{gp};p\right)_\nu}{\left(-\frac{\varepsilon_1}{g},-\varepsilon_2 p;p\right)_\nu}.$$

For $g=0$ we use the fact that $\alpha^* = -2\varepsilon(1-q) \neq 0$ to obtain

$$\frac{w(z_\nu)}{w(z_{\nu+1})} = \frac{\alpha^* p^{2\nu+1} + \beta^* p^{\nu+1}}{2fp^{\nu-1}} = \left(1+\frac{\beta^* p^{-\nu}}{\alpha^*}\right)\cdot\frac{\alpha^* p^2}{2f}\cdot p^\nu$$

with possible solution

$$w^{(II)}(z_\nu;p)\Big|_{g=0} = \frac{1}{\left(-\frac{\beta^* p^{-\nu+1}}{\alpha^*};p\right)_\nu}\left(\frac{2f}{\alpha^* p^2}\right)^\nu p^{-\binom{\nu}{2}}.$$

Case III-A. $e \neq 0$. From (11.7.1) we have for $g \neq 0$

$$\frac{w(y)}{w(q^{-1}y)} = \frac{\alpha^* y^2 + \beta^* y + gq}{q(eq^2y^2+2fqy+g)} = \frac{gq\left(1+\frac{\beta^* y}{gq}+\frac{\alpha^* y^2}{gq}\right)}{gq\left(1+\frac{2fqy}{g}+\frac{eq^2y^2}{g}\right)} = \frac{\left(1+\frac{\eta_1 y}{g}\right)\left(1+\frac{\eta_2 y}{q}\right)}{\left(1-\frac{qy}{\xi_1}\right)\left(1-\frac{qy}{\xi_2}\right)},$$

where as before $\eta_1\eta_2 = \alpha^*$, $\eta_1 q + \eta_2 g = \beta^*$, $\xi_1\xi_2 = g/e$ and $\xi_1+\xi_2 = -2f/e$, which implies that for $(\beta^*)^2 \geq 4g\alpha^* q$

$$\eta_1 = \frac{\beta^* \pm\sqrt{(\beta^*)^2-4g\alpha^* q}}{2q} \quad \text{and} \quad \eta_2 = \frac{\beta^* \mp\sqrt{(\beta^*)^2-4g\alpha^* q}}{2g},$$

and for $f^2 \geq eg$

$$\xi_1 = \frac{-f-\sqrt{f^2-eg}}{e} \quad \text{and} \quad \xi_2 = \frac{-f+\sqrt{f^2-eg}}{e}.$$

As before we easily find the solution

$$w^{(III)}(y;q^{-1}) = \frac{\left(\frac{q^2y}{\xi_1},\frac{q^2y}{\xi_2};q\right)_\infty}{\left(-\frac{\eta_1 qy}{g},-\eta_2 y;q\right)_\infty}.$$

In the case that $g=0$ we distinguish between $\beta^* \neq 0$ and $\beta^* = 0$.

For $g=0$ and $\beta^* \neq 0$ we obtain

$$\frac{w(y)}{w(q^{-1}y)} = \frac{\alpha^* y^2 + \beta^* y}{q\,(eq^2y^2 + 2fqy)} = \frac{1 + \frac{\alpha^* y}{\beta^*}}{1 + \frac{2f}{eqy}} \cdot \frac{\beta^*}{eq^3 y}$$

with possible solution

$$w^{(III)}(y;q^{-1})\Big|_{g=0,\beta^*\neq 0} = \frac{\left(-\frac{\beta^*}{eq^3y}, -\frac{eq^4y}{\beta^*};q\right)_\infty}{\left(-\frac{2f}{eqy}, -\frac{\alpha^* qy}{\beta^*};q\right)_\infty}.$$

For $g = 0$ and $\beta^* = 0$ we must have $\alpha^* \neq 0$. Assuming that $f \neq 0$ we obtain

$$\frac{w(y)}{w(q^{-1}y)} = \frac{\alpha^* y^2}{q\,(eq^2y^2 + 2fqy)} = \frac{1}{1 + \frac{eqy}{2f}} \cdot \frac{\alpha^* y}{2fq^2}$$

with possible solution

$$w^{(III)}(y;q^{-1})\Big|_{g=0,\beta^*=0} = \frac{\left(-\frac{eq^2y}{2f};q\right)_\infty}{\left(-\frac{\alpha^* y}{2fq}, -\frac{2fq^2}{\alpha^* y};q\right)_\infty}.$$

For $f = g = 0$, $\beta^* = 0$ and $\alpha^* \neq 0$ we obtain

$$\frac{w(y)}{w(q^{-1}y)} = \frac{\alpha^*}{eq^3}$$

with possible solution

$$w^{(III)}(y;q^{-1})\Big|_{f=g=0,\beta^*=0} = \frac{\left(-eq^3y, -\frac{1}{eq^2y};q\right)_\infty}{\left(-\alpha^* y, -\frac{q}{\alpha^* y};q\right)_\infty}.$$

From (11.7.2) we have for $g \neq 0$

$$\frac{w(z)}{w(pz)} = \frac{\alpha^* pz^2 + \beta^* pz + g}{ep^{-2}z^2 + 2fp^{-1}z + g} = \frac{\left(1 + \frac{\varepsilon_1 z}{g}\right)(1 + \varepsilon_2 pz)}{\left(1 - \frac{z}{\xi_1 p}\right)\left(1 - \frac{z}{\xi_2 p}\right)}$$

with possible solution

$$w^{(III)}(z;p) = \frac{\left(-\frac{\varepsilon_1 z}{g}, -\varepsilon_2 pz;p\right)_\infty}{\left(\frac{z}{\xi_1 p}, \frac{z}{\xi_2 p};p\right)_\infty}.$$

In the case that $g=0$ we distinguish between $\beta^*\neq 0$ and $\beta^*=0$.

For $g=0$ and $\beta^*\neq 0$ we obtain

$$\frac{w(z)}{w(pz)}=\frac{\alpha^* pz^2+\beta^* pz}{ep^{-2}z^2+2fp^{-1}z}=\frac{1+\frac{\alpha^* z}{\beta^*}}{1+\frac{2fp}{ez}}\cdot\frac{\beta^* p^3}{ez}$$

with possible solution

$$w^{(III)}(z;p)\Big|_{g=0,\beta^*\neq 0}=\frac{\left(-\frac{\alpha^* z}{\beta^*},-\frac{2fp^2}{ez};p\right)_\infty}{\left(-\frac{ez}{\beta^* p^3},-\frac{\beta^* p^4}{ez};p\right)_\infty}.$$

For $g=0$ and $\beta^*=0$ we must have $\alpha^*\neq 0$. Assuming that $f\neq 0$ we obtain

$$\frac{w(z)}{w(pz)}=\frac{\alpha^* pz^2}{ep^{-2}z^2+2fp^{-1}z}=\frac{1}{1+\frac{ez}{2fp}}\cdot\frac{\alpha^* p^2 z}{2f}$$

with possible solution

$$w^{(III)}(z;p)\Big|_{g=0,\beta^*=0}=\frac{\left(-\frac{\alpha^* p^2 z}{2f},-\frac{2f}{\alpha^* pz};p\right)_\infty}{\left(-\frac{ez}{2fp};p\right)_\infty}.$$

For $f=g=0$, $\beta^*=0$ and $\alpha^*\neq 0$ we obtain

$$\frac{w(z)}{w(pz)}=\frac{\alpha^* p^3}{e}$$

with possible solution

$$w^{(III)}(z;p)\Big|_{f=g=0,\beta^*=0}=\frac{\left(-\alpha^* z,-\frac{p}{\alpha^* z};p\right)_\infty}{\left(-\frac{ez}{p^3},-\frac{p^4}{ez};p\right)_\infty}.$$

Case III-B. $e\neq 0$. From (11.7.1) we have with $y_\nu=q^{-\nu}$ for $\alpha^*\neq 0$ and $g\neq 0$

$$\frac{w(y_\nu)}{w(y_{\nu+1})}=\frac{\alpha^* q^{-2\nu}+\beta^* q^{-\nu}+gq}{q(eq^{-2\nu+2}+2fq^{-\nu+1}+g)}=\frac{\left(1+\frac{gq^\nu}{\eta_1}\right)\left(1+\frac{q^{\nu+1}}{\eta_2}\right)}{(1-\xi_1 q^{\nu-1})(1-\xi_2 q^{\nu-1})}\cdot\frac{\eta_1\eta_2\xi_1\xi_2}{gq^3}$$

with possible solution

$$w^{(III)}(y_\nu;q^{-1}) = \frac{\left(\frac{\xi_1}{q},\frac{\xi_2}{q};q\right)_\nu}{\left(-\frac{g}{\eta_1},-\frac{q}{\eta_2};q\right)_\nu}\left(\frac{gq^3}{\eta_1\eta_2\xi_1\xi_2}\right)^\nu.$$

The special case $e = q^{-2}$, $2f = -\alpha - q^{-N-1}$, $g = \alpha q^{-N+1}$, $\alpha^* = \alpha\beta q$ and $\beta^* = -\alpha\beta q^2 - \alpha q^{-N+1}$, which implies that $\xi_1 = \alpha q^2$, $\xi_2 = q^{-N+1}$, $\eta_1 = -\alpha\beta q$ and $\eta_2 = -1$, leads to the weight function

$$w(y_\nu;q^{-1}) = \frac{(\alpha q,q^{-N};q)_\nu}{(\beta^{-1}q^{-N},q;q)_\nu}\left(\frac{1}{\alpha\beta}\right)^\nu$$

for the q-**Hahn** polynomials.

For $\alpha^* \neq 0$ and $g = 0$ we obtain

$$\frac{w(y_\nu)}{w(y_{\nu+1})} = \frac{\alpha^* q^{-2\nu}+\beta^* q^{-\nu}}{q(eq^{-2\nu+2}+2fq^{-\nu+1})} = \frac{1+\frac{\beta^* q^\nu}{\alpha^*}}{1+\frac{2fq^{\nu-1}}{e}}\cdot\frac{\alpha^*}{eq^3}$$

with possible solution

$$w^{(III)}(y_\nu;q^{-1})\Big|_{g=0} = \frac{\left(-\frac{2f}{eq};q\right)_\nu}{\left(-\frac{\beta^*}{\alpha^*};q\right)_\nu}\left(\frac{eq^3}{\alpha^*}\right)^\nu.$$

For $\alpha^* = 0$ we distinguish between $\beta^* \neq 0$ and $\beta^* = 0$.

For $\alpha^* = 0$ and $\beta^* \neq 0$ we obtain

$$\frac{w(y_\nu)}{w(y_{\nu+1})} = \frac{\beta^* q^{-\nu}+gq}{q(eq^{-2\nu+2}+2fq^{-\nu+1}+g)} = \frac{1+\frac{gq^{\nu+1}}{\beta^*}}{(1-\xi_1 q^{\nu-1})(1-\xi_2 q^{\nu-1})}\cdot\frac{\beta^*\xi_1\xi_2}{gq^3}\cdot q^\nu$$

with possible solution

$$w^{(III)}(y_\nu;q^{-1})\Big|_{\alpha^*=0,\beta^*\neq 0} = \frac{\left(\frac{\xi_1}{q},\frac{\xi_2}{q};q\right)_\nu}{\left(-\frac{gq}{\beta^*};q\right)_\nu}\left(\frac{gq^3}{\beta^*\xi_1\xi_2}\right)^\nu q^{-\binom{\nu}{2}}.$$

The special case $e = q^{-2}$, $2f = -b - q^{-N-1}$, $g = bq^{-N+1}$, $\alpha^* = 0$ and $\beta^* = -bq^{-N+1}$, which implies that $\xi_1 = bq^2$ and $\xi_2 = q^{-N+1}$, leads by using (1.8.18) to the weight function

$$w(y_\nu;q^{-1}) = \frac{(bq,q^{-N};q)_\nu}{(q;q)_\nu}(-b^{-1}q^N)^\nu q^{-\binom{\nu}{2}} = \frac{(bq;q)_\nu (q;q)_N}{(q;q)_\nu (q;q)_{N-\nu}} b^{-\nu}$$

for the **affine q-Krawtchouk** polynomials.

For $\alpha^* = 0$ and $\beta^* = 0$ we must have $g \neq 0$ and we obtain

$$\frac{w(y_\nu)}{w(y_{\nu+1})} = \frac{gq}{q(eq^{-2\nu+2}+2fq^{-\nu+1}+g)} = \frac{1}{(1-\xi_1 q^{\nu-1})(1-\xi_2 q^{\nu-1})} \cdot \xi_1\xi_2 q^{-2} \cdot q^{2\nu}$$

with possible solution

$$w^{(III)}(y_\nu;q^{-1})\Big|_{\alpha^*=0,\beta^*=0} = \left(\frac{\xi_1}{q},\frac{\xi_2}{q};q\right)_\nu \left(\xi_1\xi_2 q^{-2}\right)^{-\nu} q^{-2\binom{\nu}{2}}.$$

From (11.7.2) we have with $z_\nu = p^\nu$ for $g \neq 0$

$$\frac{w(z_\nu)}{w(z_{\nu+1})} = \frac{\alpha^* p^{2\nu+1}+\beta^* p^{\nu+1}+g}{ep^{2\nu-2}+2fp^{\nu-1}+g} = \frac{\left(1+\frac{\varepsilon_1 p^\nu}{g}\right)\left(1+\varepsilon_2 p^{\nu+1}\right)}{\left(1-\frac{p^{\nu-1}}{\xi_1}\right)\left(1-\frac{p^{\nu-1}}{\xi_2}\right)}$$

with possible solution

$$w^{(III)}(z_\nu;p) = \frac{\left(\frac{1}{\xi_1 p},\frac{1}{\xi_2 p};p\right)_\nu}{\left(-\frac{\varepsilon_1}{g},-\varepsilon_2 p;p\right)_\nu}.$$

For $g = 0$ we distinguish between $\beta^* \neq 0$ and $\beta^* = 0$.

For $g = 0$ and $\beta^* \neq 0$ we obtain

$$\frac{w(z_\nu)}{w(z_{\nu+1})} = \frac{\alpha^* p^{2\nu+1}+\beta^* p^{\nu+1}}{ep^{2\nu-2}+2fp^{\nu-1}} = \frac{1+\frac{\alpha^*}{\beta^*}p^\nu}{1+\frac{2f}{e}p^{-\nu+1}} \cdot \frac{\beta^* p^3}{e} \cdot p^{-\nu}$$

with possible solution

$$w^{(III)}(z_\nu;p)\Big|_{g=0,\beta^*\neq 0} = \frac{\left(-\frac{2f}{e}p^{-\nu+2};p\right)_\nu}{\left(-\frac{\alpha^*}{\beta^*};p\right)_\nu}\left(\frac{e}{\beta^* p^3}\right)^\nu p^{\binom{\nu}{2}}.$$

Note that for $f \neq 0$ we have by using (1.8.14)

$$\left(-\frac{2f}{e}p^{-\nu+2};p\right)_\nu=\left(-\frac{e}{2fp};p\right)_\nu\left(\frac{2fp}{e}\right)^\nu p^{-\binom{\nu}{2}}.$$

Then the special case $2f=-ep^{N-1}$, $g=0$, $\alpha^*=-p^{-1}$ and $\beta^*=1$ leads to the weight function

$$w(z_\nu;p)=\frac{(p^{-N};p)_\nu}{(p;p)_\nu}(-b^{-1}q)^\nu$$

for the q-**Krawtchouk** polynomials.

For $g=0$ and $\beta^*=0$ we must have $\alpha^*\neq 0$ and we obtain

$$\frac{w(z_\nu)}{w(z_{\nu+1})}=\frac{1}{1+\frac{2f}{e}p^{-\nu+1}}\cdot\frac{\alpha^*p^3}{e}$$

with possible solution

$$w^{(III)}(z_\nu;p)\Big|_{g=0,\beta^*=0}=\left(-\frac{2f}{e}p^{-\nu+2};p\right)_\nu\left(\frac{e}{\alpha^*p^3}\right)^\nu.$$

11.8 Orthogonality Relations

In the preceding section we have obtained both continuous and discrete solutions of the q^{-1}-Pearson equation (11.7.1) and the p-Pearson equation (11.7.2). In this section we will derive orthogonality relations for several cases obtained in section 11.6. We will not give explicit orthogonality relations for each different case, but we will restrict to the most important cases (either continuous or discrete).

In each different case the boundary conditions (11.4.11) or (11.4.15) should be satisfied. Therefore we have to consider

$$w(qy;q^{-1})\varphi(q^2y)\quad\text{with}\quad\varphi(y)=ey^2+2fy+g$$

and $w(y;q^{-1})$ the involving weight function which satisfies the q^{-1}-Pearson equation (11.7.1) or equivalently

$$w(p^{-1}z;p)\varphi(p^{-2}z)\quad\text{with}\quad\varphi(z)=ez^2+2fz+g$$

and $w(z;p)$ the involving weight function which satisfies the p-Pearson equation (11.7.2). These products should vanish at both ends of the interval of orthogonality.

Case I. In this case we have $e=f=0$ and $g\neq 0$. In section 11.6 we have seen that we have positive-definite orthogonality for an infinite system of polynomials in two different cases and that it is impossible to have positive-definite orthogonality for a finite system of polynomials. We will treat both cases here.

Case Ia1. $0 < q < 1$, $e = f = 0$ and $\dfrac{g}{\alpha^*} > 0$. We use the weight function

$$w^{(I)}(y;q^{-1}) = \frac{1}{\left(-\frac{\eta_1 qy}{g}, -\eta_2 y;q\right)_\infty},$$

where $\eta_1\eta_2 = \alpha^*$ and $\eta_1 q + \eta_2 g = \beta^*$. Then we have

$$w^{(I)}(qy;q^{-1})\varphi(q^2y) = \frac{g}{\left(-\frac{\eta_1 q^2 y}{g}, -\eta_2 qy;q\right)_\infty},$$

which vanishes for $y \to \pm\infty$. Now we use (1.15.13) to obtain for $q < \left|\dfrac{g}{\eta_1}\right| < 1$, $q < |\eta_2| < 1$ and $\dfrac{g}{\eta_1\eta_2} = \dfrac{g}{\alpha^*} > 0$

$$\begin{aligned} d_0 &:= \int_{-\infty}^{\infty} \frac{1}{\left(-\frac{\eta_1 qy}{g}, -\eta_2 y;q\right)_\infty} d_q y \\ &= (1-q)\frac{\left(q, -q, -1, -\frac{\eta_1\eta_2 q}{g}, -\frac{g}{\eta_1\eta_2};q\right)_\infty}{\left(-\frac{\eta_1 q}{g}, \frac{\eta_1 q}{g}, -\frac{g}{\eta_1}, \frac{g}{\eta_1}, -\eta_2, \eta_2, -\frac{q}{\eta_2}, \frac{q}{\eta_2};q\right)_\infty} > 0. \end{aligned}$$

Further we have by using (11.6.4)

$$d_n = q^{-2n}(1-q^n)\frac{g}{\eta_1\eta_2}, \quad n = 1,2,3,\ldots,$$

which implies that

$$\prod_{k=1}^{n} d_k = \left(\frac{g}{\eta_1\eta_2}\right)^n q^{-n(n+1)}(q;q)_n, \quad n = 1,2,3,\ldots.$$

This leads to the orthogonality relation

$$\begin{aligned} &\int_{-\infty}^{\infty} \frac{1}{\left(-\frac{\eta_1 qy}{g}, -\eta_2 y;q\right)_\infty} y_m^{(I)}(y;q) y_n^{(I)}(y;q)\, d_q y \\ &= (1-q)\frac{\left(q, -q, -1, -\frac{\eta_1\eta_2 q}{g}, -\frac{g}{\eta_1\eta_2};q\right)_\infty}{\left(-\frac{\eta_1 q}{g}, \frac{\eta_1 q}{g}, -\frac{g}{\eta_1}, \frac{g}{\eta_1}, -\eta_2, \eta_2, -\frac{q}{\eta_2}, \frac{q}{\eta_2};q\right)_\infty} \\ &\quad \times \left(\frac{g}{\eta_1\eta_2}\right)^n q^{-n(n+1)}(q;q)_n\, \delta_{mn}, \quad m,n = 0,1,2,\ldots, \end{aligned}$$

for $q < \left|\dfrac{g}{\eta_1}\right| < 1$, $q < |\eta_2| < 1$ and $\dfrac{g}{\eta_1\eta_2} > 0$.

The special case $e=f=0$, $g=q$, $\alpha^*=1$ and $\beta^*=0$, which implies that $\eta_1=-i$ and $\eta_2=i$, leads to the orthogonality relation

$$\int_{-\infty}^{\infty} \frac{1}{(iy,-iy;q)_\infty} y_m(y;q)y_n(y;q)\,d_qy$$
$$=(1-q)\frac{(q,-q,-1,-1,-q;q)_\infty}{(i,-i,-iq,iq,-i,i,iq,-iq;q)_\infty} q^{-n^2}(q;q)_n\,\delta_{mn}, \quad m,n=0,1,2,\ldots$$

for the **discrete q-Hermite II** polynomials.

Alternatively, we use the weight function

$$w^{(I)}(y_\nu;q^{-1})=\frac{1}{\left(-\frac{g}{\eta_1},-\frac{q}{\eta_2};q\right)_\nu}\left(\frac{gq}{\eta_1\eta_2}\right)^\nu q^{2\binom{\nu}{2}},$$

where $\eta_1\eta_2=\alpha^*$ and $\eta_1 q+\eta_2 g=\beta^*$. Then we have with $y_\nu=q^{-\nu}$

$$w^{(I)}(qy_\nu;q^{-1})\varphi(q^2y_\nu)=\frac{g}{\left(-\frac{g}{\eta_1},-\frac{q}{\eta_2};q\right)_{\nu-1}}\left(\frac{g}{\eta_1\eta_2}\right)^{\nu-1} q^{(\nu-1)^2},$$

which vanishes for $\nu=0$ if we have by using (1.8.6)

$$\frac{1}{\left(-\frac{q}{\eta_2};q\right)_{-1}}=\left(-\frac{q}{\eta_2}q^{-1};q\right)_1=1+\frac{1}{\eta_2}=0.$$

This implies that $\eta_2=-1$ and therefore $\eta_1=-\alpha^*$. Then we use (1.11.8) to obtain with $y_\nu=q^{-\nu}$ for $0<\dfrac{g}{\alpha^*}<1$

$$d_0:=\sum_{\nu=0}^{\infty} w^{(I)}(y_\nu;q^{-1})y_\nu=\sum_{\nu=0}^{\infty}\frac{q^{\nu(\nu-1)}}{\left(\frac{g}{\alpha^*},q;q\right)_\nu}\left(\frac{g}{\alpha^*}\right)^\nu=\frac{1}{\left(\frac{g}{\alpha^*};q\right)_\infty}>0.$$

Further we have by using (11.6.4)

$$d_n=q^{-2n}(1-q^n)\frac{g}{\alpha^*}, \quad n=1,2,3,\ldots,$$

which implies that

$$\prod_{k=1}^{n} d_k=\left(\frac{g}{\alpha^*}\right)^n q^{-n(n+1)}(q;q)_n, \quad n=1,2,3,\ldots.$$

This leads to the orthogonality relation

$$\sum_{\nu=0}^{\infty} \frac{q^{\nu(\nu-1)}}{\left(\frac{g}{\alpha^*},q;q\right)_\nu} \left(\frac{g}{\alpha^*}\right)^\nu y_m^{(I)}(q^{-\nu};q) y_n^{(I)}(q^{-\nu};q)$$
$$= \frac{1}{\left(\frac{g}{\alpha^*};q\right)_\infty} \left(\frac{g}{\alpha^*}\right)^n q^{-n(n+1)}(q;q)_n\,\delta_{mn}, \quad m,n=0,1,2,\ldots$$

for $0 < \dfrac{g}{\alpha^*} < 1$.

The special case $0<q<1$, $e=f=0$, $g=aq$, $\alpha^*=1$ and $\beta^*=-(a+1)q$, which implies that $\eta_1=\eta_2=-1$, leads to the orthogonality relation (for $0<aq<1$)

$$\sum_{\nu=0}^{\infty} \frac{a^\nu q^{\nu^2}}{(aq,q;q)_\nu} y_m(q^{-\nu};q) y_n(q^{-\nu};q) = \frac{1}{(aq;q)_\infty} a^n q^{-n^2}(q;q)_n\,\delta_{mn}$$

with $m,n=0,1,2,\ldots$ for the **Al-Salam-Carlitz II** polynomials.

We remark that the special case $0<q<1$, $e=f=0$, $g=-q$, $\alpha^*=1$ and $\beta^*=0$, which implies that $\eta_1=\eta_2=-1$, cannot be used here in order to obtain an orthogonality relation for the **discrete** q**-Hermite II** polynomials, since then we have $\dfrac{g}{\alpha^*}=-q<0$.

Case Ia2. $q>1$, $e=f=0$ and $\dfrac{g}{\alpha^*}<0$. Since $q>1$ we set $q=p^{-1}$ with $0<p<1$ and we use the weight function

$$w^{(I)}(z;p) = \left(-\frac{\varepsilon_1 z}{g}, -\varepsilon_2 pz;p\right)_\infty,$$

where $\varepsilon_1\varepsilon_2=\alpha^*$ and $\varepsilon_1 p^{-1}+\varepsilon_2 g=\beta^*$. Then we have

$$w^{(I)}(p^{-1}z;p)\varphi(p^{-2}z) = \left(-\frac{\varepsilon_1 z}{gp}, -\varepsilon_2 z;p\right)_\infty \cdot g,$$

which vanishes for $z=-gp/\varepsilon_1$ and for $z=-1/\varepsilon_2$. By using (1.15.11) we obtain for $\dfrac{1}{\varepsilon_2}<\dfrac{gp}{\varepsilon_1}$, $\dfrac{\varepsilon_1}{g\varepsilon_2}<1$ and $\dfrac{gp^2\varepsilon_2}{\varepsilon_1}<1$

$$d_0 := \int_{-\frac{gp}{\varepsilon_1}}^{-\frac{1}{\varepsilon_2}} \left(-\frac{\varepsilon_1 z}{g}, -\varepsilon_2 pz;p\right)_\infty d_pz$$
$$= \left(\frac{gp}{\varepsilon_1}-\frac{1}{\varepsilon_2}\right)(1-p)\left(p,\frac{\varepsilon_1}{g\varepsilon_2},\frac{gp^2\varepsilon_2}{\varepsilon_1};p\right)_\infty > 0.$$

Further we have by using (11.6.4) with $q=p^{-1}$

$$d_n = q^{-2n}(1-q^n)\frac{g}{\varepsilon_1\varepsilon_2} = -p^n(1-p^n)\frac{g}{\varepsilon_1\varepsilon_2}, \quad n = 1,2,3,\ldots,$$

which implies that

$$\prod_{k=1}^{n} d_k = \left(-\frac{g}{\varepsilon_1\varepsilon_2}\right)^n p^{\binom{n+1}{2}}(p;p)_n, \quad n = 1,2,3,\ldots.$$

This leads to the orthogonality relation

$$\begin{aligned}&\int_{-\frac{gp}{\varepsilon_1}}^{-\frac{1}{\varepsilon_2}} \left(-\frac{\varepsilon_1 z}{g}, -\varepsilon_2 pz; p\right)_\infty y_m^{(I)}(z;p^{-1}) y_n^{(I)}(z;p^{-1})\, d_pz \\ &= \left(\frac{gp}{\varepsilon_1} - \frac{1}{\varepsilon_2}\right)(1-p)\left(p, \frac{\varepsilon_1}{g\varepsilon_2}, \frac{gp^2\varepsilon_2}{\varepsilon_1}; p\right)_\infty \left(-\frac{g}{\varepsilon_1\varepsilon_2}\right)^n p^{\binom{n+1}{2}}(p;p)_n\, \delta_{mn}\end{aligned}$$

for $\dfrac{g}{\varepsilon_1\varepsilon_2} < 0$, $\dfrac{1}{\varepsilon_2} < \dfrac{gp}{\varepsilon_1}$, $\dfrac{\varepsilon_1}{g\varepsilon_2} < 1$, $\dfrac{gp^2\varepsilon_2}{\varepsilon_1} < 1$ and $m,n = 0,1,2,\ldots$.

The special case $0 < p < 1$, $e = f = 0$, $g = a/p$, $\alpha^* = 1$ and $\beta^* = -(a+1)/p$, which implies that $\varepsilon_1 = \varepsilon_2 = -1$, leads to the orthogonality relation (for $a < 0$)

$$\begin{aligned}&\int_a^1 \left(\frac{pz}{a}, pz; p\right)_\infty y_m(z;p^{-1}) y_n(z;p^{-1})\, d_pz \\ &= (1-a)(1-p)\left(p, \frac{p}{a}, ap; p\right)_\infty (-a)^n p^{\binom{n}{2}}(p;p)_n\, \delta_{mn}, \quad m,n = 0,1,2,\ldots\end{aligned}$$

for the **Al-Salam-Carlitz I** polynomials.

The special case $0 < q < 1$, $e = f = 0$, $g = -1/p$, $\alpha^* = 1$ and $\beta^* = 0$, which implies that $\varepsilon_1 = \varepsilon_2 = -1$, leads to the orthogonality relation

$$\begin{aligned}&\int_{-1}^1 (-pz, pz; p)_\infty y_m(z;p^{-1}) y_n(z;p^{-1})\, d_pz \\ &= 2(1-p)(p,-p,-p;p)_\infty p^{\binom{n}{2}}(p;p)_n\, \delta_{mn}, \quad m,n = 0,1,2,\ldots\end{aligned}$$

for the **discrete q-Hermite I** polynomials.

Case II. In this case we have $e = 0$ and $f \neq 0$. In section 11.6 we have seen that it is possible to have positive-definite orthogonality for an infinite system of polynomials in four different cases and for a finite system of polynomials in three different cases. We will only treat two infinite cases here and also one finite case.

Case IIa2. $0 < q < 1$, $e = 0$, $f \neq 0$, $(\beta^*)^2 \geq 4g\alpha^* q$, $\dfrac{\eta_1}{2f} < 1$ and $\dfrac{\eta_2 g}{2fq} < 1$. We use the weight function

$$w^{(II)}(y_\nu;q^{-1})=\frac{\left(-\frac{g}{2fq};q\right)_\nu}{\left(-\frac{g}{\eta_1},-\frac{q}{\eta_2};q\right)_\nu}\left(\frac{2fq^2}{\eta_1\eta_2}\right)^\nu q^{\binom{\nu}{2}}.$$

Then we have with $y_\nu=q^{-\nu}$

$$w^{(II)}(qy_\nu;q^{-1})\varphi(q^2y_\nu)=\frac{\left(-\frac{g}{2fq};q\right)_\nu}{\left(-\frac{g}{\eta_1},-\frac{q}{\eta_2};q\right)_{\nu-1}}\left(\frac{2f}{\eta_1\eta_2}\right)^{\nu-1}\cdot 2fq^{\nu+\binom{\nu-1}{2}},$$

which vanishes for $\nu=0$ if we have by using (1.8.6)

$$\frac{1}{\left(-\frac{q}{\eta_2};q\right)_{-1}}=\left(-\frac{q}{\eta_2}q^{-1};q\right)_1=1+\frac{1}{\eta_2}=0.$$

This implies that $\eta_2=-1$ and $\eta_1=-\alpha^*$. Then we have by using (1.11.6) with $y_\nu=q^{-\nu}$ for $\frac{2f}{\alpha^*}>0$ and $\frac{g}{\alpha^*}<1$

$$d_0:=\sum_{\nu=0}^{\infty}w^{(II)}(y_\nu;q^{-1})y_\nu=\sum_{\nu=0}^{\infty}\frac{\left(-\frac{g}{2fq};q\right)_\nu}{\left(\frac{g}{\alpha^*},q;q\right)_\nu}\left(\frac{2fq}{\alpha^*}\right)^\nu q^{\binom{\nu}{2}}=\frac{\left(-\frac{2fq}{\alpha^*};q\right)_\infty}{\left(\frac{g}{\alpha^*};q\right)_\infty}>0.$$

Further we have by using (11.6.5)

$$d_n=\left(\frac{2f}{\alpha^*}\right)^2q^{-4n+3}(1-q^n)\left(1+\frac{\alpha^*}{2f}q^{n-1}\right)\left(1+\frac{g}{2f}q^{n-2}\right),\quad n=1,2,3,\ldots,$$

which implies that

$$\prod_{k=1}^{n}d_k=\left(\frac{2f}{\alpha^*}\right)^{2n}q^{-n(2n-1)}\left(q,-\frac{\alpha^*}{2f},-\frac{g}{2fq};q\right)_n,\quad n=1,2,3,\ldots.$$

This leads to the orthogonality relation

$$\begin{aligned}&\sum_{\nu=0}^{\infty}\frac{\left(-\frac{g}{2fq};q\right)_\nu}{\left(\frac{g}{\alpha^*},q;q\right)_\nu}\left(\frac{2fq}{\alpha^*}\right)^\nu q^{\binom{\nu}{2}}y_m^{(II)}(q^{-\nu};q)y_n^{(II)}(q^{-\nu};q)\\&=\frac{\left(-\frac{2fq}{\alpha^*};q\right)_\infty}{\left(\frac{g}{\alpha^*};q\right)_\infty}\left(\frac{2f}{\alpha^*}\right)^{2n}q^{-n(2n-1)}\left(q,-\frac{\alpha^*}{2f},-\frac{g}{2fq};q\right)_n\delta_{mn},\quad m,n=0,1,2,\ldots,\end{aligned}$$

for $-\frac{g}{2fq}<1$, $\frac{2f}{\alpha^*}>0$ and $\frac{g}{\alpha^*}<1$.

The special case $2f = cq^{-1}$, $g = -bcq$, $\alpha^* = 1$ and $\beta^* = -q + bcq$, which implies that $\eta_1 = \eta_2 = -1$, leads to the orthogonality relation (for $0 \leq bq < 1$ and $c > 0$)

$$\sum_{\nu=0}^{\infty} \frac{(bq;q)_\nu}{(-bcq,q;q)_\nu} c^\nu q^{\binom{\nu}{2}} y_m(q^{-\nu};q) y_n(q^{-\nu};q)$$
$$= \frac{(-c;q)_\infty}{(-bcq;q)_\infty} c^{2n} q^{-n(2n+1)} (q,-c^{-1}q,bq;q)_n \, \delta_{mn}, \quad m,n = 0,1,2,\ldots$$

for the q-**Meixner** polynomials.

The special case $2f = aq^{-1}$, $g = 0$, $\alpha^* = 1$ and $\beta^* = -q$, which implies that $\eta_1 = \eta_2 = -1$, leads to the orthogonality relation (for $a > 0$)

$$\sum_{\nu=0}^{\infty} \frac{a^\nu}{(q;q)_\nu} q^{\binom{\nu}{2}} y_m(q^{-\nu};q) y_n(q^{-\nu};q)$$
$$= (-a;q)_\infty a^{2n} q^{-n(2n+1)} (q,-a^{-1}q;q)_n \, \delta_{mn}, \quad m,n = 0,1,2,\ldots$$

for the q-**Charlier** polynomials.

Case IIa3. $q > 1$, $e = 0$, $f \neq 0$, $(\beta^*)^2 \geq 4g\alpha^* q$, $\dfrac{\eta_1}{2f} > 1$ and $\dfrac{\eta_2 g}{2f} < 0$. Since $q > 1$ we set $q = p^{-1}$ with $0 < p < 1$. Then we have $\dfrac{\varepsilon_1}{2f} > 1$ and $\dfrac{\varepsilon_2 g}{2f} < 0$. This implies that $g \neq 0$ and we might use the weight function

$$w^{(II)}(z;p) = \frac{\left(-\frac{\varepsilon_1 z}{g}, -\varepsilon_2 pz; p\right)_\infty}{\left(-\frac{2fz}{gp}; p\right)_\infty},$$

where $\varepsilon_1 \varepsilon_2 = \alpha^* \neq 0$ and $\varepsilon_1 p^{-1} + \varepsilon_2 g = \beta^*$. Then we have

$$w^{(II)}(p^{-1}z;p)\varphi(p^{-2}z) = \frac{\left(-\frac{\varepsilon_1 z}{gp}, -\varepsilon_2 z; p\right)_\infty}{\left(-\frac{2fz}{gp^2}; p\right)_\infty} \cdot (2fp^{-2}z + g) = \frac{\left(-\frac{\varepsilon_1 z}{gp}, -\varepsilon_2 z; p\right)_\infty}{\left(-\frac{2fz}{gp}; p\right)_\infty} \cdot g,$$

which vanishes for $z = -gp/\varepsilon_1$ and for $z = -1/\varepsilon_2$. Now we use (1.15.11) to obtain for $\dfrac{1}{\varepsilon_2} < \dfrac{gp}{\varepsilon_1}$, $\dfrac{\varepsilon_1}{g\varepsilon_2} < 1$ and $\dfrac{gp^2\varepsilon_2}{\varepsilon_1} < 1$

$$d_0 := \int_{-\frac{gp}{\varepsilon_1}}^{-\frac{1}{\varepsilon_2}} \frac{\left(-\frac{\varepsilon_1 z}{g}, -\varepsilon_2 pz; p\right)_\infty}{\left(-\frac{2fz}{gp}; p\right)_\infty} d_p z = \left(\frac{gp}{\varepsilon_1} - \frac{1}{\varepsilon_2}\right)(1-p) \frac{\left(p, \frac{\varepsilon_1}{\varepsilon_2 g}, \frac{\varepsilon_2 gp^2}{\varepsilon_1}; p\right)_\infty}{\left(\frac{2f}{\varepsilon_1}, \frac{2f}{\varepsilon_2 gp}; p\right)_\infty} > 0.$$

Further we have by using (11.6.5) with $q = p^{-1}$

$$
\begin{aligned}
d_n &= \frac{4f^2}{(\alpha^*)^2} q^{-4n+3}(1-q^n)\left\{1-\frac{\beta^*}{2f}q^{n-2}+\frac{g\alpha^*}{4f^2}q^{2n-3}\right\}\\
&= -\frac{g}{\varepsilon_1\varepsilon_2}p^n(1-p^n)\left(1-\frac{2f}{\varepsilon_1}p^{n-1}\right)\left(1-\frac{2f}{\varepsilon_2 g}p^{n-2}\right), \quad n=1,2,3,\ldots,
\end{aligned}
$$

which implies that

$$
\prod_{k=1}^{n} d_k = \left(-\frac{g}{\varepsilon_1\varepsilon_2}\right)^n p^{n(n+1)/2}\left(p,\frac{2f}{\varepsilon_1},\frac{2f}{\varepsilon_2 gp};p\right)_n, \quad n=1,2,3,\ldots.
$$

This leads to the orthogonality relation

$$
\begin{aligned}
&\int_{-\frac{gp}{\varepsilon_1}}^{-\frac{1}{\varepsilon_2}} \frac{\left(-\frac{\varepsilon_1 z}{g},-\varepsilon_2 pz;p\right)_\infty}{\left(-\frac{2fz}{gp};p\right)_\infty} y_m^{(II)}(z;p)y_n^{(II)}(z;p)\,d_pz\\
&= \left(\frac{gp}{\varepsilon_1}-\frac{1}{\varepsilon_2}\right)(1-p)\frac{\left(p,\frac{\varepsilon_1}{\varepsilon_2 g},\frac{\varepsilon_2 gp^2}{\varepsilon_1};p\right)_\infty}{\left(\frac{2f}{\varepsilon_1},\frac{2f}{\varepsilon_2 gp};p\right)_\infty}\\
&\quad\times\left(-\frac{g}{\varepsilon_1\varepsilon_2}\right)^n p^{n(n+1)/2}\left(p,\frac{2f}{\varepsilon_1},\frac{2f}{\varepsilon_2 gp};p\right)_n\delta_{mn}, \quad m,n=0,1,2,\ldots
\end{aligned}
$$

for $\frac{\varepsilon_1}{2f}>1$, $\frac{\varepsilon_2 g}{2f}<0$, $\frac{1}{\varepsilon_2}<\frac{gp}{\varepsilon_1}$, $\frac{\varepsilon_1}{g\varepsilon_2}<1$ and $\frac{gp^2\varepsilon_2}{\varepsilon_1}<1$.

Case IIb1. $0<q<1$, $e=0$, $f\neq 0$, $\frac{\eta_1}{2f}>1$ with $\frac{\eta_1}{2f}q^{N_1}\leq 1<\frac{\eta_1}{2f}q^{N_1-1}$, $\frac{\eta_2 g}{2fq}>1$ with $\frac{\eta_2 g}{2f}q^{N_2-1}\leq 1<\frac{\eta_2 g}{2f}q^{N_2-2}$ and $N=\min(N_1,N_2)$. Again we use the weight function

$$
w^{(II)}(y_\nu;q^{-1}) = \frac{\left(-\frac{g}{2fq};q\right)_\nu}{\left(-\frac{g}{\eta_1},-\frac{q}{\eta_2};q\right)_\nu}\left(\frac{2fq^2}{\eta_1\eta_2}\right)^\nu q^{\binom{\nu}{2}}.
$$

Then we have as before with $y_\nu=q^{-\nu}$

$$
w^{(II)}(qy_\nu;q^{-1})\varphi(q^2y_\nu) = \frac{\left(-\frac{g}{2fq};q\right)_\nu}{\left(-\frac{g}{\eta_1},-\frac{q}{\eta_2};q\right)_{\nu-1}}\left(\frac{2f}{\eta_1\eta_2}\right)^{\nu-1}\cdot 2fq^{\nu+\binom{\nu-1}{2}},
$$

which vanishes for $\nu=0$ if $\eta_2=-1$, which implies that $\eta_1=-\alpha^*$. It also vanishes for $\nu=N+1$ if $-\frac{g}{2fq}=q^{-N}$. Then we have by using (1.11.6) with $y_\nu=q^{-\nu}$ and $g=-2fq^{-N+1}$ for $-\frac{2fq}{\alpha^*}<q^N$

$$d_0 := \sum_{\nu=0}^{N} w^{(II)}(y_\nu;q^{-1})y_\nu = \sum_{\nu=0}^{N} \frac{(q^{-N};q)_\nu}{\left(-\frac{2fq^{-N+1}}{\alpha^*},q;q\right)_\nu} \left(\frac{2fq}{\alpha^*}\right)^\nu q^{\binom{\nu}{2}}$$

$$= \frac{1}{\left(-\frac{2fq^{-N+1}}{\alpha^*};q\right)_N} > 0.$$

Now we have by using (11.6.5)

$$d_n = \left(\frac{2f}{\alpha^*}\right)^2 q^{-4n+3}(1-q^n)\left(1+\frac{\alpha^*}{2f}q^{n-1}\right)\left(1-q^{n-N-1}\right), \quad n=1,2,3,\ldots,$$

which implies that

$$\prod_{k=1}^{n} d_k = \left(\frac{2f}{\alpha^*}\right)^{2n} q^{-n(2n-1)}\left(q,-\frac{\alpha^*}{2f},q^{-N};q\right)_n, \quad n=1,2,3,\ldots.$$

This leads to the orthogonality relation

$$\sum_{\nu=0}^{N} \frac{(q^{-N};q)_\nu}{\left(-\frac{2fq^{-N+1}}{\alpha^*},q;q\right)_\nu} \left(\frac{2fq}{\alpha^*}\right)^\nu q^{\binom{\nu}{2}} y_m^{(II)}(q^{-\nu};q)y_n^{(II)}(q^{-\nu};q)$$

$$= \frac{1}{\left(-\frac{2fq^{-N+1}}{\alpha^*};q\right)_N}\left(\frac{2f}{\alpha^*}\right)^{2n} q^{-n(2n-1)}\left(q,-\frac{\alpha^*}{2f},q^{-N};q\right)_n \delta_{mn}$$

for $-\dfrac{2fq}{\alpha^*} < q^N$ and $m,n=0,1,2,\ldots,N$.

Since we have

$$w(y_\nu;q^{-1}) = \frac{(q^{-N};q)_\nu}{(b^{-1}q^{-N},q;q)_\nu}\left(-b^{-1}\right)^\nu q^{\binom{\nu}{2}} = \frac{(q;q)_N}{(bq;q)_N}\frac{(bq;q)_{N-\nu}}{(q;q)_{N-\nu}}\frac{(-1)^\nu}{(q;q)_\nu}q^{\binom{\nu}{2}},$$

the special case $2f=-b^{-1}q^{-1}$, $g=b^{-1}q^{-N}$, $\alpha^*=1$ and $\beta^*=-q-b^{-1}q^{-N}$, which implies that $\eta_1=\eta_2=-1$, leads to the orthogonality relation (for $b>q^{-N}$)

$$\sum_{\nu=0}^{N} \frac{(q;q)_N}{(bq;q)_N}\frac{(bq;q)_{N-\nu}}{(q;q)_{N-\nu}}\frac{(-1)^\nu}{(q;q)_\nu}q^{\binom{\nu}{2}} y_m(q^{-\nu};q)y_n(q^{-\nu};q)$$

$$= \frac{1}{(b^{-1}q^{-N};q)_N} b^{-2n}q^{-n(2n+1)}\left(q,bq,q^{-N};q\right)_n \delta_{mn}, \quad m,n=0,1,2,\ldots,N$$

for the **quantum q-Krawtchouk** polynomials.

Case III. In this case we have $e \neq 0$. In section 11.6 we have seen that it is possible to have positive-definite orthogonality for an infinite system of polynomials in at least nine different cases. It is also possible to have positive-definite orthogonality for finite systems of polynomials. We will only treat three finite cases here.

Case IIIb3. $0 < q < 1$, $\frac{g}{e} > 0$, $\alpha^* = 0$, $\frac{\xi_2\beta^*}{gq} < -1$ with $\frac{\xi_2\beta^*}{g}q^{N-2} < -1 \leq \frac{\xi_2\beta^*}{g}q^{N-1}$ and $\frac{\xi_1\beta^*}{gq} > -1$. We use the weight function

$$w^{(III)}(y_\nu;q^{-1})\Big|_{\alpha^*=0,\beta^*\neq 0} = \frac{\left(\frac{\xi_1}{q},\frac{\xi_2}{q};q\right)_\nu}{\left(-\frac{gq}{\beta^*};q\right)_\nu}\left(\frac{gq^3}{\beta^*\xi_1\xi_2}\right)^\nu q^{-\binom{\nu}{2}}.$$

Then we have by using $e = g/\xi_1\xi_2$ with $y_\nu = q^{-\nu}$

$$\begin{aligned}
&w^{(III)}(qy_\nu;q^{-1})\Big|_{\alpha^*=0,\beta^*\neq 0}\varphi(q^2y_\nu)\\
&= \frac{\left(\frac{\xi_1}{q},\frac{\xi_2}{q};q\right)_{\nu-1}}{\left(-\frac{gq}{\beta^*};q\right)_{\nu-1}}\left(\frac{gq^3}{\beta^*\xi_1\xi_2}\right)^{\nu-1} q^{-\binom{\nu-1}{2}}\cdot\left(eq^{-2\nu+4}+2fq^{-\nu+2}+g\right)\\
&= \frac{\left(\frac{\xi_1}{q},\frac{\xi_2}{q};q\right)_{\nu-1}}{\left(-\frac{gq}{\beta^*};q\right)_{\nu-1}}\left(\frac{gq^3}{\beta^*\xi_1\xi_2}\right)^{\nu-1} q^{-\binom{\nu-1}{2}}\cdot eq^{-2\nu+4}\left(1-\xi_1q^{\nu-2}\right)\left(1-\xi_2q^{\nu-2}\right)\\
&= \frac{\left(\frac{\xi_1}{q},\frac{\xi_2}{q};q\right)_{\nu}}{\left(-\frac{gq}{\beta^*};q\right)_{\nu-1}}\left(\frac{g}{\xi_1\xi_2}\right)^{\nu}\left(\frac{q}{\beta^*}\right)^{\nu-1} q^{2-\binom{\nu-1}{2}},
\end{aligned}$$

which vanishes for $\nu = 0$ if we have by using (1.8.6)

$$\frac{1}{\left(-\frac{gq}{\beta^*};q\right)_{-1}} = \left(-\frac{gq}{\beta^*}q^{-1};q\right)_1 = 1+\frac{g}{\beta^*} = 0.$$

This implies that $g = -\beta^*$. It also vanishes for $\nu = N+1$ if $\xi_2 = q^{-N+1}$. Then we have by using (1.11.7) with $y_\nu = q^{-\nu}$

$$d_0 := \sum_{\nu=0}^{N} w^{(III)}(y_\nu;q^{-1})\Big|_{\alpha^*=0,\beta^*\neq 0} y_\nu = \sum_{\nu=0}^{N}\frac{\left(\frac{\xi_1}{q},q^{-N};q\right)_\nu}{(q;q)_\nu}\left(-\frac{q^{N+1}}{\xi_1}\right)^\nu q^{-\binom{\nu}{2}} = \frac{q^N}{\xi_1^N}.$$

Note that this is positive for $\xi_1 > 0$. Further we have by using (11.6.6), (11.6.7) and (11.6.8)

$$d_n = -\xi_1\xi_2 q^{n-3}(1-q^n)\left(1-\xi_1q^{n-2}\right)\left(1-\xi_2q^{n-2}\right),\quad n = 1,2,3,\ldots$$

with $\xi_2 = q^{-N+1}$, which implies that

$$\prod_{k=1}^{n} d_k = \left(-\xi_1 q^{-N}\right)^n q^{n(n-3)/2}\left(q,\frac{\xi_1}{q},q^{-N};q\right)_n, \quad n=1,2,3,\ldots.$$

This leads to the orthogonality relation

$$\sum_{\nu=0}^{N}\frac{\left(\frac{\xi_1}{q},q^{-N};q\right)_\nu}{(q;q)_\nu}\left(-\frac{q^{N+1}}{\xi_1}\right)^\nu q^{-\binom{\nu}{2}} y_m^{(III)}(q^{-\nu};q)y_n^{(III)}(q^{-\nu};q)$$
$$=\frac{q^N}{\xi_1^N}\left(-\xi_1 q^{-N}\right)^n q^{n(n-3)/2}\left(q,\frac{\xi_1}{q},q^{-N};q\right)_n \delta_{mn}, \quad m,n=0,1,2,\ldots,N$$

for $\xi_1>0$.

The special case $e=q^{-2}$, $2f=-b-q^{-N-1}$, $g=bq^{-N+1}$, $\alpha^*=0$ and $\beta^*=-bq^{-N+1}$, which implies that $\xi_1=bq^2$ and $\xi_2=q^{-N+1}$, leads by using (1.8.18) and

$$\frac{(bq,q^{-N};q)_\nu}{(q;q)_\nu}(-b^{-1}q^N)^\nu q^{-\binom{\nu}{2}}=\frac{(bq;q)_\nu(q;q)_N}{(q;q)_\nu(q;q)_{N-\nu}}b^{-\nu}$$

to the orthogonality relation (for $0<bq<1$)

$$\sum_{\nu=0}^{N}\frac{(bq;q)_\nu(q;q)_N}{(q;q)_\nu(q;q)_{N-\nu}}(bq)^{-\nu}y_m(q^{-\nu};q)y_n(q^{-\nu};q)$$
$$=(-1)^n b^{n-N} q^{-N(n+1)} q^{n(n+1)/2}(q,bq,q^{-N};q)_n\,\delta_{mn}, \quad m,n=0,1,2,\ldots$$

for the **affine q-Krawtchouk** polynomials.

Case IIIb8. $q>1$, $g=0$, $\beta^*\neq 0$, $\alpha^*\leq 0<e$, $\dfrac{2fq}{\beta^*}<1$ and $0<\dfrac{2f\alpha^*}{e\beta^*}<1$ with $\dfrac{2f\alpha^*}{e\beta^*}q^N\leq 1<\dfrac{2f\alpha^*}{e\beta^*}q^{N-1}$. We use the weight function

$$w^{(III)}(z_\nu;p)\Big|_{g=0,\beta^*\neq 0}=\frac{\left(-\frac{2f}{e}p^{-\nu+2};p\right)_\nu}{\left(-\frac{\alpha^*}{\beta^*};p\right)_\nu}\left(\frac{e}{\beta^* p^3}\right)^\nu p^{\binom{\nu}{2}}.$$

Then we have with $z_\nu=p^\nu$

$$w^{(III)}(p^{-1}z_\nu;p)\Big|_{g=0,\beta^*\neq 0}\varphi(p^{-2}z_\nu) = \frac{\left(-\frac{2f}{e}p^{-\nu+3};p\right)_{\nu-1}}{\left(-\frac{\alpha^*}{\beta^*};p\right)_{\nu-1}}\left(\frac{e}{\beta^* p^3}\right)^{\nu-1} p^{\binom{\nu-1}{2}}$$
$$\times\left(ep^{2\nu-4}+2fp^{\nu-2}\right)$$
$$= \frac{\left(-\frac{2f}{e}p^{-\nu+2};p\right)_{\nu}}{\left(-\frac{\alpha^*}{\beta^*};p\right)_{\nu-1}}\left(\frac{e}{\beta^*}\right)^{\nu-1}\cdot ep^{-\nu-1+\binom{\nu-1}{2}},$$

which vanishes for $\nu=0$ if we have by using (1.8.6)

$$\frac{1}{\left(-\frac{\alpha^*}{\beta^*};p\right)_{-1}} = \left(-\frac{\alpha^*}{\beta^*}p^{-1};p\right)_1 = 1+\frac{\alpha^*}{\beta^* p} = 0.$$

This implies that $\beta^* = -\alpha^* p^{-1}$. Hence $\alpha^*\neq 0$, so that we must have $\alpha^*<0<e$. Note that for $f\neq 0$ we have by using (1.8.14)

$$\left(-\frac{2f}{e}p^{-\nu+2};p\right)_\nu = \left(-\frac{e}{2fp};p\right)_\nu\left(\frac{2fp}{e}\right)^\nu p^{-\binom{\nu}{2}},$$

which vanishes for $\nu=N+1$ if $-\dfrac{e}{2fp}=p^{-N}$. Therefore we take $2f=-ep^{N-1}$ and use (1.11.2) with $z_\nu=p^\nu$ to obtain

$$d_0 := \sum_{\nu=0}^{N} w^{(III)}(z_\nu;p)\Big|_{g=0,\alpha^*\neq 0} z_\nu = \sum_{\nu=0}^{N}\frac{(p^{-N};p)_\nu}{(p;p)_\nu}\left(\frac{ep^{N-1}}{\alpha^*}\right)^\nu = \left(\frac{e}{\alpha^* p};p\right)_N > 0,$$

since $\alpha^*<0<e$. Further we have by using (11.6.6), (11.6.7) and (11.6.9) for $2f = -ep^{N-1}$

$$d_n = -\left(\frac{\alpha^*}{e}\right)^2 p^{-3n+5}(1-p^{-n})\frac{\left(1-\frac{\alpha^*}{e}p^{-n+3}\right)\left(1-\frac{e}{\alpha^*}p^{n+N-2}\right)\left(1-p^{-n+N+1}\right)}{\left(1-\frac{\alpha^*}{e}p^{-2n+4}\right)\left(1-\frac{\alpha^*}{e}p^{-2n+3}\right)^2\left(1-\frac{\alpha^*}{e}p^{-2n+2}\right)}$$
$$= \frac{e}{\alpha^*}p^{2n+N-3}(1-p^n)\frac{\left(1-\frac{e}{\alpha^*}p^{n-3}\right)\left(1-\frac{e}{\alpha^*}p^{n+N-2}\right)\left(1-p^{n-N-1}\right)}{\left(1-\frac{e}{\alpha^*}p^{2n-4}\right)\left(1-\frac{e}{\alpha^*}p^{2n-3}\right)^2\left(1-\frac{e}{\alpha*}p^{2n-2}\right)}$$

for $n=1,2,3,\ldots$, which implies that

$$\prod_{k=1}^{n} d_k = \left(\frac{ep^{N-1}}{\alpha^*}\right)^n p^{2\binom{n}{2}}\frac{\left(p,\frac{e}{\alpha^* p^2},\frac{ep^{N-1}}{\alpha^*},p^{-N};p\right)_n}{\left(\frac{e}{\alpha^* p^2},\frac{e}{\alpha^* p};p\right)_{2n}}, \quad n=1,2,3,\ldots.$$

This leads to the orthogonality relation

$$\sum_{\nu=0}^{N}\frac{(p^{-N};p)_{\nu}}{(p;p)_{\nu}}\left(\frac{ep^{N-1}}{\alpha^*}\right)^{\nu}y_m^{(III)}(p^{\nu};p)y_n^{(III)}(p^{\nu};p)$$

$$=\left(\frac{e}{\alpha^* p};p\right)_N\left(\frac{ep^{N-1}}{\alpha^*}\right)^n p^{n(n-1)}\frac{\left(p,\frac{e}{\alpha^* p^2},\frac{ep^{N-1}}{\alpha^*},p^{-N};p\right)_n}{\left(\frac{e}{\alpha^* p^2},\frac{e}{\alpha^* p};p\right)_{2n}}\delta_{mn},\quad n=1,2,3,\ldots$$

for $\alpha^*<0<e$.

The special case $e>0$, $2f=-ep^{N-1}$, $g=0$, $\alpha^*=-p^{-1}$ and $\beta^*=1$ leads to the orthogonality relation (for $e>0$)

$$\sum_{\nu=0}^{N}\frac{(p^{-N};p)_{\nu}}{(p;p)_{\nu}}\left(-ep^N\right)^{\nu}y_m(p^{\nu};p)y_n(p^{\nu};p)$$

$$=(-e;p)_N\left(-ep^N\right)^n p^{n(n-1)}\frac{(p,-ep^N,p^{-N};p)_n}{(-ep^{-1},-e;p)_{2n}}\delta_{mn},\quad m,n=0,1,2,\ldots,N$$

for the q-**Krawtchouk** polynomials.

Case IIIb9. For $0<q<1$, $e\neq 0$, $\alpha^*\neq 0$ and $g\neq 0$ we use the weight function

$$w^{(III)}(y_{\nu};q^{-1})=\frac{\left(\frac{\xi_1}{q},\frac{\xi_2}{q};q\right)_{\nu}}{\left(-\frac{g}{\eta_1},-\frac{q}{\eta_2};q\right)_{\nu}}\left(\frac{gq^3}{\eta_1\eta_2\xi_1\xi_2}\right)^{\nu}.$$

Then we have as before with $y_{\nu}=q^{-\nu}$

$$w^{(III)}(qy_{\nu};q^{-1})\varphi(q^2y_{\nu})=\frac{\left(\frac{\xi_1}{q},\frac{\xi_2}{q};q\right)_{\nu-1}}{\left(-\frac{g}{\eta_1},-\frac{q}{\eta_2};q\right)_{\nu-1}}\left(\frac{g}{\xi_1\xi_2}\right)^{\nu}\frac{q^{\nu+1}}{(\eta_1\eta_2)^{\nu-1}},$$

which vanishes for $\nu=0$ if $\eta_2=-1$, which implies that $\eta_1=-\alpha^*$. It also vanishes for $\nu=N+1$ if $\frac{\xi_2}{q}=q^{-N}$. Then we have by using (1.11.3) with $y_{\nu}=q^{-\nu}$ and $\xi_2=q^{-N+1}$

$$d_0:=\sum_{\nu=0}^{N}w^{(III)}(y_{\nu};q^{-1})y_{\nu}=\sum_{\nu=0}^{N}\frac{\left(\frac{\xi_1}{q},q^{-N};q\right)_{\nu}}{\left(\frac{e\xi_1q^{-N+1}}{\alpha^*},q;q\right)_{\nu}}\left(\frac{eq^2}{\alpha^*}\right)^{\nu}=\frac{\left(\frac{e\xi_1q}{\alpha^*},\frac{eq^{-N+1}}{\alpha^*};q\right)_{\infty}}{\left(\frac{e\xi_1q^{-N+1}}{\alpha^*},\frac{eq^2}{\alpha^*};q\right)_{\infty}}.$$

The parameters e, ξ_1 and α^* should be chosen in such a way that $d_0>0$. Now we have as before

$$d_n = -\xi_1 q^{n-N-2}(1-q^n)\frac{1-\frac{\alpha^*}{e}q^{n-3}}{\left(1-\frac{\alpha^*}{e}q^{2n-4}\right)\left(1-\frac{\alpha^*}{e}q^{2n-3}\right)^2\left(1-\frac{\alpha^*}{e}q^{2n-2}\right)}$$
$$\times\left(1-\xi_1 q^{n-2}\right)\left(1-q^{n-N-1}\right)\left(1-\frac{\alpha^*}{e\xi_1}q^{n-1}\right)\left(1-\frac{\alpha^*}{e}q^{n+N-2}\right)$$

for $n = 1,2,3,\ldots$, which implies that

$$\prod_{k=1}^{n} d_k = \left(-\xi_1 q^{-N+1}\right)^n q^{n(n-5)/2}\frac{\left(q, \frac{\alpha^*}{eq^2}, \frac{\xi_1}{q}, q^{-N}, \frac{\alpha^*}{e\xi_1}, \frac{\alpha^* q^{N-1}}{e}; q\right)_n}{\left(\frac{\alpha^*}{eq^2}, \frac{\alpha^*}{eq}; q\right)_{2n}}, \quad n = 1,2,3,\ldots.$$

This leads to the orthogonality relation

$$\sum_{\nu=0}^{N}\frac{\left(\frac{\xi_1}{q}, q^{-N}; q\right)_\nu}{\left(\frac{e\xi_1 q^{-N+1}}{\alpha^*}, q; q\right)_\nu}\left(\frac{eq^2}{\alpha^*}\right)^\nu y_m^{(III)}(q^{-\nu};q)y_n^{(III)}(q^{-\nu};q)$$
$$= \frac{\left(\frac{e\xi_1 q}{\alpha^*}, \frac{eq^{-N+2}}{\alpha^*}; q\right)_\infty}{\left(\frac{e\xi_1 q^{-N+1}}{\alpha^*}, \frac{eq^2}{\alpha^*}; q\right)_\infty}\left(-\xi_1 q^{-N+1}\right)^n q^{n(n-5)/2}$$
$$\times\frac{\left(q, \frac{\alpha^*}{eq^2}, \frac{\xi_1}{q}, q^{-N}, \frac{\alpha^*}{e\xi_2}, \frac{\alpha^* q^{N-1}}{e}; q\right)_n}{\left(\frac{\alpha^*}{eq^2}, \frac{\alpha^*}{eq}; q\right)_{2n}}\delta_{mn}, \quad m,n = 0,1,2,\ldots,N$$

for suitable conditions on the parameters e, ξ_1 and α^*.

The special case $e = q^{-2}$, $2f = -\alpha - q^{-N-1}$, $g = \alpha q^{-N+1}$, $\alpha^* = \alpha\beta q$ and $\beta^* = -\alpha\beta q^2 - \alpha q^{-N+1}$, which implies that $\xi_1 = \alpha q^2$, $\xi_2 = q^{-N+1}$, $\eta_1 = -\alpha\beta q$ and $\eta_2 = -1$, leads to the orthogonality relation (for $0 < \alpha q < 1$ and $0 < \beta q < 1$)

$$\sum_{\nu=0}^{N}\frac{(\alpha q, q^{-N}; q)_\nu}{(\beta^{-1}q^{-N}, q; q)_\nu}\left(\frac{1}{\alpha\beta q}\right)^\nu y_m(q^{-\nu};q)y_n(q^{-\nu};q)$$
$$= \frac{(\beta^{-1}, \alpha^{-1}\beta^{-1}q^{-N-1}; q)_\infty}{(\beta^{-1}q^{-N}, \alpha^{-1}\beta^{-1}q^{-1}; q)_\infty}(-\alpha q^{-N})^n q^{n(n+1)/2}$$
$$\times\frac{(q, \alpha\beta q, \alpha q, q^{-N}, \beta q, \alpha\beta q^{N+2}; q)_n}{(\alpha\beta q, \alpha\beta q^2; q)_{2n}}\delta_{mn}, \quad m,n = 0,1,2,\ldots,N$$

for the q-**Hahn** polynomials.

Chapter 12
Orthogonal Polynomial Solutions in $q^{-x}+uq^{x}$ of Real q-Difference Equations

Classical q-Orthogonal Polynomials III

12.1 Motivation for Polynomials in $q^{-x}+uq^{x}$ Through Duality

As in chapter 7 we have the concept of duality introduced in definition 3.1. In chapter 11 we obtained a q-difference equation of the form (cf. (11.1.2))

$$A^*(x)y_n(q^{-x-1})-\{A^*(x)+B^*(x)\}y_n(q^{-x})+B^*(x)y_n(q^{-x+1})$$
$$=\lambda_n^* y_n(q^{-x}),\quad n=0,1,2,\ldots,N-1 \tag{12.1.1}$$

with $N\in\{1,2,3,\ldots\}$ or $N\to\infty$, where

$$\lambda_n^*=(q^n-1)\left\{eq(1-q^{1-n})-2\varepsilon(1-q)\right\},\quad n=0,1,2,\ldots,N-1,$$

$$A^*(x)=eq^2+2fq^{x+1}+gq^{2x}\quad\text{and}\quad B^*(x)=\alpha^*+\beta^*q^{x-1}+gq^{2x-1}$$

with

$$\alpha^*:=eq-2\varepsilon(1-q)\quad\text{and}\quad\beta^*:=2fq-\gamma(1-q),$$

where $e,f,g,\alpha^*,\beta^*\in\mathbb{R}$, $q>0$, $q\neq 1$ and $\varepsilon\neq 0$. If the regularity condition (11.2.4) holds all eigenvalues λ_n^* are different. This implies by using theorem 3.7 that there exists a sequence of dual polynomials. In this case we have

$$\lambda_n^*=(q^n-1)\left\{eq(1-q^{1-n})-2\varepsilon(1-q)\right\}\quad\text{and}\quad\kappa_n=q^{-n}$$

with $\omega=0$ and $x_0=1=q^{-0}$. Furthermore we have by using (11.2.2)

$$B^*(0)=\alpha^*+\beta^*q^{-1}+gq^{-1}=0$$

if we choose $c=-1$ in (11.2.1). Hence if

R. Koekoek et al., *Hypergeometric Orthogonal Polynomials and Their q-Analogues*, Springer Monographs in Mathematics,
DOI 10.1007/978-3-642-05014-5_12, © Springer-Verlag Berlin Heidelberg 2010

$$(A^*(n) =) eq^2 + 2fq^{n+1} + gq^{2n} \neq 0, \quad n = 0,1,2,\ldots,N-1$$

hold, the dual polynomials $\{z_m\}_{m=0}^{N}$ satisfy the three-term recurrence relation

$$\begin{aligned} &A^*(m)z_{m+1}(q^{-x}) - \{A^*(m)+B^*(m)\}z_m(q^{-x}) + B^*(m)z_{m-1}(q^{-x}) \\ &= q^{-x}z_m(q^{-x}), \quad m = 0,1,2,\ldots,N-1 \end{aligned} \tag{12.1.2}$$

with the convention that $z_{-1}(q^{-x}) := 0$. If we restrict x in (12.1.1) to $x = m$ for $m = 0,1,2,\ldots$, then we have

$$\begin{aligned} &A^*(m)y_n(q^{-m-1}) - \{A^*(m)+B^*(m)\}y_n(q^{-m}) + B^*(m)y_n(q^{-m+1}) \\ &= \lambda_n^* y_n(q^{-m}), \quad n = 0,1,2,\ldots,N-1. \end{aligned} \tag{12.1.3}$$

Since $y_n(\kappa_m) = y_n(q^{-m}) = z_m(\lambda_n)$ for all $m,n = 0,1,2,\ldots$, this implies that there exist dual polynomials with argument

$$\lambda_n^* = (q^n - 1)\left\{eq(1-q^{1-n}) - 2\varepsilon(1-q)\right\} = (q^n-1)\left(\alpha^* - eq^{2-n}\right).$$

This motivates the study of orthogonal polynomials in $q^{-x}+uq^x$ with $q > 0$ and $q \neq 1$, x a real variable and $u \in \mathbb{R}\setminus\{0\}$ a constant.

12.2 Difference Equations Having Real Polynomial Solutions with Argument $q^{-x}+uq^x$

We start with eigenvalue problems of the form

$$\widehat{\varphi}(qz)\left(\mathscr{D}_q^2\widehat{y}_n\right)(z) + \widehat{\psi}(qz)\left(\mathscr{D}_q\widehat{y}_n\right)(z) = \widehat{\lambda}_n\widehat{\rho}(qz)\widehat{y}_n(qz), \quad n = 0,1,2,\ldots \tag{12.2.1}$$

with $\mathscr{D}_q := \mathscr{A}_{q,0}$ (cf. (3.2.1)), $z \in \mathbb{C}$, $q > 0$ and $q \neq 1$. By using

$$\left(\mathscr{D}_q\widehat{y}_n\right)(z) = \frac{\widehat{y}_n(qz) - \widehat{y}_n(z)}{(q-1)z} \quad \text{and} \quad \left(\mathscr{D}_q^2\widehat{y}_n\right)(z) = \frac{\widehat{y}_n(q^2z) - (1+q)\widehat{y}_n(qz) + q\widehat{y}_n(z)}{q(q-1)^2z^2}$$

this can be written in the symmetric form

$$\widehat{C}(z)\widehat{y}_n(qz) - \left\{\widehat{C}(z) + \widehat{D}(z)\right\}\widehat{y}_n(z) + \widehat{D}(z)\widehat{y}_n(q^{-1}z) = \widehat{\lambda}_n\widehat{\rho}(z)\widehat{y}_n(z), \tag{12.2.2}$$

where

$$\widehat{C}(z) = \frac{q\widehat{\varphi}(z)}{(q-1)^2z^2} \quad \text{and} \quad \widehat{D}(z) = \frac{q^2\widehat{\varphi}(z) + q(1-q)z\widehat{\psi}(z)}{(q-1)^2z^2}.$$

For $z \in \mathbb{R}$ with $z > 0$ we may write $z = q^x$ with $x \in \mathbb{R}$, $q > 0$ and $q \neq 1$. By setting

$$\widehat{y}_n(z)=\widehat{y}_n(q^x)=y_n(x),\quad \widehat{y}_n(qz)=\widehat{y}_n(q^{x+1})=y_n(x+1),$$
$$\widehat{y}_n(q^{-1}z)=\widehat{y}_n(q^{x-1})=y_n(x-1)$$

and

$$\widehat{C}(z)=\widehat{C}(q^x)=C(x),\quad \widehat{D}(z)=\widehat{D}(q^x)=D(x)\quad\text{and}\quad \widehat{\rho}(z)=\widehat{\rho}(q^x)=\rho(x),$$

we conclude that (12.2.1) can be written in the form (cf. (2.2.12))

$$C(x)y_n(x+1)-\{C(x)+D(x)\}y_n(x)+D(x)y_n(x-1)=\lambda_n\rho(x)y_n(x) \tag{12.2.3}$$

for $n=0,1,2,\ldots$. Now we look for eigenvalues λ_n and coefficients $C(x)$, $D(x)$ and $\rho(x)$ so that for each eigenvalue λ_n there exists exactly one real polynomial solution y_n with degree$[y_n]=n$ in $q^{-x}+uq^{x}$ up to a constant factor. Since $(q^{-x}+uq^{x})^n$ can be expressed as a linear combination of $(q^{-x};q)_k(uq^x;q)_k$ for $k=0,1,2,\ldots,n$, we set

$$y_n(q^{-x}+uq^x)=\sum_{k=0}^{n}a_{n,k}\frac{(q^{-x};q)_k(uq^x;q)_k}{(q;q)_k},\quad a_{n,n}\neq 0,\quad n=0,1,2,\ldots. \tag{12.2.4}$$

Now we have

$$(q^{-x-1};q)_k(uq^{x+1};q)_k-(q^{-x};q)_k(uq^x;q)_k$$
$$=q^{-x-1}(1-q^k)(uq^{2x+1}-1)(q^{-x};q)_{k-1}(uq^{x+1};q)_{k-1}.$$

This leads to the *first simplification*

$$C(x)=q(uq^{2x-1}-1)C^*(x),\quad D(x)=(uq^{2x+1}-1)D^*(x)$$
$$\text{and}\quad \rho(x)=q^{-x}(uq^{2x-1}-1)(uq^{2x+1}-1)\rho^*(x). \tag{12.2.5}$$

For the moment we assume that $uq^{2x\pm 1}\neq 1$. Then we have

$$C^*(x)\sum_{k=1}^{n}a_{n,k}\frac{(q^{-x};q)_{k-1}(uq^{x+1};q)_{k-1}}{(q;q)_{k-1}}$$
$$-D^*(x)\sum_{k=1}^{n}a_{n,k}\frac{(q^{-x+1};q)_{k-1}(uq^{x};q)_{k-1}}{(q;q)_{k-1}}=\lambda_n\rho^*(x)y_n(x).$$

For a *second simplification* we note that

$$(q^{-x};q)_{k-1}(uq^{x+1};q)_{k-1}-(q^{-x+1};q)_{k-1}(uq^x;q)_{k-1}$$
$$=q^{-x}(1-q^{k-1})(uq^{2x}-1)(q^{-x+1};q)_{k-2}(uq^{x+1};q)_{k-2},\quad k=2,3,4,\ldots.$$

Now we define

$$C^*(x)-D^*(x)=q^{-x}(uq^{2x}-1)B(x)\quad\text{and}\quad \rho^*(x)=q^{-x}(uq^{2x}-1)\rho^{**}(x) \tag{12.2.6}$$

with the assumption that $uq^{2x} \neq 1$. Without loss of generality we may choose $\rho^{**}(x) = q$ so that we have

$$B(x)\sum_{k=1}^{n} a_{n,k}\frac{(q^{-x+1};q)_{k-1}(uq^x;q)_{k-1}}{(q;q)_{k-1}} + C^*(x)\sum_{k=2}^{n} a_{n,k}\frac{(q^{-x+1};q)_{k-2}(uq^{x+1};q)_{k-2}}{(q;q)_{k-2}}$$
$$= \lambda_n q\sum_{k=0}^{n} a_{n,k}\frac{(q^{-x};q)_k(uq^x;q)_k}{(q;q)_k}, \quad n=2,3,4,\ldots. \tag{12.2.7}$$

For $n=0$ we have $y_0(q^{-x}+uq^x) = a_{0,0}\,(\neq 0)$ which leads by using (12.2.3) to $\lambda_0 = 0$ (except for the trivial situation that $\rho^{**}(x) = 0$). For $n=1$ we have $y_1(q^{-x}+uq^x) = a_{1,0} + a_{1,1}(1-q^{-x})(1-uq^x)/(1-q)$ with $a_{1,1} \neq 0$, which leads to

$$B(x)a_{1,1} = \lambda_1 q\left\{a_{1,0} + a_{1,1}\frac{(1-q^{-x})(1-uq^x)}{1-q}\right\}.$$

Since all eigenvalues must be different, we conclude that $\lambda_1 \neq 0\,(=\lambda_0)$. Hence

$$B(x) = v + w(1-q^{-x})(1-uq^x) \quad \text{with} \quad v,w \in \mathbb{R}, \quad w \neq 0. \tag{12.2.8}$$

This form can be used as a *third simplification*. For the first term on the left-hand side of (12.2.7) we obtain

$$\sum_{k=1}^{n}\left\{vq^{k-1} + w\Big((1-u)(1-q^{k-1}) + (1-q^{-x})(1-uq^x)\Big)\right\}$$
$$\times a_{n,k}\frac{(q^{-x};q)_{k-1}(uq^x;q)_{k-1}}{(q;q)_{k-1}}$$
$$+\sum_{k=2}^{n}\left\{v - w(1-q^{-x})^2\right\}(1-uq^x)$$
$$\times a_{n,k}\frac{(q^{-x+1};q)_{k-2}(uq^{x+1};q)_{k-2}}{(q;q)_{k-2}}. \tag{12.2.9}$$

This implies that $C^*(x)$ must be of the form

$$C^*(x) = (1-uq^x)\left\{-v + w(1-q^{-x})^2 + \sigma(1-q^{-x})\right.$$
$$\left.+\tau(1-q^{-x})(1-q^{-x+1})(1-uq^{x+1})\right\}, \quad \sigma,\tau \in \mathbb{R}. \tag{12.2.10}$$

This leads to the following theorem.

Theorem 12.1. *The q-difference equation (12.2.3) only has real polynomial solutions $y_n(x^*)$ with degree$[y_n] = n$ in $x^* := q^{-x}+uq^x$ for $n = 0,1,2,\ldots$ if the coefficients $C(x)$, $D(x)$ and $\rho(x)$ have the form*

$$C(x) = q(1-uq^x)(1-uq^{2x-1}) \\ \times \left\{v - w(1-q^{-x})^2 - \sigma(1-q^{-x}) \right. \\ \left. - \tau(1-q^{-x})(1-q^{-x+1})(1-uq^{x+1})\right\},$$

$$D(x) = (1-q^{-x})(1-uq^{2x+1}) \\ \times \left\{v - w(1-uq^x)^2 - \sigma(1-uq^x) \right. \\ \left. - \tau(1-q^{-x+1})(1-uq^x)(1-uq^{x+1})\right\}$$

and

$$\rho(x) = -q^{-2x+1}(1-uq^{2x-1})(1-uq^{2x})(1-uq^{2x+1})$$

with $u, v, w, \sigma, \tau \in \mathbb{R}$, $w \neq 0$, $q > 0$ *and* $q \neq 1$.

Note that the assumptions that $uq^{2x\pm 1} \neq 1$ and $uq^{2x} \neq 1$ can be dropped.

12.3 The Basic Hypergeometric Representation

In order to find the hypergeometric representation of the polynomials in the form (12.2.4), we use (12.2.8), (12.2.10) and (12.2.9) to obtain from (12.2.7)

$$\sum_{k=1}^{n} \left\{vq^{k-1} + w\Big((1-u)(1-q^{k-1}) + (1-q^{-x})(1-uq^x)\Big)\right\} \\ \times a_{n,k}\frac{(q^{-x};q)_{k-1}(uq^x;q)_{k-1}}{(q;q)_{k-1}} \\ + \sum_{k=2}^{n} \left\{\sigma + \tau(1-q^{-x+1})(1-uq^{x+1})\right\}(1-q^{-x})(1-uq^x) \\ \times a_{n,k}\frac{(q^{-x+1};q)_{k-2}(uq^{x+1};q)_{k-2}}{(q;q)_{k-2}} \\ = \lambda_n q \sum_{k=0}^{n} a_{n,k}\frac{(q^{-x};q)_k(uq^x;q)_k}{(q;q)_k}, \quad n = 1, 2, 3, \ldots.$$

Note that this can also be written as

$$\sum_{k=0}^{n}\left\{q^{-k+1}(1-q^k)\left(w+\tau q(1-q^{k-1})\right)-\lambda_n q\right\}a_{n,k}\frac{(q^{-x};q)_k(uq^x;q)_k}{(q;q)_k}$$
$$+\sum_{k=0}^{n}\left\{vq^k-wq^{-k}(1-q^k)^2+\sigma(1-q^k)\right.$$
$$\left.-\tau q^{-k+1}(1-q^{k-1})(1-q^k)(1-uq^{k+1})\right\}$$
$$\times a_{n,k+1}\frac{(q^{-x};q)_k(uq^x;q)_k}{(q;q)_k}=0$$

for $n=1,2,3,\ldots$ with $a_{n,n+1}:=0$. By comparing the coefficients of $(q^{-x};q)_k(uq^x;q)_k$ on both sides, we find that

$$\left(\lambda_n q-q^{-k+1}(1-q^k)\left\{w+\tau q(1-q^{k-1})\right\}\right)a_{n,k}$$
$$=\left\{vq^k-wq^{-k}(1-q^k)^2+\sigma(1-q^k)\right.$$
$$\left.-\tau q^{-k+1}(1-q^{k-1})(1-q^k)(1-uq^{k+1})\right\}a_{n,k+1},$$

which holds for $k=0,1,2,\ldots,n$ with $a_{n,n+1}=0$. This leads to *eigenvalues* of the form

$$\lambda_n=(q^{-n}-1)\left\{w+\tau q(1-q^{n-1})\right\},\quad n=0,1,2,\ldots \tag{12.3.1}$$

and the *two-term recurrence relation*

$$q^{k+1}\left(w(q^{-k}-q^{-n})+\tau q\left\{q^{-k}(1+q^{2k-1})-q^{-n}(1+q^{2n-1})\right\}\right)a_{n,k}$$
$$=\left(-vq^{2k}+(1-q^k)\right.$$
$$\left.\times\left\{w(1-q^k)-\sigma q^k+\tau q(1-q^{k-1})(1-uq^{k+1})\right\}\right)a_{n,k+1} \tag{12.3.2}$$

for $k=n-1,n-2,n-3,\ldots,0$. Hence the coefficients $\{a_{n,k}\}_{k=0}^{n}$ in (12.2.4) are uniquely determined in terms of $a_{n,n}\neq 0$ if

$$(q^{-k}-q^{-n})(w+\tau q)+\tau(q^k-q^n)\neq 0 \tag{12.3.3}$$

for $k=n-1,n-2,n-3,\ldots,0$ and $n\in\{1,2,3,\ldots\}$. This condition holds if the eigenvalues in (12.3.1) are all different, since

$$(q^{-k}-q^{-n})(w+\tau q)+\tau(q^k-q^n)=\lambda_k-\lambda_n.$$

In the sequel we will always assume that this holds.

Note that the coefficients $C(x)$ and $D(x)$ can also be written as

$$\begin{aligned} q^{2x}C(x) = q(1-uq^x)(1-uq^{2x-1}) & \\ \times \{-w-\tau q + [2w+\sigma+\tau(1+q+uq^2)]\, q^x & \\ + [v-w-\sigma-\tau(1+uq+uq^2)]\, q^{2x} & \\ + \tau u q^{3x+1}\} & \end{aligned} \tag{12.3.4}$$

and

$$\begin{aligned} q^{2x}D(x) = (1-q^x)(1-uq^{2x+1}) & \\ \times \{-\tau q - [v-w-\sigma-\tau(1+uq+uq^2)]\, q^x & \\ - [2w+\sigma+\tau(1+q+uq^2)]\, uq^{2x} & \\ + (w+\tau q)u^2 q^{3x}\} & \end{aligned} \tag{12.3.5}$$

respectively. Then we have for $u \neq 0$

$$q^{2x}C(x) = -q(1-uq^x)(1-uq^{2x-1})(1-x_1q^x)(1-x_2q^x)(1-x_3q^x)$$

and

$$q^{2x}D(x) = -u^{-1}(1-q^x)(1-uq^{2x+1})(x_1-uq^x)(x_2-uq^x)(x_3-uq^x),$$

where

$$\tau = \frac{x_1x_2x_3}{uq}, \quad w = 1 - \frac{x_1x_2x_3}{u}, \quad \sigma = x_1+x_2+x_3-2-\frac{x_1x_2x_3}{uq}\left(1-q+uq^2\right)$$

and

$$v = -(1-x_1)(1-x_2)(1-x_3),$$

which leads to the following theorem:

Theorem 12.2. *The q-difference equation*

$$\begin{aligned} & q^{2x}C(x)y_n(q^{-x-1}+uq^{x+1}) \\ & \quad - q^{2x}\{C(x)+D(x)\}\, y_n(q^{-x}+uq^x) + q^{2x}D(x)y_n(q^{-x+1}+uq^{x-1}) \\ & = -q(1-uq^{2x-1})(1-uq^{2x})(1-uq^{2x+1})\lambda_n y_n(q^{-x}+uq^x) \end{aligned} \tag{12.3.6}$$

only has real polynomial solutions $y_n(x^*)$ with degree$[y_n] = n$ in $x^* = q^{-x}+uq^x$ for $n = 0,1,2,\ldots$ if the coefficients $q^{2x}C(x)$ and $q^{2x}D(x)$ and the eigenvalues λ_n have the form

$$q^{2x}C(x) = -q(1-uq^x)(1-uq^{2x-1})(1-x_1q^x)(1-x_2q^x)(1-x_3q^x), \tag{12.3.7}$$

$$q^{2x}D(x) = -u^{-1}(1-q^x)(1-uq^{2x+1})(x_1-uq^x)(x_2-uq^x)(x_3-uq^x) \tag{12.3.8}$$

and

$$\lambda_n = (q^{-n}-1)\left(1-\frac{x_1x_2x_3}{u}q^{n-1}\right), \quad n = 0,1,2,\ldots, \tag{12.3.9}$$

where $u \in \mathbb{R}\setminus\{0\}$, $q>0$, $q\neq 1$ and $x_1,x_2,x_3 \in \mathbb{R}$ or one is real and the other two are complex conjugates.

For the two-term recurrence relation (12.3.2) we obtain for $u\neq 0$

$$q(1-q^{k-n})\left(1-\frac{x_1x_2x_3}{u}q^{n+k-1}\right)a_{n,k}=(1-x_1q^k)(1-x_2q^k)(1-x_3q^k)a_{n,k+1}$$

for $k=n-1,n-2,n-3,\ldots,0$. This implies that we have

$$\begin{aligned} a_{n,k} &= \frac{(x_1q^k;q)_{n-k}(x_2q^k;q)_{n-k}(x_3q^k;q)_{n-k}}{(q^{k-n};q)_{n-k}\left(\frac{x_1x_2x_3}{u}q^{n+k-1};q\right)_{n-k}q^{n-k}}a_{n,n} \\ &= \frac{(x_1;q)_n(x_2;q)_n(x_3;q)_n}{(q^{-n};q)_n\left(\frac{x_1x_2x_3}{u}q^{n-1};q\right)_n q^n}a_{n,n}\frac{(q^{-n};q)_k\left(\frac{x_1x_2x_3}{u}q^{n-1};q\right)_k}{(x_1;q)_k(x_2;q)_k(x_3;q)_k}q^k \end{aligned}$$

for $k=0,1,2,\ldots,n$. By using (12.2.4) this leads to the representation

$$\begin{aligned} y_n(q^{-x}+uq^x) &= \frac{(x_1;q)_n(x_2;q)_n(x_3;q)_n}{(q^{-n};q)_n\left(\frac{x_1x_2x_3}{u}q^{n-1};q\right)_n q^n}a_{n,n} \\ &\quad\times\sum_{k=0}^{n}\frac{(q^{-n};q)_k\left(\frac{x_1x_2x_3}{u}q^{n-1};q\right)_k(q^{-x};q)_k(uq^x;q)_k}{(x_1;q)_k(x_2;q)_k(x_3;q)_k(q;q)_k}q^k \\ &= \frac{(x_1;q)_n(x_2;q)_n(x_3;q)_n}{(q^{-n};q)_n\left(\frac{x_1x_2x_3}{u}q^{n-1};q\right)_n q^n}a_{n,n} \\ &\quad\times{}_4\phi_3\left(\begin{matrix}q^{-n},\frac{x_1x_2x_3}{u}q^{n-1},q^{-x},uq^x\\ x_1,x_2,x_3\end{matrix};q,q\right) \end{aligned} \tag{12.3.10}$$

for the q-**Racah** polynomials with $n=0,1,2,\ldots$.

Special cases are the **dual** q-**Hahn** polynomials (for $x_3=0$) and the **dual** q-**Krawtchouk** polynomials (for $x_2=x_3=0$). Another special case is the family of **dual** q-**Charlier** polynomials (for $x_1=x_2=x_3=0$) (see [386]).

We remark that we have to choose $a_{n,n}=(q^{-n};q)_nq^n$ in order to get monic polynomials.

Remark. Note that the q-Racah, the dual q-Hahn and the dual q-Krawtchouk polynomials not only appear as *finite* systems of orthogonal polynomials. Usually one gets finite systems by setting $x_1=q^{-N}$, for instance. However, this is not necessary.

The q-Racah polynomials have a certain symmetry in n and x which leads to duality. The (not normalized) q-Racah polynomials given by (12.3.10) can be written as

$$(y_n^*(\kappa_x)=)\,y_n(q^{-x}+uq^x)={}_4\phi_3\left(\begin{matrix}q^{-n},\frac{x_1x_2x_3}{u}q^{n-1},q^{-x},uq^x\\ x_1,x_2,x_3\end{matrix};q,q\right) \tag{12.3.11}$$

for $n=0,1,2,\ldots$ with $\kappa_x=(q^{-x}-1)(1-uq^x)$. If we replace u by $x_1x_2x_3/uq$, then we have

$$(z_n^*(\lambda_x)=)\,z_n(q^{-x}+\frac{x_1x_2x_3}{uq}q^x)={}_4\phi_3\left(\begin{matrix}q^{-n},uq^n,q^{-x},\frac{x_1x_2x_3}{u}q^{x-1}\\x_1,x_2,x_3\end{matrix};q,q\right)\quad(12.3.12)$$

for $n=0,1,2,3,\ldots$ with $\lambda_x=(q^{-x}-1)(1-x_1x_2x_3q^{x-1}/u)$. Now we have $y_n^*(\kappa_m)=z_m^*(\lambda_n)$ for $m,n=0,1,2,\ldots$. In view of definition 3.1, $\{y_n(q^{-x}+uq^x)\}$ and $\{z_n(q^{-x}+\frac{x_1x_2x_3}{uq}q^x)\}$ are dual polynomial systems with respect to the sequences of eigenvalues $\{\kappa_n\}$ and $\{\lambda_n\}$. Note that we also have $y_n^*(\kappa_0)=z_0^*(\lambda_n)$. The polynomials $z_n(q^{-x}+\frac{x_1x_2x_3}{uq}q^x)$ can be considered as *dual q-Racah* polynomials. This fact will be used in the next section.

12.4 The Three-Term Recurrence Relation

In order to obtain the three-term recurrence relation, we use the concept of duality. For the q-Racah polynomials we start with the difference equation (cf. (12.2.3)) for the (not normalized) q-Racah polynomials $y_n(q^{-x}+uq^x)$ given by (12.3.11):

$$C(x)y_n^*(\kappa_{x+1})-\{C(x)+D(x)\}\,y_n^*(\kappa_x)+D(x)y_n^*(\kappa_{x-1})=\lambda_n\rho(x)y_n^*(\kappa_x).$$

Now we have $y_n^*(\kappa_x)=z_x^*(\lambda_n)$ with $\kappa_x=(q^{-x}-1)(1-uq^x)$ and $\lambda_n=(q^{-n}-1)(1-x_1x_2x_3q^{n-1}/u)$ which implies that the difference equation for the dual q-Racah polynomials $z_x^*(\lambda_n)$ given by (12.3.12) can be written as

$$C(x)z_{x+1}^*(\lambda_n)-\{C(x)+D(x)\}\,z_x^*(\lambda_n)+D(x)z_{x-1}^*(\lambda_n)=\lambda_n\rho(x)z_x^*(\lambda_n).$$

For the coefficients we have (cf. (12.3.7))

$$q^{2x}C(x)=-q(1-uq^x)(1-uq^{2x-1})(1-x_1q^x)(1-x_2q^x)(1-x_3q^x)$$

and (cf. (12.3.8))

$$q^{2x}D(x)=-u^{-1}(1-q^x)(1-uq^{2x+1})(x_1-uq^x)(x_2-uq^x)(x_3-uq^x)$$

which leads to the three-term recurrence relation

$$\begin{aligned}&q(1-uq^n)(1-uq^{2n-1})(1-x_1q^n)(1-x_2q^n)(1-x_3q^n)z_{n+1}^*(\lambda_x)\\&\quad-\big\{q(1-uq^n)(1-uq^{2n-1})(1-x_1q^n)(1-x_2q^n)(1-x_3q^n)\\&\qquad+u^{-1}(1-q^n)(1-uq^{2n+1})(x_1-uq^n)(x_2-uq^n)(x_3-uq^n)\big\}\,z_n^*(\lambda_x)\\&\quad+u^{-1}(1-q^n)(1-uq^{2n+1})(x_1-uq^n)(x_2-uq^n)(x_3-uq^n)z_{n-1}^*(\lambda_x)\\&=q(1-uq^{2n-1})(1-uq^{2n})(1-uq^{2n+1})\lambda_xz_n^*(\lambda_x),\quad n=1,2,3,\ldots\end{aligned}$$

with $z_0^*(\lambda_x)=1$ and $z_1^*(\lambda_x)=1+(1-uq)\lambda_x/(1-x_1)(1-x_2)(1-x_3)$.

By replacing u by $x_1x_2x_3/uq(=\tau)$ the polynomials $z_n^*(\lambda_n)$ given by (12.3.12) change into the polynomials $y_n^*(\kappa_x)$ given by (12.3.11). This implies that the three-term recurrence relation for the polynomials $y_n^*(\kappa_x)$ can be written as

$$\begin{aligned}
&u(u-x_1x_2x_3q^{n-1})(u-x_1x_2x_3q^{2n-2})(1-x_1q^n)(1-x_2q^n)(1-x_3q^n)y_{n+1}^*(\kappa_x)\\
&\quad-\{u(u-x_1x_2x_3q^{n-1})(u-x_1x_2x_3q^{2n-2})(1-x_1q^n)(1-x_2q^n)(1-x_3q^n)\\
&\qquad+(1-q^n)(u-x_1x_2q^{n-1})(u-x_1x_3q^{n-1})(u-x_2x_3q^{n-1})\\
&\qquad\quad\times(u-x_1x_2x_3q^{2n})\}\,y_n^*(\kappa_x)\\
&\quad+(1-q^n)(u-x_1x_2q^{n-1})(u-x_1x_3q^{n-1})(u-x_2x_3q^{n-1})\\
&\qquad\quad\times(u-x_1x_2x_3q^{2n})y_{n-1}^*(\kappa_x)\\
&=(u-x_1x_2x_3q^{2n-2})(u-x_1x_2x_3q^{2n-1})(u-x_1x_2x_3q^{2n})\kappa_x y_n^*(\kappa_x),\quad n=1,2,3,\ldots
\end{aligned}$$

with $y_0^*(\kappa_x)=1$, $y_1^*(\kappa_x)=1+(u-x_1x_2x_3)\kappa_x/u(1-x_1)(1-x_2)(1-x_3)$ and $\kappa_x=(q^{-x}-1)(1-uq^x)$. The connection with the monic q-Racah polynomials $y_n(\kappa_x)$ is given by

$$y_n^*(\kappa_x)=\frac{\left(\frac{x_1x_2x_3}{u}q^{n-1};q\right)_n}{(x_1;q)_n(x_2;q)_n(x_3;q)_n}y_n(\kappa_x),\quad n=0,1,2,\ldots.$$

Hence the three-term recurrence relation for the monic q-Racah polynomials $y_n(\kappa_x)$ can be written in the form

$$y_{n+1}(\kappa_x)=\left(\kappa_x+c_n^{(1)}+c_n^{(2)}\right)y_n(\kappa_x)-c_{n-1}^{(1)}c_n^{(2)}y_{n-1}(\kappa_x),\quad n=1,2,3,\ldots \tag{12.4.1}$$

with $y_0(\kappa_x)=1$ and $y_1(x)=\kappa_x+u(1-x_1)(1-x_2)(1-x_3)/(u-x_1x_2x_3)$, where

$$c_n^{(1)}=\frac{u(u-x_1x_2x_3q^{n-1})(1-x_1q^n)(1-x_2q^n)(1-x_3q^n)}{(u-x_1x_2x_3q^{2n-1})(u-x_1x_2x_3q^{2n})},\quad n=0,1,2,\ldots \tag{12.4.2}$$

and

$$c_n^{(2)}=\frac{(1-q^n)(u-x_1x_2q^{n-1})(u-x_1x_3q^{n-1})(u-x_2x_3q^{n-1})}{(u-x_1x_2x_3q^{2n-2})(u-x_1x_2x_3q^{2n-1})} \tag{12.4.3}$$

for $n=1,2,3,\ldots$.

12.5 Classification of the Positive-Definite Orthogonal Polynomial Solutions

Favard's theorem (theorem 3.1) can be extended for monic polynomials $y_n(q^{-x}+uq^x)$ in $q^{-x}+uq^x$ of degree $n\in\{0,1,2,\ldots\}$ with $u\in\mathbb{R}\setminus\{0\}$. The polynomials given by

$$y_{n+1}(q^{-x}+uq^x)=\left\{(q^{-x}-1)(1-uq^x)-c_n\right\}y_n(q^{-x}+uq^x)\\ -d_ny_{n-1}(q^{-x}+uq^x) \qquad (12.5.1)$$

for $n=1,2,3,\ldots$ with $y_0(q^{-x}+uq^x)=1$ and $y_1(q^{-x}+uq^x)=(q^{-x}-1)(1-uq^x)-c_0$ are orthogonal with respect to a positive-definite linear functional Λ, id est

$$\Lambda\left[y_m(q^{-x}+uq^x)y_n(q^{-x}+uq^x)\right]=\left(\prod_{k=0}^{n}d_k\right)\delta_{mn},\quad m,n=0,1,2,\ldots, \qquad (12.5.2)$$

where $\Lambda[y_0(q^{-x}+uq^x)]=d_0\in\mathbb{R}$ and $\Lambda[y_n(q^{-x}+uq^x)]=0$ for $n=1,2,3,\ldots$ iff $c_n\in\mathbb{R}$ for all $n=0,1,2,\ldots$ and $d_0,d_1,d_2,\ldots,d_n$ are positive. The proof is similar to the proof of theorem 3.1.

For the monic q-Racah polynomials given by (12.3.10) we have the three-term recurrence relation (12.4.1) with

$$c_0=-c_0^{(1)},\quad c_n=-c_n^{(1)}-c_n^{(2)},\quad n=1,2,3,\ldots,$$

$$d_n=c_{n-1}^{(1)}c_n^{(2)}=\frac{u(1-q^n)(u-x_1x_2x_3q^{n-2})}{(u-x_1x_2x_3q^{2n-3})(u-x_1x_2x_3q^{2n-2})^2(u-x_1x_2x_3q^{2n-1})}D_n$$

for $n=1,2,3,\ldots$, and

$$D_n=(1-x_1q^{n-1})(1-x_2q^{n-1})(1-x_3q^{n-1})\\ \times(u-x_1x_2q^{n-1})(u-x_1x_3q^{n-1})(u-x_2x_3q^{n-1}) \qquad (12.5.3)$$

for $n=1,2,3,\ldots$, where $u\in\mathbb{R}\setminus\{0\}$. Hence we have $c_n\in\mathbb{R}$ for all $n=0,1,2,\ldots$ if $x_1x_2x_3\in\mathbb{R}$, $x_1+x_2+x_3\in\mathbb{R}$ and $x_1x_2+x_1x_3+x_2x_3\in\mathbb{R}$. This implies that $x_1,x_2,x_3\in\mathbb{R}$ or one is real and the other two are complex conjugates.

To study the positivity of d_n for $n=1,2,3,\ldots$, we only need to consider the cases $0<q<1$ and $q>1$ in view of the argument $q^{-x}+uq^x$. Further we have $u\in\mathbb{R}\setminus\{0\}$. In all cases where $x_1x_2x_3=0$ we have

$$d_n=\frac{u(1-q^n)(u-x_1x_2x_3q^{n-2})}{(u-x_1x_2x_3q^{2n-3})(u-x_1x_2x_3q^{2n-2})^2(u-x_1x_2x_3q^{2n-1})}D_n=\frac{1-q^n}{u^2}D_n$$

for $n=1,2,3,\ldots$.

Case I. $x_1=x_2=x_3=0$. Then we have

$$d_n = u(1-q^n), \quad n = 1,2,3,\ldots.$$

This implies that we have positive-definite orthogonality in the following two infinite cases:

Case Ia1. $x_1 = x_2 = x_3 = 0$, $0 < q < 1$ and $u > 0$.

Case Ia2. $x_1 = x_2 = x_3 = 0$, $q > 1$ and $u < 0$.

In this case we have no finite systems of positive-definite orthogonal polynomials.

Case II. $x_1 \neq 0$ and $x_2 = x_3 = 0$. Then we have

$$d_n = u(1-q^n)(1-x_1q^{n-1}), \quad n = 1,2,3,\ldots.$$

This leads to positive-definite orthogonality in the following three infinite cases:

Case IIa1. $x_1 \neq 0$, $x_2 = x_3 = 0$, $0 < q < 1$, $u > 0$ and $x_1 < 1$.

Case IIa2. $x_1 \neq 0$, $x_2 = x_3 = 0$, $q > 1$, $u > 0$ and $x_1 > 1$.

Case IIa3. $x_1 \neq 0$, $x_2 = x_3 = 0$, $q > 1$, $u < 0$ and $x_1 < 0$.

It is also possible to have positive-definite orthogonality for finite systems of $N+1$ polynomials with $N \in \{1,2,3,\ldots\}$ in the following two cases:

Case IIb1. $x_1 \neq 0$, $x_2 = x_3 = 0$, $0 < q < 1$, $u < 0$ and $x_1 > 1$ with $x_1q^N \leq 1 < x_1q^{N-1}$.

Case IIb2. $x_1 \neq 0$, $x_2 = x_3 = 0$, $q > 1$, $u < 0$ and $0 < x_1 < 1$ with $x_1q^{N-1} < 1 \leq x_1q^N$.

Case III. $x_1x_2 \neq 0$ and $x_3 = 0$. Then we have

$$d_n = (1-q^n)(1-x_1q^{n-1})(1-x_2q^{n-1})(u-x_1x_2q^{n-1}), \quad n = 1,2,3,\ldots.$$

Without loss of generality we assume that $x_1 \leq x_2$. Then we have positive-definite orthogonality in the following six infinite cases:

Case IIIa1. $x_1x_2 \neq 0$, $x_3 = 0$, $0 < q < 1$, $u > 0$, $x_1 < x_2 < 1$ and $u > x_1x_2$.

Case IIIa2. $x_1x_2 \neq 0$, $x_3 = 0$, $0 < q < 1$, $x_1 = x_2$ and $u > x_1x_2$.

Case IIIa3. $x_1x_2 \neq 0$, $x_3 = 0$, $q > 1$, $x_1 < x_2 < 0$ and $u < x_1x_2$.

Case IIIa4. $x_1x_2 \neq 0$, $x_3 = 0$, $q > 1$, $x_1 < 0$, $x_2 > 1$ and $u > x_1x_2$.

Case IIIa5. $x_1x_2 \neq 0$, $x_3 = 0$, $q > 1$, $1 < x_1 < x_2$ and $u < x_1x_2$.

Case IIIa6. $x_1x_2 \neq 0$, $x_3 = 0$, $q > 1$, $x_1 = x_2$ and $u < x_1x_2$.

It is also possible to have positive-definite orthogonality for finite systems of $N+1$ polynomials with $N \in \{1,2,3,\ldots\}$ in the following seven cases:

Case IIIb1. $x_1x_2 \neq 0$, $x_3 = 0$, $0 < q < 1$, $x_1 < 0 < x_2 < 1$ and $x_1x_2q^{N-1} < u \leq x_1x_2q^N$.

Case IIIb2. $x_1x_2 \neq 0$, $x_3 = 0$, $0 < q < 1$, $x_1 < 0$, $x_2 > 1$ with $x_2q^N \leq 1 < x_2q^{N-1}$ and $u < x_1x_2$.

Case IIIb3. $x_1x_2 \neq 0$, $x_3 = 0$, $0 < q < 1$, $0 < x_1 < 1 < x_2$ with $x_2q^{N_1} \leq 1 < x_2q^{N_1-1}$ and $x_1x_2q^{N_2} \leq u < x_1x_2q^{N_2-1}$ and $N = \min(N_1, N_2)$.

Case IIIb4. $x_1x_2 \neq 0$, $x_3 = 0$, $0 < q < 1$, $1 < x_1 \leq x_2$ with $x_1q^{N_1} \leq 1 < x_1q^{N_1-1}$, $x_2q^{N_2} \leq 1 < x_2q^{N_2-1}$, $N = \min(N_1, N_2)$ and $u > x_1x_2$. Since $x_1 \leq x_2$ we have $N = N_1$ in this case.

Case IIIb5. $x_1x_2 \neq 0$, $x_3 = 0$, $q > 1$, $x_1 < 0 < x_2 < 1$ with $x_2q^{N_1-1} < 1 \leq x_2q^{N_1}$, $x_1x_2q^{N_2} \leq u < x_1x_2q^{N_2}$ and $N = \min(N_1, N_2)$.

Case IIIb6. $x_1x_2 \neq 0$, $x_3 = 0$, $q > 1$, $0 < x_1 \leq x_2 < 1$ with $x_1q^{N_1-1} < 1 \leq x_1q^{N_1}$, $x_2q^{N_2-1} < 1 \leq x_2q^{N_2}$, $N = \min(N_1, N_2)$ and $u < x_1x_2$. Since $x_1 \leq x_2$ we have $N = N_2$ in this case.

Case IIIb7. $x_1x_2 \neq 0$, $x_3 = 0$, $q > 1$, $0 < x_1 < 1 < x_2$ with $x_1q^{N_1-1} < 1 \leq x_1q^{N_1}$, $x_1x_2q^{N_2-1} < u \leq x_1x_2q^{N_2}$ and $N = \min(N_1, N_2)$.

Case IV. $x_1x_2x_3 \neq 0$. Now we write

$$d_n = u(1-q^n)D_n^{(1)}D_n^{(2)}, \quad n = 1,2,3,\ldots \tag{12.5.4}$$

with

$$D_n^{(1)} = \frac{1-\tau q^{n-1}}{(1-\tau q^{2n-2})(1-\tau q^{2n-1})^2(1-\tau q^{2n})}, \quad n = 1,2,3,\ldots \tag{12.5.5}$$

and

$$D_n^{(2)} = (1-x_1q^{n-1})(1-x_2q^{n-1})(1-x_3q^{n-1}) \times \left(1-\frac{\tau}{x_1}q^n\right)\left(1-\frac{\tau}{x_2}q^n\right)\left(1-\frac{\tau}{x_3}q^n\right) \tag{12.5.6}$$

for $n = 1,2,3,\ldots$, where

$$\tau = \frac{x_1x_2x_3}{uq}.$$

Note that $d_n \sim u$ for $n \to \infty$ both for $0 < q < 1$ and $q > 1$. This implies that $d_n > 0$ can only be true for all $n = 1,2,3,\ldots$ if $u > 0$.

q	**extra conditions**	$\mathbf{D_n^{(1)}}$	**for**
$0<q<1$	$\tau<0$	$+$	$n=1,2,3,\ldots$
	$0<\tau q<1$	$+$	$n=1,2,3,\ldots$
	$\tau q>1$ with $1<\tau q^{2N}\leq q^{-2}$	$-$	$n=1,2,3,\ldots,N$
$q>1$	$\tau<0$	$+$	$n=1,2,3,\ldots$
	$0<\tau q<1$ with $q^{-2}\leq \tau q^{2N}<1$	$+$	$n=1,2,3,\ldots,N$
	$\tau q>1$	$-$	$n=1,2,3,\ldots$

Table 12.1 sign of $D_n^{(1)}$, $q>0$ and $N \in \{1,2,3,\ldots\}$

Without loss of generality we assume that $x_1 \leq x_2 \leq x_3$. The sign of $D_n^{(1)}$ for $n = 1,2,3,\ldots$ is given by table 12.1 (cf. table 10.2). By using (12.5.4), (12.5.6) and table 12.1, we conclude that we have positive-definite orthogonality in the following four infinite cases:

Case IVa1. $x_1x_2x_3 \neq 0$, $0<q<1$, $u>0$, $\tau<0$ and $D_n^{(2)}>0$.

Case IVa2. $x_1x_2x_3 \neq 0$, $0<q<1$, $u>0$, $0<\tau q<1$ and $D_n^{(2)}>0$.

Case IVa3. $x_1x_2x_3 \neq 0$, $q>1$, $u>0$, $\tau<0$ and $D_n^{(2)}<0$.

Case IVa4. $x_1x_2x_3 \neq 0$, $q>1$, $u>0$, $\tau q>1$ and $D_n^{(2)}>0$.

It is also possible to have positive-definite orthogonality for finite systems of $N+1$ polynomials with $N \in \{1,2,3,\ldots\}$ in the following four cases:

Case IVb1. $x_1x_2x_3 \neq 0$, $0<q<1$, $u>0$, $\tau q>1$ with $1<\tau q^{2N} \leq q^{-2}$ and $D_n^{(2)}<0$.

Case IVb2. $x_1x_2x_3 \neq 0$, $0<q<1$, $u<0$, $\tau q>1$ with $1<\tau q^{2N} \leq q^{-2}$ and $D_n^{(2)}>0$.

Case IVb3. $x_1x_2x_3 \neq 0$, $q > 1$, $u > 0$, $0 < \tau q < 1$ with $q^{-2} \leq \tau q^{2N} < 1$ and $D_n^{(2)} < 0$.

Case IVb4. $x_1x_2x_3 \neq 0$, $q > 1$, $u < 0$, $0 < \tau q < 1$ with $q^{-2} \leq \tau q^{2N} < 1$ and $D_n^{(2)} > 0$.

We remark that it is also possible to have positive-definite orthogonality for finite systems of $N+1$ polynomials with $N \in \{1,2,3,\ldots\}$ in cases where $D_n^{(2)}$ has opposite sign for $n = N$ and $n = N+1$.

12.6 Solutions of the q-Pearson Equation

In this section we use the following operator[1] (cf. (2.1.1) with $\omega = 0$)

$$(\mathscr{A} f)(q^{-x}+uq^x) = \frac{f(q^{-x-1}+uq^{x+1}) - f(q^{-x}+uq^x)}{q^{-x-1}+uq^{x+1}-q^{-x}-uq^x}, \quad q>0, \quad q\neq 1, \tag{12.6.1}$$

where f is a complex-valued function in $q^{-x}+uq^x$ whose domain contains both $q^{-x}+uq^x$ and $q^{-x-1}+uq^{x+1}$ for each $x \in \mathbb{R}$. For two such functions f_1 and f_2, a product rule similar to (11.4.2) applies. In fact we have

$$\begin{aligned}(\mathscr{A}(f_1f_2))(q^{-x}+uq^x) &= (\mathscr{A} f_1)(q^{-x}+uq^x) f_2(q^{-x}+uq^x) \\ &\quad + (\mathscr{S} f_1)(q^{-x}+uq^x)(\mathscr{A} f_2)(q^{-x}+uq^x), \end{aligned} \tag{12.6.2}$$

where, analogous to (11.4.3), we now define

$$\begin{aligned}&(\mathscr{S} f)(q^{-x}+uq^x) = f(q^{-x-1}+uq^{x+1}) \\ &\text{and} \quad (\mathscr{S}^{-1} f)(q^{-x}+uq^x) = f(q^{-x+1}+uq^{x-1}).\end{aligned} \tag{12.6.3}$$

We start with a general second-order operator equation of the form

$$\begin{aligned}&\varphi(q^{-x}+uq^x)\left(\mathscr{A}^2 y_n\right)(q^{-x}+uq^x) \\ &\quad + \psi(q^{-x}+uq^x)(\mathscr{A} y_n)(q^{-x}+uq^x) = \lambda_n (\mathscr{S} y_n)(q^{-x}+uq^x)\end{aligned}$$

in terms of the operator (12.6.1) and later we will compare this to the q-difference equation (12.3.6) in theorem 12.2. If we multiply both sides of this equation by $(\mathscr{S} w)(q^{-x}+uq^x)$ we obtain

$$\begin{aligned}&(\mathscr{S} w)(q^{-x}+uq^x)\varphi(q^{-x}+uq^x)\left(\mathscr{A}^2 y_n\right)(q^{-x}+uq^x) \\ &\quad + (\mathscr{S} w)(q^{-x}+uq^x)\psi(q^{-x}+uq^x)(\mathscr{A} y_n)(q^{-x}+uq^x) \\ &= \lambda_n (\mathscr{S} w)(q^{-x}+uq^x)(\mathscr{S} y_n)(q^{-x}+uq^x),\end{aligned} \tag{12.6.4}$$

[1] It turns out that this form makes sense. However, this form no longer has the property that its action on a polynomial in $q^{-x}+uq^x$ of degree n leads to a polynomial in $q^{-x}+uq^x$ of degree $n-1$ for $n = 1,2,3,\ldots$.

which leads to the self-adjoint form

$$\begin{aligned}&\left(\mathscr{A}\left(w\left(\mathscr{S}^{-1}\varphi\right)\left(\mathscr{A}y_n\right)\right)\right)\left(q^{-x}+uq^x\right)\\&=\lambda_n\left(\mathscr{S}w\right)\left(q^{-x}+uq^x\right)\left(\mathscr{S}y_n\right)\left(q^{-x}+uq^x\right)\end{aligned}\tag{12.6.5}$$

if the Pearson operator equation

$$\left(\mathscr{A}\left(w\left(\mathscr{S}^{-1}\varphi\right)\right)\right)\left(q^{-x}+uq^x\right)=\left(\mathscr{S}w\right)\left(q^{-x}+uq^x\right)\psi(q^{-x}+uq^x)$$

holds. Furthermore, the productrule (12.6.2) leads to

$$\begin{aligned}&\left(\mathscr{A}\left(w\left(\mathscr{S}^{-1}\varphi\right)\right)\right)\left(q^{-x}+uq^x\right)\\&=\left(\mathscr{S}^{-1}\varphi\right)\left(q^{-x}+uq^x\right)\frac{w(q^{-x-1}+uq^{x+1})-w(q^{-x}+uq^x)}{q^{-x-1}+uq^{x+1}-q^{-x}-uq^x}\\&\quad+w(q^{-x-1}+uq^{x+1})\frac{\varphi(q^{-x}+uq^x)-\left(\mathscr{S}^{-1}\varphi\right)\left(q^{-x}+uq^x\right)}{q^{-x-1}+uq^{x+1}-q^{-x}-uq^x}.\end{aligned}$$

Combining the latter two equations we obtain (cf. (3.2.7) and (11.4.6))

$$\begin{aligned}&w(q^{-x}+uq^x)\varphi(q^{-x+1}+uq^{x-1})\\&=w(q^{-x-1}+uq^{x+1})\\&\quad\times\left\{\varphi(q^{-x}+uq^x)-\left(q^{-x-1}+uq^{x+1}-q^{-x}-uq^x\right)\psi(q^{-x}+uq^x)\right\}.\end{aligned}\tag{12.6.6}$$

Note that the q-difference equation (12.3.6) in theorem 12.2 can be written in the form

$$\begin{aligned}&\widetilde{C}(x)y_n(q^{-x-1}+uq^{x+1})-\left\{\widetilde{C}(x)+\widetilde{D}(x)\right\}y_n(q^{-x}+uq^x)+\widetilde{D}(x)y_n(q^{-x+1}+uq^{x-1})\\&=\lambda_n y_n(q^{-x}+uq^x),\end{aligned}\tag{12.6.7}$$

where, by using (12.3.7), (12.3.8) and (12.3.9),

$$\begin{aligned}\widetilde{C}(x)&=-\frac{q^{2x}C(x)}{q(1-uq^{2x-1})(1-uq^{2x})(1-uq^{2x+1})}\\&=\frac{(1-uq^x)(1-x_1q^x)(1-x_2q^x)(1-x_3q^x)}{(1-uq^{2x})(1-uq^{2x+1})},\end{aligned}\tag{12.6.8}$$

$$\begin{aligned}\widetilde{D}(x)&=-\frac{q^{2x}D(x)}{q(1-uq^{2x-1})(1-uq^{2x})(1-uq^{2x+1})}\\&=\frac{(1-q^x)(x_1-uq^x)(x_2-uq^x)(x_3-uq^x)}{uq(1-uq^{2x-1})(1-uq^{2x})}.\end{aligned}\tag{12.6.9}$$

and

$$\lambda_n=(q^{-n}-1)\left(1-\frac{x_1x_2x_3}{u}q^{n-1}\right),\quad n=0,1,2,\ldots.$$

Now we have by using (12.6.1)

$$(\mathscr{A}y_n)(q^{-x}+uq^x)=\frac{y_n(q^{-x-1}+uq^{x+1})-y_n(q^{-x}+uq^x)}{q^{-x-1}+uq^{x+1}-q^{-x}-uq^x} \tag{12.6.10}$$

and

$$\begin{aligned}&\left(\mathscr{A}^2y_n\right)(q^{-x}+uq^x)\\&=\frac{1}{q^{-x-1}+uq^{x+1}-q^{-x}-uq^x}\\&\quad\times\left\{\frac{y_n(q^{-x-2}+uq^{x+2})-y_n(q^{-x-1}+uq^{x+1})}{q^{-x-2}+uq^{x+2}-q^{-x-1}-uq^{x+1}}\right.\\&\qquad\left.-\frac{y_n(q^{-x-1}+uq^{x+1})-y_n(q^{-x}+uq^x)}{q^{-x-1}+uq^{x+1}-q^{-x}-uq^x}\right\}.\end{aligned} \tag{12.6.11}$$

If we apply the operator $\mathscr{S}^{-1}$ to the left of (12.6.4) we obtain

$$\begin{aligned}&w(q^{-x}+uq^x)\varphi(q^{-x+1}+uq^{x-1})\left(\mathscr{A}^2y_n\right)(q^{-x+1}+uq^{x-1})\\&\quad+w(q^{-x}+uq^x)\psi(q^{-x+1}+uq^{x-1})(\mathscr{A}y_n)(q^{-x+1}+uq^{x-1})\\&=\lambda_nw(q^{-x}+uq^x)y_n(q^{-x}+uq^x).\end{aligned}$$

Now we divide by $w(q^{-x}+uq^x)$ and use (12.6.10) and (12.6.11) to find

$$\begin{aligned}&\frac{y_n(q^{-x-1}+uq^{x+1})\varphi(q^{-x+1}+uq^{x-1})}{(q^{-x}+uq^x-q^{-x+1}-uq^{x-1})(q^{-x-1}+uq^{x+1}-q^{-x}-uq^x)}\\&\quad-\frac{y_n(q^{-x}+uq^x)}{q^{-x}+uq^x-q^{-x+1}-uq^{x-1}}\\&\qquad\times\left\{\frac{\varphi(q^{-x+1}+uq^{x-1})}{q^{-x-1}+uq^{x+1}-q^{-x}-uq^x}\right.\\&\qquad\quad\left.+\frac{\varphi(q^{-x+1}+uq^{x-1})}{q^{-x}+uq^x-q^{-x+1}-uq^{x-1}}-\psi(q^{-x+1}+uq^{x-1})\right\}\\&\quad+\frac{y_n(q^{-x+1}+uq^{x-1})}{q^{-x}+uq^x-q^{-x+1}-uq^{x-1}}\\&\qquad\times\left\{\frac{\varphi(q^{-x+1}+uq^{x-1})}{q^{-x}+uq^x-q^{-x+1}-uq^{x-1}}-\psi(q^{-x+1}+uq^{x-1})\right\}\\&=\lambda_ny_n(q^{-x}+uq^x).\end{aligned}$$

If we compare this with the q-difference equation (12.6.7) we conclude that

$$\begin{aligned}&\varphi(q^{-x+1}+uq^{x-1})\\&=\left(q^{-x}+uq^x-q^{-x+1}-uq^{x-1}\right)\left(q^{-x-1}+uq^{x+1}-q^{-x}-uq^x\right)\widetilde{C}(x)\end{aligned} \tag{12.6.12}$$

and

$$\begin{aligned}
&\psi(q^{-x+1}+uq^{x-1})\\
&=-\left(q^{-x}+uq^x-q^{-x+1}-uq^{x-1}\right)\widetilde{D}(x)+\frac{\varphi(q^{-x+1}+uq^{x-1})}{q^{-x}+uq^x-q^{-x+1}-uq^{x-1}}\\
&=-\left(q^{-x}+uq^x-q^{-x+1}-uq^{x-1}\right)\widetilde{D}(x)\\
&\qquad+\left(q^{-x-1}+uq^{x+1}-q^{-x}-uq^x\right)\widetilde{C}(x).
\end{aligned}\tag{12.6.13}$$

This implies that the q-Pearson equation (12.6.6) can be written in the form

$$\begin{aligned}
&w(q^{-x}+uq^x)\left(q^{-x}+uq^x-q^{-x+1}-uq^{x-1}\right)\widetilde{C}(x)\\
&=w(q^{-x-1}+uq^{x+1})\left(q^{-x-1}+uq^{x+1}-q^{-x}-uq^x\right)\widetilde{D}(x+1).
\end{aligned}$$

Hence we have the q-Pearson equation

$$\widetilde{w}(q^{-x}+uq^x)\widetilde{C}(x)=\widetilde{w}(q^{-x-1}+uq^{x+1})\widetilde{D}(x+1),\tag{12.6.14}$$

where

$$\widetilde{w}(q^{-x}+uq^x):=\left(q^{-x}+uq^x-q^{-x+1}-uq^{x-1}\right)w(q^{-x}+uq^x).\tag{12.6.15}$$

Finally we use (12.6.8) and (12.6.9) to conclude that we look for solutions of the q-Pearson equation

$$\begin{aligned}
&\frac{\widetilde{w}(q^{-x}+uq^x)}{\widetilde{w}(q^{-x-1}+uq^{x+1})}=\frac{\widetilde{D}(x+1)}{\widetilde{C}(x)}\\
&=\frac{(1-q^{x+1})(1-uq^{2x})(x_1-uq^{x+1})(x_2-uq^{x+1})(x_3-uq^{x+1})}{uq(1-uq^x)(1-uq^{2x+2})(1-x_1q^x)(1-x_2q^x)(1-x_3q^x)}.
\end{aligned}\tag{12.6.16}$$

As before, for $q>1$ we set $q=p^{-1}$. Then we obtain

$$\begin{aligned}
&\frac{\widetilde{w}(p^x+up^{-x})}{\widetilde{w}(p^{x+1}+up^{-x-1})}\\
&=\frac{(1-p^{-x-1})(1-up^{-2x})(x_1-up^{-x-1})(x_2-up^{-x-1})(x_3-up^{-x-1})}{up^{-1}(1-up^{-x})(1-up^{-2x-2})(1-x_1p^{-x})(1-x_2p^{-x})(1-x_3p^{-x})}.
\end{aligned}\tag{12.6.17}$$

Case I. $x_1=x_2=x_3=0$. From (12.6.16) we have

$$\frac{\widetilde{w}(q^{-x}+uq^x)}{\widetilde{w}(q^{-x-1}+uq^{x+1})}=-u^2q^{3x+2}\frac{(1-q^{x+1})(1-uq^{2x})}{(1-uq^x)(1-uq^{2x+2})}$$

with possible solution

$$w^{(I)}(q^{-x}+uq^{x})=\left(-\frac{1}{u^2}\right)^x q^{-x(3x+1)/2}(1-uq^{2x})\frac{(u;q)_x}{(q;q)_x}, \quad 0<q<1.$$

From (12.6.17) we have

$$\frac{\widetilde{w}(p^x+up^{-x})}{\widetilde{w}(p^{x+1}+up^{-x-1})}=-\frac{u(1-p^{x+1})(1-u^{-1}p^{2x})}{p^{3x+1}(1-u^{-1}p^{x})(1-u^{-1}p^{2x+2})}$$

with possible solution

$$w^{(I)}(p^x+up^{-x})=\left(-\frac{1}{u}\right)^x p^{x(3x-1)/2}(1-u^{-1}p^{2x})\frac{(u^{-1};p)_x}{(p;p)_x}, \quad 0<p<1.$$

Case II. $x_1\neq 0$ and $x_2=x_3=0$. From (12.6.16) we have

$$\frac{\widetilde{w}(q^{-x}+uq^{x})}{\widetilde{w}(q^{-x-1}+uq^{x+1})}=uq^{2x+1}\,\frac{(1-q^{x+1})(1-uq^{2x})(x_1-uq^{x+1})}{(1-uq^{x})(1-uq^{2x+2})(1-x_1q^{x})}$$

with possible solution

$$w^{(II)}(q^{-x}+uq^{x})=\left(\frac{1}{x_1uq}\right)^x q^{-2\binom{x}{2}}(1-uq^{2x})\frac{(u,x_1;q)_x}{(q,x_1^{-1}uq;q)_x}, \quad 0<q<1.$$

From (12.6.17) we have

$$\frac{\widetilde{w}(p^x+up^{-x})}{\widetilde{w}(p^{x+1}+up^{-x-1})}=\frac{u}{x_1p^{2x+1}}\,\frac{(1-p^{x+1})(1-u^{-1}p^{2x})(1-x_1u^{-1}p^{x+1})}{(1-u^{-1}p^{x})(1-u^{-1}p^{2x+2})(1-x_1^{-1}p^{x})}$$

with possible solution

$$w^{(II)}(p^x+up^{-x})=\left(\frac{x_1p}{u}\right)^x p^{2\binom{x}{2}}(1-u^{-1}p^{2x})\frac{(u^{-1},x_1^{-1};p)_x}{(p,x_1u^{-1}p;p)_x}, \quad 0<p<1.$$

Case III. $x_1x_2\neq 0$ and $x_3=0$. From (12.6.16) we have

$$\frac{\widetilde{w}(q^{-x}+uq^{x})}{\widetilde{w}(q^{-x-1}+uq^{x+1})}=-q^{x}\,\frac{(1-q^{x+1})(1-uq^{2x})(x_1-uq^{x+1})(x_2-uq^{x+1})}{(1-uq^{x})(1-uq^{2x+2})(1-x_1q^{x})(1-x_2q^{x})}$$

with possible solution

$$w^{(III)}(q^{-x}+uq^{x})=\left(-\frac{1}{x_1x_2}\right)^x q^{-\binom{x}{2}}(1-uq^{2x})\frac{(u,x_1,x_2;q)_x}{(q,x_1^{-1}uq,x_2^{-1}uq;q)_x}, \quad 0<q<1.$$

From (12.6.17) we have

$$\frac{\widetilde{w}(p^{x}+up^{-x})}{\widetilde{w}(p^{x+1}+up^{-x-1})} = -\frac{u}{x_1x_2p^{x+1}}\frac{(1-p^{x+1})(1-u^{-1}p^{2x})}{(1-u^{-1}p^{x})(1-u^{-1}p^{2x+2})} \times \frac{(1-x_1u^{-1}p^{x+1})(1-x_2u^{-1}p^{x+1})}{(1-x_1^{-1}p^{x})(1-x_2^{-1}p^{x})}$$

with possible solution

$$w^{(III)}(p^{x}+up^{-x}) = \left(-\frac{x_1x_2p}{u}\right)^{x} p^{\binom{x}{2}}(1-u^{-1}p^{2x})\frac{(u^{-1},x_1^{-1},x_2^{-1};p)_x}{(p,x_1u^{-1}p,x_2u^{-1}p;p)_x}, \quad 0<p<1.$$

Case IV. $x_1x_2x_3 \neq 0$. From (12.6.16) we have

$$\frac{\widetilde{w}(q^{-x}+uq^{x})}{\widetilde{w}(q^{-x-1}+uq^{x+1})} = \frac{x_1x_2x_3}{uq}\frac{(1-q^{x+1})(1-uq^{2x})}{(1-uq^{x})(1-uq^{2x+2})} \times \frac{(1-x_1^{-1}uq^{x+1})(1-x_2^{-1}uq^{x+1})(1-x_3^{-1}uq^{x+1})}{(1-x_1q^{x})(1-x_2q^{x})(1-x_3q^{x})}$$

with possible solution

$$w^{(IV)}(q^{-x}+uq^{x}) = \left(\frac{uq}{x_1x_2x_3}\right)^{x}(1-uq^{2x})\frac{(u,x_1,x_2,x_3;q)_x}{(q,x_1^{-1}uq,x_2^{-1}uq,x_3^{-1}uq;q)_x}, \quad 0<q<1.$$

From (12.6.17) we have

$$\frac{\widetilde{w}(p^{x}+up^{-x})}{\widetilde{w}(p^{x+1}+up^{-x-1})} = \frac{u}{x_1x_2x_3p}\frac{(1-p^{x+1})(1-u^{-1}p^{2x})}{(1-u^{-1}p^{x})(1-u^{-1}p^{2x+2})} \times \frac{(1-x_1u^{-1}p^{x+1})(1-x_2u^{-1}p^{x+1})(1-x_3u^{-1}p^{x+1})}{(1-x_1^{-1}p^{x})(1-x_2^{-1}p^{x})(1-x_3^{-1}p^{x})}$$

with possible solution

$$w^{(IV)}(p^{x}+up^{-x}) = \left(\frac{x_1x_2x_3p}{u}\right)^{x}(1-u^{-1}p^{2x}) \times \frac{(u^{-1},x_1^{-1},x_2^{-1},x_3^{-1};p)_x}{(p,x_1u^{-1}p,x_2u^{-1}p,x_3u^{-1}p;p)_x}, \quad 0<p<1.$$

12.7 Orthogonality Relations

In the preceding section we have obtained solutions of the q-Pearson equation (12.6.16) and the p-Pearson equation (12.6.17). In this section we will derive orthogonality relations for several cases obtained in section 12.5. We will not give explicit orthogonality relations for each different case, but we will restrict to the most important cases.

As in section 3.2 we now multiply (12.6.5) by $(\mathscr{S}y_m)(q^{-x}+uq^x)$ and subtract the same equation with m and n interchanged to obtain

$$\begin{aligned}&(\lambda_n-\lambda_m)(\mathscr{S}w)(q^{-x}+uq^x)(\mathscr{S}y_n)(q^{-x}+uq^x)(\mathscr{S}y_m)(q^{-x}+uq^x)\\&=\left(\mathscr{A}\left(w\left(\mathscr{S}^{-1}\varphi\right)(\mathscr{A}y_n)\right)\right)(q^{-x}+uq^x)(\mathscr{S}y_m)(q^{-x}+uq^x)\\&\qquad-\left(\mathscr{A}\left(w\left(\mathscr{S}^{-1}\varphi\right)(\mathscr{A}y_m)\right)\right)(q^{-x}+uq^x)(\mathscr{S}y_n)(q^{-x}+uq^x).\end{aligned}$$

Now we apply the operator $\mathscr{S}^{-1}$ and use the commutation rule (cf. (2.5.3) and (11.4.8))

$$\left(q^{-x}+uq^x-q^{-x+1}-uq^{x-1}\right)\mathscr{S}^{-1}\mathscr{A}=\left(q^{-x-1}+uq^{x+1}-q^{-x}-uq^x\right)\mathscr{A}\mathscr{S}^{-1},$$

which easily follows from (12.6.1) and (12.6.3), to obtain (cf. (3.2.9))

$$\begin{aligned}&(\lambda_n-\lambda_m)\widetilde{w}(q^{-x}+uq^x)y_m(q^{-x}+uq^x)y_n(q^{-x}+uq^x)\\&=\left(q^{-x-1}+uq^{x+1}-q^{-x}-uq^x\right)\\&\qquad\times\Big\{\left(\mathscr{A}\mathscr{S}^{-1}\left(w\left(\mathscr{S}^{-1}\varphi\right)(\mathscr{A}y_n)\right)\right)(q^{-x}+uq^x)y_m(q^{-x}+uq^x)\\&\qquad\quad-\left(\mathscr{A}\mathscr{S}^{-1}\left(w\left(\mathscr{S}^{-1}\varphi\right)(\mathscr{A}y_m)\right)\right)(q^{-x}+uq^x)y_n(q^{-x}+uq^x)\Big\},\end{aligned}\tag{12.7.1}$$

where $\widetilde{w}(q^{-x}+uq^x)$ is given by (12.6.15). Now the product rule (12.6.2) leads to the summations by parts formula (cf. (3.2.17))

$$\begin{aligned}&\sum_{x=0}^{N}\left(q^{-x-1}+uq^{x+1}-q^{-x}-uq^x\right)(\mathscr{A}f_1)(q^{-x}+uq^x)f_2(q^{-x}+uq^x)\\&=\Big[f_1(q^{-x}+uq^x)f_2(q^{-x}+uq^x)\Big]_{x=0}^{N+1}\\&\qquad-\sum_{x=0}^{N}\left(q^{-x-1}+uq^{x+1}-q^{-x}-uq^x\right)(\mathscr{S}f_1)(q^{-x}+uq^x)(\mathscr{A}f_2)(q^{-x}+uq^x).\end{aligned}$$

Applying this to (12.7.1) we obtain

$$\begin{aligned}(\lambda_n-\lambda_m)&\sum_{x=0}^{N}\widetilde{w}(q^{-x}+uq^x)y_m(q^{-x}+uq^x)y_n(q^{-x}+uq^x)\\&=\Big[\left(\mathscr{S}^{-1}\left(w\left(\mathscr{S}^{-1}\varphi\right)\right)\right)(q^{-x}+uq^x)\\&\quad\times\left\{\left(\mathscr{S}^{-1}\left(\mathscr{A}y_n\right)\right)(q^{-x}+uq^x)y_m(q^{-x}+uq^x)\right.\\&\qquad\left.-\left(\mathscr{S}^{-1}\left(\mathscr{A}y_m\right)\right)(q^{-x}+uq^x)y_n(q^{-x}+uq^x)\right\}\Big]_{x=0}^{N+1}.\end{aligned}$$

As in theorem 3.8 we now obtain orthogonality relations of the form

$$\sum_{x=0}^{N}\widetilde{w}(q^{-x}+uq^x)y_m(q^{-x}+uq^x)y_n(q^{-x}+uq^x)=\sigma_n\,\delta_{mn},\quad m,n=0,1,2,\ldots,N$$

with

$$\sigma_n=\prod_{k=1}^{n}d_k,\quad n=1,2,3,\ldots,N$$

and $N\in\{1,2,3,\ldots\}$ or possibly $N\to\infty$. In each different case the appropriate boundary conditions (cf. (3.2.19)) should be satisfied, id est:

$$\left(\mathscr{S}^{-1}\left(w\left(\mathscr{S}^{-1}\varphi\right)\right)\right)(q^0+uq^0)=0 \tag{12.7.2}$$

and

$$\left(\mathscr{S}^{-1}\left(w\left(\mathscr{S}^{-1}\varphi\right)\right)\right)(q^{-N-1}+uq^{N+1})=0, \tag{12.7.3}$$

where the relation between w and $\widetilde{w}$ is given by (12.6.15).

Case IVa1. $x_1x_2x_3\neq 0$, $0<q<1$, $u>0$, $\tau<0$ and $D_n^{(2)}>0$. We use the weight function

$$w^{(IV)}(q^{-x}+uq^x)=\left(\frac{uq}{x_1x_2x_3}\right)^x(1-uq^{2x})\frac{(u,x_1,x_2,x_3;q)_x}{(q,x_1^{-1}uq,x_2^{-1}uq,x_3^{-1}uq;q)_x},\quad 0<q<1.$$

Then we have by using (1.8.25) and Jackson's summation formula (1.11.10)

$$\begin{aligned}d_0:=&\sum_{x=0}^{\infty}\left(\frac{uq}{x_1x_2x_3}\right)^x(1-uq^{2x})\frac{(u,x_1,x_2,x_3;q)_x}{(q,x_1^{-1}uq,x_2^{-1}uq,x_3^{-1}uq;q)_x}\\=&\ (1-u)\,{}_6\phi_5\left(\begin{matrix}q\sqrt{u},-q\sqrt{u},u,x_1,x_2,x_3\\\sqrt{u},-\sqrt{u},x_1^{-1}uq,x_2^{-1}uq,x_3^{-1}uq\end{matrix};q,\frac{uq}{x_1x_2x_3}\right)\\=&\ (1-u)\frac{(uq,x_1^{-1}x_2^{-1}uq,x^{-1}x_3^{-1}uq,x_2^{-1}x_3^{-1}uq;q)_\infty}{(x_1^{-1}uq,x_2^{-1}uq,x_3^{-1}uq,x_1^{-1}x_2^{-1}x_3^{-1}uq;q)_\infty},\quad\left|\frac{uq}{x_1x_2x_3}\right|<1.\end{aligned}$$

Further we have

$$d_n = \frac{u(1-q^n)(1-x_1x_2x_3u^{-1}q^{n-2})(1-x_1q^{n-1})(1-x_2q^{n-1})(1-x_3q^{n-1})}{(1-x_1x_2x_3u^{-1}q^{2n-3})(1-x_1x_2x_3u^{-1}q^{2n-2})^2(1-x_1x_2x_3u^{-1}q^{2n-1})}$$
$$\times (1-x_1x_2u^{-1}q^{n-1})(1-x_1x_3u^{-1}q^{n-1})(1-x_2x_3u^{-1}q^{n-1})$$

for $n = 1,2,3,\ldots$, which implies that

$$\sigma_n = \prod_{k=1}^{n} d_k = u^n \frac{\left(q, x_1x_2x_3u^{-1}q^{-1}, x_1, x_2, x_3, x_1x_2u^{-1}, x_1x_3u^{-1}, x_2x_3u^{-1}; q\right)_n}{(x_1x_2x_3u^{-1}q^{-1}, x_1x_2x_3u^{-1}; q)_{2n}}$$

for $n = 1,2,3,\ldots$. This leads to the orthogonality relation

$$\sum_{x=0}^{\infty} \left(\frac{uq}{x_1x_2x_3}\right)^x (1-uq^{2x}) \frac{(u, x_1, x_2, x_3; q)_x}{(q, x_1^{-1}uq, x_2^{-1}uq, x_3^{-1}uq; q)_x} y_m(q^{-x}+uq^x) y_n(q^{-x}+uq^x)$$
$$= (1-u) \frac{(uq, x_1^{-1}x_2^{-1}uq, x^{-1}x_3^{-1}uq, x_2^{-1}x_3^{-1}uq; q)_\infty}{(x_1^{-1}uq, x_2^{-1}uq, x_3^{-1}uq, x_1^{-1}x_2^{-1}x_3^{-1}uq; q)_\infty}$$
$$\times u^n \frac{\left(q, x_1x_2x_3u^{-1}q^{-1}, x_1, x_2, x_3, x_1x_2u^{-1}, x_1x_3u^{-1}, x_2x_3u^{-1}; q\right)_n}{(x_1x_2x_3u^{-1}q^{-1}, x_1x_2x_3u^{-1}; q)_{2n}} \delta_{mn}$$

for $m, n = 0,1,2,\ldots$. Note that this is an orthogonality relation for an *infinite* system of q-**Racah** polynomials given by (12.3.10).

However, note that we have by using (12.6.12) and (12.6.15)

$$\left(\mathscr{S}^{-1}\left(w\left(\mathscr{S}^{-1}\varphi\right)\right)\right)(q^{-x}+uq^x) = \frac{\widetilde{w}(q^{-x+1}+uq^{x-1})\varphi(q^{-x+2}+uq^{x-2})}{q^{-x+1}+uq^{x-1}-q^{-x+2}-uq^{x-2}}$$
$$= \left(\frac{u}{x_1x_2x_3}\right)^{x-1} q^{-1}(1-q) \frac{(u, x_1, x_2, x_3; q)_x}{(q, x_1^{-1}uq, x_2^{-1}uq, x_3^{-1}uq; q)_{x-1}}.$$

Now we use (1.8.5) to obtain

$$\frac{1}{(q;q)_{-1}} = (1;q)_1 = 0 \quad \text{and} \quad \frac{1}{(x_i^{-1}uq;q)_{-1}} = (x_i^{-1}u;q)_1 = 1 - \frac{u}{x_i}, \quad x_i \neq 0.$$

This implies that the boundary condition (12.7.2) holds. If we choose x_1, x_2 or x_3 equal to q^{-N}, then the other boundary condition (12.7.3) also holds.

So, in order to find an orthogonality relation for a *finite* system of q-Racah polynomials, one of the parameters x_1, x_2 or x_3 should be set equal to q^{-N}. For instance, if we set $x_3 = q^{-N}$, then we obtain

$$w^{(IV)}(q^{-x}+uq^x) = \left(\frac{uq^{N+1}}{x_1x_2}\right)^x (1-uq^{2x}) \frac{(u, x_1, x_2, q^{-N}; q)_x}{(q, x_1^{-1}uq, x_2^{-1}uq, uq^{N+1}; q)_x}, \quad 0 < q < 1.$$

Then we have by using (1.8.25) and Jackson's summation formula (1.11.11)

$$\begin{aligned} d_0 &:= \sum_{x=0}^{N}\left(\frac{uq^{N+1}}{x_1x_2}\right)^x(1-uq^{2x})\frac{(u,x_1,x_2,q^{-N};q)_x}{(q,x_1^{-1}uq,x_2^{-1}uq,uq^{N+1};q)_x}\\ &= (1-u)\,{}_6\phi_5\left(\begin{array}{c} q\sqrt{u},-q\sqrt{u},u,x_1,x_2,q^{-N}\\ \sqrt{u},-\sqrt{u},x_1^{-1}uq,x_2^{-1}uq,uq^{N+1}\end{array};q,\frac{uq^{N+1}}{x_1x_2}\right)\\ &= (1-u)\frac{(uq,x_1^{-1}x_2^{-1}uq;q)_N}{(x_1^{-1}uq,x_2^{-1}uq;q)_N}. \end{aligned}$$

Further we have

$$\sigma_n=\prod_{k=1}^{n}d_k=u^n\frac{\left(q,x_1x_2u^{-1}q^{-N-1},q^{-N},x_1,x_2,x_1x_2u^{-1},x_1u^{-1}q^{-N},x_2u^{-1}q^{-N};q\right)_n}{(x_1x_2u^{-1}q^{-N-1},x_1x_2u^{-1}q^{-N};q)_{2n}}$$

for $n=1,2,3,\ldots,N$. This leads to the orthogonality relation

$$\begin{aligned} &\sum_{x=0}^{N}\left(\frac{uq^{N+1}}{x_1x_2}\right)^x(1-uq^{2x})\frac{(u,x_1,x_2,q^{-N};q)_x}{(q,x_1^{-1}uq,x_2^{-1}uq,uq^{N+1};q)_x}y_m(q^{-x}+uq^x)y_n(q^{-x}+uq^x)\\ &=(1-u)\frac{(uq,x_1^{-1}x_2^{-1}uq;q)_N}{(x_1^{-1}uq,x_2^{-1}uq;q)_N}\\ &\quad\times u^n\frac{\left(q,x_1x_2u^{-1}q^{-N-1},q^{-N},x_1,x_2,x_1x_2u^{-1},x_1u^{-1}q^{-N},x_2u^{-1}q^{-N};q\right)_n}{(x_1x_2u^{-1}q^{-N-1},x_1x_2u^{-1}q^{-N};q)_{2n}}\,\delta_{mn} \end{aligned}$$

for $m,n=0,1,2,\ldots,N$. Note that this is an orthogonality relation for a *finite* system of q-**Racah** polynomials.

From this general case involving the q-**Racah** polynomials we obtain orthogonality relations for the **dual** q-**Hahn** polynomials by taking the limit $x_2\to 0$. By using

$$x_2^k(x_2^{-1}uq;q)_k=(x_2-uq)(x_2-uq^2)\cdots(x_2-uq^k)\to(-1)^ku^kq^{\binom{k+1}{2}},\quad x_2\to 0$$

for $k\in\{0,1,2,\ldots\}$, we find that

$$\lim_{x_2\to 0}w^{(IV)}(q^{-x}+uq^x)=w^{(III)}(q^{-x}+uq^x).$$

We remark that this limit can only be used in the finite case. Now we have for instance

Case IIIb3. $x_1x_2\neq 0$, $x_3=0$, $0<q<1$, $0<x_1<1<x_2$ with $x_2q^{N_1}\leq 1<x_2q^{N_1-1}$ and $x_1x_2q^{N_2}\leq u<x_1x_2q^{N_2-1}$ and $N=\min(N_1,N_2)$. We use the weight function

$$w^{(III)}(q^{-x}+uq^x)=\left(-\frac{1}{x_1x_2}\right)^x q^{-\binom{x}{2}}(1-uq^{2x})\frac{(u,x_1,x_2;q)_x}{(q,x_1^{-1}uq,x_2^{-1}uq;q)_x},\quad 0<q<1.$$

Choosing $x_2=q^{-N}$ we have by using (1.8.25) and the summation formula (1.11.12)

$$d_0 := \sum_{x=0}^{N} \left(-\frac{q^N}{x_1}\right)^x q^{-\binom{x}{2}} (1-uq^{2x}) \frac{(u,x_1,q^{-N};q)_x}{(q,x_1^{-1}uq,uq^{N+1};q)_x}$$
$$= (1-u)\,{}_6\phi_4\left(\begin{array}{c} q\sqrt{u},-q\sqrt{u},u,x_1,0,q^{-N} \\ \sqrt{u},-\sqrt{u},x_1^{-1}uq,uq^{N+1} \end{array};q,\frac{q^N}{x_1}\right) = (1-u)\frac{(uq;q)_N}{x_1^N(x_1^{-1}uq;q)_N}.$$

Further we have

$$\sigma_n = \prod_{k=1}^{n} d_k = u^n \left(q,q^{-N},x_1,x_1u^{-1}q^{-N};q\right)_n, \quad n=1,2,3,\ldots,N.$$

This leads to the orthogonality relation

$$\sum_{x=0}^{N} \left(-\frac{q^N}{x_1}\right)^x q^{-\binom{x}{2}} (1-uq^{2x}) \frac{(u,x_1,q^{-N};q)_x}{(q,x_1^{-1}uq,uq^{N+1};q)_x} y_m(q^{-x}+uq^x) y_n(q^{-x}+uq^x)$$
$$= (1-u)\frac{(uq;q)_N}{x_1^N(x_1^{-1}uq;q)_N} u^n \left(q,q^{-N},x_1,x_1u^{-1}q^{-N};q\right)_n \delta_{mn}, \quad m,n=0,1,2,\ldots,N$$

for a *finite* system of **dual q-Hahn** polynomials.

In the same way as above this orthogonality relation for the **dual q-Hahn** polynomials leads to an orthogonality relation for the **dual q-Krawtchouk** polynomials by taking the limit $x_1 \to 0$. Now we have

$$\lim_{x_1\to 0} w^{(III)}(q^{-x}+uq^x) = w^{(II)}(q^{-x}+uq^x).$$

Then we have for instance

Case IIb1. $x_1 \neq 0$, $x_2 = x_3 = 0$, $0<q<1$, $u<0$ and $x_1>1$ with $x_1q^N \leq 1 < x_1q^{N-1}$. We use the weight function

$$w^{(II)}(q^{-x}+uq^x) = \left(\frac{1}{x_1uq}\right)^x q^{-2\binom{x}{2}} (1-uq^{2x}) \frac{(u,x_1;q)_x}{(q,x_1^{-1}uq;q)_x}, \quad 0<q<1.$$

Choosing $x_1 = q^{-N}$ we have by using (1.8.25) and the summation formula (1.11.13)

$$d_0 := \sum_{x=0}^{N} \left(\frac{q^{N-1}}{u}\right)^x q^{-2\binom{x}{2}} (1-uq^{2x}) \frac{(u,q^{-N};q)_x}{(q,uq^{N+1};q)_x}$$
$$= (1-u)\,{}_6\phi_3\left(\begin{array}{c} q\sqrt{u},-q\sqrt{u},u,0,0,q^{-N} \\ \sqrt{u},-\sqrt{u},uq^{N+1} \end{array};q,\frac{q^{N-1}}{u}\right)$$
$$= (1-u)(-1)^N u^{-N} q^{-\binom{N+1}{2}} (uq;q)_N.$$

Further we have

$$\sigma_n = \prod_{k=1}^{n} d_k = u^n \left(q, q^{-N}; q\right)_n, \quad n = 1, 2, 3, \ldots, N.$$

This leads to the orthogonality relation

$$\sum_{x=0}^{N} \left(\frac{q^{N-1}}{u}\right)^x q^{-2\binom{x}{2}} (1-uq^{2x}) \frac{(u, q^{-N}; q)_x}{(q, uq^{N+1}; q)_x} y_m(q^{-x}+uq^x) y_n(q^{-x}+uq^x)$$
$$= (1-u)(-1)^N u^{n-N} q^{-\binom{N+1}{2}} (uq;q)_N \left(q, q^{-N}; q\right)_n \delta_{mn}, \quad m,n = 0,1,2,\ldots,N$$

for a *finite* system of **dual q-Krawtchouk** polynomials.

This process cannot be continued in order to obtain an orthogonality relation for the **dual q-Charlier** polynomials. In that case we only have orthogonality in the case of an *infinite* family of polynomials. However, we don't know how to derive an orthogonality relation for such an *infinite* family of **dual q-Charlier** polynomials.

Chapter 13
Orthogonal Polynomial Solutions in $\frac{a}{z}+\frac{uz}{a}$ of Complex q-Difference Equations

Classical q-Orthogonal Polynomials IV

13.1 Real Polynomial Solutions in $\frac{a}{z}+\frac{uz}{a}$ with $u\in\mathbb{R}\setminus\{0\}$ and $a,z\in\mathbb{C}\setminus\{0\}$

It is also possible to obtain real polynomial solutions of the (complex) q-difference equation (12.2.1)

$$\widehat{\varphi}(qz^*)\left(\mathscr{D}_q^2\widehat{y}_n\right)(z^*)+\widehat{\psi}(qz^*)\left(\mathscr{D}_q\widehat{y}_n\right)(z^*)=\widehat{\lambda}_n\widehat{\rho}(qz^*)\widehat{y}_n(qz^*),\quad n=0,1,2,\ldots$$

with argument $z^*:=\frac{a}{z}+\frac{uz}{a}$ where $u\in\mathbb{R}\setminus\{0\}$ and $a,z\in\mathbb{C}\setminus\{0\}$. By using $z=x+iy$, $a=\alpha+i\beta$ with $x,y,\alpha,\beta\in\mathbb{R}$, we find that the imaginary part of

$$\frac{a}{z}+\frac{uz}{a}=\frac{\alpha+i\beta}{x+iy}+\frac{u(x+iy)}{\alpha+i\beta}=\frac{(\alpha+i\beta)(x-iy)}{x^2+y^2}+\frac{u(x+iy)(\alpha-i\beta)}{\alpha^2+\beta^2}$$

equals

$$\frac{(\beta x-\alpha y)(\alpha^2+\beta^2)+u(\alpha y-\beta x)(x^2+y^2)}{(x^2+y^2)(\alpha^2+\beta^2)}=\frac{(\beta x-\alpha y)\left\{\alpha^2+\beta^2-u(x^2+y^2)\right\}}{(x^2+y^2)(\alpha^2+\beta^2)}.$$

This is equal to zero for all $x\in\mathbb{R}$ and $y\in\mathbb{R}$ if

$$x^2+y^2=r^2\quad\text{and}\quad ur^2=\alpha^2+\beta^2\,(=a\bar{a}).$$

Without loss of generality we set $r=1$, which implies that z is on the unit circle. So if we define

$$z=\mathrm{e}^{i\theta},\quad a=|a|\mathrm{e}^{-i\phi},\quad \theta,\phi\in\mathbb{R},\quad |a|>0\quad\text{and}\quad u=a\bar{a}=|a|^2,$$

R. Koekoek et al., *Hypergeometric Orthogonal Polynomials and Their q-Analogues*, Springer Monographs in Mathematics,
DOI 10.1007/978-3-642-05014-5_13,

then we find that

$$\frac{a}{z}+\frac{uz}{a}=\frac{a}{z}+\bar{a}z=|a|\mathrm{e}^{-i(\theta+\phi)}+|a|\mathrm{e}^{i(\theta+\phi)}=2|a|\cos(\theta+\phi).$$

Note that if we replace $\frac{a}{z}$ by q^{-x}, then we have $\frac{a}{z}+\frac{uz}{a}\longrightarrow q^{-x}+uq^{x}$. So, similar to the situation in the previous chapter we now look for polynomial solutions of the form

$$y_n(z^*)=\sum_{k=0}^{n} b_{n,k}\frac{(\frac{a}{z};q)_k(\frac{uz}{a};q)_k}{(q;q)_k},\quad b_{n,n}\neq 0,\quad n=0,1,2,\ldots \tag{13.1.1}$$

in

$$z^*=\frac{a}{z}+\frac{uz}{a},$$

of the q-difference equation (12.2.2)

$$\widehat{C}(z^*)\widehat{y}_n(qz^*)-\left\{\widehat{C}(z^*)+\widehat{D}(z^*)\right\}\widehat{y}_n(z^*)+\widehat{D}(z^*)\widehat{y}_n(q^{-1}z^*)=\widehat{\lambda}_n\widehat{\rho}(z^*)\widehat{y}_n(z^*),$$

with $q\in\mathbb{R}\setminus\{-1,0,1\}$. Note that $y_n(z^*)$ is a polynomial with degree$[y_n]=n$ in $z^*=\frac{a}{z}+\frac{uz}{a}=2|a|\cos(\theta+\phi)$. In a similar way as in the previous chapter (cf. theorem 12.2) we now have the following theorem:

Theorem 13.1. *The q-difference equation*

$$\frac{z^2}{a^2}C(z)y_n\left(\frac{a}{qz}+\frac{uqz}{a}\right)-\frac{z^2}{a^2}\left\{C(z)+D(z)\right\}y_n\left(\frac{a}{z}+\frac{uz}{a}\right)+\frac{z^2}{a^2}D(z)y_n\left(\frac{aq}{z}+\frac{uz}{aq}\right)$$
$$=-q\left(1-\frac{\bar{a}}{aq}z^2\right)\left(1-\frac{\bar{a}}{a}z^2\right)\left(1-\frac{\bar{a}q}{a}z^2\right)\lambda_n y_n\left(\frac{a}{z}+\frac{uz}{a}\right) \tag{13.1.2}$$

only has polynomial solutions $y_n(z^*)$ with degree$[y_n]=n$ in

$$z^*=\frac{a}{z}+\frac{uz}{a}\,(=2|a|\cos(\theta+\phi))$$

for $n=0,1,2,\ldots$ if the coefficients $\frac{z^2}{a^2}C(z)$ and $\frac{z^2}{a^2}D(z)$ and the eigenvalues λ_n have the form

$$\frac{z^2}{a^2}C(z)=-q(1-\bar{a}z)\left(1-\frac{\bar{a}}{aq}z^2\right)\left(1-\frac{z_1}{a}z\right)\left(1-\frac{z_2}{a}z\right)\left(1-\frac{z_3}{a}z\right), \tag{13.1.3}$$

$$\frac{z^2}{a^2}D(z)=-\frac{1}{a\bar{a}}\left(1-\frac{z}{a}\right)\left(1-\frac{\bar{a}q}{a}z^2\right)(z_1-\bar{a}z)(z_2-\bar{a}z)(z_3-\bar{a}z) \tag{13.1.4}$$

and

$$\lambda_n=(q^{-n}-1)\left(1-\frac{z_1z_2z_3}{a\bar{a}}q^{n-1}\right),\quad n=0,1,2,\ldots, \tag{13.1.5}$$

where $a,z_1,z_2,z_3\in\mathbb{C}$, $a\neq 0$ and $q\in\mathbb{R}\setminus\{-1,0,1\}$.

By using $\frac{a}{z}=|a|\mathrm{e}^{-i(\theta+\phi)}$ and $\frac{uz}{a}=\bar{a}z=|a|\mathrm{e}^{i(\theta+\phi)}$, we obtain for $n=0,1,2,\ldots$ the representation (cf. (12.3.10))

$$y_n(z^*)=\frac{(z_1;q)_n(z_2;q)_n(z_3;q)_n}{(q^{-n};q)_n\left(\frac{z_1z_2z_3}{a\bar{a}}q^{n-1};q\right)_n q^n}b_{n,n}\times{}_4\phi_3\left(\begin{matrix}q^{-n},\frac{z_1z_2z_3}{a\bar{a}}q^{n-1},|a|\mathrm{e}^{-i(\theta+\phi)},|a|\mathrm{e}^{i(\theta+\phi)}\\ z_1,z_2,z_3\end{matrix};q,q\right)\quad(13.1.6)$$

with

$$z^*=\frac{a}{z}+\frac{uz}{a}=\frac{a}{z}+\bar{a}z=2|a|\cos(\theta+\phi)\quad(13.1.7)$$

for the **Askey-Wilson** polynomials, the **continuous q-Hahn** polynomials or the **continuous q-Jacobi** polynomials.

Special cases are the **continuous dual q-Hahn** polynomials ($z_3=0$), the **q-Meixner-Pollaczek** polynomials, the **Al-Salam-Chihara** polynomials or the **continuous q-Laguerre** polynomials ($z_2=z_3=0$) and the **continuous big q-Hermite** polynomials ($z_1=z_2=z_3=0$).

We remark that we have to choose $b_{n,n}=(q^{-n};q)_nq^n$ in order to get monic polynomials.

In view of (12.4.1), (12.4.2) and (12.4.3), we conclude that the monic Askey-Wilson, continuous q-Hahn or continuous q-Jacobi polynomials satisfy the three-term recurrence relation

$$y_{n+1}(z^*)=\left(2|a|\cos(\theta+\phi)-|a|^2-1+c_n^{(1)}+c_n^{(2)}\right)y_n(z^*)-c_{n-1}^{(1)}c_n^{(2)}y_{n-1}(z^*),\quad n=1,2,3,\ldots\quad(13.1.8)$$

with $y_0(z^*)=1$ and

$$y_1(z^*)=2|a|\cos(\theta+\phi)-|a|^2-1+|a|^2(1-z_1)(1-z_2)(1-z_3)/(|a|^2-z_1z_2z_3),$$

where

$$c_n^{(1)}=\frac{|a|^2(|a|^2-z_1z_2z_3q^{n-1})(1-z_1q^n)(1-z_2q^n)(1-z_3q^n)}{(|a|^2-z_1z_2z_3q^{2n-1})(|a|^2-z_1z_2z_3q^{2n})}\quad(13.1.9)$$

for $n=0,1,2,\ldots$, and

$$c_n^{(2)}=\frac{(1-q^n)(|a|^2-z_1z_2q^{n-1})(|a|^2-z_1z_3q^{n-1})(|a|^2-z_2z_3q^{n-1})}{(|a|^2-z_1z_2z_3q^{2n-2})(|a|^2-z_1z_2z_3q^{2n-1})}\quad(13.1.10)$$

for $n=1,2,3,\ldots$.

13.2 Classification of the Positive-Definite Orthogonal Polynomial Solutions

The classification of positive-definite orthogonal polynomial solutions can be done as in chapter 12. However, in this case we have $q \in \mathbb{R}\setminus\{-1,0,1\}$ and $u = a\bar{a} = |a|^2 > 0$. Now we have

$$c_0 = -c_0^{(1)}, \quad c_n = -c_n^{(1)} - c_n^{(2)}, \quad n = 1,2,3,\ldots,$$

$$d_n = c_{n-1}^{(1)} c_n^{(2)} = \frac{|a|^2(1-q^n)(|a|^2 - z_1z_2z_3q^{n-2})}{(|a|^2 - z_1z_2z_3q^{2n-3})(|a|^2 - z_1z_2z_3q^{2n-2})^2(|a|^2 - z_1z_2z_3q^{2n-1})} D_n$$

for $n = 1,2,3,\ldots$ and

$$\begin{aligned} D_n = {} & (1-z_1q^{n-1})(1-z_2q^{n-1})(1-z_3q^{n-1}) \\ & \times (|a|^2 - z_1z_2q^{n-1})(|a|^2 - z_1z_3q^{n-1})(|a|^2 - z_2z_3q^{n-1}) \end{aligned} \tag{13.2.1}$$

for $n = 1,2,3,\ldots$, where $q \in \mathbb{R}\setminus\{-1,0,1\}$ and $a \in \mathbb{C}\setminus\{0\}$. Hence we have $c_n \in \mathbb{R}$ for all $n = 0,1,2,\ldots$ if $z_1z_2z_3 \in \mathbb{R}$, $z_1+z_2+z_3 \in \mathbb{R}$ and $z_1z_2+z_1z_3+z_2z_3 \in \mathbb{R}$. This implies that $z_1, z_2, z_3 \in \mathbb{R}$ or one is real and the other two are complex conjugates.

As before, we start the study of the positivity of d_n for $n = 1,2,3,\ldots$ with the cases where $z_1z_2z_3 = 0$. Then we have

$$d_n = \frac{|a|^2(1-q^n)(|a|^2 - z_1z_2z_3q^{n-2})}{(|a|^2 - z_1z_2z_3q^{2n-3})(|a|^2 - z_1z_2z_3q^{2n-2})^2(|a|^2 - z_1z_2z_3q^{2n-1})} D_n = \frac{1-q^n}{|a|^4} D_n.$$

The sign of $1-q^n$ is given in table 13.1.

q	$q<-1$	$-1<q<0$	$0<q<1$	$q>1$
$\mathbf{1-q^n}$	$(-1)^{n+1}$	+	+	−

Table 13.1 sign of $1-q^n$, $n = 1,2,3,\ldots$

Case I. $z_1 = z_2 = z_3 = 0$. Then we have

$$d_n = |a|^2(1-q^n), \quad n = 1,2,3,\ldots.$$

This implies that we have positive-definite orthogonality in the following two infinite cases:

Case Ia1. $z_1 = z_2 = z_3 = 0$ and $-1 < q < 0$.

Case Ia2. $z_1 = z_2 = z_3 = 0$ and $0 < q < 1$.

In this case we have no finite systems of positive-definite orthogonal polynomials.

Case II. $z_1 \neq 0$ and $z_2 = z_3 = 0$. Then we have

$$d_n = |a|^2(1-q^n)(1-z_1q^{n-1}), \quad n = 1,2,3,\ldots.$$

In this case we must have $z_1 \in \mathbb{R}$. This leads to positive-definite orthogonality in the following four infinite cases:

Case IIa1. $z_1 \in \mathbb{R}\setminus\{0\}$, $z_2 = z_3 = 0$, $q < -1$ and the sign of $1 - z_1q^{n-1}$ equal to $(-1)^{n+1}$. This implies that $z_1 < q^{-1}$.

Case IIa2. $z_1 \in \mathbb{R}\setminus\{0\}$, $z_2 = z_3 = 0$, $-1 < q < 0$ and $1 - z_1q^{n-1} > 0$. This implies that $q^{-1} < z_1 < 1$.

Case IIa3. $z_1 \in \mathbb{R}\setminus\{0\}$, $z_2 = z_3 = 0$, $0 < q < 1$ and $1 - z_1q^{n-1} > 0$. This implies that $z_1 < 1$.

Case IIa4. $z_1 \in \mathbb{R}\setminus\{0\}$, $z_2 = z_3 = 0$, $q > 1$ and $1 - z_1q^{n-1} < 0$. This implies that $z_1 > 1$.

Also in this case we have no finite systems of positive-definite orthogonal polynomials.

Case III. $z_1z_2 \neq 0$ and $z_3 = 0$. Then we have

$$d_n = (1-q^n)(1-z_1q^{n-1})(1-z_2q^{n-1})(|a|^2 - z_1z_2q^{n-1}), \quad n = 1,2,3,\ldots.$$

This leads to positive-definite orthogonality in the following four infinite cases:

Case IIIa1. $z_1z_2 \neq 0$, $z_3 = 0$, $q < -1$ and the sign of $(1-z_1q^{n-1})(1-z_2q^{n-1})(|a|^2 - z_1z_2q^{n-1})$ equal to $(-1)^{n+1}$.

Case IIIa2. $z_1z_2 \neq 0$, $z_3 = 0$, $-1 < q < 0$ and $(1-z_1q^{n-1})(1-z_2q^{n-1})(|a|^2 - z_1z_2q^{n-1}) > 0$.

Case IIIa3. $z_1z_2 \neq 0$, $z_3 = 0$, $0 < q < 1$ and $(1-z_1q^{n-1})(1-z_2q^{n-1})(|a|^2 - z_1z_2q^{n-1}) > 0$.

Case IIIa4. $z_1z_2 \neq 0$, $z_3 = 0$, $q > 1$ and $(1-z_1q^{n-1})(1-z_2q^{n-1})(|a|^2 - z_1z_2q^{n-1}) < 0$.

It is also possible to have positive-definite orthogonality for finite systems of polynomials. However, we will not treat these finite cases.

Case IV. $z_1z_2z_3 \neq 0$. Now we write (cf. (12.5.4))

$$d_n = |a|^2(1-q^n)D_n^{(1)}D_n^{(2)}, \quad n=1,2,3,\ldots \tag{13.2.2}$$

with (cf. (12.5.5))

$$D_n^{(1)} = \frac{1-\tau q^{n-1}}{(1-\tau q^{2n-2})(1-\tau q^{2n-1})^2(1-\tau q^{2n})}, \quad n=1,2,3,\ldots \tag{13.2.3}$$

and (cf. (12.5.6))

$$D_n^{(2)} = (1-z_1q^{n-1})(1-z_2q^{n-1})(1-z_3q^{n-1}) \times\left(1-\frac{\tau}{z_1}q^n\right)\left(1-\frac{\tau}{z_2}q^n\right)\left(1-\frac{\tau}{z_3}q^n\right) \tag{13.2.4}$$

for $n=1,2,3,\ldots$, where

$$\tau = \frac{z_1z_2z_3}{|a|^2q}.$$

The sign of $1-q^n$ for $n=1,2,3,\ldots$ is given in table 13.1. The sign of $D_n^{(1)}$ for $n=1,2,3,\ldots$ is given in table 12.1 (for $q>0$) and in table 13.2 (for $q<0$).

q	**extra conditions**	$\mathbf{D_n^{(1)}}$	**for**
$q<-1$	$\tau q^2>1$	$(-1)^n$	$n=1,2,3,\ldots$
	$0<\tau q^2<1$ with $1\leq\tau q^{2N}<q^{-2}$	$+$	$n=1,2,3,\ldots,N$
	$0<\tau q<1$ with $\tau q^{2N+1}=1$	$+$	$n=1,2,3,\ldots,N$
	$0<\tau q<1$ with $q<\tau q^{2N}<q^{-1}$	$+$	$n=1,2,3,\ldots,2N+1$
	$\tau q>1$	$(-1)^{n+1}$	$n=1,2,3,\ldots$
$-1<q<0$	$\tau q^2>1$ with $1<\tau q^{2N}\leq q^{-2}$	$(-1)^n$	$n=1,2,3,\ldots,N$
	$0<\tau q^2<1$	$+$	$n=1,2,3,\ldots$
	$0<\tau q<1$	$+$	$n=1,2,3,\ldots$
	$\tau q>1$ with $\tau q^{2N+1}=1$	$(-1)^{n+1}$	$n=1,2,3,\ldots,N$
	$\tau q>1$ with $q^{-1}<\tau q^{2N}<q$	$(-1)^{n+1}$	$n=1,2,3,\ldots,2N+1$

Table 13.2 sign of $D_n^{(1)}$, $q<0$ and $N\in\{1,2,3,\ldots\}$

By using table 12.1, table 13.1 and table 13.2, we conclude that we have positive-definite orthogonality in the following eight infinite cases:

Case IVa1. $z_1z_2z_3 \neq 0$, $q < -1$, $\tau q^2 > 1$ and $D_n^{(2)} < 0$.

Case IVa2. $z_1z_2z_3 \neq 0$, $q < -1$, $\tau q > 1$ and $D_n^{(2)} > 0$.

Case IVa3. $z_1z_2z_3 \neq 0$, $-1 < q < 0$, $0 < \tau q^2 < 1$ and $D_n^{(2)} > 0$.

Case IVa4. $z_1z_2z_3 \neq 0$, $-1 < q < 0$, $0 < \tau q < 1$ and $D_n^{(2)} > 0$.

Case IVa5. $z_1z_2z_3 \neq 0$, $0 < q < 1$, $\tau < 0$ and $D_n^{(2)} > 0$.

Case IVa6. $z_1z_2z_3 \neq 0$, $0 < q < 1$, $0 < \tau q < 1$ and $D_n^{(2)} > 0$.

Case IVa7. $z_1z_2z_3 \neq 0$, $q > 1$, $\tau < 0$ and $D_n^{(2)} < 0$.

Case IVa8. $z_1z_2z_3 \neq 0$, $q > 1$, $\tau q > 1$ and $D_n^{(2)} > 0$.

It is also possible to have positive-definite orthogonality for finite systems of polynomials. However, we will not treat these finite cases.

13.3 Solutions of the q-Pearson Equation

In this section we use the following operator[1] (cf. (12.6.1))

$$(\mathscr{A}f)\left(\frac{a}{z}+\frac{uz}{a}\right)=\frac{f\left(\frac{a}{qz}+\frac{uqz}{a}\right)-f\left(\frac{a}{z}+\frac{uz}{a}\right)}{\frac{a}{qz}+\frac{uqz}{a}-\frac{a}{z}-\frac{uz}{a}},\quad q\in\mathbb{R}\setminus\{-1,0,1\},\tag{13.3.1}$$

where f is a complex-valued function in $\frac{a}{z}+\frac{uz}{a}$ whose domain contains both $\frac{a}{z}+\frac{uz}{a}$ and $\frac{a}{qz}+\frac{uqz}{a}$ for each $z \in \mathbb{C}$. For two such functions f_1 and f_2, a product rule similar to (12.6.2) applies. In fact we have

$$\begin{aligned}(\mathscr{A}(f_1f_2))\left(\frac{a}{z}+\frac{uz}{a}\right)&=(\mathscr{A}f_1)\left(\frac{a}{z}+\frac{uz}{a}\right)f_2\left(\frac{a}{z}+\frac{uz}{a}\right)\\&\quad+(\mathscr{S}f_1)\left(\frac{a}{z}+\frac{uz}{a}\right)(\mathscr{A}f_2)\left(\frac{a}{z}+\frac{uz}{a}\right),\end{aligned}\tag{13.3.2}$$

where, analogous to (12.6.3), we now define

[1] It turns out that this form makes sense. However, this form no longer has the property that its action on a polynomial in $\frac{a}{z}+\frac{uz}{a}$ of degree n leads to a polynomial in $\frac{a}{z}+\frac{uz}{a}$ of degree $n-1$ for $n = 1,2,3,\ldots$.

$$(\mathscr{S}f)\left(\frac{a}{z}+\frac{uz}{a}\right)=f\left(\frac{a}{qz}+\frac{uqz}{a}\right)$$
$$\text{and}\quad (\mathscr{S}^{-1}f)\left(\frac{a}{z}+\frac{uz}{a}\right)=f\left(\frac{aq}{z}+\frac{uz}{aq}\right). \tag{13.3.3}$$

Now we start with a general second-order operator equation of the form

$$\varphi\left(\frac{a}{z}+\frac{uz}{a}\right)(\mathscr{A}^2y_n)\left(\frac{a}{z}+\frac{uz}{a}\right)$$
$$+\psi\left(\frac{a}{z}+\frac{uz}{a}\right)(\mathscr{A}y_n)\left(\frac{a}{z}+\frac{uz}{a}\right)=\lambda_n(\mathscr{S}y_n)\left(\frac{a}{z}+\frac{uz}{a}\right)$$

in terms of the operator (13.3.1) and later we will compare this to the q-difference equation (13.1.2) in theorem 13.1. If we multiply both sides of this equation by $(\mathscr{S}w)\left(\frac{a}{z}+\frac{uz}{a}\right)$ we obtain

$$(\mathscr{S}w)\left(\frac{a}{z}+\frac{uz}{a}\right)\varphi\left(\frac{a}{z}+\frac{uz}{a}\right)(\mathscr{A}^2y_n)\left(\frac{a}{z}+\frac{uz}{a}\right)$$
$$+(\mathscr{S}w)\left(\frac{a}{z}+\frac{uz}{a}\right)\psi\left(\frac{a}{z}+\frac{uz}{a}\right)(\mathscr{A}y_n)\left(\frac{a}{z}+\frac{uz}{a}\right)$$
$$=\lambda_n(\mathscr{S}w)\left(\frac{a}{z}+\frac{uz}{a}\right)(\mathscr{S}y_n)\left(\frac{a}{z}+\frac{uz}{a}\right), \tag{13.3.4}$$

which leads to the self-adjoint form

$$\left(\mathscr{A}\left(w\left(\mathscr{S}^{-1}\varphi\right)(\mathscr{A}y_n)\right)\right)\left(\frac{a}{z}+\frac{uz}{a}\right)$$
$$=\lambda_n(\mathscr{S}w)\left(\frac{a}{z}+\frac{uz}{a}\right)(\mathscr{S}y_n)\left(\frac{a}{z}+\frac{uz}{a}\right) \tag{13.3.5}$$

if the Pearson operator equation

$$\left(\mathscr{A}\left(w\left(\mathscr{S}^{-1}\varphi\right)\right)\right)\left(\frac{a}{z}+\frac{uz}{a}\right)=(\mathscr{S}w)\left(\frac{a}{z}+\frac{uz}{a}\right)\psi\left(\frac{a}{z}+\frac{uz}{a}\right)$$

holds. Furthermore, the productrule (13.3.2) leads to

$$\begin{aligned}
&\left(\mathscr{A}\left(w\left(\mathscr{S}^{-1}\varphi\right)\right)\right)\left(\frac{a}{z}+\frac{uz}{a}\right)\\
&=\left(\mathscr{S}^{-1}\varphi\right)\left(\frac{a}{z}+\frac{uz}{a}\right)\frac{w\left(\frac{a}{qz}+\frac{uqz}{a}\right)-w\left(\frac{a}{z}+\frac{uz}{a}\right)}{\frac{a}{qz}+\frac{uqz}{a}-\frac{a}{z}-\frac{uz}{a}}\\
&\quad+w\left(\frac{a}{qz}+\frac{uqz}{a}\right)\frac{\varphi\left(\frac{a}{z}+\frac{uz}{a}\right)-\left(\mathscr{S}^{-1}\varphi\right)\left(\frac{a}{z}+\frac{uz}{a}\right)}{\frac{a}{qz}+\frac{uqz}{a}-\frac{a}{z}-\frac{uz}{a}}.
\end{aligned}$$

Combining the latter two equations we obtain (cf. (3.2.7) and (11.4.6))

$$\begin{aligned}
&w\left(\frac{a}{z}+\frac{uz}{a}\right)\varphi\left(\frac{aq}{z}+\frac{uz}{aq}\right)\\
&=w\left(\frac{a}{qz}+\frac{uqz}{a}\right)\\
&\quad\times\left\{\varphi\left(\frac{a}{z}+\frac{uz}{a}\right)-\left(\frac{a}{qz}+\frac{uqz}{a}-\frac{a}{z}-\frac{uz}{a}\right)\psi\left(\frac{a}{z}+\frac{uz}{a}\right)\right\}.
\end{aligned}\tag{13.3.6}$$

The q-difference equation (13.1.2) in theorem 13.1 can now be written in the form

$$\begin{aligned}
&\widetilde{C}(z)y_n\left(\frac{a}{qz}+\frac{uqz}{a}\right)-\left\{\widetilde{C}(z)+\widetilde{D}(z)\right\}y_n\left(\frac{a}{z}+\frac{uz}{a}\right)+\widetilde{D}(z)y_n\left(\frac{aq}{z}+\frac{uz}{aq}\right)\\
&=\lambda_n y_n\left(\frac{a}{z}+\frac{uz}{a}\right),
\end{aligned}\tag{13.3.7}$$

where, by using (13.1.3), (13.1.4) and (13.1.5),

$$\begin{aligned}
\widetilde{C}(z)&=-\frac{z^2C(z)}{a^2q\left(1-\frac{\bar{a}}{aq}z^2\right)\left(1-\frac{\bar{a}}{a}z^2\right)\left(1-\frac{\bar{a}q}{a}z^2\right)}\\
&=\frac{(1-\bar{a}z)\left(1-\frac{z_1}{a}z\right)\left(1-\frac{z_2}{a}z\right)\left(1-\frac{z_3}{a}z\right)}{\left(1-\frac{\bar{a}}{a}z^2\right)\left(1-\frac{\bar{a}q}{a}z^2\right)},
\end{aligned}\tag{13.3.8}$$

$$\begin{aligned}
\widetilde{D}(z)&=-\frac{z^2D(z)}{a^2q\left(1-\frac{\bar{a}}{aq}z^2\right)\left(1-\frac{\bar{a}}{a}z^2\right)\left(1-\frac{\bar{a}q}{a}z^2\right)}\\
&=\frac{\left(1-\frac{z}{a}\right)(z_1-\bar{a}z)(z_2-\bar{a}z)(z_3-\bar{a}z)}{a\bar{a}q\left(1-\frac{\bar{a}}{aq}z^2\right)\left(1-\frac{\bar{a}}{a}z^2\right)}
\end{aligned}\tag{13.3.9}$$

and

$$\lambda_n=(q^{-n}-1)\left(1-\frac{z_1z_2z_3}{a\bar{a}}q^{n-1}\right),\quad n=0,1,2,\ldots.$$

Note that we have by using (13.3.1)

$$(\mathscr{A}y_n)\left(\frac{a}{z}+\frac{uz}{a}\right)=\frac{y_n\left(\frac{a}{qz}+\frac{uqz}{a}\right)-y_n\left(\frac{a}{z}+\frac{uz}{a}\right)}{\frac{a}{qz}+\frac{uqz}{a}-\frac{a}{z}-\frac{uz}{a}} \tag{13.3.10}$$

and

$$(\mathscr{A}^2y_n)\left(\frac{a}{z}+\frac{uz}{a}\right)=\frac{1}{\frac{a}{qz}+\frac{uqz}{a}-\frac{a}{z}-\frac{uz}{a}}\left\{\frac{y_n\left(\frac{a}{q^2z}+\frac{uq^2z}{a}\right)-y_n\left(\frac{a}{qz}+\frac{uqz}{a}\right)}{\frac{a}{q^2z}+\frac{uq^2z}{a}-\frac{a}{qz}-\frac{uqz}{a}}\right.$$
$$\left.-\frac{y_n\left(\frac{a}{qz}+\frac{uqz}{a}\right)-y_n\left(\frac{a}{z}+\frac{uz}{a}\right)}{\frac{a}{qz}+\frac{uqz}{a}-\frac{a}{z}-\frac{uz}{a}}\right\}. \tag{13.3.11}$$

If we apply the operator $\mathscr{S}^{-1}$ to the left of (13.3.4) we obtain

$$\begin{aligned}&w\left(\frac{a}{z}+\frac{uz}{a}\right)\varphi\left(\frac{aq}{z}+\frac{uz}{aq}\right)(\mathscr{A}^2y_n)\left(\frac{aq}{z}+\frac{uz}{aq}\right)\\&\qquad+w\left(\frac{a}{z}+\frac{uz}{a}\right)\psi\left(\frac{aq}{z}+\frac{uz}{aq}\right)(\mathscr{A}y_n)\left(\frac{aq}{z}+\frac{uz}{aq}\right)\\&=\lambda_nw\left(\frac{a}{z}+\frac{uz}{a}\right)y_n\left(\frac{a}{z}+\frac{uz}{a}\right).\end{aligned}$$

Now we divide by $w\left(\frac{a}{z}+\frac{uz}{a}\right)$ and use (13.3.10) and (13.3.11) to find

$$\begin{aligned}&\frac{y_n\left(\frac{a}{qz}+\frac{uqz}{a}\right)\varphi\left(\frac{aq}{z}+\frac{uz}{aq}\right)}{\left(\frac{a}{z}+\frac{uz}{a}-\frac{aq}{z}-\frac{uz}{aq}\right)\left(\frac{a}{qz}+\frac{uqz}{a}-\frac{a}{z}-\frac{uz}{a}\right)}\\&\quad-\frac{y_n\left(\frac{a}{z}+\frac{uz}{a}\right)}{\frac{a}{z}+\frac{uz}{a}-\frac{aq}{z}-\frac{uz}{aq}}\left\{\frac{\varphi\left(\frac{aq}{z}+\frac{uz}{aq}\right)}{\frac{a}{qz}+\frac{uqz}{a}-\frac{a}{z}-\frac{uz}{a}}\right.\\&\qquad\left.+\frac{\varphi\left(\frac{aq}{z}+\frac{uz}{aq}\right)}{\frac{a}{z}+\frac{uz}{a}-\frac{aq}{z}-\frac{uz}{aq}}-\psi\left(\frac{aq}{z}+\frac{uz}{aq}\right)\right\}\\&\quad+\frac{y_n\left(\frac{aq}{z}+\frac{uz}{aq}\right)}{\frac{a}{z}+\frac{uz}{a}-\frac{aq}{z}-\frac{uz}{aq}}\left\{\frac{\varphi\left(\frac{aq}{z}+\frac{uz}{aq}\right)}{\frac{a}{z}+\frac{uz}{a}-\frac{aq}{z}-\frac{uz}{aq}}-\psi\left(\frac{aq}{z}+\frac{uz}{aq}\right)\right\}\\&=\lambda_ny_n\left(\frac{a}{z}+\frac{uz}{a}\right).\end{aligned}$$

If we compare this with the q-difference equation (13.3.7) we conclude that

$$\begin{aligned}&\varphi\left(\frac{aq}{z}+\frac{uz}{aq}\right)\\&=\left(\frac{a}{z}+\frac{uz}{a}-\frac{aq}{z}-\frac{uz}{aq}\right)\left(\frac{a}{qz}+\frac{uqz}{a}-\frac{a}{z}-\frac{uz}{a}\right)\widetilde{C}(z)\end{aligned}\tag{13.3.12}$$

and

$$\begin{aligned}&\psi\left(\frac{aq}{z}+\frac{uz}{aq}\right)\\&=-\left(\frac{a}{z}+\frac{uz}{a}-\frac{aq}{z}-\frac{uz}{aq}\right)\widetilde{D}(z)+\frac{\varphi\left(\frac{aq}{z}+\frac{uz}{aq}\right)}{\frac{a}{z}+\frac{uz}{a}-\frac{aq}{z}-\frac{uz}{aq}}\\&=-\left(\frac{a}{z}+\frac{uz}{a}-\frac{aq}{z}-\frac{uz}{aq}\right)\widetilde{D}(z)+\left(\frac{a}{qz}+\frac{uqz}{a}-\frac{a}{z}-\frac{uz}{a}\right)\widetilde{C}(z).\end{aligned}\tag{13.3.13}$$

This implies that the q-Pearson equation (13.3.6) can be written in the form

$$\begin{aligned}&w\left(\frac{a}{z}+\frac{uz}{a}\right)\left(\frac{a}{z}+\frac{uz}{a}-\frac{aq}{z}-\frac{uz}{aq}\right)\widetilde{C}(z)\\&=w\left(\frac{a}{qz}+\frac{uqz}{a}\right)\left(\frac{a}{qz}+\frac{uqz}{a}-\frac{a}{z}-\frac{uz}{a}\right)\widetilde{D}(qz).\end{aligned}$$

Hence we have the q-Pearson equation

$$\widetilde{w}\left(\frac{a}{z}+\frac{uz}{a}\right)\widetilde{C}(z)=\widetilde{w}\left(\frac{a}{qz}+\frac{uqz}{a}\right)\widetilde{D}(qz),\tag{13.3.14}$$

where

$$\widetilde{w}\left(\frac{a}{z}+\frac{uz}{a}\right):=\left(\frac{a}{z}+\frac{uz}{a}-\frac{aq}{z}-\frac{uz}{aq}\right)w\left(\frac{a}{z}+\frac{uz}{a}\right).\tag{13.3.15}$$

Finally we use (13.3.8) and (13.3.9) to conclude that we look for solutions of the q-Pearson equation

$$\begin{aligned}\frac{\widetilde{w}\left(\frac{a}{z}+\frac{uz}{a}\right)}{\widetilde{w}\left(\frac{a}{qz}+\frac{uqz}{a}\right)}&=\frac{\widetilde{D}(qz)}{\widetilde{C}(z)}=\frac{\left(1-\frac{\bar{a}}{a}z^2\right)\left(1-\frac{qz}{a}\right)(z_1-\bar{a}qz)(z_2-\bar{a}qz)(z_3-\bar{a}qz)}{a\bar{a}q\left(1-\frac{\bar{a}q^2}{a}z^2\right)(1-\bar{a}z)\left(1-\frac{z_1}{a}z\right)\left(1-\frac{z_2}{a}z\right)\left(1-\frac{z_3}{a}z\right)}\\&=\frac{\left(1-\frac{a}{qz}\right)\left(1-\frac{\bar{a}}{a}z^2\right)\left(1-\frac{\bar{a}q}{a}z^2\right)}{(1-\bar{a}z)\left(1-\frac{a}{\bar{a}q^2z^2}\right)\left(1-\frac{a}{\bar{a}qz^2}\right)}\\&\qquad\times\frac{\left(1-\frac{z_1}{\bar{a}qz}\right)\left(1-\frac{z_2}{\bar{a}qz}\right)\left(1-\frac{z_3}{\bar{a}qz}\right)}{\left(1-\frac{z_1}{a}z\right)\left(1-\frac{z_2}{a}z\right)\left(1-\frac{z_3}{a}z\right)},\end{aligned}\tag{13.3.16}$$

where

$$z=\mathrm{e}^{i\theta},\quad a=|a|\mathrm{e}^{-i\phi},\quad \theta,\phi\in\mathbb{R},\quad |a|>0\quad\text{and}\quad \frac{a}{z}+\frac{uz}{a}=2|a|\cos(\theta+\phi).$$

Note that this is equivalent to (12.6.16) with $u=|a|^2=a\bar{a}$, q^x replaced by z/a and x_i replaced by z_i for $i=1,2,3$.

Now we make the following observations:

$$w(z)=\left(\frac{\bar{a}z^2}{a},\frac{a}{\bar{a}z^2};q\right)_\infty$$
$$\Longrightarrow\quad \frac{w(z)}{w(qz)}=\frac{\left(\frac{\bar{a}z^2}{a},\frac{a}{\bar{a}z^2};q\right)_\infty}{\left(\frac{\bar{a}q^2z^2}{a},\frac{a}{\bar{a}q^2z^2};q\right)_\infty}=\frac{\left(1-\frac{\bar{a}}{a}z^2\right)\left(1-\frac{\bar{a}q}{a}z^2\right)}{\left(1-\frac{a}{\bar{a}q^2z^2}\right)\left(1-\frac{a}{\bar{a}qz^2}\right)},$$

$$w(z)=\frac{1}{\left(\bar{a}z,\frac{a}{z};q\right)_\infty}\quad\Longrightarrow\quad\frac{w(z)}{w(qz)}=\frac{\left(\bar{a}qz,\frac{a}{qz};q\right)_\infty}{\left(\bar{a}z,\frac{a}{z};q\right)_\infty}=\frac{1-\frac{a}{qz}}{1-\bar{a}z}$$

and

$$w(z)=\frac{1}{\left(\frac{z_iz}{a},\frac{z_i}{\bar{a}z};q\right)_\infty}\quad\Longrightarrow\quad\frac{w(z)}{w(qz)}=\frac{\left(\frac{z_iqz}{a},\frac{z_i}{\bar{a}qz};q\right)_\infty}{\left(\frac{z_iz}{a},\frac{z_i}{\bar{a}z};q\right)_\infty}=\frac{1-\frac{z_i}{\bar{a}qz}}{1-\frac{z_i}{a}z},\qquad i=1,2,3.$$

Case I. $z_1=z_2=z_3=0$. From (13.3.16) we have

$$\frac{\widetilde{w}(\frac{a}{z}+\frac{uz}{a})}{\widetilde{w}(\frac{a}{qz}+\frac{uqz}{a})}=\frac{\left(1-\frac{a}{qz}\right)\left(1-\frac{\bar{a}}{a}z^2\right)\left(1-\frac{\bar{a}q}{a}z^2\right)}{(1-\bar{a}z)\left(1-\frac{a}{\bar{a}q^2z^2}\right)\left(1-\frac{a}{\bar{a}qz^2}\right)}$$

with possible solution

$$w^{(I)}\left(\frac{a}{z}+\frac{uz}{a}\right)=\frac{\left(\frac{\bar{a}z^2}{a},\frac{a}{\bar{a}z^2};q\right)_\infty}{\left(\bar{a}z,\frac{a}{z};q\right)_\infty},\qquad z=\mathrm{e}^{i\theta},\quad a=|a|\mathrm{e}^{-i\phi}.$$

Case II. $z_1\in\mathbb{R}\setminus\{0\}$ and $z_2=z_3=0$. From (13.3.16) we have

$$\frac{\widetilde{w}(\frac{a}{z}+\frac{uz}{a})}{\widetilde{w}(\frac{a}{qz}+\frac{uqz}{a})}=\frac{\left(1-\frac{a}{qz}\right)\left(1-\frac{\bar{a}}{a}z^2\right)\left(1-\frac{\bar{a}q}{a}z^2\right)\left(1-\frac{z_1}{\bar{a}qz}\right)}{(1-\bar{a}z)\left(1-\frac{a}{\bar{a}q^2z^2}\right)\left(1-\frac{a}{\bar{a}qz^2}\right)\left(1-\frac{z_1}{a}z\right)}$$

with possible solution

$$w^{(II)}\left(\frac{a}{z}+\frac{uz}{a}\right)=\frac{\left(\frac{\overline{a}z^2}{a},\frac{a}{\overline{a}z^2};q\right)_\infty}{\left(\overline{a}z,\frac{a}{z},\frac{z_1z}{a},\frac{z_1}{\overline{a}z};q\right)_\infty},\qquad z=\mathrm{e}^{i\theta},\quad a=|a|\mathrm{e}^{-i\phi}.$$

Case III. $z_1z_2\neq 0$ and $z_3=0$. From (13.3.16) we have

$$\frac{\widetilde{w}(\frac{a}{z}+\frac{uz}{a})}{\widetilde{w}(\frac{a}{qz}+\frac{uqz}{a})}=\frac{\left(1-\frac{a}{qz}\right)\left(1-\frac{\overline{a}}{a}z^2\right)\left(1-\frac{\overline{a}q}{a}z^2\right)\left(1-\frac{z_1}{\overline{a}qz}\right)\left(1-\frac{z_2}{\overline{a}qz}\right)}{\left(1-\overline{a}z\right)\left(1-\frac{a}{\overline{a}q^2z^2}\right)\left(1-\frac{a}{\overline{a}qz^2}\right)\left(1-\frac{z_1}{a}z\right)\left(1-\frac{z_2}{a}z\right)}$$

with possible solution

$$w^{(III)}\left(\frac{a}{z}+\frac{uz}{a}\right)=\frac{\left(\frac{\overline{a}z^2}{a},\frac{a}{\overline{a}z^2};q\right)_\infty}{\left(\overline{a}z,\frac{a}{z},\frac{z_1z}{a},\frac{z_1}{\overline{a}z},\frac{z_2z}{a},\frac{z_2}{\overline{a}z};q\right)_\infty},\qquad z=\mathrm{e}^{i\theta},\quad a=|a|\mathrm{e}^{-i\phi}.$$

Case IV. $z_1z_2z_3\neq 0$. From (13.3.16) we have

$$\frac{\widetilde{w}(\frac{a}{z}+\frac{uz}{a})}{\widetilde{w}(\frac{a}{qz}+\frac{uqz}{a})}=\frac{\left(1-\frac{a}{qz}\right)\left(1-\frac{\overline{a}}{a}z^2\right)\left(1-\frac{\overline{a}q}{a}z^2\right)\left(1-\frac{z_1}{\overline{a}qz}\right)\left(1-\frac{z_2}{\overline{a}qz}\right)\left(1-\frac{z_3}{\overline{a}qz}\right)}{\left(1-\overline{a}z\right)\left(1-\frac{a}{\overline{a}q^2z^2}\right)\left(1-\frac{a}{\overline{a}qz^2}\right)\left(1-\frac{z_1}{a}z\right)\left(1-\frac{z_2}{a}z\right)\left(1-\frac{z_3}{a}z\right)}$$

with possible solution

$$w^{(IV)}\left(\frac{a}{z}+\frac{uz}{a}\right)=\frac{\left(\frac{\overline{a}z^2}{a},\frac{a}{\overline{a}z^2};q\right)_\infty}{\left(\overline{a}z,\frac{a}{z},\frac{z_1z}{a},\frac{z_1}{\overline{a}z},\frac{z_2z}{a},\frac{z_2}{\overline{a}z},\frac{z_3z}{a},\frac{z_3}{\overline{a}z};q\right)_\infty},\qquad z=\mathrm{e}^{i\theta},\quad a=|a|\mathrm{e}^{-i\phi}.$$

13.4 Orthogonality Relations

In the preceding section we have obtained solutions of the q-Pearson equation (13.3.16). In this section we will derive orthogonality relations for several cases obtained in section 13.2. We will not give explicit orthogonality relations for each different case, but we will restrict to the most important cases.

In order to find orthogonality relations we cannot use a similar method as in the previous chapter. However, as in theorem 3.8 we now obtain orthogonality relations of the form

$$\int_\alpha^\beta \widetilde{w}\left(\frac{a}{z}+\frac{uz}{a}\right)y_m\left(\frac{a}{z}+\frac{uz}{a}\right)y_n\left(\frac{a}{z}+\frac{uz}{a}\right)dz=\sigma_n\,\delta_{mn},\quad m,n=0,1,2,\ldots,$$

with

$$\sigma_n=\prod_{k=1}^{n} d_k,\quad n=1,2,3,\ldots.$$

In each different case the appropriate boundary conditions should be satisfied. This implies (cf. (12.7.2) and (12.7.3)) that

$$\frac{\widetilde{w}\left(\frac{aq}{z}+\frac{uz}{aq}\right)\varphi\left(\frac{aq^2}{z}+\frac{uz}{aq^2}\right)}{\frac{aq}{z}+\frac{uz}{aq}-\frac{aq^2}{z}-\frac{uz}{aq^2}}$$

should vanish for both $z=\alpha$ and $z=\beta$, the ends of the interval of orthogonality (α,β) with possibly $\alpha\to-\infty$ and/or $\beta\to\infty$.

Case Ia2. $z_1=z_2=z_3=0$ and $0<q<1$. We use the weight function

$$w^{(I)}\left(\frac{a}{z}+\frac{uz}{a}\right)=\frac{\left(\frac{\bar{a}z^2}{a},\frac{a}{\bar{a}z^2};q\right)_\infty}{\left(\bar{a}z,\frac{a}{z};q\right)_\infty},\quad z=e^{i\theta},\quad a=|a|e^{-i\phi}.$$

For the boundary conditions we should have that

$$\begin{aligned}
&\frac{\widetilde{w}\left(\frac{aq}{z}+\frac{uz}{aq}\right)\varphi\left(\frac{aq^2}{z}+\frac{uz}{aq^2}\right)}{\frac{aq}{z}+\frac{uz}{aq}-\frac{aq^2}{z}-\frac{uz}{aq^2}}\\
&=\frac{\left(\frac{\bar{a}z^2}{aq^2},\frac{aq^2}{\bar{a}z^2};q\right)_\infty}{\left(\frac{\bar{a}z}{q},\frac{aq}{z};q\right)_\infty}\left(\frac{a}{z}+\frac{uz}{a}-\frac{aq}{z}-\frac{uz}{aq}\right)\frac{\left(1-\frac{\bar{a}z}{q}\right)}{\left(1-\frac{\bar{a}z^2}{aq^2}\right)\left(1-\frac{\bar{a}z^2}{aq}\right)}\\
&=\frac{\left(\frac{\bar{a}z^2}{aq},\frac{aq^2}{\bar{a}z^2};q\right)_\infty}{\left(\bar{a}z,\frac{aq}{z};q\right)_\infty}\frac{a}{z}(1-q)
\end{aligned}$$

vanishes at both ends of the interval of orthogonality. Note that this is true for

$$z=\pm\frac{a}{|a|}\quad\Longrightarrow\quad\cos(\theta+\phi)=\pm1.$$

This implies that we should have $0\leq\theta+\phi\leq\pi$.

Note that, since $\theta\in\mathbb{R}$ and $\phi\in\mathbb{R}$, the weight function can be written in the form

$$w^{(I)}\left(2|a|\cos(\theta+\phi)\right)=\frac{\left(e^{2i(\theta+\phi)},e^{-2i(\theta+\phi)};q\right)_\infty}{\left(|a|e^{i(\theta+\phi)},|a|e^{-i(\theta+\phi)};q\right)_\infty}=\left|\frac{\left(e^{2i(\theta+\phi)};q\right)_\infty}{\left(|a|e^{i(\theta+\phi)};q\right)_\infty}\right|^2.$$

Further we have

$$d_0 := \frac{1}{2\pi}\int_{-\phi}^{-\phi+\pi}\left|\frac{\left(e^{2i(\theta+\phi)};q\right)_\infty}{\left(|a|e^{i(\theta+\phi)};q\right)_\infty}\right|^2 d\theta = \frac{1}{2\pi}\int_0^\pi \left|\frac{(e^{2i\theta};q)_\infty}{(|a|e^{i\theta};q)_\infty}\right|^2 d\theta.$$

Hence we have by using the Askey-Wilson q-beta integral (1.12.4) that

$$\begin{aligned} d_0 &= \frac{1}{2\pi}\int_0^\pi \left|\frac{(e^{2i\theta};q)_\infty}{(|a|e^{i\theta};q)_\infty}\right|^2 d\theta \\ &= \frac{1}{2\pi}\int_0^\pi \frac{(e^{2i\theta},e^{-2i\theta};q)_\infty}{(|a|e^{i\theta},|a|e^{-i\theta};q)_\infty}\,d\theta = \frac{1}{(q;q)_\infty}, \quad |a|<1. \end{aligned}$$

Further we have $d_n = |a|^2(1-q^n)$ which implies that

$$\sigma_n = \prod_{k=1}^n d_k = |a|^{2n}(q;q)_n, \quad n=1,2,3,\ldots.$$

This leads, for $0<|a|<1$, to the orthogonality relation

$$\frac{1}{2\pi}\int_0^\pi \left|\frac{(e^{2i\theta};q)_\infty}{(|a|e^{i\theta};q)_\infty}\right|^2 y_m(2|a|\cos\theta)y_n(2|a|\cos\theta)\,d\theta = \frac{|a|^{2n}(q;q)_n}{(q;q)_\infty}\,\delta_{mn}$$

with $m,n=0,1,2,\ldots$ for the **continuous big q-Hermite** polynomials. Since $|a|>0$ this orthogonality relation can be normalized to read

$$\frac{1}{2\pi}\int_0^\pi \left|\frac{(e^{2i\theta};q)_\infty}{(|a|e^{i\theta};q)_\infty}\right|^2 \widehat{y}_m(\cos\theta)\widehat{y}_n(\cos\theta)\,d\theta = \frac{(q;q)_n}{(q;q)_\infty}\,\delta_{mn}, \quad m,n=0,1,2,\ldots.$$

Then the limit case $a\to 0$ leads to the orthogonality relation

$$\frac{1}{2\pi}\int_0^\pi \left|(e^{2i\theta};q)_\infty\right|^2 y_m(\cos\theta)y_n(\cos\theta)\,d\theta = \frac{(q;q)_n}{(q;q)_\infty}\,\delta_{mn}, \quad m,n=0,1,2,\ldots$$

for the **continuous q-Hermite** polynomials.

Case IVa5. $z_1z_2z_3\neq 0$, $0<q<1$, $\tau<0$ and $D_n^{(2)}>0$. We use the weight function

$$w^{(IV)}\left(\frac{a}{z}+\frac{uz}{a}\right) = \frac{\left(\frac{\bar{a}z^2}{a},\frac{a}{\bar{a}z^2};q\right)_\infty}{\left(\bar{a}z,\frac{a}{z},\frac{z_1z}{a},\frac{z_1}{\bar{a}z},\frac{z_2z}{a},\frac{z_2}{\bar{a}z},\frac{z_3z}{a},\frac{z_3}{\bar{a}z};q\right)_\infty}, \quad z=e^{i\theta}, \quad a=|a|e^{-i\phi}.$$

Also in this case the boundary conditions hold for

$$z=\pm\frac{a}{|a|} \quad\Longrightarrow\quad \cos(\theta+\phi)=\pm 1,$$

which implies that we should have $0\leq\theta+\phi\leq\pi$.

Note that, since $\theta \in \mathbb{R}$ and $\phi \in \mathbb{R}$, the weight function can be written in the form

$$w^{(IV)}\left(2|a|\cos(\theta+\phi)\right) = \left|\frac{\left(e^{2i(\theta+\phi)};q\right)_\infty}{\left(|a|e^{i(\theta+\phi)}, \frac{z_1}{|a|}e^{i(\theta+\phi)}, \frac{z_2}{|a|}e^{i(\theta+\phi)}, \frac{z_3}{|a|}e^{i(\theta+\phi)};q\right)_\infty}\right|^2.$$

Further we have

$$d_0 := \frac{1}{2\pi}\int_{-\phi}^{-\phi+\pi}\left|\frac{\left(e^{2i(\theta+\phi)};q\right)_\infty}{\left(|a|e^{i(\theta+\phi)}, \frac{z_1}{|a|}e^{i(\theta+\phi)}, \frac{z_2}{|a|}e^{i(\theta+\phi)}, \frac{z_3}{|a|}e^{i(\theta+\phi)};q\right)_\infty}\right|^2 d\theta.$$

Hence we have as before by using the Askey-Wilson q-beta integral (1.12.4) that

$$\begin{aligned} d_0 &= \frac{1}{2\pi}\int_0^\pi\left|\frac{\left(e^{2i\theta};q\right)_\infty}{\left(|a|e^{i\theta}, \frac{z_1}{|a|}e^{i\theta}, \frac{z_2}{|a|}e^{i\theta}, \frac{z_3}{|a|}e^{i\theta};q\right)_\infty}\right|^2 d\theta \\ &= \frac{\left(\frac{z_1z_2z_3}{|a|^2};q\right)_\infty}{\left(z_1,z_2,z_3,\frac{z_1z_2}{|a|^2},\frac{z_1z_3}{|a|^2},\frac{z_2z_3}{|a|^2},q;q\right)_\infty}. \end{aligned}$$

Further we have

$$\begin{aligned} d_n &= \frac{|a|^2(1-q^n)(1-\tau q^{n-1})(1-z_1q^{n-1})(1-z_2q^{n-1})(1-z_3q^{n-1})}{(1-\tau q^{2n-2})(1-\tau q^{2n-1})^2(1-\tau q^{2n})} \\ &\quad \times\left(1-\frac{\tau}{z_1}q^n\right)\left(1-\frac{\tau}{z_2}q^n\right)\left(1-\frac{\tau}{z_3}q^n\right), \quad \tau=\frac{z_1z_2z_3}{|a|^2q} \end{aligned}$$

for $n=1,2,3,\ldots$, which implies that

$$\sigma_n = \prod_{k=1}^n d_k = |a|^{2n}\frac{\left(q,\tau,z_1,z_2,z_3,\frac{\tau q}{z_1},\frac{\tau q}{z_2},\frac{\tau q}{z_3};q\right)_n}{(\tau,\tau q;q)_{2n}}, \quad \tau=\frac{z_1z_2z_3}{|a|^2q}, \quad n=1,2,3,\ldots.$$

This leads to the orthogonality relation

$$\begin{aligned} &\frac{1}{2\pi}\int_0^\pi\left|\frac{\left(e^{2i\theta};q\right)_\infty}{\left(|a|e^{i\theta}, \frac{z_1}{|a|}e^{i\theta}, \frac{z_2}{|a|}e^{i\theta}, \frac{z_3}{|a|}e^{i\theta};q\right)_\infty}\right|^2 y_m(2|a|\cos\theta)y_n(2|a|\cos\theta)\,d\theta \\ &= \frac{\left(\frac{z_1z_2z_3}{|a|^2};q\right)_\infty}{\left(z_1,z_2,z_3,\frac{z_1z_2}{|a|^2},\frac{z_1z_3}{|a|^2},\frac{z_2z_3}{|a|^2},q;q\right)_\infty}|a|^{2n}\frac{\left(q,\frac{z_1z_2z_3}{|a|^2q},z_1,z_2,z_3,\frac{z_1z_2}{|a|^2},\frac{z_1z_3}{|a|^2},\frac{z_2z_3}{|a|^2};q\right)_n}{\left(\frac{z_1z_2z_3}{|a|^2q},\frac{z_1z_2z_3}{|a|^2};q\right)_{2n}}\delta_{mn} \end{aligned}$$

for $m,n=0,1,2,\ldots$. This is an orthogonality relation for the **Askey-Wilson** polynomials.

Similar to the situation in the preceding chapter this orthogonality relation for the **Askey-Wilson** polynomials leads to orthogonality relations for the special cases of **continuous dual q-Hahn** polynomials ($z_3 = 0$), the **q-Meixner-Pollaczek** polynomials, the **Al-Salam-Chihara** polynomials or the **continuous q-Laguerre** polynomials ($z_2 = z_3 = 0$) and the **continuous big q-Hermite** polynomials ($z_1 = z_2 = z_3 = 0$).

SCHEME
OF
BASIC HYPERGEOMETRIC ORTHOGONAL POLYNOMIALS

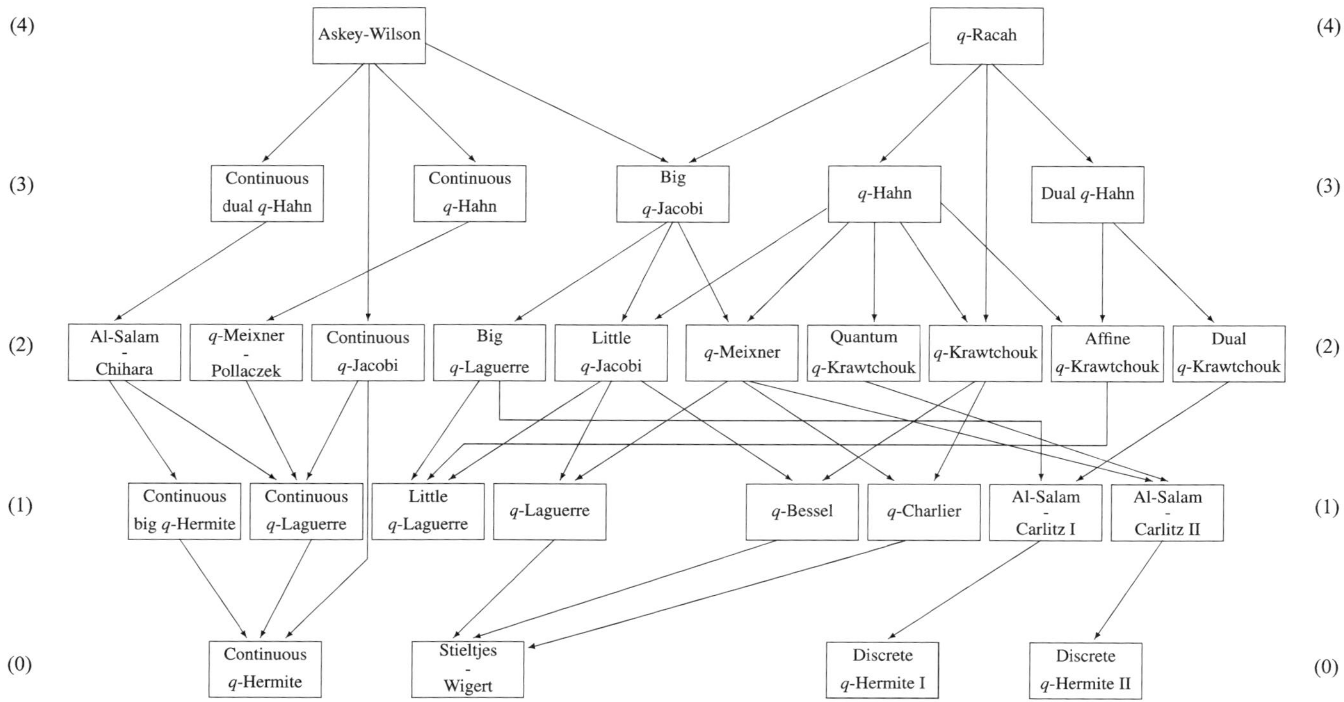

Chapter 14
Basic Hypergeometric Orthogonal Polynomials

In this chapter we deal with all families of basic hypergeometric orthogonal polynomials appearing in the q-analogue of the Askey scheme on the page 413. For each family of orthogonal polynomials we state the most important properties such as a representation as a basic hypergeometric function, orthogonality relation(s), the three-term recurrence relation, the second-order q-difference equation, the forward shift (or degree lowering) and backward shift (or degree raising) operator, a Rodrigues-type formula and some generating functions. Throughout this chapter we assume that $0 < q < 1$. In each case we use the notation which seems to be most common in the literature. Moreover, in each case we also state the limit relations between various families of q-orthogonal polynomials and the limit relations ($q \to 1$) to the classical hypergeometric orthogonal polynomials belonging to the Askey scheme on page 183. For notations the reader is referred to chapter 1.

14.1 Askey-Wilson

Basic Hypergeometric Representation

$$\begin{aligned} &\frac{a^n p_n(x;a,b,c,d|q)}{(ab,ac,ad;q)_n} \\ &= {}_4\phi_3\left(\begin{matrix} q^{-n}, abcdq^{n-1}, ae^{i\theta}, ae^{-i\theta} \\ ab, ac, ad \end{matrix}; q, q\right), \quad x = \cos\theta. \qquad (14.1.1) \end{aligned}$$

The Askey-Wilson polynomials are q-analogues of the Wilson polynomials given by (9.1.1).

R. Koekoek et al., *Hypergeometric Orthogonal Polynomials and Their q-Analogues*,
Springer Monographs in Mathematics,
DOI 10.1007/978-3-642-05014-5_14,

Orthogonality Relation

If a,b,c,d are real, or occur in complex conjugate pairs if complex, and $\max(|a|,|b|,|c|,|d|) < 1$, then we have the following orthogonality relation

$$\frac{1}{2\pi}\int_{-1}^{1}\frac{w(x)}{\sqrt{1-x^2}}p_m(x;a,b,c,d|q)p_n(x;a,b,c,d|q)\,dx = h_n\,\delta_{mn}, \tag{14.1.2}$$

where

$$w(x) := w(x;a,b,c,d|q) = \left|\frac{(e^{2i\theta};q)_\infty}{(ae^{i\theta},be^{i\theta},ce^{i\theta},de^{i\theta};q)_\infty}\right|^2 = \frac{h(x,1)h(x,-1)h(x,q^{\frac{1}{2}})h(x,-q^{\frac{1}{2}})}{h(x,a)h(x,b)h(x,c)h(x,d)},$$

with

$$h(x,\alpha) := \prod_{k=0}^{\infty}\left(1-2\alpha xq^k+\alpha^2q^{2k}\right) = \left(\alpha e^{i\theta},\alpha e^{-i\theta};q\right)_\infty, \quad x=\cos\theta$$

and

$$h_n = \frac{(abcdq^{n-1};q)_n(abcdq^{2n};q)_\infty}{(q^{n+1},abq^n,acq^n,adq^n,bcq^n,bdq^n,cdq^n;q)_\infty}.$$

If $a>1$ and b,c,d are real or one is real and the other two are complex conjugates, $\max(|b|,|c|,|d|) < 1$ and the pairwise products of a,b,c and d have absolute value less than 1, then we have another orthogonality relation given by:

$$\frac{1}{2\pi}\int_{-1}^{1}\frac{w(x)}{\sqrt{1-x^2}}p_m(x;a,b,c,d|q)p_n(x;a,b,c,d|q)\,dx + \sum_{\substack{k \\ 1<aq^k\leq a}} w_k p_m(x_k;a,b,c,d|q)p_n(x_k;a,b,c,d|q) = h_n\,\delta_{mn}, \tag{14.1.3}$$

where $w(x)$ and h_n are as before,

$$x_k = \frac{aq^k+\left(aq^k\right)^{-1}}{2}$$

and

$$w_k = \frac{(a^{-2};q)_\infty}{(q,ab,ac,ad,a^{-1}b,a^{-1}c,a^{-1}d;q)_\infty} \times \frac{(1-a^2q^{2k})(a^2,ab,ac,ad;q)_k}{(1-a^2)(q,ab^{-1}q,ac^{-1}q,ad^{-1}q;q)_k}\left(\frac{q}{abcd}\right)^k.$$

Recurrence Relation

$$2x\tilde{p}_n(x) = A_n\tilde{p}_{n+1}(x) + \left[a + a^{-1} - (A_n + C_n)\right]\tilde{p}_n(x) + C_n\tilde{p}_{n-1}(x), \qquad (14.1.4)$$

where

$$\tilde{p}_n(x) := \tilde{p}_n(x;a,b,c,d|q) = \frac{a^n p_n(x;a,b,c,d|q)}{(ab,ac,ad;q)_n}$$

and

$$\begin{cases} A_n = \dfrac{(1-abq^n)(1-acq^n)(1-adq^n)(1-abcdq^{n-1})}{a(1-abcdq^{2n-1})(1-abcdq^{2n})} \\[2ex] C_n = \dfrac{a(1-q^n)(1-bcq^{n-1})(1-bdq^{n-1})(1-cdq^{n-1})}{(1-abcdq^{2n-2})(1-abcdq^{2n-1})}. \end{cases}$$

Normalized Recurrence Relation

$$xp_n(x) = p_{n+1}(x) + \frac{1}{2}\left[a + a^{-1} - (A_n + C_n)\right]p_n(x) + \frac{1}{4}A_{n-1}C_n p_{n-1}(x), \qquad (14.1.5)$$

where

$$p_n(x;a,b,c,d|q) = 2^n(abcdq^{n-1};q)_n p_n(x).$$

q-Difference Equation

$$(1-q)^2 D_q\left[\tilde{w}(x;aq^{\frac{1}{2}},bq^{\frac{1}{2}},cq^{\frac{1}{2}},dq^{\frac{1}{2}}|q)D_q y(x)\right] + \lambda_n\tilde{w}(x;a,b,c,d|q)y(x) = 0, \quad y(x) = p_n(x;a,b,c,d|q), \qquad (14.1.6)$$

where

$$\tilde{w}(x;a,b,c,d|q) := \frac{w(x;a,b,c,d|q)}{\sqrt{1-x^2}}$$

and

$$\lambda_n = 4q^{-n+1}(1-q^n)(1-abcdq^{n-1}).$$

If we define

$$P_n(z) := \frac{(ab,ac,ad;q)_n}{a^n}\, {}_4\phi_3\left(\begin{matrix} q^{-n}, abcdq^{n-1}, az, az^{-1} \\ ab, ac, ad \end{matrix}; q, q\right)$$

then the q-difference equation can also be written in the form

$$q^{-n}(1-q^n)(1-abcdq^{n-1})P_n(z)$$
$$= A(z)P_n(qz) - \left[A(z)+A(z^{-1})\right] P_n(z) + A(z^{-1})P_n(q^{-1}z), \qquad (14.1.7)$$

where

$$A(z) = \frac{(1-az)(1-bz)(1-cz)(1-dz)}{(1-z^2)(1-qz^2)}.$$

Forward Shift Operator

$$\delta_q p_n(x;a,b,c,d|q) = -q^{-\frac{1}{2}n}(1-q^n)(1-abcdq^{n-1})(\mathrm{e}^{i\theta}-\mathrm{e}^{-i\theta})$$
$$\times\, p_{n-1}(x;aq^{\frac{1}{2}},bq^{\frac{1}{2}},cq^{\frac{1}{2}},dq^{\frac{1}{2}}|q), \qquad (14.1.8)$$

where $x=\cos\theta$, or equivalently

$$D_q p_n(x;a,b,c,d|q) = 2q^{-\frac{1}{2}(n-1)}\frac{(1-q^n)(1-abcdq^{n-1})}{1-q}$$
$$\times\, p_{n-1}(x;aq^{\frac{1}{2}},bq^{\frac{1}{2}},cq^{\frac{1}{2}},dq^{\frac{1}{2}}|q). \qquad (14.1.9)$$

Backward Shift Operator

$$\delta_q\left[\tilde{w}(x;a,b,c,d|q)p_n(x;a,b,c,d|q)\right]$$
$$= q^{-\frac{1}{2}(n+1)}(\mathrm{e}^{i\theta}-\mathrm{e}^{-i\theta})\tilde{w}(x;aq^{-\frac{1}{2}},bq^{-\frac{1}{2}},cq^{-\frac{1}{2}},dq^{-\frac{1}{2}}|q)$$
$$\times\, p_{n+1}(x;aq^{-\frac{1}{2}},bq^{-\frac{1}{2}},cq^{-\frac{1}{2}},dq^{-\frac{1}{2}}|q), \quad x=\cos\theta \qquad (14.1.10)$$

or equivalently

$$D_q\left[\tilde{w}(x;a,b,c,d|q)p_n(x;a,b,c,d|q)\right]$$
$$= -\frac{2q^{-\frac{1}{2}n}}{1-q}\tilde{w}(x;aq^{-\frac{1}{2}},bq^{-\frac{1}{2}},cq^{-\frac{1}{2}},dq^{-\frac{1}{2}}|q)$$
$$\times\, p_{n+1}(x;aq^{-\frac{1}{2}},bq^{-\frac{1}{2}},cq^{-\frac{1}{2}},dq^{-\frac{1}{2}}|q). \qquad (14.1.11)$$

Rodrigues-Type Formula

$$\tilde{w}(x;a,b,c,d|q)p_n(x;a,b,c,d|q)$$
$$= \left(\frac{q-1}{2}\right)^n q^{\frac{1}{4}n(n-1)} (\mathcal{D}_q)^n \left[\tilde{w}(x;aq^{\frac{1}{2}n},bq^{\frac{1}{2}n},cq^{\frac{1}{2}n},dq^{\frac{1}{2}n}|q)\right]. \quad (14.1.12)$$

Generating Functions

$${}_2\phi_1\left(\begin{array}{c} ae^{i\theta}, be^{i\theta} \\ ab \end{array}; q, e^{-i\theta}t\right) {}_2\phi_1\left(\begin{array}{c} ce^{-i\theta}, de^{-i\theta} \\ cd \end{array}; q, e^{i\theta}t\right)$$
$$= \sum_{n=0}^{\infty} \frac{p_n(x;a,b,c,d|q)}{(ab,cd,q;q)_n} t^n, \quad x = \cos\theta. \quad (14.1.13)$$

$${}_2\phi_1\left(\begin{array}{c} ae^{i\theta}, ce^{i\theta} \\ ac \end{array}; q, e^{-i\theta}t\right) {}_2\phi_1\left(\begin{array}{c} be^{-i\theta}, de^{-i\theta} \\ bd \end{array}; q, e^{i\theta}t\right)$$
$$= \sum_{n=0}^{\infty} \frac{p_n(x;a,b,c,d|q)}{(ac,bd,q;q)_n} t^n, \quad x = \cos\theta. \quad (14.1.14)$$

$${}_2\phi_1\left(\begin{array}{c} ae^{i\theta}, de^{i\theta} \\ ad \end{array}; q, e^{-i\theta}t\right) {}_2\phi_1\left(\begin{array}{c} be^{-i\theta}, ce^{-i\theta} \\ bc \end{array}; q, e^{i\theta}t\right)$$
$$= \sum_{n=0}^{\infty} \frac{p_n(x;a,b,c,d|q)}{(ad,bc,q;q)_n} t^n, \quad x = \cos\theta. \quad (14.1.15)$$

Limit Relations

Askey-Wilson → Continuous Dual q-Hahn

The continuous dual q-Hahn polynomials given by (14.3.1) simply follow from the Askey-Wilson polynomials given by (14.1.1) by setting $d = 0$ in (14.1.1):

$$p_n(x;a,b,c,0|q) = p_n(x;a,b,c|q). \quad (14.1.16)$$

Askey-Wilson → Continuous q-Hahn

The continuous q-Hahn polynomials given by (14.4.1) can be obtained from the Askey-Wilson polynomials given by (14.1.1) by the substitutions $\theta \to \theta + \phi$, $a \to ae^{i\phi}$, $b \to be^{i\phi}$, $c \to ce^{-i\phi}$ and $d \to de^{-i\phi}$:

$$p_n(\cos(\theta+\phi);ae^{i\phi},be^{i\phi},ce^{-i\phi},de^{-i\phi}|q) = p_n(\cos(\theta+\phi);a,b,c,d;q). \quad (14.1.17)$$

Askey-Wilson → Big q-Jacobi

The big q-Jacobi polynomials given by (14.5.1) can be obtained from the Askey-Wilson polynomials by setting $x \to \frac{1}{2}a^{-1}x$, $b = a^{-1}\alpha q$, $c = a^{-1}\gamma q$ and $d = a\beta\gamma^{-1}$ in

$$\tilde{p}_n(x;a,b,c,d|q) = \frac{a^n p_n(x;a,b,c,d|q)}{(ab,ac,ad;q)_n}$$

given by (14.1.1) and then taking the limit $a \to 0$:

$$\lim_{a\to 0} \tilde{p}_n(\tfrac{1}{2}a^{-1}x;a,a^{-1}\alpha q,a^{-1}\gamma q,a\beta\gamma^{-1}|q) = P_n(x;\alpha,\beta,\gamma;q). \quad (14.1.18)$$

Askey-Wilson → Continuous q-Jacobi

If we take $a = q^{\frac{1}{2}\alpha+\frac{1}{4}}$, $b = q^{\frac{1}{2}\alpha+\frac{3}{4}}$, $c = -q^{\frac{1}{2}\beta+\frac{1}{4}}$ and $d = -q^{\frac{1}{2}\beta+\frac{3}{4}}$ in the definition (14.1.1) of the Askey-Wilson polynomials and change the normalization we find the continuous q-Jacobi polynomials given by (14.10.1):

$$\frac{q^{(\frac{1}{2}\alpha+\frac{1}{4})n}p_n(x;q^{\frac{1}{2}\alpha+\frac{1}{4}},q^{\frac{1}{2}\alpha+\frac{3}{4}},-q^{\frac{1}{2}\beta+\frac{1}{4}},-q^{\frac{1}{2}\beta+\frac{3}{4}}|q)}{(q,-q^{\frac{1}{2}(\alpha+\beta+1)},-q^{\frac{1}{2}(\alpha+\beta+2)};q)_n} = P_n^{(\alpha,\beta)}(x|q). \quad (14.1.19)$$

Askey-Wilson → Continuous q-ultraspherical / Rogers

If we set $a = \beta^{\frac{1}{2}}$, $b = \beta^{\frac{1}{2}}q^{\frac{1}{2}}$, $c = -\beta^{\frac{1}{2}}$ and $d = -\beta^{\frac{1}{2}}q^{\frac{1}{2}}$ in the definition (14.1.1) of the Askey-Wilson polynomials and change the normalization we obtain the continuous q-ultraspherical (or Rogers) polynomials given by (14.10.17). In fact we have:

$$\frac{(\beta^2;q)_n p_n(x;\beta^{\frac{1}{2}},\beta^{\frac{1}{2}}q^{\frac{1}{2}},-\beta^{\frac{1}{2}},-\beta^{\frac{1}{2}}q^{\frac{1}{2}}|q)}{(\beta q^{\frac{1}{2}},-\beta,-\beta q^{\frac{1}{2}},q;q)_n} = C_n(x;\beta|q). \quad (14.1.20)$$

Askey-Wilson → Wilson

To find the Wilson polynomials given by (9.1.1) from the Askey-Wilson polynomials we set $a \to q^a$, $b \to q^b$, $c \to q^c$, $d \to q^d$ and $e^{i\theta} = q^{ix}$ (or $\theta = \ln q^x$) in the definition (14.1.1) and take the limit $q \to 1$:

$$\lim_{q\to 1} \frac{p_n\left(\frac{1}{2}\left(q^{ix}+q^{-ix}\right);q^a,q^b,q^c,q^d|q\right)}{(1-q)^{3n}} = W_n(x^2;a,b,c,d). \tag{14.1.21}$$

Remarks

The q-Racah polynomials given by (14.2.1) and the Askey-Wilson polynomials given by (14.1.1) are related in the following way. If we substitute $a^2 = \gamma\delta q$, $b^2 = \alpha^2\gamma^{-1}\delta^{-1}q$, $c^2 = \beta^2\gamma^{-1}\delta q$, $d^2 = \gamma\delta^{-1}q$ and $e^{2i\theta} = \gamma\delta q^{2x+1}$ in the definition (14.1.1) of the Askey-Wilson polynomials we find:

$$\begin{aligned} &R_n(\mu(x);\alpha,\beta,\gamma,\delta|q) \\ &= \frac{(\gamma\delta q)^{\frac{1}{2}n} p_n(\nu(x);\gamma^{\frac{1}{2}}\delta^{\frac{1}{2}}q^{\frac{1}{2}},\alpha\gamma^{-\frac{1}{2}}\delta^{-\frac{1}{2}}q^{\frac{1}{2}},\beta\gamma^{-\frac{1}{2}}\delta^{\frac{1}{2}}q^{\frac{1}{2}},\gamma^{\frac{1}{2}}\delta^{-\frac{1}{2}}q^{\frac{1}{2}}|q)}{(\alpha q,\beta\delta q,\gamma q;q)_n}, \end{aligned}$$

where

$$\nu(x) = \tfrac{1}{2}\gamma^{\frac{1}{2}}\delta^{\frac{1}{2}}q^{x+\frac{1}{2}} + \tfrac{1}{2}\gamma^{-\frac{1}{2}}\delta^{-\frac{1}{2}}q^{-x-\frac{1}{2}}.$$

If we replace q by q^{-1} we find

$$\tilde{p}_n(x;a,b,c,d|q^{-1}) = \tilde{p}_n(x;a^{-1},b^{-1},c^{-1},d^{-1}|q).$$

References

[3], [16], [34], [51], [72], [73], [77], [80], [82], [83], [92], [115], [116], [236], [238], [256], [257], [271], [275], [278], [282], [286], [287], [290], [294], [298], [306], [323], [326], [347], [348], [351], [371], [408], [409], [414], [416], [418], [422], [442], [443], [445], [446], [448], [449], [451], [455], [474], [495], [507], [513].

14.2 q-Racah

Basic Hypergeometric Representation

$$R_n(\mu(x);\alpha,\beta,\gamma,\delta|q) = {}_4\phi_3\left(\begin{array}{c} q^{-n},\alpha\beta q^{n+1},q^{-x},\gamma\delta q^{x+1} \\ \alpha q,\beta\delta q,\gamma q \end{array};q,q\right), \quad n=0,1,2,\ldots,N, \tag{14.2.1}$$

where

$$\mu(x) := q^{-x}+\gamma\delta q^{x+1}$$

and

$$\alpha q = q^{-N} \quad \text{or} \quad \beta\delta q = q^{-N} \quad \text{or} \quad \gamma q = q^{-N},$$

with N a nonnegative integer. Since

$$(q^{-x},\gamma\delta q^{x+1};q)_k = \prod_{j=0}^{k-1}\left(1-\mu(x)q^j+\gamma\delta q^{2j+1}\right),$$

it is clear that $R_n(\mu(x);\alpha,\beta,\gamma,\delta|q)$ is a polynomial of degree n in $\mu(x)$.

Orthogonality Relation

$$\sum_{x=0}^{N}\frac{(\alpha q,\beta\delta q,\gamma q,\gamma\delta q;q)_x}{(q,\alpha^{-1}\gamma\delta q,\beta^{-1}\gamma q,\delta q;q)_x} \times \frac{(1-\gamma\delta q^{2x+1})}{(\alpha\beta q)^x(1-\gamma\delta q)}R_m(\mu(x))R_n(\mu(x)) = h_n\,\delta_{mn}, \tag{14.2.2}$$

where

$$R_n(\mu(x)) := R_n(\mu(x);\alpha,\beta,\gamma,\delta|q)$$

and

$$h_n = \frac{(\alpha^{-1}\beta^{-1}\gamma,\alpha^{-1}\delta,\beta^{-1},\gamma\delta q^2;q)_\infty}{(\alpha^{-1}\beta^{-1}q^{-1},\alpha^{-1}\gamma\delta q,\beta^{-1}\gamma q,\delta q;q)_\infty} \times \frac{(1-\alpha\beta q)(\gamma\delta q)^n}{(1-\alpha\beta q^{2n+1})}\frac{(q,\alpha\beta\gamma^{-1}q,\alpha\delta^{-1}q,\beta q;q)_n}{(\alpha q,\alpha\beta q,\beta\delta q,\gamma q;q)_n}.$$

This implies

$$h_n = \begin{cases} \dfrac{(\beta^{-1},\gamma\delta q^2;q)_N}{(\beta^{-1}\gamma q,\delta q;q)_N}\dfrac{(1-\beta q^{-N})(\gamma\delta q)^n}{(1-\beta q^{2n-N})} \\ \quad\times\dfrac{(q,\beta q,\beta\gamma^{-1}q^{-N},\delta^{-1}q^{-N};q)_n}{(\beta q^{-N},\beta\delta q,\gamma q,q^{-N};q)_n} & \text{if}\quad \alpha q = q^{-N} \\ \dfrac{(\alpha\beta q^2,\beta\gamma^{-1};q)_N}{(\alpha\beta\gamma^{-1}q,\beta q;q)_N}\dfrac{(1-\alpha\beta q)(\beta^{-1}\gamma q^{-N})^n}{(1-\alpha\beta q^{2n+1})} \\ \quad\times\dfrac{(q,\alpha\beta q^{N+2},\alpha\beta\gamma^{-1}q,\beta q;q)_n}{(\alpha q,\alpha\beta q,\gamma q,q^{-N};q)_n} & \text{if}\quad \beta\delta q = q^{-N} \\ \dfrac{(\alpha\beta q^2,\delta^{-1};q)_N}{(\alpha\delta^{-1}q,\beta q;q)_N}\dfrac{(1-\alpha\beta q)(\delta q^{-N})^n}{(1-\alpha\beta q^{2n+1})} \\ \quad\times\dfrac{(q,\alpha\beta q^{N+2},\alpha\delta^{-1}q,\beta q;q)_n}{(\alpha q,\alpha\beta q,\beta\delta q,q^{-N};q)_n} & \text{if}\quad \gamma q = q^{-N}. \end{cases}$$

Recurrence Relation

$$\begin{aligned} &-\left(1-q^{-x}\right)\left(1-\gamma\delta q^{x+1}\right)R_n(\mu(x)) \\ &= A_n R_{n+1}(\mu(x)) - (A_n + C_n)\,R_n(\mu(x)) + C_n R_{n-1}(\mu(x)), \end{aligned} \tag{14.2.3}$$

where

$$\begin{cases} A_n = \dfrac{(1-\alpha q^{n+1})(1-\alpha\beta q^{n+1})(1-\beta\delta q^{n+1})(1-\gamma q^{n+1})}{(1-\alpha\beta q^{2n+1})(1-\alpha\beta q^{2n+2})} \\ C_n = \dfrac{q(1-q^n)(1-\beta q^n)(\gamma-\alpha\beta q^n)(\delta-\alpha q^n)}{(1-\alpha\beta q^{2n})(1-\alpha\beta q^{2n+1})}. \end{cases}$$

Normalized Recurrence Relation

$$xp_n(x) = p_{n+1}(x) + [1+\gamma\delta q - (A_n + C_n)]\,p_n(x) + A_{n-1}C_n p_{n-1}(x), \tag{14.2.4}$$

where

$$R_n(\mu(x);\alpha,\beta,\gamma,\delta|q) = \frac{(\alpha\beta q^{n+1};q)_n}{(\alpha q,\beta\delta q,\gamma q;q)_n}\,p_n(\mu(x)).$$

q-Difference Equation

$$\begin{aligned}\Delta\left[w(x-1)B(x-1)\Delta y(x-1)\right]\\ -q^{-n}(1-q^n)(1-\alpha\beta q^{n+1})w(x)y(x)=0,\end{aligned}\tag{14.2.5}$$

where

$$y(x)=R_n(\mu(x);\alpha,\beta,\gamma,\delta|q)$$

and

$$w(x):=w(x;\alpha,\beta,\gamma,\delta|q)=\frac{(\alpha q,\beta\delta q,\gamma q,\gamma\delta q;q)_x}{(q,\alpha^{-1}\gamma\delta q,\beta^{-1}\gamma q,\delta q;q)_x}\frac{(1-\gamma\delta q^{2x+1})}{(\alpha\beta q)^x(1-\gamma\delta q)}$$

and $B(x)$ as below. This q-difference equation can also be written in the form

$$\begin{aligned}&q^{-n}(1-q^n)(1-\alpha\beta q^{n+1})y(x)\\ &=B(x)y(x+1)-\left[B(x)+D(x)\right]y(x)+D(x)y(x-1),\end{aligned}\tag{14.2.6}$$

where

$$y(x)=R_n(\mu(x);\alpha,\beta,\gamma,\delta|q)$$

and

$$\begin{cases}B(x)=\dfrac{(1-\alpha q^{x+1})(1-\beta\delta q^{x+1})(1-\gamma q^{x+1})(1-\gamma\delta q^{x+1})}{(1-\gamma\delta q^{2x+1})(1-\gamma\delta q^{2x+2})}\\[2ex] D(x)=\dfrac{q(1-q^x)(1-\delta q^x)(\beta-\gamma q^x)(\alpha-\gamma\delta q^x)}{(1-\gamma\delta q^{2x})(1-\gamma\delta q^{2x+1})}.\end{cases}$$

Forward Shift Operator

$$\begin{aligned}&R_n(\mu(x+1);\alpha,\beta,\gamma,\delta|q)-R_n(\mu(x);\alpha,\beta,\gamma,\delta|q)\\ &=\frac{q^{-n-x}(1-q^n)(1-\alpha\beta q^{n+1})(1-\gamma\delta q^{2x+2})}{(1-\alpha q)(1-\beta\delta q)(1-\gamma q)}\\ &\quad\times R_{n-1}(\mu(x);\alpha q,\beta q,\gamma q,\delta|q)\end{aligned}\tag{14.2.7}$$

or equivalently

$$\begin{aligned}&\frac{\Delta R_n(\mu(x);\alpha,\beta,\gamma,\delta|q)}{\Delta\mu(x)}\\ &=\frac{q^{-n+1}(1-q^n)(1-\alpha\beta q^{n+1})}{(1-q)(1-\alpha q)(1-\beta\delta q)(1-\gamma q)}\\ &\quad\times R_{n-1}(\mu(x);\alpha q,\beta q,\gamma q,\delta|q).\end{aligned}\tag{14.2.8}$$

Backward Shift Operator

$$\begin{aligned}&(1-\alpha q^x)(1-\beta\delta q^x)(1-\gamma q^x)(1-\gamma\delta q^x)R_n(\mu(x);\alpha,\beta,\gamma,\delta|q)\\&\qquad-(1-q^x)(1-\delta q^x)(\alpha-\gamma\delta q^x)(\beta-\gamma q^x)R_n(\mu(x-1);\alpha,\beta,\gamma,\delta|q)\\&=q^x(1-\alpha)(1-\beta\delta)(1-\gamma)(1-\gamma\delta q^{2x})\\&\qquad\times R_{n+1}(\mu(x);\alpha q^{-1},\beta q^{-1},\gamma q^{-1},\delta|q)\end{aligned}\tag{14.2.9}$$

or equivalently

$$\begin{aligned}&\frac{\nabla\left[\tilde{w}(x;\alpha,\beta,\gamma,\delta|q)R_n(\mu(x);\alpha,\beta,\gamma,\delta|q)\right]}{\nabla\mu(x)}\\&=\frac{1}{(1-q)(1-\gamma\delta)}\tilde{w}(x;\alpha q^{-1},\beta q^{-1},\gamma q^{-1},\delta|q)\\&\qquad\times R_{n+1}(\mu(x);\alpha q^{-1},\beta q^{-1},\gamma q^{-1},\delta|q),\end{aligned}\tag{14.2.10}$$

where

$$\tilde{w}(x;\alpha,\beta,\gamma,\delta|q)=\frac{(\alpha q,\beta\delta q,\gamma q,\gamma\delta q;q)_x}{(q,\alpha^{-1}\gamma\delta q,\beta^{-1}\gamma q,\delta q;q)_x(\alpha\beta)^x}.$$

Rodrigues-Type Formula

$$\begin{aligned}&\tilde{w}(x;\alpha,\beta,\gamma,\delta|q)R_n(\mu(x);\alpha,\beta,\gamma,\delta|q)\\&=(1-q)^n(\gamma\delta q;q)_n\left(\nabla_\mu\right)^n\left[\tilde{w}(x;\alpha q^n,\beta q^n,\gamma q^n,\delta|q)\right],\end{aligned}\tag{14.2.11}$$

where

$$\nabla_\mu:=\frac{\nabla}{\nabla\mu(x)}.$$

Generating Functions

For $x=0,1,2,\ldots,N$ we have

$$\begin{aligned}&{}_2\phi_1\left(\begin{matrix}q^{-x},\alpha\gamma^{-1}\delta^{-1}q^{-x}\\ \alpha q\end{matrix};q,\gamma\delta q^{x+1}t\right){}_2\phi_1\left(\begin{matrix}\beta\delta q^{x+1},\gamma q^{x+1}\\ \beta q\end{matrix};q,q^{-x}t\right)\\&=\sum_{n=0}^{N}\frac{(\beta\delta q,\gamma q;q)_n}{(\beta q,q;q)_n}R_n(\mu(x);\alpha,\beta,\gamma,\delta|q)t^n,\\&\qquad\text{if}\quad\beta\delta q=q^{-N}\quad\text{or}\quad\gamma q=q^{-N}.\end{aligned}\tag{14.2.12}$$

$$
\begin{aligned}
&{}_2\phi_1\left(\begin{array}{c} q^{-x},\beta\gamma^{-1}q^{-x} \\ \beta\delta q \end{array};q,\gamma\delta q^{x+1}t\right){}_2\phi_1\left(\begin{array}{c} \alpha q^{x+1},\gamma q^{x+1} \\ \alpha\delta^{-1}q \end{array};q,q^{-x}t\right)\\
&=\sum_{n=0}^{N}\frac{(\alpha q,\gamma q;q)_n}{(\alpha\delta^{-1}q,q;q)_n}R_n(\mu(x);\alpha,\beta,\gamma,\delta|q)t^n,\\
&\qquad\qquad\text{if}\quad \alpha q=q^{-N}\quad\text{or}\quad \gamma q=q^{-N}.
\end{aligned}
\tag{14.2.13}
$$

$$
\begin{aligned}
&{}_2\phi_1\left(\begin{array}{c} q^{-x},\delta^{-1}q^{-x} \\ \gamma q \end{array};q,\gamma\delta q^{x+1}t\right){}_2\phi_1\left(\begin{array}{c} \alpha q^{x+1},\beta\delta q^{x+1} \\ \alpha\beta\gamma^{-1}q \end{array};q,q^{-x}t\right)\\
&=\sum_{n=0}^{N}\frac{(\alpha q,\beta\delta q;q)_n}{(\alpha\beta\gamma^{-1}q,q;q)_n}R_n(\mu(x);\alpha,\beta,\gamma,\delta|q)t^n,\\
&\qquad\qquad\text{if}\quad \alpha q=q^{-N}\quad\text{or}\quad \beta\delta q=q^{-N}.
\end{aligned}
\tag{14.2.14}
$$

Limit Relations

q-Racah → Big q-Jacobi

The big q-Jacobi polynomials given by (14.5.1) can be obtained from the q-Racah polynomials by setting $\delta=0$ in the definition (14.2.1):

$$
R_n(\mu(x);a,b,c,0|q)=P_n(q^{-x};a,b,c;q). \tag{14.2.15}
$$

q-Racah → q-Hahn

The q-Hahn polynomials follow from the q-Racah polynomials by the substitution $\delta=0$ and $\gamma q=q^{-N}$ in the definition (14.2.1) of the q-Racah polynomials:

$$
R_n(\mu(x);\alpha,\beta,q^{-N-1},0|q)=Q_n(q^{-x};\alpha,\beta,N|q). \tag{14.2.16}
$$

Another way to obtain the q-Hahn polynomials from the q-Racah polynomials is by setting $\gamma=0$ and $\delta=\beta^{-1}q^{-N-1}$ in the definition (14.2.1):

$$
R_n(\mu(x);\alpha,\beta,0,\beta^{-1}q^{-N-1}|q)=Q_n(q^{-x};\alpha,\beta,N|q). \tag{14.2.17}
$$

And if we take $\alpha q=q^{-N}$, $\beta\to\beta\gamma q^{N+1}$ and $\delta=0$ in the definition (14.2.1) of the q-Racah polynomials we find the q-Hahn polynomials given by (14.6.1) in the following way:

$$
R_n(\mu(x);q^{-N-1},\beta\gamma q^{N+1},\gamma,0|q)=Q_n(q^{-x};\gamma,\beta,N|q). \tag{14.2.18}
$$

Note that $\mu(x) = q^{-x}$ in each case.

q-Racah → Dual q-Hahn

To obtain the dual q-Hahn polynomials from the q-Racah polynomials we have to take $\beta = 0$ and $\alpha q = q^{-N}$ in (14.2.1):

$$R_n(\mu(x);q^{-N-1},0,\gamma,\delta|q) = R_n(\mu(x);\gamma,\delta,N|q), \tag{14.2.19}$$

with

$$\mu(x) = q^{-x} + \gamma\delta q^{x+1}.$$

We may also take $\alpha = 0$ and $\beta = \delta^{-1}q^{-N-1}$ in (14.2.1) to obtain the dual q-Hahn polynomials from the q-Racah polynomials:

$$R_n(\mu(x);0,\delta^{-1}q^{-N-1},\gamma,\delta|q) = R_n(\mu(x);\gamma,\delta,N|q), \tag{14.2.20}$$

with

$$\mu(x) = q^{-x} + \gamma\delta q^{x+1}.$$

And if we take $\gamma q = q^{-N}$, $\delta \to \alpha\delta q^{N+1}$ and $\beta = 0$ in the definition (14.2.1) of the q-Racah polynomials we find the dual q-Hahn polynomials given by (14.7.1) in the following way:

$$R_n(\mu(x);\alpha,0,q^{-N-1},\alpha\delta q^{N+1}|q) = R_n(\mu(x);\alpha,\delta,N|q), \tag{14.2.21}$$

with

$$\mu(x) = q^{-x} + \alpha\delta q^{x+1}.$$

q-Racah → q-Krawtchouk

The q-Krawtchouk polynomials given by (14.15.1) can be obtained from the q-Racah polynomials by setting $\alpha q = q^{-N}$, $\beta = -pq^N$ and $\gamma = \delta = 0$ in the definition (14.2.1) of the q-Racah polynomials:

$$R_n(q^{-x};q^{-N-1},-pq^N,0,0|q) = K_n(q^{-x};p,N;q). \tag{14.2.22}$$

Note that $\mu(x) = q^{-x}$ in this case.

q-Racah → Dual q-Krawtchouk

The dual q-Krawtchouk polynomials given by (14.17.1) easily follow from the q-Racah polynomials given by (14.2.1) by using the substitutions $\alpha = \beta = 0$, $\gamma q = q^{-N}$ and $\delta = c$:

$$R_n(\mu(x);0,0,q^{-N-1},c|q) = K_n(\lambda(x);c,N|q). \tag{14.2.23}$$

Note that

$$\mu(x) = \lambda(x) = q^{-x} + cq^{x-N}.$$

q-Racah → Racah

If we set $\alpha \to q^\alpha$, $\beta \to q^\beta$, $\gamma \to q^\gamma$, $\delta \to q^\delta$ in the definition (14.2.1) of the q-Racah polynomials and let $q \to 1$ we easily obtain the Racah polynomials given by (9.2.1):

$$\lim_{q\to 1} R_n(\mu(x);q^\alpha,q^\beta,q^\gamma,q^\delta|q) = R_n(\lambda(x);\alpha,\beta,\gamma,\delta), \tag{14.2.24}$$

where

$$\begin{cases} \mu(x) = q^{-x} + q^{x+\gamma+\delta+1} \\ \lambda(x) = x(x+\gamma+\delta+1). \end{cases}$$

Remarks

The Askey-Wilson polynomials given by (14.1.1) and the q-Racah polynomials given by (14.2.1) are related in the following way. If we substitute $\alpha = abq^{-1}$, $\beta = cdq^{-1}$, $\gamma = adq^{-1}$, $\delta = ad^{-1}$ and $q^x = a^{-1}\mathrm{e}^{-i\theta}$ in the definition (14.2.1) of the q-Racah polynomials we find:

$$\mu(x) = 2a\cos\theta$$

and

$$R_n(2a\cos\theta;abq^{-1},cdq^{-1},adq^{-1},ad^{-1}|q) = \frac{a^n p_n(x;a,b,c,d|q)}{(ab,ac,ad;q)_n}.$$

If we replace q by q^{-1} we find

$$R_n(\mu(x);\alpha,\beta,\gamma,\delta|q^{-1}) = R_n(\tilde{\mu}(x);\alpha^{-1},\beta^{-1},\gamma^{-1},\delta^{-1}|q),$$

where

$$\tilde{\mu}(x) := q^{-x} + \gamma^{-1}\delta^{-1}q^{x+1}.$$

References

[16], [27], [34], [70], [72], [80], [140], [142], [204], [233], [235], [238], [271], [301], [345], [416], [426], [442].

14.3 Continuous Dual q-Hahn

Basic Hypergeometric Representation

$$\frac{a^n p_n(x;a,b,c|q)}{(ab,ac;q)_n} = {}_3\phi_2\left(\begin{matrix} q^{-n}, ae^{i\theta}, ae^{-i\theta} \\ ab, ac \end{matrix}; q, q\right), \quad x = \cos\theta. \tag{14.3.1}$$

Orthogonality Relation

If a,b,c are real, or one is real and the other two are complex conjugates, and $\max(|a|,|b|,|c|) < 1$, then we have the following orthogonality relation

$$\frac{1}{2\pi}\int_{-1}^{1}\frac{w(x)}{\sqrt{1-x^2}}p_m(x;a,b,c|q)p_n(x;a,b,c|q)\,dx = h_n\,\delta_{mn}, \tag{14.3.2}$$

where

$$w(x) := w(x;a,b,c|q) = \left|\frac{(e^{2i\theta};q)_\infty}{(ae^{i\theta},be^{i\theta},ce^{i\theta};q)_\infty}\right|^2 = \frac{h(x,1)h(x,-1)h(x,q^{\frac{1}{2}})h(x,-q^{\frac{1}{2}})}{h(x,a)h(x,b)h(x,c)},$$

with

$$h(x,\alpha) := \prod_{k=0}^{\infty}\left(1-2\alpha x q^k + \alpha^2 q^{2k}\right) = \left(\alpha e^{i\theta}, \alpha e^{-i\theta}; q\right)_\infty, \quad x = \cos\theta$$

and

$$h_n = \frac{1}{(q^{n+1}, abq^n, acq^n, bcq^n; q)_\infty}.$$

If $a > 1$ and b and c are real or complex conjugates, $\max(|b|,|c|) < 1$ and the pairwise products of a,b and c have absolute value less than 1, then we have another orthogonality relation given by:

$$\frac{1}{2\pi}\int_{-1}^{1}\frac{w(x)}{\sqrt{1-x^2}}p_m(x;a,b,c|q)p_n(x;a,b,c|q)\,dx + \sum_{\substack{k \\ 1<aq^k\leq a}} w_k p_m(x_k;a,b,c|q)p_n(x_k;a,b,c|q) = h_n\,\delta_{mn}, \tag{14.3.3}$$

where $w(x)$ and h_n are as before,

$$x_k = \frac{aq^k + \left(aq^k\right)^{-1}}{2}$$

and

$$w_k = \frac{(a^{-2};q)_\infty}{(q,ab,ac,a^{-1}b,a^{-1}c;q)_\infty} \frac{(1-a^2q^{2k})(a^2,ab,ac;q)_k}{(1-a^2)(q,ab^{-1}q,ac^{-1}q;q)_k}(-1)^k q^{-\binom{k}{2}} \left(\frac{1}{a^2bc}\right)^k.$$

Recurrence Relation

$$2x\tilde{p}_n(x) = A_n\tilde{p}_{n+1}(x) + \left[a+a^{-1}-(A_n+C_n)\right]\tilde{p}_n(x) + C_n\tilde{p}_{n-1}(x), \tag{14.3.4}$$

where

$$\tilde{p}_n(x) := \frac{a^n p_n(x;a,b,c|q)}{(ab,ac;q)_n}$$

and

$$\begin{cases} A_n = a^{-1}(1-abq^n)(1-acq^n) \\ C_n = a(1-q^n)(1-bcq^{n-1}). \end{cases}$$

Normalized Recurrence Relation

$$\begin{aligned} xp_n(x) = {} & p_{n+1}(x) + \frac{1}{2}\left[a+a^{-1}-(A_n+C_n)\right]p_n(x) \\ & + \frac{1}{4}(1-q^n)(1-abq^{n-1}) \\ & \times (1-acq^{n-1})(1-bcq^{n-1})p_{n-1}(x), \end{aligned} \tag{14.3.5}$$

where

$$p_n(x;a,b,c|q) = 2^n p_n(x).$$

q-Difference Equation

$$\begin{aligned} & (1-q)^2 D_q\left[\tilde{w}(x;aq^{\frac{1}{2}},bq^{\frac{1}{2}},cq^{\frac{1}{2}}|q)D_q y(x)\right] \\ & + 4q^{-n+1}(1-q^n)\tilde{w}(x;a,b,c|q)y(x) = 0, \end{aligned} \tag{14.3.6}$$

where

$$y(x) = p_n(x;a,b,c|q)$$

and

$$\tilde{w}(x;a,b,c|q) := \frac{w(x;a,b,c|q)}{\sqrt{1-x^2}}.$$

If we define

$$P_n(z) := \frac{(ab,ac;q)_n}{a^n}\,{}_3\phi_2\left(\begin{matrix} q^{-n}, az, az^{-1} \\ ab, ac \end{matrix}; q, q\right)$$

then the q-difference equation can also be written in the form

$$q^{-n}(1-q^n)P_n(z) = A(z)P_n(qz) - \left[A(z)+A(z^{-1})\right]P_n(z) + A(z^{-1})P_n(q^{-1}z), \quad (14.3.7)$$

where

$$A(z) = \frac{(1-az)(1-bz)(1-cz)}{(1-z^2)(1-qz^2)}.$$

Forward Shift Operator

$$\delta_q p_n(x;a,b,c|q) = -q^{-\frac{1}{2}n}(1-q^n)(e^{i\theta}-e^{-i\theta}) \times p_{n-1}(x;aq^{\frac{1}{2}},bq^{\frac{1}{2}},cq^{\frac{1}{2}}|q), \quad x=\cos\theta \quad (14.3.8)$$

or equivalently

$$D_q p_n(x;a,b,c|q) = 2q^{-\frac{1}{2}(n-1)}\frac{1-q^n}{1-q}p_{n-1}(x;aq^{\frac{1}{2}},bq^{\frac{1}{2}},cq^{\frac{1}{2}}|q). \quad (14.3.9)$$

Backward Shift Operator

$$\delta_q\left[\tilde{w}(x;a,b,c|q)p_n(x;a,b,c|q)\right] = q^{-\frac{1}{2}(n+1)}(e^{i\theta}-e^{-i\theta})\tilde{w}(x;aq^{-\frac{1}{2}},bq^{-\frac{1}{2}},cq^{-\frac{1}{2}}|q) \times p_{n+1}(x;aq^{-\frac{1}{2}},bq^{-\frac{1}{2}},cq^{-\frac{1}{2}}|q), \quad x=\cos\theta \quad (14.3.10)$$

or equivalently

$$D_q\left[\tilde{w}(x;a,b,c|q)p_n(x;a,b,c|q)\right] = -\frac{2q^{-\frac{1}{2}n}}{1-q}\tilde{w}(x;aq^{-\frac{1}{2}},bq^{-\frac{1}{2}},cq^{-\frac{1}{2}}|q)p_{n+1}(x;aq^{-\frac{1}{2}},bq^{-\frac{1}{2}},cq^{-\frac{1}{2}}|q). \quad (14.3.11)$$

Rodrigues-Type Formula

$$\tilde{w}(x;a,b,c|q)p_n(x;a,b,c|q) = \left(\frac{q-1}{2}\right)^n q^{\frac{1}{4}n(n-1)} (D_q)^n \left[\tilde{w}(x;aq^{\frac{1}{2}n},bq^{\frac{1}{2}n},cq^{\frac{1}{2}n}|q)\right]. \tag{14.3.12}$$

Generating Functions

$$\frac{(ct;q)_\infty}{(e^{i\theta}t;q)_\infty}\,{}_2\phi_1\left(\begin{matrix} ae^{i\theta},be^{i\theta} \\ ab \end{matrix};q,e^{-i\theta}t\right) = \sum_{n=0}^{\infty}\frac{p_n(x;a,b,c|q)}{(ab,q;q)_n}t^n, \quad x=\cos\theta. \tag{14.3.13}$$

$$\frac{(bt;q)_\infty}{(e^{i\theta}t;q)_\infty}\,{}_2\phi_1\left(\begin{matrix} ae^{i\theta},ce^{i\theta} \\ ac \end{matrix};q,e^{-i\theta}t\right) = \sum_{n=0}^{\infty}\frac{p_n(x;a,b,c|q)}{(ac,q;q)_n}t^n, \quad x=\cos\theta. \tag{14.3.14}$$

$$\frac{(at;q)_\infty}{(e^{i\theta}t;q)_\infty}\,{}_2\phi_1\left(\begin{matrix} be^{i\theta},ce^{i\theta} \\ bc \end{matrix};q,e^{-i\theta}t\right) = \sum_{n=0}^{\infty}\frac{p_n(x;a,b,c|q)}{(bc,q;q)_n}t^n, \quad x=\cos\theta. \tag{14.3.15}$$

$$(t;q)_\infty\cdot{}_3\phi_2\left(\begin{matrix} ae^{i\theta},ae^{-i\theta},0 \\ ab,ac \end{matrix};q,t\right) = \sum_{n=0}^{\infty}\frac{(-1)^n a^n q^{\binom{n}{2}}}{(ab,ac,q;q)_n}p_n(x;a,b,c|q)t^n, \quad x=\cos\theta. \tag{14.3.16}$$

Limit Relations

Askey-Wilson → Continuous Dual q-Hahn

The continuous dual q-Hahn polynomials given by (14.3.1) simply follow from the Askey-Wilson polynomials given by (14.1.1) by setting $d=0$ in (14.1.1):

$$p_n(x;a,b,c,0|q) = p_n(x;a,b,c|q).$$

Continuous Dual q-Hahn → Al-Salam-Chihara

The Al-Salam-Chihara polynomials given by (14.8.1) simply follow from the continuous dual q-Hahn polynomials by taking $c=0$ in the definition (14.3.1) of the continuous dual q-Hahn polynomials:

$$p_n(x;a,b,0|q) = Q_n(x;a,b|q). \tag{14.3.17}$$

Continuous Dual q-Hahn → Continuous Dual Hahn

To find the continuous dual Hahn polynomials given by (9.3.1) from the continuous dual q-Hahn polynomials we set $a \to q^a$, $b \to q^b$, $c \to q^c$ and $\mathrm{e}^{i\theta} = q^{ix}$ (or $\theta = \ln q^x$) in the definition (14.3.1) and take the limit $q \to 1$:

$$\lim_{q\to 1} \frac{p_n(\frac{1}{2}\left(q^{ix}+q^{-ix}\right);q^a,q^b,q^c|q)}{(1-q)^{2n}} = S_n(x^2;a,b,c). \tag{14.3.18}$$

References

[73], [260].

14.4 Continuous q-Hahn

Basic Hypergeometric Representation

$$\frac{(a\mathrm{e}^{i\phi})^n p_n(x;a,b,c,d;q)}{(ab\mathrm{e}^{2i\phi},ac,ad;q)_n} = {}_4\phi_3\left(\begin{array}{c} q^{-n},abcdq^{n-1},a\mathrm{e}^{i(\theta+2\phi)},a\mathrm{e}^{-i\theta} \\ ab\mathrm{e}^{2i\phi},ac,ad \end{array};q,q\right), \quad x=\cos(\theta+\phi). \tag{14.4.1}$$

Orthogonality Relation

If $c=a$ and $d=b$ then we have, if a and b are real and $\max(|a|,|b|)<1$, or if $b=\overline{a}$ and $|a|<1$:

$$\frac{1}{4\pi}\int_{-\pi}^{\pi} w(\cos(\theta+\phi))p_m(\cos(\theta+\phi);a,b,c,d;q)p_n(\cos(\theta+\phi);a,b,c,d;q)\,d\theta$$
$$=\frac{(abcdq^{n-1};q)_n(abcdq^{2n};q)_\infty}{(q^{n+1},abq^n\mathrm{e}^{2i\phi},acq^n,adq^n,bcq^n,bdq^n,cdq^n\mathrm{e}^{-2i\phi};q)_\infty}\,\delta_{mn}, \tag{14.4.2}$$

where

$$w(x):=w(x;a,b,c,d;q)=\left|\frac{(\mathrm{e}^{2i(\theta+\phi)};q)_\infty}{(a\mathrm{e}^{i(\theta+2\phi)},b\mathrm{e}^{i(\theta+2\phi)},c\mathrm{e}^{i\theta},d\mathrm{e}^{i\theta};q)_\infty}\right|^2$$
$$=\frac{h(x,1)h(x,-1)h(x,q^{\frac{1}{2}})h(x,-q^{\frac{1}{2}})}{h(x,a\mathrm{e}^{i\phi})h(x,b\mathrm{e}^{i\phi})h(x,c\mathrm{e}^{-i\phi})h(x,d\mathrm{e}^{-i\phi})},$$

with

$$h(x,\alpha):=\prod_{k=0}^{\infty}\left(1-2\alpha xq^k+\alpha^2q^{2k}\right)=\left(\alpha\mathrm{e}^{i(\theta+\phi)},\alpha\mathrm{e}^{-i(\theta+\phi)};q\right)_\infty,\quad x=\cos(\theta+\phi).$$

Recurrence Relation

$$2x\tilde{p}_n(x)=A_n\tilde{p}_{n+1}(x)+\left[a\mathrm{e}^{i\phi}+a^{-1}\mathrm{e}^{-i\phi}-(A_n+C_n)\right]\tilde{p}_n(x)+C_n\tilde{p}_{n-1}(x), \tag{14.4.3}$$

where

$$\tilde{p}_n(x):=\tilde{p}_n(x;a,b,c,d;q)=\frac{(a\mathrm{e}^{i\phi})^n p_n(x;a,b,c,d;q)}{(ab\mathrm{e}^{2i\phi},ac,ad;q)_n}$$

and

$$\begin{cases} A_n=\dfrac{(1-ab\mathrm{e}^{2i\phi}q^n)(1-acq^n)(1-adq^n)(1-abcdq^{n-1})}{a\mathrm{e}^{i\phi}(1-abcdq^{2n-1})(1-abcdq^{2n})} \\ C_n=\dfrac{a\mathrm{e}^{i\phi}(1-q^n)(1-bcq^{n-1})(1-bdq^{n-1})(1-cd\mathrm{e}^{-2i\phi}q^{n-1})}{(1-abcdq^{2n-2})(1-abcdq^{2n-1})}. \end{cases}$$

Normalized Recurrence Relation

$$xp_n(x) = p_{n+1}(x) + \frac{1}{2}\left[ae^{i\phi} + a^{-1}e^{-i\phi} - (A_n + C_n)\right] p_n(x) + \frac{1}{4}A_{n-1}C_n p_{n-1}(x), \quad (14.4.4)$$

where

$$p_n(x;a,b,c,d;q) = 2^n (abcdq^{n-1};q)_n p_n(x).$$

q-Difference Equation

$$(1-q)^2 D_q\left[\tilde{w}(x;aq^{\frac{1}{2}},bq^{\frac{1}{2}},cq^{\frac{1}{2}},dq^{\frac{1}{2}};q) D_q y(x)\right] + \lambda_n \tilde{w}(x;a,b,c,d;q) y(x) = 0, \quad y(x) = p_n(x;a,b,c,d;q), \quad (14.4.5)$$

where

$$\tilde{w}(x;a,b,c,d;q) := \frac{w(x;a,b,c,d;q)}{\sqrt{1-x^2}}$$

and

$$\lambda_n = 4q^{-n+1}(1-q^n)(1-abcdq^{n-1}).$$

Forward Shift Operator

$$\delta_q p_n(x;a,b,c,d;q) = -q^{-\frac{1}{2}n}(1-q^n)(1-abcdq^{n-1})(e^{i(\theta+\phi)} - e^{-i(\theta+\phi)}) \times p_{n-1}(x;aq^{\frac{1}{2}},bq^{\frac{1}{2}},cq^{\frac{1}{2}},dq^{\frac{1}{2}};q), \quad x = \cos(\theta+\phi) \quad (14.4.6)$$

or equivalently

$$D_q p_n(x;a,b,c,d;q) = 2q^{-\frac{1}{2}(n-1)} \frac{(1-q^n)(1-abcdq^{n-1})}{1-q} \times p_{n-1}(x;aq^{\frac{1}{2}},bq^{\frac{1}{2}},cq^{\frac{1}{2}},dq^{\frac{1}{2}};q). \quad (14.4.7)$$

Backward Shift Operator

$$\begin{aligned}
&\delta_q\left[\tilde{w}(x;a,b,c,d;q)p_n(x;a,b,c,d;q)\right]\\
&=q^{-\frac{1}{2}(n+1)}(\mathrm{e}^{i(\theta+\phi)}-\mathrm{e}^{-i(\theta+\phi)})\tilde{w}(x;aq^{-\frac{1}{2}},bq^{-\frac{1}{2}},cq^{-\frac{1}{2}},dq^{-\frac{1}{2}};q)\\
&\qquad\times p_{n+1}(x;aq^{-\frac{1}{2}},bq^{-\frac{1}{2}},cq^{-\frac{1}{2}},dq^{-\frac{1}{2}};q),\quad x=\cos(\theta+\phi)
\end{aligned}\tag{14.4.8}$$

or equivalently

$$\begin{aligned}
&D_q\left[\tilde{w}(x;a,b,c,d;q)p_n(x;a,b,c,d;q)\right]\\
&=-\frac{2q^{-\frac{1}{2}n}}{1-q}\tilde{w}(x;aq^{-\frac{1}{2}},bq^{-\frac{1}{2}},cq^{-\frac{1}{2}},dq^{-\frac{1}{2}};q)\\
&\qquad\times p_{n+1}(x;aq^{-\frac{1}{2}},bq^{-\frac{1}{2}},cq^{-\frac{1}{2}},dq^{-\frac{1}{2}};q).
\end{aligned}\tag{14.4.9}$$

Rodrigues-Type Formula

$$\begin{aligned}
&\tilde{w}(x;a,b,c,d;q)p_n(x;a,b,c,d;q)\\
&=\left(\frac{q-1}{2}\right)^n q^{\frac{1}{4}n(n-1)}\left(D_q\right)^n\left[\tilde{w}(x;aq^{\frac{1}{2}n},bq^{\frac{1}{2}n},cq^{\frac{1}{2}n},dq^{\frac{1}{2}n};q)\right].
\end{aligned}\tag{14.4.10}$$

Generating Functions

$$\begin{aligned}
&{}_2\phi_1\left(\begin{matrix}a\mathrm{e}^{i(\theta+2\phi)},b\mathrm{e}^{i(\theta+2\phi)}\\ ab\mathrm{e}^{2i\phi}\end{matrix};q,\mathrm{e}^{-i(\theta+\phi)}t\right)\\
&\qquad\times{}_2\phi_1\left(\begin{matrix}c\mathrm{e}^{-i(\theta+2\phi)},d\mathrm{e}^{-i(\theta+2\phi)}\\ cd\mathrm{e}^{-2i\phi}\end{matrix};q,\mathrm{e}^{i(\theta+\phi)}t\right)\\
&=\sum_{n=0}^{\infty}\frac{p_n(x;a,b,c,d;q)t^n}{(ab\mathrm{e}^{2i\phi},cd\mathrm{e}^{-2i\phi},q;q)_n},\quad x=\cos(\theta+\phi).
\end{aligned}\tag{14.4.11}$$

$$\begin{aligned}
&{}_2\phi_1\left(\begin{matrix}a\mathrm{e}^{i(\theta+2\phi)},c\mathrm{e}^{i\theta}\\ ac\end{matrix};q,\mathrm{e}^{-i(\theta+\phi)}t\right){}_2\phi_1\left(\begin{matrix}b\mathrm{e}^{-i\theta},d\mathrm{e}^{-i(\theta+2\phi)}\\ bd\end{matrix};q,\mathrm{e}^{i(\theta+\phi)}t\right)\\
&=\sum_{n=0}^{\infty}\frac{p_n(x;a,b,c,d;q)}{(ac,bd,q;q)_n}t^n,\quad x=\cos(\theta+\phi).
\end{aligned}\tag{14.4.12}$$

$$
{}_2\phi_1\left(\begin{array}{c} a\mathrm{e}^{i(\theta+2\phi)}, d\mathrm{e}^{i\theta} \\ ad \end{array}; q, \mathrm{e}^{-i(\theta+\phi)}t\right) {}_2\phi_1\left(\begin{array}{c} b\mathrm{e}^{-i\theta}, c\mathrm{e}^{-i(\theta+2\phi)} \\ bc \end{array}; q, \mathrm{e}^{i(\theta+\phi)}t\right)
$$
$$
= \sum_{n=0}^{\infty} \frac{p_n(x;a,b,c,d;q)}{(ad,bc,q;q)_n} t^n, \quad x = \cos(\theta+\phi). \tag{14.4.13}
$$

Limit Relations

Askey-Wilson → Continuous q-Hahn

The continuous q-Hahn polynomials given by (14.4.1) can be obtained from the Askey-Wilson polynomials given by (14.1.1) by the substitutions $\theta \to \theta+\phi$, $a \to a\mathrm{e}^{i\phi}$, $b \to b\mathrm{e}^{i\phi}$, $c \to c\mathrm{e}^{-i\phi}$ and $d \to d\mathrm{e}^{-i\phi}$:

$$
p_n(\cos(\theta+\phi); a\mathrm{e}^{i\phi}, b\mathrm{e}^{i\phi}, c\mathrm{e}^{-i\phi}, d\mathrm{e}^{-i\phi}|q) = p_n(\cos(\theta+\phi); a,b,c,d;q).
$$

Continuous q-Hahn → q-Meixner-Pollaczek

The q-Meixner-Pollaczek polynomials given by (14.9.1) simply follow from the continuous q-Hahn polynomials if we set $d = a$ and $b = c = 0$ in the definition (14.4.1) of the continuous q-Hahn polynomials:

$$
\frac{p_n(\cos(\theta+\phi); a,0,0,a;q)}{(q;q)_n} = P_n(\cos(\theta+\phi); a|q). \tag{14.4.14}
$$

Continuous q-Hahn → Continuous Hahn

If we set $a \to q^a$, $b \to q^b$, $c \to q^c$, $d \to q^d$ and $\mathrm{e}^{-i\theta} = q^{ix}$ (or $\theta = \ln q^{-x}$) in the definition (14.4.1) of the continuous q-Hahn polynomials and take the limit $q \to 1$ we find the continuous Hahn polynomials given by (9.4.1) in the following way:

$$
\lim_{q\to 1} \frac{p_n(\cos(\ln q^{-x}+\phi); q^a, q^b, q^c, q^d; q)}{(1-q)^n (q;q)_n} = (-2\sin\phi)^n p_n(x;a,b,c,d). \tag{14.4.15}
$$

Remark

If we replace q by q^{-1} we find

$$\tilde{p}_n(x;a,b,c,d;q^{-1}) = \tilde{p}_n(x;a^{-1},b^{-1},c^{-1},d^{-1};q).$$

References

[34], [72], [238].

14.5 Big q-Jacobi

Basic Hypergeometric Representation

$$P_n(x;a,b,c;q) = {}_3\phi_2\left(\begin{matrix} q^{-n}, abq^{n+1}, x \\ aq, cq \end{matrix} ; q, q\right). \tag{14.5.1}$$

Orthogonality Relation

For $0 < aq < 1$, $0 \leq bq < 1$ and $c < 0$ we have

$$\begin{aligned}
&\int_{cq}^{aq} \frac{(a^{-1}x, c^{-1}x;q)_\infty}{(x, bc^{-1}x;q)_\infty} P_m(x;a,b,c;q) P_n(x;a,b,c;q)\, d_qx \\
&= aq(1-q)\frac{(q, abq^2, a^{-1}c, ac^{-1}q;q)_\infty}{(aq, bq, cq, abc^{-1}q;q)_\infty} \\
&\qquad \times \frac{(1-abq)}{(1-abq^{2n+1})} \frac{(q, bq, abc^{-1}q;q)_n}{(aq, abq, cq;q)_n} (-acq^2)^n q^{\binom{n}{2}}\, \delta_{mn}.
\end{aligned} \tag{14.5.2}$$

Recurrence Relation

$$\begin{aligned}
(x-1)P_n(x;a,b,c;q) &= A_n P_{n+1}(x;a,b,c;q) - (A_n + C_n) P_n(x;a,b,c;q) \\
&\quad + C_n P_{n-1}(x;a,b,c;q),
\end{aligned} \tag{14.5.3}$$

where

$$\begin{cases} A_n = \dfrac{(1-aq^{n+1})(1-abq^{n+1})(1-cq^{n+1})}{(1-abq^{2n+1})(1-abq^{2n+2})} \\[2ex] C_n = -acq^{n+1}\dfrac{(1-q^n)(1-abc^{-1}q^n)(1-bq^n)}{(1-abq^{2n})(1-abq^{2n+1})}. \end{cases}$$

Normalized Recurrence Relation

$$xp_n(x) = p_{n+1}(x) + [1-(A_n+C_n)]\,p_n(x) + A_{n-1}C_n p_{n-1}(x), \tag{14.5.4}$$

where

$$P_n(x;a,b,c;q) = \frac{(abq^{n+1};q)_n}{(aq,cq;q)_n}\,p_n(x).$$

q-Difference Equation

$$\begin{aligned} &q^{-n}(1-q^n)(1-abq^{n+1})x^2y(x) \\ &= B(x)y(qx) - [B(x)+D(x)]\,y(x) + D(x)y(q^{-1}x), \end{aligned} \tag{14.5.5}$$

where

$$y(x) = P_n(x;a,b,c;q)$$

and

$$\begin{cases} B(x) = aq(x-1)(bx-c) \\[1ex] D(x) = (x-aq)(x-cq). \end{cases}$$

Forward Shift Operator

$$\begin{aligned} &P_n(x;a,b,c;q) - P_n(qx;a,b,c;q) \\ &= \frac{q^{-n+1}(1-q^n)(1-abq^{n+1})}{(1-aq)(1-cq)}xP_{n-1}(qx;aq,bq,cq;q) \end{aligned} \tag{14.5.6}$$

or equivalently

$$\mathscr{D}_q P_n(x;a,b,c;q) = \frac{q^{-n+1}(1-q^n)(1-abq^{n+1})}{(1-q)(1-aq)(1-cq)}P_{n-1}(qx;aq,bq,cq;q). \tag{14.5.7}$$

Backward Shift Operator

$$
\begin{aligned}
&(x-a)(x-c)P_n(x;a,b,c;q)-a(x-1)(bx-c)P_n(qx;a,b,c;q)\\
&=(1-a)(1-c)xP_{n+1}(x;aq^{-1},bq^{-1},cq^{-1};q)
\end{aligned}
\tag{14.5.8}
$$

or equivalently

$$
\begin{aligned}
&\mathscr{D}_q\left[w(x;a,b,c;q)P_n(x;a,b,c;q)\right]\\
&=\frac{(1-a)(1-c)}{ac(1-q)}w(x;aq^{-1},bq^{-1},cq^{-1};q)\\
&\quad\times P_{n+1}(x;aq^{-1},bq^{-1},cq^{-1};q),
\end{aligned}
\tag{14.5.9}
$$

where

$$
w(x;a,b,c;q)=\frac{(a^{-1}x,c^{-1}x;q)_\infty}{(x,bc^{-1}x;q)_\infty}.
$$

Rodrigues-Type Formula

$$
\begin{aligned}
&w(x;a,b,c;q)P_n(x;a,b,c;q)\\
&=\frac{a^nc^nq^{n(n+1)}(1-q)^n}{(aq,cq;q)_n}\left(\mathscr{D}_q\right)^n\left[w(x;aq^n,bq^n,cq^n;q)\right].
\end{aligned}
\tag{14.5.10}
$$

Generating Functions

$$
\begin{aligned}
&{}_2\phi_1\left(\begin{matrix}aqx^{-1},0\\ aq\end{matrix};q,xt\right)\,{}_1\phi_1\left(\begin{matrix}bc^{-1}x\\ bq\end{matrix};q,cqt\right)\\
&=\sum_{n=0}^{\infty}\frac{(cq;q)_n}{(bq,q;q)_n}P_n(x;a,b,c;q)t^n.
\end{aligned}
\tag{14.5.11}
$$

$$
\begin{aligned}
&{}_2\phi_1\left(\begin{matrix}cqx^{-1},0\\ cq\end{matrix};q,xt\right)\,{}_1\phi_1\left(\begin{matrix}bc^{-1}x\\ abc^{-1}q\end{matrix};q,aqt\right)\\
&=\sum_{n=0}^{\infty}\frac{(aq;q)_n}{(abc^{-1}q,q;q)_n}P_n(x;a,b,c;q)t^n.
\end{aligned}
\tag{14.5.12}
$$

Limit Relations

Askey-Wilson → Big q-Jacobi

The big q-Jacobi polynomials given by (14.5.1) can be obtained from the Askey-Wilson polynomials by setting $x \rightarrow \frac{1}{2}a^{-1}x$, $b = a^{-1}\alpha q$, $c = a^{-1}\gamma q$ and $d = a\beta\gamma^{-1}$ in

$$\tilde{p}_n(x;a,b,c,d|q) = \frac{a^n p_n(x;a,b,c,d|q)}{(ab,ac,ad;q)_n}$$

given by (14.1.1) and then taking the limit $a \rightarrow 0$:

$$\lim_{a\rightarrow 0} \tilde{p}_n(\tfrac{1}{2}a^{-1}x;a,a^{-1}\alpha q,a^{-1}\gamma q,a\beta\gamma^{-1}|q) = P_n(x;\alpha,\beta,\gamma;q).$$

q-Racah → Big q-Jacobi

The big q-Jacobi polynomials given by (14.5.1) can be obtained from the q-Racah polynomials by setting $\delta = 0$ in the definition (14.2.1):

$$R_n(\mu(x);a,b,c,0|q) = P_n(q^{-x};a,b,c;q).$$

Big q-Jacobi → Big q-Laguerre

If we set $b = 0$ in the definition (14.5.1) of the big q-Jacobi polynomials we obtain the big q-Laguerre polynomials given by (14.11.1):

$$P_n(x;a,0,c;q) = P_n(x;a,c;q). \tag{14.5.13}$$

Big q-Jacobi → Little q-Jacobi

The little q-Jacobi polynomials given by (14.12.1) can be obtained from the big q-Jacobi polynomials by the substitution $x \rightarrow cqx$ in the definition (14.5.1) and then by the limit $c \rightarrow -\infty$:

$$\lim_{c\rightarrow -\infty} P_n(cqx;a,b,c;q) = p_n(x;a,b|q). \tag{14.5.14}$$

Big q-Jacobi → q-Meixner

If we set $b = -a^{-1}cd^{-1}$ (with $d > 0$) in the definition (14.5.1) of the big q-Jacobi polynomials and take the limit $c \rightarrow -\infty$ we obtain the q-Meixner polynomials given by (14.13.1):

$$\lim_{c\to-\infty} P_n(q^{-x};a,-a^{-1}cd^{-1},c;q) = M_n(q^{-x};a,d;q). \tag{14.5.15}$$

Big q-Jacobi → Jacobi

If we set $c = 0$, $a = q^\alpha$ and $b = q^\beta$ in the definition (14.5.1) of the big q-Jacobi polynomials and let $q \to 1$ we find the Jacobi polynomials given by (9.8.1):

$$\lim_{q\to 1} P_n(x;q^\alpha,q^\beta,0;q) = \frac{P_n^{(\alpha,\beta)}(2x-1)}{P_n^{(\alpha,\beta)}(1)}. \tag{14.5.16}$$

If we take $c = -q^\gamma$ for arbitrary real γ instead of $c = 0$ we find

$$\lim_{q\to 1} P_n(x;q^\alpha,q^\beta,-q^\gamma;q) = \frac{P_n^{(\alpha,\beta)}(x)}{P_n^{(\alpha,\beta)}(1)}. \tag{14.5.17}$$

Remarks

The big q-Jacobi polynomials with $c = 0$ and the little q-Jacobi polynomials given by (14.12.1) are related in the following way:

$$P_n(x;a,b,0;q) = \frac{(bq;q)_n}{(aq;q)_n}(-1)^n a^n q^{n+\binom{n}{2}} p_n(a^{-1}q^{-1}x;b,a|q).$$

Sometimes the big q-Jacobi polynomials are defined in terms of four parameters instead of three. In fact the polynomials given by the definition

$$P_n(x;a,b,c,d;q) = {}_3\phi_2\left(\begin{matrix} q^{-n}, abq^{n+1}, ac^{-1}qx \\ aq, -ac^{-1}dq \end{matrix}; q, q\right)$$

are orthogonal on the interval $[-d,c]$ with respect to the weight function

$$\frac{(c^{-1}qx, -d^{-1}qx;q)_\infty}{(ac^{-1}qx, -bd^{-1}qx;q)_\infty} d_q x.$$

These polynomials are not really different from those given by (14.5.1) since we have

$$P_n(x;a,b,c,d;q) = P_n(ac^{-1}qx;a,b,-ac^{-1}d;q)$$

and

$$P_n(x;a,b,c;q) = P_n(x;a,b,aq,-cq;q).$$

References

[12], [16], [34], [79], [80], [157], [211], [238], [256], [259], [261], [271], [295], [298], [305], [320], [345], [348], [351], [408], [416], [419], [420], [421], [474], [482].

Special Case

14.5.1 Big q-Legendre

Basic Hypergeometric Representation

The big q-Legendre polynomials are big q-Jacobi polynomials with $a = b = 1$:

$$P_n(x;c;q) = {}_3\phi_2\left(\begin{matrix} q^{-n}, q^{n+1}, x \\ q, cq \end{matrix}; q, q\right). \tag{14.5.18}$$

Orthogonality Relation

$$\int_{cq}^{q} P_m(x;c;q)P_n(x;c;q)\,d_qx$$
$$= q(1-c)\frac{(1-q)}{(1-q^{2n+1})}\frac{(c^{-1}q;q)_n}{(cq;q)_n}(-cq^2)^n q^{\binom{n}{2}}\,\delta_{mn}, \quad c<0. \tag{14.5.19}$$

Recurrence Relation

$$(x-1)P_n(x;c;q) = A_nP_{n+1}(x;c;q) - (A_n+C_n)\,P_n(x;c;q) + C_nP_{n-1}(x;c;q), \tag{14.5.20}$$

where

$$\begin{cases} A_n = \dfrac{(1-q^{n+1})(1-cq^{n+1})}{(1+q^{n+1})(1-q^{2n+1})} \\[2ex] C_n = -cq^{n+1}\dfrac{(1-q^n)(1-c^{-1}q^n)}{(1+q^n)(1-q^{2n+1})}. \end{cases}$$

Normalized Recurrence Relation

$$xp_n(x) = p_{n+1}(x) + [1 - (A_n + C_n)]\, p_n(x) + A_{n-1}C_n p_{n-1}(x), \qquad (14.5.21)$$

where

$$P_n(x;c;q) = \frac{(q^{n+1};q)_n}{(q,cq;q)_n} p_n(x).$$

q-Difference Equation

$$\begin{aligned} &q^{-n}(1-q^n)(1-q^{n+1})x^2y(x) \\ &= B(x)y(qx) - [B(x)+D(x)]\, y(x) + D(x)y(q^{-1}x), \end{aligned} \qquad (14.5.22)$$

where

$$y(x) = P_n(x;c;q)$$

and

$$\begin{cases} B(x) = q(x-1)(x-c) \\ D(x) = (x-q)(x-cq). \end{cases}$$

Rodrigues-Type Formula

$$\begin{aligned} P_n(x;c;q) &= \frac{c^n q^{n(n+1)}(1-q)^n}{(q,cq;q)_n} (\mathscr{D}_q)^n \left[(q^{-n}x, c^{-1}q^{-n}x;q)_n\right] \qquad (14.5.23) \\ &= \frac{(1-q)^n}{(q,cq;q)_n} (\mathscr{D}_q)^n \left[(qx^{-1}, cqx^{-1};q)_n x^{2n}\right]. \end{aligned}$$

Generating Functions

$${}_2\phi_1\left(\begin{matrix} qx^{-1}, 0 \\ q \end{matrix}; q, xt\right) {}_1\phi_1\left(\begin{matrix} c^{-1}x \\ q \end{matrix}; q, cqt\right) = \sum_{n=0}^{\infty} \frac{(cq;q)_n}{(q,q;q)_n} P_n(x;c;q)t^n. \qquad (14.5.24)$$

$${}_2\phi_1\left(\begin{matrix} cqx^{-1}, 0 \\ cq \end{matrix}; q, xt\right) {}_1\phi_1\left(\begin{matrix} c^{-1}x \\ c^{-1}q \end{matrix}; q, qt\right) = \sum_{n=0}^{\infty} \frac{P_n(x;c;q)}{(c^{-1}q;q)_n} t^n. \qquad (14.5.25)$$

Limit Relations

Big q-Legendre → Legendre / Spherical

If we set $c = 0$ in the definition (14.5.18) of the big q-Legendre polynomials and let $q \to 1$ we simply obtain the Legendre (or spherical) polynomials given by (9.8.62):

$$\lim_{q\to 1} P_n(x;0;q) = P_n(2x-1). \tag{14.5.26}$$

If we take $c = -q^{\gamma}$ for arbitrary real γ instead of $c = 0$ we find

$$\lim_{q\to 1} P_n(x;-q^{\gamma};q) = P_n(x). \tag{14.5.27}$$

References

[324], [345].

14.6 q-Hahn

Basic Hypergeometric Representation

$$Q_n(q^{-x};\alpha,\beta,N|q) = {}_3\phi_2\left(\begin{matrix} q^{-n},\alpha\beta q^{n+1},q^{-x} \\ \alpha q,q^{-N} \end{matrix} ;q,q\right), \quad n = 0,1,2,\ldots,N. \tag{14.6.1}$$

Orthogonality Relation

For $0 < \alpha q < 1$ and $0 < \beta q < 1$, or for $\alpha > q^{-N}$ and $\beta > q^{-N}$, we have

$$\begin{aligned}&\sum_{x=0}^{N} \frac{(\alpha q,q^{-N};q)_x}{(q,\beta^{-1}q^{-N};q)_x}(\alpha\beta q)^{-x}Q_m(q^{-x};\alpha,\beta,N|q)Q_n(q^{-x};\alpha,\beta,N|q)\\ &= \frac{(\alpha\beta q^2;q)_N}{(\beta q;q)_N(\alpha q)^N}\frac{(q,\alpha\beta q^{N+2},\beta q;q)_n}{(\alpha q,\alpha\beta q,q^{-N};q)_n}\frac{(1-\alpha\beta q)(-\alpha q)^n}{(1-\alpha\beta q^{2n+1})}q^{\binom{n}{2}-Nn}\,\delta_{mn}.\end{aligned} \tag{14.6.2}$$

Recurrence Relation

$$-\left(1-q^{-x}\right)Q_n(q^{-x}) = A_nQ_{n+1}(q^{-x}) - (A_n+C_n)\,Q_n(q^{-x}) \\ +C_nQ_{n-1}(q^{-x}), \tag{14.6.3}$$

where

$$Q_n(q^{-x}) := Q_n(q^{-x};\alpha,\beta,N|q)$$

and

$$\begin{cases} A_n = \dfrac{(1-q^{n-N})(1-\alpha q^{n+1})(1-\alpha\beta q^{n+1})}{(1-\alpha\beta q^{2n+1})(1-\alpha\beta q^{2n+2})} \\ \\ C_n = -\dfrac{\alpha q^{n-N}(1-q^n)(1-\alpha\beta q^{n+N+1})(1-\beta q^n)}{(1-\alpha\beta q^{2n})(1-\alpha\beta q^{2n+1})}. \end{cases}$$

Normalized Recurrence Relation

$$xp_n(x) = p_{n+1}(x) + [1-(A_n+C_n)]\,p_n(x) + A_{n-1}C_np_{n-1}(x), \tag{14.6.4}$$

where

$$Q_n(q^{-x};\alpha,\beta,N|q) = \frac{(\alpha\beta q^{n+1};q)_n}{(\alpha q,q^{-N};q)_n}p_n(q^{-x}).$$

q-Difference Equation

$$q^{-n}(1-q^n)(1-\alpha\beta q^{n+1})y(x) \\ = B(x)y(x+1) - [B(x)+D(x)]\,y(x) + D(x)y(x-1), \tag{14.6.5}$$

where

$$y(x) = Q_n(q^{-x};\alpha,\beta,N|q)$$

and

$$\begin{cases} B(x) = (1-q^{x-N})(1-\alpha q^{x+1}) \\ \\ D(x) = \alpha q(1-q^x)(\beta - q^{x-N-1}). \end{cases}$$

Forward Shift Operator

$$
\begin{aligned}
&Q_n(q^{-x-1};\alpha,\beta,N|q)-Q_n(q^{-x};\alpha,\beta,N|q)\\
&=\frac{q^{-n-x}(1-q^n)(1-\alpha\beta q^{n+1})}{(1-\alpha q)(1-q^{-N})}Q_{n-1}(q^{-x};\alpha q,\beta q,N-1|q)
\end{aligned}
\tag{14.6.6}
$$

or equivalently

$$
\begin{aligned}
\frac{\Delta Q_n(q^{-x};\alpha,\beta,N|q)}{\Delta q^{-x}}=\frac{q^{-n+1}(1-q^n)(1-\alpha\beta q^{n+1})}{(1-q)(1-\alpha q)(1-q^{-N})}&\\
\times Q_{n-1}(q^{-x};\alpha q,\beta q,N-1|q).&
\end{aligned}
\tag{14.6.7}
$$

Backward Shift Operator

$$
\begin{aligned}
&(1-\alpha q^x)(1-q^{x-N-1})Q_n(q^{-x};\alpha,\beta,N|q)\\
&\quad-\alpha(1-q^x)(\beta-q^{x-N-1})Q_n(q^{-x+1};\alpha,\beta,N|q)\\
&=q^x(1-\alpha)(1-q^{-N-1})Q_{n+1}(q^{-x};\alpha q^{-1},\beta q^{-1},N+1|q)
\end{aligned}
\tag{14.6.8}
$$

or equivalently

$$
\begin{aligned}
&\frac{\nabla\left[w(x;\alpha,\beta,N|q)Q_n(q^{-x};\alpha,\beta,N|q)\right]}{\nabla q^{-x}}\\
&=\frac{1}{1-q}w(x;\alpha q^{-1},\beta q^{-1},N+1|q)Q_{n+1}(q^{-x};\alpha q^{-1},\beta q^{-1},N+1|q),
\end{aligned}
\tag{14.6.9}
$$

where

$$
w(x;\alpha,\beta,N|q)=\frac{(\alpha q,q^{-N};q)_x}{(q,\beta^{-1}q^{-N};q)_x}(\alpha\beta)^{-x}.
$$

Rodrigues-Type Formula

$$
\begin{aligned}
&w(x;\alpha,\beta,N|q)Q_n(q^{-x};\alpha,\beta,N|q)\\
&=(1-q)^n\left(\nabla_q\right)^n\left[w(x;\alpha q^n,\beta q^n,N-n|q)\right],
\end{aligned}
\tag{14.6.10}
$$

where

$$
\nabla_q:=\frac{\nabla}{\nabla q^{-x}}.
$$

Generating Functions

For $x = 0, 1, 2, \ldots, N$ we have

$$
{}_1\phi_1\left(\begin{matrix} q^{-x} \\ \alpha q \end{matrix}; q, \alpha q t\right) {}_2\phi_1\left(\begin{matrix} q^{x-N}, 0 \\ \beta q \end{matrix}; q, q^{-x} t\right)
= \sum_{n=0}^{N} \frac{(q^{-N};q)_n}{(\beta q, q;q)_n} Q_n(q^{-x};\alpha,\beta,N|q) t^n. \tag{14.6.11}
$$

$$
{}_2\phi_1\left(\begin{matrix} q^{-x}, \beta q^{N+1-x} \\ 0 \end{matrix}; q, -\alpha q^{x-N+1} t\right) {}_2\phi_0\left(\begin{matrix} q^{x-N}, \alpha q^{x+1} \\ - \end{matrix}; q, -q^{-x} t\right)
= \sum_{n=0}^{N} \frac{(\alpha q, q^{-N};q)_n}{(q;q)_n} q^{-\binom{n}{2}} Q_n(q^{-x};\alpha,\beta,N|q) t^n. \tag{14.6.12}
$$

Limit Relations

q-Racah → q-Hahn

The q-Hahn polynomials follow from the q-Racah polynomials by the substitution $\delta = 0$ and $\gamma q = q^{-N}$ in the definition (14.2.1) of the q-Racah polynomials:

$$R_n(\mu(x);\alpha,\beta,q^{-N-1},0|q) = Q_n(q^{-x};\alpha,\beta,N|q).$$

Another way to obtain the q-Hahn polynomials from the q-Racah polynomials is by setting $\gamma = 0$ and $\delta = \beta^{-1} q^{-N-1}$ in the definition (14.2.1):

$$R_n(\mu(x);\alpha,\beta,0,\beta^{-1}q^{-N-1}|q) = Q_n(q^{-x};\alpha,\beta,N|q).$$

And if we take $\alpha q = q^{-N}$, $\beta \to \beta\gamma q^{N+1}$ and $\delta = 0$ in the definition (14.2.1) of the q-Racah polynomials we find the q-Hahn polynomials given by (14.6.1) in the following way:

$$R_n(\mu(x);q^{-N-1},\beta\gamma q^{N+1},\gamma,0|q) = Q_n(q^{-x};\gamma,\beta,N|q).$$

Note that $\mu(x) = q^{-x}$ in each case.

q-Hahn → Little q-Jacobi

If we set $x \to N - x$ in the definition (14.6.1) of the q-Hahn polynomials and take the limit $N \to \infty$ we find the little q-Jacobi polynomials:

$$\lim_{N\to\infty} Q_n(q^{x-N};\alpha,\beta,N|q) = p_n(q^x;\alpha,\beta|q), \tag{14.6.13}$$

where $p_n(q^x;\alpha,\beta|q)$ is given by (14.12.1).

q-Hahn $\to$ q-Meixner

The q-Meixner polynomials given by (14.13.1) can be obtained from the q-Hahn polynomials by setting $\alpha = b$ and $\beta = -b^{-1}c^{-1}q^{-N-1}$ in the definition (14.6.1) of the q-Hahn polynomials and letting $N \to \infty$:

$$\lim_{N\to\infty} Q_n(q^{-x};b,-b^{-1}c^{-1}q^{-N-1},N|q) = M_n(q^{-x};b,c;q). \tag{14.6.14}$$

q-Hahn $\to$ Quantum q-Krawtchouk

The quantum q-Krawtchouk polynomials given by (14.14.1) simply follow from the q-Hahn polynomials by setting $\beta = p$ in the definition (14.6.1) of the q-Hahn polynomials and taking the limit $\alpha \to \infty$:

$$\lim_{\alpha\to\infty} Q_n(q^{-x};\alpha,p,N|q) = K_n^{qtm}(q^{-x};p,N;q). \tag{14.6.15}$$

q-Hahn $\to$ q-Krawtchouk

If we set $\beta = -\alpha^{-1}q^{-1}p$ in the definition (14.6.1) of the q-Hahn polynomials and then let $\alpha \to 0$ we obtain the q-Krawtchouk polynomials given by (14.15.1):

$$\lim_{\alpha\to 0} Q_n(q^{-x};\alpha,-\alpha^{-1}q^{-1}p,N|q) = K_n(q^{-x};p,N;q). \tag{14.6.16}$$

q-Hahn $\to$ Affine q-Krawtchouk

The affine q-Krawtchouk polynomials given by (14.16.1) can be obtained from the q-Hahn polynomials by the substitution $\alpha = p$ and $\beta = 0$ in (14.6.1):

$$Q_n(q^{-x};p,0,N|q) = K_n^{Aff}(q^{-x};p,N;q). \tag{14.6.17}$$

q-Hahn $\to$ Hahn

The Hahn polynomials given by (9.5.1) simply follow from the q-Hahn polynomials given by (14.6.1), after setting $\alpha \to q^{\alpha}$ and $\beta \to q^{\beta}$, in the following way:

$$\lim_{q\to 1} Q_n(q^{-x};q^{\alpha},q^{\beta},N|q) = Q_n(x;\alpha,\beta,N). \tag{14.6.18}$$

Remark

The q-Hahn polynomials given by (14.6.1) and the dual q-Hahn polynomials given by (14.7.1) are related in the following way:

$$Q_n(q^{-x};\alpha,\beta,N|q) = R_x(\mu(n);\alpha,\beta,N|q),$$

with

$$\mu(n) = q^{-n} + \alpha\beta q^{n+1}$$

or

$$R_n(\mu(x);\gamma,\delta,N|q) = Q_x(q^{-n};\gamma,\delta,N|q),$$

where

$$\mu(x) = q^{-x} + \gamma\delta q^{x+1}.$$

References

[16], [30], [34], [70], [72], [80], [186], [205], [235], [238], [261], [305], [326], [329], [343], [345], [416], [419], [442], [486], [488], [489].

14.7 Dual q-Hahn

Basic Hypergeometric Representation

$$R_n(\mu(x);\gamma,\delta,N|q) = {}_3\phi_2\left(\begin{matrix} q^{-n}, q^{-x}, \gamma\delta q^{x+1} \\ \gamma q, q^{-N} \end{matrix}; q, q\right), \quad n = 0,1,2,\ldots,N, \tag{14.7.1}$$

where

$$\mu(x) := q^{-x} + \gamma\delta q^{x+1}.$$

Orthogonality Relation

For $0 < \gamma q < 1$ and $0 < \delta q < 1$, or for $\gamma > q^{-N}$ and $\delta > q^{-N}$, we have

$$\sum_{x=0}^{N} \frac{(\gamma q, \gamma\delta q, q^{-N};q)_x}{(q, \gamma\delta q^{N+2}, \delta q;q)_x} \frac{(1-\gamma\delta q^{2x+1})}{(1-\gamma\delta q)(-\gamma q)^x} q^{Nx-\binom{x}{2}} \times R_m(\mu(x);\gamma,\delta,N|q) R_n(\mu(x);\gamma,\delta,N|q) = \frac{(\gamma\delta q^2;q)_N}{(\delta q;q)_N} (\gamma q)^{-N} \frac{(q,\delta^{-1}q^{-N};q)_n}{(\gamma q, q^{-N};q)_n} (\gamma\delta q)^n \, \delta_{mn}. \tag{14.7.2}$$

Recurrence Relation

$$-\left(1-q^{-x}\right)\left(1-\gamma\delta q^{x+1}\right) R_n(\mu(x)) = A_n R_{n+1}(\mu(x)) - (A_n + C_n) R_n(\mu(x)) + C_n R_{n-1}(\mu(x)), \tag{14.7.3}$$

where

$$R_n(\mu(x)) := R_n(\mu(x);\gamma,\delta,N|q)$$

and

$$\begin{cases} A_n = \left(1-q^{n-N}\right)\left(1-\gamma q^{n+1}\right) \\ C_n = \gamma q\left(1-q^n\right)\left(\delta - q^{n-N-1}\right). \end{cases}$$

Normalized Recurrence Relation

$$x p_n(x) = p_{n+1}(x) + [1 + \gamma\delta q - (A_n + C_n)] \, p_n(x) + \gamma q(1-q^n)(1-\gamma q^n) \times (1-q^{n-N-1})(\delta - q^{n-N-1}) p_{n-1}(x), \tag{14.7.4}$$

where

$$R_n(\mu(x);\gamma,\delta,N|q) = \frac{1}{(\gamma q, q^{-N};q)_n} p_n(\mu(x)).$$

q-Difference Equation

$$q^{-n}(1-q^n) y(x) = B(x) y(x+1) - [B(x) + D(x)] \, y(x) + D(x) y(x-1), \tag{14.7.5}$$

where

$$y(x) = R_n(\mu(x);\gamma,\delta,N|q)$$

and

$$\begin{cases} B(x) = \dfrac{(1-q^{x-N})(1-\gamma q^{x+1})(1-\gamma\delta q^{x+1})}{(1-\gamma\delta q^{2x+1})(1-\gamma\delta q^{2x+2})} \\ \\ D(x) = -\dfrac{\gamma q^{x-N}(1-q^x)(1-\gamma\delta q^{x+N+1})(1-\delta q^x)}{(1-\gamma\delta q^{2x})(1-\gamma\delta q^{2x+1})}. \end{cases}$$

Forward Shift Operator

$$\begin{aligned} &R_n(\mu(x+1);\gamma,\delta,N|q) - R_n(\mu(x);\gamma,\delta,N|q) \\ &= \frac{q^{-n-x}(1-q^n)(1-\gamma\delta q^{2x+2})}{(1-\gamma q)(1-q^{-N})} R_{n-1}(\mu(x);\gamma q,\delta,N-1|q) \end{aligned} \tag{14.7.6}$$

or equivalently

$$\begin{aligned} &\frac{\Delta R_n(\mu(x);\gamma,\delta,N|q)}{\Delta\mu(x)} \\ &= \frac{q^{-n+1}(1-q^n)}{(1-q)(1-\gamma q)(1-q^{-N})} R_{n-1}(\mu(x);\gamma q,\delta,N-1|q). \end{aligned} \tag{14.7.7}$$

Backward Shift Operator

$$\begin{aligned} &(1-\gamma q^x)(1-\gamma\delta q^x)(1-q^{x-N-1})R_n(\mu(x);\gamma,\delta,N|q) \\ &\quad + \gamma q^{x-N-1}(1-q^x)(1-\gamma\delta q^{x+N+1})(1-\delta q^x)R_n(\mu(x-1);\gamma,\delta,N|q) \\ &= q^x(1-\gamma)(1-q^{-N-1})(1-\gamma\delta q^{2x})R_{n+1}(\mu(x);\gamma q^{-1},\delta,N+1|q) \end{aligned} \tag{14.7.8}$$

or equivalently

$$\begin{aligned} &\frac{\nabla\left[w(x;\gamma,\delta,N|q)R_n(\mu(x);\gamma,\delta,N|q)\right]}{\nabla\mu(x)} \\ &= \frac{1}{(1-q)(1-\gamma\delta)} w(x;\gamma q^{-1},\delta,N+1|q) \\ &\quad \times R_{n+1}(\mu(x);\gamma q^{-1},\delta,N+1|q), \end{aligned} \tag{14.7.9}$$

where

$$w(x;\gamma,\delta,N|q) = \frac{(\gamma q,\gamma\delta q,q^{-N};q)_x}{(q,\gamma\delta q^{N+2},\delta q;q)_x}(-\gamma^{-1})^x q^{Nx-\binom{x}{2}}.$$

Rodrigues-Type Formula

$$\begin{aligned} & w(x;\gamma,\delta,N|q)R_n(\mu(x);\gamma,\delta,N|q) \\ & = (1-q)^n(\gamma\delta q;q)_n\left(\nabla_\mu\right)^n\left[w(x;\gamma q^n,\delta,N-n|q)\right], \end{aligned} \tag{14.7.10}$$

where

$$\nabla_\mu := \frac{\nabla}{\nabla\mu(x)}.$$

Generating Functions

For $x = 0,1,2,\ldots,N$ we have

$$\begin{aligned} & (q^{-N}t;q)_{N-x}\cdot {}_2\phi_1\left(\begin{array}{c} q^{-x},\delta^{-1}q^{-x} \\ \gamma q \end{array};q,\gamma\delta q^{x+1}t\right) \\ & = \sum_{n=0}^{N}\frac{(q^{-N};q)_n}{(q;q)_n}R_n(\mu(x);\gamma,\delta,N|q)t^n. \end{aligned} \tag{14.7.11}$$

$$\begin{aligned} & (\gamma\delta qt;q)_x\cdot {}_2\phi_1\left(\begin{array}{c} q^{x-N},\gamma q^{x+1} \\ \delta^{-1}q^{-N} \end{array};q,q^{-x}t\right) \\ & = \sum_{n=0}^{N}\frac{(q^{-N},\gamma q;q)_n}{(\delta^{-1}q^{-N},q;q)_n}R_n(\mu(x);\gamma,\delta,N|q)t^n. \end{aligned} \tag{14.7.12}$$

Limit Relations

q-Racah → Dual q-Hahn

To obtain the dual q-Hahn polynomials from the q-Racah polynomials we have to take $\beta = 0$ and $\alpha q = q^{-N}$ in (14.2.1):

$$R_n(\mu(x);q^{-N-1},0,\gamma,\delta|q) = R_n(\mu(x);\gamma,\delta,N|q),$$

with

$$\mu(x) = q^{-x} + \gamma\delta q^{x+1}.$$

We may also take $\alpha = 0$ and $\beta = \delta^{-1}q^{-N-1}$ in (14.2.1) to obtain the dual q-Hahn polynomials from the q-Racah polynomials:

$$R_n(\mu(x);0,\delta^{-1}q^{-N-1},\gamma,\delta|q) = R_n(\mu(x);\gamma,\delta,N|q),$$

with

$$\mu(x) = q^{-x} + \gamma\delta q^{x+1}.$$

And if we take $\gamma q = q^{-N}$, $\delta \to \alpha\delta q^{N+1}$ and $\beta = 0$ in the definition (14.2.1) of the q-Racah polynomials we find the dual q-Hahn polynomials given by (14.7.1) in the following way:

$$R_n(\mu(x);\alpha,0,q^{-N-1},\alpha\delta q^{N+1}|q) = R_n(\mu(x);\alpha,\delta,N|q),$$

with

$$\mu(x) = q^{-x} + \alpha\delta q^{x+1}.$$

Dual q-Hahn → Affine q-Krawtchouk

The affine q-Krawtchouk polynomials given by (14.16.1) can be obtained from the dual q-Hahn polynomials by the substitution $\gamma = p$ and $\delta = 0$ in (14.7.1):

$$R_n(\mu(x);p,0,N|q) = K_n^{Aff}(q^{-x};p,N;q). \tag{14.7.13}$$

Note that $\mu(x) = q^{-x}$ in this case.

Dual q-Hahn → Dual q-Krawtchouk

The dual q-Krawtchouk polynomials given by (14.17.1) can be obtained from the dual q-Hahn polynomials by setting $\delta = c\gamma^{-1}q^{-N-1}$ in (14.7.1) and letting $\gamma \to 0$:

$$\lim_{\gamma\to 0} R_n(\mu(x);\gamma,c\gamma^{-1}q^{-N-1},N|q) = K_n(\lambda(x);c,N|q). \tag{14.7.14}$$

Dual q-Hahn → Dual Hahn

The dual Hahn polynomials given by (9.6.1) follow from the dual q-Hahn polynomials by simply taking the limit $q \to 1$ in the definition (14.7.1) of the dual q-Hahn polynomials after applying the substitution $\gamma \to q^{\gamma}$ and $\delta \to q^{\delta}$:

$$\lim_{q\to 1} R_n(\mu(x);q^{\gamma},q^{\delta},N|q) = R_n(\lambda(x);\gamma,\delta,N), \tag{14.7.15}$$

where

$$\begin{cases} \mu(x) = q^{-x} + q^{x+\gamma+\delta+1} \\ \lambda(x) = x(x+\gamma+\delta+1). \end{cases}$$

Remark

The dual q-Hahn polynomials given by (14.7.1) and the q-Hahn polynomials given by (14.6.1) are related in the following way:

$$Q_n(q^{-x};\alpha,\beta,N|q) = R_x(\mu(n);\alpha,\beta,N|q),$$

with

$$\mu(n) = q^{-n} + \alpha\beta q^{n+1}$$

or

$$R_n(\mu(x);\gamma,\delta,N|q) = Q_x(q^{-n};\gamma,\delta,N|q),$$

where

$$\mu(x) = q^{-x} + \gamma\delta q^{x+1}.$$

References

[31], [34], [70], [72], [80], [238], [329], [416], [488].

14.8 Al-Salam-Chihara

Basic Hypergeometric Representation

$$\begin{aligned} Q_n(x;a,b|q) &= \frac{(ab;q)_n}{a^n}\,{}_3\phi_2\left(\begin{matrix} q^{-n}, ae^{i\theta}, ae^{-i\theta} \\ ab, 0 \end{matrix}; q, q\right) \\ &= (ae^{i\theta};q)_n e^{-in\theta}\,{}_2\phi_1\left(\begin{matrix} q^{-n}, be^{-i\theta} \\ a^{-1}q^{-n+1}e^{-i\theta} \end{matrix}; q, a^{-1}qe^{i\theta}\right) \\ &= (be^{-i\theta};q)_n e^{in\theta}\,{}_2\phi_1\left(\begin{matrix} q^{-n}, ae^{i\theta} \\ b^{-1}q^{-n+1}e^{i\theta} \end{matrix}; q, b^{-1}qe^{-i\theta}\right), \quad x = \cos\theta. \end{aligned} \tag{14.8.1}$$

Orthogonality Relation

If a and b are real or complex conjugates and $\max(|a|,|b|) < 1$, then we have the following orthogonality relation

$$\frac{1}{2\pi}\int_{-1}^{1}\frac{w(x)}{\sqrt{1-x^2}}Q_m(x;a,b|q)Q_n(x;a,b|q)\,dx = \frac{\delta_{mn}}{(q^{n+1},abq^n;q)_\infty}, \tag{14.8.2}$$

where

$$w(x) := w(x;a,b|q) = \left|\frac{(e^{2i\theta};q)_\infty}{(ae^{i\theta},be^{i\theta};q)_\infty}\right|^2 = \frac{h(x,1)h(x,-1)h(x,q^{\frac{1}{2}})h(x,-q^{\frac{1}{2}})}{h(x,a)h(x,b)},$$

with

$$h(x,\alpha) := \prod_{k=0}^{\infty}\left(1-2\alpha x q^k+\alpha^2 q^{2k}\right) = \left(\alpha e^{i\theta},\alpha e^{-i\theta};q\right)_\infty, \quad x=\cos\theta.$$

If $a>1$ and $|ab|<1$, then we have another orthogonality relation given by:

$$\frac{1}{2\pi}\int_{-1}^{1}\frac{w(x)}{\sqrt{1-x^2}}Q_m(x;a,b|q)Q_n(x;a,b|q)\,dx + \sum_{\substack{k\\1<aq^k\leq a}} w_k Q_m(x_k;a,b|q)Q_n(x_k;a,b|q) = \frac{\delta_{mn}}{(q^{n+1},abq^n;q)_\infty}, \tag{14.8.3}$$

where $w(x)$ is as before,

$$x_k = \frac{aq^k+\left(aq^k\right)^{-1}}{2}$$

and

$$w_k = \frac{(a^{-2};q)_\infty}{(q,ab,a^{-1}b;q)_\infty}\frac{(1-a^2q^{2k})(a^2,ab;q)_k}{(1-a^2)(q,ab^{-1}q;q)_k}q^{-k^2}\left(\frac{1}{a^3b}\right)^k.$$

Recurrence Relation

$$2xQ_n(x) = Q_{n+1}(x)+(a+b)q^nQ_n(x)+(1-q^n)(1-abq^{n-1})Q_{n-1}(x). \tag{14.8.4}$$

Normalized Recurrence Relation

$$xp_n(x) = p_{n+1}(x)+\frac{1}{2}(a+b)q^np_n(x)+\frac{1}{4}(1-q^n)(1-abq^{n-1})p_{n-1}(x), \tag{14.8.5}$$

where

$$Q_n(x;a,b|q) = 2^np_n(x).$$

q-Difference Equation

$$(1-q)^2 D_q\left[\tilde{w}(x;aq^{\frac{1}{2}},bq^{\frac{1}{2}}|q)D_q y(x)\right] + 4q^{-n+1}(1-q^n)\tilde{w}(x;a,b|q)y(x) = 0, \tag{14.8.6}$$

where

$$y(x) = Q_n(x;a,b|q)$$

and

$$\tilde{w}(x;a,b|q) := \frac{w(x;a,b|q)}{\sqrt{1-x^2}}.$$

If we define

$$P_n(z) := \frac{(ab;q)_n}{a^n}\,{}_3\phi_2\left(\begin{matrix} q^{-n}, az, az^{-1} \\ ab, 0 \end{matrix}; q, q\right)$$

then the q-difference equation can also be written in the form

$$q^{-n}(1-q^n)P_n(z) = A(z)P_n(qz) - \left[A(z)+A(z^{-1})\right]P_n(z) + A(z^{-1})P_n(q^{-1}z), \tag{14.8.7}$$

where

$$A(z) = \frac{(1-az)(1-bz)}{(1-z^2)(1-qz^2)}.$$

Forward Shift Operator

$$\delta_q Q_n(x;a,b|q) = -q^{-\frac{1}{2}n}(1-q^n)(e^{i\theta}-e^{-i\theta})Q_{n-1}(x;aq^{\frac{1}{2}},bq^{\frac{1}{2}}|q), \quad x=\cos\theta \tag{14.8.8}$$

or equivalently

$$D_q Q_n(x;a,b|q) = 2q^{-\frac{1}{2}(n-1)}\frac{1-q^n}{1-q}Q_{n-1}(x;aq^{\frac{1}{2}},bq^{\frac{1}{2}}|q). \tag{14.8.9}$$

Backward Shift Operator

$$\delta_q\left[\tilde{w}(x;a,b|q)Q_n(x;a,b|q)\right] = q^{-\frac{1}{2}(n+1)}(e^{i\theta}-e^{-i\theta})\tilde{w}(x;aq^{-\frac{1}{2}},bq^{-\frac{1}{2}}|q) \times Q_{n+1}(x;aq^{-\frac{1}{2}},bq^{-\frac{1}{2}}|q), \quad x=\cos\theta \tag{14.8.10}$$

or equivalently

$$\begin{aligned}&D_q\left[\tilde{w}(x;a,b|q)Q_n(x;a,b|q)\right]\\&=-\frac{2q^{-\frac{1}{2}n}}{1-q}\tilde{w}(x;aq^{-\frac{1}{2}},bq^{-\frac{1}{2}}|q)Q_{n+1}(x;aq^{-\frac{1}{2}},bq^{-\frac{1}{2}}|q).\end{aligned}\tag{14.8.11}$$

Rodrigues-Type Formula

$$\begin{aligned}&\tilde{w}(x;a,b|q)Q_n(x;a,b|q)\\&=\left(\frac{q-1}{2}\right)^n q^{\frac{1}{4}n(n-1)}\left(\mathcal{D}_q\right)^n\left[\tilde{w}(x;aq^{\frac{1}{2}n},bq^{\frac{1}{2}n}|q)\right].\end{aligned}\tag{14.8.12}$$

Generating Functions

$$\frac{(at,bt;q)_\infty}{(\mathrm{e}^{i\theta}t,\mathrm{e}^{-i\theta}t;q)_\infty}=\sum_{n=0}^{\infty}\frac{Q_n(x;a,b|q)}{(q;q)_n}t^n,\quad x=\cos\theta.\tag{14.8.13}$$

$$\frac{1}{(\mathrm{e}^{i\theta}t;q)_\infty}\,{}_2\phi_1\left(\begin{matrix}a\mathrm{e}^{i\theta},b\mathrm{e}^{i\theta}\\ab\end{matrix};q,\mathrm{e}^{-i\theta}t\right)=\sum_{n=0}^{\infty}\frac{Q_n(x;a,b|q)}{(ab,q;q)_n}t^n,\quad x=\cos\theta.\tag{14.8.14}$$

$$\begin{aligned}&(t;q)_\infty\cdot{}_2\phi_1\left(\begin{matrix}a\mathrm{e}^{i\theta},a\mathrm{e}^{-i\theta}\\ab\end{matrix};q,t\right)\\&=\sum_{n=0}^{\infty}\frac{(-1)^n a^n q^{\binom{n}{2}}}{(ab,q;q)_n}Q_n(x;a,b|q)t^n,\quad x=\cos\theta.\end{aligned}\tag{14.8.15}$$

$$\begin{aligned}&\frac{(\gamma\mathrm{e}^{i\theta}t;q)_\infty}{(\mathrm{e}^{i\theta}t;q)_\infty}\,{}_3\phi_2\left(\begin{matrix}\gamma,a\mathrm{e}^{i\theta},b\mathrm{e}^{i\theta}\\ab,\gamma\mathrm{e}^{i\theta}t\end{matrix};q,\mathrm{e}^{-i\theta}t\right)\\&=\sum_{n=0}^{\infty}\frac{(\gamma;q)_n}{(ab,q;q)_n}Q_n(x;a,b|q)t^n,\quad x=\cos\theta,\quad\gamma\text{ arbitrary}.\end{aligned}\tag{14.8.16}$$

Limit Relations

Continuous Dual q-Hahn → Al-Salam-Chihara

The Al-Salam-Chihara polynomials given by (14.8.1) simply follow from the continuous dual q-Hahn polynomials by taking $c = 0$ in the definition (14.3.1) of the continuous dual q-Hahn polynomials:

$$p_n(x;a,b,0|q) = Q_n(x;a,b|q).$$

Al-Salam-Chihara → Continuous Big q-Hermite

If we take $b = 0$ in the definition (14.8.1) of the Al-Salam-Chihara polynomials we simply obtain the continuous big q-Hermite polynomials given by (14.18.1):

$$Q_n(x;a,0|q) = H_n(x;a|q). \tag{14.8.17}$$

Al-Salam-Chihara → Continuous q-Laguerre

The continuous q-Laguerre polynomials given by (14.19.1) can be obtained from the Al-Salam-Chihara polynomials given by (14.8.1) by taking $a = q^{\frac{1}{2}\alpha+\frac{1}{4}}$ and $b = q^{\frac{1}{2}\alpha+\frac{3}{4}}$:

$$Q_n(x;q^{\frac{1}{2}\alpha+\frac{1}{4}},q^{\frac{1}{2}\alpha+\frac{3}{4}}|q) = \frac{(q;q)_n}{q^{(\frac{1}{2}\alpha+\frac{1}{4})n}}P_n^{(\alpha)}(x|q). \tag{14.8.18}$$

Al-Salam-Chihara → Meixner-Pollaczek

If we set $a = q^{\lambda}\mathrm{e}^{-i\phi}$, $b = q^{\lambda}\mathrm{e}^{i\phi}$ and $\mathrm{e}^{i\theta} = q^{ix}\mathrm{e}^{i\phi}$ in the definition (14.8.1) of the Al-Salam-Chihara polynomials and take the limit $q \to 1$ we obtain the Meixner-Pollaczek polynomials given by (9.7.1) in the following way:

$$\lim_{q\to 1} \frac{Q_n(\cos(\ln q^x + \phi);q^{\lambda}\mathrm{e}^{i\phi},q^{\lambda}\mathrm{e}^{-i\phi}|q)}{(q;q)_n} = P_n^{(\lambda)}(x;\phi). \tag{14.8.19}$$

References

[16], [20], [21], [64], [73], [86], [118], [148], [151], [160], [209], [292], [295], [320], [328].

14.9 q-Meixner-Pollaczek

Basic Hypergeometric Representation

$$P_n(x;a|q) = a^{-n}\mathrm{e}^{-in\phi}\frac{(a^2;q)_n}{(q;q)_n}\,{}_3\phi_2\left(\begin{matrix}q^{-n},a\mathrm{e}^{i(\theta+2\phi)},a\mathrm{e}^{-i\theta}\\ a^2,0\end{matrix};q,q\right) \tag{14.9.1}$$
$$= \frac{(a\mathrm{e}^{-i\theta};q)_n}{(q;q)_n}\mathrm{e}^{in(\theta+\phi)}\,{}_2\phi_1\left(\begin{matrix}q^{-n},a\mathrm{e}^{i\theta}\\ a^{-1}q^{-n+1}\mathrm{e}^{i\theta}\end{matrix};q,qa^{-1}\mathrm{e}^{-i(\theta+2\phi)}\right)$$

with $x=\cos(\theta+\phi)$.

Orthogonality Relation

$$\frac{1}{2\pi}\int_{-\pi}^{\pi}w(\cos(\theta+\phi);a|q)P_m(\cos(\theta+\phi);a|q)P_n(\cos(\theta+\phi);a|q)\,d\theta$$
$$=\frac{\delta_{mn}}{(q;q)_n(q,a^2q^n;q)_\infty},\quad 0<a<1, \tag{14.9.2}$$

where

$$w(x;a|q)=\left|\frac{(\mathrm{e}^{2i(\theta+\phi)};q)_\infty}{(a\mathrm{e}^{i(\theta+2\phi)},a\mathrm{e}^{i\theta};q)_\infty}\right|^2=\frac{h(x,1)h(x,-1)h(x,q^{\frac{1}{2}})h(x,-q^{\frac{1}{2}})}{h(x,a\mathrm{e}^{i\phi})h(x,a\mathrm{e}^{-i\phi})},$$

with

$$h(x,\alpha):=\prod_{k=0}^{\infty}\left(1-2\alpha xq^k+\alpha^2q^{2k}\right)=\left(\alpha\mathrm{e}^{i(\theta+\phi)},\alpha\mathrm{e}^{-i(\theta+\phi)};q\right)_\infty,\quad x=\cos(\theta+\phi).$$

Recurrence Relation

$$2xP_n(x;a|q)=(1-q^{n+1})P_{n+1}(x;a|q)+2aq^n\cos\phi P_n(x;a|q)$$
$$+(1-a^2q^{n-1})P_{n-1}(x;a|q). \tag{14.9.3}$$

Normalized Recurrence Relation

$$xp_n(x) = p_{n+1}(x) + aq^n\cos\phi\, p_n(x) + \frac{1}{4}(1-q^n)(1-a^2q^{n-1})p_{n-1}(x), \qquad (14.9.4)$$

where

$$P_n(x;a|q) = \frac{2^n}{(q;q)_n}p_n(x).$$

q-Difference Equation

$$(1-q)^2D_q\left[\tilde{w}(x;aq^{\frac{1}{2}}|q)D_qy(x)\right] + 4q^{-n+1}(1-q^n)\tilde{w}(x;a|q)y(x) = 0, \qquad (14.9.5)$$

where

$$y(x) = P_n(x;a|q)$$

and

$$\tilde{w}(x;a|q) := \frac{w(x;a|q)}{\sqrt{1-x^2}}.$$

Forward Shift Operator

$$\delta_qP_n(x;a|q) = -q^{-\frac{1}{2}n}(\mathrm{e}^{i(\theta+\phi)} - \mathrm{e}^{-i(\theta+\phi)})P_{n-1}(x;aq^{\frac{1}{2}}|q), \quad x = \cos\theta \qquad (14.9.6)$$

or equivalently

$$D_qP_n(x;a|q) = \frac{2q^{-\frac{1}{2}(n-1)}}{1-q}P_{n-1}(x;aq^{\frac{1}{2}}|q). \qquad (14.9.7)$$

Backward Shift Operator

$$\begin{aligned}&\delta_q\left[\tilde{w}(x;a|q)P_n(x;a|q)\right]\\ &= q^{-\frac{1}{2}(n+1)}(1-q^{n+1})(\mathrm{e}^{i\theta} - \mathrm{e}^{-i\theta})\\ &\quad \times \tilde{w}(x;aq^{-\frac{1}{2}}|q)P_{n+1}(x;aq^{-\frac{1}{2}}|q), \quad x = \cos\theta \end{aligned} \qquad (14.9.8)$$

or equivalently

$$D_q\left[\tilde{w}(x;a|q)P_n(x;a|q)\right] = -2q^{-\frac{1}{2}n}\frac{1-q^{n+1}}{1-q}\tilde{w}(x;aq^{-\frac{1}{2}}|q)P_{n+1}(x;aq^{-\frac{1}{2}}|q). \quad (14.9.9)$$

Rodrigues-Type Formula

$$\tilde{w}(x;a|q)P_n(x;a|q)=\left(\frac{q-1}{2}\right)^n q^{\frac{1}{4}n(n-1)}\frac{1}{(q;q)_n}(D_q)^n\left[\tilde{w}(x;aq^{\frac{1}{2}n}|q)\right]. \quad (14.9.10)$$

Generating Functions

$$\left|\frac{(ae^{i\phi}t;q)_\infty}{(e^{i(\theta+\phi)}t;q)_\infty}\right|^2 =\frac{(ae^{i\phi}t,ae^{-i\phi}t;q)_\infty}{(e^{i(\theta+\phi)}t,e^{-i(\theta+\phi)}t;q)_\infty}=\sum_{n=0}^{\infty}P_n(x;a|q)t^n, \quad x=\cos(\theta+\phi). \quad (14.9.11)$$

$$\frac{1}{(e^{i(\theta+\phi)}t;q)_\infty}\,{}_2\phi_1\left(\begin{array}{c} ae^{i(\theta+2\phi)},ae^{i\theta} \\ a^2 \end{array};q,\,e^{-i(\theta+\phi)}t\right) =\sum_{n=0}^{\infty}\frac{P_n(x;a|q)}{(a^2;q)_n}t^n, \quad x=\cos(\theta+\phi). \quad (14.9.12)$$

Limit Relations

Continuous q-Hahn → q-Meixner-Pollaczek

The q-Meixner-Pollaczek polynomials given by (14.9.1) simply follow from the continuous q-Hahn polynomials if we set $d=a$ and $b=c=0$ in the definition (14.4.1) of the continuous q-Hahn polynomials:

$$\frac{p_n(\cos(\theta+\phi);a,0,0,a;q)}{(q;q)_n}=P_n(\cos(\theta+\phi);a|q).$$

q-Meixner-Pollaczek → Continuous q-ultraspherical / Rogers

If we take $\theta=0$ and $a=\beta$ in the definition (14.9.1) of the q-Meixner-Pollaczek polynomials we obtain the continuous q-ultraspherical (or Rogers) polynomials given by (14.10.17):

$$P_n(\cos\phi;\beta|q)=C_n(\cos\phi;\beta|q). \quad (14.9.13)$$

q-Meixner-Pollaczek → Continuous q-Laguerre

If we take $e^{i\phi} = q^{-\frac{1}{4}}$, $a = q^{\frac{1}{2}\alpha+\frac{1}{2}}$ and $e^{i\theta} \to q^{\frac{1}{4}}e^{i\theta}$ in the definition (14.9.1) of the q-Meixner-Pollaczek polynomials we obtain the continuous q-Laguerre polynomials given by (14.19.1):

$$P_n(\cos(\theta+\phi); q^{\frac{1}{2}\alpha+\frac{1}{2}}|q) = q^{-(\frac{1}{2}\alpha+\frac{1}{4})n} P_n^{(\alpha)}(\cos\theta|q). \tag{14.9.14}$$

q-Meixner-Pollaczek → Meixner-Pollaczek

To find the Meixner-Pollaczek polynomials given by (9.7.1) from the q-Meixner-Pollaczek polynomials we substitute $a = q^{\lambda}$ and $e^{i\theta} = q^{-ix}$(or $\theta = \ln q^{-x}$) in the definition (14.9.1) of the q-Meixner-Pollaczek polynomials and take the limit $q \to 1$ to find:

$$\lim_{q\to 1} P_n(\cos(\ln q^{-x}+\phi); q^{\lambda}|q) = P_n^{(\lambda)}(x; -\phi). \tag{14.9.15}$$

References

[16], [21], [64], [72], [135], [270].

14.10 Continuous q-Jacobi

Basic Hypergeometric Representation

If we take $a = q^{\frac{1}{2}\alpha+\frac{1}{4}}$, $b = q^{\frac{1}{2}\alpha+\frac{3}{4}}$, $c = -q^{\frac{1}{2}\beta+\frac{1}{4}}$ and $d = -q^{\frac{1}{2}\beta+\frac{3}{4}}$ in the definition (14.1.1) of the Askey-Wilson polynomials we find after renormalizing

$$\begin{aligned} &P_n^{(\alpha,\beta)}(x|q)\\ &= \frac{(q^{\alpha+1};q)_n}{(q;q)_n}\, {}_4\phi_3\left(\begin{array}{c} q^{-n}, q^{n+\alpha+\beta+1}, q^{\frac{1}{2}\alpha+\frac{1}{4}}e^{i\theta}, q^{\frac{1}{2}\alpha+\frac{1}{4}}e^{-i\theta} \\ q^{\alpha+1}, -q^{\frac{1}{2}(\alpha+\beta+1)}, -q^{\frac{1}{2}(\alpha+\beta+2)} \end{array}; q, q\right) \end{aligned} \tag{14.10.1}$$

with $x = \cos\theta$.

Orthogonality Relation

For $\alpha \geq -\frac{1}{2}$ and $\beta \geq -\frac{1}{2}$ we have

$$\frac{1}{2\pi}\int_{-1}^{1}\frac{w(x)}{\sqrt{1-x^2}}P_m^{(\alpha,\beta)}(x|q)P_n^{(\alpha,\beta)}(x|q)\,dx$$
$$=\frac{(q^{\frac{1}{2}(\alpha+\beta+2)},q^{\frac{1}{2}(\alpha+\beta+3)};q)_\infty}{(q,q^{\alpha+1},q^{\beta+1},-q^{\frac{1}{2}(\alpha+\beta+1)},-q^{\frac{1}{2}(\alpha+\beta+2)};q)_\infty}\frac{1-q^{\alpha+\beta+1}}{1-q^{2n+\alpha+\beta+1}}$$
$$\times\frac{(q^{\alpha+1},q^{\beta+1},-q^{\frac{1}{2}(\alpha+\beta+3)};q)_n}{(q,q^{\alpha+\beta+1},-q^{\frac{1}{2}(\alpha+\beta+1)};q)_n}q^{(\alpha+\frac{1}{2})n}\,\delta_{mn}, \tag{14.10.2}$$

where

$$w(x):=w(x;q^\alpha,q^\beta|q)=\left|\frac{(e^{2i\theta};q)_\infty}{(q^{\frac{1}{2}\alpha+\frac{1}{4}}e^{i\theta},q^{\frac{1}{2}\alpha+\frac{3}{4}}e^{i\theta},-q^{\frac{1}{2}\beta+\frac{1}{4}}e^{i\theta},-q^{\frac{1}{2}\beta+\frac{3}{4}}e^{i\theta};q)_\infty}\right|^2$$
$$=\left|\frac{(e^{i\theta},-e^{i\theta};q^{\frac{1}{2}})_\infty}{(q^{\frac{1}{2}\alpha+\frac{1}{4}}e^{i\theta},-q^{\frac{1}{2}\beta+\frac{1}{4}}e^{i\theta};q^{\frac{1}{2}})_\infty}\right|^2$$
$$=\frac{h(x,1)h(x,-1)h(x,q^{\frac{1}{2}})h(x,-q^{\frac{1}{2}})}{h(x,q^{\frac{1}{2}\alpha+\frac{1}{4}})h(x,q^{\frac{1}{2}\alpha+\frac{3}{4}})h(x,-q^{\frac{1}{2}\beta+\frac{1}{4}})h(x,-q^{\frac{1}{2}\beta+\frac{3}{4}})},$$

with

$$h(x,\alpha):=\prod_{k=0}^{\infty}\left(1-2\alpha xq^k+\alpha^2q^{2k}\right)=\left(\alpha e^{i\theta},\alpha e^{-i\theta};q\right)_\infty,\quad x=\cos\theta.$$

Recurrence Relation

$$2x\tilde{P}_n(x|q)=A_n\tilde{P}_{n+1}(x|q)+\left[q^{\frac{1}{2}\alpha+\frac{1}{4}}+q^{-\frac{1}{2}\alpha-\frac{1}{4}}-(A_n+C_n)\right]\tilde{P}_n(x|q)$$
$$+C_n\tilde{P}_{n-1}(x|q), \tag{14.10.3}$$

where

$$\tilde{P}_n(x|q):=\frac{(q;q)_n}{(q^{\alpha+1};q)_n}P_n^{(\alpha,\beta)}(x|q)$$

and

$$\begin{cases} A_n = \dfrac{(1-q^{n+\alpha+1})(1-q^{n+\alpha+\beta+1})(1+q^{n+\frac{1}{2}(\alpha+\beta+1)})(1+q^{n+\frac{1}{2}(\alpha+\beta+2)})}{q^{\frac{1}{2}\alpha+\frac{1}{4}}(1-q^{2n+\alpha+\beta+1})(1-q^{2n+\alpha+\beta+2})} \\ C_n = \dfrac{q^{\frac{1}{2}\alpha+\frac{1}{4}}(1-q^n)(1-q^{n+\beta})(1+q^{n+\frac{1}{2}(\alpha+\beta)})(1+q^{n+\frac{1}{2}(\alpha+\beta+1)})}{(1-q^{2n+\alpha+\beta})(1-q^{2n+\alpha+\beta+1})}. \end{cases}$$

Normalized Recurrence Relation

$$xp_n(x) = p_{n+1}(x) + \frac{1}{2}\left[q^{\frac{1}{2}\alpha+\frac{1}{4}} + q^{-\frac{1}{2}\alpha-\frac{1}{4}} - (A_n + C_n)\right] p_n(x) + \frac{1}{4}A_{n-1}C_n p_{n-1}(x), \tag{14.10.4}$$

where

$$P_n^{(\alpha,\beta)}(x|q) = \frac{2^n q^{(\frac{1}{2}\alpha+\frac{1}{4})n}(q^{n+\alpha+\beta+1};q)_n}{(q,-q^{\frac{1}{2}(\alpha+\beta+1)},-q^{\frac{1}{2}(\alpha+\beta+2)};q)_n} p_n(x).$$

q-Difference Equation

$$(1-q)^2 D_q\left[\tilde{w}(x;q^{\alpha+1},q^{\beta+1}|q)D_q y(x)\right] + \lambda_n \tilde{w}(x;q^\alpha,q^\beta|q)y(x) = 0, \tag{14.10.5}$$

where

$$y(x) = P_n^{(\alpha,\beta)}(x|q)$$

and

$$\tilde{w}(x;q^\alpha,q^\beta|q) := \frac{w(x;q^\alpha,q^\beta|q)}{\sqrt{1-x^2}},$$

$$\lambda_n = 4q^{-n+1}(1-q^n)(1-q^{n+\alpha+\beta+1}).$$

Forward Shift Operator

$$\delta_q P_n^{(\alpha,\beta)}(x|q) = -\frac{q^{-n+\frac{1}{2}\alpha+\frac{3}{4}}(1-q^{n+\alpha+\beta+1})(e^{i\theta}-e^{-i\theta})}{(1+q^{\frac{1}{2}(\alpha+\beta+1)})(1+q^{\frac{1}{2}(\alpha+\beta+2)})} \times P_{n-1}^{(\alpha+1,\beta+1)}(x|q), \quad x=\cos\theta \tag{14.10.6}$$

or equivalently

$$D_qP_n^{(\alpha,\beta)}(x|q)=\frac{2q^{-n+\frac{1}{2}\alpha+\frac{5}{4}}(1-q^{n+\alpha+\beta+1})}{(1-q)(1+q^{\frac{1}{2}(\alpha+\beta+1)})(1+q^{\frac{1}{2}(\alpha+\beta+2)})}\times P_{n-1}^{(\alpha+1,\beta+1)}(x|q). \tag{14.10.7}$$

Backward Shift Operator

$$\delta_q\left[\tilde{w}(x;q^\alpha,q^\beta|q)P_n^{(\alpha,\beta)}(x|q)\right]=q^{-\frac{1}{2}\alpha-\frac{1}{4}}(1-q^{n+1})(1+q^{\frac{1}{2}(\alpha+\beta-1)})(1+q^{\frac{1}{2}(\alpha+\beta)})(e^{i\theta}-e^{-i\theta})\times\tilde{w}(x;q^{\alpha-1},q^{\beta-1}|q)P_{n+1}^{(\alpha-1,\beta-1)}(x|q),\quad x=\cos\theta \tag{14.10.8}$$

or equivalently

$$D_q\left[\tilde{w}(x;q^\alpha,q^\beta|q)P_n^{(\alpha,\beta)}(x|q)\right]=-2q^{-\frac{1}{2}\alpha+\frac{1}{4}}\frac{(1-q^{n+1})(1+q^{\frac{1}{2}(\alpha+\beta-1)})(1+q^{\frac{1}{2}(\alpha+\beta)})}{1-q}\times\tilde{w}(x;q^{\alpha-1},q^{\beta-1}|q)P_{n+1}^{(\alpha-1,\beta-1)}(x|q). \tag{14.10.9}$$

Rodrigues-Type Formula

$$\tilde{w}(x;q^\alpha,q^\beta|q)P_n^{(\alpha,\beta)}(x|q)=\left(\frac{q-1}{2}\right)^n\frac{q^{\frac{1}{4}n^2+\frac{1}{2}n\alpha}}{(q,-q^{\frac{1}{2}(\alpha+\beta+1)},-q^{\frac{1}{2}(\alpha+\beta+2)};q)_n}\times(D_q)^n\left[\tilde{w}(x;q^{\alpha+n},q^{\beta+n}|q)\right]. \tag{14.10.10}$$

Generating Functions

$${}_2\phi_1\left(\begin{matrix}q^{\frac{1}{2}\alpha+\frac{1}{4}}e^{i\theta},q^{\frac{1}{2}\alpha+\frac{3}{4}}e^{i\theta}\\ q^{\alpha+1}\end{matrix};q,e^{-i\theta}t\right){}_2\phi_1\left(\begin{matrix}-q^{\frac{1}{2}\beta+\frac{1}{4}}e^{-i\theta},-q^{\frac{1}{2}\beta+\frac{3}{4}}e^{-i\theta}\\ q^{\beta+1}\end{matrix};q,e^{i\theta}t\right)=\sum_{n=0}^{\infty}\frac{(-q^{\frac{1}{2}(\alpha+\beta+1)},-q^{\frac{1}{2}(\alpha+\beta+2)};q)_n}{(q^{\alpha+1},q^{\beta+1};q)_n}\frac{P_n^{(\alpha,\beta)}(x|q)}{q^{(\frac{1}{2}\alpha+\frac{1}{4})n}}t^n,\quad x=\cos\theta. \tag{14.10.11}$$

$$
{}_2\phi_1\left(\begin{array}{c} q^{\frac{1}{2}\alpha+\frac{1}{4}}e^{i\theta}, -q^{\frac{1}{2}\beta+\frac{1}{4}}e^{i\theta} \\ -q^{\frac{1}{2}(\alpha+\beta+1)} \end{array}; q, e^{-i\theta}t\right) {}_2\phi_1\left(\begin{array}{c} q^{\frac{1}{2}\alpha+\frac{3}{4}}e^{-i\theta}, -q^{\frac{1}{2}\beta+\frac{3}{4}}e^{-i\theta} \\ -q^{\frac{1}{2}(\alpha+\beta+3)} \end{array}; q, e^{i\theta}t\right)
$$
$$
= \sum_{n=0}^{\infty} \frac{(-q^{\frac{1}{2}(\alpha+\beta+2)};q)_n}{(-q^{\frac{1}{2}(\alpha+\beta+3)};q)_n} \frac{P_n^{(\alpha,\beta)}(x|q)}{q^{(\frac{1}{2}\alpha+\frac{1}{4})n}} t^n, \quad x = \cos\theta. \tag{14.10.12}
$$

$$
{}_2\phi_1\left(\begin{array}{c} q^{\frac{1}{2}\alpha+\frac{1}{4}}e^{i\theta}, -q^{\frac{1}{2}\beta+\frac{3}{4}}e^{i\theta} \\ -q^{\frac{1}{2}(\alpha+\beta+2)} \end{array}; q, e^{-i\theta}t\right) {}_2\phi_1\left(\begin{array}{c} q^{\frac{1}{2}\alpha+\frac{3}{4}}e^{-i\theta}, -q^{\frac{1}{2}\beta+\frac{1}{4}}e^{-i\theta} \\ -q^{\frac{1}{2}(\alpha+\beta+2)} \end{array}; q, e^{i\theta}t\right)
$$
$$
= \sum_{n=0}^{\infty} \frac{(-q^{\frac{1}{2}(\alpha+\beta+1)};q)_n}{(-q^{\frac{1}{2}(\alpha+\beta+2)};q)_n} \frac{P_n^{(\alpha,\beta)}(x|q)}{q^{(\frac{1}{2}\alpha+\frac{1}{4})n}} t^n, \quad x = \cos\theta. \tag{14.10.13}
$$

Limit Relations

Askey-Wilson → Continuous q-Jacobi

If we take $a = q^{\frac{1}{2}\alpha+\frac{1}{4}}$, $b = q^{\frac{1}{2}\alpha+\frac{3}{4}}$, $c = -q^{\frac{1}{2}\beta+\frac{1}{4}}$ and $d = -q^{\frac{1}{2}\beta+\frac{3}{4}}$ in the definition (14.1.1) of the Askey-Wilson polynomials and change the normalization we find the continuous q-Jacobi polynomials given by (14.10.1):

$$
\frac{q^{(\frac{1}{2}\alpha+\frac{1}{4})n} p_n(x; q^{\frac{1}{2}\alpha+\frac{1}{4}}, q^{\frac{1}{2}\alpha+\frac{3}{4}}, -q^{\frac{1}{2}\beta+\frac{1}{4}}, -q^{\frac{1}{2}\beta+\frac{3}{4}}|q)}{(q, -q^{\frac{1}{2}(\alpha+\beta+1)}, -q^{\frac{1}{2}(\alpha+\beta+2)};q)_n} = P_n^{(\alpha,\beta)}(x|q).
$$

Continuous q-Jacobi → Continuous q-Laguerre

The continuous q-Laguerre polynomials given by (14.19.1) follow simply from the continuous q-Jacobi polynomials given by (14.10.1) by taking the limit $\beta \to \infty$:

$$
\lim_{\beta\to\infty} P_n^{(\alpha,\beta)}(x|q) = P_n^{(\alpha)}(x|q). \tag{14.10.14}
$$

Continuous q-Jacobi → Jacobi

If we take the limit $q \to 1$ in the definitions (14.10.1) of the continuous q-Jacobi polynomials we simply find the Jacobi polynomials given by (9.8.1):

$$
\lim_{q\to 1} P_n^{(\alpha,\beta)}(x|q) = P_n^{(\alpha,\beta)}(x). \tag{14.10.15}
$$

Remarks

In [441] M. Rahman takes $a = q^{\frac{1}{2}}$, $b = q^{\alpha+\frac{1}{2}}$, $c = -q^{\beta+\frac{1}{2}}$ and $d = -q^{\frac{1}{2}}$ in the definition (14.1.1) of the Askey-Wilson polynomials to obtain after renormalizing

$$P_n^{(\alpha,\beta)}(x;q) = \frac{(q^{\alpha+1},-q^{\beta+1};q)_n}{(q,-q;q)_n} \times {}_4\phi_3\left(\begin{matrix} q^{-n}, q^{n+\alpha+\beta+1}, q^{\frac{1}{2}}e^{i\theta}, q^{\frac{1}{2}}e^{-i\theta} \\ q^{\alpha+1}, -q^{\beta+1}, -q \end{matrix} ; q, q\right) \quad (14.10.16)$$

with $x = \cos\theta$. These two q-analogues of the Jacobi polynomials are not really different, since they are connected by the quadratic transformation:

$$P_n^{(\alpha,\beta)}(x|q^2) = \frac{(-q;q)_n}{(-q^{\alpha+\beta+1};q)_n} q^{n\alpha} P_n^{(\alpha,\beta)}(x;q).$$

The continuous q-Jacobi polynomials given by (14.10.16) and the continuous q-ultraspherical (or Rogers) polynomials given by (14.10.17) are connected by the quadratic transformations:

$$C_{2n}(x;q^\lambda|q) = \frac{(q^\lambda,-q;q)_n}{(q^{\frac{1}{2}},-q^{\frac{1}{2}};q)_n} q^{-\frac{1}{2}n} P_n^{(\lambda-\frac{1}{2},-\frac{1}{2})}(2x^2-1;q)$$

and

$$C_{2n+1}(x;q^\lambda|q) = \frac{(q^\lambda,-1;q)_{n+1}}{(q^{\frac{1}{2}},-q^{\frac{1}{2}};q)_{n+1}} q^{-\frac{1}{2}n} x P_n^{(\lambda-\frac{1}{2},\frac{1}{2})}(2x^2-1;q).$$

If we replace q by q^{-1} we find

$$P_n^{(\alpha,\beta)}(x|q^{-1}) = q^{-n\alpha} P_n^{(\alpha,\beta)}(x|q) \quad \text{and} \quad P_n^{(\alpha,\beta)}(x;q^{-1}) = q^{-n(\alpha+\beta)} P_n^{(\alpha,\beta)}(x;q).$$

References

[72], [208], [236], [238], [288], [290], [293], [351], [414], [416], [441], [443], [444], [446], [474], [495].

Special Cases

14.10.1 Continuous q-Ultraspherical / Rogers

Basic Hypergeometric Representation

If we set $a=\beta^{\frac{1}{2}}$, $b=\beta^{\frac{1}{2}}q^{\frac{1}{2}}$, $c=-\beta^{\frac{1}{2}}$ and $d=-\beta^{\frac{1}{2}}q^{\frac{1}{2}}$ in the definition (14.1.1) of the Askey-Wilson polynomials and change the normalization we obtain the continuous q-ultraspherical (or Rogers) polynomials:

$$\begin{aligned} C_n(x;\beta|q) &= \frac{(\beta^2;q)_n}{(q;q)_n}\beta^{-\frac{1}{2}n}\,{}_4\phi_3\left(\begin{matrix} q^{-n},\beta^2q^n,\beta^{\frac{1}{2}}\mathrm{e}^{i\theta},\beta^{\frac{1}{2}}\mathrm{e}^{-i\theta} \\ \beta q^{\frac{1}{2}},-\beta,-\beta q^{\frac{1}{2}} \end{matrix};q,q\right) \qquad (14.10.17)\\ &= \frac{(\beta^2;q)_n}{(q;q)_n}\beta^{-n}\mathrm{e}^{-in\theta}\,{}_3\phi_2\left(\begin{matrix} q^{-n},\beta,\beta\mathrm{e}^{2i\theta} \\ \beta^2,0 \end{matrix};q,q\right)\\ &= \frac{(\beta;q)_n}{(q;q)_n}\mathrm{e}^{in\theta}\,{}_2\phi_1\left(\begin{matrix} q^{-n},\beta \\ \beta^{-1}q^{-n+1} \end{matrix};q,\beta^{-1}q\mathrm{e}^{-2i\theta}\right), \quad x=\cos\theta. \end{aligned}$$

Orthogonality Relation

$$\begin{aligned} &\frac{1}{2\pi}\int_{-1}^{1}\frac{w(x)}{\sqrt{1-x^2}}C_m(x;\beta|q)C_n(x;\beta|q)\,dx\\ &= \frac{(\beta,\beta q;q)_\infty}{(\beta^2,q;q)_\infty}\frac{(\beta^2;q)_n}{(q;q)_n}\frac{(1-\beta)}{(1-\beta q^n)}\delta_{mn}, \quad |\beta|<1, \qquad (14.10.18) \end{aligned}$$

where

$$\begin{aligned} w(x):=w(x;\beta|q) &= \left|\frac{(\mathrm{e}^{2i\theta};q)_\infty}{(\beta^{\frac{1}{2}}\mathrm{e}^{i\theta},\beta^{\frac{1}{2}}q^{\frac{1}{2}}\mathrm{e}^{i\theta},-\beta^{\frac{1}{2}}\mathrm{e}^{i\theta},-\beta^{\frac{1}{2}}q^{\frac{1}{2}}\mathrm{e}^{i\theta};q)_\infty}\right|^2\\ &= \left|\frac{(\mathrm{e}^{2i\theta};q)_\infty}{(\beta\mathrm{e}^{2i\theta};q)_\infty}\right|^2 = \frac{h(x,1)h(x,-1)h(x,q^{\frac{1}{2}})h(x,-q^{\frac{1}{2}})}{h(x,\beta^{\frac{1}{2}})h(x,\beta^{\frac{1}{2}}q^{\frac{1}{2}})h(x,-\beta^{\frac{1}{2}})h(x,-\beta^{\frac{1}{2}}q^{\frac{1}{2}})}, \end{aligned}$$

with

$$h(x,\alpha):=\prod_{k=0}^{\infty}\left(1-2\alpha xq^k+\alpha^2q^{2k}\right)=\left(\alpha\mathrm{e}^{i\theta},\alpha\mathrm{e}^{-i\theta};q\right)_\infty, \quad x=\cos\theta.$$

Recurrence Relation

$$2(1-\beta q^n)xC_n(x;\beta|q) = (1-q^{n+1})C_{n+1}(x;\beta|q) + (1-\beta^2q^{n-1})C_{n-1}(x;\beta|q). \quad (14.10.19)$$

Normalized Recurrence Relation

$$xp_n(x) = p_{n+1}(x) + \frac{(1-q^n)(1-\beta^2q^{n-1})}{4(1-\beta q^{n-1})(1-\beta q^n)}p_{n-1}(x), \quad (14.10.20)$$

where

$$C_n(x;\beta|q) = \frac{2^n(\beta;q)_n}{(q;q)_n}p_n(x).$$

q-Difference Equation

$$(1-q)^2D_q\left[\tilde{w}(x;\beta q|q)D_qy(x)\right] + \lambda_n\tilde{w}(x;\beta|q)y(x) = 0, \quad (14.10.21)$$

where

$$y(x) = C_n(x;\beta|q),$$

$$\tilde{w}(x;\beta|q) := \frac{w(x;\beta|q)}{\sqrt{1-x^2}}$$

and

$$\lambda_n = 4q^{-n+1}(1-q^n)(1-\beta^2q^n).$$

Forward Shift Operator

$$\delta_qC_n(x;\beta|q) = -q^{-\frac{1}{2}n}(1-\beta)(e^{i\theta}-e^{-i\theta})C_{n-1}(x;\beta q|q), \quad x=\cos\theta \quad (14.10.22)$$

or equivalently

$$D_qC_n(x;\beta|q) = 2q^{-\frac{1}{2}(n-1)}\frac{1-\beta}{1-q}C_{n-1}(x;\beta q|q). \quad (14.10.23)$$

Backward Shift Operator

$$\delta_q\left[\tilde{w}(x;\beta|q)C_n(x;\beta|q)\right]$$
$$=q^{-\frac{1}{2}(n+1)}\frac{(1-q^{n+1})(1-\beta^2q^{n-1})}{(1-\beta q^{-1})}(e^{i\theta}-e^{-i\theta})$$
$$\times\tilde{w}(x;\beta q^{-1}|q)C_{n+1}(x;\beta q^{-1}|q),\quad x=\cos\theta \tag{14.10.24}$$

or equivalently

$$D_q\left[\tilde{w}(x;\beta|q)C_n(x;\beta|q)\right]$$
$$=-\frac{2q^{-\frac{1}{2}n}(1-q^{n+1})(1-\beta^2q^{n-1})}{(1-q)(1-\beta q^{-1})}$$
$$\times\tilde{w}(x;\beta q^{-1}|q)C_{n+1}(x;\beta q^{-1}|q). \tag{14.10.25}$$

Rodrigues-Type Formula

$$\tilde{w}(x;\beta|q)C_n(x;\beta|q)$$
$$=\left(\frac{q-1}{2}\right)^n q^{\frac{1}{4}n(n-1)}\frac{(\beta;q)_n}{(q,\beta^2q^n;q)_n}(D_q)^n\left[\tilde{w}(x;\beta q^n|q)\right]. \tag{14.10.26}$$

Generating Functions

$$\left|\frac{(\beta e^{i\theta}t;q)_\infty}{(e^{i\theta}t;q)_\infty}\right|^2=\frac{(\beta e^{i\theta}t,\beta e^{-i\theta}t;q)_\infty}{(e^{i\theta}t,e^{-i\theta}t;q)_\infty}=\sum_{n=0}^{\infty}C_n(x;\beta|q)t^n,\quad x=\cos\theta. \tag{14.10.27}$$

$$\frac{1}{(e^{i\theta}t;q)_\infty}\,{}_2\phi_1\left(\begin{matrix}\beta,\beta e^{2i\theta}\\ \beta^2\end{matrix};q,e^{-i\theta}t\right)=\sum_{n=0}^{\infty}\frac{C_n(x;\beta|q)}{(\beta^2;q)_n}t^n,\quad x=\cos\theta. \tag{14.10.28}$$

$$(e^{-i\theta}t;q)_\infty\cdot{}_2\phi_1\left(\begin{matrix}\beta,\beta e^{2i\theta}\\ \beta^2\end{matrix};q,e^{-i\theta}t\right)$$
$$=\sum_{n=0}^{\infty}\frac{(-1)^n\beta^nq^{\binom{n}{2}}}{(\beta^2;q)_n}C_n(x;\beta|q)t^n,\quad x=\cos\theta. \tag{14.10.29}$$

$$
{}_2\phi_1\left(\begin{array}{c}\beta^{\frac{1}{2}}e^{i\theta},\beta^{\frac{1}{2}}q^{\frac{1}{2}}e^{i\theta}\\ \beta q^{\frac{1}{2}}\end{array};q,e^{-i\theta}t\right){}_2\phi_1\left(\begin{array}{c}-\beta^{\frac{1}{2}}e^{-i\theta},-\beta^{\frac{1}{2}}q^{\frac{1}{2}}e^{-i\theta}\\ \beta q^{\frac{1}{2}}\end{array};q,e^{i\theta}t\right)
$$
$$
=\sum_{n=0}^{\infty}\frac{(-\beta,-\beta q^{\frac{1}{2}};q)_n}{(\beta^2,\beta q^{\frac{1}{2}};q)_n}C_n(x;\beta|q)t^n,\quad x=\cos\theta. \tag{14.10.30}
$$

$$
{}_2\phi_1\left(\begin{array}{c}\beta^{\frac{1}{2}}e^{i\theta},-\beta^{\frac{1}{2}}e^{i\theta}\\ -\beta\end{array};q,e^{-i\theta}t\right){}_2\phi_1\left(\begin{array}{c}\beta^{\frac{1}{2}}q^{\frac{1}{2}}e^{-i\theta},-\beta^{\frac{1}{2}}q^{\frac{1}{2}}e^{-i\theta}\\ -\beta q\end{array};q,e^{i\theta}t\right)
$$
$$
=\sum_{n=0}^{\infty}\frac{(\beta q^{\frac{1}{2}},-\beta q^{\frac{1}{2}};q)_n}{(\beta^2,-\beta q;q)_n}C_n(x;\beta|q)t^n,\quad x=\cos\theta. \tag{14.10.31}
$$

$$
{}_2\phi_1\left(\begin{array}{c}\beta^{\frac{1}{2}}e^{i\theta},-\beta^{\frac{1}{2}}q^{\frac{1}{2}}e^{i\theta}\\ -\beta q^{\frac{1}{2}}\end{array};q,e^{-i\theta}t\right){}_2\phi_1\left(\begin{array}{c}\beta^{\frac{1}{2}}q^{\frac{1}{2}}e^{-i\theta},-\beta^{\frac{1}{2}}e^{-i\theta}\\ -\beta q^{\frac{1}{2}}\end{array};q,e^{i\theta}t\right)
$$
$$
=\sum_{n=0}^{\infty}\frac{(-\beta,\beta q^{\frac{1}{2}};q)_n}{(\beta^2,-\beta q^{\frac{1}{2}};q)_n}C_n(x;\beta|q)t^n,\quad x=\cos\theta. \tag{14.10.32}
$$

$$
\frac{(\gamma e^{i\theta}t;q)_\infty}{(e^{i\theta}t;q)_\infty}\,{}_3\phi_2\left(\begin{array}{c}\gamma,\beta,\beta e^{2i\theta}\\ \beta^2,\gamma e^{i\theta}t\end{array};q,e^{-i\theta}t\right)
$$
$$
=\sum_{n=0}^{\infty}\frac{(\gamma;q)_n}{(\beta^2;q)_n}C_n(x;\beta|q)t^n,\quad x=\cos\theta,\quad \gamma\text{ arbitrary}. \tag{14.10.33}
$$

Limit Relations

Askey-Wilson → Continuous q-ultraspherical / Rogers

If we set $a=\beta^{\frac{1}{2}}$, $b=\beta^{\frac{1}{2}}q^{\frac{1}{2}}$, $c=-\beta^{\frac{1}{2}}$ and $d=-\beta^{\frac{1}{2}}q^{\frac{1}{2}}$ in the definition (14.1.1) of the Askey-Wilson polynomials and change the normalization we obtain the continuous q-ultraspherical (or Rogers) polynomials given by (14.10.17). In fact we have:

$$
\frac{(\beta^2;q)_n p_n(x;\beta^{\frac{1}{2}},\beta^{\frac{1}{2}}q^{\frac{1}{2}},-\beta^{\frac{1}{2}},-\beta^{\frac{1}{2}}q^{\frac{1}{2}}|q)}{(\beta q^{\frac{1}{2}},-\beta,-\beta q^{\frac{1}{2}},q;q)_n}=C_n(x;\beta|q).
$$

q-Meixner-Pollaczek → Continuous q-ultraspherical / Rogers

If we take $\theta=0$ and $a=\beta$ in the definition (14.9.1) of the q-Meixner-Pollaczek polynomials we obtain the continuous q-ultraspherical (or Rogers) polynomials given by

(14.10.17):

$$P_n(\cos\phi;\beta|q) = C_n(\cos\phi;\beta|q).$$

Continuous q-ultraspherical / Rogers → Continuous q-Hermite

The continuous q-Hermite polynomials given by (14.26.1) can be obtained from the continuous q-ultraspherical (or Rogers) polynomials given by (14.10.17) by taking the limit $\beta \to 0$. In fact we have

$$\lim_{\beta\to 0} C_n(x;\beta|q) = \frac{H_n(x|q)}{(q;q)_n}. \tag{14.10.34}$$

Continuous q-ultraspherical / Rogers → Gegenbauer / Ultraspherical

If we set $\beta = q^{\lambda}$ in the definition (14.10.17) of the continuous q-ultraspherical (or Rogers) polynomials and let q tend to 1 we obtain the Gegenbauer (or ultraspherical) polynomials given by (9.8.19):

$$\lim_{q\to 1} C_n(x;q^{\lambda}|q) = C_n^{(\lambda)}(x). \tag{14.10.35}$$

Remarks

The continuous q-ultraspherical (or Rogers) polynomials can also be written as:

$$C_n(x;\beta|q) = \sum_{k=0}^{n} \frac{(\beta;q)_k(\beta;q)_{n-k}}{(q;q)_k(q;q)_{n-k}} e^{i(n-2k)\theta}, \quad x = \cos\theta.$$

They can be obtained from the continuous q-Jacobi polynomials given by (14.10.1) in the following way. Set $\beta = \alpha$ in the definition (14.10.1) and replace $q^{\alpha+\frac{1}{2}}$ by β and we find the continuous q-ultraspherical (or Rogers) polynomials with a different normalization. We have

$$P_n^{(\alpha,\alpha)}(x|q) \overset{q^{\alpha+\frac{1}{2}}\to\beta}{\longrightarrow} \frac{(\beta q^{\frac{1}{2}};q)_n}{(\beta^2;q)_n} \beta^{\frac{1}{2}n} C_n(x;\beta|q).$$

If we set $\beta = q^{\alpha+\frac{1}{2}}$ in the definition (14.10.17) of the q-ultraspherical (or Rogers) polynomials we find the continuous q-Jacobi polynomials given by (14.10.1) with $\beta = \alpha$. In fact we have

$$C_n(x;q^{\alpha+\frac{1}{2}}|q) = \frac{(q^{2\alpha+1};q)_n}{(q^{\alpha+1};q)_n q^{(\frac{1}{2}\alpha+\frac{1}{4})n}} P_n^{(\alpha,\alpha)}(x|q).$$

If we replace q by q^{-1} we find

$$C_n(x;\beta|q^{-1}) = (\beta q)^n C_n(x;\beta^{-1}|q).$$

The special case $\beta = q$ of the continuous q-ultraspherical (or Rogers) polynomials equals the Chebyshev polynomials of the second kind given by (9.8.36). In fact we have

$$C_n(x;q|q) = \frac{\sin(n+1)\theta}{\sin\theta} = U_n(x), \quad x = \cos\theta.$$

The limit case $\beta \to 1$ leads to the Chebyshev polynomials of the first kind given by (9.8.35) in the following way:

$$\lim_{\beta\to 1} \frac{1-q^n}{2(1-\beta)} C_n(x;\beta|q) = \cos n\theta = T_n(x), \quad x = \cos\theta, \quad n = 1,2,3,\ldots.$$

The continuous q-Jacobi polynomials given by (14.10.16) and the continuous q-ultraspherical (or Rogers) polynomials given by (14.10.17) are connected by the quadratic transformations:

$$C_{2n}(x;q^\lambda|q) = \frac{(q^\lambda,-q;q)_n}{(q^{\frac{1}{2}},-q^{\frac{1}{2}};q)_n} q^{-\frac{1}{2}n} P_n^{(\lambda-\frac{1}{2},-\frac{1}{2})}(2x^2-1;q)$$

and

$$C_{2n+1}(x;q^\lambda|q) = \frac{(q^\lambda,-1;q)_{n+1}}{(q^{\frac{1}{2}},-q^{\frac{1}{2}};q)_{n+1}} q^{-\frac{1}{2}n} x P_n^{(\lambda-\frac{1}{2},\frac{1}{2})}(2x^2-1;q).$$

Finally we remark that the continuous q-ultraspherical (or Rogers) polynomials are related to the continuous q-Legendre polynomials given by (14.10.36) in the following way:

$$C_n(x;q^{\frac{1}{2}}|q) = q^{-\frac{1}{4}n} P_n(x|q).$$

References

[13], [14], [16], [34], [51], [52], [53], [62], [63], [64], [66], [72], [80], [114], [119], [120], [210], [230], [231], [232], [234], [236], [237], [238], [271], [288], [294], [295], [299], [325], [326], [344], [350], [351], [395], [414], [416], [421], [446], [448], [452], [453], [454], [460], [461], [462], [473].

14.10.2 Continuous q-Legendre

Basic Hypergeometric Representation

The continuous q-Legendre polynomials are continuous q-Jacobi polynomials with $\alpha = \beta = 0$:

$$P_n(x|q) = {}_4\phi_3\left(\begin{matrix} q^{-n}, q^{n+1}, q^{\frac{1}{4}}e^{i\theta}, q^{\frac{1}{4}}e^{-i\theta} \\ q, -q^{\frac{1}{2}}, -q \end{matrix}; q, q\right), \quad x = \cos\theta. \tag{14.10.36}$$

Orthogonality Relation

$$\frac{1}{2\pi}\int_{-1}^{1}\frac{w(x;1|q)}{\sqrt{1-x^2}}P_m(x|q)P_n(x|q)\,dx = \frac{(q^{\frac{1}{2}};q)_\infty}{(q,q,-q^{\frac{1}{2}},-q;q)_\infty}\frac{q^{\frac{1}{2}n}}{1-q^{n+\frac{1}{2}}}\,\delta_{mn}, \tag{14.10.37}$$

where

$$\begin{aligned} w(x;a|q) &= \left|\frac{(e^{2i\theta};q)_\infty}{(aq^{\frac{1}{4}}e^{i\theta}, aq^{\frac{3}{4}}e^{i\theta}, -aq^{\frac{1}{4}}e^{i\theta}, -aq^{\frac{3}{4}}e^{i\theta};q)_\infty}\right|^2 \\ &= \left|\frac{(e^{i\theta}, -e^{i\theta};q^{\frac{1}{2}})_\infty}{(aq^{\frac{1}{4}}e^{i\theta}, -aq^{\frac{1}{4}}e^{i\theta};q^{\frac{1}{2}})_\infty}\right|^2 = \left|\frac{(e^{2i\theta};q)_\infty}{(a^2q^{\frac{1}{2}}e^{2i\theta};q)_\infty}\right|^2 \\ &= \frac{h(x,1)h(x,-1)h(x,q^{\frac{1}{2}})h(x,-q^{\frac{1}{2}})}{h(x,aq^{\frac{1}{4}})h(x,aq^{\frac{3}{4}})h(x,-aq^{\frac{1}{4}})h(x,-aq^{\frac{3}{4}})}, \end{aligned}$$

with

$$h(x,\alpha) := \prod_{k=0}^{\infty}\left(1-2\alpha xq^k+\alpha^2q^{2k}\right) = \left(\alpha e^{i\theta}, \alpha e^{-i\theta};q\right)_\infty, \quad x = \cos\theta.$$

Recurrence Relation

$$2(1-q^{n+\frac{1}{2}})xP_n(x|q) = q^{-\frac{1}{4}}(1-q^{n+1})P_{n+1}(x|q) + q^{\frac{1}{4}}(1-q^n)P_{n-1}(x|q). \tag{14.10.38}$$

Normalized Recurrence Relation

$$xp_n(x) = p_{n+1}(x) + \frac{(1-q^n)^2}{4(1-q^{n-\frac{1}{2}})(1-q^{n+\frac{1}{2}})} p_{n-1}(x), \tag{14.10.39}$$

where

$$P_n(x|q) = \frac{2^n q^{\frac{1}{4}n}(q^{\frac{1}{2}};q)_n}{(q;q)_n} p_n(x).$$

q-Difference Equation

$$(1-q)^2 D_q \left[\tilde{w}(x;q^{\frac{1}{2}}|q) D_q y(x)\right] + \lambda_n \tilde{w}(x;1|q) y(x) = 0, \quad y(x) = P_n(x|q), \tag{14.10.40}$$

where

$$\lambda_n = 4q^{-n+1}(1-q^n)(1-q^{n+1})$$

and

$$\tilde{w}(x;a|q) := \frac{w(x;a|q)}{\sqrt{1-x^2}}.$$

Rodrigues-Type Formula

$$\tilde{w}(x;1|q) P_n(x|q) = \left(\frac{q-1}{2}\right)^n \frac{q^{\frac{1}{4}n^2}}{(q,-q^{\frac{1}{2}},-q;q)_n} (D_q)^n \left[\tilde{w}(x;q^{\frac{1}{2}n}|q)\right]. \tag{14.10.41}$$

Generating Functions

$$\left|\frac{(q^{\frac{1}{2}}e^{i\theta}t;q)_\infty}{(e^{i\theta}t;q)_\infty}\right|^2 = \frac{(q^{\frac{1}{2}}e^{i\theta}t, q^{\frac{1}{2}}e^{-i\theta}t;q)_\infty}{(e^{i\theta}t, e^{-i\theta}t;q)_\infty} = \sum_{n=0}^{\infty} \frac{P_n(x|q)}{q^{\frac{1}{4}n}} t^n, \quad x = \cos\theta. \tag{14.10.42}$$

$$\frac{1}{(e^{i\theta}t;q)_\infty} {}_2\phi_1\left(\begin{matrix} q^{\frac{1}{2}}, q^{\frac{1}{2}}e^{2i\theta} \\ q \end{matrix}; q, e^{-i\theta}t\right) = \sum_{n=0}^{\infty} \frac{P_n(x|q)}{(q;q)_n q^{\frac{1}{4}n}} t^n, \quad x = \cos\theta. \tag{14.10.43}$$

$$(e^{-i\theta}t;q)_\infty \cdot {}_2\phi_1\left(\begin{matrix} q^{\frac{1}{2}}, q^{\frac{1}{2}}e^{2i\theta} \\ q \end{matrix} ; q, e^{-i\theta}t\right) = \sum_{n=0}^{\infty} \frac{(-1)^n q^{\frac{1}{4}n+\binom{n}{2}}}{(q;q)_n} P_n(x|q)t^n, \quad x = \cos\theta. \tag{14.10.44}$$

$${}_2\phi_1\left(\begin{matrix} q^{\frac{1}{4}}e^{i\theta}, q^{\frac{3}{4}}e^{i\theta} \\ q \end{matrix} ; q, e^{-i\theta}t\right) {}_2\phi_1\left(\begin{matrix} -q^{\frac{1}{4}}e^{-i\theta}, -q^{\frac{3}{4}}e^{-i\theta} \\ q \end{matrix} ; q, e^{i\theta}t\right) = \sum_{n=0}^{\infty} \frac{(-q^{\frac{1}{2}}, -q;q)_n}{(q,q;q)_n} \frac{P_n(x|q)}{q^{\frac{1}{4}n}} t^n, \quad x = \cos\theta. \tag{14.10.45}$$

$${}_2\phi_1\left(\begin{matrix} q^{\frac{1}{4}}e^{i\theta}, -q^{\frac{1}{4}}e^{i\theta} \\ -q^{\frac{1}{2}} \end{matrix} ; q, e^{-i\theta}t\right) {}_2\phi_1\left(\begin{matrix} q^{\frac{3}{4}}e^{-i\theta}, -q^{\frac{3}{4}}e^{-i\theta} \\ -q^{\frac{3}{2}} \end{matrix} ; q, e^{i\theta}t\right) = \sum_{n=0}^{\infty} \frac{(-q;q)_n}{(-q^{\frac{3}{2}};q)_n} \frac{P_n(x|q)}{q^{\frac{1}{4}n}} t^n, \quad x = \cos\theta. \tag{14.10.46}$$

$${}_2\phi_1\left(\begin{matrix} q^{\frac{1}{4}}e^{i\theta}, -q^{\frac{3}{4}}e^{i\theta} \\ -q \end{matrix} ; q, e^{-i\theta}t\right) {}_2\phi_1\left(\begin{matrix} q^{\frac{3}{4}}e^{-i\theta}, -q^{\frac{1}{4}}e^{-i\theta} \\ -q \end{matrix} ; q, e^{i\theta}t\right) = \sum_{n=0}^{\infty} \frac{(-q^{\frac{1}{2}};q)_n}{(-q;q)_n} \frac{P_n(x|q)}{q^{\frac{1}{4}n}} t^n, \quad x = \cos\theta. \tag{14.10.47}$$

$$\frac{(\gamma e^{i\theta}t;q)_\infty}{(e^{i\theta}t;q)_\infty} {}_3\phi_2\left(\begin{matrix} \gamma, q^{\frac{1}{2}}, q^{\frac{1}{2}}e^{2i\theta} \\ q, \gamma e^{i\theta}t \end{matrix} ; q, e^{-i\theta}t\right) = \sum_{n=0}^{\infty} \frac{(\gamma;q)_n}{(q;q)_n} \frac{P_n(x|q)}{q^{\frac{1}{4}n}} t^n, \quad x = \cos\theta, \quad \gamma \text{ arbitrary}. \tag{14.10.48}$$

Limit Relations

Continuous q-Legendre → Legendre / Spherical

The Legendre (or spherical) polynomials given by (9.8.62) easily follow from the continuous q-Legendre polynomials given by (14.10.36) by taking the limit $q \to 1$:

$$\lim_{q\to 1} P_n(x|q) = P_n(x). \tag{14.10.49}$$

Remarks

The continuous q-Legendre polynomials can also be written as:

$$P_n(x|q) = q^{\frac{1}{4}n} \sum_{k=0}^{n} \frac{(q^{\frac{1}{2}};q)_k (q^{\frac{1}{2}};q)_{n-k}}{(q;q)_k (q;q)_{n-k}} e^{i(n-2k)\theta}, \quad x = \cos\theta.$$

If we set $\alpha = \beta = 0$ in (14.10.16) we find

$$P_n(x;q) = {}_4\phi_3 \left(\begin{matrix} q^{-n}, q^{n+1}, q^{\frac{1}{2}} e^{i\theta}, q^{\frac{1}{2}} e^{-i\theta} \\ q, -q, -q \end{matrix} ; q, q \right), \quad x = \cos\theta,$$

but these are not really different from those given by (14.10.36) in view of the quadratic transformation

$$P_n(x|q^2) = P_n(x;q).$$

If we replace q by q^{-1} we find

$$P_n(x|q^{-1}) = P_n(x|q).$$

The continuous q-Legendre polynomials are related to the continuous q-ultraspherical (or Rogers) polynomials given by (14.10.17) in the following way:

$$P_n(x|q) = q^{\frac{1}{4}n} C_n(x; q^{\frac{1}{2}}|q).$$

References

[320], [323], [341], [345], [348].

14.11 Big q-Laguerre

Basic Hypergeometric Representation

$$P_n(x;a,b;q) = {}_3\phi_2 \left(\begin{matrix} q^{-n}, 0, x \\ aq, bq \end{matrix} ; q, q \right) \tag{14.11.1}$$
$$= \frac{1}{(b^{-1}q^{-n};q)_n} {}_2\phi_1 \left(\begin{matrix} q^{-n}, aqx^{-1} \\ aq \end{matrix} ; q, \frac{x}{b} \right).$$

Orthogonality Relation

For $0 < aq < 1$ and $b < 0$ we have

$$\int_{bq}^{aq} \frac{(a^{-1}x, b^{-1}x;q)_\infty}{(x;q)_\infty} P_m(x;a,b;q)P_n(x;a,b;q)\,d_qx$$
$$= aq(1-q)\frac{(q,a^{-1}b,ab^{-1}q;q)_\infty}{(aq,bq;q)_\infty}\frac{(q;q)_n}{(aq,bq;q)_n}(-abq^2)^n q^{\binom{n}{2}}\,\delta_{mn}. \quad (14.11.2)$$

Recurrence Relation

$$(x-1)P_n(x;a,b;q) = A_nP_{n+1}(x;a,b;q) - (A_n+C_n)P_n(x;a,b;q) + C_nP_{n-1}(x;a,b;q), \quad (14.11.3)$$

where

$$\begin{cases} A_n = (1-aq^{n+1})(1-bq^{n+1}) \\ C_n = -abq^{n+1}(1-q^n). \end{cases}$$

Normalized Recurrence Relation

$$xp_n(x) = p_{n+1}(x) + [1-(A_n+C_n)]\,p_n(x) - abq^{n+1}(1-q^n)(1-aq^n)(1-bq^n)p_{n-1}(x), \quad (14.11.4)$$

where

$$P_n(x;a,b;q) = \frac{1}{(aq,bq;q)_n}p_n(x).$$

q-Difference Equation

$$q^{-n}(1-q^n)x^2y(x) = B(x)y(qx) - [B(x)+D(x)]\,y(x) + D(x)y(q^{-1}x), \quad (14.11.5)$$

where

$$y(x) = P_n(x;a,b;q)$$

and

$$\begin{cases} B(x) = abq(1-x) \\ D(x) = (x-aq)(x-bq). \end{cases}$$

Forward Shift Operator

$$P_n(x;a,b;q)-P_n(qx;a,b;q)=\frac{q^{-n+1}(1-q^n)}{(1-aq)(1-bq)}xP_{n-1}(qx;aq,bq;q) \tag{14.11.6}$$

or equivalently

$$\mathscr{D}_qP_n(x;a,b;q)=\frac{q^{-n+1}(1-q^n)}{(1-q)(1-aq)(1-bq)}P_{n-1}(qx;aq,bq;q). \tag{14.11.7}$$

Backward Shift Operator

$$\begin{aligned}&(x-a)(x-b)P_n(x;a,b;q)-ab(1-x)P_n(qx;a,b;q)\\&=(1-a)(1-b)xP_{n+1}(x;aq^{-1},bq^{-1};q)\end{aligned} \tag{14.11.8}$$

or equivalently

$$\begin{aligned}&\mathscr{D}_q\left[w(x;a,b;q)P_n(x;a,b;q)\right]\\&=\frac{(1-a)(1-b)}{ab(1-q)}w(x;aq^{-1},bq^{-1};q)P_{n+1}(x;aq^{-1},bq^{-1};q),\end{aligned} \tag{14.11.9}$$

where

$$w(x;a,b;q)=\frac{(a^{-1}x,b^{-1}x;q)_\infty}{(x;q)_\infty}.$$

Rodrigues-Type Formula

$$\begin{aligned}&w(x;a,b;q)P_n(x;a,b;q)\\&=\frac{a^nb^nq^{n(n+1)}(1-q)^n}{(aq,bq;q)_n}(\mathscr{D}_q)^n\left[w(x;aq^n,bq^n;q)\right].\end{aligned} \tag{14.11.10}$$

Generating Functions

$$(bqt;q)_\infty\cdot{}_2\phi_1\left(\begin{matrix}aqx^{-1},0\\aq\end{matrix};q,xt\right)=\sum_{n=0}^{\infty}\frac{(bq;q)_n}{(q;q)_n}P_n(x;a,b;q)t^n. \tag{14.11.11}$$

$$(aqt;q)_\infty \cdot {}_2\phi_1\left(\begin{matrix} bqx^{-1},0 \\ bq \end{matrix} ; q, xt\right) = \sum_{n=0}^{\infty} \frac{(aq;q)_n}{(q;q)_n} P_n(x;a,b;q)t^n. \tag{14.11.12}$$

$$(t;q)_\infty \cdot {}_3\phi_2\left(\begin{matrix} 0,0,x \\ aq,bq \end{matrix} ; q, t\right) = \sum_{n=0}^{\infty} \frac{(-1)^n q^{\binom{n}{2}}}{(q;q)_n} P_n(x;a,b;q)t^n. \tag{14.11.13}$$

Limit Relations

Big q-Jacobi → Big q-Laguerre

If we set $b=0$ in the definition (14.5.1) of the big q-Jacobi polynomials we obtain the big q-Laguerre polynomials given by (14.11.1):

$$P_n(x;a,0,c;q) = P_n(x;a,c;q).$$

Big q-Laguerre → Little q-Laguerre / Wall

The little q-Laguerre (or Wall) polynomials given by (14.20.1) can be obtained from the big q-Laguerre polynomials by taking $x \to bqx$ in (14.11.1) and letting $b \to -\infty$:

$$\lim_{b\to-\infty} P_n(bqx;a,b;q) = p_n(x;a|q). \tag{14.11.14}$$

Big q-Laguerre → Al-Salam-Carlitz I

If we set $x \to aqx$ and $b \to ab$ in the definition (14.11.1) of the big q-Laguerre polynomials and take the limit $a \to 0$ we obtain the Al-Salam-Carlitz I polynomials given by (14.24.1):

$$\lim_{a\to 0} \frac{P_n(aqx;a,ab;q)}{a^n} = q^n U_n^{(b)}(x;q). \tag{14.11.15}$$

Big q-Laguerre → Laguerre

The Laguerre polynomials given by (9.12.1) can be obtained from the big q-Laguerre polynomials by the substitution $a=q^\alpha$ and $b=(1-q)^{-1}q^\beta$ in the definition (14.11.1) of the big q-Laguerre polynomials and the limit $q \to 1$:

$$\lim_{q\to 1} P_n(x;q^\alpha,(1-q)^{-1}q^\beta;q) = \frac{L_n^{(\alpha)}(x-1)}{L_n^{(\alpha)}(0)}. \tag{14.11.16}$$

Remark

The big q-Laguerre polynomials given by (14.11.1) and the affine q-Krawtchouk polynomials given by (14.16.1) are related in the following way:

$$K_n^{Aff}(q^{-x};p,N;q) = P_n(q^{-x};p,q^{-N-1};q).$$

References

[16], [28], [74], [157], [284].

14.12 Little q-Jacobi

Basic Hypergeometric Representation

$$p_n(x;a,b|q) = {}_2\phi_1\left(\begin{matrix} q^{-n}, abq^{n+1} \\ aq \end{matrix}; q, qx\right). \tag{14.12.1}$$

Orthogonality Relation

For $0 < aq < 1$ and $bq < 1$ we have

$$\begin{aligned}&\sum_{k=0}^{\infty} \frac{(bq;q)_k}{(q;q)_k}(aq)^k p_m(q^k;a,b|q)p_n(q^k;a,b|q)\\ &= \frac{(abq^2;q)_\infty}{(aq;q)_\infty}\frac{(1-abq)(aq)^n}{(1-abq^{2n+1})}\frac{(q,bq;q)_n}{(aq,abq;q)_n}\delta_{mn}.\end{aligned} \tag{14.12.2}$$

Recurrence Relation

$$\begin{aligned}-xp_n(x;a,b|q) &= A_n p_{n+1}(x;a,b|q) - (A_n + C_n)\,p_n(x;a,b|q)\\ &\quad + C_n p_{n-1}(x;a,b|q),\end{aligned} \tag{14.12.3}$$

where

$$\begin{cases} A_n = q^n \dfrac{(1-aq^{n+1})(1-abq^{n+1})}{(1-abq^{2n+1})(1-abq^{2n+2})} \\ \\ C_n = aq^n \dfrac{(1-q^n)(1-bq^n)}{(1-abq^{2n})(1-abq^{2n+1})}. \end{cases}$$

Normalized Recurrence Relation

$$xp_n(x) = p_{n+1}(x) + (A_n + C_n)p_n(x) + A_{n-1}C_n p_{n-1}(x), \tag{14.12.4}$$

where

$$p_n(x;a,b|q) = \frac{(-1)^n q^{-\binom{n}{2}}(abq^{n+1};q)_n}{(aq;q)_n} p_n(x).$$

q-Difference Equation

$$\begin{aligned} &q^{-n}(1-q^n)(1-abq^{n+1})xy(x) \\ &= B(x)y(qx) - [B(x)+D(x)]\,y(x) + D(x)y(q^{-1}x), \end{aligned} \tag{14.12.5}$$

where

$$y(x) = p_n(x;a,b|q)$$

and

$$\begin{cases} B(x) = a(bqx-1) \\ \\ D(x) = x-1. \end{cases}$$

Forward Shift Operator

$$\begin{aligned} &p_n(x;a,b|q) - p_n(qx;a,b|q) \\ &= -\frac{q^{-n+1}(1-q^n)(1-abq^{n+1})}{(1-aq)} xp_{n-1}(x;aq,bq|q) \end{aligned} \tag{14.12.6}$$

or equivalently

$$\mathcal{D}_q p_n(x;a,b|q) = -\frac{q^{-n+1}(1-q^n)(1-abq^{n+1})}{(1-q)(1-aq)} p_{n-1}(x;aq,bq|q). \tag{14.12.7}$$

Backward Shift Operator

$$a(bx-1)p_n(x;a,b|q)-(x-1)p_n(q^{-1}x;a,b|q)$$
$$=(1-a)p_{n+1}(x;aq^{-1},bq^{-1}|q) \qquad (14.12.8)$$

or equivalently

$$\mathscr{D}_{q^{-1}}\left[w(x;\alpha,\beta|q)p_n(x;q^\alpha,q^\beta|q)\right]$$
$$=\frac{1-q^\alpha}{q^{\alpha-1}(1-q)}w(x;\alpha-1,\beta-1|q)p_{n+1}(x;q^{\alpha-1},q^{\beta-1}|q), \qquad (14.12.9)$$

where

$$w(x;\alpha,\beta|q)=\frac{(qx;q)_\infty}{(q^{\beta+1}x;q)_\infty}x^\alpha.$$

Rodrigues-Type Formula

$$w(x;\alpha,\beta|q)p_n(x;q^\alpha,q^\beta|q)$$
$$=\frac{q^{n\alpha+\binom{n}{2}}(1-q)^n}{(q^{\alpha+1};q)_n}\left(\mathscr{D}_{q^{-1}}\right)^n[w(x;\alpha+n,\beta+n|q)]. \qquad (14.12.10)$$

Generating Function

$${}_0\phi_1\left(\begin{matrix}-\\aq\end{matrix};q,aqxt\right){}_2\phi_1\left(\begin{matrix}x^{-1},0\\bq\end{matrix};q,xt\right)=\sum_{n=0}^\infty\frac{(-1)^nq^{\binom{n}{2}}}{(bq,q;q)_n}p_n(x;a,b|q)t^n. \qquad (14.12.11)$$

Limit Relations

Big q-Jacobi → Little q-Jacobi

The little q-Jacobi polynomials given by (14.12.1) can be obtained from the big q-Jacobi polynomials by the substitution $x\to cqx$ in the definition (14.5.1) and then by the limit $c\to-\infty$:

$$\lim_{c\to-\infty}P_n(cqx;a,b,c;q)=p_n(x;a,b|q).$$

q-Hahn $\rightarrow$ Little q-Jacobi

If we set $x \rightarrow N-x$ in the definition (14.6.1) of the q-Hahn polynomials and take the limit $N \rightarrow \infty$ we find the little q-Jacobi polynomials:

$$\lim_{N\rightarrow\infty} Q_n(q^{x-N};\alpha,\beta,N|q) = p_n(q^x;\alpha,\beta|q),$$

where $p_n(q^x;\alpha,\beta|q)$ is given by (14.12.1).

Little q-Jacobi $\rightarrow$ Little q-Laguerre / Wall

The little q-Laguerre (or Wall) polynomials given by (14.20.1) are little q-Jacobi polynomials with $b=0$. So if we set $b=0$ in the definition (14.12.1) of the little q-Jacobi polynomials we obtain the little q-Laguerre (or Wall) polynomials:

$$p_n(x;a,0|q) = p_n(x;a|q). \tag{14.12.12}$$

Little q-Jacobi $\rightarrow$ q-Laguerre

If we substitute $a=q^\alpha$ and $x \rightarrow -b^{-1}q^{-1}x$ in the definition (14.12.1) of the little q-Jacobi polynomials and then take the limit $b \rightarrow -\infty$ we find the q-Laguerre polynomials given by (14.21.1):

$$\lim_{b\rightarrow-\infty} p_n(-b^{-1}q^{-1}x;q^\alpha,b|q) = \frac{(q;q)_n}{(q^{\alpha+1};q)_n} L_n^{(\alpha)}(x;q). \tag{14.12.13}$$

Little q-Jacobi $\rightarrow$ q-Bessel

If we set $b \rightarrow -a^{-1}q^{-1}b$ in the definition (14.12.1) of the little q-Jacobi polynomials and then take the limit $a \rightarrow 0$ we obtain the q-Bessel polynomials given by (14.22.1):

$$\lim_{a\rightarrow 0} p_n(x;a,-a^{-1}q^{-1}b|q) = y_n(x;b;q). \tag{14.12.14}$$

Little q-Jacobi $\rightarrow$ Jacobi / Laguerre

The Jacobi polynomials given by (9.8.1) simply follow from the little q-Jacobi polynomials given by (14.12.1) in the following way:

$$\lim_{q\rightarrow 1} p_n(x;q^\alpha,q^\beta|q) = \frac{P_n^{(\alpha,\beta)}(1-2x)}{P_n^{(\alpha,\beta)}(1)}. \tag{14.12.15}$$

If we take $a = q^\alpha$, $b = -q^\beta$ for arbitrary real β and $x \to \frac{1}{2}(1-q)x$ in the definition (14.12.1) of the little q-Jacobi polynomials and then take the limit $q \to 1$ we obtain the Laguerre polynomials given by (9.12.1):

$$\lim_{q\to 1} p_n(\tfrac{1}{2}(1-q)x; q^\alpha, -q^\beta|q) = \frac{L_n^{(\alpha)}(x)}{L_n^{(\alpha)}(0)}. \tag{14.12.16}$$

Remarks

The little q-Jacobi polynomials given by (14.12.1) and the big q-Jacobi polynomials given by (14.5.1) are related in the following way:

$$p_n(x;a,b|q) = \frac{(bq;q)_n}{(aq;q)_n}(-1)^n b^{-n} q^{-n-\binom{n}{2}} P_n(bqx;b,a,0;q).$$

The little q-Jacobi polynomials and the q-Meixner polynomials given by (14.13.1) are related in the following way:

$$M_n(q^{-x};b,c;q) = p_n(-c^{-1}q^n;b,b^{-1}q^{-n-x-1}|q).$$

References

[12], [16], [23], [24], [33], [34], [51], [79], [80], [157], [212], [213], [214], [235], [238], [256], [261], [271], [287], [298], [304], [326], [329], [343], [345], [346], [348], [402], [408], [416], [442], [480], [482], [485].

Special Case

14.12.1 Little q-Legendre

Basic Hypergeometric Representation

The little q-Legendre polynomials are little q-Jacobi polynomials with $a = b = 1$:

$$p_n(x|q) = {}_2\phi_1\left(\begin{matrix} q^{-n}, q^{n+1} \\ q \end{matrix}; q, qx\right). \tag{14.12.17}$$

Orthogonality Relation

$$\int_0^1 p_m(x|q)p_n(x|q)\,d_qx = (1-q)\sum_{k=0}^{\infty} q^k p_m(q^k|q)p_n(q^k|q)$$
$$= \frac{(1-q)q^n}{(1-q^{2n+1})}\,\delta_{mn}. \tag{14.12.18}$$

Recurrence Relation

$$-xp_n(x|q) = A_n p_{n+1}(x|q) - (A_n + C_n)\,p_n(x|q) + C_n p_{n-1}(x|q), \tag{14.12.19}$$

where

$$\begin{cases} A_n = q^n \dfrac{(1-q^{n+1})}{(1+q^{n+1})(1-q^{2n+1})} \\ C_n = q^n \dfrac{(1-q^n)}{(1+q^n)(1-q^{2n+1})}. \end{cases}$$

Normalized Recurrence Relation

$$xp_n(x) = p_{n+1}(x) + (A_n + C_n)p_n(x) + A_{n-1}C_n p_{n-1}(x), \tag{14.12.20}$$

where

$$p_n(x|q) = \frac{(-1)^n q^{-\binom{n}{2}}(q^{n+1};q)_n}{(q;q)_n}\,p_n(x).$$

q-Difference Equation

$$q^{-n}(1-q^n)(1-q^{n+1})xy(x)$$
$$= B(x)y(qx) - [B(x) + D(x)]\,y(x) + D(x)y(q^{-1}x), \tag{14.12.21}$$

where

$$y(x) = p_n(x|q)$$

and

$$\begin{cases} B(x) = qx - 1 \\ D(x) = x - 1. \end{cases}$$

Rodrigues-Type Formula

$$p_n(x|q) = \frac{q^{\binom{n}{2}}(1-q)^n}{(q;q)_n}\left(\mathscr{D}_{q^{-1}}\right)^n\left[(qx;q)_n x^n\right]. \tag{14.12.22}$$

Generating Function

$${}_0\phi_1\left(\begin{matrix}-\\q\end{matrix};q,qxt\right){}_2\phi_1\left(\begin{matrix}x^{-1},0\\q\end{matrix};q,xt\right) = \sum_{n=0}^{\infty}\frac{(-1)^n q^{\binom{n}{2}}}{(q,q;q)_n}p_n(x|q)t^n. \tag{14.12.23}$$

Limit Relation

Little q-Legendre $\rightarrow$ Legendre / Spherical

If we take the limit $q \rightarrow 1$ in the definition (14.12.17) of the little q-Legendre polynomials we simply find the Legendre (or spherical) polynomials given by (9.8.62):

$$\lim_{q\rightarrow 1} p_n(x|q) = P_n(1-2x). \tag{14.12.24}$$

References

[345], [346], [447], [504].

14.13 q-Meixner

Basic Hypergeometric Representation

$$M_n(q^{-x};b,c;q) = {}_2\phi_1\left(\begin{matrix}q^{-n},q^{-x}\\bq\end{matrix};q,-\frac{q^{n+1}}{c}\right). \tag{14.13.1}$$

Orthogonality Relation

$$\sum_{x=0}^{\infty}\frac{(bq;q)_x}{(q,-bcq;q)_x}c^x q^{\binom{x}{2}}M_m(q^{-x};b,c;q)M_n(q^{-x};b,c;q)$$
$$=\frac{(-c;q)_\infty}{(-bcq;q)_\infty}\frac{(q,-c^{-1}q;q)_n}{(bq;q)_n}q^{-n}\,\delta_{mn},\quad 0\leq bq<1,\quad c>0. \quad (14.13.2)$$

Recurrence Relation

$$\begin{aligned}&q^{2n+1}(1-q^{-x})M_n(q^{-x})\\&=c(1-bq^{n+1})M_{n+1}(q^{-x})\\&\qquad-\left[c(1-bq^{n+1})+q(1-q^n)(c+q^n)\right]M_n(q^{-x})\\&\qquad\qquad+q(1-q^n)(c+q^n)M_{n-1}(q^{-x}),\end{aligned} \quad (14.13.3)$$

where

$$M_n(q^{-x}):=M_n(q^{-x};b,c;q).$$

Normalized Recurrence Relation

$$\begin{aligned}xp_n(x)&=p_{n+1}(x)+\left[1+q^{-2n-1}\left\{c(1-bq^{n+1})+q(1-q^n)(c+q^n)\right\}\right]p_n(x)\\&\qquad+cq^{-4n+1}(1-q^n)(1-bq^n)(c+q^n)p_{n-1}(x),\end{aligned} \quad (14.13.4)$$

where

$$M_n(q^{-x};b,c;q)=\frac{(-1)^n q^{n^2}}{(bq;q)_n c^n}p_n(q^{-x}).$$

q-Difference Equation

$$-(1-q^n)y(x)=B(x)y(x+1)-[B(x)+D(x)]y(x)+D(x)y(x-1), \quad (14.13.5)$$

where

$$y(x)=M_n(q^{-x};b,c;q)$$

and

$$\begin{cases} B(x) = cq^x(1-bq^{x+1}) \\ D(x) = (1-q^x)(1+bcq^x). \end{cases}$$

Forward Shift Operator

$$\begin{aligned} &M_n(q^{-x-1};b,c;q) - M_n(q^{-x};b,c;q) \\ &= -\frac{q^{-x}(1-q^n)}{c(1-bq)} M_{n-1}(q^{-x};bq,cq^{-1};q) \end{aligned} \tag{14.13.6}$$

or equivalently

$$\frac{\Delta M_n(q^{-x};b,c;q)}{\Delta q^{-x}} = -\frac{q(1-q^n)}{c(1-q)(1-bq)} M_{n-1}(q^{-x};bq,cq^{-1};q). \tag{14.13.7}$$

Backward Shift Operator

$$\begin{aligned} &cq^x(1-bq^x)M_n(q^{-x};b,c;q) - (1-q^x)(1+bcq^x)M_n(q^{-x+1};b,c;q) \\ &= cq^x(1-b)M_{n+1}(q^{-x};bq^{-1},cq;q) \end{aligned} \tag{14.13.8}$$

or equivalently

$$\begin{aligned} &\frac{\nabla\left[w(x;b,c;q)M_n(q^{-x};b,c;q)\right]}{\nabla q^{-x}} \\ &= \frac{1}{1-q} w(x;bq^{-1},cq;q)M_{n+1}(q^{-x};bq^{-1},cq;q), \end{aligned} \tag{14.13.9}$$

where

$$w(x;b,c;q) = \frac{(bq;q)_x}{(q,-bcq;q)_x} c^x q^{\binom{x+1}{2}}.$$

Rodrigues-Type Formula

$$w(x;b,c;q)M_n(q^{-x};b,c;q) = (1-q)^n \left(\nabla_q\right)^n \left[w(x;bq^n,cq^{-n};q)\right], \tag{14.13.10}$$

where

$$\nabla_q := \frac{\nabla}{\nabla q^{-x}}.$$

Generating Functions

$$\frac{1}{(t;q)_\infty}\,{}_1\phi_1\left(\begin{matrix}q^{-x}\\ bq\end{matrix};q,-c^{-1}qt\right)=\sum_{n=0}^{\infty}\frac{M_n(q^{-x};b,c;q)}{(q;q)_n}t^n. \tag{14.13.11}$$

$$\frac{1}{(t;q)_\infty}\,{}_1\phi_1\left(\begin{matrix}-b^{-1}c^{-1}q^{-x}\\ -c^{-1}q\end{matrix};q,bqt\right)$$
$$=\sum_{n=0}^{\infty}\frac{(bq;q)_n}{(-c^{-1}q,q;q)_n}M_n(q^{-x};b,c;q)t^n. \tag{14.13.12}$$

Limit Relations

Big q-Jacobi → q-Meixner

If we set $b=-a^{-1}cd^{-1}$ (with $d>0$) in the definition (14.5.1) of the big q-Jacobi polynomials and take the limit $c\to-\infty$ we obtain the q-Meixner polynomials given by (14.13.1):

$$\lim_{c\to-\infty}P_n(q^{-x};a,-a^{-1}cd^{-1},c;q)=M_n(q^{-x};a,d;q). \tag{14.13.13}$$

q-Hahn → q-Meixner

The q-Meixner polynomials given by (14.13.1) can be obtained from the q-Hahn polynomials by setting $\alpha=b$ and $\beta=-b^{-1}c^{-1}q^{-N-1}$ in the definition (14.6.1) of the q-Hahn polynomials and letting $N\to\infty$:

$$\lim_{N\to\infty}Q_n(q^{-x};b,-b^{-1}c^{-1}q^{-N-1},N|q)=M_n(q^{-x};b,c;q).$$

q-Meixner → q-Laguerre

The q-Laguerre polynomials given by (14.21.1) can be obtained from the q-Meixner polynomials given by (14.13.1) by setting $b=q^\alpha$ and $q^{-x}\to cq^\alpha x$ in the definition (14.13.1) of the q-Meixner polynomials and then taking the limit $c\to\infty$:

$$\lim_{c\to\infty}M_n(cq^\alpha x;q^\alpha,c;q)=\frac{(q;q)_n}{(q^{\alpha+1};q)_n}L_n^{(\alpha)}(x;q). \tag{14.13.14}$$

q-Meixner → q-Charlier

The q-Charlier polynomials given by (14.23.1) can easily be obtained from the q-Meixner given by (14.13.1) by setting $b = 0$ in the definition (14.13.1) of the q-Meixner polynomials:

$$M_n(x;0,c;q) = C_n(x;c;q). \tag{14.13.15}$$

q-Meixner → Al-Salam-Carlitz II

The Al-Salam-Carlitz II polynomials given by (14.25.1) can be obtained from the q-Meixner polynomials given by (14.13.1) by setting $b = -ac^{-1}$ in the definition (14.13.1) of the q-Meixner polynomials and then taking the limit $c \to 0$:

$$\lim_{c\to 0} M_n(x;-ac^{-1},c;q) = \left(-\frac{1}{a}\right)^n q^{\binom{n}{2}} V_n^{(a)}(x;q). \tag{14.13.16}$$

q-Meixner → Meixner

To find the Meixner polynomials given by (9.10.1) from the q-Meixner polynomials given by (14.13.1) we set $b = q^{\beta-1}$ and $c \to (1-c)^{-1}c$ and let $q \to 1$:

$$\lim_{q\to 1} M_n(q^{-x};q^{\beta-1},(1-c)^{-1}c;q) = M_n(x;\beta,c). \tag{14.13.17}$$

Remarks

The q-Meixner polynomials given by (14.13.1) and the little q-Jacobi polynomials given by (14.12.1) are related in the following way:

$$M_n(q^{-x};b,c;q) = p_n(-c^{-1}q^n;b,b^{-1}q^{-n-x-1}|q).$$

The q-Meixner polynomials and the quantum q-Krawtchouk polynomials given by (14.14.1) are related in the following way:

$$K_n^{qtm}(q^{-x};p,N;q) = M_n(q^{-x};q^{-N-1},-p^{-1};q).$$

References

[16], [27], [28], [30], [74], [80], [125], [238], [261], [276], [416], [472].

14.14 Quantum q-Krawtchouk

Basic Hypergeometric Representation

$$K_n^{qtm}(q^{-x};p,N;q) = {}_2\phi_1\left(\begin{matrix} q^{-n}, q^{-x} \\ q^{-N} \end{matrix}; q, pq^{n+1}\right), \quad n = 0,1,2,\ldots,N. \quad (14.14.1)$$

Orthogonality Relation

$$\sum_{x=0}^{N} \frac{(pq;q)_{N-x}}{(q;q)_x(q;q)_{N-x}}(-1)^{N-x}q^{\binom{x}{2}}K_m^{qtm}(q^{-x};p,N;q)K_n^{qtm}(q^{-x};p,N;q)$$
$$= \frac{(-1)^n p^N (q;q)_{N-n}(q,pq;q)_n}{(q,q;q)_N} q^{\binom{N+1}{2}-\binom{n+1}{2}+Nn}\,\delta_{mn}, \quad p > q^{-N}. \quad (14.14.2)$$

Recurrence Relation

$$\begin{aligned} &-pq^{2n+1}(1-q^{-x})K_n^{qtm}(q^{-x}) \\ &= (1-q^{n-N})K_{n+1}^{qtm}(q^{-x}) \\ &\quad - \left[(1-q^{n-N}) + q(1-q^n)(1-pq^n)\right] K_n^{qtm}(q^{-x}) \\ &\quad\quad + q(1-q^n)(1-pq^n)K_{n-1}^{qtm}(q^{-x}), \end{aligned} \quad (14.14.3)$$

where

$$K_n^{qtm}(q^{-x}) := K_n^{qtm}(q^{-x};p,N;q).$$

Normalized Recurrence Relation

$$\begin{aligned} xp_n(x) &= p_{n+1}(x) + \left[1 - p^{-1}q^{-2n-1}\left\{(1-q^{n-N}) + q(1-q^n)(1-pq^n)\right\}\right] p_n(x) \\ &\quad + p^{-2}q^{-4n+1}(1-q^n)(1-pq^n)(1-q^{n-N-1})p_{n-1}(x), \end{aligned} \quad (14.14.4)$$

where

$$K_n^{qtm}(q^{-x};p,N;q) = \frac{p^n q^{n^2}}{(q^{-N};q)_n} p_n(q^{-x}).$$

q-Difference Equation

$$-p(1-q^n)y(x) = B(x)y(x+1) - [B(x)+D(x)]y(x) + D(x)y(x-1), \quad (14.14.5)$$

where

$$y(x) = K_n^{qtm}(q^{-x};p,N;q)$$

and

$$\begin{cases} B(x) = -q^x(1-q^{x-N}) \\ D(x) = (1-q^x)(p-q^{x-N-1}). \end{cases}$$

Forward Shift Operator

$$\begin{aligned} &K_n^{qtm}(q^{-x-1};p,N;q) - K_n^{qtm}(q^{-x};p,N;q) \\ &= \frac{pq^{-x}(1-q^n)}{1-q^{-N}} K_{n-1}^{qtm}(q^{-x};pq,N-1;q) \end{aligned} \quad (14.14.6)$$

or equivalently

$$\frac{\Delta K_n^{qtm}(q^{-x};p,N;q)}{\Delta q^{-x}} = \frac{pq(1-q^n)}{(1-q)(1-q^{-N})} K_{n-1}^{qtm}(q^{-x};pq,N-1;q). \quad (14.14.7)$$

Backward Shift Operator

$$\begin{aligned} &(1-q^{x-N-1})K_n^{qtm}(q^{-x};p,N;q) \\ &\quad + q^{-x}(1-q^x)(p-q^{x-N-1})K_n^{qtm}(q^{-x+1};p,N;q) \\ &= (1-q^{-N-1})K_{n+1}^{qtm}(q^{-x};pq^{-1},N+1;q) \end{aligned} \quad (14.14.8)$$

or equivalently

$$\begin{aligned} &\frac{\nabla\left[w(x;p,N;q)K_n^{qtm}(q^{-x};p,N;q)\right]}{\nabla q^{-x}} \\ &= \frac{1}{1-q} w(x;pq^{-1},N+1;q) K_{n+1}^{qtm}(q^{-x};pq^{-1},N+1;q), \end{aligned} \quad (14.14.9)$$

where

$$w(x;p,N;q) = \frac{(q^{-N};q)_x}{(q,p^{-1}q^{-N};q)_x}(-p)^{-x} q^{\binom{x+1}{2}}.$$

Rodrigues-Type Formula

$$w(x;p,N;q)K_n^{qtm}(q^{-x};p,N;q) = (1-q)^n \left(\nabla_q\right)^n \left[w(x;pq^n,N-n;q)\right], \quad (14.14.10)$$

where

$$\nabla_q := \frac{\nabla}{\nabla q^{-x}}.$$

Generating Functions

For $x = 0,1,2,\ldots,N$ we have

$$(q^{x-N}t;q)_{N-x} \cdot {}_2\phi_1\left(\begin{matrix} q^{-x}, pq^{N+1-x} \\ 0 \end{matrix}; q, q^{x-N}t\right)$$
$$= \sum_{n=0}^{N} \frac{(q^{-N};q)_n}{(q;q)_n} K_n^{qtm}(q^{-x};p,N;q)t^n. \quad (14.14.11)$$

$$(q^{-x}t;q)_x \cdot {}_2\phi_1\left(\begin{matrix} q^{x-N}, 0 \\ pq \end{matrix}; q, q^{-x}t\right)$$
$$= \sum_{n=0}^{N} \frac{(q^{-N};q)_n}{(pq,q;q)_n} K_n^{qtm}(q^{-x};p,N;q)t^n. \quad (14.14.12)$$

Limit Relations

q-Hahn → Quantum q-Krawtchouk

The quantum q-Krawtchouk polynomials given by (14.14.1) simply follow from the q-Hahn polynomials by setting $\beta = p$ in the definition (14.6.1) of the q-Hahn polynomials and taking the limit $\alpha \to \infty$:

$$\lim_{\alpha\to\infty} Q_n(q^{-x};\alpha,p,N|q) = K_n^{qtm}(q^{-x};p,N;q).$$

Quantum q-Krawtchouk → Al-Salam-Carlitz II

If we set $p = a^{-1}q^{-N-1}$ in the definition (14.14.1) of the quantum q-Krawtchouk polynomials and let $N \to \infty$ we obtain the Al-Salam-Carlitz II polynomials given by (14.25.1). In fact we have

$$\lim_{N\to\infty} K_n^{qtm}(x;a^{-1}q^{-N-1},N;q) = \left(-\frac{1}{a}\right)^n q^{\binom{n}{2}} V_n^{(a)}(x;q). \tag{14.14.13}$$

Quantum q-Krawtchouk → Krawtchouk

The Krawtchouk polynomials given by (9.11.1) easily follow from the quantum q-Krawtchouk polynomials given by (14.14.1) in the following way:

$$\lim_{q\to 1} K_n^{qtm}(q^{-x};p,N;q) = K_n(x;p^{-1},N). \tag{14.14.14}$$

Remarks

The quantum q-Krawtchouk polynomials given by (14.14.1) and the q-Meixner polynomials given by (14.13.1) are related in the following way:

$$K_n^{qtm}(q^{-x};p,N;q) = M_n(q^{-x};q^{-N-1},-p^{-1};q).$$

The quantum q-Krawtchouk polynomials are related to the affine q-Krawtchouk polynomials given by (14.16.1) by the transformation $q \leftrightarrow q^{-1}$ in the following way:

$$K_n^{qtm}(q^x;p,N;q^{-1}) = (p^{-1}q;q)_n \left(-\frac{p}{q}\right)^n q^{-\binom{n}{2}} K_n^{Aff}(q^{x-N};p^{-1},N;q).$$

References

[238], [343], [345], [472].

14.15 q-Krawtchouk

Basic Hypergeometric Representation

For $n = 0, 1, 2, \ldots, N$ we have

$$K_n(q^{-x};p,N;q) = {}_3\phi_2\left(\begin{matrix} q^{-n}, q^{-x}, -pq^n \\ q^{-N}, 0 \end{matrix}; q, q\right) \tag{14.15.1}$$
$$= \frac{(q^{x-N};q)_n}{(q^{-N};q)_n q^{nx}}\, {}_2\phi_1\left(\begin{matrix} q^{-n}, q^{-x} \\ q^{N-x-n+1} \end{matrix}; q, -pq^{n+N+1}\right).$$

Orthogonality Relation

$$\sum_{x=0}^{N} \frac{(q^{-N};q)_x}{(q;q)_x}(-p)^{-x}K_m(q^{-x};p,N;q)K_n(q^{-x};p,N;q)$$
$$= \frac{(q,-pq^{N+1};q)_n}{(-p,q^{-N};q)_n}\frac{(1+p)}{(1+pq^{2n})}$$
$$\times(-pq;q)_N p^{-N} q^{-\binom{N+1}{2}}\left(-pq^{-N}\right)^n q^{n^2}\,\delta_{mn}, \quad p>0. \quad (14.15.2)$$

Recurrence Relation

$$-\left(1-q^{-x}\right)K_n(q^{-x}) = A_nK_{n+1}(q^{-x}) - (A_n+C_n)K_n(q^{-x}) + C_nK_{n-1}(q^{-x}), \quad (14.15.3)$$

where

$$K_n(q^{-x}) := K_n(q^{-x};p,N;q)$$

and

$$\begin{cases} A_n = \dfrac{(1-q^{n-N})(1+pq^n)}{(1+pq^{2n})(1+pq^{2n+1})} \\[2ex] C_n = -pq^{2n-N-1}\dfrac{(1+pq^{n+N})(1-q^n)}{(1+pq^{2n-1})(1+pq^{2n})}. \end{cases}$$

Normalized Recurrence Relation

$$xp_n(x) = p_{n+1}(x) + [1-(A_n+C_n)]\,p_n(x) + A_{n-1}C_np_{n-1}(x), \quad (14.15.4)$$

where

$$K_n(q^{-x};p,N;q) = \frac{(-pq^n;q)_n}{(q^{-N};q)_n}p_n(q^{-x}).$$

q-Difference Equation

$$q^{-n}(1-q^n)(1+pq^n)y(x)$$
$$= (1-q^{x-N})y(x+1) - \left[(1-q^{x-N}) - p(1-q^x)\right]y(x)$$
$$-p(1-q^x)y(x-1), \quad (14.15.5)$$

where

$$y(x) = K_n(q^{-x}; p, N; q).$$

Forward Shift Operator

$$\begin{aligned} &K_n(q^{-x-1}; p, N; q) - K_n(q^{-x}; p, N; q) \\ &= \frac{q^{-n-x}(1-q^n)(1+pq^n)}{1-q^{-N}} K_{n-1}(q^{-x}; pq^2, N-1; q) \end{aligned} \tag{14.15.6}$$

or equivalently

$$\frac{\Delta K_n(q^{-x}; p, N; q)}{\Delta q^{-x}} = \frac{q^{-n+1}(1-q^n)(1+pq^n)}{(1-q)(1-q^{-N})} K_{n-1}(q^{-x}; pq^2, N-1; q). \tag{14.15.7}$$

Backward Shift Operator

$$\begin{aligned} &(1-q^{x-N-1})K_n(q^{-x}; p, N; q) + pq^{-1}(1-q^x)K_n(q^{-x+1}; p, N; q) \\ &= q^x(1-q^{-N-1})K_{n+1}(q^{-x}; pq^{-2}, N+1; q) \end{aligned} \tag{14.15.8}$$

or equivalently

$$\begin{aligned} &\frac{\nabla \left[w(x; p, N; q) K_n(q^{-x}; p, N; q)\right]}{\nabla q^{-x}} \\ &= \frac{1}{1-q} w(x; pq^{-2}, N+1; q) K_{n+1}(q^{-x}; pq^{-2}, N+1; q), \end{aligned} \tag{14.15.9}$$

where

$$w(x; p, N; q) = \frac{(q^{-N}; q)_x}{(q; q)_x} \left(-\frac{q}{p}\right)^x.$$

Rodrigues-Type Formula

$$w(x; p, N; q) K_n(q^{-x}; p, N; q) = (1-q)^n \left(\nabla_q\right)^n \left[w(x; pq^{2n}, N-n; q)\right], \tag{14.15.10}$$

where

$$\nabla_q := \frac{\nabla}{\nabla q^{-x}}.$$

Generating Function

For $x = 0,1,2,\ldots,N$ we have

$$
{}_1\phi_1\left(\begin{matrix} q^{-x} \\ 0 \end{matrix}; q, pqt\right) {}_2\phi_0\left(\begin{matrix} q^{x-N}, 0 \\ - \end{matrix}; q, -q^{-x}t\right)
= \sum_{n=0}^{N} \frac{(q^{-N};q)_n}{(q;q)_n} q^{-\binom{n}{2}} K_n(q^{-x};p,N;q) t^n. \tag{14.15.11}
$$

Limit Relations

q-Racah → q-Krawtchouk

The q-Krawtchouk polynomials given by (14.15.1) can be obtained from the q-Racah polynomials by setting $\alpha q = q^{-N}$, $\beta = -pq^N$ and $\gamma = \delta = 0$ in the definition (14.2.1) of the q-Racah polynomials:

$$
R_n(q^{-x};q^{-N-1},-pq^N,0,0|q) = K_n(q^{-x};p,N;q).
$$

Note that $\mu(x) = q^{-x}$ in this case.

q-Hahn → q-Krawtchouk

If we set $\beta = -\alpha^{-1}q^{-1}p$ in the definition (14.6.1) of the q-Hahn polynomials and then let $\alpha \to 0$ we obtain the q-Krawtchouk polynomials given by (14.15.1):

$$
\lim_{\alpha\to 0} Q_n(q^{-x};\alpha,-\alpha^{-1}q^{-1}p,N|q) = K_n(q^{-x};p,N;q).
$$

q-Krawtchouk → q-Bessel

If we set $x \to N-x$ in the definition (14.15.1) of the q-Krawtchouk polynomials and then take the limit $N \to \infty$ we obtain the q-Bessel polynomials given by (14.22.1):

$$
\lim_{N\to\infty} K_n(q^{x-N};p,N;q) = y_n(q^x;p;q). \tag{14.15.12}
$$

q-Krawtchouk → q-Charlier

By setting $p = a^{-1}q^{-N}$ in the definition (14.15.1) of the q-Krawtchouk polynomials and then taking the limit $N \to \infty$ we obtain the q-Charlier polynomials given by

(14.23.1):

$$\lim_{N\to\infty} K_n(q^{-x};a^{-1}q^{-N},N;q) = C_n(q^{-x};a;q). \tag{14.15.13}$$

q-Krawtchouk → Krawtchouk

If we take the limit $q \to 1$ in the definition (14.15.1) of the q-Krawtchouk polynomials we simply find the Krawtchouk polynomials given by (9.11.1) in the following way:

$$\lim_{q\to 1} K_n(q^{-x};p,N;q) = K_n(x;(p+1)^{-1},N). \tag{14.15.14}$$

Remark

The q-Krawtchouk polynomials given by (14.15.1) and the dual q-Krawtchouk polynomials given by (14.17.1) are related in the following way:

$$K_n(q^{-x};p,N;q) = K_x(\lambda(n);-pq^N,N|q)$$

with

$$\lambda(n) = q^{-n} - pq^n$$

or

$$K_n(\lambda(x);c,N|q) = K_x(q^{-n};-cq^{-N},N;q)$$

with

$$\lambda(x) = q^{-x} + cq^{x-N}.$$

References

[30], [70], [80], [125], [238], [416], [421], [487], [488].

14.16 Affine q-Krawtchouk

Basic Hypergeometric Representation

$$K_n^{Aff}(q^{-x};p,N;q) = {}_3\phi_2\left(\begin{matrix} q^{-n},0,q^{-x} \\ pq,q^{-N} \end{matrix};q,q\right) \tag{14.16.1}$$
$$= \frac{(-pq)^n q^{\binom{n}{2}}}{(pq;q)_n}\,{}_2\phi_1\left(\begin{matrix} q^{-n},q^{x-N} \\ q^{-N} \end{matrix};q,\frac{q^{-x}}{p}\right), \quad n=0,1,2,\ldots,N.$$

Orthogonality Relation

$$\sum_{x=0}^{N}\frac{(pq;q)_x(q;q)_N}{(q;q)_x(q;q)_{N-x}}(pq)^{-x}K_m^{Aff}(q^{-x};p,N;q)K_n^{Aff}(q^{-x};p,N;q)$$
$$= (pq)^{n-N}\frac{(q;q)_n(q;q)_{N-n}}{(pq;q)_n(q;q)_N}\,\delta_{mn}, \quad 0<pq<1. \tag{14.16.2}$$

Recurrence Relation

$$-(1-q^{-x})K_n^{Aff}(q^{-x})$$
$$= (1-q^{n-N})(1-pq^{n+1})K_{n+1}^{Aff}(q^{-x})$$
$$- \left[(1-q^{n-N})(1-pq^{n+1}) - pq^{n-N}(1-q^n)\right]K_n^{Aff}(q^{-x})$$
$$- pq^{n-N}(1-q^n)K_{n-1}^{Aff}(q^{-x}), \tag{14.16.3}$$

where

$$K_n^{Aff}(q^{-x}) := K_n^{Aff}(q^{-x};p,N;q).$$

Normalized Recurrence Relation

$$xp_n(x) = p_{n+1}(x) + \left[1-\left\{(1-q^{n-N})(1-pq^{n+1}) - pq^{n-N}(1-q^n)\right\}\right]p_n(x)$$
$$- pq^{n-N}(1-q^n)(1-pq^n)(1-q^{n-N-1})p_{n-1}(x), \tag{14.16.4}$$

where

$$K_n^{Aff}(q^{-x};p,N;q) = \frac{1}{(pq,q^{-N};q)_n} p_n(q^{-x}).$$

q-Difference Equation

$$q^{-n}(1-q^n)y(x) = B(x)y(x+1) - [B(x)+D(x)]\,y(x) + D(x)y(x-1), \quad (14.16.5)$$

where

$$y(x) = K_n^{Aff}(q^{-x};p,N;q)$$

and

$$\begin{cases} B(x) = (1-q^{x-N})(1-pq^{x+1}) \\ D(x) = -p(1-q^x)q^{x-N}. \end{cases}$$

Forward Shift Operator

$$K_n^{Aff}(q^{-x-1};p,N;q) - K_n^{Aff}(q^{-x};p,N;q) = \frac{q^{-n-x}(1-q^n)}{(1-pq)(1-q^{-N})} K_{n-1}^{Aff}(q^{-x};pq,N-1;q) \quad (14.16.6)$$

or equivalently

$$\frac{\Delta K_n^{Aff}(q^{-x};p,N;q)}{\Delta q^{-x}} = \frac{q^{-n+1}(1-q^n)}{(1-q)(1-pq)(1-q^{-N})} K_{n-1}^{Aff}(q^{-x};pq,N-1;q). \quad (14.16.7)$$

Backward Shift Operator

$$(1-pq^x)(1-q^{-x+N+1})K_n^{Aff}(q^{-x};p,N;q) - p(1-q^x)K_n^{Aff}(q^{-x+1};p,N;q) = (1-p)(1-q^{N+1})K_{n+1}^{Aff}(q^{-x};pq^{-1},N+1;q) \quad (14.16.8)$$

or equivalently

$$\frac{\nabla\left[w(x;p,N;q)K_n^{Aff}(q^{-x};p,N;q)\right]}{\nabla q^{-x}}$$
$$=\frac{1-q^{N+1}}{1-q}w(x;pq^{-1},N+1;q)K_{n+1}^{Aff}(q^{-x};pq^{-1},N+1;q), \quad (14.16.9)$$

where

$$w(x;p,N;q)=\frac{(pq;q)_x}{(q;q)_x(q;q)_{N-x}}p^{-x}.$$

Rodrigues-Type Formula

$$w(x;p,N;q)K_n^{Aff}(q^{-x};p,N;q)$$
$$=\frac{(-1)^n q^{-Nn+\binom{n}{2}}(1-q)^n}{(q^{-N};q)_n}(\nabla_q)^n\left[w(x;pq^n,N-n;q)\right], \quad (14.16.10)$$

where

$$\nabla_q:=\frac{\nabla}{\nabla q^{-x}}.$$

Generating Functions

For $x=0,1,2,\ldots,N$ we have

$$(q^{-N}t;q)_{N-x}\cdot{}_1\phi_1\left(\begin{matrix}q^{-x}\\ pq\end{matrix};q,\,pqt\right)=\sum_{n=0}^{N}\frac{(q^{-N};q)_n}{(q;q)_n}K_n^{Aff}(q^{-x};p,N;q)t^n. \quad (14.16.11)$$

$$(-pq^{-N+1}t;q)_x\cdot{}_2\phi_0\left(\begin{matrix}q^{x-N},pq^{x+1}\\ -\end{matrix};q,\,-q^{-x}t\right)$$
$$=\sum_{n=0}^{N}\frac{(pq,q^{-N};q)_n}{(q;q)_n}q^{-\binom{n}{2}}K_n^{Aff}(q^{-x};p,N;q)t^n. \quad (14.16.12)$$

Limit Relations

q-Hahn $\rightarrow$ Affine q-Krawtchouk

The affine q-Krawtchouk polynomials given by (14.16.1) can be obtained from the q-Hahn polynomials by the substitution $\alpha = p$ and $\beta = 0$ in (14.6.1):

$$Q_n(q^{-x}; p, 0, N|q) = K_n^{Aff}(q^{-x}; p, N; q).$$

Dual q-Hahn $\rightarrow$ Affine q-Krawtchouk

The affine q-Krawtchouk polynomials given by (14.16.1) can be obtained from the dual q-Hahn polynomials by the substitution $\gamma = p$ and $\delta = 0$ in (14.7.1):

$$R_n(\mu(x); p, 0, N|q) = K_n^{Aff}(q^{-x}; p, N; q).$$

Note that $\mu(x) = q^{-x}$ in this case.

Affine q-Krawtchouk $\rightarrow$ Little q-Laguerre / Wall

If we set $x \rightarrow N - x$ in the definition (14.16.1) of the affine q-Krawtchouk polynomials and take the limit $N \rightarrow \infty$ we simply obtain the little q-Laguerre (or Wall) polynomials given by (14.20.1):

$$\lim_{N\to\infty} K_n^{Aff}(q^{x-N}; p, N; q) = p_n(q^x; p; q). \tag{14.16.13}$$

Affine q-Krawtchouk $\rightarrow$ Krawtchouk

If we let $q \rightarrow 1$ in the definition (14.16.1) of the affine q-Krawtchouk polynomials we obtain:

$$\lim_{q\to 1} K_n^{Aff}(q^{-x}; p, N|q) = K_n(x; 1-p, N), \tag{14.16.14}$$

where $K_n(x; 1-p, N)$ is the Krawtchouk polynomial given by (9.11.1).

Remarks

The affine q-Krawtchouk polynomials given by (14.16.1) and the big q-Laguerre polynomials given by (14.11.1) are related in the following way:

$$K_n^{Aff}(q^{-x}; p, N; q) = P_n(q^{-x}; p, q^{-N-1}; q).$$

The affine q-Krawtchouk polynomials are related to the quantum q-Krawtchouk polynomials given by (14.14.1) by the transformation $q \leftrightarrow q^{-1}$ in the following way:

$$K_n^{Aff}(q^x;p,N;q^{-1}) = \frac{1}{(p^{-1}q;q)_n} K_n^{qtm}(q^{x-N};p^{-1},N;q).$$

References

[80], [141], [161], [162], [186], [214], [238], [488].

14.17 Dual q-Krawtchouk

Basic Hypergeometric Representation

$$\begin{aligned} K_n(\lambda(x);c,N|q) &= {}_3\phi_2\left(\begin{array}{c} q^{-n},q^{-x},cq^{x-N} \\ q^{-N},0 \end{array};q,q\right) \qquad (14.17.1) \\ &= \frac{(q^{x-N};q)_n}{(q^{-N};q)_n q^{nx}}\, {}_2\phi_1\left(\begin{array}{c} q^{-n},q^{-x} \\ q^{N-x-n+1} \end{array};q,cq^{x+1}\right), \quad n=0,1,2,\ldots,N, \end{aligned}$$

where

$$\lambda(x) := q^{-x} + cq^{x-N}.$$

Orthogonality Relation

$$\sum_{x=0}^{N} \frac{(cq^{-N},q^{-N};q)_x}{(q,cq;q)_x} \frac{(1-cq^{2x-N})}{(1-cq^{-N})} c^{-x} q^{x(2N-x)} K_m(\lambda(x))K_n(\lambda(x))$$
$$= (c^{-1};q)_N \frac{(q;q)_n}{(q^{-N};q)_n} (cq^{-N})^n \,\delta_{mn}, \quad c<0, \qquad (14.17.2)$$

where

$$K_n(\lambda(x)) := K_n(\lambda(x);c,N|q).$$

Recurrence Relation

$$\begin{aligned}
&-(1-q^{-x})(1-cq^{x-N})K_n(\lambda(x))\\
&=(1-q^{n-N})K_{n+1}(\lambda(x))\\
&\quad-\left[(1-q^{n-N})+cq^{-N}(1-q^n)\right]K_n(\lambda(x))\\
&\quad\quad+cq^{-N}(1-q^n)K_{n-1}(\lambda(x)),
\end{aligned} \tag{14.17.3}$$

where

$$K_n(\lambda(x)) := K_n(\lambda(x);c,N|q).$$

Normalized Recurrence Relation

$$\begin{aligned}
xp_n(x) = p_{n+1}(x)+(1+c)q^{n-N}p_n(x)\\
+cq^{-N}(1-q^n)(1-q^{n-N-1})p_{n-1}(x),
\end{aligned} \tag{14.17.4}$$

where

$$K_n(\lambda(x);c,N|q) = \frac{1}{(q^{-N};q)_n}p_n(\lambda(x)).$$

q-Difference Equation

$$q^{-n}(1-q^n)y(x) = B(x)y(x+1)-[B(x)+D(x)]y(x)+D(x)y(x-1), \tag{14.17.5}$$

where

$$y(x) = K_n(\lambda(x);c,N|q)$$

and

$$\begin{cases}
B(x) = \dfrac{(1-q^{x-N})(1-cq^{x-N})}{(1-cq^{2x-N})(1-cq^{2x-N+1})}\\[2ex]
D(x) = cq^{2x-2N-1}\dfrac{(1-q^x)(1-cq^x)}{(1-cq^{2x-N-1})(1-cq^{2x-N})}.
\end{cases}$$

Forward Shift Operator

$$
\begin{aligned}
&K_n(\lambda(x+1);c,N|q)-K_n(\lambda(x);c,N|q)\\
&=\frac{q^{-n-x}(1-q^n)(1-cq^{2x-N+1})}{1-q^{-N}}K_{n-1}(\lambda(x);c,N-1|q) \qquad (14.17.6)
\end{aligned}
$$

or equivalently

$$
\frac{\Delta K_n(\lambda(x);c,N|q)}{\Delta\lambda(x)}=\frac{q^{-n+1}(1-q^n)}{(1-q)(1-q^{-N})}K_{n-1}(\lambda(x);c,N-1|q). \qquad (14.17.7)
$$

Backward Shift Operator

$$
\begin{aligned}
&(1-q^{x-N-1})(1-cq^{x-N-1})K_n(\lambda(x);c,N|q)\\
&\quad -cq^{2(x-N-1)}(1-q^x)(1-cq^x)K_n(\lambda(x-1);c,N|q)\\
&=q^x(1-q^{-N-1})(1-cq^{2x-N-1})K_{n+1}(\lambda(x);c,N+1|q) \qquad (14.17.8)
\end{aligned}
$$

or equivalently

$$
\begin{aligned}
&\frac{\nabla\left[w(x;c,N|q)K_n(\lambda(x);c,N|q)\right]}{\nabla\lambda(x)}\\
&=\frac{1}{(1-q)(1-cq^{-N-1})}w(x;c,N+1|q)K_{n+1}(\lambda(x);c,N+1|q), \qquad (14.17.9)
\end{aligned}
$$

where

$$
w(x;c,N|q)=\frac{(q^{-N},cq^{-N};q)_x}{(q,cq;q)_x}c^{-x}q^{2Nx-x(x-1)}.
$$

Rodrigues-Type Formula

$$
\begin{aligned}
&w(x;c,N|q)K_n(\lambda(x);c,N|q)\\
&=(1-q)^n(cq^{-N};q)_n\left(\nabla_\lambda\right)^n\left[w(x;c,N-n|q)\right], \qquad (14.17.10)
\end{aligned}
$$

where

$$
\nabla_\lambda:=\frac{\nabla}{\nabla\lambda(x)}.
$$

Generating Function

For $x = 0, 1, 2, \ldots, N$ we have

$$(cq^{-N}t;q)_x \cdot (q^{-N}t;q)_{N-x} = \sum_{n=0}^{N} \frac{(q^{-N};q)_n}{(q;q)_n} K_n(\lambda(x);c,N|q)t^n. \tag{14.17.11}$$

Limit Relations

q-Racah → Dual q-Krawtchouk

The dual q-Krawtchouk polynomials given by (14.17.1) easily follow from the q-Racah polynomials given by (14.2.1) by using the substitutions $\alpha = \beta = 0$, $\gamma q = q^{-N}$ and $\delta = c$:

$$R_n(\mu(x);0,0,q^{-N-1},c|q) = K_n(\lambda(x);c,N|q).$$

Note that

$$\mu(x) = \lambda(x) = q^{-x} + cq^{x-N}.$$

Dual q-Hahn → Dual q-Krawtchouk

The dual q-Krawtchouk polynomials given by (14.17.1) can be obtained from the dual q-Hahn polynomials by setting $\delta = c\gamma^{-1}q^{-N-1}$ in (14.7.1) and letting $\gamma \to 0$:

$$\lim_{\gamma \to 0} R_n(\mu(x);\gamma,c\gamma^{-1}q^{-N-1},N|q) = K_n(\lambda(x);c,N|q).$$

Dual q-Krawtchouk → Al-Salam-Carlitz I

If we set $c = a^{-1}$ in the definition (14.17.1) of the dual q-Krawtchouk polynomials and take the limit $N \to \infty$ we simply obtain the Al-Salam-Carlitz I polynomials given by (14.24.1):

$$\lim_{N \to \infty} K_n(\lambda(x);a^{-1},N|q) = \left(-\frac{1}{a}\right)^n q^{-\binom{n}{2}} U_n^{(a)}(q^x;q). \tag{14.17.12}$$

Note that $\lambda(x) = q^{-x} + a^{-1}q^{x-N}$.

Dual q-Krawtchouk → Krawtchouk

If we set $c = 1 - p^{-1}$ in the definition (14.17.1) of the dual q-Krawtchouk polynomials and take the limit $q \to 1$ we simply find the Krawtchouk polynomials given by (9.11.1):

$$\lim_{q\to 1} K_n(\lambda(x); 1-p^{-1}, N|q) = K_n(x; p, N). \qquad (14.17.13)$$

Remark

The dual q-Krawtchouk polynomials given by (14.17.1) and the q-Krawtchouk polynomials given by (14.15.1) are related in the following way:

$$K_n(q^{-x}; p, N; q) = K_x(\lambda(n); -pq^N, N|q)$$

with

$$\lambda(n) = q^{-n} - pq^n$$

or

$$K_n(\lambda(x); c, N|q) = K_x(q^{-n}; -cq^{-N}, N; q)$$

with

$$\lambda(x) = q^{-x} + cq^{x-N}.$$

References

[141], [326], [345], [348].

14.18 Continuous Big q-Hermite

Basic Hypergeometric Representation

$$\begin{aligned} H_n(x; a|q) &= a^{-n} {}_3\phi_2\left(\begin{matrix} q^{-n}, ae^{i\theta}, ae^{-i\theta} \\ 0, 0 \end{matrix}; q, q\right) \qquad (14.18.1) \\ &= e^{in\theta} {}_2\phi_0\left(\begin{matrix} q^{-n}, ae^{i\theta} \\ - \end{matrix}; q, q^n e^{-2i\theta}\right), \quad x = \cos\theta. \end{aligned}$$

Orthogonality Relation

If a is real and $|a| < 1$, then we have the following orthogonality relation

$$\frac{1}{2\pi}\int_{-1}^{1}\frac{w(x)}{\sqrt{1-x^2}}H_m(x;a|q)H_n(x;a|q)\,dx=\frac{\delta_{mn}}{(q^{n+1};q)_\infty}, \tag{14.18.2}$$

where

$$w(x):=w(x;a|q)=\left|\frac{(\mathrm{e}^{2i\theta};q)_\infty}{(a\mathrm{e}^{i\theta};q)_\infty}\right|^2=\frac{h(x,1)h(x,-1)h(x,q^{\frac{1}{2}})h(x,-q^{\frac{1}{2}})}{h(x,a)},$$

with

$$h(x,\alpha):=\prod_{k=0}^{\infty}\left(1-2\alpha xq^k+\alpha^2q^{2k}\right)=\left(\alpha\mathrm{e}^{i\theta},\alpha\mathrm{e}^{-i\theta};q\right)_\infty,\quad x=\cos\theta.$$

If $a > 1$, then we have another orthogonality relation given by:

$$\frac{1}{2\pi}\int_{-1}^{1}\frac{w(x)}{\sqrt{1-x^2}}H_m(x;a|q)H_n(x;a|q)\,dx + \sum_{\substack{k \\ 1<aq^k\leq a}} w_kH_m(x_k;a|q)H_n(x_k;a|q)=\frac{\delta_{mn}}{(q^{n+1};q)_\infty}, \tag{14.18.3}$$

where $w(x)$ is as before,

$$x_k=\frac{aq^k+\left(aq^k\right)^{-1}}{2}$$

and

$$w_k=\frac{(a^{-2};q)_\infty}{(q;q)_\infty}\frac{(1-a^2q^{2k})(a^2;q)_k}{(1-a^2)(q;q)_k}q^{-\frac{3}{2}k^2-\frac{1}{2}k}\left(-\frac{1}{a^4}\right)^k.$$

Recurrence Relation

$$2xH_n(x;a|q)=H_{n+1}(x;a|q)+aq^nH_n(x;a|q)+(1-q^n)H_{n-1}(x;a|q). \tag{14.18.4}$$

Normalized Recurrence Relation

$$xp_n(x)=p_{n+1}(x)+\frac{1}{2}aq^np_n(x)+\frac{1}{4}(1-q^n)p_{n-1}(x), \tag{14.18.5}$$

where

$$H_n(x;a|q) = 2^n p_n(x).$$

q-Difference Equations

$$(1-q)^2 D_q \left[\tilde{w}(x;aq^{\frac{1}{2}}|q) D_q y(x)\right] + 4q^{-n+1}(1-q^n)\tilde{w}(x;a|q)y(x) = 0, \quad (14.18.6)$$

where

$$y(x) = H_n(x;a|q)$$

and

$$\tilde{w}(x;a|q) := \frac{w(x;a|q)}{\sqrt{1-x^2}}.$$

If we define

$$P_n(z) := a^{-n} {}_3\phi_2\left(\begin{matrix} q^{-n}, az, az^{-1} \\ 0, 0 \end{matrix}; q, q\right)$$

then the q-difference equation can also be written in the form

$$q^{-n}(1-q^n)P_n(z) = A(z)P_n(qz) - \left[A(z) + A(z^{-1})\right] P_n(z) + A(z^{-1})P_n(q^{-1}z), \quad (14.18.7)$$

where

$$A(z) = \frac{(1-az)}{(1-z^2)(1-qz^2)}.$$

Forward Shift Operator

$$\delta_q H_n(x;a|q) = -q^{-\frac{1}{2}n}(1-q^n)(e^{i\theta} - e^{-i\theta})H_{n-1}(x;aq^{\frac{1}{2}}|q), \quad x = \cos\theta \quad (14.18.8)$$

or equivalently

$$D_q H_n(x;a|q) = \frac{2q^{-\frac{1}{2}(n-1)}(1-q^n)}{1-q} H_{n-1}(x;aq^{\frac{1}{2}}|q). \quad (14.18.9)$$

Backward Shift Operator

$$\delta_q \left[\tilde{w}(x;a|q)H_n(x;a|q)\right] = q^{-\frac{1}{2}(n+1)}(e^{i\theta} - e^{-i\theta}) \times \tilde{w}(x;aq^{-\frac{1}{2}}|q)H_{n+1}(x;aq^{-\frac{1}{2}}|q), \quad x = \cos\theta \quad (14.18.10)$$

or equivalently

$$D_q\left[\tilde{w}(x;a|q)H_n(x;a|q)\right]=-\frac{2q^{-\frac{1}{2}n}}{1-q}\tilde{w}(x;aq^{-\frac{1}{2}}|q)H_{n+1}(x;aq^{-\frac{1}{2}}|q). \tag{14.18.11}$$

Rodrigues-Type Formula

$$w(x;a|q)H_n(x;a|q)=\left(\frac{q-1}{2}\right)^n q^{\frac{1}{4}n(n-1)}\left(D_q\right)^n\left[w(x;aq^{\frac{1}{2}n}|q)\right]. \tag{14.18.12}$$

Generating Functions

$$\frac{(at;q)_\infty}{(e^{i\theta}t,e^{-i\theta}t;q)_\infty}=\sum_{n=0}^{\infty}\frac{H_n(x;a|q)}{(q;q)_n}t^n,\quad x=\cos\theta. \tag{14.18.13}$$

$$\begin{aligned}&(e^{i\theta}t;q)_\infty\cdot{}_1\phi_1\left(\begin{matrix}ae^{i\theta}\\ e^{i\theta}t\end{matrix};q,e^{-i\theta}t\right)\\&=\sum_{n=0}^{\infty}\frac{(-1)^nq^{\binom{n}{2}}}{(q;q)_n}H_n(x;a|q)t^n,\quad x=\cos\theta.\end{aligned} \tag{14.18.14}$$

$$\begin{aligned}&\frac{(\gamma e^{i\theta}t;q)_\infty}{(e^{i\theta}t;q)_\infty}\,{}_2\phi_1\left(\begin{matrix}\gamma,ae^{i\theta}\\ \gamma e^{i\theta}t\end{matrix};q,e^{-i\theta}t\right)\\&=\sum_{n=0}^{\infty}\frac{(\gamma;q)_n}{(q;q)_n}H_n(x;a|q)t^n,\quad x=\cos\theta,\quad \gamma\text{ arbitrary}.\end{aligned} \tag{14.18.15}$$

Limit Relations

Al-Salam-Chihara → Continuous Big q-Hermite

If we take $b=0$ in the definition (14.8.1) of the Al-Salam-Chihara polynomials we simply obtain the continuous big q-Hermite polynomials given by (14.18.1):

$$Q_n(x;a,0|q)=H_n(x;a|q).$$

Continuous Big q-Hermite $\rightarrow$ Continuous q-Hermite

The continuous q-Hermite polynomials given by (14.26.1) can easily be obtained from the continuous big q-Hermite polynomials given by (14.18.1) by taking $a=0$:

$$H_n(x;0|q)=H_n(x|q). \tag{14.18.16}$$

Continuous Big q-Hermite $\rightarrow$ Hermite

If we set $a=0$ and $x\rightarrow x\sqrt{\frac{1}{2}(1-q)}$ in the definition (14.18.1) of the continuous big q-Hermite polynomials and let q tend to 1, we obtain the Hermite polynomials given by (9.15.1) in the following way:

$$\lim_{q\rightarrow 1}\frac{H_n(x\sqrt{\frac{1}{2}(1-q)};0|q)}{\left(\frac{1-q}{2}\right)^{\frac{n}{2}}}=H_n(x). \tag{14.18.17}$$

If we take $a\rightarrow a\sqrt{2(1-q)}$ and $x\rightarrow x\sqrt{\frac{1}{2}(1-q)}$ in the definition (14.18.1) of the continuous big q-Hermite polynomials and take the limit $q\rightarrow 1$ we find the Hermite polynomials given by (9.15.1) with shifted argument:

$$\lim_{q\rightarrow 1}\frac{H_n(x\sqrt{\frac{1}{2}(1-q)};a\sqrt{2(1-q)}|q)}{\left(\frac{1-q}{2}\right)^{\frac{n}{2}}}=H_n(x-a). \tag{14.18.18}$$

References

[73], [85], [86], [207].

14.19 Continuous q-Laguerre

Basic Hypergeometric Representation

The continuous q-Laguerre polynomials can be obtained from the continuous q-Jacobi polynomials given by (14.10.1) by taking the limit $\beta \to \infty$:

$$P_n^{(\alpha)}(x|q) = \frac{(q^{\alpha+1};q)_n}{(q;q)_n}\,{}_3\phi_2\left(\begin{matrix} q^{-n}, q^{\frac{1}{2}\alpha+\frac{1}{4}}\mathrm{e}^{i\theta}, q^{\frac{1}{2}\alpha+\frac{1}{4}}\mathrm{e}^{-i\theta} \\ q^{\alpha+1}, 0 \end{matrix}; q, q\right) \tag{14.19.1}$$
$$= \frac{(q^{\frac{1}{2}\alpha+\frac{3}{4}}\mathrm{e}^{-i\theta};q)_n}{(q;q)_n} q^{(\frac{1}{2}\alpha+\frac{1}{4})n}\mathrm{e}^{in\theta}$$
$$\times\,{}_2\phi_1\left(\begin{matrix} q^{-n}, q^{\frac{1}{2}\alpha+\frac{1}{4}}\mathrm{e}^{i\theta} \\ q^{-\frac{1}{2}\alpha+\frac{1}{4}-n}\mathrm{e}^{i\theta} \end{matrix}; q, q^{-\frac{1}{2}\alpha+\frac{1}{4}}\mathrm{e}^{-i\theta}\right), \quad x = \cos\theta.$$

Orthogonality Relation

For $\alpha \geq -\frac{1}{2}$ we have

$$\frac{1}{2\pi}\int_{-1}^{1} \frac{w(x)}{\sqrt{1-x^2}} P_m^{(\alpha)}(x|q) P_n^{(\alpha)}(x|q)\,dx$$
$$= \frac{1}{(q, q^{\alpha+1};q)_\infty} \frac{(q^{\alpha+1};q)_n}{(q;q)_n} q^{(\alpha+\frac{1}{2})n}\,\delta_{mn}, \tag{14.19.2}$$

where

$$w(x) := w(x;q^\alpha|q) = \left|\frac{(\mathrm{e}^{2i\theta};q)_\infty}{(q^{\frac{1}{2}\alpha+\frac{1}{4}}\mathrm{e}^{i\theta}, q^{\frac{1}{2}\alpha+\frac{3}{4}}\mathrm{e}^{i\theta};q)_\infty}\right|^2 = \left|\frac{(\mathrm{e}^{i\theta}, -\mathrm{e}^{i\theta};q^{\frac{1}{2}})_\infty}{(q^{\frac{1}{2}\alpha+\frac{1}{4}}\mathrm{e}^{i\theta};q^{\frac{1}{2}})_\infty}\right|^2$$
$$= \frac{h(x,1)h(x,-1)h(x,q^{\frac{1}{2}})h(x,-q^{\frac{1}{2}})}{h(x,q^{\frac{1}{2}\alpha+\frac{1}{4}})h(x,q^{\frac{1}{2}\alpha+\frac{3}{4}})},$$

with

$$h(x,\alpha) := \prod_{k=0}^{\infty}\left(1 - 2\alpha x q^k + \alpha^2 q^{2k}\right) = \left(\alpha\mathrm{e}^{i\theta}, \alpha\mathrm{e}^{-i\theta};q\right)_\infty, \quad x = \cos\theta.$$

Recurrence Relation

$$\begin{aligned}2xP_n^{(\alpha)}(x|q) &= q^{-\frac{1}{2}\alpha-\frac{1}{4}}(1-q^{n+1})P_{n+1}^{(\alpha)}(x|q)\\ &\quad + q^{n+\frac{1}{2}\alpha+\frac{1}{4}}(1+q^{\frac{1}{2}})P_n^{(\alpha)}(x|q)\\ &\quad + q^{\frac{1}{2}\alpha+\frac{1}{4}}(1-q^{n+\alpha})P_{n-1}^{(\alpha)}(x|q). \qquad (14.19.3)\end{aligned}$$

Normalized Recurrence Relation

$$\begin{aligned}xp_n(x) &= p_{n+1}(x) + \frac{1}{2}q^{n+\frac{1}{2}\alpha+\frac{1}{4}}(1+q^{\frac{1}{2}})p_n(x)\\ &\quad + \frac{1}{4}(1-q^n)(1-q^{n+\alpha})p_{n-1}(x), \qquad (14.19.4)\end{aligned}$$

where

$$P_n^{(\alpha)}(x|q) = \frac{2^n q^{(\frac{1}{2}\alpha+\frac{1}{4})n}}{(q;q)_n}p_n(x).$$

q-Difference Equations

$$(1-q)^2 D_q\left[\tilde{w}(x;q^{\alpha+1}|q)D_q y(x)\right] + 4q^{-n+1}(1-q^n)\tilde{w}(x;q^{\alpha}|q)y(x) = 0, \quad (14.19.5)$$

where

$$y(x) = P_n^{(\alpha)}(x|q)$$

and

$$\tilde{w}(x;q^{\alpha}|q) := \frac{w(x;q^{\alpha}|q)}{\sqrt{1-x^2}}.$$

Forward Shift Operator

$$\delta_q P_n^{(\alpha)}(x|q) = -q^{-n+\frac{1}{2}\alpha+\frac{3}{4}}(\mathrm{e}^{i\theta}-\mathrm{e}^{-i\theta})P_{n-1}^{(\alpha+1)}(x|q), \quad x=\cos\theta \qquad (14.19.6)$$

or equivalently

$$D_q P_n^{(\alpha)}(x|q) = \frac{2q^{-n+\frac{1}{2}\alpha+\frac{5}{4}}}{1-q}P_{n-1}^{(\alpha+1)}(x|q). \qquad (14.19.7)$$

Backward Shift Operator

$$\begin{aligned}
&\delta_q\left[\tilde{w}(x;q^{\alpha}|q)P_n^{(\alpha)}(x|q)\right]\\
&=q^{-\frac{1}{2}\alpha-\frac{1}{4}}(1-q^{n+1})(\mathrm{e}^{i\theta}-\mathrm{e}^{-i\theta})\\
&\qquad\times\tilde{w}(x;q^{\alpha-1}|q)P_{n+1}^{(\alpha-1)}(x|q),\quad x=\cos\theta
\end{aligned}\tag{14.19.8}$$

or equivalently

$$\begin{aligned}
&D_q\left[\tilde{w}(x;q^{\alpha}|q)P_n^{(\alpha)}(x|q)\right]\\
&=-2q^{-\frac{1}{2}\alpha+\frac{1}{4}}\frac{1-q^{n+1}}{1-q}\tilde{w}(x;q^{\alpha-1}|q)P_{n+1}^{(\alpha-1)}(x|q).
\end{aligned}\tag{14.19.9}$$

Rodrigues-Type Formula

$$\tilde{w}(x;q^{\alpha}|q)P_n^{(\alpha)}(x|q)=\left(\frac{q-1}{2}\right)^n\frac{q^{\frac{1}{4}n^2+\frac{1}{2}n\alpha}}{(q;q)_n}(D_q)^n\left[\tilde{w}(x;q^{\alpha+n}|q)\right].\tag{14.19.10}$$

Generating Functions

$$\frac{(q^{\alpha+\frac{1}{2}}t,q^{\alpha+1}t;q)_{\infty}}{(q^{\frac{1}{2}\alpha+\frac{1}{4}}\mathrm{e}^{i\theta}t,q^{\frac{1}{2}\alpha+\frac{1}{4}}\mathrm{e}^{-i\theta}t;q)_{\infty}}=\sum_{n=0}^{\infty}P_n^{(\alpha)}(x|q)t^n,\quad x=\cos\theta.\tag{14.19.11}$$

$$\begin{aligned}
&\frac{1}{(\mathrm{e}^{i\theta}t;q)_{\infty}}\,{}_2\phi_1\left(\begin{matrix}q^{\frac{1}{2}\alpha+\frac{1}{4}}\mathrm{e}^{i\theta},q^{\frac{1}{2}\alpha+\frac{3}{4}}\mathrm{e}^{i\theta}\\ q^{\alpha+1}\end{matrix};q,\mathrm{e}^{-i\theta}t\right)\\
&=\sum_{n=0}^{\infty}\frac{P_n^{(\alpha)}(x|q)t^n}{(q^{\alpha+1};q)_nq^{(\frac{1}{2}\alpha+\frac{1}{4})n}},\quad x=\cos\theta.
\end{aligned}\tag{14.19.12}$$

$$\begin{aligned}
&(t;q)_{\infty}\cdot{}_2\phi_1\left(\begin{matrix}q^{\frac{1}{2}\alpha+\frac{1}{4}}\mathrm{e}^{i\theta},q^{\frac{1}{2}\alpha+\frac{1}{4}}\mathrm{e}^{-i\theta}\\ q^{\alpha+1}\end{matrix};q,t\right)\\
&=\sum_{n=0}^{\infty}\frac{(-1)^nq^{\binom{n}{2}}}{(q^{\alpha+1};q)_n}P_n^{(\alpha)}(x|q)t^n,\quad x=\cos\theta.
\end{aligned}\tag{14.19.13}$$

$$\frac{(\gamma e^{i\theta}t;q)_\infty}{(e^{i\theta}t;q)_\infty} {}_3\phi_2\left(\begin{matrix}\gamma, q^{\frac{1}{2}\alpha+\frac{1}{4}}e^{i\theta}, q^{\frac{1}{2}\alpha+\frac{3}{4}}e^{i\theta} \\ q^{\alpha+1}, \gamma e^{i\theta}t\end{matrix}; q, e^{-i\theta}t\right)$$
$$= \sum_{n=0}^{\infty} \frac{(\gamma;q)_n}{(q^{\alpha+1};q)_n} \frac{P_n(x|q)}{q^{(\frac{1}{2}\alpha+\frac{1}{4})n}} t^n, \quad x = \cos\theta, \quad \gamma \text{ arbitrary}. \quad (14.19.14)$$

Limit Relations

Al-Salam-Chihara → Continuous q-Laguerre

The continuous q-Laguerre polynomials given by (14.19.1) can be obtained from the Al-Salam-Chihara polynomials given by (14.8.1) by taking $a = q^{\frac{1}{2}\alpha+\frac{1}{4}}$ and $b = q^{\frac{1}{2}\alpha+\frac{3}{4}}$:

$$Q_n(x; q^{\frac{1}{2}\alpha+\frac{1}{4}}, q^{\frac{1}{2}\alpha+\frac{3}{4}}|q) = \frac{(q;q)_n}{q^{(\frac{1}{2}\alpha+\frac{1}{4})n}} P_n^{(\alpha)}(x|q).$$

q-Meixner-Pollaczek → Continuous q-Laguerre

If we take $e^{i\phi} = q^{-\frac{1}{4}}$, $a = q^{\frac{1}{2}\alpha+\frac{1}{2}}$ and $e^{i\theta} \to q^{\frac{1}{4}}e^{i\theta}$ in the definition (14.9.1) of the q-Meixner-Pollaczek polynomials we obtain the continuous q-Laguerre polynomials given by (14.19.1):

$$P_n(\cos(\theta+\phi); q^{\frac{1}{2}\alpha+\frac{1}{2}}|q) = q^{-(\frac{1}{2}\alpha+\frac{1}{4})n} P_n^{(\alpha)}(\cos\theta|q).$$

Continuous q-Jacobi → Continuous q-Laguerre

The continuous q-Laguerre polynomials given by (14.19.1) follow simply from the continuous q-Jacobi polynomials given by (14.10.1) by taking the limit $\beta \to \infty$:

$$\lim_{\beta\to\infty} P_n^{(\alpha,\beta)}(x|q) = P_n^{(\alpha)}(x|q).$$

Continuous q-Laguerre → Continuous q-Hermite

The continuous q-Hermite polynomials given by (14.26.1) can be obtained from the continuous q-Laguerre polynomials given by (14.19.1) by taking the limit $\alpha \to \infty$ in the following way:

$$\lim_{\alpha\to\infty} \frac{P_n^{(\alpha)}(x|q)}{q^{(\frac{1}{2}\alpha+\frac{1}{4})n}} = \frac{H_n(x|q)}{(q;q)_n}. \quad (14.19.15)$$

Continuous q-Laguerre → Laguerre

If we set $x \to q^x$ in the definitions (14.19.1) of the continuous q-Laguerre polynomials and take the limit $q \to 1$ we find the Laguerre polynomials given by (9.12.1). In fact we have:

$$\lim_{q\to 1} P_n^{(\alpha)}(q^x|q) = L_n^{(\alpha)}(2x). \tag{14.19.16}$$

Remark

If we let β tend to infinity in (14.10.16) and renormalize we obtain

$$P_n^{(\alpha)}(x;q) = \frac{(q^{\alpha+1};q)_n}{(q;q)_n}\, {}_3\phi_2\left(\begin{matrix} q^{-n}, q^{\frac{1}{2}}e^{i\theta}, q^{\frac{1}{2}}e^{-i\theta} \\ q^{\alpha+1}, -q \end{matrix}; q, q\right), \quad x = \cos\theta. \tag{14.19.17}$$

These two q-analogues of the Laguerre polynomials are connected by the following quadratic transformation:

$$P_n^{(\alpha)}(x|q^2) = q^{n\alpha} P_n^{(\alpha)}(x;q).$$

Reference

[72].

14.20 Little q-Laguerre / Wall

Basic Hypergeometric Representation

$$\begin{aligned} p_n(x;a|q) &= {}_2\phi_1\left(\begin{matrix} q^{-n}, 0 \\ aq \end{matrix}; q, qx\right) \\ &= \frac{1}{(a^{-1}q^{-n};q)_n}\, {}_2\phi_0\left(\begin{matrix} q^{-n}, x^{-1} \\ - \end{matrix}; q, \frac{x}{a}\right). \end{aligned} \tag{14.20.1}$$

Orthogonality Relation

$$\sum_{k=0}^{\infty}\frac{(aq)^k}{(q;q)_k}p_m(q^k;a|q)p_n(q^k;a|q)=\frac{(aq)^n}{(aq;q)_\infty}\frac{(q;q)_n}{(aq;q)_n}\,\delta_{mn},\quad 0<aq<1. \quad (14.20.2)$$

Recurrence Relation

$$-xp_n(x;a|q)=A_np_{n+1}(x;a|q)-(A_n+C_n)\,p_n(x;a|q)+C_np_{n-1}(x;a|q), \quad (14.20.3)$$

where

$$\begin{cases} A_n=q^n(1-aq^{n+1}) \\ C_n=aq^n(1-q^n). \end{cases}$$

Normalized Recurrence Relation

$$xp_n(x)=p_{n+1}(x)+(A_n+C_n)p_n(x)+aq^{2n-1}(1-q^n)(1-aq^n)p_{n-1}(x), \quad (14.20.4)$$

where

$$p_n(x;a|q)=\frac{(-1)^nq^{-\binom{n}{2}}}{(aq;q)_n}p_n(x).$$

q-Difference Equation

$$-q^{-n}(1-q^n)xy(x)=ay(qx)+(x-a-1)y(x)+(1-x)y(q^{-1}x), \quad (14.20.5)$$

where

$$y(x)=p_n(x;a|q).$$

Forward Shift Operator

$$p_n(x;a|q)-p_n(qx;a|q)=-\frac{q^{-n+1}(1-q^n)}{1-aq}xp_{n-1}(x;aq|q) \quad (14.20.6)$$

or equivalently

$$\mathscr{D}_qp_n(x;a|q)=-\frac{q^{-n+1}(1-q^n)}{(1-q)(1-aq)}p_{n-1}(x;aq|q). \quad (14.20.7)$$

Backward Shift Operator

$$ap_n(x;a|q)-(1-x)p_n(q^{-1}x;a|q)=(a-1)p_{n+1}(x;q^{-1}a|q) \qquad (14.20.8)$$

or equivalently

$$\mathscr{D}_{q^{-1}}\left[w(x;\alpha|q)p_n(x;q^{\alpha}|q)\right]$$
$$=\frac{1-q^{\alpha}}{q^{\alpha-1}(1-q)}w(x;\alpha-1|q)p_{n+1}(x;q^{\alpha-1}|q), \qquad (14.20.9)$$

where

$$w(x;\alpha|q)=(qx;q)_{\infty}x^{\alpha}.$$

Rodrigues-Type Formula

$$w(x;\alpha|q)p_n(x;q^{\alpha}|q)=\frac{q^{n\alpha+\binom{n}{2}}(1-q)^n}{(q^{\alpha+1};q)_n}\left(\mathscr{D}_{q^{-1}}\right)^n\left[w(x;\alpha+n|q)\right]. \qquad (14.20.10)$$

Generating Function

$$\frac{(t;q)_{\infty}}{(xt;q)_{\infty}}\,{}_0\phi_1\left(\begin{matrix}-\\ aq\end{matrix};q,aqxt\right)=\sum_{n=0}^{\infty}\frac{(-1)^nq^{\binom{n}{2}}}{(q;q)_n}p_n(x;a|q)t^n. \qquad (14.20.11)$$

Limit Relations

Big q-Laguerre → Little q-Laguerre / Wall

The little q-Laguerre (or Wall) polynomials given by (14.20.1) can be obtained from the big q-Laguerre polynomials by taking $x\rightarrow bqx$ in (14.11.1) and letting $b\rightarrow-\infty$:

$$\lim_{b\rightarrow-\infty}P_n(bqx;a,b;q)=p_n(x;a|q).$$

Little q-Jacobi → Little q-Laguerre / Wall

The little q-Laguerre (or Wall) polynomials given by (14.20.1) are little q-Jacobi polynomials with $b=0$. So if we set $b=0$ in the definition (14.12.1) of the little q-Jacobi polynomials we obtain the little q-Laguerre (or Wall) polynomials:

$$p_n(x;a,0|q) = p_n(x;a|q).$$

Affine q-Krawtchouk → Little q-Laguerre / Wall

If we set $x \rightarrow N - x$ in the definition (14.16.1) of the affine q-Krawtchouk polynomials and take the limit $N \rightarrow \infty$ we simply obtain the little q-Laguerre (or Wall) polynomials given by (14.20.1):

$$\lim_{N\rightarrow\infty} K_n^{Aff}(q^{x-N};p,N;q) = p_n(q^x;p;q).$$

Little q-Laguerre / Wall → Laguerre / Charlier

If we set $a = q^\alpha$ and $x \rightarrow (1-q)x$ in the definition (14.20.1) of the little q-Laguerre (or Wall) polynomials and let q tend to 1, we obtain the Laguerre polynomials given by (9.12.1):

$$\lim_{q\rightarrow 1} p_n((1-q)x;q^\alpha|q) = \frac{L_n^{(\alpha)}(x)}{L_n^{(\alpha)}(0)}. \tag{14.20.12}$$

If we set $a \rightarrow (q-1)a$ and $x \rightarrow q^x$ in the definition (14.20.1) of the little q-Laguerre (or Wall) polynomials and take the limit $q \rightarrow 1$ we obtain the Charlier polynomials given by (9.14.1) in the following way:

$$\lim_{q\rightarrow 1} \frac{p_n(q^x;(q-1)a|q)}{(1-q)^n} = \frac{C_n(x;a)}{a^n}. \tag{14.20.13}$$

Remark

If we set $a = q^\alpha$ and replace q by q^{-1} we find the q-Laguerre polynomials given by (14.21.1) in the following way:

$$p_n(x;q^{-\alpha}|q^{-1}) = \frac{(q;q)_n}{(q^{\alpha+1};q)_n} L_n^{(\alpha)}(-x;q).$$

References

[16], [26], [84], [144], [146], [147], [157], [214], [238], [345], [346], [416], [504], [509].

14.21 q-Laguerre

Basic Hypergeometric Representation

$$L_n^{(\alpha)}(x;q) = \frac{(q^{\alpha+1};q)_n}{(q;q)_n} \, {}_1\phi_1\left(\begin{matrix} q^{-n} \\ q^{\alpha+1} \end{matrix}; q, -q^{n+\alpha+1}x\right) \tag{14.21.1}$$
$$= \frac{1}{(q;q)_n} \, {}_2\phi_1\left(\begin{matrix} q^{-n}, -x \\ 0 \end{matrix}; q, q^{n+\alpha+1}\right).$$

Orthogonality Relation

The q-Laguerre polynomials satisfy two kinds of orthogonality relations, an absolutely continuous one and a discrete one. These orthogonality relations are given by, respectively:

$$\int_0^\infty \frac{x^\alpha}{(-x;q)_\infty} L_m^{(\alpha)}(x;q) L_n^{(\alpha)}(x;q)\, dx$$
$$= \frac{(q^{-\alpha};q)_\infty}{(q;q)_\infty} \frac{(q^{\alpha+1};q)_n}{(q;q)_n q^n} \Gamma(-\alpha)\Gamma(\alpha+1)\, \delta_{mn}, \quad \alpha > -1 \tag{14.21.2}$$

and

$$\sum_{k=-\infty}^{\infty} \frac{q^{k\alpha+k}}{(-cq^k;q)_\infty} L_m^{(\alpha)}(cq^k;q) L_n^{(\alpha)}(cq^k;q)$$
$$= \frac{(q, -cq^{\alpha+1}, -c^{-1}q^{-\alpha};q)_\infty}{(q^{\alpha+1}, -c, -c^{-1}q;q)_\infty} \frac{(q^{\alpha+1};q)_n}{(q;q)_n q^n} \delta_{mn}, \quad \alpha > -1, \quad c > 0. \tag{14.21.3}$$

For $c = 1$ the latter orthogonality relation can also be written as

$$\int_0^\infty \frac{x^\alpha}{(-x;q)_\infty} L_m^{(\alpha)}(x;q) L_n^{(\alpha)}(x;q)\, d_q x$$
$$= \frac{1-q}{2} \frac{(q, -q^{\alpha+1}, -q^{-\alpha};q)_\infty}{(q^{\alpha+1}, -q, -q;q)_\infty} \frac{(q^{\alpha+1};q)_n}{(q;q)_n q^n} \delta_{mn}, \quad \alpha > -1. \tag{14.21.4}$$

Recurrence Relation

$$-q^{2n+\alpha+1}xL_n^{(\alpha)}(x;q) = (1-q^{n+1})L_{n+1}^{(\alpha)}(x;q) \\ - \left[(1-q^{n+1})+q(1-q^{n+\alpha})\right]L_n^{(\alpha)}(x;q) \\ + q(1-q^{n+\alpha})L_{n-1}^{(\alpha)}(x;q). \qquad (14.21.5)$$

Normalized Recurrence Relation

$$xp_n(x) = p_{n+1}(x) + q^{-2n-\alpha-1}\left[(1-q^{n+1})+q(1-q^{n+\alpha})\right]p_n(x) \\ + q^{-4n-2\alpha+1}(1-q^n)(1-q^{n+\alpha})p_{n-1}(x), \qquad (14.21.6)$$

where

$$L_n^{(\alpha)}(x;q) = \frac{(-1)^n q^{n(n+\alpha)}}{(q;q)_n}p_n(x).$$

q-Difference Equation

$$-q^{\alpha}(1-q^n)xy(x) = q^{\alpha}(1+x)y(qx) - [1+q^{\alpha}(1+x)]y(x) + y(q^{-1}x), \quad (14.21.7)$$

where

$$y(x) = L_n^{(\alpha)}(x;q).$$

Forward Shift Operator

$$L_n^{(\alpha)}(x;q) - L_n^{(\alpha)}(qx;q) = -q^{\alpha+1}xL_{n-1}^{(\alpha+1)}(qx;q) \qquad (14.21.8)$$

or equivalently

$$\mathcal{D}_q L_n^{(\alpha)}(x;q) = -\frac{q^{\alpha+1}}{1-q}L_{n-1}^{(\alpha+1)}(qx;q). \qquad (14.21.9)$$

Backward Shift Operator

$$L_n^{(\alpha)}(x;q) - q^{\alpha}(1+x)L_n^{(\alpha)}(qx;q) = (1-q^{n+1})L_{n+1}^{(\alpha-1)}(x;q) \qquad (14.21.10)$$

or equivalently

$$\mathscr{D}_q\left[w(x;\alpha;q)L_n^{(\alpha)}(x;q)\right]=\frac{1-q^{n+1}}{1-q}w(x;\alpha-1;q)L_{n+1}^{(\alpha-1)}(x;q), \tag{14.21.11}$$

where

$$w(x;\alpha;q)=\frac{x^{\alpha}}{(-x;q)_\infty}.$$

Rodrigues-Type Formula

$$w(x;\alpha;q)L_n^{(\alpha)}(x;q)=\frac{(1-q)^n}{(q;q)_n}\left(\mathscr{D}_q\right)^n\left[w(x;\alpha+n;q)\right]. \tag{14.21.12}$$

Generating Functions

$$\frac{1}{(t;q)_\infty}\,{}_1\phi_1\left(\begin{matrix}-x\\ 0\end{matrix};q,q^{\alpha+1}t\right)=\sum_{n=0}^{\infty}L_n^{(\alpha)}(x;q)t^n. \tag{14.21.13}$$

$$\frac{1}{(t;q)_\infty}\,{}_0\phi_1\left(\begin{matrix}-\\ q^{\alpha+1}\end{matrix};q,-q^{\alpha+1}xt\right)=\sum_{n=0}^{\infty}\frac{L_n^{(\alpha)}(x;q)}{(q^{\alpha+1};q)_n}t^n. \tag{14.21.14}$$

$$(t;q)_\infty\cdot{}_0\phi_2\left(\begin{matrix}-\\ q^{\alpha+1},t\end{matrix};q,-q^{\alpha+1}xt\right)=\sum_{n=0}^{\infty}\frac{(-1)^nq^{\binom{n}{2}}}{(q^{\alpha+1};q)_n}L_n^{(\alpha)}(x;q)t^n. \tag{14.21.15}$$

$$\begin{aligned}&\frac{(\gamma t;q)_\infty}{(t;q)_\infty}\,{}_1\phi_2\left(\begin{matrix}\gamma\\ q^{\alpha+1},\gamma t\end{matrix};q,-q^{\alpha+1}xt\right)\\&=\sum_{n=0}^{\infty}\frac{(\gamma;q)_n}{(q^{\alpha+1};q)_n}L_n^{(\alpha)}(x;q)t^n,\quad \gamma\text{ arbitrary}.\end{aligned} \tag{14.21.16}$$

Limit Relations

Little q-Jacobi $\rightarrow$ q-Laguerre

If we substitute $a=q^{\alpha}$ and $x\rightarrow -b^{-1}q^{-1}x$ in the definition (14.12.1) of the little q-Jacobi polynomials and then take the limit $b\rightarrow-\infty$ we find the q-Laguerre polynomials given by (14.21.1):

$$\lim_{b\rightarrow-\infty}p_n(-b^{-1}q^{-1}x;q^{\alpha},b|q)=\frac{(q;q)_n}{(q^{\alpha+1};q)_n}L_n^{(\alpha)}(x;q).$$

q-Meixner $\rightarrow$ q-Laguerre

The q-Laguerre polynomials given by (14.21.1) can be obtained from the q-Meixner polynomials given by (14.13.1) by setting $b = q^{\alpha}$ and $q^{-x} \rightarrow cq^{\alpha}x$ in the definition (14.13.1) of the q-Meixner polynomials and then taking the limit $c \rightarrow \infty$:

$$\lim_{c\rightarrow\infty} M_n(cq^{\alpha}x; q^{\alpha}, c; q) = \frac{(q;q)_n}{(q^{\alpha+1};q)_n} L_n^{(\alpha)}(x;q).$$

q-Laguerre $\rightarrow$ Stieltjes-Wigert

If we set $x \rightarrow xq^{-\alpha}$ in the definition (14.21.1) of the q-Laguerre polynomials and take the limit $\alpha \rightarrow \infty$ we simply obtain the Stieltjes-Wigert polynomials given by (14.27.1):

$$\lim_{\alpha\rightarrow\infty} L_n^{(\alpha)}\left(xq^{-\alpha};q\right) = S_n(x;q). \tag{14.21.17}$$

q-Laguerre $\rightarrow$ Laguerre / Charlier

If we set $x \rightarrow (1-q)x$ in the definition (14.21.1) of the q-Laguerre polynomials and take the limit $q \rightarrow 1$ we obtain the Laguerre polynomials given by (9.12.1):

$$\lim_{q\rightarrow 1} L_n^{(\alpha)}((1-q)x;q) = L_n^{(\alpha)}(x). \tag{14.21.18}$$

If we set $x \rightarrow -q^{-x}$ and $q^{\alpha} = a^{-1}(q-1)^{-1}$ (or $\alpha = -(\ln q)^{-1}\ln(q-1)a$) in the definition (14.21.1) of the q-Laguerre polynomials, multiply by $(q;q)_n$, and take the limit $q \rightarrow 1$ we obtain the Charlier polynomials given by (9.14.1):

$$\lim_{q\rightarrow 1} (q;q)_n L_n^{(\alpha)}(-q^{-x};q) = C_n(x;a), \tag{14.21.19}$$

where

$$q^{\alpha} = \frac{1}{a(q-1)} \quad \text{or} \quad \alpha = -\frac{\ln(q-1)a}{\ln q}.$$

Remarks

The q-Laguerre polynomials are sometimes called the generalized Stieltjes-Wigert polynomials.

If we replace q by q^{-1} we obtain the little q-Laguerre (or Wall) polynomials given by (14.20.1) in the following way:

$$L_n^{(\alpha)}(x;q^{-1}) = \frac{(q^{\alpha+1};q)_n}{(q;q)_n q^{n\alpha}} p_n(-x;q^\alpha|q).$$

The q-Laguerre polynomials given by (14.21.1) and the q-Bessel polynomials given by (14.22.1) are related in the following way:

$$\frac{y_n(q^x;a;q)}{(q;q)_n} = L_n^{(x-n)}(aq^n;q).$$

The q-Laguerre polynomials given by (14.21.1) and the q-Charlier polynomials given by (14.23.1) are related in the following way:

$$\frac{C_n(-x;-q^{-\alpha};q)}{(q;q)_n} = L_n^{(\alpha)}(x;q).$$

Since the Stieltjes and Hamburger moment problems corresponding to the q-Laguerre polynomials are indeterminate there exist many different weight functions.

References

[12], [16], [49], [51], [72], [84], [139], [144], [146], [147], [150], [199], [238], [256], [276], [291], [302], [411].

14.22 q-Bessel

Basic Hypergeometric Representation

$$\begin{aligned} y_n(x;a;q) &= {}_2\phi_1\left(\begin{matrix} q^{-n}, -aq^n \\ 0 \end{matrix}; q, qx\right) \qquad (14.22.1) \\ &= (q^{-n+1}x;q)_n \cdot {}_1\phi_1\left(\begin{matrix} q^{-n} \\ q^{-n+1}x \end{matrix}; q, -aq^{n+1}x\right) \\ &= (-aq^n x)^n \cdot {}_2\phi_1\left(\begin{matrix} q^{-n}, x^{-1} \\ 0 \end{matrix}; q, -\frac{q^{-n+1}}{a}\right). \end{aligned}$$

Orthogonality Relation

$$\sum_{k=0}^{\infty}\frac{a^k}{(q;q)_k}q^{\binom{k+1}{2}}y_m(q^k;a;q)y_n(q^k;a;q)$$
$$=(q;q)_n(-aq^n;q)_{\infty}\frac{a^nq^{\binom{n+1}{2}}}{(1+aq^{2n})}\delta_{mn},\quad a>0. \tag{14.22.2}$$

Recurrence Relation

$$-xy_n(x;a;q)=A_ny_{n+1}(x;a;q)-(A_n+C_n)y_n(x;a;q)+C_ny_{n-1}(x;a;q), \tag{14.22.3}$$

where

$$\begin{cases} A_n=q^n\dfrac{(1+aq^n)}{(1+aq^{2n})(1+aq^{2n+1})} \\ C_n=aq^{2n-1}\dfrac{(1-q^n)}{(1+aq^{2n-1})(1+aq^{2n})}. \end{cases}$$

Normalized Recurrence Relation

$$xp_n(x)=p_{n+1}(x)+(A_n+C_n)p_n(x)+A_{n-1}C_np_{n-1}(x), \tag{14.22.4}$$

where

$$y_n(x;a;q)=(-1)^nq^{-\binom{n}{2}}(-aq^n;q)_np_n(x).$$

q-Difference Equation

$$-q^{-n}(1-q^n)(1+aq^n)xy(x)$$
$$=axy(qx)-(ax+1-x)y(x)+(1-x)y(q^{-1}x), \tag{14.22.5}$$

where

$$y(x)=y_n(x;a;q).$$

Forward Shift Operator

$$y_n(x;a;q) - y_n(qx;a;q) = -q^{-n+1}(1-q^n)(1+aq^n)xy_{n-1}(x;aq^2;q) \qquad (14.22.6)$$

or equivalently

$$\mathscr{D}_q y_n(x;a;q) = -\frac{q^{-n+1}(1-q^n)(1+aq^n)}{1-q} y_{n-1}(x;aq^2;q). \qquad (14.22.7)$$

Backward Shift Operator

$$aq^{x-1}y_n(q^x;a;q) - (1-q^x)y_n(q^{x-1}x;a;q) = -y_{n+1}(q^x;aq^{-2};q) \qquad (14.22.8)$$

or equivalently

$$\frac{\nabla\left[w(x;a;q)y_n(q^x;a;q)\right]}{\nabla q^x} = \frac{q^2}{a(1-q)} w(x;aq^{-2};q)y_{n+1}(q^x;aq^{-2};q), \qquad (14.22.9)$$

where

$$w(x;a;q) = \frac{a^x q^{\binom{x}{2}}}{(q;q)_x}.$$

Rodrigues-Type Formula

$$w(x;a;q)y_n(q^x;a;q) = a^n(1-q)^n q^{n(n-1)} \left(\nabla_q\right)^n \left[w(x;aq^{2n};q)\right], \qquad (14.22.10)$$

where

$$\nabla_q := \frac{\nabla}{\nabla q^x}.$$

Generating Functions

$$\begin{aligned} &{}_0\phi_1\left(\begin{matrix} - \\ 0 \end{matrix}; q, -aq^{x+1}t\right) {}_2\phi_0\left(\begin{matrix} q^{-x}, 0 \\ - \end{matrix}; q, q^x t\right) \\ &= \sum_{n=0}^{\infty} \frac{y_n(q^x;a;q)}{(q;q)_n} t^n, \quad x = 0,1,2,\ldots. \end{aligned} \qquad (14.22.11)$$

$$\frac{(t;q)_\infty}{(xt;q)_\infty} {}_1\phi_3\left(\begin{matrix} xt \\ 0,0,t \end{matrix}; q, -aqxt\right) = \sum_{n=0}^{\infty} \frac{(-1)^n q^{\binom{n}{2}}}{(q;q)_n} y_n(x;a;q)t^n. \qquad (14.22.12)$$

Limit Relations

Little q-Jacobi $\rightarrow$ q-Bessel

If we set $b \rightarrow -a^{-1}q^{-1}b$ in the definition (14.12.1) of the little q-Jacobi polynomials and then take the limit $a \rightarrow 0$ we obtain the q-Bessel polynomials given by (14.22.1):

$$\lim_{a\rightarrow 0} p_n(x;a,-a^{-1}q^{-1}b|q) = y_n(x;b;q).$$

q-Krawtchouk $\rightarrow$ q-Bessel

If we set $x \rightarrow N-x$ in the definition (14.15.1) of the q-Krawtchouk polynomials and then take the limit $N \rightarrow \infty$ we obtain the q-Bessel polynomials given by (14.22.1):

$$\lim_{N\rightarrow\infty} K_n(q^{x-N};p,N;q) = y_n(q^x;p;q).$$

q-Bessel $\rightarrow$ Stieltjes-Wigert

The Stieltjes-Wigert polynomials given by (14.27.1) can be obtained from the q-Bessel polynomials by setting $x \rightarrow a^{-1}x$ in the definition (14.22.1) of the q-Bessel polynomials and then taking the limit $a \rightarrow \infty$. In fact we have

$$\lim_{a\rightarrow\infty} y_n(a^{-1}x;a;q) = (q;q)_n S_n(x;q). \tag{14.22.13}$$

q-Bessel $\rightarrow$ Bessel

If we set $x \rightarrow -\frac{1}{2}(1-q)^{-1}x$ and $a \rightarrow -q^{a+1}$ in the definition (14.22.1) of the q-Bessel polynomials and take the limit $q \rightarrow 1$ we find the Bessel polynomials given by (9.13.1):

$$\lim_{q\rightarrow 1} y_n(-\tfrac{1}{2}(1-q)^{-1}x;-q^{a+1};q) = y_n(x;a). \tag{14.22.14}$$

q-Bessel $\rightarrow$ Charlier

If we set $x \rightarrow q^x$ and $a \rightarrow a(1-q)$ in the definition (14.22.1) of the q-Bessel polynomials and take the limit $q \rightarrow 1$ we find the Charlier polynomials given by (9.14.1):

$$\lim_{q\rightarrow 1} \frac{y_n(q^x;a(1-q);q)}{(q-1)^n} = a^n C_n(x;a). \tag{14.22.15}$$

Remark

In [318] and [319] these q-Bessel polynomials were called *alternative q-Charlier polynomials*.
The q-Bessel polynomials given by (14.22.1) and the q-Laguerre polynomials given by (14.21.1) are related in the following way:

$$\frac{y_n(q^x;a;q)}{(q;q)_n} = L_n^{(x-n)}(aq^n;q).$$

Reference

[157].

14.23 q-Charlier

Basic Hypergeometric Representation

$$\begin{aligned} C_n(q^{-x};a;q) &= {}_2\phi_1\left(\begin{matrix} q^{-n}, q^{-x} \\ 0 \end{matrix}; q, -\frac{q^{n+1}}{a}\right) \qquad (14.23.1)\\ &= (-a^{-1}q;q)_n \cdot {}_1\phi_1\left(\begin{matrix} q^{-n} \\ -a^{-1}q \end{matrix}; q, -\frac{q^{n+1-x}}{a}\right). \end{aligned}$$

Orthogonality Relation

$$\begin{aligned} &\sum_{x=0}^{\infty} \frac{a^x}{(q;q)_x} q^{\binom{x}{2}} C_m(q^{-x};a;q) C_n(q^{-x};a;q) \\ &= q^{-n}(-a;q)_\infty(-a^{-1}q,q;q)_n\,\delta_{mn}, \quad a > 0. \qquad (14.23.2) \end{aligned}$$

Recurrence Relation

$$\begin{aligned} & q^{2n+1}(1-q^{-x})C_n(q^{-x}) \\ & = aC_{n+1}(q^{-x}) - [a+q(1-q^n)(a+q^n)]\, C_n(q^{-x}) \\ & \qquad + q(1-q^n)(a+q^n)C_{n-1}(q^{-x}), \end{aligned} \tag{14.23.3}$$

where

$$C_n(q^{-x}) := C_n(q^{-x};a;q).$$

Normalized Recurrence Relation

$$\begin{aligned} xp_n(x) = p_{n+1}(x) + \left[1 + q^{-2n-1}\left\{a+q(1-q^n)(a+q^n)\right\}\right] p_n(x) \\ + aq^{-4n+1}(1-q^n)(a+q^n)p_{n-1}(x), \end{aligned} \tag{14.23.4}$$

where

$$C_n(q^{-x};a;q) = \frac{(-1)^n q^{n^2}}{a^n} p_n(q^{-x}).$$

q-Difference Equation

$$q^n y(x) = aq^x y(x+1) - q^x(a-1)y(x) + (1-q^x)y(x-1), \tag{14.23.5}$$

where

$$y(x) = C_n(q^{-x};a;q).$$

Forward Shift Operator

$$C_n(q^{-x-1};a;q) - C_n(q^{-x};a;q) = -a^{-1}q^{-x}(1-q^n)C_{n-1}(q^{-x};aq^{-1};q) \tag{14.23.6}$$

or equivalently

$$\frac{\Delta C_n(q^{-x};a;q)}{\Delta q^{-x}} = -\frac{q(1-q^n)}{a(1-q)} C_{n-1}(q^{-x};aq^{-1};q). \tag{14.23.7}$$

Backward Shift Operator

$$C_n(q^{-x};a;q)-a^{-1}q^{-x}(1-q^x)C_n(q^{-x+1};a;q)=C_{n+1}(q^{-x};aq;q) \tag{14.23.8}$$

or equivalently

$$\frac{\nabla\left[w(x;a;q)C_n(q^{-x};a;q)\right]}{\nabla q^{-x}}=\frac{1}{1-q}w(x;aq;q)C_{n+1}(q^{-x};aq;q), \tag{14.23.9}$$

where

$$w(x;a;q)=\frac{a^x q^{\binom{x+1}{2}}}{(q;q)_x}.$$

Rodrigues-Type Formula

$$w(x;a;q)C_n(q^{-x};a;q)=(1-q)^n\left(\nabla_q\right)^n\left[w(x;aq^{-n};q)\right], \tag{14.23.10}$$

where

$$\nabla_q:=\frac{\nabla}{\nabla q^{-x}}.$$

Generating Functions

$$\frac{1}{(t;q)_\infty}\,{}_1\phi_1\left(\begin{matrix}q^{-x}\\0\end{matrix};q,-a^{-1}qt\right)=\sum_{n=0}^{\infty}\frac{C_n(q^{-x};a;q)}{(q;q)_n}t^n. \tag{14.23.11}$$

$$\frac{1}{(t;q)_\infty}\,{}_0\phi_1\left(\begin{matrix}-\\-a^{-1}q\end{matrix};q,-a^{-1}q^{-x+1}t\right)=\sum_{n=0}^{\infty}\frac{C_n(q^{-x};a;q)}{(-a^{-1}q,q;q)_n}t^n. \tag{14.23.12}$$

Limit Relations

q-Meixner $\rightarrow$ q-Charlier

The q-Charlier polynomials given by (14.23.1) can easily be obtained from the q-Meixner given by (14.13.1) by setting $b=0$ in the definition (14.13.1) of the q-Meixner polynomials:

$$M_n(x;0,c;q)=C_n(x;c;q). \tag{14.23.13}$$

q-Krawtchouk → q-Charlier

By setting $p = a^{-1}q^{-N}$ in the definition (14.15.1) of the q-Krawtchouk polynomials and then taking the limit $N \to \infty$ we obtain the q-Charlier polynomials given by (14.23.1):

$$\lim_{N\to\infty} K_n(q^{-x}; a^{-1}q^{-N}, N; q) = C_n(q^{-x}; a; q).$$

q-Charlier → Stieltjes-Wigert

If we set $q^{-x} \to ax$ in the definition (14.23.1) of the q-Charlier polynomials and take the limit $a \to \infty$ we obtain the Stieltjes-Wigert polynomials given by (14.27.1) in the following way:

$$\lim_{a\to\infty} C_n(ax; a; q) = (q;q)_n S_n(x;q). \tag{14.23.14}$$

q-Charlier → Charlier

If we set $a \to a(1-q)$ in the definition (14.23.1) of the q-Charlier polynomials and take the limit $q \to 1$ we obtain the Charlier polynomials given by (9.14.1):

$$\lim_{q\to 1} C_n(q^{-x}; a(1-q); q) = C_n(x; a). \tag{14.23.15}$$

Remark

The q-Charlier polynomials given by (14.23.1) and the q-Laguerre polynomials given by (14.21.1) are related in the following way:

$$\frac{C_n(-x; -q^{-\alpha}; q)}{(q;q)_n} = L_n^{(\alpha)}(x;q).$$

References

[30], [80], [238], [261], [328], [416], [524].

14.24 Al-Salam-Carlitz I

Basic Hypergeometric Representation

$$U_n^{(a)}(x;q) = (-a)^n q^{\binom{n}{2}} {}_2\phi_1\left(\begin{matrix} q^{-n}, x^{-1} \\ 0 \end{matrix}; q, \frac{qx}{a}\right). \tag{14.24.1}$$

Orthogonality Relation

$$\int_a^1 (qx, a^{-1}qx; q)_\infty U_m^{(a)}(x;q) U_n^{(a)}(x;q)\, d_q x$$
$$= (-a)^n (1-q)(q;q)_n (q, a, a^{-1}q; q)_\infty q^{\binom{n}{2}}\, \delta_{mn}, \quad a < 0. \tag{14.24.2}$$

Recurrence Relation

$$xU_n^{(a)}(x;q) = U_{n+1}^{(a)}(x;q) + (a+1)q^n U_n^{(a)}(x;q) - aq^{n-1}(1-q^n)U_{n-1}^{(a)}(x;q). \tag{14.24.3}$$

Normalized Recurrence Relation

$$xp_n(x) = p_{n+1}(x) + (a+1)q^n p_n(x) - aq^{n-1}(1-q^n)p_{n-1}(x), \tag{14.24.4}$$

where

$$U_n^{(a)}(x;q) = p_n(x).$$

q-Difference Equation

$$(1-q^n)x^2 y(x) = aq^{n-1}y(qx) - \left[aq^{n-1} + q^n(1-x)(a-x)\right] y(x)$$
$$+ q^n(1-x)(a-x)y(q^{-1}x), \tag{14.24.5}$$

where

$$y(x) = U_n^{(a)}(x;q).$$

Forward Shift Operator

$$U_n^{(a)}(x;q)-U_n^{(a)}(qx;q)=(1-q^n)xU_{n-1}^{(a)}(x;q) \tag{14.24.6}$$

or equivalently

$$\mathscr{D}_qU_n^{(a)}(x;q)=\frac{1-q^n}{1-q}U_{n-1}^{(a)}(x;q). \tag{14.24.7}$$

Backward Shift Operator

$$aU_n^{(a)}(x;q)-(1-x)(a-x)U_n^{(a)}(q^{-1}x;q)=-q^{-n}xU_{n+1}^{(a)}(x;q) \tag{14.24.8}$$

or equivalently

$$\mathscr{D}_{q^{-1}}\left[w(x;a;q)U_n^{(a)}(x;q)\right]=\frac{q^{-n+1}}{a(1-q)}w(x;a;q)U_{n+1}^{(a)}(x;q), \tag{14.24.9}$$

where

$$w(x;a;q)=(qx,a^{-1}qx;q)_\infty.$$

Rodrigues-Type Formula

$$w(x;a;q)U_n^{(a)}(x;q)=a^nq^{\frac{1}{2}n(n-3)}(1-q)^n\left(\mathscr{D}_{q^{-1}}\right)^n[w(x;a;q)]. \tag{14.24.10}$$

Generating Function

$$\frac{(t,at;q)_\infty}{(xt;q)_\infty}=\sum_{n=0}^\infty\frac{U_n^{(a)}(x;q)}{(q;q)_n}t^n. \tag{14.24.11}$$

Limit Relations

Big q-Laguerre $\rightarrow$ Al-Salam-Carlitz I

If we set $x\rightarrow aqx$ and $b\rightarrow ab$ in the definition (14.11.1) of the big q-Laguerre polynomials and take the limit $a\rightarrow 0$ we obtain the Al-Salam-Carlitz I polynomials given by (14.24.1):

$$\lim_{a \to 0} \frac{P_n(aqx;a,ab;q)}{a^n} = q^n U_n^{(b)}(x;q).$$

Dual q-Krawtchouk → Al-Salam-Carlitz I

If we set $c = a^{-1}$ in the definition (14.17.1) of the dual q-Krawtchouk polynomials and take the limit $N \to \infty$ we simply obtain the Al-Salam-Carlitz I polynomials given by (14.24.1):

$$\lim_{N \to \infty} K_n(\lambda(x);a^{-1},N|q) = \left(-\frac{1}{a}\right)^n q^{-\binom{n}{2}} U_n^{(a)}(q^x;q).$$

Note that $\lambda(x) = q^{-x} + a^{-1}q^{x-N}$.

Al-Salam-Carlitz I → Discrete q-Hermite I

The discrete q-Hermite I polynomials given by (14.28.1) can easily be obtained from the Al-Salam-Carlitz I polynomials given by (14.24.1) by the substitution $a = -1$:

$$U_n^{(-1)}(x;q) = h_n(x;q). \tag{14.24.12}$$

Al-Salam-Carlitz I → Charlier / Hermite

If we set $a \to a(q-1)$ and $x \to q^x$ in the definition (14.24.1) of the Al-Salam-Carlitz I polynomials and take the limit $q \to 1$ after dividing by $a^n(1-q)^n$ we obtain the Charlier polynomials given by (9.14.1):

$$\lim_{q \to 1} \frac{U_n^{(a(q-1))}(q^x;q)}{(1-q)^n} = a^n C_n(x;a). \tag{14.24.13}$$

If we set $x \to x\sqrt{1-q^2}$ and $a \to a\sqrt{1-q^2}-1$ in the definition (14.24.1) of the Al-Salam-Carlitz I polynomials, divide by $(1-q^2)^{\frac{n}{2}}$, and let q tend to 1 we obtain the Hermite polynomials given by (9.15.1) with shifted argument. In fact we have

$$\lim_{q \to 1} \frac{U_n^{(a\sqrt{1-q^2}-1)}(x\sqrt{1-q^2};q)}{(1-q^2)^{\frac{n}{2}}} = \frac{H_n(x-a)}{2^n}. \tag{14.24.14}$$

Remark

The Al-Salam-Carlitz I polynomials are related to the Al-Salam-Carlitz II polynomials given by (14.25.1) in the following way:

$$U_n^{(a)}(x;q^{-1}) = V_n^{(a)}(x;q).$$

References

[16], [18], [20], [68], [80], [144], [146], [157], [160], [180], [238], [269], [289], [312], [524].

14.25 Al-Salam-Carlitz II

Basic Hypergeometric Representation

$$V_n^{(a)}(x;q) = (-a)^n q^{-\binom{n}{2}} {}_2\phi_0\left(\begin{matrix} q^{-n}, x \\ - \end{matrix} ; q, \frac{q^n}{a} \right). \tag{14.25.1}$$

Orthogonality Relation

$$\begin{aligned} &\sum_{k=0}^{\infty} \frac{q^{k^2} a^k}{(q;q)_k (aq;q)_k} V_m^{(a)}(q^{-k};q) V_n^{(a)}(q^{-k};q) \\ &= \frac{(q;q)_n a^n}{(aq;q)_\infty q^{n^2}} \delta_{mn}, \quad 0 < aq < 1. \end{aligned} \tag{14.25.2}$$

Recurrence Relation

$$\begin{aligned} xV_n^{(a)}(x;q) = {} & V_{n+1}^{(a)}(x;q) + (a+1)q^{-n} V_n^{(a)}(x;q) \\ & + aq^{-2n+1}(1-q^n) V_{n-1}^{(a)}(x;q). \end{aligned} \tag{14.25.3}$$

Normalized Recurrence Relation

$$xp_n(x) = p_{n+1}(x) + (a+1)q^{-n}p_n(x) + aq^{-2n+1}(1-q^n)p_{n-1}(x), \tag{14.25.4}$$

where

$$V_n^{(a)}(x;q) = p_n(x).$$

q-Difference Equation

$$\begin{aligned}&-(1-q^n)x^2y(x)\\&= (1-x)(a-x)y(qx) - [(1-x)(a-x)+aq]y(x) + aqy(q^{-1}x),\end{aligned} \tag{14.25.5}$$

where

$$y(x) = V_n^{(a)}(x;q).$$

Forward Shift Operator

$$V_n^{(a)}(x;q) - V_n^{(a)}(qx;q) = q^{-n+1}(1-q^n)xV_{n-1}^{(a)}(qx;q) \tag{14.25.6}$$

or equivalently

$$\mathscr{D}_q V_n^{(a)}(x;q) = \frac{q^{-n+1}(1-q^n)}{1-q}V_{n-1}^{(a)}(qx;q). \tag{14.25.7}$$

Backward Shift Operator

$$aV_n^{(a)}(x;q) - (1-x)(a-x)V_n^{(a)}(qx;q) = -q^n xV_{n+1}^{(a)}(x;q) \tag{14.25.8}$$

or equivalently

$$\mathscr{D}_q\left[w(x;a;q)V_n^{(a)}(x;q)\right] = -\frac{q^n}{a(1-q)}w(x;a;q)V_{n+1}^{(a)}(x;q), \tag{14.25.9}$$

where

$$w(x;a;q) = \frac{1}{(x,a^{-1}x;q)_\infty}.$$

Rodrigues-Type Formula

$$w(x;a;q)V_n^{(a)}(x;q) = a^n(q-1)^n q^{-\binom{n}{2}} (\mathscr{D}_q)^n \left[w(x;a;q)\right]. \tag{14.25.10}$$

Generating Functions

$$\frac{(xt;q)_\infty}{(t,at;q)_\infty} = \sum_{n=0}^{\infty} \frac{(-1)^n q^{\binom{n}{2}}}{(q;q)_n} V_n^{(a)}(x;q)t^n. \tag{14.25.11}$$

$$(at;q)_\infty \cdot {}_1\phi_1\left(\begin{array}{c} x \\ at \end{array}; q, t\right) = \sum_{n=0}^{\infty} \frac{q^{n(n-1)}}{(q;q)_n} V_n^{(a)}(x;q)t^n. \tag{14.25.12}$$

Limit Relations

q-Meixner → Al-Salam-Carlitz II

The Al-Salam-Carlitz II polynomials given by (14.25.1) can be obtained from the q-Meixner polynomials given by (14.13.1) by setting $b = -ac^{-1}$ in the definition (14.13.1) of the q-Meixner polynomials and then taking the limit $c \to 0$:

$$\lim_{c\to 0} M_n(x;-ac^{-1},c;q) = \left(-\frac{1}{a}\right)^n q^{\binom{n}{2}} V_n^{(a)}(x;q).$$

Quantum q-Krawtchouk → Al-Salam-Carlitz II

If we set $p = a^{-1}q^{-N-1}$ in the definition (14.14.1) of the quantum q-Krawtchouk polynomials and let $N \to \infty$ we obtain the Al-Salam-Carlitz II polynomials given by (14.25.1). In fact we have

$$\lim_{N\to\infty} K_n^{qtm}(x;a^{-1}q^{-N-1},N;q) = \left(-\frac{1}{a}\right)^n q^{\binom{n}{2}} V_n^{(a)}(x;q).$$

Al-Salam-Carlitz II → Discrete q-Hermite II

The discrete q-Hermite II polynomials given by (14.29.1) follow from the Al-Salam-Carlitz II polynomials given by (14.25.1) by the substitution $a = -1$ in the following way:

$$i^{-n}V_n^{(-1)}(ix;q) = \tilde{h}_n(x;q). \tag{14.25.13}$$

Al-Salam-Carlitz II → Charlier / Hermite

If we set $a \to a(1-q)$ and $x \to q^{-x}$ in the definition (14.25.1) of the Al-Salam-Carlitz II polynomials and taking the limit $q \to 1$ we find

$$\lim_{q\to 1} \frac{V_n^{(a(1-q))}(q^{-x};q)}{(q-1)^n} = a^n C_n(x;a). \tag{14.25.14}$$

If we set $x \to ix\sqrt{1-q^2}$ and $a \to ia\sqrt{1-q^2}-1$ in the definition (14.25.1) of the Al-Salam-Carlitz II polynomials, divide by $i^n(1-q^2)^{\frac{n}{2}}$, and let q tend to 1 we obtain the Hermite polynomials given by (9.15.1) with shifted argument. In fact we have

$$\lim_{q\to 1} \frac{V_n^{(ia\sqrt{1-q^2}-1)}(ix\sqrt{1-q^2};q)}{i^n(1-q^2)^{\frac{n}{2}}} = \frac{H_n(x-a)}{2^n}. \tag{14.25.15}$$

Remark

The Al-Salam-Carlitz II polynomials are related to the Al-Salam-Carlitz I polynomials given by (14.24.1) in the following way:

$$V_n^{(a)}(x;q^{-1}) = U_n^{(a)}(x;q).$$

References

[16], [18], [20], [67], [101], [143], [144], [146], [160], [214], [269], [276].

14.26 Continuous q-Hermite

Basic Hypergeometric Representation

$$H_n(x|q) = e^{in\theta}\,{}_2\phi_0\left(\begin{matrix} q^{-n}, 0 \\ - \end{matrix}\,;\, q, q^n e^{-2i\theta}\right), \quad x = \cos\theta. \tag{14.26.1}$$

Orthogonality Relation

$$\frac{1}{2\pi}\int_{-1}^{1}\frac{w(x|q)}{\sqrt{1-x^2}}H_m(x|q)H_n(x|q)\,dx=\frac{\delta_{mn}}{(q^{n+1};q)_\infty}, \tag{14.26.2}$$

where

$$w(x|q)=\left|\left(e^{2i\theta};q\right)_\infty\right|^2=h(x,1)h(x,-1)h(x,q^{\frac{1}{2}})h(x,-q^{\frac{1}{2}}),$$

with

$$h(x,\alpha):=\prod_{k=0}^{\infty}\left(1-2\alpha xq^k+\alpha^2q^{2k}\right)=\left(\alpha e^{i\theta},\alpha e^{-i\theta};q\right)_\infty,\quad x=\cos\theta.$$

Recurrence Relation

$$2xH_n(x|q)=H_{n+1}(x|q)+(1-q^n)H_{n-1}(x|q). \tag{14.26.3}$$

Normalized Recurrence Relation

$$xp_n(x)=p_{n+1}(x)+\frac{1}{4}(1-q^n)p_{n-1}(x), \tag{14.26.4}$$

where

$$H_n(x|q)=2^np_n(x).$$

q-Difference Equation

$$(1-q)^2D_q\left[\tilde{w}(x|q)D_qy(x)\right]+4q^{-n+1}(1-q^n)\tilde{w}(x|q)y(x)=0, \tag{14.26.5}$$

where

$$y(x)=H_n(x|q)$$

and

$$\tilde{w}(x|q):=\frac{w(x|q)}{\sqrt{1-x^2}}.$$

Forward Shift Operator

$$\delta_qH_n(x|q)=-q^{-\frac{1}{2}n}(1-q^n)(e^{i\theta}-e^{-i\theta})H_{n-1}(x|q),\quad x=\cos\theta \tag{14.26.6}$$

or equivalently

$$D_qH_n(x|q)=\frac{2q^{-\frac{1}{2}(n-1)}(1-q^n)}{1-q}H_{n-1}(x|q). \tag{14.26.7}$$

Backward Shift Operator

$$\delta_q\left[\tilde{w}(x|q)H_n(x|q)\right]=q^{-\frac{1}{2}(n+1)}(\mathrm{e}^{i\theta}-\mathrm{e}^{-i\theta})\tilde{w}(x|q)H_{n+1}(x|q),\quad x=\cos\theta \tag{14.26.8}$$

or equivalently

$$D_q\left[\tilde{w}(x|q)H_n(x|q)\right]=-\frac{2q^{-\frac{1}{2}n}}{1-q}\tilde{w}(x|q)H_{n+1}(x|q). \tag{14.26.9}$$

Rodrigues-Type Formula

$$\tilde{w}(x|q)H_n(x|q)=\left(\frac{q-1}{2}\right)^n q^{\frac{1}{4}n(n-1)}\left(D_q\right)^n\left[\tilde{w}(x|q)\right]. \tag{14.26.10}$$

Generating Functions

$$\frac{1}{|(\mathrm{e}^{i\theta}t;q)_\infty|^2}=\frac{1}{(\mathrm{e}^{i\theta}t,\mathrm{e}^{-i\theta}t;q)_\infty}=\sum_{n=0}^{\infty}\frac{H_n(x|q)}{(q;q)_n}t^n,\quad x=\cos\theta. \tag{14.26.11}$$

$$\begin{aligned}&(\mathrm{e}^{i\theta}t;q)_\infty\cdot{}_1\phi_1\left(\begin{matrix}0\\ \mathrm{e}^{i\theta}t\end{matrix};q,\mathrm{e}^{-i\theta}t\right)\\ &=\sum_{n=0}^{\infty}\frac{(-1)^nq^{\binom{n}{2}}}{(q;q)_n}H_n(x|q)t^n,\quad x=\cos\theta.\end{aligned} \tag{14.26.12}$$

$$\begin{aligned}&\frac{(\gamma\mathrm{e}^{i\theta}t;q)_\infty}{(\mathrm{e}^{i\theta}t;q)_\infty}{}_2\phi_1\left(\begin{matrix}\gamma,0\\ \gamma\mathrm{e}^{i\theta}t\end{matrix};q,\mathrm{e}^{-i\theta}t\right)\\ &=\sum_{n=0}^{\infty}\frac{(\gamma;q)_n}{(q;q)_n}H_n(x|q)t^n,\quad x=\cos\theta,\quad \gamma\text{ arbitrary}.\end{aligned} \tag{14.26.13}$$

Limit Relations

Continuous Big q-Hermite → Continuous q-Hermite

The continuous q-Hermite polynomials given by (14.26.1) can easily be obtained from the continuous big q-Hermite polynomials given by (14.18.1) by taking $a = 0$:

$$H_n(x;0|q) = H_n(x|q).$$

Continuous q-Laguerre → Continuous q-Hermite

The continuous q-Hermite polynomials given by (14.26.1) can be obtained from the continuous q-Laguerre polynomials given by (14.19.1) by taking the limit $\alpha \to \infty$ in the following way:

$$\lim_{\alpha\to\infty} \frac{P_n^{(\alpha)}(x|q)}{q^{(\frac{1}{2}\alpha+\frac{1}{4})n}} = \frac{H_n(x|q)}{(q;q)_n}.$$

Continuous q-Hermite → Hermite

The Hermite polynomials given by (9.15.1) can be obtained from the continuous q-Hermite polynomials given by (14.26.1) by setting $x \to x\sqrt{\frac{1}{2}(1-q)}$. In fact we have

$$\lim_{q\to 1} \frac{H_n(x\sqrt{\frac{1}{2}(1-q)}|q)}{\left(\frac{1-q}{2}\right)^{\frac{n}{2}}} = H_n(x). \tag{14.26.14}$$

Remark

The continuous q-Hermite polynomials can also be written as:

$$H_n(x|q) = \sum_{k=0}^{n} \frac{(q;q)_n}{(q;q)_k(q;q)_{n-k}} \mathrm{e}^{i(n-2k)\theta}, \quad x = \cos\theta.$$

References

[8], [16], [17], [25], [34], [51], [52], [62], [63], [72], [73], [75], [76], [78], [80], [86], [87], [100], [113], [119], [122], [210], [234], [238], [272], [276], [285], [294], [296], [416], [460], [461], [462], [481].

14.27 Stieltjes-Wigert

Basic Hypergeometric Representation

$$S_n(x;q) = \frac{1}{(q;q)_n}\,{}_1\phi_1\left(\begin{matrix} q^{-n} \\ 0 \end{matrix};q,-q^{n+1}x\right). \tag{14.27.1}$$

Orthogonality Relation

$$\int_0^\infty \frac{S_m(x;q)S_n(x;q)}{(-x,-qx^{-1};q)_\infty}\,dx = -\frac{\ln q}{q^n}\frac{(q;q)_\infty}{(q;q)_n}\,\delta_{mn}. \tag{14.27.2}$$

Recurrence Relation

$$\begin{aligned}&-q^{2n+1}xS_n(x;q)\\&= (1-q^{n+1})S_{n+1}(x;q) - [1+q-q^{n+1}]S_n(x;q) + qS_{n-1}(x;q).\end{aligned} \tag{14.27.3}$$

Normalized Recurrence Relation

$$\begin{aligned}xp_n(x) &= p_{n+1}(x) + q^{-2n-1}\left[1+q-q^{n+1}\right]p_n(x)\\&\quad + q^{-4n+1}(1-q^n)p_{n-1}(x),\end{aligned} \tag{14.27.4}$$

where

$$S_n(x;q) = \frac{(-1)^n q^{n^2}}{(q;q)_n}p_n(x).$$

q-Difference Equation

$$-x(1-q^n)y(x) = xy(qx) - (x+1)y(x) + y(q^{-1}x), \quad y(x) = S_n(x;q). \tag{14.27.5}$$

Forward Shift Operator

$$S_n(x;q)-S_n(qx;q)=-qxS_{n-1}(q^2x;q) \tag{14.27.6}$$

or equivalently

$$\mathscr{D}_qS_n(x;q)=-\frac{q}{1-q}S_{n-1}(q^2x;q). \tag{14.27.7}$$

Backward Shift Operator

$$S_n(x;q)-xS_n(qx;q)=(1-q^{n+1})S_{n+1}(q^{-1}x;q), \tag{14.27.8}$$

or equivalently

$$\mathscr{D}_q\left[w(x;q)S_n(x;q)\right]=\frac{1-q^{n+1}}{1-q}q^{-1}w(q^{-1}x;q)S_{n+1}(q^{-1}x;q), \tag{14.27.9}$$

where

$$w(x;q)=\frac{1}{(-x,-qx^{-1};q)_\infty}.$$

Rodrigues-Type Formula

$$w(x;q)S_n(x;q)=\frac{q^n(1-q)^n}{(q;q)_n}\left((\mathscr{D}_q)^n w\right)(q^nx;q). \tag{14.27.10}$$

Generating Functions

$$\frac{1}{(t;q)_\infty}\,{}_0\phi_1\left(\begin{matrix}-\\0\end{matrix};q,-qxt\right)=\sum_{n=0}^{\infty}S_n(x;q)t^n. \tag{14.27.11}$$

$$(t;q)_\infty\cdot{}_0\phi_2\left(\begin{matrix}-\\0,t\end{matrix};q,-qxt\right)=\sum_{n=0}^{\infty}(-1)^nq^{\binom{n}{2}}S_n(x;q)t^n. \tag{14.27.12}$$

$$\frac{(\gamma t;q)_\infty}{(t;q)_\infty}\,{}_1\phi_2\left(\begin{matrix}\gamma\\0,\gamma t\end{matrix};q,-qxt\right)=\sum_{n=0}^{\infty}(\gamma;q)_nS_n(x;q)t^n,\quad \gamma\text{ arbitrary}. \tag{14.27.13}$$

Limit Relations

q-Laguerre → Stieltjes-Wigert

If we set $x \to xq^{-\alpha}$ in the definition (14.21.1) of the q-Laguerre polynomials and take the limit $\alpha \to \infty$ we simply obtain the Stieltjes-Wigert polynomials given by (14.27.1):

$$\lim_{\alpha\to\infty} L_n^{(\alpha)}\left(xq^{-\alpha};q\right) = S_n(x;q).$$

q-Bessel → Stieltjes-Wigert

The Stieltjes-Wigert polynomials given by (14.27.1) can be obtained from the q-Bessel polynomials by setting $x \to a^{-1}x$ in the definition (14.22.1) of the q-Bessel polynomials and then taking the limit $a \to \infty$. In fact we have

$$\lim_{a\to\infty} y_n(a^{-1}x;a;q) = (q;q)_n S_n(x;q).$$

q-Charlier → Stieltjes-Wigert

If we set $q^{-x} \to ax$ in the definition (14.23.1) of the q-Charlier polynomials and take the limit $a \to \infty$ we obtain the Stieltjes-Wigert polynomials given by (14.27.1) in the following way:

$$\lim_{a\to\infty} C_n(ax;a;q) = (q;q)_n S_n(x;q).$$

Stieltjes-Wigert → Hermite

The Hermite polynomials given by (9.15.1) can be obtained from the Stieltjes-Wigert polynomials given by (14.27.1) by setting $x \to q^{-1}x\sqrt{2(1-q)}+1$ and taking the limit $q \to 1$ in the following way:

$$\lim_{q\to 1} \frac{(q;q)_n S_n(q^{-1}x\sqrt{2(1-q)}+1;q)}{\left(\frac{1-q}{2}\right)^{\frac{n}{2}}} = (-1)^n H_n(x). \tag{14.27.14}$$

Remark

Since the Stieltjes and Hamburger moment problems corresponding to the Stieltjes-Wigert polynomials are indeterminate there exist many different weight functions. For instance, they are also orthogonal with respect to the weight function

$$w(x)=\frac{\gamma}{\sqrt{\pi}}\exp\left(-\gamma^2\ln^2 x\right),\quad x>0,\quad \text{with}\quad \gamma^2=-\frac{1}{2\ln q}.$$

References

[49], [51], [86], [145], [146], [160], [276], [416], [490], [493], [503], [511].

14.28 Discrete q-Hermite I

Basic Hypergeometric Representation

The discrete q-Hermite I polynomials are Al-Salam-Carlitz I polynomials with $a=-1$:

$$h_n(x;q)=U_n^{(-1)}(x;q)=q^{\binom{n}{2}}\,{}_2\phi_1\left(\begin{matrix}q^{-n},x^{-1}\\0\end{matrix};q,-qx\right) \tag{14.28.1}$$
$$=x^n\,{}_2\phi_0\left(\begin{matrix}q^{-n},q^{-n+1}\\-\end{matrix}\,\middle|\,q^2;\frac{q^{2n-1}}{x^2}\right).$$

Orthogonality Relation

$$\int_{-1}^{1}(qx,-qx;q)_\infty h_m(x;q)h_n(x;q)\,d_qx$$
$$=(1-q)(q;q)_n(q,-1,-q;q)_\infty q^{\binom{n}{2}}\,\delta_{mn}. \tag{14.28.2}$$

Recurrence Relation

$$xh_n(x;q)=h_{n+1}(x;q)+q^{n-1}(1-q^n)h_{n-1}(x;q). \tag{14.28.3}$$

Normalized Recurrence Relation

$$xp_n(x)=p_{n+1}(x)+q^{n-1}(1-q^n)p_{n-1}(x), \tag{14.28.4}$$

where

$$h_n(x;q) = p_n(x).$$

q-Difference Equation

$$-q^{-n+1}x^2y(x) = y(qx) - (1+q)y(x) + q(1-x^2)y(q^{-1}x), \tag{14.28.5}$$

where

$$y(x) = h_n(x;q).$$

Forward Shift Operator

$$h_n(x;q) - h_n(qx;q) = (1-q^n)xh_{n-1}(x;q) \tag{14.28.6}$$

or equivalently

$$\mathscr{D}_q h_n(x;q) = \frac{1-q^n}{1-q}h_{n-1}(x;q). \tag{14.28.7}$$

Backward Shift Operator

$$h_n(x;q) - (1-x^2)h_n(q^{-1}x;q) = q^{-n}xh_{n+1}(x;q) \tag{14.28.8}$$

or equivalently

$$\mathscr{D}_{q^{-1}}\left[w(x;q)h_n(x;q)\right] = -\frac{q^{-n+1}}{1-q}w(x;q)h_{n+1}(x;q), \tag{14.28.9}$$

where

$$w(x;q) = (qx, -qx;q)_\infty.$$

Rodrigues-Type Formula

$$w(x;q)h_n(x;q) = (q-1)^n q^{\frac{1}{2}n(n-3)}\left(\mathscr{D}_{q^{-1}}\right)^n\left[w(x;q)\right]. \tag{14.28.10}$$

Generating function

$$\frac{(t^2;q^2)_\infty}{(xt;q)_\infty} = \sum_{n=0}^{\infty}\frac{h_n(x;q)}{(q;q)_n}t^n. \tag{14.28.11}$$

Limit Relations

Al-Salam-Carlitz I → Discrete q-Hermite I

The discrete q-Hermite I polynomials given by (14.28.1) can easily be obtained from the Al-Salam-Carlitz I polynomials given by (14.24.1) by the substitution $a = -1$:

$$U_n^{(-1)}(x;q) = h_n(x;q).$$

Discrete q-Hermite I → Hermite

The Hermite polynomials given by (9.15.1) can be found from the discrete q-Hermite I polynomials given by (14.28.1) in the following way:

$$\lim_{q\to 1} \frac{h_n(x\sqrt{1-q^2};q)}{(1-q^2)^{\frac{n}{2}}} = \frac{H_n(x)}{2^n}. \tag{14.28.12}$$

Remark

The discrete q-Hermite I polynomials are related to the discrete q-Hermite II polynomials given by (14.29.1) in the following way:

$$h_n(ix;q^{-1}) = i^n \tilde{h}_n(x;q).$$

References

[16], [18], [80], [100], [119], [238], [261], [349].

14.29 Discrete q-Hermite II

Basic Hypergeometric Representation

The discrete q-Hermite II polynomials are Al-Salam-Carlitz II polynomials with $a = -1$:

$$\tilde{h}_n(x;q) = i^{-n}V_n^{(-1)}(ix;q) = i^{-n}q^{-\binom{n}{2}}\,{}_2\phi_0\left(\begin{matrix} q^{-n}, ix \\ - \end{matrix}\,;q,-q^n\right) \tag{14.29.1}$$
$$= x^n\,{}_2\phi_1\left(\begin{matrix} q^{-n}, q^{-n+1} \\ 0 \end{matrix}\,\middle|\,q^2;-\frac{q^2}{x^2}\right).$$

Orthogonality Relation

$$\sum_{k=-\infty}^{\infty}\left[\tilde{h}_m(cq^k;q)\tilde{h}_n(cq^k;q)+\tilde{h}_m(-cq^k;q)\tilde{h}_n(-cq^k;q)\right]w(cq^k;q)q^k$$
$$= 2\frac{(q^2,-c^2q,-c^{-2}q;q^2)_\infty}{(q,-c^2,-c^{-2}q^2;q^2)_\infty}\frac{(q;q)_n}{q^{n^2}}\,\delta_{mn}, \quad c>0, \tag{14.29.2}$$

where

$$w(x;q) = \frac{1}{(ix,-ix;q)_\infty} = \frac{1}{(-x^2;q^2)_\infty}.$$

For $c = 1$ this orthogonality relation can also be written as

$$\int_{-\infty}^{\infty}\frac{\tilde{h}_m(x;q)\tilde{h}_n(x;q)}{(-x^2;q^2)_\infty}\,d_qx = \frac{(q^2,-q,-q;q^2)_\infty}{(q^3,-q^2,-q^2;q^2)_\infty}\frac{(q;q)_n}{q^{n^2}}\,\delta_{mn}. \tag{14.29.3}$$

Recurrence Relation

$$x\tilde{h}_n(x;q) = \tilde{h}_{n+1}(x;q) + q^{-2n+1}(1-q^n)\tilde{h}_{n-1}(x;q). \tag{14.29.4}$$

Normalized Recurrence Relation

$$xp_n(x) = p_{n+1}(x) + q^{-2n+1}(1-q^n)p_{n-1}(x), \tag{14.29.5}$$

where

$$\tilde{h}_n(x;q) = p_n(x).$$

q-Difference Equation

$$\begin{aligned}&-(1-q^n)x^2\tilde{h}_n(x;q)\\&=(1+x^2)\tilde{h}_n(qx;q)-(1+x^2+q)\tilde{h}_n(x;q)+q\tilde{h}_n(q^{-1}x;q).\end{aligned} \quad (14.29.6)$$

Forward Shift Operator

$$\tilde{h}_n(x;q)-\tilde{h}_n(qx;q)=q^{-n+1}(1-q^n)x\tilde{h}_{n-1}(qx;q) \quad (14.29.7)$$

or equivalently

$$\mathscr{D}_q\tilde{h}_n(x;q)=\frac{q^{-n+1}(1-q^n)}{1-q}\tilde{h}_{n-1}(qx;q). \quad (14.29.8)$$

Backward Shift Operator

$$\tilde{h}_n(x;q)-(1+x^2)\tilde{h}_n(qx;q)=-q^nx\tilde{h}_{n+1}(x;q) \quad (14.29.9)$$

or equivalently

$$\mathscr{D}_q\left[w(x;q)\tilde{h}_n(x;q)\right]=-\frac{q^n}{1-q}w(x;q)\tilde{h}_{n+1}(x;q). \quad (14.29.10)$$

Rodrigues-Type Formula

$$w(x;q)\tilde{h}_n(x;q)=(q-1)^nq^{-\binom{n}{2}}\left(\mathscr{D}_q\right)^n\left[w(x;q)\right]. \quad (14.29.11)$$

Generating Functions

$$\frac{(-xt;q)_\infty}{(-t^2;q^2)_\infty}=\sum_{n=0}^{\infty}\frac{q^{\binom{n}{2}}}{(q;q)_n}\tilde{h}_n(x;q)t^n. \quad (14.29.12)$$

$$(-it;q)_\infty\cdot{}_1\phi_1\left(\begin{matrix}ix\\-it\end{matrix};q,it\right)=\sum_{n=0}^{\infty}\frac{(-1)^nq^{n(n-1)}}{(q;q)_n}\tilde{h}_n(x;q)t^n. \quad (14.29.13)$$

Limit Relations

Al-Salam-Carlitz II → Discrete q-Hermite II

The discrete q-Hermite II polynomials given by (14.29.1) follow from the Al-Salam-Carlitz II polynomials given by (14.25.1) by the substitution $a = -1$ in the following way:

$$i^{-n}V_n^{(-1)}(ix;q) = \tilde{h}_n(x;q).$$

Discrete q-Hermite II → Hermite

The Hermite polynomials given by (9.15.1) can be found from the discrete q-Hermite II polynomials given by (14.29.1) in the following way:

$$\lim_{q\to 1} \frac{\tilde{h}_n(x\sqrt{1-q^2};q)}{(1-q^2)^{\frac{n}{2}}} = \frac{H_n(x)}{2^n}. \tag{14.29.14}$$

Remark

The discrete q-Hermite II polynomials are related to the discrete q-Hermite I polynomials given by (14.28.1) in the following way:

$$\tilde{h}_n(x;q^{-1}) = i^{-n}h_n(ix;q).$$

References

[100], [349].

Bibliography

1. N. ABDUL-HALIM and W.A. AL-SALAM, *A characterization of the Laguerre polynomials.* Rendiconti del Seminario Matematico della Università di Padova **34**, 1964, 176–179.
2. M. ABRAMOWITZ and I.A. STEGUN (eds.), *Handbook of mathematical functions (with formulas, graphs, and mathematical tables).* Dover Publications, New York, 1970.
3. L.D. ABREU and J. BUSTOZ, *Turán inequalities for symmetric Askey-Wilson polynomials.* The Rocky Mountain Journal of Mathematics **30**, 2000, 401–409.
4. S. AHMED, A. LAFORGIA and M.E. MULDOON, *On the spacing of the zeros of some classical orthogonal polynomials.* Journal of the London Mathematical Society (2) **25**, 1982, 246–252.
5. S. AHMED, M.E. MULDOON and R. SPIGLER, *Inequalities and numerical bounds for zeros of ultraspherical polynomials.* SIAM Journal on Mathematical Analysis **17**, 1986, 1000–1007.
6. K. ALLADI and M.L. ROBINSON, *Legendre polynomials and irrationality.* Journal für die Reine und Angewandte Mathematik **318**, 1980, 137–155.
7. WM.R. ALLAWAY, *The representation of orthogonal polynomials in terms of a differential operator.* Journal of Mathematical Analysis and Applications **56**, 1976, 288–293.
8. WM.R. ALLAWAY, *Some properties of the q-Hermite polynomials.* Canadian Journal of Mathematics **32**, 1980, 686–694.
9. WM.R. ALLAWAY, *Convolution orthogonality and the Jacobi polynomials.* Canadian Mathematical Bulletin **32**, 1989, 298–308.
10. WM.R. ALLAWAY, *Convolution shift, c-orthogonality preserving maps, and the Laguerre polynomials.* Journal of Mathematical Analysis and Applications **157**, 1991, 284–299.
11. N.A. AL-SALAM, *Orthogonal polynomials of hypergeometric type.* Duke Mathematical Journal **33**, 1966, 109–121.
12. N.A. AL-SALAM, *On some q-operators with applications.* Indagationes Mathematicae **51**, 1989, 1–13.
13. W. AL-SALAM, WM.R. ALLAWAY and R. ASKEY, *A characterization of the continuous q-ultraspherical polynomials.* Canadian Mathematical Bulletin **27**, 1984, 329–336.
14. W. AL-SALAM, W.R. ALLAWAY and R. ASKEY, *Sieved ultraspherical polynomials.* Transactions of the American Mathematical Society **284**, 1984, 39–55.
15. W.A. AL-SALAM, *Operational representations for the Laguerre and other polynomials.* Duke Mathematical Journal **31**, 1964, 127–142.
16. W.A. AL-SALAM, *Characterization theorems for orthogonal polynomials.* In: Orthogonal Polynomials: Theory and Practice (ed. P. Nevai), Kluwer Academic Publishers, Dordrecht, 1990, 1–24.
17. W.A. AL-SALAM, *A characterization of the Rogers q-Hermite polynomials.* International Journal of Mathematics and Mathematical Sciences **18**, 1995, 641–647.

R. Koekoek et al., *Hypergeometric Orthogonal Polynomials and Their q-Analogues*, Springer Monographs in Mathematics,
DOI 10.1007/978-3-642-05014-5, © Springer-Verlag Berlin Heidelberg 2010

18. W.A. AL-SALAM and L. CARLITZ, *Some orthogonal q-polynomials.* Mathematische Nachrichten **30**, 1965, 47–61.
19. W.A. AL-SALAM and T.S. CHIHARA, *Another characterization of the classical orthogonal polynomials.* SIAM Journal of Mathematical Analysis **3**, 1972, 65–70.
20. W.A. AL-SALAM and T.S. CHIHARA, *Convolutions of orthonormal polynomials.* SIAM Journal of Mathematical Analysis **7**, 1976, 16–28.
21. W.A. AL-SALAM and T.S. CHIHARA, *q-Pollaczek polynomials and a conjecture of Andrews and Askey.* SIAM Journal of Mathematical Analysis **18**, 1987, 228–242.
22. W.A. AL-SALAM and M.E.H. ISMAIL, *Polynomials orthogonal with respect to discrete convolution.* Journal of Mathematical Analysis and Applications **55**, 1976, 125–139.
23. W.A. AL-SALAM and M.E.H. ISMAIL, *Reproducing kernels for q-Jacobi polynomials.* Proceedings of the American Mathematical Society **67**, 1977, 105–110.
24. W.A. AL-SALAM and M.E.H. ISMAIL, *Orthogonal polynomials associated with the Rogers-Ramanujan continued fraction.* Pacific Journal of Mathematics **104**, 1983, 269–283.
25. W.A. AL-SALAM and M.E.H. ISMAIL, *q-Beta integrals and the q-Hermite polynomials.* Pacific Journal of Mathematics **135**, 1988, 209–221.
26. W.A. AL-SALAM and A. VERMA, *On an orthogonal polynomial set.* Indagationes Mathematicae **44**, 1982, 335–340.
27. W.A. AL-SALAM and A. VERMA, *Some remarks on q-beta integral.* Proceedings of the American Mathematical Society **85**, 1982, 360–362.
28. W.A. AL-SALAM and A. VERMA, *On the Geronimus polynomial sets.* In: Orthogonal Polynomials and Their Applications (eds. M. Alfaro et al.). Lecture Notes in Mathematics **1329**, Springer-Verlag, New York, 1988, 193–202.
29. M. ÁLVAREZ DE MORALES, T.E. PÉREZ, M.A. PIÑAR and A. RONVEAUX, *Non-standard orthogonality for Meixner polynomials. Orthogonal polynomials: numerical and symbolic algorithms, Leganés 1998,* Electronic Transactions on Numerical Analysis **9**, 1999, 1–25.
30. R. ÁLVAREZ-NODARSE and A. RONVEAUX, *Recurrence relations for connection coefficients between Q-orthogonal polynomials of discrete variables in the non-uniform lattice* $x(s) = q^{2s}$. Journal of Physics A. Mathematical and General **29**, 1996, 7165–7175.
31. R. ÁLVAREZ-NODARSE and YU.F. SMIRNOV, *The dual Hahn q-polynomials in the lattice* $x(s) = [s]_q[s+1]_q$ *and the q-algebras* $SU_q(2)$ *and* $SU_q(1,1)$. Journal of Physics A. Mathematical and General **29**, 1996, 1435–1451.
32. E.X.L. DE ANDRADE, D.K. DIMITROV and A. SRI RANGA, *Characterization of generalized Bessel polynomials in terms of polynomial inequalities.* Journal of Mathematical Analysis and Applications **221**, 1998, 538–543.
33. G.E. ANDREWS and R. ASKEY, *Enumeration of partitions: the role of Eulerian series and q-orthogonal polynomials.* In: Higher Combinatorics (ed. M. Aigner), Reidel Publicatons, Boston, 1977, 3–26.
34. G.E. ANDREWS and R. ASKEY, *Classical orthogonal polynomials.* In: Polynômes Orthogonaux et Applications (eds. C. Brezinski et al.). Lecture Notes in Mathematics **1171**, Springer-Verlag, New York, 1985, 36–62.
35. G.E. ANDREWS, R. ASKEY and R. ROY, *Special Functions.* Encyclopedia of Mathematics and Its Applications **71**, Cambridge University Press, Cambridge, 1999.
36. T.K. ARAAYA, *Linearization and connection problems for the symmetric Meixner-Pollaczek polynomials.* International Journal of Pure and Applied Mathematics **17**, 2004, 409–422.
37. T.K. ARAAYA, *The symmetric Meixner-Pollaczek polynomials with real parameter.* Journal of Mathematical Analysis and Applications **305**, 2005, 411–423.
38. I. AREA, D.K. DIMITROV, E. GODOY and A. RONVEAUX, *Zeros of Gegenbauer and Hermite polynomials and connection coefficients.* Mathematics of Computation **73**, 2004, 1937–1951.
39. I. AREA, E. GODOY, A. RONVEAUX and A. ZARZO, *Minimal recurrence relations for connection coefficients between classical orthogonal polynomials: discrete case.* Journal of Computational and Applied Mathematics **87**, 1997, 321–337.
40. R. ASKEY, *Product of ultraspherical polynomials.* The American Mathematical Monthly **74**, 1967, 1221–1222.

41. R. ASKEY, *Dual equations and classical orthogonal polynomials.* Journal of Mathematical Analysis and Applications **24**, 1968, 677–685.
42. R. ASKEY, *An inequality for the classical polynomials.* Indagationes Mathematicae **32**, 1970, 22–25.
43. R. ASKEY, *Positive Jacobi polynomial sums.* The Tôhoku Mathematical Journal (2) **24**, 1972, 109–119.
44. R. ASKEY, *Summability of Jacobi series.* Transactions of the American Mathematical Society **179**, 1973, 71–84.
45. R. ASKEY, *Jacobi polynomials I. New proofs of Koornwinder's Laplace type integral representation and Bateman's bilinear sum.* SIAM Journal on Mathematical Analysis **5**, 1974, 119–124.
46. R. ASKEY, *Orthogonal polynomials and special functions.* CBMS Regional Conference Series, Volume 21. Society for Industrial and Applied Mathematics, Philadelphia, 1975.
47. R. ASKEY, *Jacobi's generating function for Jacobi polynomials.* Proceedings of the American Mathematical Society **71**, 1978, 243–246.
48. R. ASKEY, *Continuous Hahn polynomials.* Journal of Physics A. Mathematical and General **18**, 1985, L1017–L1019.
49. R. ASKEY, *Limits of some q-Laguerre polynomials.* Journal of Approximation Theory **46**, 1986, 213–216.
50. R. ASKEY, *An integral of Ramanujan and orthogonal polynomials.* Journal of the Indian Mathematical Society **51**, 1987, 27–36.
51. R. ASKEY, *Beta integrals and the associated orthogonal polynomials.* In: Number Theory (ed. K. Alladi). Lecture Notes in Mathematics **1395**, Springer-Verlag, New York, 1989, 84–121.
52. R. ASKEY, *Continuous q-Hermite polynomials when $q > 1$.* In: q-Series and Partitions (ed. D. Stanton). The IMA Volumes in Mathematics and Its Applications **18**, Springer-Verlag, New York, 1989, 151–158.
53. R. ASKEY, *Divided difference operators and classical orthogonal polynomials.* The Rocky Mountain Journal of Mathematics **19**, 1989, 33–37.
54. R. ASKEY, *Duality for classical orthogonal polynomials.* Journal of Computational and Applied Mathematics **178**, 2005, 37–43.
55. R. ASKEY and J. FITCH, *Integral representations for Jacobi polynomials and some applications.* Journal of Mathematical Analysis and Applications **26**, 1969, 411–437.
56. R. ASKEY and G. GASPER, *Jacobi polynomial expansions of Jacobi polynomials with non-negative coefficients.* Proceedings of the Cambridge Philosophical Society **70**, 1971, 243–255.
57. R. ASKEY and G. GASPER, *Linearization of the product of Jacobi polynomials III.* Canadian Journal of Mathematics **23**, 1971, 332–338.
58. R. ASKEY and G. GASPER, *Positive Jacobi polynomial sums II.* American Journal of Mathematics **98**, 1976, 709–737.
59. R. ASKEY and G. GASPER, *Convolution structures for Laguerre polynomials.* Journal d'Analyse Mathématique **31**, 1977, 48–68.
60. R. ASKEY, G. GASPER and L.A. HARRIS, *An inequality for Tchebycheff polynomials and extensions.* Journal of Approximation Theory **14**, 1975, 1–11.
61. R. ASKEY and M.E.H. ISMAIL, *Permutation problems and special functions.* Canadian Journal of Mathematics **28**, 1976, 853–874.
62. R. ASKEY and M.E.H. ISMAIL, *The Rogers q-ultraspherical polynomials.* In: Approximation Theory III, Academic Press, New York, 1980, 175–182.
63. R. ASKEY and M.E.H. ISMAIL, *A generalization of ultraspherical polynomials.* In: Studies in Pure Mathematics (ed. P. Erdős), Birkhäuser Verlag, Basel, 1983, 55–78.
64. R. ASKEY and M.E.H. ISMAIL, *Recurrence relations, continued fractions and orthogonal polynomials.* Memoirs of the American Mathematical Society **300**, Providence, Rhode Island, 1984.
65. R. ASKEY, M.E.H. ISMAIL and T. KOORNWINDER, *Weighted permutation problems and Laguerre polynomials.* Journal of Combinatorial Theory A **25**, 1978, 277–287.

66. R. ASKEY, T.H. KOORNWINDER and M. RAHMAN, *An integral of products of ultraspherical functions and a q-extension.* Journal of the London Mathematical Society (2) **33**, 1986, 133–148.
67. R. ASKEY and S.K. SUSLOV, *The q-harmonic oscillator and an analogue of the Charlier polynomials.* Journal of Physics A. Mathematical and General **26**, 1993, L693–L698.
68. R. ASKEY and S.K. SUSLOV, *The q-harmonic oscillator and the Al-Salam and Carlitz polynomials.* Letters in Mathematical Physics **29**, 1993, 123–132.
69. R. ASKEY and S. WAINGER, *A convolution structure for Jacobi series.* American Journal of Mathematics **91**, 1969, 463–485.
70. R. ASKEY and J. WILSON, *A set of orthogonal polynomials that generalize the Racah coefficients or* $6-j$ *symbols.* SIAM Journal on Mathematical Analysis **10**, 1979, 1008–1016.
71. R. ASKEY and J. WILSON, *A set of hypergeometric orthogonal polynomials.* SIAM Journal on Mathematical Analysis **13**, 1982, 651–655.
72. R. ASKEY and J.A. WILSON, *Some basic hypergeometric orthogonal polynomials that generalize Jacobi polynomials.* Memoirs of the American Mathematical Society **319**, Providence, Rhode Island, 1985.
73. R.A. ASKEY, M. RAHMAN and S.K. SUSLOV, *On a general q-Fourier transformation with nonsymmetric kernels.* Journal of Computational and Applied Mathematics **68**, 1996, 25–55.
74. M.N. ATAKISHIYEV, N.M. ATAKISHIYEV and A.U. KLIMYK, *Big q-Laguerre and q-Meixner polynomials and representations of the quantum algebra* $U_q(\mathrm{su}_{1,1})$. Journal of Physics. A. Mathematical and General **36**, 2003, 10335–10347.
75. M.N. ATAKISHIYEV and A.U. KLIMYK, *The factorization of a q-difference equation for continuous q-Hermite polynomials.* Journal of Physics. A. Mathematical and Theoretical **40**, 2007, 9311–9317.
76. N.M. ATAKISHIYEV, *On the Fourier-Gauss transforms of some q-exponential and q-trigonometric functions.* Journal of Physics A. Mathematical and General **29**, 1996, 7177–7181.
77. N.M. ATAKISHIYEV, *Fourier-Gauss transforms of the Askey-Wilson polynomials.* Journal of Physics A. Mathematical and General **30**, 1997, L815–L820.
78. N.M. ATAKISHIYEV and P. FEINSILVER, *On the coherent states for the q-Hermite polynomials and related Fourier transformation.* Journal of Physics A. Mathematical and General **29**, 1996, 1659–1664.
79. N.M. ATAKISHIYEV and A.U. KLIMYK, *On q-orthogonal polynomials, dual to little and big q-Jacobi polynomials.* Journal of Mathematical Analysis and Applications **294**, 2004, 246–257.
80. N.M. ATAKISHIYEV, M. RAHMAN and S.K. SUSLOV, *On classical orthogonal polynomials.* Constructive Approximation **11**, 1995, 181–226.
81. N.M. ATAKISHIYEV and S.K. SUSLOV, *The Hahn and Meixner polynomials of an imaginary argument and some of their applications.* Journal of Physics A. Mathematical and General **18**, 1985, 1583–1596.
82. N.M. ATAKISHIYEV and S.K. SUSLOV, *Continuous orthogonality property for some classical polynomials of a discrete variable.* Revista Mexicana de Fisica **34**, 1988, 541–563.
83. N.M. ATAKISHIYEV and S.K. SUSLOV, *On the Askey-Wilson polynomials.* Constructive Approximation **8**, 1992, 363–369.
84. M.K. ATAKISHIYEVA and N.M. ATAKISHIYEV, *q-Laguerre and Wall polynomials are related by the Fourier-Gauss transform.* Journal of Physics A. Mathematical and General **30**, 1997, L429–L432.
85. M.K. ATAKISHIYEVA and N.M. ATAKISHIYEV, *Fourier-Gauss transforms of the continuous big q-Hermite polynomials.* Journal of Physics A. Mathematical and General **30**, 1997, L559–L565.
86. M.K. ATAKISHIYEVA and N.M. ATAKISHIYEV, *Fourier-Gauss transforms of the Al-Salam-Chihara polynomials.* Journal of Physics A. Mathematical and General **30**, 1997, L655–L661.
87. M.K. ATAKISHIYEVA and N.M. ATAKISHIYEV, *On continuous q-Hermite polynomials and the classical Fourier transform.* Journal of Physics A. Mathematical and Theoretical **41**, 2008, 125201.

88. R. AZOR, J. GILLIS and J.D. VICTOR, *Combinatorial applications of Hermite polynomials.* SIAM Journal on Mathematical Analysis **13**, 1982, 879–890.
89. E. BADERTSCHER and T.H. KOORNWINDER, *Continuous Hahn polynomials of differential operator argument and analysis on Riemannian symmetric spaces of constant curvature.* Canadian Journal of Mathematics **44**, 1992, 750–773.
90. W.N. BAILEY, *Generalized Hypergeometric Series.* Cambridge Tracts in Mathematics and Mathematical Physics **32**, Cambridge University Press, 1935.
91. W.N. BAILEY, *The generating function of Jacobi polynomials.* Journal of the London Mathematical Society **13**, 1938, 8–12.
92. G. BANGEREZAKO, *The factorization method for the Askey-Wilson polynomials.* Journal of Computational and Applied Mathematics **107**, 1999, 219–232.
93. W.W. BARRETT, *An alternative approach to Laguerre polynomial identities in combinatorics.* Canadian Journal of Mathematics **31**, 1979, 312–319.
94. H. BAVINCK, *On the zeros of certain linear combinations of Chebyshev polynomials.* Journal of Computational and Applied Mathematics **65**, 1995, 19–26.
95. H. BAVINCK, *A new result for Laguerre polynomials.* Journal of Physics A. Mathematical and General **29**, 1996, L277–L279.
96. H. BAVINCK, *Some new results for Charlier and Meixner polynomials.* Journal of Computational and Applied Mathematics **90**, 1998, 51–56.
97. H. BAVINCK and H. VAN HAERINGEN, *Difference equations for generalized Meixner polynomials.* Journal of Mathematical Analysis and Applications **184**, 1994, 453–463.
98. H. BAVINCK and R. KOEKOEK, *On a difference equation for generalizations of Charlier polynomials.* Journal of Approximation Theory **81**, 1995, 195–206.
99. C. BERG, *Integrals involving Gegenbauer and Hermite polynomials on the imaginary axis.* Aequationes Mathematicae **40**, 1990, 83–88.
100. C. BERG and M.E.H. ISMAIL, *q-Hermite polynomials and classical orthogonal polynomials.* Canadian Journal of Mathematics **48**, 1996, 43–63.
101. C. BERG and G. VALENT, *Nevanlinna extremal measures for some orthogonal polynomials related to birth and death processes.* Journal of Computational and Applied Mathematics **57**, 1995, 29–43.
102. C. BERG and C. VIGNAT, *Linearization coefficients of Bessel polynomials and properties of student t-distributions.* Constructive Approximation **27**, 2008, 15–32.
103. B.R. BHONSLE, *On a series of Rainville involving Legendre polynomials.* Proceedings of the American Mathematical Society **8**, 1957, 10–14.
104. G.G. BILODEAU, *Gegenbauer polynomials and fractional derivatives.* The American Mathematical Monthly **71**, 1964, 1026–1028.
105. G.G. BILODEAU, *On the summability of series of Hermite polynomials.* Journal of Mathematical Analysis and Applications **8**, 1964, 406–422.
106. S. BOCHNER, *Über Sturm-Liouvillesche Polynomsysteme.* Mathematische Zeitschrift **29**, 1929, 730–736.
107. B. BOJANOV and G. NIKOLOV, *Duffin and Schaeffer type inequality for ultraspherical polynomials.* Journal of Approximation Theory **84**, 1996, 129–138.
108. A. BORODIN and G. OLSHANSKI, *Infinite random matrices and ergodic measures.* Communications in Mathematical Physics **223**, 2001, 87–123.
109. F. BRAFMAN, *Generating functions of Jacobi and related polynomials.* Proceedings of the American Mathematical Society **2**, 1951, 942–949.
110. F. BRAFMAN, *An ultraspherical generating function.* Pacific Journal of Mathematics **7**, 1957, 1319–1323.
111. F. BRAFMAN, *Some generating functions for Laguerre and Hermite polynomials.* Canadian Journal of Mathematics **9**, 1957, 180–187.
112. W.C. BRENKE, *On generating functions of polynomial systems.* The American Mathematical Monthly **52**, 1945, 297–301.
113. D.M. BRESSOUD, *A simple proof of Mehler's formula for q-Hermite polynomials.* Indiana University Mathematics Journal **29**, 1980, 577–580.

114. D.M. BRESSOUD, *Linearization and related formulas for q-ultraspherical polynomials.* SIAM Journal on Mathematical Analysis **12**, 1981, 161–168.
115. B.M. BROWN, W.D. EVANS and M.E.H. ISMAIL, *The Askey-Wilson polynomials and q-Sturm-Liouville problems.* Mathematical Proceedings of the Cambridge Philosophical Society **119**, 1996, 1–16.
116. B.M. BROWN and M.E.H. ISMAIL, *A right inverse of the Askey-Wilson operator.* Proceedings of the American Mathematical Society **123**, 1995, 2071–2079.
117. J.W. BROWN, *New generating functions for classical polynomials.* Proceedings of the American Mathematical Society **21**, 1969, 263–268.
118. W. BRYC, W. MATYSIAK and P.J. SZABŁOWSKI, *Probabilistic aspects of Al-Salam-Chihara polynomials.* Proceedings of the American Mathematical Society **133**, 2005, 1127–1134.
119. J. BUSTOZ and M.E.H. ISMAIL, *The associated ultraspherical polynomials and their q-analogues.* Canadian Journal of Mathematics **34**, 1982, 718–736.
120. J. BUSTOZ and M.E.H. ISMAIL, *Turán inequalities for ultraspherical and continuous q-ultraspherical polynomials.* SIAM Journal on Mathematical Analysis **14**, 1983, 807–818.
121. J. BUSTOZ and M.E.H. ISMAIL, *A positive trigonometric sum.* SIAM Journal on Mathematical Analysis **20**, 1989, 176–181.
122. J. BUSTOZ and M.E.H. ISMAIL, *Turán inequalities for symmetric orthogonal polynomials.* International Journal of Mathematics and Mathematical Sciences **20**, 1997, 1–7.
123. J. BUSTOZ and N. SAVAGE, *Inequalities for ultraspherical and Laguerre polynomials.* SIAM Journal on Mathematical Analysis **10**, 1979, 902–912.
124. J. BUSTOZ and N. SAVAGE, *Inequalities for ultraspherical and Laguerre polynomials II.* SIAM Journal on Mathematical Analysis **11**, 1980, 876–884.
125. C. CAMPIGOTTO, YU.F. SMIRNOV and S.G. ENIKEEV, *q-Analogue of the Krawtchouk and Meixner orthogonal polynomials.* Journal of Computational and Applied Mathematics **57**, 1995, 87–97.
126. L. CARLITZ, *A note on the Bessel polynomials.* Duke Mathematical Journal **24**, 1957, 151–162.
127. L. CARLITZ, *Some arithmetic properties of the Legendre polynomials.* Proceedings of the Cambridge Philosophical Society **53**, 1957, 265–268.
128. L. CARLITZ, *A note on the Laguerre polynomials.* The Michigan Mathematical Journal **7**, 1960, 219–223.
129. L. CARLITZ, *On the product of two Laguerre polynomials.* Journal of the London Mathematical Society **36**, 1961, 399–402.
130. L. CARLITZ, *Some generating functions of Weisner.* Duke Mathematical Journal **28**, 1961, 523–529.
131. L. CARLITZ, *A characterization of the Laguerre polynomials.* Monatshefte für Mathematik **66**, 1962, 389–392.
132. L. CARLITZ, *The generating function for the Jacobi polynomial.* Rendiconti del Seminario Matematico della Università di Padova **38**, 1967, 86–88.
133. L. CARLITZ, *Some generating functions for Laguerre polynomials.* Duke Mathematical Journal **35**, 1968, 825–827.
134. L. CARLITZ and H.M. SRIVASTAVA, *Some new generating functions for the Hermite polynomials.* Journal of Mathematical Analysis and Applications **149**, 1990, 513–520.
135. J.A. CHARRIS and M.E.H. ISMAIL, *On sieved orthogonal polynomials. V: sieved Pollaczek polynomials.* SIAM Journal on Mathematical Analysis **18**, 1987, 1177–1218.
136. K.-Y. CHEN and H.M. SRIVASTAVA, *A limit relationship between Laguerre and Hermite polynomials.* Integral Transforms and Special Functions **16**, 2005, 75–80.
137. L.-C. CHEN and M.E.H. ISMAIL, *On asymptotics of Jacobi polynomials.* SIAM Journal on Mathematical Analysis **22**, 1991, 1442–1449.
138. Y. CHEN and M.E.H. ISMAIL, *Asymptotics of extreme zeros of the Meixner-Pollaczek polynomials.* Journal of Computational and Applied Mathematics **82**, 1997, 59–78.
139. Y. CHEN, M.E.H. ISMAIL and K.A. MUTTALIB, *Asymptotics of basic Bessel functions and q-Laguerre polynomials.* Journal of Computational and Applied Mathematics **54**, 1994, 263–272.

140. L. CHIHARA, *On the zeros of the Askey-Wilson polynomials, with applications to coding theory.* SIAM Journal on Mathematical Analysis **18**, 1987, 191–207.
141. L. CHIHARA and D. STANTON, *Zeros of generalized Krawtchouk polynomials.* Journal of Approximation Theory **60**, 1990, 43–57.
142. L.M. CHIHARA, *Askey-Wilson polynomials, kernel polynomials and association schemes.* Graphs and Combinatorics **9**, 1993, 213–223.
143. T.S. CHIHARA, *On indeterminate Hamburger moment problems.* Pacific Journal of Mathematics **27**, 1968, 475–484.
144. T.S. CHIHARA, *Orthogonal polynomials with Brenke type generating functions.* Duke Mathematical Journal **35**, 1968, 505–517.
145. T.S. CHIHARA, *A characterization and a class of distribution functions for the Stieltjes-Wigert polynomials.* Canadian Mathematical Bulletin **13**, 1970, 529–532.
146. T.S. CHIHARA, *An introduction to orthogonal polynomials.* Gordon and Breach, New York, 1978.
147. T.S. CHIHARA, *On generalized Stieltjes-Wigert and related orthogonal polynomials.* Journal of Computational and Applied Mathematics **5**, 1979, 291–297.
148. T.S. CHIHARA and M.E.H. ISMAIL, *Extremal measures for a system of orthogonal polynomials.* Constructive Approximation **9**, 1993, 111–119.
149. Y. CHOW, L. GATTESCHI and R. WONG, *A Bernstein-type inequality for the Jacobi polynomial.* Proceedings of the American Mathematical Society **121**, 1994, 703–709.
150. J.S. CHRISTIANSEN, *The moment problem associated with the q-Laguerre polynomials.* Constructive Approximation **19**, 2003, 1–22.
151. J.S. CHRISTIANSEN and M.E.H. ISMAIL, *A moment problem and a family of integral evaluations.* Transactions of the American Mathematical Society **358**, 2006, 4071–4097.
152. Z. CIESIELSKI, *Explicit formula relating the Jacobi, Hahn and Bernstein polynomials.* SIAM Journal on Mathematical Analysis **18**, 1987, 1573–1575.
153. M.E. COHEN, *Some classes of generating functions for the Laguerre and Hermite polynomials.* Mathematics of Computation **31**, 1977, 511–518.
154. A.K. COMMON, *Uniform inequalities for ultraspherical polynomials and Bessel functions of fractional order.* Journal of Approximation Theory **49**, 1987, 331–339.
155. R.D. COOPER, M.R. HOARE and M. RAHMAN, *Stochastic processes and special functions: On the probabilistic origin of some positive kernels associated with classical orthogonal polynomials.* Journal of Mathematical Analysis and Applications **61**, 1977, 262–291.
156. A.E. DANESE, *Some identities and inequalities involving ultraspherical polynomials.* Duke Mathematical Journal **26**, 1959, 349–359.
157. S. DATTA and J. GRIFFIN, *A characterization of some q-orthogonal polynomials.* Ramanujan Journal **12**, 2006, 425–437.
158. G. DATTOLI, H.M. SRIVASTAVA and C. CESARANO, *The Laguerre and Legendre polynomials from an operational point of view.* Applied Mathematics and Computation **124**, 2001, 117–127.
159. G. DATTOLI, H.M. SRIVASTAVA and D. SACCHETTI, *A note on the Bessel polynomials.* Mathematica Macedonica **1**, 2003, 1–7.
160. J.S. DEHESA, *On a general system of orthogonal q-polynomials.* Journal of Computational and Applied Mathematics **5**, 1979, 37–45.
161. P. DELSARTE, *Association schemes and t-designs in regular semilattices.* Journal of Combinatorial Theory A **20**, 1976, 230–243.
162. P. DELSARTE and J.M. GOETHALS, *Alternating bilinear forms over $GF(q)$.* Journal of Combinatorial Theory A **19**, 1975, 26–50.
163. H. DETTE, *Extremal properties of ultraspherical polynomials.* Journal of Approximation Theory **76**, 1994, 246–273.
164. H. DETTE, *New bounds for Hahn and Krawtchouk polynomials.* SIAM Journal on Mathematical Analysis **26**, 1995, 1647–1659.
165. H. DETTE and W.J. STUDDEN, *On a new characterization of the classical orthogonal polynomials.* Journal of Approximation Theory **71**, 1992, 3–17.

166. H. Dette and W.J. Studden, *Some new asymptotic properties for the zeros of Jacobi, Laguerre, and Hermite polynomials.* Constructive Approximation **11**, 1995, 227–238.

167. A. Dijksma and T.H. Koornwinder, *Spherical harmonics and the product of two Jacobi polynomials.* Indagationes Mathematicae **33**, 1971, 191–196.

168. K. Dilcher and K.B. Stolarsky, *Zeros of the Wronskian of Chebyshev and ultraspherical polynomials.* The Rocky Mountain Journal of Mathematics **23**, 1993, 49–65.

169. D.K. Dimitrov, *On a conjecture concerning monotonicity of zeros of ultraspherical polynomials.* Journal of Approximation Theory **85**, 1996, 88–97.

170. D.K. Dimitrov, *Convexity of the extreme zeros of Gegenbauer and Laguerre polynomials.* Journal of Computational and Applied Mathematics **153**, 2003, 171–180.

171. D.K. Dimitrov and F.R. Rafaeli, *Monotonicity of zeros of Jacobi polynomials.* Journal of Approximation Theory **149**, 2007, 15–29.

172. D.K. Dimitrov and R.O. Rodrigues, *On the behaviour of zeros of Jacobi polynomials.* Journal of Approximation Theory **116**, 2002, 224–239.

173. E.H. Doha, *On the coefficients of differentiated expansions and derivatives of Jacobi polynomials.* Journal of Physics. A. Mathematical and General **35**, 2002, 3467–3478.

174. E.H. Doha, *On the coefficients of integrated expansions and integrals of ultraspherical polynomials and their applications for solving differential equations.* Journal of Computational and Applied Mathematics **139**, 2002, 275–298.

175. E.H. Doha, *Explicit formulae for the coefficients of integrated expansions of Jacobi polynomials and their integrals.* Integral Transforms and Special Functions **14**, 2003, 69–86.

176. E.H. Doha, *On the connection coefficients and recurrence relations arising from expansions in series of Laguerre polynomials.* Journal of Physics. A. Mathematical and General **36**, 2003, 5449–5462.

177. E.H. Doha, *On the connection coefficients and recurrence relations arising from expansions in series of Hermite polynomials.* Integral Transforms and Special Functions **15**, 2004, 13–29.

178. E.H. Doha, *On the construction of recurrence relations for the expansion and connection coefficients in series of Jacobi polynomials.* Journal of Physics. A. Mathematical and General **37**, 2004, 657–675.

179. E.H. Doha and H.M. Ahmed, *Recurrences and explicit formulae for the expansion and connection coefficients in series of Bessel polynomials.* Journal of Physics. A. Mathematical and General **37**, 2004, 8045–8063.

180. E.H. Doha and H.M. Ahmed, *Efficient algorithms for construction of recurrence relations for the expansion and connection coefficients in series of Al-Salam-Carlitz I polynomials.* Journal of Physics. A. Mathematical and General **38**, 2005, 10107–10121.

181. E.H. Doha and H.M. Ahmed, *On the coefficients of integrated expansions of Bessel polynomials.* Journal of Computational and Applied Mathematics **187**, 2006, 58–71.

182. D. Dominici, *Asymptotic analysis of the Krawtchouk polynomials by the WKB method.* Ramanujan Journal **15**, 2008, 303–338.

183. K. Driver and P. Duren, *Zeros of ultraspherical polynomials and the Hilbert-Klein formulas.* Journal of Computational and Applied Mathematics **135**, 2001, 293–301.

184. C.F. Dunkl, *A Krawtchouk polynomial addition theorem and wreath products of symmetric groups.* Indiana University Mathematics Journal **25**, 1976, 335–358.

185. C.F. Dunkl, *An addition theorem for Hahn polynomials: the spherical functions.* SIAM Journal on Mathematical Analysis **9**, 1978, 627–637.

186. C.F. Dunkl, *An addition theorem for some q-Hahn polynomials.* Monatshefte für Mathematik **85**, 1978, 5–37.

187. C.F. Dunkl, *Orthogonal polynomials with symmetry of order three.* Canadian Journal of Mathematics **36**, 1984, 685–717.

188. C.F. Dunkl and D.E. Ramirez, *Krawtchouk polynomials and the symmetrization of hypergroups.* SIAM Journal on Mathematical Analysis **5**, 1974, 351–366.

189. G.K. Eagleson, *A duality relation for discrete orthogonal systems.* Studia Scientiarum Mathematicarum Hungarica **3**, 1968, 127–136.

190. A. Elbert and A. Laforgia, *A note to the paper of Ahmed, Muldoon and Spigler.* SIAM Journal on Mathematical Analysis **17**, 1986, 1008–1009.

191. A. Elbert and A. Laforgia, *Some monotonicity properties of the zeros of ultraspherical polynomials.* Acta Mathematica Hungarica **48**, 1986, 155–159.
192. A. Elbert and A. Laforgia, *Monotonicity results on the zeros of generalized Laguerre polynomials.* Journal of Approximation Theory **51**, 1987, 168–174.
193. A. Elbert and A. Laforgia, *New properties of the zeros of a Jacobi polynomial in relation to their centroid.* SIAM Journal on Mathematical Analysis **18**, 1987, 1563–1572.
194. A. Elbert and A. Laforgia, *Upper bounds for the zeros of ultraspherical polynomials.* Journal of Approximation Theory **61**, 1990, 88–97.
195. A. Elbert and A. Laforgia, *An inequality for Legendre polynomials.* Journal of Mathematical Physics **35**, 1994, 1348–1360.
196. A. Elbert, A. Laforgia and L.G. Rodonò, *On the zeros of Jacobi polynomials.* Acta Mathematica Hungarica **64**, 1994, 351–359.
197. A. Erdélyi, *Asymptotic forms for Laguerre polynomials.* Journal of the Indian Mathematical Society **24**, 1960, 235–250.
198. A. Erdélyi et al., *Higher Transcendental Functions.* Volume II of the Bateman Manuscript Project. McGraw-Hill Book Company, New York, 1953.
199. H. Exton, *Basic Laguerre polynomials.* Pure and Applied Mathematika Sciences **5**, 1977, 19–24.
200. H. Exton, *New generating functions for Gegenbauer polynomials.* Journal of Computational and Applied Mathematics **67**, 1996, 191–193.
201. H. Exton, *Summation formulae involving the Laguerre polynomial.* Journal of Computational and Applied Mathematics **100**, 1998, 225–227.
202. J. Faldey and W. Gawronski, *On the limit distributions of the zeros of Jonquière polynomials and generalized classical orthogonal polynomials.* Journal of Approximation Theory **81**, 1995, 231–249.
203. P. Feinsilver and R. Schott, *Operator calculus approach to orthogonal polynomial expansions.* Journal of Computational and Applied Mathematics **66**, 1996, 185–199.
204. I. Fischer, *A Rodrigues-type formula for the q-Racah polynomials and some related results.* Contemporary Mathematics **169**, 1994, 253–259.
205. I. Fischer, *Discrete orthogonal polynomial expansions of averaged functions.* Journal of Mathematical Analysis and Applications **194**, 1995, 934–947.
206. M. Flensted-Jensen, *Spherical functions on a simply connected semisimple Lie group. II. The Paley-Wiener theorem for the rank one case.* Mathematische Annalen **228**, 1977, 65–92.
207. R. Floreanini, J. LeTourneux and L. Vinet, *An algebraic interpretation of the continuous big q-Hermite polynomials.* Journal of Mathematical Physics **36**, 1995, 5091–5097.
208. R. Floreanini, J. LeTourneux and L. Vinet, *A q-deformed $e(4)$ and continuous q-Jacobi polynomials.* Journal of Mathematical Physics **37**, 1996, 4135–4149.
209. R. Floreanini, J. LeTourneux and L. Vinet, *Symmetry techniques for the Al-Salam-Chihara polynomials.* Journal of Physics A. Mathematical and General **30**, 1997, 3107–3114.
210. R. Floreanini and L. Vinet, *A model for the continuous q-ultraspherical polynomials.* Journal of Mathematical Physics **36**, 1995, 3800–3813.
211. R. Floreanini and L. Vinet, *Quantum algebra approach to q-Gegenbauer polynomials.* Journal of Computational and Applied Mathematics **57**, 1995, 123–133.
212. P.G.A. Floris, *A noncommutative discrete hypergroup associated with q-disk polynomials.* Journal of Computational and Applied Mathematics **68**, 1996, 69–78.
213. P.G.A. Floris, *Addition formula for q-disk polynomials.* Compositio Mathematica **108**, 1997, 123–149.
214. P.G.A. Floris and H.T. Koelink, *A commuting q-analogue of the addition formula for disk polynomials.* Constructive Approximation **13**, 1997, 511–535.
215. D. Foata and J. Labelle, *Modèles combinatoires pour les polynômes de Meixner.* European Journal of Combinatorics **4**, 1983, 305–311.
216. D. Foata and P. Leroux, *Polynômes de Jacobi, interprétation combinatoire et fonction génératrice.* Proceedings of the American Mathematical Society **87**, 1983, 47–53.
217. B. Gabutti, *Some characteristic property of the Meixner polynomials.* Journal of Mathematical Analysis and Applications **95**, 1983, 265–277.

218. B. GABUTTI and M.L. MATHIS, *A characterization of the Meixner polynomials.* Annals of Numerical Mathematics **2**, 1995, 247–253.
219. G. GASPER, *Nonnegative sums of cosine, ultraspherical and Jacobi polynomials.* Journal of Mathematical Analysis and Applications **26**, 1969, 60–68.
220. G. GASPER, *Linearization of the product of Jacobi polynomials I.* Canadian Journal of Mathematics **22**, 1970, 171–175.
221. G. GASPER, *Linearization of the product of Jacobi polynomials II.* Canadian Journal of Mathematics **22**, 1970, 582–593.
222. G. GASPER, *On the extension of Turan's inequality to Jacobi polynomials.* Duke Mathematical Journal **38**, 1971, 415–428.
223. G. GASPER, *Positivity and the convolution structure for Jacobi series.* Annals of Mathematics **93**, 1971, 112–118.
224. G. GASPER, *An inequality of Turan type for Jacobi polynomials.* Proceedings of the American Mathematical Society **32**, 1972, 435–439.
225. G. GASPER, *Banach algebras for Jacobi series and positivity of a kernel.* Annals of Mathematics **95**, 1972, 261–280.
226. G. GASPER, *Nonnegativity of a discrete Poisson kernel for the Hahn polynomials.* Journal of Mathematical Analysis and Applications **42**, 1973, 438–451.
227. G. GASPER, *On two conjectures of Askey concerning normalized Hankel determinants for the classical polynomials.* SIAM Journal on Mathematical Analysis **4**, 1973, 508–513.
228. G. GASPER, *Projection formulas for orthogonal polynomials of a discrete variable.* Journal of Mathematical Analysis and Applications **45**, 1974, 176–198.
229. G. GASPER, *Positive sums of the classical orthogonal polynomials.* SIAM Journal on Mathematical Analysis **8**, 1977, 423–447.
230. G. GASPER, *Orthogonality of certain functions with respect to complex valued weights.* Canadian Journal of Mathematics **33**, 1981, 1261–1270.
231. G. GASPER, *Rogers' linearization formula for the continuous q-ultraspherical polynomials and quadratic transformation formulas.* SIAM Journal on Mathematical Analysis **16**, 1985, 1061–1071.
232. G. GASPER, *q-Extensions of Clausen's formula and of the inequalities used by De Branges in his proof of the Bieberbach, Robertson, and Milin conjectures.* SIAM Journal on Mathematical Analysis **20**, 1989, 1019–1034.
233. G. GASPER and M. RAHMAN, *Nonnegative kernels in product formulas for q-Racah polynomials I.* Journal of Mathematical Analysis and Applications **95**, 1983, 304–318.
234. G. GASPER and M. RAHMAN, *Positivity of the Poisson kernel for the continuous q-ultraspherical polynomials.* SIAM Journal on Mathematical Analysis **14**, 1983, 409–420.
235. G. GASPER and M. RAHMAN, *Product formulas of Watson, Bailey and Bateman types and positivity of the Poisson kernel for q-Racah polynomials.* SIAM Journal on Mathematical Analysis **15**, 1984, 768–789.
236. G. GASPER and M. RAHMAN, *Positivity of the Poisson kernel for the continuous q-Jacobi polynomials and some quadratic transformation formulas for basic hypergeometric series.* SIAM Journal on Mathematical Analysis **17**, 1986, 970–999.
237. G. GASPER and M. RAHMAN, *A nonterminating q-Clausen formula and some related product formulas.* SIAM Journal on Mathematical Analysis **20**, 1989, 1270–1282.
238. G. GASPER and M. RAHMAN, *Basic Hypergeometric Series.* Encyclopedia of Mathematics and Its Applications **35**, Cambridge University Press, Cambridge, 1990.
239. L. GATTESCHI, *New inequalities for the zeros of Jacobi polynomials.* SIAM Journal on Mathematical Analysis **18**, 1987, 1549–1562.
240. L. GATTESCHI, *Asymptotics and bounds for the zeros of Laguerre polynomials: a survey.* Journal of Computational and Applied Mathematics **144**, 2002, 7–27.
241. W. GAUTSCHI, *On a conjectured inequality for the largest zero of Jacobi polynomials.* Numerical Algorithms **49**, 2008, 195–198.
242. W. GAUTSCHI, *On conjectured inequalities for zeros of Jacobi polynomials.* Numerical Algorithms **50**, 2009, 93–96.

243. W. GAUTSCHI, *New conjectured inequalities for zeros of Jacobi polynomials.* Numerical Algorithms **50**, 2009, 293–296.
244. W. GAUTSCHI and P. LEOPARDI, *Conjectured inequalities for Jacobi polynomials and their largest zeros.* Numerical Algorithms **45**, 2007, 217–230.
245. W. GAWRONSKI, *On the asymptotic distribution of the zeros of Hermite, Laguerre, and Jonquière polynomials.* Journal of Approximation Theory **50**, 1987, 214–231.
246. W. GAWRONSKI, *Strong asymptotics and the asymptotic zero distributions of Laguerre polynomials* $L_n^{(an+\alpha)}$ *and Hermite polynomials* $H_n^{(an+\alpha)}$. Analysis **13**, 1993, 29–67.
247. W. GAWRONSKI and B. SHAWYER, *Strong asymptotics and the limit distribution of the zeros of Jacobi polynomials* $P_n^{(an+\alpha,bn+\beta)}$. In: Progress in Approximation Theory, Academic Press, Boston, 1991, 379–404.
248. J. GILLIS, *Integrals of products of Laguerre polynomials.* SIAM Journal on Mathematical Analysis **6**, 1975, 318–339.
249. J. GILLIS, M.E.H. ISMAIL and T. OFFER, *An asymptotic problem in derangement theory.* SIAM Journal on Mathematical Analysis **21**, 1990, 262–269.
250. J. GILLIS, J. JEDWAB and D. ZEILBERGER, *A combinatorial interpretation of the integral of the products of Legendre polynomials.* SIAM Journal on Mathematical Analysis **19**, 1988, 1455–1461.
251. E. GODOY, A. RONVEAUX, A. ZARZO and I. AREA, *Minimal recurrence relations for connection coefficients between classical orthogonal polynomials: continuous case.* Journal of Computational and Applied Mathematics **84**, 1997, 257–275.
252. W.M.Y. GOH, *Plancherel-Rotach asymptotics for the Charlier polynomials.* Constructive Approximation **14**, 1998, 151–168.
253. I.S. GRADSHTEYN and I.M. RYZHIK, *Table of integrals, series, and products.* Corrected and enlarged edition. Academic Press, Orlando, Florida, 1980.
254. YA.I. GRANOVSKII, I.M. LUTZENKO and A.S. ZHEDANOV, *Mutual integrability, quadratic algebras, and dynamical symmetry.* Annales of Physics **217**, 1992, 1–20.
255. E. GROSSWALD, *The Bessel Polynomials.* Lecture Notes in Mathematics **698**, Springer-Verlag, New York, 1978.
256. F.A. GRÜNBAUM and L. HAINE, *The q-version of a theorem of Bochner.* Journal of Computational and Applied Mathematics **68**, 1996, 103–114.
257. F.A. GRÜNBAUM and L. HAINE, *Some functions that generalize the Askey-Wilson polynomials.* Communications in Mathematical Physics **184**, 1997, 173–202.
258. D.P. GUPTA, M.E.H. ISMAIL and D.R. MASSON, *Associated continuous Hahn polynomials.* Canadian Journal of Mathematics **43**, 1991, 1263–1280.
259. D.P. GUPTA, M.E.H. ISMAIL and D.R. MASSON, *Contiguous relations, basic hypergeometric functions, and orthogonal polynomials II. Associated big q-Jacobi polynomials.* Journal of Mathematical Analysis and Applications **171**, 1992, 477–497.
260. D.P. GUPTA, M.E.H. ISMAIL and D.R. MASSON, *Contiguous relations, basic hypergeometric functions, and orthogonal polynomials III. Associated continuous dual q-Hahn polynomials.* Journal of Computational and Applied Mathematics **68**, 1996, 115–149.
261. W. HAHN, *Über Orthogonalpolynome, die q-Differenzengleichungen genügen.* Mathematische Nachrichten **2**, 1949, 4–34.
262. M. HAJMIRZAAHMAD, *Jacobi polynomial expansions.* Journal of Mathematical Analysis and Applications **181**, 1994, 35–61.
263. M. HAJMIRZAAHMAD, *Laguerre polynomial expansions.* Journal of Computational and Applied Mathematics **59**, 1995, 25–37.
264. T. HARTMANN and E.P. STEPHAN, *Rates of convergence for collocation with Jacobi polynomials for the airfoil equation.* Journal of Computational and Applied Mathematics **51**, 1994, 179–191.
265. M.R. HOARE and M. RAHMAN, *Cumulative Bernoulli trials and Krawtchouk processes.* Stochastic Processes and their Applications **16**, 1984, 113–139.
266. R. HORTON, *Jacobi polynomials IV. A family of variation diminishing kernels.* SIAM Journal on Mathematical Analysis **6**, 1975, 544–550.

267. M.E.H. ISMAIL, *On obtaining generating functions of Boas and Buck type for orthogonal polynomials.* SIAM Journal on Mathematical Analysis **5**, 1974, 202–208.
268. M.E.H. ISMAIL, *Connection relations and bilinear formulas for the classical orthogonal polynomials.* Journal of Mathematical Analysis and Applications **57**, 1977, 487–496.
269. M.E.H. ISMAIL, *A queueing model and a set of orthogonal polynomials.* Journal of Mathematical Analysis and Applications **108**, 1985, 575–594.
270. M.E.H. ISMAIL, *On sieved orthogonal polynomials, I: Symmetric Pollaczek analogues.* SIAM Journal on Mathematical Analysis **16**, 1985, 1093–1113.
271. M.E.H. ISMAIL, *Asymptotics of the Askey-Wilson and q-Jacobi polynomials.* SIAM Journal on Mathematical Analysis **17**, 1986, 1475–1482.
272. M.E.H. ISMAIL, *On sieved orthogonal polynomials, IV: Generating functions.* Journal of Approximation Theory **46**, 1986, 284–296.
273. M.E.H. ISMAIL, *Asymptotics of Pollaczek polynomials and their zeros.* SIAM Journal on Mathematical Analysis **25**, 1994, 462–473.
274. M.E.H. ISMAIL, *Relativistic orthogonal polynomials are Jacobi polynomials.* Journal of Physics A. Mathematical and General **29**, 1996, 3199–3202.
275. M.E.H. ISMAIL, *A generalization of a theorem of Bochner.* Journal of Computational and Applied Mathematics **159**, 2003, 319–324.
276. M.E.H. ISMAIL, *Asymptotics of q-orthogonal polynomials and a q-Airy function.* International Mathematics Research Notices **18**, 2005, 1063–1088.
277. M.E.H. ISMAIL, *Classical and Quantum Orthogonal Polynomials in One Variable.* Encyclopedia of Mathematics and Its Applications **98**, Cambridge University Press, Cambridge, 2005.
278. M.E.H. ISMAIL, J. LETESSIER, D.R. MASSON and G. VALENT, *Birth and death processes and orthogonal polynomials.* In: Orthogonal Polynomials: Theory and Practice (ed. P. Nevai), Kluwer Academic Publishers, Dordrecht, 1990, 229–255.
279. M.E.H. ISMAIL, J. LETESSIER and G. VALENT, *Linear birth and death models and associated Laguerre and Meixner polynomials.* Journal of Approximation Theory **55**, 1988, 337–348.
280. M.E.H. ISMAIL, J. LETESSIER and G. VALENT, *Quadratic birth and death processes and associated continuous dual Hahn polynomials.* SIAM Journal on Mathematical Analysis **20**, 1989, 727–737.
281. M.E.H. ISMAIL, J. LETESSIER, G. VALENT and J. WIMP, *Two families of associated Wilson polynomials.* Canadian Journal of Mathematics **42**, 1990, 659–695.
282. M.E.H. ISMAIL, J. LETESSIER, G. VALENT and J. WIMP, *Some results on associated Wilson polynomials.* In: Orthogonal Polynomials and their Applications (eds. C. Brezinski, L. Gori and A. Ronveaux). IMACS Annals on Computing and Applied Mathematics **9**, J.C. Baltzer Scientific Publishing Company, Basel, 1991, 293–298.
283. M.E.H. ISMAIL and X. LI, *Bound on the extreme zeros of orthogonal polynomials.* Proceedings of the American Mathematical Society **115**, 1992, 131–140.
284. M.E.H. ISMAIL and C.A. LIBIS, *Contiguous relations, basic hypergeometric functions, and orthogonal polynomials I.* Journal of Mathematical Analysis and Applications **141**, 1989, 349–372.
285. M.E.H. ISMAIL and D.R. MASSON, *q-Hermite polynomials, biorthogonal rational functions, and q-beta integrals.* Transactions of the American Mathematical Society **346**, 1994, 63–116.
286. M.E.H. ISMAIL and D.R. MASSON, *Generalized orthogonality and continued fractions.* Journal of Approximation Theory **83**, 1995, 1–40.
287. M.E.H. ISMAIL, D.R. MASSON and M. RAHMAN, *Complex weight functions for classical orthogonal polynomials.* Canadian Journal of Mathematics **43**, 1991, 1294–1308.
288. M.E.H. ISMAIL, D.R. MASSON and S.K. SUSLOV, *Some generating functions for q-polynomials.* In: Special Functions, q-Series and Related Topics (eds. M.E.H. Ismail, D.R. Masson and M. Rahman). Fields Institute Communications **14**, 1997, 91–108.
289. M.E.H. ISMAIL and M.E. MULDOON, *A discrete approach to monotonicity of zeros of orthogonal polynomials.* Transactions of the American Mathematical Society **323**, 1991, 65–78.

290. M.E.H. ISMAIL and M. RAHMAN, *The associated Askey-Wilson polynomials.* Transactions of the American Mathematical Society **328**, 1991, 201–237.
291. M.E.H. ISMAIL and M. RAHMAN, *The q-Laguerre polynomials and related moment problems.* Journal of Mathematical Analysis and Applications **218**, 1998, 155–174.
292. M.E.H. ISMAIL, M. RAHMAN and S.K. SUSLOV, *Some summation theorems and transformations for q-series.* Canadian Journal of Mathematics **49**, 1997, 543–567.
293. M.E.H. ISMAIL, M. RAHMAN and R. ZHANG, *Diagonalization of certain integral operators II.* Journal of Computational and Applied Mathematics **68**, 1996, 163–196.
294. M.E.H. ISMAIL and D. STANTON, *On the Askey-Wilson and Rogers polynomials.* Canadian Journal of Mathematics **40**, 1988, 1025–1045.
295. M.E.H. ISMAIL and D. STANTON, *Classical orthogonal polynomials as moments.* Canadian Journal of Mathematics **49**, 1997, 520–542.
296. M.E.H. ISMAIL, D. STANTON and G. VIENNOT, *The combinatorics of q-Hermite polynomials and the Askey-Wilson integral.* European Journal of Combinatorics **8**, 1987, 379–392.
297. M.E.H. ISMAIL and M.V. TAMHANKAR, *A combinatorial approach to some positivity problems.* SIAM Journal on Mathematical Analysis **10**, 1979, 478–485.
298. M.E.H. ISMAIL and J.A. WILSON, *Asymptotic and generating relations for the q-Jacobi and* ${}_4\phi_3$ *polynomials.* Journal of Approximation Theory **36**, 1982, 43–54.
299. M.E.H. ISMAIL and R. ZHANG, *Diagonalization of certain integral operators.* Advances in Mathematics **109**, 1994, 1–33.
300. D.M. JACKSON, *Laguerre polynomials and derangements.* Mathematical Proceedings of the Cambridge Philosophical Society **80**, 1976, 213–214.
301. V.K. JAIN and H.M. SRIVISTAVA, *New results involving a certain class of q-orthogonal polynomials.* Journal of Mathematical Analysis and Applications **166**, 1992, 331–344.
302. V.K. JAIN and H.M. SRIVISTAVA, *Some families of multilinear q-generating functions and combinatorial q-series identities.* Journal of Mathematical Analysis and Applications **192**, 1995, 413–438.
303. X.-S. JIN and R. WONG, *Uniform asymptotic expansions for Meixner polynomials.* Constructive Approximation **14**, 1998, 113–150.
304. K.W.J. KADELL, *The little q-Jacobi functions of complex order.* In: Theory and Applications of Special Functions. Developments in Mathematics **13**, Springer, New York, 2005, 301–338.
305. E.G. KALNINS and W. MILLER JR., *q-Series and orthogonal polynomials associated with Barnes' first lemma.* SIAM Journal on Mathematical Analysis **19**, 1988, 1216–1231.
306. E.G. KALNINS and W. MILLER JR., *Symmetry techniques for q-series: Askey-Wilson polynomials.* The Rocky Mountain Journal of Mathematics **19**, 1989, 223–230.
307. S. KARLIN and J. MCGREGOR, *Linear growth; birth and death processes.* Journal of Mathematics and Mechanics **7**, 1958, 643–662.
308. S. KARLIN and J.L. MCGREGOR, *The Hahn polynomials, formulas and an application.* Scripta Mathematicae **26**, 1961, 33–46.
309. A.W. KEMP, *Heine-Euler extensions of the Poisson distribution.* Communications in Statistics, Theory and Methods **21**, 1992, 571–588.
310. A.W. KEMP, *Steady-state Markov chain models for the Heine and Euler distributions.* Journal of Applied Probability **29**, 1992, 869–876.
311. A.W. KEMP, *Characterizations involving* $U|(U+V=m)$ *for certain discrete distributions.* Journal of Statistical Planning and Inference **109**, 2003, 31–41.
312. D. KIM, *On combinatorics of Al-Salam Carlitz polynomials.* European Journal of Combinatorics **18**, 1997, 295–302.
313. E. KOCHNEFF, *Expansions in Laguerre polynomials of negative order.* Journal of Approximation Theory **81**, 1995, 332–346.
314. E. KOCHNEFF, *Expansions in Jacobi polynomials of negative order.* Constructive Approximation **13**, 1997, 435–446.
315. E. KOCHNEFF, *Weighted norm inequalities for fractional integrals with an application to mean convergence of Laguerre series.* Journal of Approximation Theory **91**, 1997, 84–102.
316. J. KOEKOEK and R. KOEKOEK, *The Jacobi inversion formula.* Complex Variables. Theory and Application **39**, 1999, 1–18.

317. R. KOEKOEK, *Inversion formulas involving orthogonal polynomials and some of their applications.* In: Special functions, Proceedings of the International Workshop, Hong Kong 1999, World Scientific Publishing, Singapore, 2000, 166–180.
318. R. KOEKOEK and R.F. SWARTTOUW, *The Askey-scheme of hypergeometric orthogonal polynomials and its q-analogue.* Delft University of Technology, Report no. **94-05**, 1994.
319. R. KOEKOEK and R.F. SWARTTOUW, *The Askey-scheme of hypergeometric orthogonal polynomials and its q-analogue.* Delft University of Technology, Report no. **98-17**, 1998.
320. E. KOELINK, *Addition formulas for q-special functions.* In: Special Functions, q-Series and Related Topics (eds. M.E.H. Ismail, D.R. Masson and M. Rahman). Fields Institute Communications **14**, 1997, 109–129.
321. E. KOELINK and J.V. STOKMAN, *The Askey-Wilson function transform.* International Mathematics Research Notices **22**, 2001, 1203–1227.
322. E. KOELINK and J.V. STOKMAN, *The Askey-Wilson function transform scheme.* In: Special Functions 2000: Current Perspective and Future Directions (eds. J. Bustoz, M.E.H. Ismail and S.K. Suslov). NATO Science Series II **30**, Kluwer, Dordrecht, 2001, 221–241.
323. H.T. KOELINK, *The addition formula for continuous q-Legendre polynomials and associated spherical elements on the* $SU(2)$ *quantum group related to Askey-Wilson polynomials.* SIAM Journal on Mathematical Analysis **25**, 1994, 197–217.
324. H.T. KOELINK, *Addition formula for big q-Legendre polynomials from the quantum* $SU(2)$ *group.* Canadian Journal of Mathematics **47**, 1995, 436–448.
325. H.T. KOELINK, *Identities for q-ultraspherical polynomials and Jacobi functions.* Proceedings of the American Mathematical Society **123**, 1995, 2479–2487.
326. H.T. KOELINK, *Askey-Wilson polynomials and the quantum* $SU(2)$ *group: survey and applications.* Acta Applicandae Mathematicae **44**, 1996, 295–352.
327. H.T. KOELINK, *On Jacobi and continuous Hahn polynomials.* Proceedings of the American Mathematical Society **124**, 1996, 887–898.
328. H.T. KOELINK, *Yet another basic analogue of Graf's addition formula.* Journal of Computational and Applied Mathematics **68**, 1996, 209–220.
329. H.T. KOELINK and T.H. KOORNWINDER, *The Clebsch-Gordan coefficients for the quantum group* $S_\mu U(2)$ *and q-Hahn polynomials.* Indagationes Mathematicae **51**, 1989, 443–456.
330. T. KOORNWINDER, *The addition formula for Jacobi polynomials and spherical harmonics.* SIAM Journal on Applied Mathematics **25**, 1973, 236–246.
331. T. KOORNWINDER, *Jacobi polynomials II. An analytic proof of the product formula.* SIAM Journal on Mathematical Analysis **5**, 1974, 125–137.
332. T. KOORNWINDER, *Jacobi polynomials III. An analytic proof of the addition formula.* SIAM Journal on Mathematical Analysis **6**, 1975, 533–543.
333. T. KOORNWINDER, *The addition formula for Laguerre polynomials.* SIAM Journal on Mathematical Analysis **8**, 1977, 535–540.
334. T. KOORNWINDER, *Yet another proof of the addition formula for Jacobi polynomials.* Journal of Mathematical Analysis and Applications **61**, 1977, 136–141.
335. T. KOORNWINDER, *Positivity proofs for linearization and connection coefficients of orthogonal polynomials satisfying an addition formula.* Journal of the London Mathematical Society (2) **18**, 1978, 101–114.
336. T.H. KOORNWINDER, *The addition formula for Jacobi polynomials I. Summary of results.* Indagationes Mathematicae **34**, 1972, 188–191.
337. T.H. KOORNWINDER, *Clebsch-Gordan coefficients for* $SU(2)$ *and Hahn polynomials.* Nieuw Archief voor Wiskunde **29**, 1981, 140–155.
338. T.H. KOORNWINDER, *Krawtchouk polynomials, a unification of two different group theoretic interpretations.* SIAM Journal on Mathematical Analysis **13**, 1982, 1011–1023.
339. T.H. KOORNWINDER, *Special orthogonal polynomial systems mapped onto each other by the Fourier-Jacobi transform.* In: Polynômes Orthogonaux et Applications (eds. C. Brezinski et al.). Lecture Notes in Mathematics **1171**, Springer-Verlag, New York, 1985, 174–183.
340. T.H. KOORNWINDER, *Group theoretic interpretations of Askey's scheme of hypergeometric orthogonal polynomials.* In: Orthogonal Polynomials and Their Applications (eds. M. Alfaro et al.). Lecture Notes in Mathematics **1329**, Springer-Verlag, New York, 1988, 46–72.

341. T.H. KOORNWINDER, *Continuous q-Legendre polynomials as spherical matrix elements of irreducible representations of the quantum $SU(2)$ group.* C.W.I. Quarterly **2**, 1989, 171–173.
342. T.H. KOORNWINDER, *Meixner-Pollaczek polynomials and the Heisenberg algebra.* Journal of Mathematical Physics **30**, 1989, 767–769.
343. T.H. KOORNWINDER, *Representations of the twisted $SU(2)$ quantum group and some q-hypergeometric orthogonal polynomials.* Indagationes Mathematicae **51**, 1989, 97–117.
344. T.H. KOORNWINDER, *Jacobi functions as limit cases of q-ultraspherical polynomials.* Journal of Mathematical Analysis and Applications **148**, 1990, 44–54.
345. T.H. KOORNWINDER, *Orthogonal polynomials in connection with quantum groups.* In: Orthogonal Polynomials: Theory and Practice (ed. P. Nevai), Kluwer, Dordrecht, 1990, 257–292.
346. T.H. KOORNWINDER, *The addition formula for little q-Legendre polynomials and the $SU(2)$ quantum group.* SIAM Journal on Mathematical Analysis **22**, 1991, 295–301.
347. T.H. KOORNWINDER, *Askey-Wilson polynomials for root systems of type BC.* Contemporary Mathematics **138**, 1992, 189–204.
348. T.H. KOORNWINDER, *Askey-Wilson polynomials as zonal spherical functions on the $SU(2)$ quantum group.* SIAM Journal on Mathematical Analysis **24**, 1993, 795–813.
349. T.H. KOORNWINDER, *Special functions and q-commuting variables.* In: Special Functions, *q*-Series and Related Topics (eds. M.E.H. Ismail, D.R. Masson and M. Rahman). Fields Institute Communications **14**, 1997, 131–166.
350. T.H. KOORNWINDER, *A second addition formula for continuous q-ultraspherical polynomials.* In: Theory and Applications of Special Functions. Developments in Mathematics **13**, Springer, New York, 2005, 339–360.
351. T.H. KOORNWINDER, *The structure relation for Askey-Wilson polynomials.* Journal of Computational and Applied Mathematics **207**, 2007, 214–226.
352. H.L. KRALL and O. FRINK, *A new class of orthogonal polynomials: The Bessel polynomials.* Transactions of the American Mathematical Society **65**, 1949, 100–115.
353. I. KRASIKOV, *Bounds for zeros of the Charlier polynomials.* Methods and Applications of Analysis **9**, 2002, 599–610.
354. I. KRASIKOV, *Bounds for zeros of the Laguerre polynomials.* Journal of Approximation Theory **121**, 2003, 287–291.
355. I. KRASIKOV, *New bounds on the Hermite polynomials.* East Journal on Approximations **10**, 2004, 355–362.
356. I. KRASIKOV, *Inequalities for Laguerre polynomials.* East Journal on Approximations **11**, 2005, 257–268.
357. A.B.J. KUIJLAARS and A. MARTÍNEZ-FINKELSHTEIN, *Strong asymptotics for Jacobi polynomials with varying nonstandard parameters.* Journal d'Analyse Mathématique **94**, 2004, 195–234.
358. A.B.J. KUIJLAARS, A. MARTÍNEZ-FINKELSHTEIN and R. ORIVE, *Orthogonality of Jacobi polynomials with general parameters.* Electronic Transactions on Numerical Analysis **19**, 2005, 1–17.
359. A.B.J. KUIJLAARS and K.T.-R. MCLAUGHLIN, *Asymptotic zero behavior of Laguerre polynomials with negative parameter.* Constructive Approximation **20**, 2004, 497–523.
360. K.H. KWON and L.L. LITTLEJOHN, *The orthogonality of the Laguerre polynomials $\{L_n^{(-k)}(x)\}$ for positive integers k.* Annals of Numerical Mathematics **2**, 1995, 289–303.
361. K.H. KWON, B.H. YOO and G.J. YOON, *A characterization of Hermite polynomials.* Journal of Computational and Applied Mathematics **78**, 1997, 295–299.
362. J. LABELLE, *Tableau d'Askey.* In: Polynômes Orthogonaux et Applications, (eds. C. Brezinski et al.). Lecture Notes in Mathematics **1171**, Springer-Verlag, New York, 1985, xxxvi–xxxvii.
363. J. LABELLE and Y.N. YEH, *Combinatorial proofs of some limit formulas involving orthogonal polynomials.* Discrete Mathematics **79**, 1989, 77–93.
364. J. LABELLE and Y.N. YEH, *Combinatorial proofs of symmetry formulas for the generalized hypergeometric series.* Journal of Mathematical Analysis and Applications **139**, 1989, 36–48.

365. J. LABELLE and Y.N. YEH, *The combinatorics of Laguerre, Charlier, and Hermite polynomials.* Studies in Applied Mathematics **80**, 1989, 25–36.
366. A. LAFORGIA, *A monotonic property for the zeros of ultraspherical polynomials.* Proceedings of the American Mathematical Society **83**, 1981, 757–758.
367. T.P. LAINE, *Projection formulas and a new proof of the addition formula for the Jacobi polynomials.* SIAM Journal on Mathematical Analysis **13**, 1982, 324–330.
368. P.A. LEE, *Explicit formula for the multiple generating function of product of generalized Laguerre polynomials.* Journal of Physics A. Mathematical and General **30**, 1997, L183–L186.
369. P.A. LEE, *Markov processes and a multiple generating function of product of generalized Laguerre polynomials.* Journal of Physics A. Mathematical and General **30**, 1997, L373–L377.
370. P.A. LEE, *On an integral of product of Laguerre polynomials.* International Journal of Computer Mathematics **64**, 1997, 303–307.
371. D.A. LEONARD, *Orthogonal polynomials, duality and association schemes.* SIAM Journal on Mathematical Analysis **13**, 1982, 656–663.
372. P. LESKY, *Über Polynomsysteme, die Sturm-Liouvilleschen Differenzengleichungen genügen.* Mathematische Zeitschrift **78**, 1962, 439–445.
373. P. LESKY, *Der Satz von Favard für Endliche Orthogonalsysteme.* Österreichische Akademie der Wissenschaften. Mathematisch-Naturwissenschaftliche Klasse. Anzeiger **123**, 1986, 45–49.
374. P. LESKY, *Eine Vervollständigung der Hahnpolynome.* Österreichische Akademie der Wissenschaften. Mathematisch-Naturwissenschaftliche Klasse. Sitzungsberichte. Abteilung II. Mathematische, Physikalische und Technische Wissenschaften **197**, 1988, 379–389.
375. P. LESKY, *Die Vervollständigung der diskreten klassischen Orthogonalpolynome.* Österreichische Akademie der Wissenschaften. Mathematisch-Naturwissenschaftliche Klasse. Sitzungsberichte. Abteilung II. Mathematische, Physikalische und Technische Wissenschaften **198**, 1989, 295–315.
376. P.A. LESKY, *Racahpolynome und duale Hahnpolynome als Lösungen von Eigenwertproblemen.* Österreichische Akademie der Wissenschaften Mathematisch-Naturwissenschaftliche Klasse. Sitzungsberichte. Abteilung II. Mathematische, Physikalische und Technische Wissenschaften **202**, 1993, 163–172.
377. P.A. LESKY, *Dualität bei Polynomlösungen von reellen Differenzengleichungen zweiter Ordnung.* Österreichische Akademie der Wissenschaften Mathematisch-Naturwissenschaftliche Klasse. Sitzungsberichte. Abteilung II. Mathematische, Physikalische und Technische Wissenschaften **203**, 1994, 63–80.
378. P.A. LESKY, *Dualität bei Polynomlösungen von komplexen Differenzengleichungen zweiter Ordnung.* Österreichische Akademie der Wissenschaften Mathematisch-Naturwissenschaftliche Klasse. Sitzungsberichte. Abteilung II. Mathematische, Physikalische und Technische Wissenschaften **203**, 1994, 81–100.
379. P.A. LESKY, *Zweigliedrige Rekursionen für die Koeffizienten von Polynomlösungen Sturm-Liouvillescher q-Differenzengleichungen.* Zeitschrift für Angewandte Mathematik und Mechanik **74**, 1994, 497–500.
380. P.A. LESKY, *Orthogonalität von dualen Hahnpolynomen und continuous dualen Hahnpolynomen.* Österreichische Akademie der Wissenschaften Mathematisch-Naturwissenschaftliche Klasse. Sitzungsberichte. Abteilung II. Mathematische, Physikalische und Technische Wissenschaften **204**, 1995, 81–95.
381. P.A. LESKY, *Vervollständigung der klassischen Orthogonalpolynome durch Ergänzungen zum Askey-Schema der hypergeometrischen orthogonalen Polynome.* Österreichische Akademie der Wissenschaften Mathematisch-Naturwissenschaftliche Klasse. Sitzungsberichte. Abteilung II. Mathematische, Physikalische und Technische Wissenschaften **204**, 1995, 151–166.
382. P.A. LESKY, *Endliche und undendliche Systeme von kontinuierlichen klassischen Orthogonalpolynomen.* Zeitschrift für Angewandte Mathematik und Mechanik **76**, 1996, 181–184.

383. P.A. LESKY, *Unendliche und endliche Orthogonalsysteme von continuous Hahnpolynomen.* Results in Mathematics **31**, 1997, 127–135.

384. P.A. LESKY, *Einordnung der Polynome von Romanovski-Bessel in das Askey-Tableau.* Zeitschrift für Angewandte Mathematik und Mechanik **78**, 1998, 646–648.

385. P.A. LESKY, *Orthogonalpolynome in x and q^{-x} als Lösungen von reellen q-Operatorgleichungen zweiter Ordnung.* Monatshefte für Mathematik **132**, 2001, 123–140.

386. P.A. LESKY, *Reelle Polynome in $\frac{a}{z}+\frac{uz}{a}$ $(a,z \in \mathbb{C},\ u \in \mathbb{R})$ als Lösungen von q-Operatorgleichungen zweiter Ordnung.* Österreichische Akademie der Wissenschaften Mathematisch-Naturwissenschaftliche Klasse. Sitzungsberichte. Abteilung II. Mathematische, Physikalische und Technische Wissenschaften **210**, 2001, 33–59.

387. P.A. LESKY and B.M. WAIBEL, *Orthogonalität von Racahpolynomen und Wilsonpolynomen.* Results in Mathematics **35**, 1999, 119–133.

388. J. LETESSIER and G. VALENT, *The generating function method for quadratic asymptotically symmetric birth and death processes.* SIAM Journal on Applied Mathematics **44**, 1984, 773–783.

389. J. LETESSIER and G. VALENT, *Dual birth and death processes and orthogonal polynomials.* SIAM Journal on Applied Mathematics **46**, 1986, 393–405.

390. S. LEWANOWICZ, *Second-order recurrence relation for the linearization coefficients of the classical orthogonal polynomials.* Journal of Computational and Applied Mathematics **69**, 1996, 159–170.

391. S. LEWANOWICZ, *Recurrence relations for the connection coefficients of orthogonal polynomials of a discrete variable.* Journal of Computational and Applied Mathematics **76**, 1996, 213–229.

392. X. LI and R. WONG, *On the asymptotics of the Meixner-Pollaczek polynomials and their zeros.* Constructive Approximation **17**, 2001, 59–90.

393. Z. LI, *Conjugate Jacobi series and conjugate functions.* Journal of Approximation Theory **86**, 1996, 179–196.

394. Z. LI, *On the Cesàro means of conjugate Jacobi series.* Journal of Approximation Theory **91**, 1997, 103–116.

395. Z. LIU, *The q-ultraspherical polynomials and some identities of the Rogers-Ramanujan type.* Chinese Quarterly Journal of Mathematics **15**, 2000, 1–5.

396. J.L. LÓPEZ and N.M. TEMME, *Hermite polynomials in asymptotic representations of generalized Bernoulli, Euler, Bessel, and Buchholz polynomials.* Journal of Mathematical Analysis and Applications **239**, 1999, 457–477.

397. J.L. LÓPEZ and N.M. TEMME, *Convergent asymptotic expansions of Charlier, Laguerre and Jacobi polynomials.* Proceedings of the Royal Society of Edinburgh **134**, 2004, 537–555.

398. L. LORCH, *Inequalities for ultraspherical polynomials and the gamma function.* Journal of Approximation Theory **40**, 1984, 115–120.

399. Y.L. LUKE, *The special functions and their approximations. Volume 1.* Mathematics in Science and Engineering, Volume 53. Academic Press, New York, 1969.

400. A. MARTÍNEZ-FINKELSHTEIN, P. MARTÍNEZ-GONZÁLEZ and R. ORIVE, *Zeros of Jacobi polynomials with varying non-classical parameters.* In: Special functions, Proceedings of the International Workshop, Hong Kong 1999, World Scientific Publishing, Singapore, 2000, 98–113.

401. D.R. MASSON, *Associated Wilson polynomials.* Constructive Approximation **7**, 1991, 521–534.

402. T. MASUDA, K. MIMACHI, Y. NAKAGAMI, M. NOUMI and K. UENO, *Representations of the quantum group $SU_q(2)$ and the little q-Jacobi polynomials.* Journal of Functional Analysis **99**, 1991, 357–386.

403. A.M. MATHAI, *A Handbook of Generalized Special Functions for Statistical and Physical Sciences.* Clarendon Press, Oxford, 1993.

404. P. MEER and I. WEISS, *Smoothed differentiation filters for images.* Journal of Visual Communication and Image Representation **3**, 1992, 58–72.

405. H.G. MEIJER, *Asymptotic expansion of Jacobi polynomials.* In: Polynômes Orthogonaux et Applications (eds. C. Brezinski et al.). Lecture Notes in Mathematics **1171**, Springer-Verlag, New York, 1985, 380–389.
406. J. MEIXNER, *Orthogonale Polynomsysteme mit einer besonderen Gestalt der erzeugenden Funktion.* Journal of the London Mathematical Society **9**, 1934, 6–13.
407. W. MILLER JR., *A note on Wilson polynomials.* SIAM Journal on Mathematical Analysis **18**, 1987, 1221–1226.
408. W. MILLER JR., *Symmetry techniques and orthogonality for q-series.* In: q-Series and Partitions (ed. D. Stanton). The IMA Volumes in Mathematics and Its Applications **18**, Springer-Verlag, New York, 1989, 191–212.
409. K. MIMACHI, *Askey-Wilson polynomials by means of a q-Selberg type integral.* Advances in Mathematics **147**, 1999, 315–327.
410. K. MIMACHI, *Integral representations of the Wilson polynomials and the continuous dual Hahn polynomials.* Advances in Applied Mathematics **23**, 1999, 340–359.
411. D.S. MOAK, *The q-analogue of the Laguerre polynomials.* Journal of Mathematical Analysis and Applications **81**, 1981, 20–47.
412. D.S. MOAK, E.B. SAFF and R.S. VARGA, *On the zeros of Jacobi polynomials* $P_n^{(\alpha_n,\beta_n)}(x)$. Transactions of the American Mathematical Society **249**, 1979, 159–162.
413. B. NAGEL, *The relativistic Hermite polynomial is a Gegenbauer polynomial.* Journal of Mathematical Physics **35**, 1994, 1549–1554.
414. B. NASSRALLAH and M. RAHMAN, *Projection formulas, a reproducing kernel and a generating function for q-Wilson polynomials.* SIAM Journal on Mathematical Analysis **16**, 1985, 186–197.
415. E. NEUMAN, *On Hahn polynomials and continuous dual Hahn polynomials.* Journal of Computational and Applied Mathematics **8**, 2006, 229–248.
416. A.F. NIKIFOROV, S.K. SUSLOV and V.B. UVAROV, *Classical orthogonal polynomials of a discrete variable.* Springer Series in Computational Physics, Springer-Verlag, New York, 1991.
417. A.F. NIKIFOROV and V.B. UVAROV, *Special Functions of Mathematical Physics.* Birkhäuser Verlag, Basel, 1988.
418. M. NOUMI and K. MIMACHI, *Askey-Wilson polynomials and the quantum group* $SU_q(2)$. Proceedings of the Japan Academy, Series A. Mathematical Sciences **66**, 1990, 146–149.
419. M. NOUMI and K. MIMACHI, *Big q-Jacobi polynomials, q-Hahn polynomials, and a family of quantum* 3*-spheres.* Letters in Mathematical Physics **19**, 1990, 299–305.
420. M. NOUMI and K. MIMACHI, *Quantum* 2*-spheres and big q-Jacobi polynomials.* Communications in Mathematical Physics **128**, 1990, 521–531.
421. M. NOUMI and K. MIMACHI, *Roger's q-ultraspherical polynomials on a quantum* 2*-sphere.* Duke Mathematical Journal **63**, 1991, 65–80.
422. M. NOUMI and K. MIMACHI, *Askey-Wilson polynomials as spherical functions on* $SU_q(2)$. In: Quantum groups (Leningrad, 1990). Lecture Notes in Mathematics **1510**, Springer-Verlag, New York, 1992, 98–103.
423. M. NOUMI and J.V. STOKMAN, *Askey-Wilson polynomials: an affine Hecke algebraic approach.* In: Laredo Lectures on Orthogonal Polynomials and Special Functions (eds. R. Álvarez-Nodarse, F. Marcellán and W. Van Assche). Nova Science Publishers, New York, 2004, 111–144.
424. F.W.J. OLVER, *Asymptotics and special functions.* Computer Science and Applied Mathematics, Academic Press, New York, 1974.
425. T.E. PÉREZ and M.A. PIÑAR, *On Sobolev orthogonality for the generalized Laguerre polynomials.* Journal of Approximation Theory **86**, 1996, 278–285.
426. M. PERLSTADT, *A property of orthogonal polynomial families with polynomial duals.* SIAM Journal on Mathematical Analysis **15**, 1984, 1043–1054.
427. G. PITTALUGA and L. SACRIPANTE, *Bounds for the zeros of Hermite polynomials.* Annals of Numerical Mathematics **2**, 1995, 371–379.
428. J. PRASAD and H. HAYASHI, *On the uniform approximation of smooth functions by Jacobi polynomials.* Canadian Journal of Mathematics **25**, 1973, 216–223.

429. W.-Y. QIU and R. WONG, *Asymptotic expansion of the Krawtchouk polynomials and their zeros.* Computational Methods and Function Theory **4**, 2004, 189–226.
430. M. RAHMAN, *Construction of a family of positive kernels from Jacobi polynomials.* SIAM Journal on Mathematical Analysis **7**, 1976, 92–116.
431. M. RAHMAN, *A five-parameter family of positive kernels from Jacobi polynomials.* SIAM Journal on Mathematical Analysis **7**, 1976, 386–413.
432. M. RAHMAN, *Some positive kernels and bilinear sums for Hahn polynomials.* SIAM Journal on Mathematical Analysis **7**, 1976, 414–435.
433. M. RAHMAN, *On a generalization of the Poisson kernel for Jacobi polynomials.* SIAM Journal on Mathematical Analysis **8**, 1977, 1014–1031.
434. M. RAHMAN, *A generalization of Gasper's kernel for Hahn polynomials: application to Pollaczek polynomials.* Canadian Journal of Mathematics **30**, 1978, 133–146.
435. M. RAHMAN, *A positive kernel for Hahn-Eberlein polynomials.* SIAM Journal on Mathematical Analysis **9**, 1978, 891–905.
436. M. RAHMAN, *An elementary proof of Dunkl's addition theorem for Krawtchouk polynomials.* SIAM Journal on Mathematical Analysis **10**, 1979, 438–445.
437. M. RAHMAN, *A product formula and a non-negative Poisson kernel for Racah-Wilson polynomials.* Canadian Journal of Mathematics **32**, 1980, 1501–1517.
438. M. RAHMAN, *A non-negative representation of the linearization coefficients of the product of Jacobi polynomials.* Canadian Journal of Mathematics **33**, 1981, 915–928.
439. M. RAHMAN, *A stochastic matrix and bilinear sums for Racah-Wilson polynomials.* SIAM Journal on Mathematical Analysis **12**, 1981, 145–160.
440. M. RAHMAN, *Families of biorthogonal rational functions in a discrete variable.* SIAM Journal on Mathematical Analysis **12**, 1981, 355–367.
441. M. RAHMAN, *The linearization of the product of continuous q-Jacobi polynomials.* Canadian Journal of Mathematics **33**, 1981, 961–987.
442. M. RAHMAN, *Reproducing kernels and bilinear sums for q-Racah and q-Wilson polynomials.* Transactions of the American Mathematical Society **273**, 1982, 483–508.
443. M. RAHMAN, *A q-extension of Feldheim's bilinear sum for Jacobi polynomials and some applications.* Canadian Journal of Mathematics **37**, 1985, 551–578.
444. M. RAHMAN, *A product formula for the continuous q-Jacobi polynomials.* Journal of Mathematical Analysis and Applications **118**, 1986, 309–322.
445. M. RAHMAN, *q-Wilson functions of the second kind.* SIAM Journal on Mathematical Analysis **17**, 1986, 1280–1286.
446. M. RAHMAN, *A projection formula for the Askey-Wilson polynomials and an application.* Proceedings of the American Mathematical Society **103**, 1988, 1099–1107.
447. M. RAHMAN, *A simple proof of Koornwinder's addition formula for the little q-Legendre polynomials.* Proceedings of the American Mathematical Society **107**, 1989, 373–381.
448. M. RAHMAN, *Askey-Wilson functions of the first and second kinds: series and integral representations of* $C_n^2(x;\beta|q)+D_n^2(x;\beta|q)$. Journal of Mathematical Analysis and Applications **164**, 1992, 263–284.
449. M. RAHMAN, *Some generating functions for the associated Askey-Wilson polynomials.* Journal of Computational and Applied Mathematics **68**, 1996, 287–296.
450. M. RAHMAN and M.J. SHAH, *An infinite series with products of Jacobi polynomials and Jacobi functions of the second kind.* SIAM Journal on Mathematical Analysis **16**, 1985, 859–875.
451. M. RAHMAN and S.K. SUSLOV, *Singular analogue of the Fourier transformation for the Askey-Wilson polynomials.* In: Symmetries and integrability of difference equations (Estérel, PQ, 1994). Centre de Recherches Mathématiques. CRM Proceedings and Lecture Notes **9**, 1996, 289–302.
452. M. RAHMAN and A. VERMA, *A q-integral representation of Rogers' q-ultraspherical polynomials and some applications.* Constructive Approximation **2**, 1986, 1–10.
453. M. RAHMAN and A. VERMA, *Product and addition formulas for the continuous q-ultraspherical polynomials.* SIAM Journal on Mathematical Analysis **17**, 1986, 1461–1474.

454. M. RAHMAN and A. VERMA, *Infinite sums of product of continuous q-ultraspherical functions.* The Rocky Mountain Journal of Mathematics **17**, 1987, 371–384.
455. M. RAHMAN and A. VERMA, *Positivity of the Poisson kernel for the Askey-Wilson polynomials.* Indian Journal of Mathematics **33**, 1991, 287–306.
456. E.D. RAINVILLE, *Special functions.* The Macmillan Company, New York, 1960.
457. M.O. RAYES, V. TREVISAN and P.S. WANG, *Factorization properties of Chebyshev polynomials.* Computers and Mathematics with Applications **50**, 2005, 1231–1240.
458. M. REIMER, *Uniform inequalities for Gegenbauer polynomials.* Acta Mathematica Hungarica **70**, 1996, 13–26.
459. T.J. RIVLIN, *The Chebyshev polynomials.* Pure and Applied Mathematics. A Wiley Interscience series of Texts, Monographs and Tracts. John Wiley and Sons, New York, 1974.
460. L.J. ROGERS, *On the expansion of some infinite products.* Proceedings of the London Mathematical Society **24**, 1893, 337–352.
461. L.J. ROGERS, *Second memoir on the expansion of certain infinite products.* Proceedings of the London Mathematical Society **25**, 1894, 318–343.
462. L.J. ROGERS, *Third memoir on the expansion of certain infinite products.* Proceedings of the London Mathematical Society **26**, 1895, 15–32.
463. V. ROMANOVSKI, *Sur quelques classes nouvelles de polynômes orthogonaux.* Comptes Rendus de l'Académie des Sciences de Paris **188**, 1929, 1023–1025.
464. P. RUSEV, *A note on classical generating function for the Jacobi polynomials.* Fractional Calculus and Applied Analysis **8**, 2005, 63–66.
465. P. SABLONNIÈRE, *Discrete Bernstein bases and Hahn polynomials.* Journal of Computational and Applied Mathematics **49**, 1993, 233–241.
466. M. DE SAINTE-CATHERINE and G. VIENNOT, *Combinatorial interpretation of integrals of products of Hermite, Laguerre and Tchebycheff polynomials.* In: Polynômes Orthogonaux et Applications (eds. C. Brezinski et al.). Lecture Notes in Mathematics **1171**, Springer-Verlag, New York, 1985, 120–128.
467. F. SARTORETTO and R. SPIGLER, *Computing expansion coefficients in orthogonal bases of ultraspherical polynomials.* Journal of Computational and Applied Mathematics **56**, 1994, 295–303.
468. A. SAVITZKY and M.J.E. GOLAY, *Smoothing and differentiation of data by simplified least squares procedures.* Analytical Chemistry **36**, 1964, 1627–1639.
469. W. SCHOUTENS, *Stochastic Processes and Orthogonal Polynomials.* Lecture Notes in Stochastics **146**, Springer-Verlag, New York, 2000.
470. Y.G. SHI, *Regularity and explicit representation of* $(0,1,\ldots,m-2,m)$*-interpolation on the zeros of the Jacobi polynomials.* Journal of Approximation Theory **76**, 1994, 274–285.
471. L.J. SLATER, *Generalized Hypergeometric Functions.* Cambridge University Press, 1966.
472. Y.F. SMIRNOV and C. CAMPIGOTTO, *The quantum q-Krawtchouk and q-Meixner polynomials and their related D-functions for the quantum groups* $\mathrm{SU}_q(2)$ *and* $\mathrm{SU}_q(1,1)$. Proceedings of the 10th International Congress on Computational and Applied Mathematics, ICCAM-2002. Journal of Computational and Applied Mathematics 164/165, 2004, 643–660.
473. V. SPIRIDONOV and A. ZHEDANOV, *q-Ultraspherical polynomials for q a root of unity.* Letters in Mathematical Physics **37**, 1996, 173–180.
474. V. SPIRIDONOV and A. ZHEDANOV, *Zeros and orthogonality of the Askey-Wilson polynomials for q a root of unity.* Duke Mathematical Journal **89**, 1997, 283–305.
475. H.M. SRIVASTAVA, *Note on generalized Laguerre polynomials.* Bulletin of the Calcutta Mathematical Society **58**, 1966, 81–82.
476. H.M. SRIVASTAVA, *A note on certain dual series equations involving Laguerre polynomials.* Pacific Journal of Mathematics **30**, 1969, 525–527.
477. H.M. SRIVASTAVA, *Generating functions for Jacobi and Laguerre polynomials.* Proceedings of the American Mathematical Society **23**, 1969, 590–595.
478. H.M. SRIVASTAVA, *Dual series relations involving generalized Laguerre polynomials.* Journal of Mathematical Analysis and Applications **31**, 1970, 587–594.
479. H.M. SRIVASTAVA, *A class of generating functions for generalized hypergeometric polynomials.* Journal of Mathematical Analysis and Applications **35**, 1971, 230–235.

480. H.M. SRIVASTAVA, *An elementary proof of Bailey's bilinear generating function for Jacobi polynomials.* IMA Journal of Applied Mathematics **29**, 1982, 275–280.
481. H.M. SRIVASTAVA and V.K. JAIN, *Some multilinear generating functions for q-Hermite polynomials.* Journal of Mathematical Analysis and Applications **144**, 1989, 147–157.
482. H.M. SRIVASTAVA and V.K. JAIN, *Some formulas involving q-Jacobi and related polynomials.* Annali di Matematica Pura ed Applicata (4) **157**, 1990, 63–75.
483. H.M. SRIVASTAVA and J.P. SINGHAL, *New generating functions for Jacobi and related polynomials.* Journal of Mathematical Analysis and Applications **41**, 1973, 748–752.
484. R. SRIVASTAVA and K.K. MATHUR, *Weighted* $(0;0,2)$*-interpolation on the roots of Hermite polynomials.* Acta Mathematica Hungarica **70**, 1996, 57–73.
485. D. STANTON, *A short proof of a generating function for Jacobi polynomials.* Proceedings of the American Mathematical Society **80**, 1980, 398–400.
486. D. STANTON, *Product formulas for q-Hahn polynomials.* SIAM Journal on Mathematical Analysis **11**, 1980, 100–107.
487. D. STANTON, *Some q-Krawtchouk polynomials on Chevalley groups.* American Journal of Mathematics **102**, 1980, 625–662.
488. D. STANTON, *Orthogonal polynomials and Chevalley groups.* In: Special Functions: Group Theoretical Aspects and Applications (eds. R.A. Askey et al.), D. Reidel Publishing Company, Dordrecht, 1984, 87–128.
489. D. STANTON, *An introduction to group representations and orthogonal polynomials.* In: Orthogonal Polynomials: Theory and Practice (ed. P. Nevai), Kluwer Academic Publishers, Dordrecht, 1990, 419–433.
490. T.J. STIELTJES, *Recherches sur les fractions continues.* Annales de la Faculté des Sciences de Toulouse **8**, 1894, J1-122; **9**, 1895, A5-47; Œuvres Complètes, Volume 2, 398–566.
491. J.V. STOKMAN, *On BC type basic hypergeometric orthogonal polynomials.* Transactions of the American Mathematical Society **352**, 2000, 1527–1579.
492. F.H. SZAFRANIEC, *Charlier polynomials and translational invariance in the quantum harmonic oscillator.* Mathematische Nachrichten **241**, 2002, 163–169.
493. G. SZEGŐ, *Orthogonal polynomials.* American Mathematical Society, Colloquium Publications **23**, Providence, Rhode Island, Fourth edition, 1975.
494. G. SZEGŐ, *An outline of the history of orthogonal polynomials.* With four pages of comments and references by the editor. In: Collected Papers, Volume 3 (ed. R. Askey). Contemporary Mathematicians, Birkhäuser, Boston, 1982, 855–869.
495. R. SZWARC, *Nonnegative linearization and quadratic transformation of Askey-Wilson polynomials.* Canadian Mathematical Bulletin **39**, 1996, 241–249.
496. N.M. TEMME, *Special Functions: An Introduction to the Classical Functions of Mathematical Physics.* John Wiley and Sons, New York, 1996.
497. N.M. TEMME and J.L. LÓPEZ, *The role of Hermite polynomials in asymptotic analysis.* In: Special functions, Proceedings of the International Workshop, Hong Kong 1999, World Scientific Publishing, Singapore, 2000, 339–350.
498. N.M. TEMME and J.L. LÓPEZ, *The Askey scheme for hypergeometric orthogonal polynomials viewed from asymptotic analysis.* In: Proceedings of the Fifth International Symposium on Orthogonal Polynomials, Special Functions and their Applications (Patras, 1999). Journal of Computational and Applied Mathematics **133**, 2001, 623–633.
499. P. TERWILLIGER, *Two linear transformations each tridiagonal with respect to an eigenbasis of the other: comments on the split decomposition.* Journal of Computational and Applied Mathematics **178**, 2005, 437–452.
500. P. TERWILLIGER, *An algebraic approach to the Askey scheme of orthogonal polynomials.* In: Orthogonal Polynomials and Special Functions. Lecture Notes in Mathematics **1883**, 2006, 255–330.
501. S.B. TRIČKOVIĆ and M.S. STANKOVIĆ, *A new approach to the orthogonality of the Laguerre and Hermite polynomials.* Integral Transforms and Special Functions **17**, 2006, 661–672.
502. T.N. TRIVEDI and P.N. PATHAK, *A note on an integral involving Laguerre polynomials.* International Journal of Mathematical Sciences **3**, 2004, 369–373.

503. G. Valent and W. Van Assche, *The impact of Stieltjes' work on continued fractions and orthogonal polynomials: additional material.* Journal of Computational and Applied Mathematics **65**, 1995, 419–447.
504. W. Van Assche and T.H. Koornwinder, *Asymptotic behaviour for Wall polynomials and the addition formula for little q-Legendre polynomials.* SIAM Journal on Mathematical Analysis **22**, 1991, 302–311.
505. P. Vértesi, *On the zeros of generalized Jacobi polynomials.* Annals of Numerical Mathematics **4**, 1997, 561–577.
506. G. Viennot, *A combinatorial theory for general orthogonal polynomials with extensions and applications.* In: Polynômes Orthogonaux et Applications (eds. C. Brezinski et al.). Lecture Notes in Mathematics **1171**, Springer-Verlag, New York, 1985, 139–157.
507. L. Vinet and A. Zhedanov, *Generalized Bochner theorem: characterization of the Askey-Wilson polynomials.* Journal of Computational and Applied Mathematics **211**, 2008, 45–56.
508. B. Viswanathan, *Generating functions for ultraspherical functions.* Canadian Journal of Mathematics **20**, 1968, 120–134.
509. H.S. Wall, *A continued fraction related to some partition formulas of Euler.* The American Mathematical Monthly **48**, 1941, 102–108.
510. L. Weisner, *Generating functions for Hermite functions.* Canadian Journal of Mathematics **11**, 1959, 141–147.
511. S. Wigert, *Sur les polynomes orthogonaux et l'approximation des fonctions continues.* Arkiv för Matematik, Astronomi och Fysik **17**, 1923, 1–15.
512. J.A. Wilson, *Some hypergeometric orthogonal polynomials.* SIAM Journal on Mathematical Analysis **11**, 1980, 690–701.
513. J.A. Wilson, *Asymptotics for the* ${}_4F_3$ *polynomials.* Journal of Approximation Theory **66**, 1991, 58–71.
514. M.W. Wilson, *Nonnegative expansions of polynomials.* Proceedings of the American Mathematical Society **24**, 1970, 100–102.
515. M.W. Wilson, *On the Hahn polynomials.* SIAM Journal on Mathematical Analysis **1**, 1970, 131–139.
516. J. Wimp, *Explicit formulas for the associated Jacobi polynomials and some applications.* Canadian Journal of Mathematics **39**, 1987, 983–1000.
517. J. Wimp, *Pollaczek polynomials and Padé approximants: some closed-form expressions.* Journal of Computational and Applied Mathematics **32**, 1990, 301–310.
518. R. Wong and J.-M. Zhang, *Asymptotic monotonicity of the relative extrema of Jacobi polynomials.* Canadian Journal of Mathematics **46**, 1994, 1318–1337.
519. R. Wong and W. Zhang, *Uniform asymptotics for Jacobi polynomials with varying large negative parameters – a Riemann-Hilbert approach.* Transactions of the American Mathematical Society **358**, 2006, 2663–2694.
520. M. Wyman, *The asymptotic behaviour of the Hermite polynomials.* Canadian Journal of Mathematics **15**, 1963, 332–349.
521. A. Zarzo, I. Area, E. Godoy and A. Ronveaux, *Results for some inversion problems for classical continuous and discrete orthogonal polynomials.* Journal of Physics A. Mathematical and General **30**, 1997, L35–L40.
522. A.I. Zayed, *Jacobi polynomials as generalized Faber polynomials.* Transactions of the American Mathematical Society **321**, 1990, 363–378.
523. J. Zeng, *Linéarisation de produits de polynômes de Meixner, Krawtchouk, et Charlier.* SIAM Journal on Mathematical Analysis **21**, 1990, 1349–1368.
524. J. Zeng, *The q-Stirling numbers, continued fractions and the q-Charlier and q-Laguerre polynomials.* Journal of Computational and Applied Mathematics **57**, 1995, 413–424.
525. W. Zhang, *On Chebyshev polynomials and Fibonacci numbers.* The Fibonacci Quarterly **40**, 2002, 424–428.
526. Z. Zhang and J. Wang, *On some identities involving the Chebyshev polynomials.* The Fibonacci Quarterly **42**, 2004, 245–249.

Index

R. Koekoek et al., *Hypergeometric Orthogonal Polynomials and Their q-Analogues*, Springer Monographs in Mathematics,
DOI 10.1007/978-3-642-05014-5,

Printed by Printforce, the Netherlands